Handbook of Generalized Convexity and Generalized Monotonicity

Handbook of Generalized Convexity and Generalized Monotonicity

Edited by

NICOLAS HADJISAVVAS
University of Aegean

SÁNDOR KOMLÓSI
University of Pécs

SIEGFRIED SCHAIBLE
University of California at Riverside

Springer

Library of Congress Cataloging-in-Publication Data

A C.I.P. Catalogue record for this book is available
from the Library of Congress.

ISBN 0-387-23255-9 e-ISBN 0-387- 23393-8 Printed on acid-free paper.

Printed in the United States of America.

9 8 7 6 5 4 3 2 1 SPIN 11319351

springeronline.com

Contents

Preface

Convexity of functions plays a central role in many branches of applied mathematics. One of the reasons is that it is very well fitted to extremum problems. For instance, some necessary conditions for the existence of a minimum become also sufficient in the presence of convexity. However, by far not all real-life problems can be described by a convex mathematical model. In many cases, nonconvex functions provide a much more accurate representation of reality. It turns out that often these functions, in spite of being nonconvex, retain some of the nice properties and characteristics of convex functions. For instance, their presence may ensure that necessary conditions for a minimum are also sufficient or that a local minimum is also a global one. This led to the introduction of several generalizations of the classical concept of a convex function. It happened in various fields such as economics, engineering, management science, probability theory and various applied sciences for example, mostly during the second half of the last (20th) century.

For instance, a well-known and useful property of a convex function is that its sublevel sets are convex. Many simple nonconvex functions also have this property. If we consider the class of all functions whose sublevel sets are convex, we obtain what is called the class of quasiconvex functions. It is much larger than the class of convex functions. The introduction of quasiconvex functions is usually attributed to de Finetti, in 1949. However quasiconvexity played already a crucial role in the 1928 minimax theorem of John von Neumann where it was introduced as a technical hypothesis, not as a new type of a function[1].

Many other classes of generalized convex functions have been introduced since then. In recent years, besides real-valued generalized convex functions, also vector-valued as well as multi-valued generalized convex

[1] See A. Guerraggio and E. Molho, "The origins of quasi-concavity: a development between mathematics and economics", Historia Mathematica 31, 2004, 62–75.

functions have been investigated intensively. During the last forty years, a significant increase of research activities in this field has been witnessed.

A well-known feature of convexity is that it is closely related to monotonicity: a differentiable function is convex if and only if its gradient is a monotone map. This can be extended to nondifferentiable functions as well, through the use of generalized derivatives, subdifferentials and multi-valued maps. Similar connections were discovered between generalized convex functions and certain classes of maps, generically called generalized monotone. For instance, a differentiable function is quasiconvex if and only if its gradient has a property that, quite naturally, was called quasimonotonicity.

Generalized monotone maps are not new in the literature. It is very interesting to observe that the first appearance of generalized monotonicity (well before the birth of this terminology) dates back to 1936. Even more remarkable, it occurred independently and almost at the same time in two seminal articles: the one, by Georgescu-Roegen (1936), dealt with the concept of a local preference in consumer theory of economics, and the other one, by Wald (1936), contained the first rigorous proof of the existence of a competitive general equilibrium. It is also remarkable that various axioms on revealed preferences in consumer theory are in fact generalized monotonicity conditions.

The recognition of close links between the already well established field of generalized convexity and the relatively undeveloped field of generalized monotonicity gave a boost to both and led to a major increase of research activities. Today generalized monotonicity is frequently used in complementarity problems, variational inequalities and equilibrium problems.

The first volume devoted exclusively to generalized convexity were the Proceedings of a NATO Advanced Study Institute organized by Avriel, Schaible and Ziemba in Vancouver, Canada on August 4-15, 1980; "Generalized Concavity in Optimization and Economics" by Schaible and Ziemba (1981, Academic Press). It was followed by the monograph "Generalized Concavity" by Avriel, Diewert, Schaible, Zang (1988, Plenum Publishing Corporation). Both volumes maintain the close relationship between the mathematics of generalized convex functions and its relevance to applications, especially in economic theory which described the field from the beginning.

It was Mordecai Avriel who initiated the first monograph solely devoted to generalizations of convexity back in 1978. Shortly after that he also proposed to organize an international conference on this topic. He often emphasized the importance of generalizing convexity with applications in mind. This gives the research area its interdisciplinary nature.

The extensive amount of new knowledge accumulated since the initial two volumes, exclusively contained in research papers, gave birth to the idea of presenting the main research results in both generalized convexity and generalized monotonicity in the form of a handbook.

The present volume consists of fourteen chapters, written by leading experts of various areas of Generalized Convexity and Generalized Monotonicity. Each chapter starts from the very basics and goes to the state of the art of its subject. Given the amount of material in the field, there are some topics that were not included. We trust that the comprehensive bibliography provided by each chapter can help to bridge the gap.

The chapters are grouped into two parts of the book, depending on whether their main focus is generalized convexity or generalized monotonicity.

Chapter 1, by *Frenk and Kassay*, provides an introduction to convex and quasiconvex analysis in a finite-dimensional setting. Since this analysis is heavily based on the study of certain sets, the most important algebraic and topological properties of linear subspaces, affine sets, convex and evenly convex sets and cones are included here. The well-known separation results are also presented and then used to derive the so-called dual representations for (evenly) convex sets. The next part is devoted fundamentally to the study of convex, quasiconvex and evenly quasiconvex functions. It is shown that this study can be reduced to the study of the sets considered in the preceding part. As such, the equivalent formulation of a separation result for convex sets is now given by the dual representation of a function. The key result in this respect is the Fenchel-Moreau theorem within convex analysis and its generalization to evenly quasiconvex functions. The remainder of the chapter presents some important applications of convex and quasiconvex analysis to optimization theory, game theory and the study of positively homogeneous, evenly quasiconvex functions.

Chapter 2, written by *Crouzeix*, is devoted to first and second order characterizations of generalized convex functions and first order criteria for generalized monotonicity. First order characterizations of differentiable generalized convex functions (quasiconvex, pseudoconvex, etc.) include characterizations in terms of generalized monotonicity (quasimonotonicity, pseudomonotonicity, etc.) of the gradient map. Second order characterizations of generalized convexity are inherently linked with first order characterizations of generalized monotonicity. In this study the positive (semi-)definiteness of the restriction of a quadratic form to a linear subspace is of great importance. An exhaustive study is presented on this subject. The chapter concludes with some important applications on Cobb-Douglas functions, conditions for a function to be convexifiable, generalized convex quadratic functions, generalized monotone affine maps, interior point methods and additive separability.

It is well known from convex analysis that a "nice" geometrical structure of convex functions, the convexity of the epigraph, has strong implications in terms of continuity and directional differentiability. In the case of quasiconvex functions the epigraph is no longer convex, but the sublevel sets are. In parallel to convex functions, here it is the geometrical structure of the sublevel sets which implies important continuity and differentiability properties for quasiconvex functions. *Chapter 3*, by *Crouzeix*, provides a deep insight into quasiconvex analysis discussing quasiconvex regularizations, cone-increasingness, as well as continuity and differentiability properties of quasiconvex functions. In the absence of a directional derivative, generalized derivatives like the lower and upper Dini derivatives can be used. In certain cases these derivatives are linked with important geometrical objects: the normal cone to sublevel sets and the quasi-subdifferential of a quasiconvex function at a given point. The role and usefulness of these objects are thoroughly investigated in this part.

It is well known that in classical scalar optimization theory, convexity plays a fundamental role since it guarantees important properties such as: a local minimizer is also a global one; a stationary point is a global minimizer (the first order necessary optimality conditions are sufficient). The results obtained in the scalar case have had a major influence on the area of vector optimization, which has been well developed during recent decades. In vector optimization the concept of a minimum is usually translated by means of an ordering cone in the image space of the objective function. This cone induces only a partial order which is the main reason why there are several ways of extending the notion of generalized convexity in vector optimization. *Chapter 4*, by *Cambini and Martein*, discusses the role of generalized convexity properties in scalar and vector optimization problems in finite dimensions. In the scalar-valued case extremal problems with quasiconvex, quasilinear, pseudoconvex, pseudolinear, preinvex and invex functions are studied. In the vector-valued case optimality conditions are derived for certain classes of differentiable vector-valued generalized convex functions including C-quasiconvex, $(C^\sharp, C^\S)$-pseudoconvex, $(C^\sharp, C^\S)$-pseudolinear, (C, η)-invex functions where C denotes the ordering cone of the image space.

Chapter 5, by *Luc*, also deals with vector optimization problems, but in a more general setting, namely in a real Hausdorff topological vector space and in terms of nonlinear analysis. Different classes of generalized convex vector functions are introduced and studied. Existence results for efficient solutions are given and structural properties such as compactness and connectedness of solution sets are investigated. Optimality conditions are provided in terms of three kinds of derivatives: classical derivatives, contingent derivatives and approximate Jacobians. The last part of the chapter discusses scalarization methods for solving vector optimization problems.

A central topic in optimization is convex duality theory. Given a primal convex minimization problem, one embeds it into a family of perturbed optimization problems and then, relative to these perturbations, one associates it with a so-called dual problem. The deep relations existing between the primal and the dual are helpful for analyzing properties of the original problem and, in particular, for obtaining optimality conditions. They are also used to devise numerical algorithms. In case of problems arising in applications in various sciences, particularly in economics, dual problems usually have a nice interpretation that yields a new motivation for analyzing them. In *Chapter 6*, by *Martínez–Legaz*, generalized convex duality and its economic applications are discussed. The extension of convex duality theory to the generalized convex case is based on the Fenchel-Moreau generalized duality scheme and also on the general theory of conjugations. In quasiconvex analysis the so-called level sets conjugations are very useful. The Fenchel-Moreau general duality scheme and these conjugations are discussed in details in this chapter. The remainder of the chapter is devoted to economic applications: duality between direct and indirect utility functions in consumer theory, monotonicity of demand functions and consumer duality theory in the absence of a utility function.

The notion of a subdifferential of a convex function at a given point of its domain plays a central role in convex analysis. In fact the subdifferential plays two different roles. One of them is local: it furnishes a local approximation to a convex function in a neighborhood of a given point. The other one is global: it is a tool for constructing supporting hyperplanes to the epigraph of a convex function. Generalizations of the first line of thought of the subdifferential leads to nonsmooth analysis, while the second one leads to abstract convexity. Abstract convexity is the topic of *Chapter 7*, written by *Rubinov and Dutta*. One of the fundamental results of convex analysis is that each lower semicontinuous convex function is the upper envelope (pointwise supremum) of all affine functions majorized by the function. The main motivation for the development of abstract convexity lies in the fact that the envelope representation is very convenient, even when we consider the upper envelope of sets of nonaffine functions. Two kinds of objects are studied in the framework of abstract convexity: abstract convex functions and abstract convex sets. Abstract convex functions can be represented via the envelope representation of not necessary linear or affine elementary functions. Abstract convex sets are defined via the following property: a point not belonging to an abstract convex set can be separated from it by an elementary function. This chapter presents the main definitions related to abstract convexity and provides numerous examples of abstract convex functions based on various sets of elementary functions, including abstract quasiconvex functions. Some applications of abstract convexity can also be found in it: a Minkowski duality scheme, a Fenchel-Moreau conjugation, and Hadamard type inequalities for quasiconvex functions.

Chapter 8, by *Frenk and Schaible*, is devoted to Fractional Programming. It concludes the first part of the book which focuses on generalized convexity. Fractional programs are ratio-optimization problems involving one or several ratios in the objective function. Such objective functions are usually not convex, but often generalized convex in case of convex, concave or affine numerators and denominators. An extensive literature comparable to that for other major classes of optimization problems has been developed. During the last forty years fractional programming has benefited from advances in generalized convexity and vice versa.

The chapter begins with a survey of a large variety of applications of single-ratio, min-max and sum-of-ratios fractional programs. Recent applications are emphasized. Some case studies are mentioned too. This survey of fractional programming applications reinforces the relevance of generalized convexity to applied disciplines, in particular to management science. With that it complements the presentation of generalized convexity applications in economic theory in Chapter 6. The survey of fractional programming applications is followed by a detailed analysis of a min-max fractional program which includes the single-ratio and the more restricted, classical min-max fractional program as special cases. Basis of the analysis is a parametric approach. First a primal Dinkelbach-type algorithm is proposed and its convergence properties are analyzed. Then duality is introduced giving rise to a dual max-min fractional program and duality relations are established. Finally a dual Dinkelbach-type algorithm is presented and its convergence properties are derived.

The first chapter of the second part of the book is *Chapter 9*, by *Hadjisavvas and Schaible*. It starts with an introductory presentation of nine kinds of generalized monotone maps and shows the links between these maps and differentiable generalized convex functions: a function belongs to one of nine classes of generalized convex functions if and only if its gradient belongs to a corresponding class of generalized monotone maps. The remainder of the chapter is devoted to criteria for the generalized monotonicity of maps. This is done first for differentiable maps and then for nondifferentiable maps (locally Lipschitz or just continuous). Finally, the special case of affine maps is discussed in considerable detail in view of its relevance to generalized convex quadratic programs and generalized monotone linear complementarity problems.

Generalized convex functions are not necessarily differentiable. In this case, the gradient is usually replaced by an appropriate generalized derivative. *Chapter 10*, by *Komlósi*, is devoted to the study of nonsmooth generalized convex functions with the help of special classes of generalized derivatives. Several results are presented on the links between generalized monotonicity of generalized derivatives and generalized convexity of the functions under discussion. The abundance of different notions of generalized derivatives has motivated an axiomatic treatment resulting, among others, in the abstract concept of first order approximations. The usefulness of quasiconvex first order approximations in optimization theory is investigated. In particular, generalized upper quasidifferentiable functions are studied, and quasiconvex Farkas-type theorems and KKT-type optimality conditions are established.

The relationship between generalized convex functions and generalized monotone maps can also be approached by using subdifferentials instead of generalized derivatives. For nonconvex functions, the standard Fenchel-Moreau subdifferential is not suitable. This led Clarke, Rockafellar and others to the introduction of many different subdifferentials. In *Chapter 11*, by *Hadjisavvas*, the relationship of generalized convexity of lower semicontinuous functions to the generalized monotonicity of their subdifferential is presented. This unifying approach includes most subdifferentials. Some new kinds of generalized monotonicity, such as cyclic generalized monotonicity and proper quasimonotonicity are introduced. The latter notion has proved to be closely linked to Minty variational inequalities. Finally, some recent results on the maximality of pseudomonotone operators and its relationship to continuity are presented.

Generalized monotonicity is not exclusively a "dual" concept of generalized convexity. The concept of pseudomonotonicity, for instance, appeared first in complementarity problems. *Chapter 12*, by *Yao and Chadli*, demonstrates the importance of this concept in complementarity problems and variational inequalities. In the first part of the chapter recent results on existence and uniqueness of solutions for complementarity problems and variational inequalities in infinite-dimensional spaces under pseudomonotonicity are reported. An application of the complementarity problem to the post-critical equilibrium state of a thin elastic plate is given. In the next part topological pseudomonotonicity is presented, and again existence results are derived and some applications are given. The notion of topological pseudomonotonicity was introduced by Brézis in 1968 for nonlinear operators in order to obtain existence theorems for quasi-linear elliptic and parabolic partial differential equations. In the final part some possible extensions of complementarity and variational inequality problems are given, and the equivalence of several problems including complementarity problems, least element problems and variational inequality problems is discussed.

Equilibrium problems provide the most general and unified framework for optimization, complementarity problems and variational inequalities. Traditionally, most works on various aspects of equilibrium problems were limited to monotone problems. However, the monotonicity assumption turned out to be too restrictive for many applied problems, especially in economics and management science. During the last decade generalized monotone equilibrium problems were investigated rather intensively and much progress has been made in different directions. *Chapter 13*, by *Konnov*, presents basic results in the theory and for the construction of solution methods for generalized monotone equilibrium problems and variational inequalities. Vector-valued equilibrium problems are also considered.

Probably the first use of generalized monotonicity (well before the birth of this terminology) was made already sixty eight years ago in economics. As said before, it occurred independently in an article by Georgescu-Roegen (1936), dealing with the concept of a local preference in consumer theory, and in an article by Wald (1936), containing the first rigorous proof of the existence of a competitive general equilibrium. These significant advances are presented in *Chapter 14*, by John. This chapter provides further important examples of the use of generalized convexity and generalized monotonicity in economics. The first example is consumer theory. The utility function approach puts emphasis on the quasiconcavity and pseudoconcavity of that function. The demand relation approach displays generalized monotonicity properties that are, in the economic literature, well known as axioms in revealed preference theory. In case of convex (respectively semistrictly or strictly convex) nontransitive preferences, various generalized monotonicity properties of demand arise quite naturally. The second interesting example is general equilibrium theory. The relevance of pseudomonotone excess demand to the stability of an equilibrium is recognized by its representability as a solution of a Minty variational inequality problem.

Research in generalized convexity and generalized monotonicity continues, and scientific results are numerous. They are presented regularly in special session clusters at many major conferences each year, in particular at the International Symposium on Generalized Convexity and Monotonicity which is organized usually every three years. Until today, there have been seven such symposia. The last three were organized by the "Working Group on Generalized Convexity". This is a growing association of several hundred researchers, whose website (`http://www.genconv.org`) provides information on past and future conferences, proceedings, publications etc.

We express our sincere gratitude to Dr. John Martindale, Kluwer Academic Publishers for his outstanding, continuous support in the production of this volume.

NICOLAS HADJISAVVAS

SANDOR KOMLOSI

SIEGFRIED SCHAIBLE

This book is dedicated
to Mordecai Avriel,
who initiated both the
first monograph and
the first international
conference in this
growing research area
twenty five years ago.

Contributing Authors

Alberto Cambini is professor of Mathematics for Economic Decisions at the Faculty of Economy, University of Pisa, 56124 Pisa, Italy. His research has focused on nonlinear programming and, in particular, on fractional programming as well as generalized convexity and optimality in scalar and vector optimization. He was the chair of the Organizing Committee for the third International Symposium on Generalized Convexity. He received from the University of Pisa the "Ordine del Cherubino" (highest research award) in 1991.
E-mail acambini@ec.unipi.it.
Personal page http://www.ec.unipi.it/acambini

Ouayl Chadli is assistant professor at the Department of Economics of the University Ibn Zohr of Agadir, B.P 8658 Hay Dakhla, Agadir 80000 Morocco. His research interests are: variational and hemivariational inequalities, equilibrium problems and their applications, vector optimization and mathematical economics.
E-mail ochadli@esta.ac.ma

Jean-Pierre Crouzeix is professor at the Blaise Pascal University in Clermont-Ferrand, 63170 Aubière, France. His interests in research concern, for a large part, (generalized) convexity, (generalized) monotonicity and their applications.
E-mail jp.crouzeix@math.univ-bpclermont.fr
Personal page http://wwwlma.univ-bpclermont.fr/~crouzeix/

Joydeep Dutta is an Assistant Professor of Mathematics at the Indian Institute of Technology, Kanpur, 208016, India. He received his Ph.D. in 1998 from the Indian Institute of Technology, Kharagpur. Subsequently he did his post-doctoral studies at the Indian Statistical Institute, New Delhi and the Universidad Autonoma de Barcelona. His main research

interests are in nonsmooth optimization and abstract convexity. He has published in both of these areas.
E-mail jdutta@iitk.ac.in

Hans Frenk is associate professor of Operations Research at the Econometric Institute of the Erasmus University, 3000 DR Rotterdam, The Netherlands. His research is in applied stochastic processes, optimization theory, equilibrium theory, convex/quasiconvex analysis and the application of deterministic and stochastic techniques to problems in Management Science. He published both in applied and theoretical journals and is a research fellow of the Erasmus Research Institute of Management (ERIM).
E-mail frenk@few.eur.nl.
Personal page http://www.few.eur.nl/few/people/frenk/

Nicolas Hadjisavvas is professor of Optimization at the Department of Product and Systems Design Engineering of the University of the Aegean, 84100 Hermoupolis, Greece. He is currently the chair of the international "Working Group on Generalized Convexity". He published several papers on the theory of generalized monotone maps and their applications to variational inequalities and equilibrium problems.
E-mail nhad@aegean.gr
Personal page http://www.syros.aegean.gr/users/nhad/

Reinhard John is Lecturer at the Department of Economics at the University of Bonn, D-53113 Bonn, Germany. He published several papers that have focused on the use of generalized monotonicity concepts in consumer theory and general equilibrium theory.
E-mail rjohn@uni-bonn.de.

Gábor Kassay is an associate professor at the Department of Analysis and Optimization of the Babes-Bolyai University, 3400 Cluj, Romania. He published several papers on generalized convexity/monotonicity and a book concerned with the equilibrium problem and its applications.
E-mail kassay@math.ubbcluj.ro

Sándor Komlósi is professor of Management Science at the Faculty of Business and Economics, University of Pécs, 7622 Pécs, Hungary. His research has focused on optimization theory, in particular, on generalized convexity and generalized monotonicity in nonsmooth optimization. He

is a past chair of the "Working Group on Generalized Convexity" and a former President of the Hungarian Operational Research Society.
E-mail komlosi@ktk.pte.hu
Personal page http://www.ktk.pte.hu/main//main.php?id=41

Igor Konnov is professor at the Department of Applied Mathematics of the Kazan University, Kazan 420008, Russia. His fields of scientific interests are the theory and solution methods for nonsmooth optimization, equilibrium problems and variational inequalities, including vector and generalized monotone problems, and their applications in Mathematical Physics, Economics and Transportation. He is the author of three books in these fields and coordinator of the Working Group "Equilibrium Models in Complex Systems" (EMICS).
E-mail Igor.Konnov@ksu.ru
Personal page
http://www.kcn.ru/tat_en/university/persons/9116.en.html

Dinh The Luc is currently a professor at the Department of Mathematics, University of Avignon, 33 rue Louis Pasteur, 84000 Avignon, France. His research activities include nonsmooth analysis, optimization and variational inequalities.
E-mail dtluc@univ-avignon.fr
Personal page
http://www.lia.univ-avignon.fr/equipes/RO/?lang=&rubrique=pageluc

Laura Martein is professor of Mathematics for Economic Decisions at the Faculty of Economy, University of Pisa, 56124 Pisa, Italy. Her research has focused on nonlinear programming and, in particular, on fractional programming as well as generalized convexity and optimality in scalar and vector optimization. She received from the University of Pisa the "Ordine del Cherubino" (highest research award) in 2003.
E-mail lmartein@ec.unipi.it
Personal page http://www.ec.unipi.it/lmartein/

Juan Enrique Martínez-Legaz is professor in the Department of Economics and Economics History and a researcher in the Center for the Study of Organizations and Decisions in Economics (CODE) at the University Autonoma de Barcelona, 08193 Bellaterra, Spain. He is currently a member of the Scientific Committee of the "Working Group on Generalized Convexity". His research has focused on generalized convexity

and its applications to mathematical economics.
E-mail JuanEnrique.Martinez@uab.es
Personal page http://idea.uab.es/jemartinez/

Alexander Rubinov is Professor of Mathematics in the School of Information Technology and Mathematical Sciences, University of Ballarat, 3353 Victoria, Australia. He is also Director of the Centre for Informatics and Applied Optimization at this university. He published several papers and two books on abstract convexity, monotonic analysis and their applications in different areas, including generalized Lagrange-type functions and numerical global optimization.
E-mail a.rubinov@ballarat.edu.au.
Personal page
http://uob-community.ballarat.edu.au/~arubinov/index.html

Siegfried Schaible is a Professor in the A.G. Anderson Graduate School of Management, University of California, Riverside, CA 92521, USA. His research has focused on fractional programming as well as generalizations of convexity and monotonicity for optimization, variational inequalities and equilibrium problems. He is founding chair of the international "Working Group on Generalized Convexity" (1994).
E-mail siegfried.schaible@ucr.edu.
Personal page http://www.agsm.ucr.edu/faculty/pages/schaible.html

Jen-Chih Yao is professor of the Department of Applied Mathematics, National Sun Yat-sen University, Kaohsiung, Taiwan 80424, ROC. His research interests are (vector) variational inequalities, (vector) equilibrium problems, (vector) complementarity problems, solution methods for variational inequality and equilibrium problems.
E-mail yaojc@math.nsysu.edu.tw

I

GENERALIZED CONVEXITY

Chapter 1

INTRODUCTION TO CONVEX AND QUASICONVEX ANALYSIS

Johannes B.G. Frenk
Department of Economics, Econometric Institute
Erasmus University
3000DR Rotterdam, The Netherlands
frenk@few.eur.nl

Gábor Kassay
Faculty of Mathematics and Computer Science
Babes Bolyai University
3400 Cluj, Romania
kassay@math.ubbcluj.ro

Abstract In the first chapter of this book the basic results within convex and quasiconvex analysis are presented. In Section 2 we consider in detail the algebraic and topological properties of convex sets within $\mathbb{R}^n$ together with their primal and dual representations. In Section 3 we apply the results for convex sets to convex and quasiconvex functions and show how these results can be used to give primal and dual representations of the functions considered in this field. As such, most of the results are well known with the exception of Subsection 3.4 dealing with dual representations of quasiconvex functions. In Section 3 we consider applications of convex analysis to noncooperative game and minimax theory, Lagrangian duality in optimization and the properties of positively homogeneous evenly quasiconvex functions. Among these result an elementary proof of the well-known Sion's minimax theorem concerning quasiconvex-quasiconcave bifunctions is presented, thereby avoiding the less elementary fixed point arguments. Most of the results are proved in detail and the authors have tried to make these proofs as transparent as possible. Remember that convex analysis deals with the study of convex cones and convex sets and these objects are generalizations of linear subspaces and affine sets, thereby extending the field of linear algebra. Although some of the proofs are technical, it is possible to give a clear geometrical interpretation of the main ideas of convex analysis. Finally in Section 5 we list a short and probably incomplete overview on the history of convex and quasiconvex analysis.

Keywords: Convex Analysis, Quasiconvex Analysis, Noncooperative games, Minimax, Optimization theory.

1. Introduction

In this chapter the fundamental questions studied within the field of convex and quasiconvex analysis are discussed. Although some of these questions can also be answered within infinite dimensional real topological vector spaces, our universe will be the finite dimensional real linear space $\mathbb{R}^n$ equipped with the well-known Euclidean norm $\|.\|$. Since convex and quasiconvex analysis can be seen as the study of certain sets, we consider in Section 2 the basic sets studied in this field and list with or without proof the most important algebraic and topological properties of those sets. In this section a proof based on elementary calculus of the important separation result for disjoint convex sets in $\mathbb{R}^n$ will be given. In Section 3 we introduce the so-called convex and quasiconvex functions and show that the study of these functions can be reduced to the study of the sets considered in Section 2. As such, the formulation of the separation result for disjoint convex sets is now given by the dual representation of a convex or quasiconvex function. In Section 4 we will discuss important applications of convex and quasiconvex analysis to optimization theory, game theory and the study of positively homogeneous evenly quasiconvex functions. Finally in Section 5 we consider some of the historical developments within the field of convex and quasiconvex analysis.

2. Sets studied within convex and quasiconvex analysis

In this section the basic sets studied within convex and quasiconvex analysis in $\mathbb{R}^n$ are discussed and their most important properties listed. Since in some cases these properties are well-known we often mention them without any proof. We introduce in Subsection 2.1 the definition of a linear subspace, an affine set, a cone and a convex set in $\mathbb{R}^n$ together with their so-called primal representation. Also the important concept of a hull operation applied to an arbitrary set is considered. In Subsection 2.2 the topological properties of the sets considered in Subsection 2.1 are listed and in Subsection 2.3 we prove the well-known separation result for disjoint convex sets. Finally in Subsection 2.4 this separation result is applied to derive the so-called dual representation of a closed convex set. In case proofs are included we have tried to make these proofs as transparent and simple as possible. Also in some cases these proofs can be easily adapted, if our universe is an infinite dimensional real topological vector space. Most of the material in this section together with the proofs can be found in Lancaster and Tismenetsky (cf. [47]) for the linear algebra part, while for the convex analysis part the reader is referred to Rockafellar (cf. [63]) and Hiriart-Urruty and Lemaréchal (cf. [34], [35]).

2.1 Algebraic properties of sets

As already observed our universe will always be the n-dimensional Euclidean space $\mathbb{R}^n$ and any element of $\mathbb{R}^n$ is denoted by the vector $\mathbf{x} = (x_1, ..., x_n)^\top, x_i \in \mathbb{R}$ or $\mathbf{y} = (y_1, ..., y_n)^\top, y_i \in \mathbb{R}$. The inner product $< .,. >: \mathbb{R}^n \times \mathbb{R}^n \to \mathbb{R}$ is then given by

$$< \mathbf{x}, \mathbf{y} > := \sum\nolimits_{i=1}^{n} x_i y_i = \mathbf{x}^\top \mathbf{y},$$

while the Euclidean norm $\|.\|$ is defined by

$$\|\mathbf{x}\| := \sqrt[2]{< \mathbf{x}, \mathbf{x} >}.$$

To simplify the notation, we also introduce for the sets $A, B \subseteq \mathbb{R}^n$ and $\alpha, \beta \in \mathbb{R}$ the *Minkowsky sum* $\alpha A + \beta B$ given by

$$\alpha A + \beta B := \{\alpha \mathbf{x} + \beta \mathbf{y} : \mathbf{x} \in A, \mathbf{y} \in B\}.$$

The first sets to be introduced are the main topic of study within *linear algebra* (cf. [47]).

Definition 1.1 *A set $L \subseteq \mathbb{R}^n$ is called a linear subspace if L is nonempty and $\alpha L + \beta L \subseteq L$ for every α, $\beta \in \mathbb{R}$. Moreover, a set $M \subseteq \mathbb{R}^n$ is called affine if $\alpha M + (1 - \alpha) M \subseteq M$ for every $\alpha \in \mathbb{R}$.*

The empty set $\emptyset$ and $\mathbb{R}^n$ are extreme examples of an affine set. Also it can be shown that the set M is affine and $\mathbf{0} \in M$ if and only if M is a linear subspace and for each nonempty affine set M there exists a unique linear subspace L_M satisfying

$$M = L_M + \mathbf{x} \tag{1.1}$$

for any given $\mathbf{x} \in M$ (cf. [63]).

Since $\mathbb{R}^n$ is a linear subspace, we can apply to any nonempty set $S \subseteq \mathbb{R}^n$ the so-called *linear hull operation* and construct the set

$$lin(S) := \cap\{L : S \subseteq L \text{ and } L \text{ a linear subspace}\}. \tag{1.2}$$

For any collection of linear subspaces $L_i, i \in I$ containing S it is obvious that the intersection $\cap_{i \in I} L_i$ is again a linear subspace containing S and this shows that the set $lin(S)$ is the smallest linear subspace containing S. The set $lin(S)$ is called the *linear hull generated by the set* S and if S has a finite number of elements the linear hull is called *finitely generated.* By a similar argument one can construct, using the so-called *affine hull operation,* the smallest affine set containing S. This set, denoted by $aff(S)$, is called the *affine hull generated by the set* S and is given by

$$aff(S) := \cap\{M : S \subseteq M \text{ and } M \text{ an affine set}\}. \tag{1.3}$$

If the set S has a finite number of elements, the affine hull is called *finitely generated.* Since any linear subspace is an affine set, it is clear that $aff(S) \subseteq lin(S)$. To give a so-called *primal representation* of these sets we introduce the next definition.

Definition 1.2 *A vector* $\mathbf{x}$ *is a linear combination of the vectors* $\mathbf{x}_1, ..., \mathbf{x}_k$ *if*

$$\mathbf{x} = \sum_{i=1}^{k} \alpha_i \mathbf{x}_i, \alpha_i \in \mathbb{R}, 1 \leq i \leq k.$$

A vector $\mathbf{x}$ *is an affine combination of the vectors* $\mathbf{x}_1, ..., \mathbf{x}_k$ *if*

$$\mathbf{x} = \sum_{i=1}^{k} \alpha_i \mathbf{x}_i, \alpha_i \in \mathbb{R}, 1 \leq i \leq k \text{ and } \sum_{i=1}^{k} \alpha_i = 1.$$

A linear combination of the nonempty set S *is given by the set* $\sum_{i=1}^{k} \alpha_i S$ *with* $\alpha_i \in \mathbb{R}, 1 \leq i \leq k$, *while an affine combination of the same set is given by the set* $\sum_{i=1}^{k} \alpha_i S$ *with* $\sum_{i=1}^{k} \alpha_i = 1$ *and* $\alpha_i \in \mathbb{R}, 1 \leq i \leq k$.

A trivial consequence of Definitions 1.1 and 1.2 is given by the next result which also holds in infinite dimensional linear spaces.

Lemma 1.1 *A nonempty set* $L \subseteq \mathbb{R}^n$ *is a linear subspace if and only if it contains all linear combinations of the set* L. *Moreover, a nonempty set* $M \subseteq \mathbb{R}^n$ *is an affine set if and only if it contains all affine combinations of the set* M.

The result in Lemma 1.1 yields a primal representation of a linear subspace and an affine set. In particular, we obtain from Lemma 1.1

that the set $lin(S)$ ($aff(S)$) with $S \subseteq \mathbb{R}^n$ nonempty equals all linear (affine) combinations of the set S. This means

$$lin(S) = \cup_{k=1}^{\infty}\{\sum_{i=1}^{k} \alpha_i S : \alpha_i \in \mathbb{R}\} \tag{1.4}$$

and

$$aff(S) = \cup_{k=1}^{\infty}\{\sum_{i=1}^{k} \alpha_i S : \alpha_i \in \mathbb{R} \text{ and } \sum_{i=1}^{k} \alpha_i = 1\}. \tag{1.5}$$

For any nonempty sets $S_1 \subseteq \mathbb{R}^n$ and $S_2 \subseteq \mathbb{R}^m$ one can now show using relation (1.4) that

$$lin(S_1 \times S_2) = lin(S_1) \times lin(S_2) \tag{1.6}$$

and using relation (1.5) that

$$aff(S_1 \times S_2) = aff(S_1) \times aff(S_2). \tag{1.7}$$

Also, for $A : \mathbb{R}^n \to \mathbb{R}^m$ a linear mapping, it is easy to verify that

$$A(lin(S)) = lin(A(S)) \tag{1.8}$$

and for $A : \mathbb{R}^n \to \mathbb{R}^m$an affine mapping, that

$$A(aff(S)) = aff(A(S)). \tag{1.9}$$

Recall a mapping $A : \mathbb{R}^n \to \mathbb{R}^m$ is called linear if

$$A(\alpha\mathbf{x} + \beta\mathbf{y}) = \alpha A(\mathbf{x}) + \beta A(\mathbf{y})$$

for every $\alpha, \beta \in \mathbb{R}$ and $\mathbf{x}, \mathbf{y} \in \mathbb{R}^n$ and it is called affine if

$$A(\alpha\mathbf{x} + (1-\alpha)\mathbf{y}) = \alpha A(\mathbf{x}) + (1-\alpha)A(\mathbf{y})$$

for every $\alpha \in \mathbb{R}$ and $\mathbf{x}, \mathbf{y} \in \mathbb{R}^n$. Moreover, in case we apply relation (1.7) to the affine mapping $A : \mathbb{R}^{2n} \to \mathbb{R}^n$, given by $A(\mathbf{x}, \mathbf{y}) = \alpha\mathbf{x} + \beta\mathbf{y}$, with $\alpha, \beta \in \mathbb{R}$ and use relation (1.9) the following rule for the affine hull of the sum of sets is easy to verify.

Lemma 1.2 *For any nonempty sets $S_1, S_2 \subseteq \mathbb{R}^n$ and $\alpha, \beta \in \mathbb{R}$ it follows that*

$$aff(\alpha S_1 + \beta S_2) = \alpha aff(S_1) + \beta aff(S_2).$$

Another application of relations (1.4) and (1.5) yields the next result.

Lemma 1.3 *For any nonempty set $S \subseteq \mathbb{R}^n$ and $\mathbf{x}_0$ belonging to $aff(S)$ it follows that $aff(S) = \mathbf{x}_0 + lin(S - \mathbf{x}_0)$.*

An improvement of Lemma 1.1 is given by the observation that any linear subspace (affine set) of $\mathbb{R}^n$ can be written as the linear or affine hull of a *finite* subset $S \subseteq \mathbb{R}^n$. To show this improvement one needs to introduce the next definition (cf. [47]).

Definition 1.3 *The vectors $\mathbf{x}_1, ..., \mathbf{x}_k$ are called linearly independent if*

$$\sum_{i=1}^{k} \alpha_i \mathbf{x}_i = \mathbf{0} \text{ and } \alpha_i \in \mathbb{R} \Rightarrow \alpha_i = 0,\ 1 \leq i \leq k.$$

Moreover, the vectors $\mathbf{x}_1, ..., \mathbf{x}_k$ are called affinely independent if

$$\sum_{i=1}^{k} \alpha_i \mathbf{x}_i = \mathbf{0} \text{ and } \sum_{i=1}^{k} \alpha_i = 0 \Rightarrow \alpha_i = 0,\ 1 \leq i \leq k.$$

For $k \geq 2$ an equivalent characterization of affinely independent vectors is given by the observation that the vectors $\mathbf{x}_1, ..., \mathbf{x}_k$ are affinely independent if and only if the vectors $\mathbf{x}_2 - \mathbf{x}_1, ..., \mathbf{x}_k - \mathbf{x}_1$ are linear independent (cf. [34]). To explain the name linearly and affinely independent we observe that the vectors $\mathbf{x}_1, ..., \mathbf{x}_k$ are linearly independent if and only if any vector $\mathbf{x}$ belonging to the linear hull $lin(\{\mathbf{x}_1, ..., \mathbf{x}_k\})$ can be written as a unique linear combination of the vectors $\mathbf{x}_1, ..., \mathbf{x}_k$. Moreover, the vectors $\mathbf{x}_1, ..., \mathbf{x}_k$ are affinely independent if and only if any vector $\mathbf{x}$ belonging to the affine hull $aff(\{\mathbf{x}_1, ..., \mathbf{x}_k\})$ can be written as a unique affine combination of the vectors $\mathbf{x}_1, ..., \mathbf{x}_k$. The improvement of Lemma 1.1 is given by the following result well-known within linear algebra (cf. [47]).

Lemma 1.4 *For any linear subspace $L \subseteq \mathbb{R}^n$ containing nonzero elements there exists a set of linearly independent vectors $\mathbf{x}_1, ..., \mathbf{x}_k$, $k \leq n$ satisfying $lin(\{\mathbf{x}_1, ..., \mathbf{x}_k\}) = L$. Also for any nonempty affine set $M \subseteq \mathbb{R}^n$ there exists a set of affinely independent vectors $\mathbf{x}_0, ..., \mathbf{x}_k, k \leq n$ satisfying $aff(\{\mathbf{x}_0, ..., \mathbf{x}_k\}) = M$.*

By Lemma 1.4 any linear subspace $L \subseteq \mathbb{R}^n$ containing nonzero elements can be represented as the linear hull of $k \leq n$ linearly independent vectors. If this holds, the dimension $\dim(L)$ of the linear subspace L is given by k. Since any $\mathbf{x}$ belonging to L can be written as a unique linear combination of linearly independent vectors this shows (cf. [47]) that $\dim(L)$ is well defined for L containing nonzero elements. If $L = \{\mathbf{0}\}$ the dimension $\dim(L)$ is by definition equal to 0. To extend this to affine sets we observe by relation (1.1) that any nonempty affine set M is parallel to its unique subspace L_M and the dimension $\dim(M)$ of a nonempty affine set M is now given by $\dim(L_M)$. By definition the dimension of the empty set $\varnothing$ equals -1. Finally, the dimension $\dim(S)$ of an arbitrary set $S \subseteq \mathbb{R}^n$ is given by $\dim(\mathrm{aff}(S))$. In the next definition we will introduce the sets which are the main objects of study within the field of convex and quasiconvex analysis.

Definition 1.4 *A set $C \subseteq \mathbb{R}^n$ is called convex if $\alpha C + (1-\alpha)C \subseteq C$ for every $0 < \alpha < 1$. Moreover, a set $K \subseteq \mathbb{R}^n$ is called a cone if $\alpha K \subseteq K$ for every $\alpha > 0$.*

The empty set $\varnothing$ is an extreme example of a convex set and a cone. An affine set is clearly a convex set but it is obvious that not every convex set is an affine set. This shows that convex analysis is an extension of linear algebra. Moreover, it is easy to show for every cone K that

$$K \text{ convex} \Leftrightarrow K + K \subseteq K. \tag{1.10}$$

Finally, for $A : \mathbb{R}^n \to \mathbb{R}^m$ an affine mapping and $C \subseteq \mathbb{R}^n$ a nonempty convex set it follows that the set $A(C)$ is convex, while for $A : \mathbb{R}^n \to \mathbb{R}^m$ a linear mapping and $K \subseteq \mathbb{R}^n$ a nonempty cone the set $A(K)$ is a cone.

To relate convex sets to convex cones we observe for $\mathbb{R}_+ := [0, \infty)$ and any nonempty set $S \subseteq \mathbb{R}^n$ that the set

$$\mathbb{R}_+(S \times \{1\}) := \{(\alpha\mathbf{x}, \alpha) : \alpha \geq 0, \mathbf{x} \in S\} \subseteq \mathbb{R}^{n+1}$$

is a cone. This implies by relation (1.10) that the set $\mathbb{R}_+(C \times \{1\})$ is a convex cone for any convex set $C \subseteq \mathbb{R}^n$. It is now clear for any nonempty set $S \subseteq \mathbb{R}^n$ that

$$\mathbb{R}_+(S \times \{1\}) \cap (\mathbb{R}^n \times \{1\}) = S \times \{1\} \tag{1.11}$$

and so any convex set C can be seen as an intersection of the convex cone $\mathbb{R}_+(C \times \{1\})$ and the affine set $\mathbb{R}^n \times \{1\}$. This shows that convex sets are closely related to convex cones and by relation (1.11) one can study convex sets by only studying affine sets and convex cones containing $\mathbf{0}$. We will not pursue this approach but only remark that the above relation is sometimes useful. Introducing an important subclass of convex sets, let $\mathbf{a}$ be a nonzero vector belonging to $\mathbb{R}^n$ and $b \in \mathbb{R}$ and

$$H^{<}(\mathbf{a}, b) := \{\mathbf{x} \in \mathbb{R}^n : \mathbf{a}^\intercal \mathbf{x} < b\}. \tag{1.12}$$

The set $H^{<}(\mathbf{a}, b)$ is called a halfspace and clearly this halfspace is a convex set. Moreover, the set $H^{\leq}(\mathbf{a}, b) := \{\mathbf{x} \in \mathbb{R}^n : \mathbf{a}^\intercal \mathbf{x} \leq b\}$ is also called a halfspace and this set is also a convex set. Another important subclass of convex sets useful within the study of quasiconvex functions is given by the following definition (cf. [23]).

Definition 1.5 *A set $C_e \subseteq \mathbb{R}^n$ is called evenly convex if $C_e = \mathbb{R}^n$ or C_e is the intersection of a collection of halfspaces $H^{<}(\mathbf{a}, b)$.*

Clearly the empty set $\varnothing$ is evenly convex and since any halfspace $H^{\leq}(\mathbf{a}, b)$ can be obtained by intersecting the halfspaces $H^{<}(\mathbf{a}, b+\frac{1}{n}), n \geq 1$ it also follows that any halfspace $H^{\leq}(\mathbf{a}, b)$ is evenly convex. In Subsection 2.3 it will be shown that any closed or open convex set is evenly convex. However, there exist convex sets which are not evenly convex.

Example 1.1 If $C := \{(x_1, x_2) : 0 \leq x_1 \leq 1, 0 \leq x_2 < 1\} \cup \{(1,1)\}$, then it follows that C is convex but not evenly convex.

Since $\mathbb{R}^n$ is a convex set, we can apply to any nonempty set $S \subseteq \mathbb{R}^n$ the so-called *convex hull operation* and construct the nonempty set

$$co(S) := \cap\{C : S \subseteq C \text{ and } C \text{ a convex set}\}. \tag{1.13}$$

For any collection of convex sets $C_i, i \in I$ containing S it is obvious that the intersection $\cap_{i \in I} C_i$ is again a convex set containing S and this shows that the set co(S) is the smallest convex set containing S. The set $co(S)$ is called the *convex hull generated by the set* S and if S has a finite number of elements the convex hull is called *finitely generated.* Since $\mathbb{R}^n$ is by definition evenly convex one can construct by a similar argument using the so-called *evenly convex hull operation* the smallest evenly convex set containing the nonempty set S. This set, denoted by $ec(S)$, is called the *evenly convex hull generated by the set* S and is given by

$$ec(S) := \cap \{C_e : S \subseteq C_e \text{ and } C_e \text{ an evenly convex set}\}. \tag{1.14}$$

Since any evenly convex set is convex it follows that $co(S) \subseteq ec(S)$.

By the so-called *canonic hull operation* one can also construct the smallest convex cone containing the nonempty set S, and the smallest convex cone containing $S \cup \{\mathbf{0}\}$. The last set is given by

$$cone(S) := \cap\{K : S \cup \{\mathbf{0}\} \subseteq K \text{ and } K \text{ a convex cone}\}. \tag{1.15}$$

Unfortunately this set is called the *convex cone generated by* S (cf. [63]). Clearly the set $cone(S)$ is in general not equal to the smallest convex cone containing S unless the zero element belongs to S. To give an alternative characterization of the above sets we introduce the next definition.

Definition 1.6 *A vector* $\mathbf{x}$ *is a canonical combination of the vectors* $\mathbf{x}_1, ..., \mathbf{x}_k$ *if*

$$\mathbf{x} = \sum_{i=1}^{k} \alpha_i \mathbf{x}_i, \alpha_i \geq 0, 1 \leq i \leq k.$$

The vector $\mathbf{x}$ *is called a strict canonical combination of the same vectors if* $\alpha_i > 0, 1 \leq i \leq k$. *A vector* $\mathbf{x}$ *is a convex combination of the vectors* $\mathbf{x}_1, ..., \mathbf{x}_k$ *if*

$$\mathbf{x} = \sum_{i=1}^{k} \alpha_i \mathbf{x}_i \text{ and } \sum_{i=1}^{k} \alpha_i = 1, \alpha_i > 0.$$

A canonical combination of the nonempty set S *is given by the set* $\sum_{i=1}^{k} \alpha_i S$ *with* $\alpha_i \geq 0, 1 \leq i \leq k$, *while a strict canonical combination of the same set is given by* $\sum_{i=1}^{k} \alpha_i S$ *with* $\alpha_i > 0, 1 \leq i \leq k$. *Finally a convex combination of the set* S *is given by the set* $\sum_{i=1}^{k} \alpha_i S$ *with* $\sum_{i=1}^{k} \alpha_i = 1, \alpha_i > 0, 1 \leq i \leq k$.

A trivial consequence of Definitions 1.4 and 1.6 is given by the next result which also holds in infinite dimensional linear spaces.

Lemma 1.5 *A nonempty set* $K \subseteq \mathbb{R}^n$ *is a convex cone (convex cone containing* $\mathbf{0}$*) if and only if it contains all strict canonical (canonical) combinations of the set* K. *Moreover, a nonempty set* $C \subseteq \mathbb{R}^n$ *is a convex set if and only if it contains all convex combinations of the set* C.

The result in Lemma 1.5 yields a primal representation of a convex cone and a convex set. In particular, we obtain from Lemma 1.5 that the set *cone*(S) (*co*(S)) with $S \subseteq \mathbb{R}^n$ nonempty equals all canonical (convex) combinations of the set S. This means

$$cone(S) = \cup_{k=1}^{\infty} \{\sum_{i=1}^{k} \alpha_i S : \alpha_i \geq 0\} \tag{1.16}$$

and

$$co(S) = \cup_{k=1}^{\infty} \{\sum_{i=1}^{k} \alpha_i S : \sum_{i=1}^{k} \alpha_i = 1, \alpha_i > 0\}. \tag{1.17}$$

We observe that the representations of *cone*(S) and *co*(S), listed in relations (1.16) and (1.17), are the "convex equivalences" of the representation of $lin(S)$ and $aff(S)$ given by relations (1.4) and (1.5). Moreover, to relate the above representations, it is easy to see that

$$cone(S) = \mathbb{R}_+(co(S)). \tag{1.18}$$

Since by relations (1.16) and (1.17) a convex cone containing $\mathbf{0}$ (convex set) can be seen as a generalization of a linear subspace (affine set) one might wonder whether a similar result as in Lemma 1.4 holds. Hence we wonder whether any convex cone containing $\mathbf{0}$ (convex set) can be seen as a canonical (convex) combination of a finite set S.

Example 1.2 Contrary to linear subspaces it is not true that any convex cone containing $\mathbf{0}$ is a canonical combination of a finite set. An example is given by the so-called L^2 or ice-cream cone $K = \{(\mathbf{x}, t) : \|\mathbf{x}\| \leq t\} \subseteq \mathbb{R}^{n+1}$.

Despite this negative result it is possible in finite dimensional linear spaces to improve for canonical hulls and convex hulls the representation given by relations (1.16) and (1.17). In the next result it is shown that any element belonging to cone(S) with S containing nonzero elements can be written as a canonical combination of at most n linearly independent vectors belonging to S. This is called *Caratheodory's theorem* for canonical hulls. Using this result and relation (1.11) a related result holds for convex hulls and in this case linearly independent is replaced by affinely independent and at most n is replaced by at most $n+1$. Clearly this result (cf. [63]) is the "convex equivalence" of Lemma 1.4.

Lemma 1.6 *If $S \subseteq \mathbb{R}^n$ is a set containing nonzero elements, then for any $\mathbf{x}$ belonging to cone(S) there exists a set of linearly independent vectors $\mathbf{x}_1, ..., \mathbf{x}_k, k \leq n$ belonging to S such that $\mathbf{x}$ can be written as a canonical combination of these vectors. Moreover, for any $\mathbf{x} \in co(S)$ there exists a set of affinely independent vectors $\mathbf{x}_1, ..., \mathbf{x}_k, k \leq n+1$ belonging to S such that $\mathbf{x}$ can be written as a convex combination of these vectors.*

Proof. Clearly for $\mathbf{0} \in cone(S)$ the desired result holds and so $\mathbf{x} \in cone(S)$ should be nonzero. By relation (1.16) there exists some finite set $\{\mathbf{x}_1, ..., \mathbf{x}_k\}$ and $\alpha_i > 0, 1 \leq i \leq k$ satisfying $\mathbf{x} = \sum_{i=1}^k \alpha_i \mathbf{x}_i$. If the vectors $\mathbf{x}_i, 1 \leq i \leq k$ are linearly independent, then clearly $k \leq n$ and we are done. Otherwise, there exists a nonzero sequence $\beta_i, 1 \leq i \leq k$ satisfying $\mathbf{0} = \sum_{i=1}^k \beta_i \mathbf{x}_i$ and without loss of generality we may assume that the set $I := \{1 \leq i \leq k : \beta_i > 0\}$ is nonempty. If $\epsilon := \min\{\frac{\alpha_i}{\beta_i} : i \in I\} > 0$ and $i^* := \arg\min\{\frac{\alpha_i}{\beta_i} : i \in I\}$ we obtain that

$$\mathbf{x} = \sum_{i=1}^{k} (\alpha_i - \epsilon\beta_i)\mathbf{x}_i = \sum_{i=1, i \neq i^*}^{k} (\alpha_i - \epsilon\beta_i)\mathbf{x}_i$$

and so $\mathbf{x}$ can be written as a strict canonical combination of at most $k-1$ vectors. Applying now the same procedure again until we have identified a subset of $\{\mathbf{x}_1, ..., \mathbf{x}_k\}$ consisting of linearly independent vectors the first part follows. To show the result for convex hulls it follows for any $\mathbf{x} \in co(S)$ that $(\mathbf{x}, 1)$ belongs to $co(S) \times \{1\} \subseteq \mathbb{R}^+(co(S) \times \{1\}\} \subseteq \mathbb{R}^{n+1}$.

By relation (1.18) the set $\mathbb{R}^+(co(S) \times \{1\}$ is the convex cone generated by $S \times \{1\}$ and by the first part the vector $(\mathbf{x}, 1)$ can be written as a canonical combination of at most $n+1$ linearly independent vectors $(\mathbf{x}_i, 1) \in S \times \{1\}$. Hence one can find positive scalars α_i satisfying $\mathbf{x} = \sum_{i=1}^k \alpha_i \mathbf{x}_i, k \leq n+1$ and $\sum_{i=1}^k \alpha_i = 1$ and since the vectors $(\mathbf{x}_i, 1) \in S \times \{1\}$ are linearly independent if and only if the vectors $\mathbf{x}_i \in S$ are affinely independent the desired result follows. □

Although in the above lemma $k \leq n$ for cones and $k \leq n+1$ for convex hulls it is easy to see that n can be replaced by $\dim(S) \leq n$. This concludes our discussion on algebraic properties of linear subspaces, affine sets, convex sets, and convex cones. In the next subsection we investigate topological properties of these sets.

2.2 Topological properties of sets

In this subsection we focus on the topological properties of the different classes of sets used within linear algebra and convex analysis. To start with affine sets one can show the following result. This result can be easily verified using Lemma 1.4 (cf. [46]).

Lemma 1.7 *Any affine set $M \subseteq \mathbb{R}^n$ is closed.*

An important consequence of Lemma 1.7 is given by the following observation. For a given set $S \subseteq \mathbb{R}^n$ let $int(S)$ and $cl(S)$ denote the *interior*, respectively the *closure* of the set S. By Lemma 1.7 we obtain $cl(S) \subseteq aff(S) \subseteq lin(S)$ and this yields by the monotonicity of the hull operation that

$$aff(cl((S))) = aff(S) \text{ and } lin(cl(S)) = lin(S). \tag{1.19}$$

Opposed to affine sets it is not true that convex cones and convex sets are closed. However, as will be shown later, the algebraic property *convexity* and the topological property *closed* are necessary and sufficient to give a so-called *dual representation* of a set. Due to this important representation one needs beforehand easy sufficient conditions on a convex set to be closed. Recall that every affine set can be seen as the *affine hull* of a *finite set of affinely independent vectors* and this property implies that every affine set is closed. By this observation it seems reasonable to consider convex sets which are the *convex hull* of a smaller set and identify which property on the smaller set S one needs to guarantee that the convex set $co(S)$ is closed. Looking at the following counterexample it is not sufficient to impose that the set S is closed and this implies that we need a stronger property on S.

Example 1.3 If $S = \{\mathbf{0}\} \cup \{(x,1) : x \geq 0\}$, then S is closed and its convex hull given by $co(S) = \{(x_1, x_2) : 0 < x_2 \leq 1, x_1 \geq 0\} \cup \{\mathbf{0}\}$ is clearly not closed.

In the above counterexample the closed set S is unbounded and this prevents $co(S)$ to be closed. Imposing now the additional property that the closed set S is bounded or equivalently compact, one can show that $co(S)$ is compact and hence closed. Using relation (1.18) this also yields a way to identify for which sets S the set $cone(S)$ is closed. So finiteness of the generator S for affine sets should be replaced by compactness of S for convex hulls. To prove the next result we first introduce the so-called *unit simplex*

$$\Delta_{n+1} := \{\alpha : \sum_{i=1}^{n+1} \alpha_i = 1 \text{ and } \alpha_i \geq 0\} \subseteq \mathbb{R}^{n+1}.$$

If the function $f : \Delta_{n+1} \times S^{n+1} \to \mathbb{R}^n$ with S^k denoting the k-fold Cartesian product of the set $S \subseteq \mathbb{R}^n$ is given by

$$f(\alpha, \mathbf{x}_1, ..., \mathbf{x}_{n+1}) = \sum_{i=1}^{n+1} \alpha_i \mathbf{x}_i,$$

then by Lemma 1.6 it follows that

$$co(S) = f(\Delta_{n+1} \times S^{n+1}). \tag{1.20}$$

Using relation (1.20) one can now show the following result (cf. [34]).

Lemma 1.8 *If the nonempty set $S \subseteq \mathbb{R}^n$ is compact, then the set $co(S)$ is compact. Moreover, if S is compact and $\mathbf{0}$ does not belong to $co(S)$, then the set $cone(S)$ is closed.*

Proof. It is well known, that the set $\Delta_{n+1} \times S^{n+1}$ is compact (cf. [64]) and this shows by relation (1.20) and f a continuous function that $co(S)$ is compact. To verify the second part we observe by relation (1.18) that $cone(S) = \mathbb{R}_+(co(S))$ and so we need to show that the set $\mathbb{R}_+(co(S))$ is closed. Consider now an arbitrary sequence $t_n\mathbf{x}_n, n \in \mathbb{N}$ belonging to $\mathbb{R}_+(co(S))$ satisfying $\lim_{n\uparrow\infty} t_n\mathbf{x}_n = \mathbf{y}$. This implies $\lim_{n\uparrow\infty} t_n\|\mathbf{x}_n\| = \|\mathbf{y}\|$ and since $\mathbf{0} \notin co(S)$ and $co(S)$ is compact there exists a subsequence $N_0 \subseteq \mathbb{N}$ with $\lim_{n\in N_0\uparrow\infty} \mathbf{x}_n = \mathbf{x}_\infty \in co(S)$ and $\mathbf{x}_\infty \neq \mathbf{0}$. Hence we obtain

$$0 \leq \lim_{n\in N_0\uparrow\infty} t_n = \lim_{n\in N_0\uparrow\infty} \frac{t_n \parallel \mathbf{x}_n \parallel}{\parallel \mathbf{x}_n \parallel} = \frac{\|\mathbf{y}\|}{\parallel \mathbf{x}_\infty \parallel} := t_\infty < \infty.$$

and so $\mathbf{y} = t_\infty \mathbf{x}_\infty \in \mathbb{R}_+(\mathrm{co}(S))$, showing the desired result. □

The following example shows that the condition $\mathbf{0} \notin co(S)$ cannot be omitted in Lemma 1.8.

Example 1.4 If the condition $\mathbf{0} \notin S$ is omitted in Lemma 1.8, then the set $cone(S)$ might not be closed as shown by the following example. Let $S = \{(x_1, x_2) : (x_1 - 1)^2 + x_2^2 \leq 1\}$. Clearly S is compact and $\mathbf{0} \in S$. Moreover, by relation (1.18) it follows that $cone(S) = \{(x_1, x_2) : x_1 > 0\} \cup \{\mathbf{0}\}$ and this set is not closed.

An immediate consequence of Caratheodory's theorem (Lemma 1.6) and Lemma 1.8 is given by the next result for convex cones generated by some nonempty set S.

Lemma 1.9 *If the set $S \subseteq \mathbb{R}^n$ contains a finite number of elements, then the set $cone(S)$ is closed.*

Proof. For the finite set S we consider the finite set $V := \{I : I \subseteq S$ and the set I consists of linearly independent vectors$\}$. By Lemma 1.6 it follows that $cone(S) = \cup_{I \in V} cone(I)$. Since each I belonging to V is a finite set of linearly independent vectors the set I is compact and $\mathbf{0}$ does not belong to $\text{co}(I)$. This shows by Lemma 1.8 that $cone(I)$ is closed for every I belonging to V and since V is a finite set the result follows. □

Next we introduce within a finite dimensional linear space the definition of a relative interior point, generalizing the notion of an interior point. A similar notion can also be defined within a so-called (infinite dimensional) locally convex topological vector space (cf. [58]).

Definition 1.7 *If $E := \{\mathbf{x} \in \mathbb{R}^n : \|\mathbf{x}\| < 1\}$, a vector $\mathbf{x} \in \mathbb{R}^n$ is called a relative interior point of the set $S \subseteq \mathbb{R}^n$ if $\mathbf{x}$ belongs to $aff(S)$ and there exists some $\epsilon > 0$ such that*

$$(\mathbf{x} + \epsilon E) \cap aff(S) \subseteq S.$$

The relative interior $ri(S)$ of any set S is given by $ri(S) := \{\mathbf{x} \in \mathbb{R}^n : \mathbf{x}$ is a relative interior point of $S\}$. The set $S \subseteq \mathbb{R}^n$ is called relatively open if S equals $ri(S)$ and it is called regular if $ri(S)$ is nonempty.

As shown by the next example it is quite natural to assume that $\mathbf{x}$ belongs to $aff(S)$. This assumption implies that $ri(S) \subseteq S$.

Example 1.5 Consider the set $S = \{0\} \times [-1, 1] \subseteq \mathbb{R}^2$ and let $\mathbf{x} = (1, 0)$. Clearly the set $aff(S)$ is given by $\{0\} \times \mathbb{R}$ and for $\epsilon = 1$ it follows that $(\mathbf{x} + E) \cap aff(S) \subseteq S$. If one would delete in the definition of a relative interior point the condition that $\mathbf{x}$ must belong to $aff(S)$, then according to this, the vector $(1, 0)$ would be a relative interior point of the set S. However, the vector $(1, 0)$ is not an element of S and so this definition is not natural.

By Definition 1.7 it is clear for $S \subseteq \mathbb{R}^n$ full dimensional or equivalently $aff(S) = \mathbb{R}^n$ that relative interior means interior and hence relative refers to relative with respect to $aff(S)$. By the same definition, we also obtain that every affine set is relatively open. Moreover, since by Lemma 1.7 the set $aff(S)$ is closed it follows that $cl(S) \subseteq aff(S)$ and so it is useless to introduce closure relative to the affine hull of a given set S. Contrary to the different hull operations the relative interior operator is not a monotone operator. This means that $S_1 \subseteq S_2$ does not imply that $ri(S_1) \subseteq ri(S_2)$.

Example 1.6 If $C_1 = \{0\}$ and $C_2 = [0,1]$, then both sets are convex and $ri(C_1) = \{0\}$ and $ri(C_2) = (0,1)$. This shows $C_1 \subseteq C_2$ and $ri(C_1) \not\subseteq ri(C_2)$.

To guarantee that the relative interior operator is monotone we need to impose the additional condition that $aff(S_1) = aff(S_2)$. If this holds it is easy to check that

$$S_1 \subseteq S_2 \Rightarrow ri(S_1) \subseteq ri(S_2). \tag{1.21}$$

By the above observation it is important to know which different sets cannot be distinguished by the affine hull operator. The next lemma shows that this holds for the sets S, $cl(S)$, $co(S)$ and $cl(co(S))$. This result can be easily verified using $cl(co(S)) \subseteq aff(S)$.

Lemma 1.10 *It follows for every nonempty set $S \subseteq \mathbb{R}^n$ that*

$$aff(S) = aff(cl(S)) = aff(co(S)) = aff(cl(co(S))).$$

By relation (1.21) and Lemma 1.10 we obtain $ri(S) \subseteq ri(cl(S)) \subseteq ri(cl(co(S)))$ and $ri(S) \subseteq ri(co(S))$ for arbitrary sets $S \subseteq \mathbb{R}^n$. Moreover, by relation (1.7) it is easy to verify that

$$ri(S_1 \times S_2) = ri(S_1) \times ri(S_2). \tag{1.22}$$

Since we also like to show $aff(ri(S)) = aff(S)$ an alternative definition of a relative interior point is given by the next lemma.

Lemma 1.11 *If the set $S \subseteq \mathbb{R}^n$ is regular, then the vector $\mathbf{x}$ is a relative interior point of the set S if and only if $\mathbf{x}$ belongs to $aff(S)$ and there exists some $\epsilon > 0$ such that $(\mathbf{x} + \epsilon E) \cap aff(S) \subseteq ri(S)$.*

Proof. We only need to verify the if implication. Let $\mathbf{x}$ be a relative interior point of the set S. This means $\mathbf{x} \in aff(S)$ and there exists some $\epsilon > 0$ such that $(\mathbf{x}+\epsilon E) \cap aff(S) \subseteq S$. Since $\mathbf{x} \in aff(S)$ we obtain that

$(\mathbf{x} + \delta E) \cap aff(S)$ is nonempty for every $\delta > 0$ and so we may consider any point $\mathbf{y}$ belonging to $(\mathbf{x} + \frac{\epsilon}{2}E) \cap aff(S)$. Clearly $\mathbf{y} \in aff(S)$ and $\mathbf{y} + \frac{\epsilon}{2}E \subseteq \mathbf{x} + \epsilon E$ and this shows $(\mathbf{y} + \frac{\epsilon}{2}E) \cap aff(S) \subseteq S$. Hence $\mathbf{y}$ belongs to $ri(S)$ and we have verified that $(\mathbf{x} + \frac{\epsilon}{2}E) \cap aff(S) \subseteq ri(S)$. □

The next result shows for regular sets $S \subseteq \mathbb{R}^n$ that the affine hull operation cannot distinguish the sets $ri(S)$ and S and so this lemma can be seen as an extension of Lemma 1.10.

Lemma 1.12 *If the set $S \subseteq \mathbb{R}^n$ is regular, then it follows that*

$$aff(ri(S)) = aff(S).$$

Proof. It is clear that $aff(ri(S)) \subseteq aff(S)$ and to show the converse inclusion it is sufficient to verify that $S \backslash ri(S) \subseteq aff(ri(S))$. Let $\mathbf{x} \in S \backslash ri(S)$. Since the set S is regular one can find some $\mathbf{y} \in ri(S) \subseteq S$ and so by Lemma 1.11 there exists some $\epsilon > 0$ satisfying

$$(\mathbf{y} + \epsilon E) \cap aff(S) \subseteq ri(S). \tag{1.23}$$

Clearly the set $[\mathbf{y}, \mathbf{x}] := \{(1 - \alpha)\mathbf{y} + \alpha\mathbf{x} : 0 \leq \alpha \leq 1\}$ belongs to $co(S) \subseteq aff(S)$ and this implies by relation (1.23) that $(\mathbf{y} + \epsilon E) \cap [\mathbf{y}, \mathbf{x}] \subseteq ri(S)$. This means that the halfline starting in $\mathbf{y}$ and passing through $\mathbf{x}_1 \in (\mathbf{y} + \epsilon E) \cap [\mathbf{y}, \mathbf{x}] \subseteq ri(S)$ is a subset of $aff(ri(S))$ and contains $\mathbf{x}$. Hence $\mathbf{x}$ belongs to $aff(ri(S))$ and so $S \backslash ri(S) \subseteq aff(ri(S))$. □

An immediate consequence of Lemmas 1.11 and 1.12 is given by the observation that for any regular set $S \subseteq \mathbb{R}^n$ it follows that $\mathbf{x}$ is a relative interior point of S if and only if $\mathbf{x}$ belongs to $aff(ri(S))$ and there exists some $\epsilon > 0$ satisfying $(\mathbf{x} + \epsilon E) \cap aff(ri(S)) \subseteq ri(S)$. This implies for every regular set $S \subseteq \mathbb{R}^n$ that $ri(ri(S)) = ri(S)$, and since by definition $ri(\varnothing) = \varnothing$, we obtain for any set $S \subseteq \mathbb{R}^n$ that

$$ri(ri(S)) = ri(S). \tag{1.24}$$

Keeping in mind the close relationship between affine hulls and convex sets and the observation that nonempty affine sets are regular (in fact $ri(M) = M$!) we might wonder whether convex sets are regular. This is indeed the case as the following result shows (cf. [63]).

Lemma 1.13 *Every nonempty convex set $C \subseteq \mathbb{R}^n$ is regular.*

Although convexity is not a necessary condition for a set to be regular it follows by the definition of a regular set that at least around any relative interior point the set must be "locally" convex. A set, which clearly violates this condition, is the set $\mathbb{Q}$ of rational numbers and this set is therefore not regular. Besides convexity of the set C the proof of Lemma 1.13 uses also that C is a subset of a finite dimensional linear space. If the last condition does not hold and C is an infinite dimensional convex subset of a locally convex topological vector space, then the above result might not hold. We will now list some important properties of relative interiors. To start with this, we first verify the following technical result.

Lemma 1.14 *If $S_1, S_2 \subseteq \mathbb{R}^n$ are nonempty sets, then it follows for every $0 < \alpha < 1$ that*

$$(\alpha S_1 + (1-\alpha)S_2) \cap aff(S_1) \subseteq \alpha S_1 + (1-\alpha)(S_2 \cap aff(S_1)).$$

Proof. Consider for $0 < \alpha < 1$ the vector $\mathbf{y} = \alpha \mathbf{x}_1 + (1-\alpha)\mathbf{x}_2$ with $\mathbf{x}_i \in S_i$, $i = 1, 2$ and $\mathbf{y} \in aff(S_1)$. It is now necessary to verify that $\mathbf{x}_2$ belongs to $S_2 \cap aff(S_1)$. By the definition of $\mathbf{y}$ and $0 < \alpha < 1$ we obtain that

$$\mathbf{x}_2 = \frac{1}{1-\alpha}\mathbf{y} - \frac{\alpha}{1-\alpha}\mathbf{x}_1 \in \frac{1}{1-\alpha} aff(S_1) - \frac{\alpha}{1-\alpha} S_1,$$

and so it follows that $\mathbf{x}_2$ belongs to $aff(S_1)$. Hence the vector $\mathbf{x}_2$ belongs to $S_2 \cap aff(S_1)$ and this shows the desired result. □

Applying now Lemma 1.14, the next important result for convex sets can be shown. This result will play an prominent role in verifying the topological properties of convex sets.

Lemma 1.15 *If $C \subseteq \mathbb{R}^n$ is a nonempty convex set, then it follows for every $0 \leq \alpha < 1$ that*

$$\alpha cl(C) + (1-\alpha) ri(C) \subseteq ri(C).$$

Proof. To prove the above result it is sufficient to show that $\alpha cl(C) + (1-\alpha)\mathbf{x}_2 \subseteq ri(C)$ for any $\mathbf{x}_2 \in ri(C)$ and $0 < \alpha < 1$. Clearly this set is a subset of $aff(C)$ and since $\mathbf{x}_2$ belongs to $ri(C) \subseteq C$ there exists some $\epsilon > 0$ satisfying

$$(\mathbf{x}_2 + \frac{(1+\alpha)\epsilon}{1-\alpha} E) \cap aff(C) \subseteq C. \tag{1.25}$$

Moreover, since $cl(C) \subseteq \cap_{\epsilon > 0}(C + \epsilon E)$ it follows that $cl(C) \subseteq C + \epsilon E$, and this implies

$$\alpha cl(C) + (1-\alpha)\mathbf{x}_2 + \epsilon E \subseteq \alpha C + (1-\alpha)(\mathbf{x}_2 + \frac{(1+\alpha)\epsilon}{1-\alpha} E).$$

Applying now Lemma 1.14 and relation (1.25) we obtain by the convexity of the set C that

$$(\alpha cl(C) + (1-\alpha)\mathbf{x}_2 + \epsilon E) \cap aff(C) \subseteq \alpha C + (1-\alpha)C \subseteq C$$

and this shows the result. □

By Lemmas 1.13 and 1.15 it follows for any nonempty convex set C that the set $ri(C)$ is nonempty and convex. Also, since $cl(C) = \cap_{\epsilon>0}(C + \epsilon E)$ we obtain that $cl(C)$ is a convex set. An easy and important consequence of Lemma 1.15 is given by the observation that the relative interior operator cannot distinguish the convex sets C and $cl(C)$. A similar observation holds for the closure operator applied to the convex sets $ri(C)$ and C. The next result also plays an important role in the proof of the weak separation result to be discussed in Subsection 2.3.

Lemma 1.16 *If $C \subseteq \mathbb{R}^n$ is a nonempty convex set, then it follows that*

$$cl(ri(C)) = cl(C) \text{ and } ri(C) = ri(cl(C)).$$

Proof. To prove the first formula we only need to check that $cl(C) \subseteq cl(ri(C))$. To verify this we consider $\mathbf{x} \in cl(C)$ and select some $\mathbf{y}$ belonging to ri(C). By Lemma 1.15 the half-open line segment $[\mathbf{y}, \mathbf{x})$ belongs to $ri(C)$ and this implies that the vector $\mathbf{x}$ belongs to $cl(ri(C))$. Hence $cl(C) \subseteq cl(ri(C))$ and the first formula is verified. To prove the second formula, it follows immediately by relation (1.21) that $ri(C) \subseteq ri(cl(C))$. To verify $ri(cl(C)) \subseteq ri(C)$ consider an arbitrary $\mathbf{x}$ belonging to $ri(cl(C))$ and so one can find some $\epsilon > 0$ satisfying

$$(\mathbf{x} + \epsilon E) \cap aff(cl(C)) \subseteq cl(C). \tag{1.26}$$

Moreover, since $ri(C)$ is nonempty, construct for some $\mathbf{y} \in ri(C)$ the line $T := \{(1-t)\mathbf{x} + t\mathbf{y} : t \in \mathbb{R}\}$ through the points $\mathbf{x}$ and $\mathbf{y}$. Since $\mathbf{x} \in ri(cl(C))$ and $\mathbf{y} \in ri(C)$ it follows that $T \subseteq aff(cl(C))$ and so by relation (1.26) there exists some $\mu < 0$ satisfying $\mathbf{y}_1 := (1-\mu)\mathbf{x} + \mu\mathbf{y} \in cl(C)$. This shows

$$\mathbf{x} = \frac{1}{1-\mu}\mathbf{y}_1 - \frac{\mu}{1-\mu}\mathbf{y}, \tag{1.27}$$

and since $\mathbf{y}_1 \in cl(C)$ and $\mathbf{y} \in ri(C)$ this implies by Lemma 1.15 and relation (1.27) that $\mathbf{x} \in ri(C)$. Hence it follows that $ri(cl(C)) \subseteq ri(C)$, and this proves the second formula. □

In the above lemma one might wonder whether the convexity of the set C is necessary. In the following example we present a regular set

S with $ri(S)$ and $cl(S)$ convex and S not convex and this set does not satisfy the result of Lemma 1.16.

Example 1.7 Let $S = [0,1] \cup ((1,2] \cap \mathbb{Q})$. This set is clearly not convex and $ri(S) = (0,1)$ while $cl(S) = [0,2]$. Moreover, $ri(cl(S)) \neq ri(S)$ and $cl(ri(S)) \neq cl(S)$.

We will now give a primal representation of the relative interior of a convex set S (cf. [63]).

Lemma 1.17 *If $S \subseteq \mathbb{R}^n$ is a nonempty convex set, then it follows that*

$$ri(S) = \{\mathbf{x} \in \mathbb{R}^n : \forall_{\mathbf{y} \in cl(S)}\ \exists_{\mu<0} \text{ such that } (1-\mu)\mathbf{x} + \mu\mathbf{y} \in S\}.$$

The above result is equivalent to the geometrically obvious fact that for S a convex set and any $\mathbf{x} \in ri(S)$ and $\mathbf{y} \in S$ the line segment $[\mathbf{y}, \mathbf{x}]$ can be extended beyond $\mathbf{x}$ without leaving S. Also, by relation (1.24) and Lemma 1.16 another primal representation of $ri(S)$ with S a convex set is given by

$$ri(S) = \{\mathbf{x} \in \mathbb{R}^n : \forall_{\mathbf{y} \in cl(S)}\ \exists_{\mu<0} \text{ such that } (1-\mu)\mathbf{x} + \mu\mathbf{y} \in ri(S)\}.$$

Since affine mappings preserve convexity it is also of interest to know how the relative interior operator behaves under an affine mapping. Using Lemma 1.17 one can show the next result (cf. [63]).

Lemma 1.18 *If $A : \mathbb{R}^n \to \mathbb{R}^m$ is an affine mapping and $C \subseteq \mathbb{R}^n$ is a nonempty convex set, then it follows that $A(ri(C)) = ri(A(C))$. Moreover, if $C \subseteq \mathbb{R}^m$ is a nonempty convex set satisfying $A^{-1}(ri(C)) := \{\mathbf{x} \in \mathbb{R}^n : A(\mathbf{x}) \in ri(C)\}$ is nonempty, then $ri(A^{-1}(C)) = A^{-1}(ri(C))$.*

As shown by the following counterexample the condition $A^{-1}(ri(C))$ is nonempty cannot be omitted in the previous lemma.

Example 1.8 Let $A : \mathbb{R} \to \mathbb{R}$ given by $A(x) = 1$ for all $x \in \mathbb{R}$ and let $C := [0,1] \subset \mathbb{R}$. Then clearly $ri(C) = (0,1)$, $A^{-1}(ri(C)) = \emptyset$ and $ri(A^{-1}(C)) = \mathbb{R}$.

An immediate consequence of Lemma 1.18 is given by the observation that

$$ri(\alpha S_1 + \beta S_2) = \alpha ri(S_1) + \beta ri(S_2), \tag{1.28}$$

for any $\alpha, \beta \in \mathbb{R}$ and $S_i \subseteq \mathbb{R}^n, i = 1,2$ convex sets. To conclude our discussion on topological properties for sets we finally mention the following result (cf. [63]).

Lemma 1.19 *If the sets C_i, $i \in I$ are convex and $\cap_{i\in I} ri(C_i)$ is nonempty, then it follows that $cl(\cap_{i\in I} C_i) = \cap_{i\in I} cl(C_i)$. Moreover, if the set I is finite, we obtain $ri\,(\cap_{i\in I} C_i) = \cap_{i\in I} ri(C_i)$.*

As shown by the next counterexample it is necessary to assume in Lemma 1.19 that the intersection $\cap_{i\in I} ri(C_i)$ is nonempty.

Example 1.9 Let $C_1 = \{\mathbf{x} \in \mathbb{R}^2 : x_1 > 0, x_2 > 0\} \cup \{\mathbf{0}\}$ and $C_2 = \{\mathbf{x} \in \mathbb{R}^2 : x_2 = 0\}$. It is obvious that $ri(C_1) = \{\mathbf{x} : x_1 > 0, x_2 > 0\}$ and $ri(C_2) = C_2$, and so we obtain $ri(C_1) \cap ri(C_2) = \varnothing$ and $ri(C_1 \cap C_2) \neq ri(C_1) \cap ri(C_2)$. For the same example it is also easy to see that $cl(C_1 \cap C_2) \neq cl(C_1) \cap cl(C_2)$.

In the following counterexample we show that the second result listed in Lemma 1.19 does not hold if the set I is not finite.

Example 1.10 Let $I = (0, \infty)$ and $C_\alpha = [0, 1+\alpha], \alpha > 0$. For this example it follows $ri(\cap_{\alpha>0} C_\alpha) = ri([0,1]) = (0,1)$, and since $ri(C_\alpha) = (0, 1+\alpha)$ for each $\alpha > 0$, we obtain $\cap_{\alpha>0} ri(C_\alpha) = (0,1]$.

This last example concludes our discussion of topological properties of convex sets. In the next subsection we will discuss basic separation results for those sets.

2.3 Separation of convex sets

For a nonempty convex set $C \subseteq \mathbb{R}^n$ consider for any $\mathbf{y} \in \mathbb{R}^n$ the so-called *minimum norm problem* given by

$$v(\mathbf{y}) := \inf\{\| \mathbf{x} - \mathbf{y} \|^2 : \mathbf{x} \in C\}. \tag{$P(\mathrm{y})$}$$

If additionally C is closed, a standard application of the Weierstrass theorem (cf. [64]) shows that for every $\mathbf{y}$ the optimal objective value $v(\mathbf{y})$ in the above optimization problem is attained. To verify that the minimum norm problem has a unique solution, observe for any $\mathbf{z}_1, \mathbf{z}_2$ belonging to $\mathbb{R}^n$ that

$$\|\mathbf{z}_1 + \mathbf{z}_2\|^2 + \|\mathbf{z}_1 - \mathbf{z}_2\|^2 = 2\|\mathbf{z}_1\|^2 + 2\|\mathbf{z}_2\|^2. \tag{1.29}$$

For every $\mathbf{x}_1 \neq \mathbf{x}_2$ belonging to C it follows by relation (1.29) with $\mathbf{z}_i$ replaced by $\mathbf{x}_i - \mathbf{y}$ for $i = 1, 2$ that

$$\|\frac{1}{2}(\mathbf{x}_1 + \mathbf{x}_2) - \mathbf{y}\|^2 < \frac{1}{2}\|\mathbf{x}_1 - \mathbf{y}\|^2 + \frac{1}{2}\|\mathbf{x}_2 - \mathbf{y}\|^2,$$

and so for $\mathbf{x}_i, i = 1, 2$ different optimal solutions of the minimum norm problem $(P(\mathrm{y}))$ we obtain that $\|\frac{1}{2}(\mathbf{x}_1 + \mathbf{x}_2) - \mathbf{y}\|^2 < v(\mathbf{y})$. Since the set C

is convex and hence $\frac{1}{2}(\mathbf{x}_1 + \mathbf{x}_2)$ belongs to C, this yields a contradiction and the optimal solution is therefore unique. Denoting now this optimal solution by $p_C(\mathbf{y})$ one can show the following result (cf. [34]).

Lemma 1.20 *For any $\mathbf{y} \in \mathbb{R}^n$ and $C \subseteq \mathbb{R}^n$ a nonempty closed convex set it follows that*

$$\mathbf{z} = p_C(\mathbf{y}) \Leftrightarrow \mathbf{z} \in C \text{ and } (\mathbf{z} - \mathbf{y})^\intercal(\mathbf{x} - \mathbf{z}) \geq 0 \text{ for every } \mathbf{x} \in C.$$

Moreover, for every $\mathbf{x} \in C$ the triangle inequality

$$\|\mathbf{x} - p_C(\mathbf{y})\|^2 + \|p_C(\mathbf{y}) - \mathbf{y}\|^2 \leq \|\mathbf{x} - \mathbf{y}\|^2$$

holds.

Proof. To show the only if implication we observe that

$$0 \leq (\mathbf{z} - \mathbf{y})^\intercal(\mathbf{x} - \mathbf{z}) = -\|\mathbf{z} - \mathbf{y}\|^2 + (\mathbf{z} - \mathbf{y})^\top(\mathbf{x} - \mathbf{y})$$

and this shows by the Cauchy-Schwarz inequality (cf. [46])

$$0 \leq (\mathbf{z} - \mathbf{y})^\intercal(\mathbf{x} - \mathbf{z}) \leq -\|\mathbf{z} - \mathbf{y}\|^2 + \|\mathbf{z} - \mathbf{y}\|\|\mathbf{x} - \mathbf{y}\| \tag{1.30}$$

for every $\mathbf{x} \in C$. If $\mathbf{y} \in C$ we obtain, substituting $\mathbf{x} = \mathbf{y}$ in relation (1.30), that $0 \leq -\|\mathbf{z} - \mathbf{y}\|^2$ and by the nonnegativity of $\|.\|^2$ this yields $0 = \|\mathbf{z} - \mathbf{y}\|^2$. Also, using $\mathbf{y} \in C$, we obtain $\mathbf{y} = p_C(\mathbf{y})$ and so $\mathbf{z} = \mathbf{y} = p_C(\mathbf{y})$. Moreover, if $\mathbf{y} \notin C$, then $\|\mathbf{z} - \mathbf{y}\| > 0$ and this implies by relation (1.30) that $\|\mathbf{z} - \mathbf{y}\| \leq \|\mathbf{x} - \mathbf{y}\|$ for every $\mathbf{x} \in C$. Hence $\mathbf{z}$ is an optimal solution and by the uniqueness of this solution we obtain $\mathbf{z} = p_C(\mathbf{y})$. To verify the if implication, it follows for $\mathbf{z} = p_C(\mathbf{y})$ that $\mathbf{z} \in C$ and since C is convex this shows

$$\|\mathbf{z} - \mathbf{y}\|^2 \leq \|\alpha\mathbf{x} + (1 - \alpha)\mathbf{z} - \mathbf{y}\|^2 = \|\mathbf{z} - \mathbf{y} + \alpha(\mathbf{x} - \mathbf{z})\|^2 \tag{1.31}$$

for every $\mathbf{x} \in C$ and $0 < \alpha < 1$. Rewriting relation (1.31) we obtain for every $0 < \alpha < 1$ that $2(\mathbf{z} - \mathbf{y})^\top(\mathbf{x} - \mathbf{z}) + \alpha\|\mathbf{x} - \mathbf{z}\|^2 \geq 0$ and letting $\alpha \downarrow 0$ the desired inequality follows. To show the triangle inequality, we observe using $\|\mathbf{z}_1\|^2 - \|\mathbf{z}_2\|^2 =< \mathbf{z}_1 - \mathbf{z}_2, \mathbf{z}_1 + \mathbf{z}_2 >$ for every $\mathbf{z}_1, \mathbf{z}_2$ that

$$\|\mathbf{x} - p_C(\mathbf{y})\|^2 - \|\mathbf{x} - \mathbf{y}\|^2 =< \mathbf{y} - p_C(\mathbf{y}), 2\mathbf{x} - \mathbf{y} - p_C(\mathbf{y}) > .$$

The last term equals $-\|p_C(\mathbf{y}) - \mathbf{y})\|^2 + 2 < \mathbf{y} - p_C(\mathbf{y}), \mathbf{x} - p_C(\mathbf{y}) >$ and applying now the first part yields the desired inequality. □

Actually the above result is nothing else than the first order necessary and sufficient condition for a minimum of a convex function on a closed convex set. We will now prove one of the most fundamental results in convex analysis. This result has an obvious geometric interpretation and serves as a basic tool in deriving dual representations. In infinite dimensional locally convex topological vector spaces the next result is also known as the Hahn-Banach theorem (cf. [65]).

Theorem 1.1 *If $C \subseteq \mathbb{R}^n$ is a nonempty convex set and $\mathbf{y}$ does not belong to the set $cl(C)$, then there exists some nonzero vector $\mathbf{y}_0 \in \mathbb{R}^n$ and $\epsilon > 0$ with $\mathbf{y}_0^\top \mathbf{x} \geq \mathbf{y}_0^\top \mathbf{y} + \epsilon$ for every $\mathbf{x}$ belonging to $cl(C)$. In particular, the vector $\mathbf{y}_0 := p_{cl(C)}(\mathbf{y}) - \mathbf{y}$ satisfies this inequality.*

Proof. By Lemma 1.20 we obtain for every $\mathbf{x} \in cl(C)$ and the nonzero vector $\mathbf{y}_0 := p_{cl(C)}(\mathbf{y}) - \mathbf{y}$ that $\mathbf{y}_0^\top \mathbf{x} \geq {\mathbf{y}_0}^\top p_{cl(C)}(\mathbf{y})$. This shows

$$\mathbf{y}_0^\top \mathbf{x} \geq \|\mathbf{y}_0\|^2 + \mathbf{y}_0^\top \mathbf{y} \tag{1.32}$$

and since $\mathbf{y}_0 \neq \mathbf{0}$ the desired result follows. □

The nonzero vector $\mathbf{y}_0$ belonging to $cl(C) - \mathbf{y}$ is called the *normal vector* of the *separating hyperplane*

$$H^{=}(\mathbf{a}, a) := \{\mathbf{x} \in \mathbb{R}^n : \mathbf{a}^\top \mathbf{x} = a\},$$

$\mathbf{a} = \mathbf{y}_0$ and $a = {\mathbf{y}_0}^\top \mathbf{y} + \frac{\epsilon}{2}$, and this hyperplane strongly separates the closed convex set $cl(C)$ and $\mathbf{y}$. Since $\mathbf{y}_0 \neq \mathbf{0}$ we may take as a normal vector of the hyperplane the vector $\mathbf{y}_0 \|\mathbf{y}_0\|^{-1}$ and this vector has norm 1 and belongs to $cone(cl(C) - \mathbf{y})$.

The strong separation result of Theorem 1.1 can be used to prove the following "weaker" separation result valid under a weaker condition on the point $\mathbf{y}$. Instead of $\mathbf{y}$ does not belong to $cl(C)$ we assume that $\mathbf{y}$ does not belong to $\mathrm{ri}(C)$. By Theorem 1.1 it is clear that we may assume without loss of generality that $\mathbf{y}$ belongs to the *relative boundary* $rbd(C) := cl(C) \backslash ri(C)$ of the convex set $C \subseteq \mathbb{R}^n$.

Theorem 1.2 *If $C \subseteq \mathbb{R}^n$ is a nonempty convex set and $\mathbf{y}$ does not belong to $ri(C)$, then there exists some nonzero vector $\mathbf{y}_0$ belonging to the unique linear subspace $L_{aff(C)}$ satisfying $\mathbf{y}_0^\top \mathbf{x} \geq \mathbf{y}_0^\top \mathbf{y}$ for every $\mathbf{x} \in C$. Moreover, for the vector $\mathbf{y}_0$ there exists some $\mathbf{x}_0 \in C$ such that $\mathbf{y}_0^\top \mathbf{x}_0 > \mathbf{y}_0^\top \mathbf{y}$.*

Proof. Consider for every $n \in \mathbb{N}$ the set $(\mathbf{y} + n^{-1}E) \cap aff(cl(C))$. By Lemma 1.16 it follows that $\mathbf{y}$ does not belong to $ri(cl(C))$ and so there exists some vector $\mathbf{y}_n$ satisfying

$$\mathbf{y}_n \notin cl(C) \text{ and } \mathbf{y}_n \in (\mathbf{y} + n^{-1}E) \cap aff(cl(C)). \tag{1.33}$$

The set $cl(C)$ is a closed convex set and by relation (1.33) and Theorem 1.1 one can find some vector $\mathbf{y}_n^* \in \mathbb{R}^n$ satisfying

$$\| \mathbf{y}_n^* \| = 1,\ \mathbf{y}_n^* \in cone(cl(C) - \mathbf{y}_n) \subseteq L_{aff(C)} \text{ and } \mathbf{y}_n^{*\intercal}\mathbf{x} \geq \mathbf{y}_n^{*\intercal}\mathbf{y}_n \quad (1.34)$$

for every $\mathbf{x} \in cl(C)$. The sequence $\{\mathbf{y}_n^* : n \in \mathbb{N}\}$ belongs to a compact set and so there exists a convergent subsequence $\{\mathbf{y}_n^* : n \in N_0\}$ with

$$\lim_{n \in N_0 \uparrow \infty} \mathbf{y}_n^* = \mathbf{y}_0. \quad (1.35)$$

This implies by relations (1.33), (1.34) and (1.35) that

$$\mathbf{y}_0^\intercal \mathbf{x} = \lim_{n \in N_0 \uparrow \infty} \mathbf{y}_n^{*\intercal}\mathbf{x} \geq \lim_{n \in N_0 \uparrow \infty} \mathbf{y}_n^{*\top} \mathbf{y}_n = \mathbf{y}_0^\intercal \mathbf{y} \quad (1.36)$$

for every $\mathbf{x} \in cl(C)$ and

$$\mathbf{y}_0 \in L_{aff(C)} \text{ and } \| \mathbf{y}_0 \| = 1. \quad (1.37)$$

Suppose now that there does not exist some $\mathbf{x}_0 \in C$ satisfying $\mathbf{y}_0^\intercal \mathbf{x}_0 > \mathbf{y}_0^\intercal \mathbf{y}$. By relation (1.36) this implies that $\mathbf{y}_0^\intercal(\mathbf{x} - \mathbf{y}) = 0$ for every $\mathbf{x} \in C$ and since $\mathbf{y}$ belongs to $\text{cl}(C) \subseteq \text{aff}(C)$ we obtain by relation (1.4) and Lemma 1.3 that $\mathbf{y}_0^\intercal \mathbf{z} = 0$ for every $\mathbf{z}$ belonging to $L_{aff(C)}$. Since by relation (1.37) the vector $\mathbf{y}_0$ belongs to $L_{aff(C)}$ this implies $\| \mathbf{y}_0 \|^2 = 0$ and so we contradict $\| \mathbf{y}_0 \| = 1$. Hence it must follow that there exists some $\mathbf{x}_0 \in C$ satisfying $\mathbf{y}_0^\intercal \mathbf{x}_0 > \mathbf{y}_0^\top \mathbf{y}$ and this proves the desired result.

□

The separation of Theorem 1.2 is called a *proper separation* between the set C and the vector $\mathbf{y}$. One can also introduce proper separation between two convex sets.

Definition 1.8 *The convex sets $C_1, C_2 \subseteq \mathbb{R}^n$ are called properly separated if there exist some $\mathbf{y}_0 \in \mathbb{R}^n$ satisfying*

$$\inf_{x \in C_1} \mathbf{y}_0^\top \mathbf{x} \geq \sup_{\mathbf{x} \in C_2} \mathbf{y}_0^\top \mathbf{x} \text{ and } \mathbf{y}_0^\top \mathbf{x}_1 > \mathbf{y}_0^\top \mathbf{x}_2$$

for some $\mathbf{x}_1 \in C_1$ and $\mathbf{x}_2 \in C_2$.

An immediate consequence of Theorem 1.2 is given by the next result.

Theorem 1.3 *If the convex sets $C_1, C_2 \subseteq \mathbb{R}^n$ satisfy $ri(C_1) \cap ri(C_2) = \varnothing$, then the two sets can be properly separated.*

Proof. By relation (1.28) we obtain for $\alpha = 1$ and $\beta = -1$ that $ri(C_1 - C_2) = ri(C_1) - ri(C_2)$, and this shows $ri(C_1) \cap ri(C_2) = \varnothing$ if and only if

$\mathbf{0} \notin ri(C_1 - C_2)$. Applying now Theorem 1.2 with $\mathbf{y} = \mathbf{0}$ and the convex set given by $C_1 - C_2$, the result follows. □

The above separation results are the corner stones of convex and quasiconvex analysis. Observe in infinite dimensional locally convex topological vector spaces one can show similar separation results under stronger assumptions on the convex sets C_1 and C_2 (cf. [65],[17],[58]). An easy consequence of the separation results is given by the observation that closed convex sets and relatively open convex sets are evenly convex. These convex sets play an important role in duality theory for quasiconvex functions.

Lemma 1.21 *If the nonempty convex set $C \subseteq \mathbb{R}^n$ is closed or relatively open, then C is evenly convex.*

Proof. If $C = \mathbb{R}^n$, the result follows by definition and so we may suppose that the closed set C is a proper subset of $\mathbb{R}^n$. Hence there exists some $\mathbf{y} \notin C$ and this implies by Theorem 1.1 that there exists some $\mathbf{a} \in \mathbb{R}^n$ and $b \in \mathbb{R}$ satisfying $C \subseteq H^{<}(\mathbf{a}, b)$. This shows that the set $\mathcal{H}_C$ of all open halfspaces H satisfying $C \subseteq H$ is nonempty and by the definition of $\mathcal{H}_C$ it is clear that $C \subseteq \cap\{H : H \in \mathcal{H}_C\}$. Again by Theorem 1.1 one can show using contradiction that C equals $\cap\{H : H \in \mathcal{H}_C\}$ and this shows that every closed convex set is evenly convex. To verify the second result, we observe $\mathcal{H}_{\text{cl}(C)} \subseteq \mathcal{H}_C$ and since $\mathcal{H}_{\text{cl}(C)}$ is nonempty by the first part, it follows that $\mathcal{H}_C$ is nonempty and $C \subseteq \cap\{H : H \in \mathcal{H}_C\}$. To show that $C = \cap\{H : H \in \mathcal{H}_C\}$ we assume by contradiction that there exists some $\mathbf{y} \notin C$ with $\mathbf{y} \in H$ for every $H \in \mathcal{H}_C$. Due to $\mathbf{y} \notin C$ it follows by Theorem 1.2 that there exists some nonzero $\mathbf{y}_0 \in L_{aff(C)}$ satisfying

$$\mathbf{y}_0^\top \mathbf{x} \geq \mathbf{y}_0^\top \mathbf{y} \tag{1.38}$$

for every $\mathbf{x} \in C$. Since the convex set C is relatively open there exists for every $\mathbf{x} \in C$ some $\epsilon > 0$ satisfying $\mathbf{x} - \epsilon\mathbf{y}_0 \in C$ and so by relation (1.38) we obtain for every $\mathbf{x} \in C$ that $\mathbf{y}_0^\top \mathbf{x} = \mathbf{y}_0^\top(\mathbf{x} - \epsilon\mathbf{y}_0) + \epsilon\|\mathbf{y}_0\|^2 > \mathbf{y}_0^\top \mathbf{y}$. Hence the open halfspace $H^{<}(\mathbf{a}, b) := \{\mathbf{x} \in \mathbb{R}^n : \mathbf{a}^\top \mathbf{x} < b\}$ with $\mathbf{a} := -\mathbf{y}_0$ and $b := -\mathbf{y}_0^\top \mathbf{y}$ belongs to $\mathcal{H}_C$ and since $\mathbf{y} \notin H^{<}(\mathbf{a}, b)$ this contradicts $\mathbf{y} \in H$ for every H belonging to $\mathcal{H}_C$. □

This concludes our discussion of separation results of convex sets. In the next subsection we will use these separation results to derive dual representations for convex sets.

2.4 Dual representations of convex sets

In contrast to the primal representation of a linear subspace, affine set, convex cone and convex set discussed in Subsection 2.1 we can also give a so-called *dual representation* of these sets. From a geometrical point of view a primal representation is a representation from "within" the set, while a dual representation turns out to be a representation from "outside" the set. Such a characterization can be seen as an improvement of the hull operation given by relations (1.2), (1.3), (1.16) and (1.17). We start with linear subspaces or affine sets (cf. [47]).

Definition 1.9 *If $S \subseteq \mathbb{R}^n$ is some nonempty set, then the nonempty set $S^{\perp} \subseteq \mathbb{R}^n$ given by $S^{\perp} := \{\mathbf{x}^* \in \mathbb{R}^n : \mathbf{x}^{\intercal}\mathbf{x}^* = 0 \text{ for every } \mathbf{x} \in S\}$ is called the orthogonal complement of the set S.*

It is easy to verify that the orthogonal complement $S^{\perp}$ of the set S is a linear subspace. Moreover, a basic result (cf. [47]) in linear algebra is given by the following.

Lemma 1.22 *For any linear subspace L it follows that $(L^{\perp})^{\perp} = L$.*

By Lemma 1.22 a so-called dual representation of any linear hull $lin(S)$ with S nonempty can be constructed. Using $S \subseteq lin(S)$ it follows by Lemma 1.22 that $(S^{\perp})^{\perp} \subseteq (lin(S)^{\perp})^{\perp} = lin(S)$. Since $lin(S)$ is the smallest linear subspace containing S and $(S^{\perp})^{\perp}$ is clearly a linear subspace containing S the previous inclusion implies

$$(S^{\perp})^{\perp} = lin(S). \tag{1.39}$$

The alternative representation of $lin(S)$ in relation (1.39) is called a dual representation. To construct a dual representation for an affine hull we observe by Lemma 1.3 and the dual representation of a linear hull that $aff(S) = \mathbf{x}_0 + ((S - \mathbf{x}_0)^{\perp})^{\perp}$ for $\mathbf{x}_0$ belonging to $aff(S)$. Since it is easy to verify that $(S - \mathbf{x}_0)^{\perp} = (S - \mathbf{x}_1)^{\perp}$ for every $\mathbf{x}_1 \in aff(S)$ we obtain for affine hulls the dual representation

$$aff(S) = \mathbf{x}_0 + ((S - \mathbf{x}_1)^{\perp})^{\perp} \tag{1.40}$$

for every $\mathbf{x}_0, \mathbf{x}_1 \in aff(S)$.

Next we discuss the dual representation of a closed convex set containing $\mathbf{0}$ and a closed convex cone. This dual representation will be verified by means of the strong separation result listed in Theorem 1.1. Recall first the definition of a support function.

Definition 1.10 *If $S \subseteq \mathbb{R}^n$ is some nonempty set, then the function $\sigma_S : \mathbb{R}^n \to (-\infty, \infty]$ given by $\sigma_S(\mathbf{s}) := \sup\{\mathbf{s}^\intercal \mathbf{x} : \mathbf{x} \in S\}$ is called the support function of the set S.*

An equivalent formulation of Theorem 1.1 involving the support function of the closed convex set C is given by the following result.

Theorem 1.4 *If $C \subseteq \mathbb{R}^n$ is a proper nonempty convex set, then it follows that $\mathbf{x}_0 \in cl(C)$ if and only if $\mathbf{s}^\top \mathbf{x}_0 \leq \sigma_{cl(C)}(\mathbf{s})$ for every $\mathbf{s} \in \mathbb{R}^n$.*

Proof. Clearly $\mathbf{x}_0 \in cl(C)$ implies that $\mathbf{s}^\intercal \mathbf{x}_0 \leq \sigma_{cl(C)}(\mathbf{s})$ for every $\mathbf{s}$ belonging to $\mathbb{R}^n$. To show the reverse implication let $\mathbf{s}^\intercal \mathbf{x}_0 \leq \sigma_{cl(C)}(\mathbf{s})$ for every $\mathbf{s} \in \mathbb{R}^n$ and suppose by contradiction that $\mathbf{x}_0 \notin cl(C)$. By Theorem 1.1 there exists some nonzero vector $\mathbf{y}_0 \in \mathbb{R}^n$ and $\epsilon > 0$ satisfying $-\mathbf{y}_0^\top \mathbf{x} \leq -\mathbf{y}_0^\top \mathbf{x}_0 - \epsilon$ for every $\mathbf{x}$ belonging to $cl(C)$. This implies $\sigma_{cl(C)}(-\mathbf{y}_0) \leq -\mathbf{y}_0^\top \mathbf{x}_0 - \epsilon < -\mathbf{y}_0^\top \mathbf{x}_0$, contradicting our initial assumption and so it must follow that $\mathbf{x}_0$ belongs to $cl(C)$. □

To generalize the dual representation of linear subspaces in Lemma 1.22 to the larger class of closed convex sets containing $\mathbf{0}$ we need to generalize the orthogonality relation given in Definition 1.9.

Definition 1.11 *If $S \subseteq \mathbb{R}^n$ is a nonempty set, then the set S^0, given by $S^0 := \{\mathbf{x}^* \in \mathbb{R}^n : \mathbf{x}^\intercal \mathbf{x}^* \leq 1 \text{ for every } \mathbf{x} \in S\}$, is called the polar of the set S. Moreover, the bipolar S^{00} of the set S is defined by $S^{00} := (S^0)^0$.*

The polar S^0 of a nonempty set $S \subseteq \mathbb{R}^n$ is a nonempty closed convex set and satisfies $S^0 = (cl(S))^0$. If the nonempty set $K \subseteq \mathbb{R}^n$ is a convex cone, then it is easy to show that $K^0 = \{\mathbf{x}^* \in \mathbb{R}^n : \mathbf{x}^\intercal \mathbf{x}^* \leq 0 \text{ for every } \mathbf{x} \in K\}$ and K^0 is a closed convex cone, while for L a linear subspace it follows that $L^0 = L^\perp$. Hence the *polar* operator applied to a linear subspace reduces to the *orthogonal* operator and can therefore be seen as a generalization of this operator. To prove a generalization of Lemma 1.22 it is convenient to introduce the so-called Minkowski functional (cf. [65]). Recall in the next definition that $\inf\{\emptyset\} := \infty$.

Definition 1.12 *The Minkowski functional or gauge of the nonempty set $S \subseteq \mathbb{R}^n$ is given by the function $\gamma_S : \mathbb{R}^n \to [0, \infty]$ defined by*

$$\gamma_S(\mathbf{s}) := \inf\{t > 0 : \mathbf{s} \in tS\}.$$

As shown by the next result the support function of any set S containing the zero vector $\mathbf{0}$ equals the gauge of the closed convex polar S^0.

Lemma 1.23 *If $S \subseteq \mathbb{R}^n$ is a nonempty set containing $\mathbf{0}$, then it follows that $\sigma_{cl(S)}(\mathbf{s}) = \gamma_{S^0}(\mathbf{s})$ for every $\mathbf{s} \in \mathbb{R}^n$.*

Proof. Since $\mathbf{0}$ belongs to $cl(S)$, it follows that the support function $\sigma_{cl(S)}$ of the set $cl(S)$ is nonnegative. Consider now the following two cases. If $\sigma_{cl(S)}(\mathbf{s}_0) = 0$, we obtain $t^{-1}\mathbf{s}_0^\top \mathbf{x} \leq 0$ for every $t > 0$ and $\mathbf{x} \in S$. This shows $t^{-1}\mathbf{s}_0 \in S^0$ for every $t > 0$ and so $\gamma_{S^0}(\mathbf{s}_0) = 0 = \sigma_{cl(S)}(\mathbf{s}_0)$. Moreover, if $\sigma_{cl(S)}(\mathbf{s}_0) > 0$, we obtain using $\sigma_{cl(S)} = \sigma_S$ that

$$0 < \sigma_S(\mathbf{s}_0) = \inf\{t > 0 : \mathbf{s}_0^\top \mathbf{x} \leq t, \mathbf{x} \in S\} = \inf\{t > 0 : \frac{\mathbf{s}_0}{t} \in S^0\}$$

and this shows the desired result. □

Finally we can prove the so-called *bipolar* theorem for closed convex sets containing $\mathbf{0}$, generalizing Lemma 1.22. This representation can be seen as a so-called dual representation of a closed convex set containing $\mathbf{0}$.

Theorem 1.5 *If $C \subseteq \mathbb{R}^n$ is a nonempty convex set with $\mathbf{0} \in cl(C)$, then it follows that $C^{00} = cl(C)$.*

Proof. It is obvious that $cl(C) \subseteq C^{00}$ and so we only need to verify the reverse inclusion. Since for any $\mathbf{s} \in \mathbb{R}^n$ satisfying $\gamma_{C^0}(\mathbf{s}) < \infty$ it follows that

$$(\gamma_{C^0}(\mathbf{s}) + \epsilon)^{-1}\mathbf{s} \in C^0$$

for every $\epsilon > 0$, we obtain for every $\mathbf{x}_0 \in C^{00}$ that $\mathbf{s}^\top \mathbf{x}_0 \leq \gamma_{C^0}(\mathbf{s}) + \epsilon$. This implies $\mathbf{s}^\top \mathbf{x}_0 \leq \gamma_{C^0}(\mathbf{s})$ and since this inequality trivially holds for $\gamma_{C^0}(\mathbf{s}) = \infty$ we obtain by Lemma 1.23 that $\mathbf{s}^\top \mathbf{x}_0 \leq \sigma_{cl(C)}(\mathbf{s})$ for every $\mathbf{s}$. Applying now Theorem 1.4 shows $\mathbf{x}_0 \in cl(C)$ and we have checked that $C^{00} \subseteq cl(C)$. □

By a similar approach as used after Lemma 1.22 it is easy to construct a dual representation of the convex set $co(S \cup \{\mathbf{0}\})$ with S a nonempty set. First we observe by the definition of the polar operator and using Theorem 1.5 that $S^{00} \subseteq (co(S \cup \{\mathbf{0}\})^{00} = cl(co(S \cup \{\mathbf{0}\}))$. Since S^{00} is a closed convex set containing $S \cup \{\mathbf{0}\}$ and $cl(co(S \cup \{\mathbf{0}\}))$ is the smallest closed convex set containing $S \cup \{\mathbf{0}\}$ we obtain by the previous inclusion the general formula

$$S^{00} = cl(co(S \cup \{\mathbf{0}\}\}). \tag{1.41}$$

The formula, listed in relation (1.41), is called the *bipolar theorem* for arbitrary sets $S \subseteq \mathbb{R}^n$. Replacing Theorem 1.4 by its equivalent version valid in locally convex topological vector spaces one can verify using a similar proof the bipolar theorem (cf. [10], [37]) in locally convex topological vector spaces.

An important special case of Theorem 1.5 is given by $K^{00} = cl(K)$ with K a convex cone. By means of similar proof techniques (cf. [67]) it is also possible to give a dual representation of the relative interior $ri(K)$ of a convex cone K. Without proof we now list the following result. For related results, valid in infinite dimensional topological vector spaces, the reader should consult [38].

Theorem 1.6 *For any nonempty convex cone $K \subseteq \mathbb{R}^n$ it follows that*

$$\mathbf{x} \in ri(K) \Leftrightarrow \mathbf{x} \in (K^{\perp})^{\perp} \text{ and } \mathbf{x}^{*\top}\mathbf{x} < 0 \text{ for } \mathbf{x}^* \in K^0 \cap (K^{\perp})^{\perp} \backslash \{0\}.$$

This concludes our section on sets. In the next section we will consider functions studied within convex and quasiconvex analysis.

3. Functions studied within convex and quasiconvex analysis

In this section we first introduce in Subsection 3.1 the different classes of functions studied within convex and quasiconvex analysis and derive their algebraic properties. These algebraic properties are an easy consequence of two important relations between functions and sets and the properties of sets derived in Subsection 2.1. Also from Subsection 2.1 we know how to apply hull operations to sets and using this it is also possible to construct so-called hull functions. These different hull functions are also introduced in Subsection 3.1 and their properties will be derived. In Subsection 3.2 topological properties of functions are introduced together with some of the "topological" hull functions. It will turn out that especially the class of lower semicontinuous functions is extremely important in this field. Finally in Subsections 3.3 and 3.4 dual characterizations of the considered functions will be derived. The key results in these sections are the Fenchel-Moreau theorem within convex analysis and its generalization to the so-called evenly quasiconvex and lower semicontinuous quasiconvex functions.

3.1 Algebraic properties of functions

In this subsection we relate functions to sets and use the algebraic properties of sets given in Subsection 2.1 to derive algebraic properties of functions. To start with this approach, let $f : \mathbb{R}^n \to [-\infty, \infty]$ be an extended real valued function and associate with f its so-called *epigraph*

$$epi(f) := \{(\mathbf{x}, r) \in \mathbb{R}^{n+1} : f(\mathbf{x}) \leq r\}. \tag{1.42}$$

A related set is the strict *epigraph*

$$\widetilde{epi}(f) := \{(\mathbf{x}, r) \in \mathbb{R}^{n+1} : f(\mathbf{x}) < r\}. \tag{1.43}$$

Within convex analysis it is now useful to represent a function f by the obvious relation (cf. [63])

$$f(\mathbf{x}) = \inf\{r : (\mathbf{x}, r) \in epi(f)\}. \tag{1.44}$$

By definition $\inf\{\emptyset\} = \infty$ and this only happens if the vector $\mathbf{x}$ does not belong to the so-called *effective domain*

$$dom(f) := \{\mathbf{x} \in \mathbb{R}^n : f(\mathbf{x}) < \infty\} \tag{1.45}$$

of the function f. By this observation it follows that $dom(f)$ is nonempty if and only if $epi(f)$ is nonempty and if this holds we obtain

$$dom(f) = A(epi(f)) \tag{1.46}$$

with A the projection of $\mathbb{R}^{n+1}$ onto $\mathbb{R}^n$ given by $A(\mathbf{x}, r) = \mathbf{x}$. As shown by the following definition, the representation of the function f given by relation (1.44) is useful in the study of convex functions.

Definition 1.13 *The function $f : \mathbb{R}^n \to [-\infty, \infty]$ is called convex if the set $epi(f)$ is convex. Moreover, the function $f : \mathbb{R}^n \to [-\infty, \infty]$ is called positively homogeneous if the set $epi(f)$ is a cone.*

An equivalent definition of a convex function is given by the next result, which is easy to verify.

Lemma 1.24 *A function $f : \mathbb{R}^n \to [-\infty, \infty]$ is convex if and only if the set $\widetilde{epi}(f)$ is convex.*

Using Lemma 1.24 we obtain that a function $f : \mathbb{R}^n \to [-\infty, \infty]$ is convex if and only if for every $0 < \alpha < 1$

$$f(\alpha\mathbf{x}_1 + (1-\alpha)\mathbf{x}_2) < \alpha r_1 + (1-\alpha) r_2 \tag{1.47}$$

whenever $f(\mathbf{x}_i) < r_i \in \mathbb{R}$. In case we know additionally that $f > -\infty$ we obtain by relation (1.44) that f is convex if and only if for every $0 < \alpha < 1$

$$f(\alpha\mathbf{x}_1 + (1-\alpha)\mathbf{x}_2) \leq \alpha f(\mathbf{x}_1) + (1-\alpha) f(\mathbf{x}_2) \tag{1.48}$$

and so we recover the more familiar definition of a convex function. An important special case satisfying relation (1.48) is given by $f > -\infty$ and

$dom(f)$ is nonempty. If this holds the function f is called *proper.* Also the next result is easy to verify.

Lemma 1.25 *The function* $f : \mathbb{R}^n \to [-\infty, \infty]$ *is positively homogeneous if and only if* $f(\alpha \mathbf{x}) = \alpha f(\mathbf{x})$ *for every* $\mathbf{x} \in \mathbb{R}^n$ *and* $\alpha > 0$.

To investigate under which operations on convex functions this property is preserved we observe for any collection of functions f_i, $i \in I$ that

$$epi(\sup{}_{i\in I} f_i) = \cap_{i\in I} epi(f_i). \tag{1.49}$$

Since the intersection of convex sets is again convex we obtain by relation (1.49) that the function $\sup_{i\in I} f_i$ is convex if f_i is convex for every $i \in I$. Moreover, by relation (1.48), it follows that any strict canonical combination of the convex functions $f_i > -\infty$, $i = 1, 2$ is again convex.

In case we use the representation of a function f, given by relation (1.44), and the various hull operations on a set defined in Subsection 2.1 it is easy to introduce the various so-called hull functions of f. The first hull function is given by the next definition (cf. [63]). In this volume the various hull functions, given in this subsection and the next, are also discussed by Crouzeix (cf. [11]).

Definition 1.14 *For any function* $f : \mathbb{R}^n \to [-\infty, \infty]$ *the function* $f_c : \mathbb{R}^n \to [-\infty, \infty]$, *given by* $f_c(\mathbf{x}) := \inf\{r : (\mathbf{x}, r) \in co(epi(f))\}$, *is called the convex hull function of the function* f.

The next result yields an interpretation of the convex hull function of a function f. Recall that the convex hull of the empty set is again the empty set.

Lemma 1.26 *For any function* $f : \mathbb{R}^n \to [-\infty, \infty]$ *the convex hull function* f_c *is the greatest convex function majorized by* f. *Moreover, it follows that* $\widetilde{epi}(f_c)) \subseteq co(epi(f)) \subseteq epi(f_c)$ *and* $dom(f_c) = co(dom(f))$.

Proof. Without loss of generality we may assume that $epi(f)$ or equivalently $dom(f)$ is nonempty. Since $co(epi(f))$ is a convex set we obtain by Definition 1.14 for every $r_i > f_c(\mathbf{x}_i), i = 1, 2$ that $f_c(\alpha\mathbf{x}_1 + (1-\alpha)\mathbf{x}_2) < \alpha r_1 + (1-\alpha)r_2$ for every $0 < \alpha < 1$. This shows by relation (1.47) that the function f_c is convex. Moreover, if $h \leq f$ and h is convex, then $co(epi(f)) \subseteq co(epi(h)) = epi(h)$ and so f_c is the greatest convex function majorized by f. Using again Definition 1.14 it is also easy to verify that $\widetilde{epi}(f_c) \subseteq co(epi(f)) \subseteq epi(f_c)$. To show the last part of this lemma, let $\mathbf{x} \in dom(f_c)$ and so $(\mathbf{x}, r) \in co(epi(f))$ for every $r > f_c(\mathbf{x})$. This implies by relation (1.46) that $\mathbf{x} \in A(co(epi(f)) = co(A(epi(f)) =$

$co(dom(f))$ with A the projection of $\mathbb{R}^{n+1}$ onto $\mathbb{R}^n$ and we have verified $dom(f_c) \subseteq co(dom(f))$. Also, for $\mathbf{x} \in co(dom(f))$, we obtain by relation (1.46) that $\mathbf{x} \in A(co(epi(f)))$ and so $\mathbf{x}$ belongs to $dom(f_c)$ showing the reverse inclusion. □

In general it follows that $epi(f_c) \neq co(epi(f))$. A direct consequence of Lemma 1.26 and the fact that $\sup_{i \in I} f_i$ is convex for f_i, $i \in I$ a collection of convex functions, is the often used representation of the function f_c given by

$$f_c(\mathbf{x}) = \sup\{h(\mathbf{x}) : h \leq f \text{ and } h \text{ is a convex function}\}. \tag{1.50}$$

Next to the epigraph of a function $f : \mathbb{R}^n \to [-\infty, \infty]$ one also considers the so-called *lower-level set* $L(f, r), r \in \mathbb{R}$ of the function f given by

$$L(f, r) := \{\mathbf{x} \in \mathbb{R}^n : f(\mathbf{x}) \leq r\}. \tag{1.51}$$

A related set is the *strict lower-level set* of the function f of level r represented by

$$\widetilde{L}(f, r) := \{\mathbf{x} \in \mathbb{R}^n : f(\mathbf{x}) < r\}. \tag{1.52}$$

Within quasiconvex analysis it is now useful to represent a function f by the obvious relation (cf. [15])

$$f(\mathbf{x}) = \inf\{r : \mathbf{x} \in L(f, r)\}. \tag{1.53}$$

As shown by the following definition, the representation of the function f, given by relation (1.53), is useful in the study of quasiconvex functions.

Definition 1.15 *The function $f : \mathbb{R}^n \to [-\infty, \infty]$ is called quasiconvex if for every $r \in \mathbb{R}$ the lower-level set $L(f, r)$ is convex. Moreover, the function f is called evenly quasiconvex if for every $r \in \mathbb{R}$ the lower level set $L(f, r)$ is evenly convex.*

To derive the relation between convex and quasiconvex functions we observe that $epi(f) \cap (\mathbb{R}^n \times \{r\}) = L(f, r) \times \{r\}$ for every $r \in \mathbb{R}$. This implies that a convex function is also a quasiconvex function. Since each monotonic (increasing or decreasing) function $f : \mathbb{R} \to \mathbb{R}$ is quasiconvex, but not necessarily convex, the converse is not true. For quasiconvex functions a similar result as in Lemma 1.24 can be easily verified.

Lemma 1.27 *A function $f : \mathbb{R}^n \to [-\infty, \infty]$ is quasiconvex if and only if the set $\widetilde{L}(f, r)$ is convex for every $r \in \mathbb{R}$.*

To recover a more familiar representation of a quasiconvex function it can be shown easily (cf. [2]) that a function $f : \mathbb{R}^n \to [-\infty, \infty]$ is quasiconvex if and only if $f(\alpha \mathbf{x}_1 + (1-\alpha)\mathbf{x}_2) \leq \max\{f(\mathbf{x}_1), f(\mathbf{x}_2)\}$, for every $0 < \alpha < 1$.

As for convex functions, one is interested under which operations on quasiconvex functions this property is preserved. Clearly for any collection of functions $f_i, i \in I$ it follows that

$$L(\sup{}_{i\in I} f_i, r) = \cap_{i \in I} L(f_i, r) \tag{1.54}$$

and this shows that the function $\sup_{i\in I} f_i$ is quasiconvex if f_i is quasiconvex for every $i \in I$. Opposed to convex functions, it is not true that a strict canonical combination of quasiconvex functions is quasiconvex and this is shown by the following example.

Example 1.11 Let $f_i : \mathbb{R} \to \mathbb{R}, i = 1, 2$ be given by $f_1(x) = x$ and

$$f_2(x) = x^2 \text{ for } |x| \leq 1 \text{ and } f_2(x) = 1 \text{ otherwise.}$$

These functions are quasiconvex, but it is easy to verify by means of a picture that the sum of the two functions is not quasiconvex.

Using relation (1.53), one can apply the different hull operations to the lower level set. The first hull function constructed in this way is listed in the next definition (cf. [15], [11]).

Definition 1.16 *For any function* $f : \mathbb{R}^n \to [-\infty, \infty]$ *the function* $f_q : \mathbb{R}^n \to [-\infty, \infty]$, *given by* $f_q(\mathbf{x}) := \inf\{r : \mathbf{x} \in co(L(f, r))\}$, *is called the quasiconvex hull function of the function* f.

The next result (cf. [15]) yields an interpretation of the quasiconvex hull function of a function f.

Lemma 1.28 *For any function* $f : \mathbb{R}^n \to [-\infty, \infty]$ *the quasiconvex hull function* f_q *is the greatest quasiconvex function majorized by* f. *Moreover, it follows that* $L(f_q, r) = \cap_{\beta > r} co(L(f, \beta))$ *for every* $r \in \mathbb{R}$.

Proof. Again we may assume without loss of generality that $dom(f)$ is nonempty. By Definition 1.16 it follows that $L(f_q, r) \subseteq \cap_{\beta > r} co(L(f, \beta))$. Since it is obvious that the reverse inclusion holds, we obtain $L(f_q, r) = \cap_{\beta > r} co(L(f, \beta))$. By this relation it is clear that the function f_q is quasiconvex and applying a similar argument as in Lemma 1.26 to lower level sets it can be shown that this function is the greatest quasiconvex function majorized by the function f. □

A direct consequence of Lemma 1.28 and the fact that $\sup_{i\in I} f_i$ is quasiconvex for f_i, $i \in I$ a collection of quasiconvex functions, is the often used representation of f_q given by

$$f_q(\mathbf{x}) = \sup\{h(\mathbf{x}) : h \leq f \text{ and } h \text{ is a quasiconvex function}\}. \quad (1.55)$$

To conclude this subsection, we consider a hull function based on evenly convex sets (cf. [55], [11]). It will turn out that this function plays an important role in duality theory for quasiconvex functions.

Definition 1.17 *For any function $f : \mathbb{R}^n \to [-\infty, \infty]$ the function $f_{ec} : \mathbb{R}^n \to [-\infty, \infty]$, given by $f_{ec}(\mathbf{x}) := \inf\{r : \mathbf{x} \in ec(L(f, r))\}$, is called the evenly quasiconvex hull function of the function f.*

As done for the quasiconvex hull function one can show by a similar proof the following result (cf. [55]).

Lemma 1.29 *For any function $f : \mathbb{R}^n \to [-\infty, \infty]$ the evenly quasiconvex hull function f_{ec} is the greatest evenly quasiconvex function majorized by f. Moreover, it follows that $L(f_{ec}, r) = \cap_{\beta > r} ec(L(f, \beta))$ for every $r \in \mathbb{R}$.*

A direct consequence of Lemma 1.29 and the fact that $\sup_{i \in I} f_i$ is evenly quasiconvex for f_i, $i \in I$ a collection of evenly quasiconvex functions, is the often used representation of f_{ec} given by

$$f_{ec}(\mathbf{x}) = \sup\{h(\mathbf{x}) : h \leq f, \ h \text{ evenly quasiconvex function}\}. \quad (1.56)$$

Since an evenly quasiconvex function is clearly a quasiconvex function it holds that $f_{ec} \leq f_q$. This concludes our discussion of algebraic properties of convex and quasiconvex functions. In the next subsection we will consider topological properties of functions.

3.2 Topological properties of functions

In this subsection we first introduce the class of lower semicontinuous functions. These functions play an important role within the theory of convex functions.

Definition 1.18 *If $f : \mathbb{R}^n \to [-\infty, \infty]$ is some function, then this function is called lower semicontinuous at $\mathbf{x} \in \mathbb{R}^n$ if $\liminf_{\mathbf{y} \to \mathbf{x}} f(\mathbf{y}) = f(\mathbf{x})$ with*

$$\liminf_{\mathbf{y} \to \mathbf{x}} f(\mathbf{y}) := \sup_{\epsilon > 0} \inf\{f(\mathbf{y}) : \mathbf{y} \in \mathbf{x} + \epsilon E\}. \quad (1.57)$$

Moreover, the function $f : \mathbb{R}^n \to [-\infty, \infty]$ is called upper semicontinuous at $\mathbf{x} \in \mathbb{R}^n$ if the function $-f$ is lower semicontinuous at $\mathbf{x}$ and it is *called continuous at* $\mathbf{x}$ *if it is both lower and upper semicontinuous at* $\mathbf{x}$. *The function* $f : \mathbb{R}^n \to [-\infty, \infty]$ *is called lower semicontinuous (upper semicontinuous) if* f *is lower semicontinuous (upper semicontinuous) at every* $\mathbf{x} \in \mathbb{R}^n$ *and it is called continuous if it is both upper and lower semicontinuous.*

We mostly abbreviate lower semicontinuous by l.s.c.. To relate the above definition of liminf to the liminf of a sequence we observe for every sequence $\mathbf{y}_k, k \in \mathbb{N}$ that $\liminf_{k\uparrow\infty} f(\mathbf{y}_k) := \lim_{n\uparrow\infty} \inf_{k\geq n} f(\mathbf{y}_k)$. Using this definition one can easily show the following result.

Lemma 1.30 *The function* $f : \mathbb{R}^n \to [-\infty, \infty]$ *is l.s.c. at* $\mathbf{x} \in \mathbb{R}^n$ *if and only if* $\liminf_{k\uparrow\infty} f(\mathbf{y}_k) \geq f(\mathbf{x})$ *for every sequence* $\mathbf{y}_k, k \in \mathbb{N}$ *satisfying* $\lim_{k\uparrow\infty} \mathbf{y}_k = \mathbf{x} \in \mathbb{R}^n$.

Using Lemma 1.30 the following important characterization of l.s.c. functions can be proved (cf. [63], [1]).

Theorem 1.7 *If* $f : \mathbb{R}^n \to [-\infty, \infty]$ *is an extended real valued function, then the following conditions are equivalent:*

1. *The function* f *is l.s.c..*
2. *The set epi*(f) *is closed.*
3. *The set* $L(f, r)$ *is closed for every* $r \in \mathbb{R}$.

It is useful to know under which operations on l.s.c. functions this property is preserved. Since $epi(\sup_{i\in I} f_i) = \cap_{i\in I} epi(f_i)$ and the intersection of closed sets is again a closed set we obtain by Theorem 1.7 that the function $\sup_{i\in I} f_i$ is l.s.c. if each function $f_i, i \in I$ is l.s.c.. Also it follows for every finite set I that $epi(\min_{i\in I} f_i) = \cup_{i\in I} epi(f_i)$ and this shows by Theorem 1.7 and the fact that a finite union of closed sets is closed, that the function $\min_{i\in I} f_i$ is l.s.c. if each f_i, $i \in I$ is l.s.c.. Finally, for arbitrary functions $f_i : \mathbb{R}^n \to (-\infty, \infty], i = 1, 2$ we obtain that

$$L(f_1 + f_2, r)^c = \cup_{q\in\mathbb{Q}}(L(f_1, r - q)^c \cup L(f_2, q)^c)$$

with A^c denoting the complement of the set $A \subseteq \mathbb{R}^n$ and this implies using Theorem 1.7 that the function $\alpha f_1 + \beta f_2$ is l.s.c. for every $\alpha, \beta \geq 0$, if the functions $f_i > -\infty, i = 1, 2$ are l.s.c..

To verify the next theorem we introduce for any function $f : \mathbb{R}^n \to [-\infty, \infty]$ the (possibly empty) set of continuous real valued minorants $\mathcal{C}_f$ of f given by

$$\mathcal{C}_f := \{h : h \leq f \text{ and } h \text{ is a real valued continuous function}\}.$$

In the next result it is now shown that any l.s.c. function can be seen as a pointwise limit of an increasing sequence of real valued continuous functions.

Theorem 1.8 *For any function $f : \mathbb{R}^n \to (-\infty, \infty]$ the following conditions are equivalent:*

1 The function f is l.s.c..

2 There exists an increasing sequence of continuous functions $(h_m)_{m \in \mathbb{N}}$ satisfying $f(\mathbf{x}) = \lim_{m \uparrow \infty} h_m(\mathbf{x})$ for every $\mathbf{x} \in \mathbb{R}^n$.

3 $f(\mathbf{x}) = \sup\{h(\mathbf{x}) : h \in \mathcal{C}_f\}$ with $\mathcal{C}_f$ nonempty.

Proof. We only give a proof of $1 \Rightarrow 2$ since the other implications are obvious. We first show the desired result for a nonnegative uniformly bounded function f. Actually, if the function f is nonnegative and uniformly bounded, then the sequence $f_m : \mathbb{R}^n \to [0, \infty), m \in \mathbb{N}$ given by $f_m(\mathbf{x}) := \inf\{f(\mathbf{z}) + m\|\mathbf{x} - \mathbf{z}\| : \mathbf{z} \in \mathbb{R}^n\}$ is increasing, converges pointwise to $f(\mathbf{x})$ and each f_m is continuous (actually Lipschitz continuous with Lipschitz constant m!). To reduce the general case of a proper l.s.c. function f to this special case, replace the proper l.s.c. function f by the nonnegative uniformly bounded l.s.c. function $g = k \circ f$, where $k(x) = \frac{1}{2}\pi + \arctan(x)$ and apply the first part. Hence there exists an increasing sequence g_m of continuous functions converging pointwise to g. Use now that the function $k : (-\infty, \infty] \to (0, \pi]$ is one-to-one, strictly increasing and continuous with a continuous inverse $k^{\leftarrow}$and select the sequence $h_m := k^{\leftarrow} \circ g_m$. □

By Theorem 1.8 we obtain that the set of l.s.c. functions is the smallest set of functions, which are closed under taking the sup operation to any collection of functions belonging to this set and which contain the set of continuous real valued functions.

As in the previous subsection, we are going to introduce hull operations related to functions. In this case topological properties will be involved. First we consider the so-called l.s.c. hull function of a function f (cf. [63], [11]).

Definition 1.19 *For any function $f : \mathbb{R}^n \to [-\infty, \infty]$ the function $\overline{f} : \mathbb{R}^n \to [-\infty, \infty]$, given by $\overline{f}(\mathbf{x}) := \inf\{r : (\mathbf{x}, r) \in cl(epi(f))\}$, is called the l.s.c. hull function of the function f.*

In the next result an interpretation of the l.s.c. hull function of a function f is given.

Lemma 1.31 *For any function $f : \mathbb{R}^n \to [-\infty, \infty]$ the l.s.c. hull function $\overline{f}$ is the greatest l.s.c. function majorized by f. Moreover, its epigraph equals $cl(epi(f))$ and $dom(f) \subseteq dom(\overline{f}) \subseteq cl(dom(f))$. If additionally $dom(f)$ is a convex set, it follows that $ri(dom(\overline{f})) = ri(dom(f))$.*

Proof. By Definition 1.19 we obtain $(\mathbf{x}, r) \in epi(\overline{f}) \Leftrightarrow \forall_{\epsilon>0}\ (\mathbf{x}, r+\epsilon) \in cl(epi(f)) \Leftrightarrow (\mathbf{x}, r) \in cl(epi(f))$. This means that $epi(\overline{f})$ equals $cl(epi(f))$ and by Theorem 1.7 the function $\overline{f}$ is l.s.c.. Moreover, if $h \leq f$ and h is l.s.c., then by Theorem 1.7 we obtain $cl(epi(f)) \subseteq cl(epi(h)) = epi(h)$ and so it follows that $h \leq \overline{f}$. To verify the last part we may assume without loss of generality that $dom(f)$ is nonempty. Since $\overline{f} \leq f$ it follows that $dom(f) \subseteq dom(\overline{f})$ and by relation (1.46) we obtain $dom(\overline{f}) = A(cl(epi(f)) \subseteq cl(A(epi(f)) = cl(dom(f))$. Finally, if $dom(f)$ is a nonempty convex set it follows by Lemma 1.16 that $ri(dom(f)) = ri(cl(dom(f)))$ and since $dom(f) \subseteq dom(\overline{f}) \subseteq cl(dom(f))$ we obtain $ri(dom(\overline{f})) = ri(dom(f))$. □

A direct consequence of Lemma 1.31 and the fact that $\sup_{i\in I} f_i$ is l.s.c. for f_i, $i \in I$ a collection of l.s.c. functions, is the often used representation of $\overline{f}$ given by

$$\overline{f}(\mathbf{x}) = \sup\{h(\mathbf{x}) : h \leq f \text{ and } h \text{ is a l.s.c. function }\}. \tag{1.58}$$

For nondecreasing functions $f : \mathbb{R} \to [-\infty, \infty]$ it is possible to give a more detailed description of the l.s.c. hull function $\overline{f}$ of f. To show this result we first introduce the next definition.

Definition 1.20 *For any function $f : \mathbb{R} \to [-\infty, \infty]$ the function $f^{\diamond} : \mathbb{R} \to [-\infty, \infty]$ is given by*

$$f^{\diamond}(t) := \sup_{s<t} f(s).$$

The next result is needed in the proof of a dual representation of a l.s.c. quasiconvex function.

Lemma 1.32 *For any nondecreasing function $f : \mathbb{R} \to [-\infty, \infty]$ it follows that $\overline{f}(t) = f^{\diamond}(t)$ for every $t \in \mathbb{R}$.*

Proof. Since the function f is nondecreasing, it is easy to verify that $f^{\diamond}$ is nondecreasing and $f^{\diamond} \leq f$. We now verify that the function $f^{\diamond}$ is l.s.c. and so by Theorem 1.7 we need to check that the lower-level set $L(f^{\diamond}, r)$ is closed for every $r \in \mathbb{R}$. Assume now by contradiction that there exists some $r_0 \in \mathbb{R}$ such that the set $L(f^{\diamond}, r_0)$ is not closed. Hence there exists a sequence $\{t_n : n \in \mathbb{N}\} \subseteq L(f^{\diamond}, r_0)$ with $t_\infty := \lim_{n\uparrow\infty} t_n$ and t_∞ does not belong to $L(f^{\diamond}, r_0)$. Since $f^{\diamond}$ is nondecreasing and $f^{\diamond}(t_\infty) > r_0$ it follows that $t_n < t_\infty$ for every $n \in \mathbb{N}$ and by Definition 1.20 one can find some $s_0 < t_\infty$ satisfying $f(s_0) > r_0$. This implies that there exists some t_n satisfying $s_0 < t_n < t_\infty$ and so $f^\diamond(t_n) \geq f(s_0) > r_0$ contradicting t_n belongs to $L(f^\diamond, r_0)$. Therefore $f^\diamond$ is l.s.c. and using $f^\diamond \leq f$ it follows by relation (1.58) that $f^\diamond \leq \overline{f}$. Suppose now by contradiction that $f^\diamond(t_0) < \overline{f}(t_0)$ for some t_0. By relation (1.57) and $\overline{f}$ is l.s.c. this implies that there exists some $\epsilon > 0$ satisfying $\overline{f}(t) > f^\diamond(t_0)$ for every $t_0 - \epsilon \leq t \leq t_0 + \epsilon$ and so

$$f(t_0 - \epsilon) \geq \overline{f}(t_0 - \epsilon) > f^\diamond(t_0) \geq f(t_0 - \epsilon).$$

This yields a contradiction and the result is proved. □

The next result relates $\overline{f}$ to f and this result is nothing else than a "function value translation" of the original definition of the l.s.c. hull function $\overline{f}$ of f.

Lemma 1.33 *For any function $f : \mathbb{R}^n \to [-\infty, \infty]$ and $\mathbf{x} \in \mathbb{R}^n$ it follows that $\overline{f}(\mathbf{x}) = \liminf_{\mathbf{y}\to\mathbf{x}} f(\mathbf{y})$.*

Proof. Since $\overline{f} \leq f$ and $\overline{f}$ is a l.s.c. function we obtain that $\overline{f}(\mathbf{x}) = \liminf_{\mathbf{y}\to\mathbf{x}} \overline{f}(\mathbf{y}) \leq \liminf_{\mathbf{y}\to\mathbf{x}} f(\mathbf{y})$. Suppose now by contradiction that $\overline{f}(\mathbf{x}) < \liminf_{\mathbf{y}\to\mathbf{x}} f(\mathbf{y})$. If this holds, then clearly $\overline{f}(\mathbf{x}) < \infty$ and by the definition of liminf there exists some finite γ and $\epsilon > 0$ satisfying $f(\mathbf{x}+\mathbf{y}) > \gamma > \overline{f}(\mathbf{x})$ for every $\mathbf{y} \in \epsilon E$. This implies that the open set $(\mathbf{x}+\epsilon E) \times (-\infty, \gamma)$ containing the point $(\mathbf{x}, \overline{f}(\mathbf{x}))$ has an empty intersection with epi(f). However, by Lemma 1.31 it follows that $(\mathbf{x}, \overline{f}(\mathbf{x}))$ belongs to *cl*(*epi*(f)) and so every open set containing $(\mathbf{x}, \overline{f}(\mathbf{x}))$ must have a nonempty intersection with *epi*(f). Hence we obtain a contradiction and so the result is proved. □

By Lemma 1.33 and Definition 1.18 it follows immediately that

$$f \text{ is l.s.c. at } \mathbf{x} \Leftrightarrow \overline{f}(\mathbf{x}) = f(\mathbf{x}). \tag{1.59}$$

Using Theorem 1.7 and Lemmas 1.31 and 1.33 one can show that the l.s.c. hull operation applied to functions preserves the convexity and quasiconvexity property.

Lemma 1.34 *If the function $f : \mathbb{R}^n \to [-\infty, \infty)$ is convex (quasiconvex), then also the l.s.c. hull function $\overline{f}$ of f is convex (quasiconvex).*

Proof. If the function f is convex, then *epi*(f) is a convex set and hence also *cl*(*epi*(f)) is a convex set. Since by Lemma 1.31 the epigraph of $\overline{f}$ is given by *cl*(*epi*(f)) this shows that $\overline{f}$ is a convex function. To verify that $\overline{f}$ is quasiconvex for f quasiconvex we need to verify by Lemma 1.27 that the set $\widetilde{L}(\overline{f}, r)$ is convex for every $r \in \mathbb{R}$. If the vectors $\mathbf{x}_i, i = 1, 2$ belong to $\widetilde{L}(\overline{f}, r)$ it follows by Lemma 1.33 that

$$\inf\{f(\mathbf{y}) : \mathbf{y} \in \mathbf{x}_i + \epsilon E\} \leq \overline{f}(\mathbf{x}_i) < r$$

for every $\epsilon > 0$ and $i = 1, 2$. This implies for every $\epsilon > 0$ and $i = 1, 2$ that there exists some vector $\mathbf{y}_{i,\epsilon} \in \mathbf{x}_i + \epsilon E$ satisfying

$$f(\mathbf{y}_{i,\epsilon}) \leq r_0 := \frac{1}{2}(\max\{\overline{f}(\mathbf{x}_1), \overline{f}(\mathbf{x}_2)\} + r) < r$$

Applying now the quasiconvexity of the function f we obtain for every $0 < \alpha < 1$ that

$$f(\alpha\mathbf{y}_{1,\epsilon} + (1 - \alpha)\mathbf{y}_{2,\epsilon}) \leq \max\{f(\mathbf{y}_{1,\epsilon}), f(\mathbf{y}_{2,\epsilon})\} \leq r_0$$

and since the vector $\alpha\mathbf{y}_{1,\epsilon} + (1 - \alpha)\mathbf{y}_{2,\epsilon}$ belongs to the set $\alpha\mathbf{x}_1 + (1 - \alpha)\mathbf{x}_2 + \epsilon E$ this yields

$$\inf\{f(\mathbf{y}) : \mathbf{y} \in \alpha\mathbf{x}_i + (1 - \alpha)\mathbf{x}_2 + \epsilon E\} \leq r_0 < r$$

for every $\epsilon > 0$. Using again Lemma 1.33 we obtain

$$\overline{f}(\alpha\mathbf{x}_1 + (1 - \alpha)\mathbf{x}_2) = \liminf_{\mathbf{y} \to \alpha\mathbf{x}_1 + (1-\alpha)\mathbf{x}_2} f(\mathbf{y}) \leq r_0 < r$$

and it follows that $\alpha\mathbf{x}_1 + (1 - \alpha)\mathbf{x}_2$ belongs to $\widetilde{L}(\overline{f}, r)$. □

To improve Lemma 1.33 for convex functions f we need to give a representation of the relative interior of the epigraph of a convex function. This representation is an immediate consequence of the following observation. If $f : \mathbb{R}^n \to [-\infty, \infty]$ is a convex function and $f(\mathbf{x})$ is finite for some $\mathbf{x}$, then clearly $\{\mathbf{x}\} \times (f(\mathbf{x}), \infty) = ri(\{\mathbf{x}\} \times [f(\mathbf{x}), \infty))$ and so

$$\{\mathbf{x}\} \times (f(\mathbf{x}), \infty) = ri((\{\mathbf{x}\} \times \mathbb{R}) \cap epi(f)). \tag{1.60}$$

A similar observation also holds for $f(\mathbf{x}) = -\infty$ and this shows that relation (1.60) is valid for every $\mathbf{x} \in dom(f)$. Also by relation (1.46) and Lemma 1.18 we obtain

$$ri(dom(f)) = ri(A(epi(f)) = A(ri(epi(f))) \tag{1.61}$$

with $A : \mathbb{R}^{n+1} \to \mathbb{R}^n$ the projection on $\mathbb{R}^n$ and so it follows by relation (1.61) that

$$\mathbf{x} \in ri(dom(f)) \Leftrightarrow (\{\mathbf{x}\} \times \mathbb{R}) \cap ri(epi(f)) \neq \varnothing. \tag{1.62}$$

Since the set $\{\mathbf{x}\} \times \mathbb{R}$ is affine and therefore relatively open we obtain by relation (1.62) that the conditions of Lemma 1.19 hold and hence by relation (1.60) we obtain

$$\{\mathbf{x}\} \times (f(\mathbf{x}), \infty) = (\{\mathbf{x}\} \times \mathbb{R}) \cap ri(epi(f)) \tag{1.63}$$

for every $\mathbf{x} \in ri(dom(f))$. Using this equality the next representation is easy to verify.

Lemma 1.35 *If the function $f : \mathbb{R}^n \to [-\infty, \infty]$ is convex and $dom(f)$ is nonempty, then the set $ri(epi(f))$ is nonempty and*

$$ri(epi(f)) = \{(\mathbf{x}, r) \in \mathbb{R}^{n+1} : f(\mathbf{x}) < r, \mathbf{x} \in ri(dom(f))\}.$$

Proof. If $\mathbf{x}$ belongs to $ri(dom(f))$ and $f(\mathbf{x}) < r$ it follows by relation (1.63) that $(\mathbf{x}, r) \in ri(epi(f))$. To show the reverse inclusion we proceed as follows. If $(\mathbf{x}, r)$ belongs to $ri(epi(f))$ then by relation (1.61) we obtain $\mathbf{x} \in ri(dom(f))$. Applying now relation (1.63) yields $f(\mathbf{x}) < r$. □

In case f is a convex function with $dom(f)$ nonempty, the result of Lemma 1.33 can be improved as follows.

Lemma 1.36 *If the function $f : \mathbb{R}^n \to [-\infty, \infty]$ is convex and $dom(f)$ is nonempty, then $\overline{f}(\mathbf{x}) = \lim_{t \downarrow 0} f(\mathbf{x}+t(\mathbf{y}-\mathbf{x}))$ for every $\mathbf{y} \in ri(dom(f))$. Moreover, if $\mathbf{x} \in ri(dom(f))$, then it follows that $\overline{f}(\mathbf{x}) = f(\mathbf{x})$.*

Proof. By Lemma 1.33 it is obvious that $\overline{f}(\mathbf{x}) \leq \liminf_{t \downarrow 0} f(\mathbf{x}+t(\mathbf{y}-\mathbf{x}))$. If $\overline{f}(\mathbf{x}) = \infty$ then the result holds by the previous inequality and so we assume $\overline{f}(\mathbf{x}) < \infty$. This implies that $(\mathbf{x}, r) \in epi(\overline{f}) = cl(epi(f))$ for every $r > \overline{f}(\mathbf{x})$ and since $\mathbf{y} \in ri(dom(f))$ it follows by Lemma 1.35 that $(\mathbf{y}, r_1) \in ri(epi(f))$ for every $r_1 > f(\mathbf{y})$. Applying now Lemma 1.15 we obtain for every $0 < t < 1$ that $((1-t)\mathbf{x}+t\mathbf{y}, (1-t)r+tr_1) \in epi(f)$ and this shows $f(\mathbf{x} + t(\mathbf{y} - \mathbf{x})) = f(t\mathbf{y} + (1-t)\mathbf{x}) \leq tr_1 + (1-t)r$. Hence it follows that $\limsup_{t \downarrow 0} f(\mathbf{x} + t(\mathbf{y} - \mathbf{x})) \leq r$ and since $r > \overline{f}(\mathbf{x})$ we obtain $\limsup_{t \downarrow 0} f(\mathbf{x}+t(\mathbf{y}-\mathbf{x})) \leq \overline{f}(\mathbf{x})$. This proves the first part and to verify the second part we first observe that the convex set $dom(f)$ is nonempty and so by Lemma 1.13 the set $ri(dom(f))$ is nonempty. By Lemma 1.31 and 1.35 and $\overline{f}$ is convex it now follows that

$$\{(\mathbf{x}, r) \in \mathbb{R}^{n+1} : \overline{f}(\mathbf{x}) < r, \mathbf{x} \in ri(dom(f))\} = ri(cl(epi(f))).$$

This implies using Lemma 1.16 and f is convex that

$$\{(\mathbf{x}, r) \in \mathbb{R}^{n+1} : \overline{f}(\mathbf{x}) < r, \mathbf{x} \in ri(dom(f))\} \subseteq epi(f), \tag{1.64}$$

and by contradiction we obtain $\overline{f}(\mathbf{x}) \geq f(\mathbf{x})$ for every $\mathbf{x} \in ri(dom(f))$. Since always $\overline{f}(\mathbf{x}) \leq f(\mathbf{x})$ the proof is completed. □

We now introduce the most important hull function used within the field of convex analysis (cf. [63], [11]).

Definition 1.21 *For any function $f : \mathbb{R}^n \to [-\infty, \infty]$ the function $f_{\overline{c}} : \mathbb{R}^n \to [-\infty, \infty]$, given by $f_{\overline{c}}(\mathbf{x}) := \inf\{r : (\mathbf{x}, r) \in cl(co(epi(f)))\}$, is called the l.s.c. convex hull function of the function f.*

Using now a similar approach as in Lemma 1.31 one can prove the following result.

Lemma 1.37 *For any function $f : \mathbb{R}^n \to [-\infty, \infty]$ the l.s.c. convex hull function $f_{\overline{c}}$ is the greatest l.s.c. convex function majorized by f. Moreover, it follows that $epi(f_{\overline{c}}) = cl(co(epi(f)))$, $dom(f_c) \subseteq dom(f_{\overline{c}})) \subseteq cl(dom(f_c))$ and $ri(dom(f_c)) = ri(dom(f_{\overline{c}}))$.*

A direct consequence of Lemma 1.37 and the fact that $\sup_{i \in I} f_i$ is a l.s.c. convex function for f_i, $i \in I$ a collection of l.s.c. convex functions, is the often used representation of $f_{\overline{c}}$ given by

$$f_{\overline{c}}(\mathbf{x}) = \sup\{h(\mathbf{x}) : h \leq f \text{ and } h \text{ is a l.s.c. convex function}\}. \tag{1.65}$$

To relate the various hull functions based on relation (1.44) we observe by Lemmas 1.26 and 1.34 that the function $\overline{f_c}$ is convex and l.s.c. Since $\overline{f_c} \leq \overline{f} \leq f$ this shows by Lemma 1.37 that $\overline{f_c} \leq f_{\overline{c}}$. Also by Lemmas 1.26 and 1.37 it holds that the l.s.c. function $f_{\overline{c}}$ is bounded from above by f_c. This implies by Lemma 1.31 that $f_{\overline{c}} \leq \overline{f_c}$ and combining both inequalities yields

$$f_{\overline{c}}(\mathbf{x}) = \overline{f_c}(\mathbf{x}) \tag{1.66}$$

for every $\mathbf{x} \in \mathbb{R}^n$. An immediate consequence of relation (1.66) is now given by the chain of inequalities

$$f_{\overline{c}}(\mathbf{x}) \leq f_c(\mathbf{x}) \leq f(\mathbf{x}) \text{ and } f_{\overline{c}}(\mathbf{x}) \leq \overline{f}(\mathbf{x}) \leq f(\mathbf{x}). \tag{1.67}$$

for every $\mathbf{x} \in \mathbb{R}^n$. We finally consider hull functions based on the lower level set (cf. [15],[11]).

Definition 1.22 *For any function $f : \mathbb{R}^n \to [-\infty, \infty]$ the function $f_{\overline{q}} : \mathbb{R}^n \to [-\infty, \infty]$, given by $f_{\overline{q}}(\mathbf{x}) := \inf\{r : \mathbf{x} \in cl(co(L(f, r)))\}$, is called the l.s.c. quasiconvex hull function of the function f.*

Using a similar approach as in Lemma 1.28 one can show the following result.

Lemma 1.38 *For any function $f : \mathbb{R}^n \to [-\infty, \infty]$ the l.s.c. quasiconvex hull function $f_{\overline{q}}$ is the greatest l.s.c. quasiconvex function majorized by f. Moreover, it follows that $L(f_{\overline{q}}, r) = \cap_{\beta > r} cl(co(L(f, \beta)))$ for every $r \in \mathbb{R}$.*

A direct consequence of Lemma 1.38 and the fact that $\sup_{i \in I} f_i$ is a l.s.c. quasiconvex function for $f_i, i \in I$ a collection of l.s.c. quasiconvex functions, is the often used representation of $f_{\overline{q}}$ given by

$$f_{\overline{q}}(\mathbf{x}) = \sup\{h(\mathbf{x}) : h \leq f \text{ and } h \text{ is a l.s.c. quasiconvex function}\}. \tag{1.68}$$

To relate the various hull functions based on relation (1.53) we first observe by Lemma 1.21 that every closed convex set is evenly convex and so it follows that

$$f_{\overline{q}}(\mathbf{x}) \leq f_{ec}(\mathbf{x}) \leq f_q(\mathbf{x}) \leq f(\mathbf{x}). \tag{1.69}$$

for every $\mathbf{x} \in \mathbb{R}^n$. Moreover, using $\overline{f_q} \leq \overline{f} \leq f$, relation (1.68) and Lemma 1.34 we obtain $\overline{f_q} \leq f_{\overline{q}}$ and since by relation (1.69) and Lemma 1.38 also $f_{\overline{q}} = \overline{f_{\overline{q}}} \leq \overline{f_{ec}} \leq \overline{f_q}$, this finally yields

$$f_{\overline{q}}(\mathbf{x}) = \overline{f_{ec}}(\mathbf{x}) = \overline{f_q}(\mathbf{x}) \tag{1.70}$$

for every $\mathbf{x} \in \mathbb{R}^n$. The above representations of the hull functions do not depend on the fact that the domain is finite dimensional and so we can also introduce the same hull functions in linear topological vector spaces (cf. [56]). In the next two subsections we consider the dual representations of some of the hull functions.

3.3 Dual representations of convex functions

In this subsection we will consider in detail properties of convex functions, which can be derived using the strong and weak separation results for nonempty convex sets. In particular, we will discuss a dual representation of a l.s.c. convex function f satisfying $f > -\infty$. As always in mathematics one likes to approximate complicated functions by simpler functions. For convex functions these simpler functions are given by the so-called affine minorants.

Definition 1.23 *For any function $f : \mathbb{R} \to [-\infty, \infty]$ the affine function $a : \mathbb{R}^n \to \mathbb{R}$, given by $a(\mathbf{x}) = \mathbf{a}^\intercal \mathbf{x} + \alpha$, with $\mathbf{a} \in \mathbb{R}^n$ and $\alpha \in \mathbb{R}$ is called an affine minorant of the function f if $f(\mathbf{x}) \geq a(\mathbf{x})$ for every $\mathbf{x}$ belonging to $\mathbb{R}^n$. Moreover, the possibly empty set of affine minorants of the function f is denoted by $\mathcal{A}_f$.*

Since any affine minorant a of a function f is continuous and convex it is easy to verify the following result.

Lemma 1.39 *For any function* $f : \mathbb{R}^n \to [-\infty, \infty]$ *it follows that* $\mathcal{A}_f = \mathcal{A}_{f_c} = \mathcal{A}_{f_{\overline{c}}} = \mathcal{A}_{\overline{f_c}}$.

Proof. We only give a proof of the above result for $\mathcal{A}_f$ nonempty. Since by relations (1.67) and (1.66) we know that $f_{\overline{c}} = \overline{f_c} \leq f_c \leq f$ it follows immediately that $\mathcal{A}_{f_{\overline{c}}} = \mathcal{A}_{\overline{f_c}} \subseteq \mathcal{A}_{f_c} \subseteq \mathcal{A}_f$. Moreover, if the function a belongs to $\mathcal{A}_f$, then clearly $a \leq f$ and a is continuous and convex. This implies by relation (1.65) that $a \leq f_{\overline{c}}$ and hence the affine function a belongs to $\mathcal{A}_{f_{\overline{c}}}$. □

Since an affine function is always finite valued the set $\mathcal{A}_f$ is empty if there exists some $\mathbf{x} \in \mathbb{R}^n$ satisfying $f(\mathbf{x}) = -\infty$ and so it is necessary to consider functions $f : \mathbb{R}^n \to (-\infty, \infty]$. In Theorem 1.9 necessary and sufficient conditions are given for $\mathcal{A}_f$ to be nonempty. To prove this result we first need to verify the next important lemma.

Lemma 1.40 *If* $f : \mathbb{R} \to [-\infty, \infty]$ *is an arbitrary function and* $f_{\overline{c}}(\mathbf{x}_0)$ *is finite for some* $\mathbf{x}_0$, *then the set* $\mathcal{A}_f$ *is nonempty.*

Proof. It follows that the vector $(\mathbf{x}_0, f_{\overline{c}}(\mathbf{x}_0)-1)$ does not belong to the set $epi(f_{\overline{c}})$. By Lemma 1.37 the nonempty set $epi(f_{\overline{c}})$ is convex and closed and applying Theorem 1.1, there exists some nonzero vector $(\mathbf{y}_0, \beta)$ satisfying

$$\mathbf{y}_0^\top \mathbf{x} + \beta r > \mathbf{y}_0^\top \mathbf{x}_0 + \beta(f_{\overline{c}}(\mathbf{x}_0) - 1)$$

for every $(\mathbf{x},r) \in epi(f_{\overline{c}})$. Since $(\mathbf{x}_0, f_{\overline{c}}(\mathbf{x}_0))$ belongs to $epi(f_{\overline{c}})$ this implies $\beta > 0$ and so for every $(\mathbf{x},r) \in epi(f_{\overline{c}})$ the inequality

$$r > -\beta^{-1}\mathbf{y}_0^\top(\mathbf{x} - \mathbf{x}_0) + f_{\overline{c}}(\mathbf{x}_0) - 1 \tag{1.71}$$

holds. By relation (1.71) it follows by contradiction that $f_{\overline{c}}(\mathbf{x}) > -\infty$ for every $\mathbf{x} \in dom(f_{\overline{c}})$ and this yields using $dom(f) \subseteq dom(f_{\overline{c}})$ that $(\mathbf{x}, f_{\overline{c}}(\mathbf{x})) \in epi(f_{\overline{c}})$ for every $\mathbf{x} \in dom(f)$. Substituting this into relation (1.71) we obtain

$$f(\mathbf{x}) \geq f_{\overline{c}}(\mathbf{x}) > -\beta^{-1}\mathbf{y}_0^\top(\mathbf{x} - \mathbf{x}_0) + f_{\overline{c}}(\mathbf{x}_0) - 1$$

for every $\mathbf{x} \in dom(f)$. Since the previous inequality trivially holds for $\mathbf{x} \notin dom(f)$ the function $a(\mathbf{x}) := -\beta^{-1}\mathbf{y}_0^\top(\mathbf{x} - \mathbf{x}_0) + f_{\overline{c}}(\mathbf{x}_0) - 1$ is an affine minorant of f and the desired result is proved. □

Using Lemma 1.40 one can show the following theorem.

Theorem 1.9 *For any function $f : \mathbb{R}^n \to [-\infty, \infty]$ the following conditions are equivalent:*

1 The set $\mathcal{A}_f$ is nonempty.

2 $f_c > -\infty$.

3 $f_{\overline{c}} > -\infty$.

Proof. If the set $\mathcal{A}_f$ is nonempty then for any $a \in \mathcal{A}_f$ we obtain by relation (1.65) that $f_{\overline{c}}(\mathbf{x}) \geq a(\mathbf{x})$ for every $\mathbf{x} \in \mathbb{R}^n$, and this shows the implication $1 \Rightarrow 3$. Due to $f_{\overline{c}} \leq f_c$ the implication $3 \Rightarrow 2$ is obvious. To show the implication $2 \Rightarrow 1$ consider some f satisfying $f_c > -\infty$. In case $dom(f_c)$ is empty it follows that $f \equiv \infty$ and so trivially $\mathcal{A}_f$ is nonempty. Therefore assume that $dom(f_c)$ is nonempty. By Lemma 1.26 this is a nonempty convex set and so by Lemma 1.13 one can find some $\mathbf{x}_0 \in ri(dom(f_c))$. Since $f_c > -\infty$ is a convex function it follows by Lemma 1.36 that $-\infty < f_c(\mathbf{x}_0) = \overline{f_c}(\mathbf{x}_0) = f_{\overline{c}}(\mathbf{x}_0) < \infty$ and so we have found some $\mathbf{x}_0$ satisfying $f_{\overline{c}}(\mathbf{x}_0)$ is finite. Applying now Lemma 1.40 yields $\mathcal{A}_f$ is nonempty and the result is proved. □

As shown by the following example it is not true that $\mathcal{A}_f$ is nonempty for $f > -\infty$.

Example 1.12 For the concave function $f : \mathbb{R} \to \mathbb{R}$ given by $f(x) = -x^2$ it is easy to verify that $co(epi(f)) = \mathbb{R}^2$ and $f > -\infty$. Hence we obtain that $\mathcal{A}_{f_c}$ is empty and this yields by Lemma 1.39 that $\mathcal{A}_f$ is empty.

To prove an important representation for a subclass of convex functions we introduce the following definition.

Definition 1.24 *The function $f : \mathbb{R}^n \to [-\infty, \infty]$ belongs to the set $\Gamma(\mathbb{R}^n)$ if f is convex and l.s.c. and $f > -\infty$.*

It is now possible to prove the following representation for the set $\Gamma(\mathbb{R}^n)$. This result is known as Minkowski's theorem.

Theorem 1.10 *For any function $f : \mathbb{R}^n \to [-\infty, \infty]$ it follows that*

$$f \in \Gamma(\mathbb{R}^n) \Leftrightarrow f(\mathbf{x}) = \sup\{a(\mathbf{x}) : a \in \mathcal{A}_f\} \text{ and the set } \mathcal{A}_f \text{ is nonempty.}$$

Proof. If the function $f : \mathbb{R}^n \to [-\infty, \infty]$ has the representation $f(\mathbf{x}) = \sup\{a(\mathbf{x}) : a \in \mathcal{A}_f\}$ and the set $\mathcal{A}_f$ is nonempty, then clearly the function f is l.s.c., convex and $f > -\infty$ and so f belongs to $\Gamma(\mathbb{R}^n)$. To prove the reverse implication, we observe for $f \in \Gamma(\mathbb{R}^n)$ that $f_c = f > -\infty$ and this shows by Theorem 1.9 that the set $\mathcal{A}_f$ is nonempty and hence $f(\mathbf{x}) \geq \sup\{a(\mathbf{x}) : a \in \mathcal{A}_f\} > -\infty$. Suppose now by contradiction that $f(\mathbf{x}_0) > \sup\{a(\mathbf{x}_0) : a \in \mathcal{A}_f\}$ for some $\mathbf{x}_0 \in \mathbb{R}^n$. Hence one can find some $\gamma \in \mathbb{R}$ satisfying

$$f(\mathbf{x}_0) > \gamma > \sup\{a(\mathbf{x}_0) : a \in \mathcal{A}_f\}, \tag{1.72}$$

and so $(\mathbf{x}_0, \gamma) \notin epi(f)$. If $epi(f)$ is empty, then the affine function $a(\mathbf{x}) = \gamma$ is an affine minorant of f and this contradicts relation (1.72). Therefore we assume that $epi(f)$ is nonempty and since this set is closed and convex there exists by Theorem 1.1 a nonzero vector $(\mathbf{y}_0, \beta)$ and $\epsilon > 0$ satisfying

$$\mathbf{y}_0^\top \mathbf{x} + \beta r \geq \mathbf{y}_0^\top \mathbf{x}_0 + \beta\gamma + \epsilon \tag{1.73}$$

for every $(\mathbf{x}, r) \in epi(f)$. Since for $(\mathbf{x}, r) \in epi(f)$ and $h > 0$ the vector $(\mathbf{x}, r + h)$ belongs to $epi(f)$ it follows by relation (1.73) that $\beta \geq 0$. Consider now the two cases $f(\mathbf{x}_0) < \infty$ and $f(\mathbf{x}_0) = \infty$. If $f(\mathbf{x}_0) < \infty$ we obtain by relation (1.73) replacing $(\mathbf{x}, r)$ by $(\mathbf{x}_0, f(\mathbf{x}_0))$ that $\beta(f(\mathbf{x}_0) - \gamma) \geq \epsilon$ and this implies using relation (1.72) that $\beta > 0$. Hence by relation (1.73) it holds that

$$f(\mathbf{x}) \geq a(\mathbf{x}) := -\beta^{-1}\mathbf{y}_0^\top(\mathbf{x} - \mathbf{x}_0) + \gamma$$

for every $\mathbf{x}$ belonging to $dom(f)$ and we have found some $a \in \mathcal{A}_f$ satisfying $a(\mathbf{x}_0) = \gamma$ contradicting relation (1.72). If $f(\mathbf{x}_0) = \infty$ and $\beta > 0$ in relation (1.73), then by the same proof we obtain a contradiction and so we consider the last case $f(\mathbf{x}_0) = \infty$ and $\beta = 0$. Introduce now the affine function $a_0 : \mathbb{R}^n \to \mathbb{R}$, given by

$$a_0(\mathbf{x}) = -\mathbf{y}_0^\top(\mathbf{x} - \mathbf{x}_0) + \epsilon.$$

By relation (1.73) $a_0(\mathbf{x}) \leq 0$ for every $\mathbf{x} \in dom(f)$ and $a_0(\mathbf{x}_0) > 0$. Since $\mathcal{A}_f$ is nonempty, select some $a \in \mathcal{A}_f$ and by relation (1.72) it follows that $\lambda_0 := a_0(\mathbf{x}_0)^{-1}(\gamma - a(\mathbf{x}_0)) > 0$. Introducing now the affine function $a_{\lambda_0} : \mathbb{R}^n \to \mathbb{R}$ given by

$$a_{\lambda_0}(\mathbf{x}) := a(\mathbf{x}) + \lambda_0 a_0(\mathbf{x})$$

we obtain $a_{\lambda_0}(\mathbf{x}_0) = \gamma$ and since $a_0(\mathbf{x}) \leq 0$ for every $\mathbf{x} \in dom(f)$ and $a \in \mathcal{A}_f$ we also obtain $a_{\lambda_0} \in \mathcal{A}_f$. Hence a_{λ_0} is an affine minorant of f satisfying $a_{\lambda_0}(\mathbf{x}_0) = \gamma$ and this contradicts relation (1.72) showing the desired result. □

An immediate consequence of Minkowski's theorem and Lemma 1.39 is listed in the next result.

Theorem 1.11 *If* $f : \mathbb{R}^n \to [-\infty, \infty]$ *is a function satisfying* $f_{\overline{c}} > -\infty$, *then it follows that* $\overline{f_c}(\mathbf{x}) = f_{\overline{c}}(\mathbf{x}) = \sup\{a(\mathbf{x}) : a \in \mathcal{A}_f\}$ *and the set* $\mathcal{A}_f$ *is nonempty.*

Proof. By relation (1.66) and Theorem 1.10 we obtain that $\overline{f_c}(\mathbf{x}) = f_{\overline{c}}(\mathbf{x}) = \sup\{a(\mathbf{x}) : a \in \mathcal{A}_{f_{\overline{c}}}\}$ with the set $\mathcal{A}_{f_{\overline{c}}}$ is nonempty. Applying now Lemma 1.39 the desired result follows. □

In Theorem 1.11 we only guarantee that any function $f_{\overline{c}} > -\infty$ can be approximated from below by affine functions. However, it is sometimes useful to derive an approximation formula in terms of the original function f. This formula was first constructed in its general form by Fenchel (cf. [21]) and it has an easy geometrical interpretation (cf. [27]).

Definition 1.25 *For any function* $f : \mathbb{R}^n \to [-\infty, \infty]$ *the function* $f^* : \mathbb{R}^n \to [-\infty, \infty]$ *given by* $f^*(\mathbf{a}) := \sup\{\mathbf{a}^\top\mathbf{x} - f(\mathbf{x}) : \mathbf{x} \in \mathbb{R}^n\}$ *is called the conjugate function of the function* f. *The function* $f^{**} : \mathbb{R}^n \to [-\infty, \infty]$ *given by* $f^{**}(\mathbf{x}) := \sup\{\mathbf{a}^\top\mathbf{x} - f^*(\mathbf{a}) : \mathbf{a} \in \mathbb{R}^n\}$ *is called the biconjugate function of* f.

By the above definition it is immediately clear that the conjugate function f^* is convex and l.s.c.. Moreover, if the function $f : \mathbb{R}^n \to [-\infty, \infty]$ is proper and the set $\mathcal{A}_f$ of affine minorants is nonempty, then it is easy to verify that the function f^* is also proper. As shown by the next result the biconjugate function has a clear geometrical interpretation.

Lemma 1.41 *If* $f : \mathbb{R}^n \to [-\infty, \infty]$ *is an arbitrary function satisfying* $\mathcal{A}_f$ *is nonempty, then it follows that* $(\mathbf{a}, r) \in epi(f^*)$ *if and only if* $a \in \mathcal{A}_f$ *with* $a(\mathbf{x}) = \mathbf{a}^\top\mathbf{x} - r$. *Additionally, it holds that* $f^{**}(\mathbf{x}) = \sup\{a(\mathbf{x}) : a \in \mathcal{A}_f\}$ *for every* $\mathbf{x} \in \mathbb{R}^n$.

Proof. To verify the equivalence relation we observe for $a(\mathbf{x}) = \mathbf{a}^\top\mathbf{x} - r \leq f(\mathbf{x})$ for every $\mathbf{x} \in \mathbb{R}^n$ that $r \geq f^*(\mathbf{a}) = \sup\{\mathbf{a}^\top\mathbf{x} - f(\mathbf{x}) : \mathbf{x} \in \mathbb{R}^n\}$ or $(\mathbf{a}, r) \in epi(f^*)$. Moreover, if $(\mathbf{a}, r) \in epi(f^*)$ we obtain $r \geq f^*(\mathbf{a})$ and this implies for every $\mathbf{x} \in \mathbb{R}^n$ that $a(\mathbf{x}) = \mathbf{a}^\top\mathbf{x} - r \leq f(\mathbf{x})$. To prove the relation for the biconjugate function it follows by the definition of $epi(f^*)$ that $f^{**}(\mathbf{x}) = \sup\{\mathbf{a}^\top\mathbf{x} - r : (\mathbf{a}, r) \in epi(f^*)\}$. Since by the first part $(\mathbf{a}, r) \in epi(f^*)$ if and only if $a(\mathbf{x}) = \mathbf{a}^\top\mathbf{x} - r$ is an affine minorant of f, this shows that $f^{**}(\mathbf{x}) = \sup\{a(\mathbf{x}) : a \in \mathcal{A}_f\}$ for every $\mathbf{x} \in \mathbb{R}^n$ and hence the equality for the biconjugate function is verified. □

To prove one of the most important theorems in convex analysis we introduce the definition of the closure of the function f.

Definition 1.26 *If $f : \mathbb{R}^n \to [-\infty, \infty]$ is an arbitrary function, then the closure $cl(f) : \mathbb{R}^n \to [-\infty, \infty]$ of the function f is given by*

$$cl(f) := \begin{cases} \overline{f} & \text{if } \overline{f} > -\infty \\ -\infty & \text{otherwise} \end{cases} .$$

Clearly the function $cl(f)$ is l.s.c. and satisfies $cl(f) \leq \overline{f}$. Also it is easy to verify by Lemma 1.41, Theorem 1.9 and using $\mathcal{A}_f = \mathcal{A}_{\overline{f}}$ that

$$cl(f)^* = f^* \tag{1.74}$$

for any convex function f. The next result is known as the Fenchel-Moreau theorem and is one of the most important results in convex analysis.

Theorem 1.12 *For any function $f : \mathbb{R}^n \to [-\infty, \infty]$ it follows that $f^{**}(\mathbf{x}) = cl(f_c)(\mathbf{x})$ for every $\mathbf{x} \in \mathbb{R}^n$.*

Proof. If $f_{\overline{c}}(\mathbf{x}_0) = -\infty$ for some $\mathbf{x}_0 \in \mathbb{R}^n$ then $f^* \equiv \infty$. To show this, suppose by contradiction that $f^*(\mathbf{a}_0) < \infty$ for some $\mathbf{a}_0$. This implies the existence of some $r \in \mathbb{R}$ satisfying $r \geq \mathbf{a}_0^\top \mathbf{x} \quad f(\mathbf{x})$ for every $\mathbf{x} \in \mathbb{R}^n$ and so the function $a(\mathbf{x}) = \mathbf{a}_0^\top \mathbf{x} - r$ is an affine minorant of f. Hence by relation (1.65) we obtain that $f_{\overline{c}}(\mathbf{x}_0) > -\infty$ and this contradicts our initial assumption. Since $f^* \equiv \infty$ we obtain $f^{**} \equiv -\infty$ and by Definition 1.26 we obtain $f^{**} = cl(f_c)$. In case $f_{\overline{c}} > -\infty$ the result follows by Theorem 1.11 and Lemma 1.41. □

An important consequence of the Fenchel-Moreau theorem is given by the following result. Recall a function is *sublinear*, if it is positively homogeneous and convex.

Lemma 1.42 *Any l.s.c. sublinear function $f : \mathbb{R}^n \to (-\infty, \infty]$ has the representation*

$$f(\mathbf{x}) = \sup\{\mathbf{a}^\top \mathbf{x} : \mathbf{a} \in C\}$$

with $C = \{\mathbf{a} \in \mathbb{R}^n : f^*(\mathbf{a}) \leq 0\}$ a nonempty closed convex set.

Proof. By the Fenchel Moreau theorem it follows that

$$f(\mathbf{x}) = f^{**}(\mathbf{x}) = \sup_{\mathbf{a} \in \mathbb{R}^n}\{\mathbf{a}^\top \mathbf{x} - f^*(\mathbf{a})\}.$$

Since f is positively homogeneous we obtain by Lemma 1.25 that

$$\alpha f^*(\mathbf{a}) = \sup_{\mathbf{x} \in \mathbb{R}^n}\{\mathbf{a}^\top(\alpha \mathbf{x}) - f(\alpha \mathbf{x})\} = f^*(\mathbf{a})$$

for every $\alpha > 0$ and $\mathbf{a} \in \mathbb{R}^n$ and this shows that $f^*(\mathbf{a}) \in \{\infty, -\infty, 0\}$. If $f^*(\mathbf{a}) = \infty$ for every $\mathbf{a} \in \mathbb{R}^n$, then $f^{**}(\mathbf{x}) = -\infty$ for every $\mathbf{x}$ and this shows by the Fenchel Moreau theorem that $f(\mathbf{x}) = -\infty$ for every $\mathbf{x}$, contradicting $f > -\infty$. Therefore f^* is not identically ∞ and this yields that the set C is not empty. Again by the Fenchel Moreau theorem we obtain

$$f(\mathbf{x}) = f^{**}(\mathbf{x}) = \sup_{\mathbf{a}\in\mathbb{R}^n}\{\mathbf{a}^\top\mathbf{x} - f^*(\mathbf{a})\} = \sup_{\mathbf{a}\in C}\mathbf{a}^\top\mathbf{x}.$$

and since the function f^* is l.s.c. and convex the nonempty set C is closed and convex. □

Finally we introduce the so-called subgradient set of a function at a point.

Definition 1.27 *For any function $f : \mathbb{R}^n \to [-\infty, \infty]$ and $\mathbf{x}_0 \in \mathbb{R}^n$ the subset of $\mathbb{R}^n$ consisting of those vectors $\mathbf{a}_0$ satisfying $f(\mathbf{x}) \geq f(\mathbf{x}_0) + \mathbf{a}_0^\top(\mathbf{x} - \mathbf{x}_0)$ for every $\mathbf{x} \in \mathbb{R}^n$ is called the subgradient set of the function f at the point $\mathbf{x}_0$. This set is denoted by $\partial f(\mathbf{x}_0)$ and its elements are called subgradients.*

If $f(\mathbf{x}_0) = -\infty$, then clearly $\partial f(\mathbf{x}_0) = \mathbb{R}^n$ and so it is sufficient to consider those $\mathbf{x}_0 \in \mathbb{R}^n$ satisfying $f(\mathbf{x}_0) > -\infty$. Moreover, if $f(\mathbf{x}_0) > -\infty$ and dom(f) is empty, then again $\partial f(\mathbf{x}_0) = \mathbb{R}^n$ and hence we only need to consider $f(\mathbf{x}_0) > -\infty$ and dom(f) is not empty. If $\mathbf{x}_0 \notin dom(f)$ or $f(\mathbf{x}_0) = \infty$, then this implies, using dom(f) is nonempty, that $\partial f(\mathbf{x}_0) = \emptyset$ and so the only interesting case which remains is given by $f(\mathbf{x}_0)$ finite. It is now relatively easy to prove for $f(\mathbf{x}_0)$ finite that $\partial f(\mathbf{x}_0) \neq \emptyset$ is equivalent to another condition related to the conjugate function.

Lemma 1.43 *If $f : \mathbb{R}^n \to [-\infty, \infty]$ is an arbitrary function satisfying $f(\mathbf{x}_0)$ is finite for some $\mathbf{x}_0$, then it follows that $\mathbf{a}_0 \in \partial f(\mathbf{x}_0)$ if and only if $f(\mathbf{x}_0) + f^*(\mathbf{a}_0) = \mathbf{a}_0^\top\mathbf{x}_0$.*

Proof. If $\mathbf{a}_0 \in \partial f(\mathbf{x}_0)$ then by definition $f(\mathbf{x}) \geq f(\mathbf{x}_0) + \mathbf{a}_0^\top(\mathbf{x} - \mathbf{x}_0)$ for every $\mathbf{x}$ and this implies using $f(\mathbf{x}_0)$ is finite that $\mathbf{a}_0^\top\mathbf{x}_0 - f(\mathbf{x}_0) \geq \mathbf{a}_0^\top\mathbf{x} - f(\mathbf{x})$ for every $\mathbf{x}$. Hence we obtain that $\mathbf{a}_0^\top\mathbf{x}_0 - f(\mathbf{x}_0) = f^*(\mathbf{a}_0)$ and this shows the equality. To verify the reverse implication is trivial and so we omit its proof. □

Up to now we did not show any existence result for the subgradient set of f at $\mathbf{x}_0$ in case $f(\mathbf{x}_0)$ is finite. Such a result will be given by the next theorem.

Theorem 1.13 *If the function $f : \mathbb{R}^n \to [-\infty, \infty]$ is convex and $f(\mathbf{x}_0)$ is finite for some $\mathbf{x}_0 \in ri(dom(f))$, then the set $\partial f(\mathbf{x}_0)$ is nonempty.*

Proof. If $\mathbf{x}_0 \in ri(dom(f))$ and $f(\mathbf{x}_0)$ is finite we obtain by Lemma 1.35 that $(\mathbf{x}_0, f(\mathbf{x}_0)) \notin ri(epi(f))$. This implies by the convexity of the set $epi(f)$ and Theorem 1.2 that there exists some nonzero vector $(\mathbf{y}_0, \beta) \in L_{aff(epi(f))}$ satisfying

$$\mathbf{y}_0^\intercal \mathbf{x} + \beta r \geq \mathbf{y}_0^\intercal \mathbf{x}_0 + \beta f(\mathbf{x}_0) \tag{1.75}$$

for $(\mathbf{x}, r) \in epi(f)$. Moreover, using $(\mathbf{x}_0, f(\mathbf{x}_0) + h)$ belongs to epi(f) for every $h \geq 0$, we obtain $\beta \geq 0$ and to show that $\beta > 0$ assume by contradiction that $\beta = 0$. Hence it follows by relation (1.75) that

$$\mathbf{y}_0^\intercal \mathbf{x} \geq \mathbf{y}_0^\intercal \mathbf{x}_0 \tag{1.76}$$

for every $(\mathbf{x}, r) \in epi(f)$. Since $aff(epi(f)) = aff(dom(f)) \times \mathbb{R}$ and so

$$L_{aff(epi(f))} = L_{aff(dom(f))} \times \mathbb{R}$$

we know that $\mathbf{y}_0$ belongs to $L_{aff(dom(f))}$. This implies, using $\mathbf{x}_0$ belongs to $ri(dom(f))$, that there exists some $\epsilon > 0$ satisfying $\mathbf{x}_0 - \epsilon \mathbf{y}_0 \in dom(f)$ and applying now relation (1.76) with $\mathbf{x}$ replaced by $\mathbf{x}_0 - \epsilon \mathbf{y}_0$ yields $-\epsilon \|\mathbf{y}_0\|^2 \geq 0$. Hence it follows that $(\mathbf{y}_0, \beta) = \mathbf{0}$ and we obtain a contradiction. Therefore it must hold that $\beta > 0$ and dividing now the inequality in relation (1.75) by $\beta > 0$ and using that $f(\mathbf{x})$ is finite for every $\mathbf{x} \in dom(f)$ yields

$$f(\mathbf{x}) \geq f(\mathbf{x}_0) - \beta^{-1} \mathbf{y}_0^\intercal (\mathbf{x} - \mathbf{x}_0)$$

for every $\mathbf{x} \in dom(f)$. This shows that the vector $\mathbf{a}_0 = -\beta^{-1}\mathbf{y}_0$ is a subgradient of the function f at the point $\mathbf{x}_0$ and so $\partial f(\mathbf{x}_0)$ is a nonempty set. □

In case $\mathbf{x}_0$ does not belong to $ri(dom(f))$ for some convex function f it might happen that f does not have a subgradient at the point $\mathbf{x}_0$. This is shown by the following example.

Example 1.13 Consider the convex function $f : \mathbb{R} \to (-\infty, \infty]$ given by $f(x) = -\sqrt{x}$ for $x \geq 0$ and $f(x) = \infty$ otherwise. Clearly 0 belongs to the relative boundary of $dom(f)$ but $\partial f(0)$ is empty.

In case the function $f > -\infty$ is a sublinear function one can show the following improvement of Theorem 1.13 replacing the condition $\mathbf{0} \in ri(dom(f))$ by the condition $\mathbf{0} \in dom(f)$.

Theorem 1.14 *If the function $f : \mathbb{R}^n \to (-\infty, \infty]$ is sublinear and $\mathbf{0} \in dom(f)$, then the set $\partial f(\mathbf{0})$ is nonempty and $\partial f(\mathbf{0}) = \{\mathbf{a} \in \mathbb{R}^n : f^*(\mathbf{a}) \leq 0\}$.*

Proof. Since f is convex it follows that $f_c = f > -\infty$ and this implies by Theorem 1.9 that $\mathcal{A}_f$ is nonempty and so $\overline{f}$ is a proper function. Since by Definition 1.13 and Lemma 1.31 the function $\overline{f}$ is also sublinear one may apply Lemma 1.42 and this shows $\overline{f}(\mathbf{x}) = \sup\{\mathbf{a}^\top\mathbf{x} : \mathbf{a} \in C\}$ with $C = \{\mathbf{a} \in \mathbb{R}^n : \overline{f}^*(\mathbf{a}) \leq 0\}$ a nonempty closed convex set. By relation (1.74) and $\overline{f} = cl(f)$ it follows that $\overline{f}^* = f^*$ and so $C = \{\mathbf{a} \in \mathbb{R}^n : f^*(\mathbf{a}) \leq 0\}$. We will now verify that $\partial f(\mathbf{0}) = C$. By the definition of f^* we obtain for $\mathbf{a} \in C$ that $f(\mathbf{x}) \geq \mathbf{a}^\top\mathbf{x}$ for every $\mathbf{x} \in \mathbb{R}^n$. Since $f(0)$ is finite and f positively homogeneous it follows that $f(0) = 0$ and so it follows that $\mathbf{a} \in \partial f(\mathbf{0})$. This shows $C \subseteq \partial f(\mathbf{0})$ and to verify the reverse inclusion we observe for every $\mathbf{a} \in \partial f(\mathbf{0})$ that $f(\mathbf{x}) \geq \mathbf{a}^\top\mathbf{x}$ for every $\mathbf{x}$. This implies $f^*(\mathbf{a}) \leq 0$ and so $\mathbf{a}$ belongs to C. Hence $C = \partial f(\mathbf{0})$ is nonempty and the proof is completed. □

In Theorem 1.14 we actually show for $f : \mathbb{R}^n \to (-\infty, \infty]$ sublinear and $\mathbf{0} \in dom(f)$ that

$$\overline{f}(\mathbf{x}) = \sup\{\mathbf{a}^\top\mathbf{x} : \mathbf{a} \in \partial f(\mathbf{0})\} \text{ and } \partial f(\mathbf{0}) \neq \varnothing. \tag{1.77}$$

A nice implication of Theorem 1.13 is the observation that convex functions have remarkable continuity properties. Before showing this result we need the following technical lemmas.

Lemma 1.44 *If the vectors $\mathbf{z}_i \in \mathbb{R}^n, 1 \leq i \leq k \leq n$ form an orthonormal system and the set P is the convex hull generated by the set $S = \{\mathbf{z}_1, ..., \mathbf{z}_k, -\mathbf{z}_1, ..., -\mathbf{z}_k\}$, then it follows that*

$$k^{-\frac{1}{2}} E \cap lin(\{\mathbf{z}_1, ..., \mathbf{z}_k\}) \subseteq P.$$

Proof. Since the vectors $\mathbf{z}_i, 1 \leq i \leq k$ form an orthonormal system we obtain for any vector $\alpha^\top = (\alpha_1, ..., \alpha_k) \in \mathbb{R}^k$ that

$$\|\sum_{i=1}^k \alpha_i \mathbf{z}_i\|^2 = \|\alpha\|^2. \tag{1.78}$$

Applying now the Cauchy-Schwartz inequality to the inner product of the vectors $|\alpha|^\top := (|\alpha_1|, ..., |\alpha_k|)$ and $\mathbf{e}^\top = (1, ..., 1)$ it follows that

$$\sum_{i=1}^k |\alpha_i| =< |\alpha|, \mathbf{e} >\leq \|\alpha\| k^{\frac{1}{2}}$$

and this implies by relation (1.78) that

$$\|\sum_{i=1}^k \alpha_i \mathbf{z}_i\| \geq k^{-\frac{1}{2}} \sum_{i=1}^k |\alpha_i|. \tag{1.79}$$

Consider now an arbitrary vector $\mathbf{y}$ belonging to $k^{-\frac{1}{2}}E \cap lin(\{\mathbf{z}_1, ..., \mathbf{z}_k\})$. Since the vectors $\mathbf{z}_i, 1 \leq i \leq k$ are independent, there exists a unique vector $\alpha^\top = (\alpha_1, ..., \alpha_k) \in \mathbb{R}^k$ such that

$$\mathbf{y} = \sum_{i \in I} \alpha_i \mathbf{z}_i + \sum_{i \notin I} -\alpha_i(-\mathbf{z}_i)$$

with $I := \{1 \leq i \leq k : \alpha_i > 0\}$. Applying now the inequality in relation (1.79) it follows that

$$k^{-\frac{1}{2}} \geq \|\mathbf{y}\| \geq k^{-\frac{1}{2}} \left(\sum_{i \in I} \alpha_i + \sum_{i \notin I} -\alpha_i\right)$$

and this shows that the vector $\mathbf{y}$ belongs to P. □

Another result which is needed in the proof of Theorem 1.15 is given by the following lemma.

Lemma 1.45 *If the function $f : \mathbb{R}^n \to (-\infty, \infty]$ is convex and for some $\mathbf{x}_0$ and $\delta > 0$ there exists some finite constants m, M satisfying $m \leq f(\mathbf{x}) \leq M$ for every $\mathbf{x}$ belonging to $(\mathbf{x}_0 + 2\delta E) \cap dom(f)$, then one can find some $L > 0$ satisfying*

$$|f(\mathbf{x}_1) - f(\mathbf{x}_2)| \leq L\|\mathbf{x}_1 - \mathbf{x}_2\|$$

for every $\mathbf{x}_1, \mathbf{x}_2$ belonging to $(\mathbf{x}_0 + \delta E) \cap ri(dom(f))$.

Proof. Let $\mathbf{x}_1$ and $\mathbf{x}_2$ be two different vectors belonging to $(\mathbf{x}_0 + \delta E) \cap ri(dom(f))$. This yields that the vector $\mathbf{x}_1 - \mathbf{x}_2$ belongs to $L_{aff(dom(f))}$ and since $\mathbf{x}_1$ is a relative interior point of the convex set $dom(f)$ one can find some $0 < \epsilon < \delta$ satisfying

$$\mathbf{x}_3 := \mathbf{x}_1 + \epsilon\|\mathbf{x}_1 - \mathbf{x}_2\|^{-1}(\mathbf{x}_1 - \mathbf{x}_2) \in dom(f). \tag{1.80}$$

Hence the vector $\mathbf{x}_3$ belongs to $(\mathbf{x}_0 + 2\delta E) \cap dom(f)$ and by relation (1.80) we obtain

$$\mathbf{x}_1 = \frac{\|\mathbf{x}_1 - \mathbf{x}_2\|}{\|\mathbf{x}_1 - \mathbf{x}_2\| + \epsilon}\mathbf{x}_3 + \frac{\epsilon}{\|\mathbf{x}_1 - \mathbf{x}_2\| + \epsilon}\mathbf{x}_2.$$

Using now relation (1.48) and the fact that the function f is bounded from above and below on $(\mathbf{x}_0 + 2\delta E) \cap dom(f)$ it follows for $L := \epsilon^{-1}(M - m)$ that

$$f(\mathbf{x}_1) - f(\mathbf{x}_2) \leq \frac{\|\mathbf{x}_1 - \mathbf{x}_2\|}{\|\mathbf{x}_1 - \mathbf{x}_2\| + \epsilon}(f(\mathbf{x}_3) - f(\mathbf{x}_2)) \leq L\|\mathbf{x}_1 - \mathbf{x}_2\|.$$

Reversing the roles of $\mathbf{x}_1$ and $\mathbf{x}_2$ yields a similar bound for $f(\mathbf{x}_2) - f(\mathbf{x}_1)$ and the desired inequality is verified. □

The above property of the function f is called *Lipschitz continuity* on the set $(\mathbf{x}_0 + \delta E) \cap ri(dom(f))$. Using Lemmas 1.45, 1.44 and Theorem 1.13 one can now show the next result, which is an improvement of Lemma 1.36.

Theorem 1.15 *If $f : \mathbb{R}^n \rightarrow (-\infty, \infty]$ is a convex function, then it follows that f is continuous on $ri(dom(f))$ and Lipschitz continuous on every compact subset of $ri(dom(f))$.*

Proof. If $\mathbf{x}_0 \in ri(dom(f))$ one can find some $\epsilon > 0$ satisfying

$$(\mathbf{x}_0 + 2\epsilon E) \cap aff(dom(f)) \subseteq dom(f). \tag{1.81}$$

To give a more detailed characterization of $aff(dom(f))$ we observe by Lemma 1.4, that there exists a set of $k \leq n$ linearly independent vectors $\mathbf{z}_1, ..., \mathbf{z}_k$ satisfying $L_{aff(dom(f))} = lin(\{\mathbf{z}_1, ..., \mathbf{z}_k\})$ and so

$$aff(dom(f)) = \mathbf{x}_0 + lin(\{\mathbf{z}_1, ..., \mathbf{z}_k\}). \tag{1.82}$$

Without loss of generality (Use the well-known Gram-Schmidt orthogonalization process (cf. [47])) we may assume that the set $\{\mathbf{z}_1, ..., \mathbf{z}_k\}$ is an orthonormal system. By relations (1.82) and (1.81) and dom(f) a convex set it follows that the set $\mathbf{x}_0 + P$ with P the convex hull generated by the set $S = \{\epsilon\mathbf{z}_1, ..., \epsilon\mathbf{z}_k, -\epsilon\mathbf{z}_1, ..., -\epsilon\mathbf{z}_k\}$ belongs to dom(f). Also by the convexity of the function f and relation (1.48) we obtain that

$$f(\mathbf{x}) \leq \max\{f(\mathbf{x}_0 + \epsilon\mathbf{z}_i), f(\mathbf{x}_0 - \epsilon\mathbf{z}_i), 1 \leq i \leq k\} < \infty$$

for every $\mathbf{x} \in \mathbf{x}_0 + P$. Since by Lemma 1.44 there exists some $\gamma > 0$ satisfying

$$(\mathbf{x}_0 + \gamma E) \cap aff(dom(f)) \subseteq P.$$

this shows that the function f is bounded from above on $(\mathbf{x}_0 + \gamma E) \cap dom(f)$. Using Theorem 1.13 we also obtain that the function f is bounded from below on $(\mathbf{x}_0 + \gamma E) \cap dom(f)$ and applying now Lemma 1.45 with 2δ replaced by γ yields the desired result. □

This concludes our discussion on dual representations and conjugation for convex functions. In the next subsection we consider the same topic for quasiconvex functions.

3.4 Dual representations of quasiconvex functions

In this section we study dual representations of evenly quasiconvex and l.s.c. quasiconvex functions. Most of the results of this section can be found in [56]. Unfortunately in [56] no geometrical interpretation of the results are given and for such an interpretation the reader should consult [27]. In [56] it is shown, that one can use the same approach as in convex analysis and this results in proving that certain subsets of quasiconvex functions can be approximated from below by so-called c-affine functions with $c : \mathbb{R} \to [-\infty, \infty]$ belonging to a given class $\mathcal{C}$ of extended real valued univariate functions. Recall that a function is called univariate if its domain is given by $\mathbb{R}$. As in convex analysis the used approximations and the generalized biconjugate functions have a clear geometrical interpretation (cf. [27]). To start with this approach we introduce in the next definition the class of c-affine functions. More general classes of so-called coupling functions a are discussed in this volume by Martínez-Legaz (cf. [49]).

Definition 1.28 *For a given univariate function $c : \mathbb{R} \to [-\infty, \infty]$ the function $a : \mathbb{R}^n \to [-\infty, \infty]$ is called a c-affine function, if there exist some $\mathbf{a} \in \mathbb{R}^n$ and $r \in \mathbb{R}$ such that $a(\mathbf{x}) = c(\mathbf{a}^\top \mathbf{x}) + r$ for every $\mathbf{x} \in \mathbb{R}^n$. If $\mathcal{C}$ denotes a subset of the set of extended real valued univariate functions the function a is called a $\mathcal{C}$-affine function, if for some $c \in \mathcal{C}$ the function a is a c-affine function. The function a is called a $\mathcal{C}$-affine minorant of the function $f : \mathbb{R}^n \to [-\infty, \infty]$ if $a(\mathbf{x}) \leq f(\mathbf{x})$ for every $\mathbf{x} \in \mathbb{R}^n$ and a is a $\mathcal{C}$-affine function. The set $\mathcal{C}\mathcal{A}_f$ denotes now the (possibly empty) set of $\mathcal{C}$-affine minorants of f.*

To specify the set $\mathcal{C}$ we first consider the set $\mathcal{C}_0$ of extended real valued nondecreasing univariate functions $c : \mathbb{R} \to [-\infty, \infty]$ and the proper subset $\mathcal{C}_1 \subseteq \mathcal{C}_0$ of extended real valued nondecreasing l.s.c. univariate functions. Since for any $c \in \mathcal{C}_i, i = 0, 1$ and $r \in \mathbb{R}$ also the function $c^* : \mathbb{R} \to [-\infty, \infty]$, given by $c^*(t) = c(t) + r$, belongs to $\mathcal{C}_i$, $i = 0, 1$, we observe for these classes of extended real valued univariate functions that the class of $\mathcal{C}_i$-affine functions, $i = 0, 1$ reduces to the set of functions $a : \mathbb{R}^n \to [-\infty, \infty]$, given by $a(\mathbf{x}) = c(\mathbf{a}^\top \mathbf{x})$ for some $\mathbf{a} \in \mathbb{R}^n$ and $c \in \mathcal{C}_i$. Clearly $\mathcal{C}_1\mathcal{A}_f \subseteq \mathcal{C}_0\mathcal{A}_f$ and since the function $a : \mathbb{R}^n \to [-\infty, \infty]$ with $a(\mathbf{x}) = -\infty$ for every $\mathbf{x} \in \mathbb{R}^n$ belongs to the set $\mathcal{C}_1$, we obtain that $\mathcal{C}_1\mathcal{A}_f$ is nonempty for every $f : \mathbb{R}^n \to [-\infty, \infty]$. This is a major difference with the set of affine minorants of a function f, since this set might be empty. Observe in Theorem 1.9 we showed that this set is nonempty if and only if $f_c > -\infty$. One can now show the following result for $\mathcal{C}$-affine functions with $\mathcal{C}$ either equal to $\mathcal{C}_1$ or $\mathcal{C}_0$.

Lemma 1.46 *If $a : \mathbb{R}^n \to [-\infty, \infty]$ is $\mathcal{C}_0$-affine, then the function a is evenly quasiconvex. Moreover, if a is $\mathcal{C}_1$-affine, then the function a is l.s.c. and quasiconvex.*

Proof. If a is a $\mathcal{C}_0$-affine function, then there exists some $c \in \mathcal{C}_0$ and $\mathbf{a} \in \mathbb{R}^n$ such that $L(a, r) = \{\mathbf{x} \in \mathbb{R}^n : \mathbf{a}^\top \mathbf{x} \in L(c, r)\}$ for every $r \in \mathbb{R}$ with $L(c, r)$ the lower level set of the function c. Since c is nondecreasing, this lower level set is either empty or an interval given by $(-\infty, \beta_r)$ or $(-\infty, \beta_r]$ with $\beta_r := \sup\{t \in \mathbb{R} : c(t) \leq r\}$. Hence the set $L(a, r)$ is either empty or an open or closed halfspace and this shows that $L(a, r)$ is evenly convex. Similarly for $c \in \mathcal{C}_1$ we obtain, using Theorem 1.7, that $L(c, r)$ is empty or $(-\infty, \beta_r]$ and hence $L(a, r)$ is empty or a closed halfspace. This shows that the function a is quasiconvex and by Theorem 1.7 it is also l.s.c.. □

By Lemma 1.28, 1.29, 1.38 and 1.46 and $f_{\overline{q}} = \overline{f_q} \leq f_{ec} \leq f_q \leq f$ (see relations (1.69) and 1.70).) one can show, applying a similar proof as in Lemma 1.39, that the following result holds.

Lemma 1.47 *For any function $f : \mathbb{R}^n \to [-\infty, \infty]$ it follows that $\mathcal{C}_0\mathcal{A}_f = \mathcal{C}_0\mathcal{A}_{f_{ec}}$ and $\mathcal{C}_1\mathcal{A}_f = \mathcal{C}_1\mathcal{A}_{f_q} = \mathcal{C}_1\mathcal{A}_{\overline{f_q}} = \mathcal{C}_1\mathcal{A}_{f_{\overline{q}}}$.*

Contrary to functions studied in convex analysis, we do not have to determine for which extended real valued functions the sets $\mathcal{C}_i\mathcal{A}_f, i = 1, 2$ are nonempty and so we can start generalizing Minkowsky's theorem (see Theorem 1.10) to evenly quasiconvex and l.s.c. quasiconvex functions. In the proof of this generalization and in the remainder of this subsection an important role is played by the following functions.

Definition 1.29 *For any function $f : \mathbb{R}^n \to [-\infty, \infty]$ and $\mathbf{a} \in \mathbb{R}^n$, let $c_{\mathbf{a}} : \mathbb{R} \to [-\infty, \infty]$ denote the function $c_{\mathbf{a}}(t) := \inf\{f(\mathbf{y}) : \mathbf{a}^\top \mathbf{y} \geq t\}$.*

It is now possible to show the following result.

Theorem 1.16 *If $f : \mathbb{R}^n \to [-\infty, \infty]$ is an evenly quasiconvex function, then $f(\mathbf{x}) = \sup\{a(\mathbf{x}) : a \in \mathcal{C}_0\mathcal{A}_f\}$ for every $\mathbf{x} \in \mathbb{R}^n$. Moreover, if f is an l.s.c. quasiconvex function, then $f(\mathbf{x}) = \sup\{a(\mathbf{x}) : a \in \mathcal{C}_1\mathcal{A}_f\}$ for every $\mathbf{x} \in \mathbb{R}^n$.*

Proof. Since the set $\mathcal{C}_0\mathcal{A}_f$ is nonempty, we obtain by the definition of $\mathcal{C}_0\mathcal{A}_f$ that $f(\mathbf{x}) \geq \sup\{a(\mathbf{x}) : a \in \mathcal{C}_0\mathcal{A}_f\}$ for every $\mathbf{x} \in \mathbb{R}^n$. Suppose now by contradiction that $f(\mathbf{x}_0) > \sup\{a(\mathbf{x}_0) : a \in \mathcal{C}_0\mathcal{A}_f\}$ for some $\mathbf{x}_0$ and so there exists some $\gamma \in \mathbb{R}$ satisfying

$$f(\mathbf{x}_0) > \gamma > \sup\{a(\mathbf{x}_0) : a \in \mathcal{C}_0\mathcal{A}_f\}. \tag{1.83}$$

If the set $L(f,\gamma)$ is empty, it follows that $f(\mathbf{x}) > \gamma$ for every $\mathbf{x} \in \mathbb{R}^n$ and choosing $c(t) = \gamma$ for every $t \in \mathbb{R}$ and $a(\mathbf{x}) = c(\mathbf{a}^\top \mathbf{x})$ with $\mathbf{a} \in \mathbb{R}^n$ arbitrary, we obtain that $a \in \mathcal{C}_1\mathcal{A}_f \subseteq \mathcal{C}_0\mathcal{A}_f$ contradicting relation (1.83). Therefore the set $L(f,\gamma)$ is nonempty and since the function f is evenly quasiconvex one can find a collection of vectors $(\mathbf{a}_i, b_i)_{i\in I}$ satisfying

$$L(f,\gamma) = \cap_{i\in I} H^{<}(\mathbf{a}_i, b_i). \tag{1.84}$$

By relation (1.83) the vector $\mathbf{x}_0$ does not belong to $L(f,\gamma)$ and this shows by relation (1.84) that there exists some $i \in I$ with a nonzero $\mathbf{a}_i$ satisfying $\mathbf{a}_i^\top \mathbf{x}_0 \geq b_i$. This implies again by relation (1.84) that

$$\{\mathbf{y} \in \mathbb{R}^n : \mathbf{a}_i^\top \mathbf{y} \geq \mathbf{a}_i^\top \mathbf{x}_0\} \subseteq \{\mathbf{y} \in \mathbb{R}^n : f(\mathbf{y}) > \gamma\}. \tag{1.85}$$

Since the vector $\mathbf{a}_i$ is nonzero, the function $c_{\mathbf{a}_i}$, given in Definition 1.29, is nondecreasing and so the function $a(\mathbf{x}) := c_{\mathbf{a}_i}(\mathbf{a}_i^\top \mathbf{x})$ is $\mathcal{C}_0$-affine and by relation (1.85) it satisfies $a(\mathbf{x}_0) \geq \gamma$. Also for every $\mathbf{x} \in \mathbb{R}^n$ we obtain that $a(\mathbf{x}) \leq f(\mathbf{x})$ and so we have constructed a $\mathcal{C}_0$-affine minorant a of the function f satisfying $a(\mathbf{x}_0) \geq \gamma$. This contradicts relation (1.83) and hence we have shown that $f(\mathbf{x}) = \sup\{a(\mathbf{x}) : a \in \mathcal{C}_0\mathcal{A}_f\}$ for every $\mathbf{x} \in \mathbb{R}^n$. To verify the representation for f quasiconvex and l.s.c. we again assume by contradiction that there exists some $\gamma \in \mathbb{R}$ satisfying

$$f(\mathbf{x}_0) > \gamma > \sup\{a(\mathbf{x}_0) : a \in \mathcal{C}_1\mathcal{A}_f\} \tag{1.86}$$

for some $\mathbf{x}_0$. If the convex set $L(f,\gamma)$ is empty then as in the first part we obtain a contradiction. Therefore the closed convex set $L(f,\gamma)$ is nonempty and since by relation (1.86) it holds that $\mathbf{x}_0$ does not belong to $L(f,\gamma)$, there exist by Theorem 1.1 some nonzero vector $\mathbf{a}_0 \in \mathbb{R}^n$ and $\beta \in \mathbb{R}$ satisfying $\mathbf{a}_0^\top \mathbf{x} < \beta < \mathbf{a}_0^\top \mathbf{x}_0$ for every $\mathbf{x} \in L(f,\gamma)$. This implies for every $\mathbf{y}$ satisfying $\mathbf{a}_0^\top \mathbf{y} \geq \beta$ that $f(\mathbf{y}) > \gamma$ and so $c_{\mathbf{a}_0}(\beta) \geq \gamma$. Introducing now the function $a(\mathbf{x}) := c^{\Diamond}_{\mathbf{a}_0}(\mathbf{a}_0^\top \mathbf{x})$ with $c^{\Diamond}_{\mathbf{a}_0}(t)$ listed in Definition 1.20 this implies

$$a(\mathbf{x}_0) = c^{\Diamond}_{\mathbf{a}_0}(\mathbf{a}_0^\top \mathbf{x}_0) = \sup_{s<\mathbf{a}_0^\top \mathbf{x}_0} c_{\mathbf{a}_0}(s) \geq c_{\mathbf{a}_0}(\beta) \geq \gamma.$$

By Lemma 1.32 the function $c^{\Diamond}_{\mathbf{a}_0}$ is l.s.c. and $c^{\Diamond}_{\mathbf{a}_0}(\mathbf{a}_0^\top \mathbf{x}) \leq c_{\mathbf{a}_0}(\mathbf{a}_0^\top \mathbf{x}) \leq f(\mathbf{x})$ for every $\mathbf{x}$. Hence we have constructed a $\mathcal{C}_1$-affine minorant a of the function f satisfying $a(\mathbf{x}_0) \geq \gamma$ and this contradicts relation (1.86).

Therefore $f(\mathbf{x}) = \sup\{a(\mathbf{x}) : a \in \mathcal{C}_1\mathcal{A}_f\}$ for every $\mathbf{x} \in \mathbb{R}^n$ and the proof is completed. □

By Theorem 1.16 it is clear that the set of $\mathcal{C}_1$-affine functions ($\mathcal{C}_0$-affine functions) play the same role for l.s.c. quasiconvex functions (evenly quasiconvex functions) as the affine functions do for l.s.c. convex functions. However, besides this observation, it is also interesting to investigate the question whether these sets of $\mathcal{C}$-affine minorants are the smallest possible class satisfying the above property. In this section we will also pay attention to this question. An immediate consequence of Theorem 1.16 and Lemma 1.47 is given by the next result.

Theorem 1.17 *For any function $f : \mathbb{R}^n \to [-\infty, \infty]$ it follows that $f_{ec}(\mathbf{x}) = \sup\{a(\mathbf{x}) : a \in \mathcal{C}_0\mathcal{A}_f\}$ and $f_{\overline{q}}(\mathbf{x}) = \overline{f_q}(\mathbf{x}) = \sup\{a(\mathbf{x}) : a \in \mathcal{C}_1\mathcal{A}_f\}$ for every $\mathbf{x} \in \mathbb{R}^n$.*

Proof. By Theorem 1.16 we obtain $f_{ec}(\mathbf{x}) = \sup\{a(\mathbf{x}) : a \in \mathcal{C}_0\mathcal{A}_{f_{ec}}\}$ and since by Lemma 1.47 it holds that $\mathcal{C}_0\mathcal{A}_f = \mathcal{C}_0\mathcal{A}_{f_{ec}}$ the first formula follows. The second formula can be verified similarly. □

Studying the proof of Theorem 1.16 for evenly quasiconvex functions one can actually show the following improvement of Theorem 1.17.

Theorem 1.18 *If $f : \mathbb{R}^n \to [-\infty, \infty]$ is an arbitrary function, then it follows for every $\mathbf{x} \in \mathbb{R}^n$ that*

$$f_{ec}(\mathbf{x}) = \sup_{\mathbf{a} \in \mathbb{R}^n} c_{\mathbf{a}}(\mathbf{a}^\top \mathbf{x})$$

with the function $c_{\mathbf{a}}$ given in Definition 1.29.

Proof. It follows for every $\mathbf{a}$ and $\mathbf{x} \in \mathbb{R}^n$ that $c_{\mathbf{a}}(\mathbf{a}^\top \mathbf{x}) \le f(\mathbf{x})$. Since $c_{\mathbf{a}} \in \mathcal{C}_0$ this implies by Lemma 1.46 that the function $\mathbf{x} \to c_{\mathbf{a}}(\mathbf{a}^\top \mathbf{x})$ is evenly quasiconvex and so by Lemma 1.29 we obtain for every $\mathbf{x} \in \mathbb{R}^n$ that $f_{ec}(\mathbf{x}) \ge \sup_{\mathbf{a} \in \mathbb{R}^n} c_{\mathbf{a}}(\mathbf{a}^\top \mathbf{x})$. Suppose now by contradiction that $f_{ec}(\mathbf{x}_0) > \sup_{\mathbf{a} \in \mathbb{R}^n} c_{\mathbf{a}}(\mathbf{a}^\top \mathbf{x}_0)$ for some $\mathbf{x}_0$ and so there exists some $\gamma \in \mathbb{R}$ satisfying

$$f_{ec}(\mathbf{x}_0) > \gamma > \sup_{\mathbf{a} \in \mathbb{R}^n} c_{\mathbf{a}}(\mathbf{a}^\top \mathbf{x}_0). \tag{1.87}$$

If the set $L(f_{ec}, \gamma)$ is empty we obtain $f(\mathbf{x}) \ge f_{ec}(\mathbf{x}) > \gamma$ for every $\mathbf{x} \in \mathbb{R}^n$ and this implies $c_{\mathbf{a}}(\mathbf{a}^\top \mathbf{x}_0) \ge \gamma$ for every $\mathbf{a} \in \mathbb{R}^n$ contradicting relation (1.87). Therefore the set $L(f_{ec}, \gamma)$ is nonempty and since by Lemma 1.29 the function f_{ec} is evenly quasiconvex one can find a collection of vectors $(\mathbf{a}_i, b_i)_{i \in I}$ satisfying

$$L(f_{ec}, \gamma) = \cap_{i \in I} H^{<}(\mathbf{a}_i, b_i). \tag{1.88}$$

By relation (1.87) we know $\mathbf{x}_0$ does not belong to $L(f_{ec}, \gamma)$ and so by relation (1.88) there exists some $i \in I$ and a nonzero vector $\mathbf{a}_i$ satisfying $\mathbf{a}_i^\top \mathbf{x}_0 \geq b_i$. This implies using $f(\mathbf{x}) \geq f_{ec}(\mathbf{x})$ for every $\mathbf{x} \in \mathbb{R}^n$ and relation (1.88) that

$$\{\mathbf{y} \in \mathbb{R}^n : \mathbf{a}_i^\top \mathbf{y} \geq \mathbf{a}_i^\top \mathbf{x}_0\} \subseteq \{\mathbf{y} \in \mathbb{R}^n : f(\mathbf{y}) > \gamma\}$$

and so it follows that $c_{\mathbf{a}_i}(\mathbf{a}_i^\top \mathbf{x}_0) \geq \gamma$. This yields $\sup_{\mathbf{a} \in \mathbb{R}^n} c_{\mathbf{a}}(\mathbf{a}^\top \mathbf{x_0}) \geq c_{\mathbf{a}_i}(\mathbf{a}_i^\top \mathbf{x}_0) \geq \gamma$ contradicting relation (1.87). This shows the desired representation and our proof is completed. □

Also for l.s.c. quasiconvex functions one can show the following improvement of Theorem 1.16. Observe this formula is more complicated than the corresponding formula for evenly quasiconvex functions.

Theorem 1.19 *If* $f : \mathbb{R}^n \to [-\infty, \infty]$ *is an arbitrary function, then it follows for every* $\mathbf{x} \in \mathbb{R}^n$ *that*

$$f_{\overline{q}}(\mathbf{x}) = \overline{f_q}(\mathbf{x}) = \sup_{\mathbf{a} \in \mathbb{R}^n} \overline{c_{\mathbf{a}}}(\mathbf{a}^\top \mathbf{x}) = \sup_{\mathbf{a} \in \mathbb{R}^n} c_{\mathbf{a}}^{\diamondsuit}(\mathbf{a}^\top \mathbf{x})$$

with $\overline{c_{\mathbf{a}}}$ *denoting the l.s.c. hull of the function* $c_{\mathbf{a}}$ *and* $c_{\mathbf{a}}^{\diamondsuit}$ *listed in Definition 1.20.*

Proof. By Lemma 1.32 and relation 1.70 it is sufficient to show for every $\mathbf{x} \in \mathbb{R}^n$ that $\overline{f_q}(\mathbf{x}) = \sup_{\mathbf{a} \in \mathbb{R}^n} c_{\mathbf{a}}^{\diamondsuit}(\mathbf{a}^\top \mathbf{x})$. To verify this we first observe for every $\mathbf{a}$ and $\mathbf{x} \in \mathbb{R}^n$ that $c_{\mathbf{a}}(\mathbf{a}^\top \mathbf{x}) \leq f(\mathbf{x})$ and so we obtain $c_{\mathbf{a}}^{\diamondsuit}(\mathbf{a}^\top \mathbf{x}) \leq f(\mathbf{x})$ for every $\mathbf{x}$. By Lemma 1.32 the function $c_{\mathbf{a}}^{\diamondsuit} : \mathbb{R} \to [-\infty, \infty]$ is l.s.c. and nondecreasing and this implies by Lemma 1.46 that $\mathbf{x} \to c_{\mathbf{a}}^{\diamondsuit}(\mathbf{a}^\top \mathbf{x})$ is quasiconcex and l.s.c.. Therefore we obtain for every $\mathbf{x}$ that

$$\overline{f_q}(\mathbf{x}) \geq \sup_{\mathbf{a} \in \mathbb{R}^n} c_{\mathbf{a}}^{\diamondsuit}(\mathbf{a}^\top \mathbf{x}).$$

Suppose now by contradiction that $\overline{f_q}(\mathbf{x}_0) > \sup_{\mathbf{a} \in \mathbb{R}^n} c_{\mathbf{a}}^{\diamondsuit}(\mathbf{a}^\top \mathbf{x}_0)$ for some $\mathbf{x}_0$ and so there exists some $\gamma \in \mathbb{R}$ satisfying

$$\overline{f_q}(\mathbf{x}_0) > \gamma > \sup_{\mathbf{a} \in \mathbb{R}^n} c_{\mathbf{a}}^{\diamondsuit}(\mathbf{a}^\top \mathbf{x}_0). \tag{1.89}$$

If the set $L(\overline{f_q}, \gamma)$ is empty we obtain $f(\mathbf{x}) \geq \overline{f_q}(\mathbf{x}) > \gamma$ and we obtain as in Theorem 1.18 a contradiction with relation (1.89). Therefore, the closed convex set $L(\overline{f_q}, \gamma)$ is nonempty and since by relation (1.89) it holds that $\mathbf{x}_0$ does not belong to $L(\overline{f_q}, \gamma)$ there exist by Theorem 1.1 some nonzero vector $\mathbf{a}_0 \in \mathbb{R}^n$ and $\beta \in \mathbb{R}$ satisfying $\mathbf{a}_0^\top \mathbf{x} < \beta < \mathbf{a}_0^\top \mathbf{x}_0$ for every $\mathbf{x} \in L(\overline{f_q}, \gamma)$. Hence it follows for every $\mathbf{y}$ satisfying $\mathbf{a}_0^\top \mathbf{y} \geq \beta$ that $f(\mathbf{y}) \geq \overline{f_q}(\mathbf{y}) > \gamma$ and this yields $c_{\mathbf{a}_0}(\beta) \geq \gamma$. Using this observation we obtain

$$c_{\mathbf{a}_0}^{\diamondsuit}(\mathbf{a}_0^\top \mathbf{x}_0) = \sup_{s < \mathbf{a}_0^\top \mathbf{x}_0} c_{\mathbf{a}_0}(s) \geq c_{\mathbf{a}_0}(\beta) \geq \gamma$$

and this contradicts relation (1.89) completing the proof. □

It is also possible to show for every $\mathbf{a} \in \mathbb{R}^n$ that the function $c_{\mathbf{a}}^{\lozenge}$ is actually the inverse of another function.

Lemma 1.48 *If $f : \mathbb{R}^n \to [-\infty, \infty]$ is a function with dom(f) nonempty and the function $h_{\mathbf{a}} : \mathbb{R} \to [-\infty, \infty], \mathbf{a} \in \mathbb{R}^n$ is given by $h_{\mathbf{a}}(\alpha) := \sup\{\mathbf{a}^\top \mathbf{y} : \mathbf{y} \in L(f, \alpha)\}$, then it follows for every $t \in \mathbb{R}$ that*

$$c_{\mathbf{a}}^{\lozenge}(t) = \inf\{\alpha \in \mathbb{R} : h_{\mathbf{a}}(\alpha) \geq t\}.$$

Proof. Since dom(f) is nonempty, there exists some $\alpha \in \mathbb{R}$ satisfying $L(f, \alpha)$ is nonempty. If for some $\alpha_0 \in \mathbb{R}$ it follows that $h_{\mathbf{a}}(\alpha_0) \geq t$, then for every $s < t$ there exists some $\mathbf{y}_0$ satisfying $f(\mathbf{y}_0) \leq \alpha_0$ and $\mathbf{a}^\top \mathbf{y}_0 \geq s$. This implies $\alpha_0 \geq f(\mathbf{y}_0) \geq c_{\mathbf{a}}(s)$ and hence $\inf\{\alpha \in \mathbb{R} : h_{\mathbf{a}}(\alpha) \geq t\} \geq c_{\mathbf{a}}(s)$. Since $s < t$ we obtain $\inf\{\alpha \in \mathbb{R} : h_{\mathbf{a}}(\alpha) \geq t\} \geq \sup_{s<t} c_{\mathbf{a}}(s) = c_{\mathbf{a}}^{\lozenge}(t)$ and to show equality we assume by contradiction that there exists some t_0 satisfying

$$\inf\{\alpha \in \mathbb{R} : h_{\mathbf{a}}(\alpha) \geq t_0\} > c_{\mathbf{a}}^{\lozenge}(t_0).$$

If this holds one can find some α_0 satisfying $\alpha_0 > c_{\mathbf{a}}^{\lozenge}(t_0)$ and $h_{\mathbf{a}}(\alpha_0) < t_0$. Hence there exists some $\epsilon > 0$ satisfying $\alpha_0 > c_{\mathbf{a}}^{\lozenge}(t_0)$ and $h_{\mathbf{a}}(\alpha_0) < t_0 - \epsilon$. Since $h_{\mathbf{a}}(\alpha_0) < t_0 - \epsilon$ we obtain for every $\mathbf{y}$ satisfying $\mathbf{a}^\top \mathbf{y} \geq t_0 - \epsilon$ that $f(\mathbf{y}) > \alpha_0$. This implies $c_{\mathbf{a}}(t_0 - \epsilon) \geq \alpha_0$ and it follows $\alpha_0 > c_{\mathbf{a}}^{\lozenge}(t_0) \geq c_{\mathbf{a}}(t_0 - \epsilon) \geq \alpha_0$. This is clearly a contradiction and the proof is completed. □

In case dom(f) is empty and so $f \equiv \infty$ and we use the well-known convention that $\sup\{\varnothing\} = -\infty$ and $\inf\{\varnothing\} = \infty$ then it is easy to verify that the above relation still holds. The next result first verified in [15] is an immediate consequence of Lemma 1.48 and Theorem 1.19.

Theorem 1.20 *If $f : \mathbb{R}^n \to [-\infty, \infty]$ is an arbitrary function, then it follows that*

$$\overline{f_q}(\mathbf{x}) = f_{\overline{q}}(\mathbf{x}) = \sup_{\mathbf{a} \in \mathbb{R}^n} \inf\{\alpha \in \mathbb{R} : \sup_{\mathbf{y} \in L(f,\, \alpha)} \mathbf{a}^\top \mathbf{y} \geq \mathbf{a}^\top \mathbf{x}\}$$

for every $\mathbf{x} \in \mathbb{R}^n$.

Actually the result in Theorem 1.18 and 1.19 can be seen as a generalization of the Fenchel-Moreau theorem for l.s.c. convex hulls. To show this we need to generalize the notion of conjugate and biconjugate functions used within convex analysis. Since we are dealing with extended real valued functions we use the convention that $(-\infty) + (+\infty) = (+\infty) + (-\infty) = -\infty$ and $-(-\infty) = \infty$.

Definition 1.30 *Let $\mathcal{C}$ be a nonempty collection of extended real valued univariate functions. For any function $f : \mathbb{R}^n \to [-\infty, \infty]$ and $c \in \mathcal{C}$ the function $f^c(\mathbf{a}) := \sup\{c(\mathbf{a}^\top \mathbf{x}) - f(\mathbf{x}) : \mathbf{x} \in \mathbb{R}^n\}$ is called the c-conjugate function of the function f. The function $f^{\mathcal{CC}}(\mathbf{x}) := \sup\{c(\mathbf{x}^\top \mathbf{a}) - f^c(\mathbf{a}) : \mathbf{a} \in \mathbb{R}^n, c \in \mathcal{C}\}$ is called the bi-$\mathcal{C}$-conjugate function of f.*

By a similar proof as in Lemma 1.41 it is easy to give a geometrical interpretation of the biconjugate function.

Lemma 1.49 *For $\mathcal{C}$ a nonempty collection of extended real valued univariate functions and $f : \mathbb{R}^n \to [-\infty, \infty]$ an arbitrary function it follows that $(\mathbf{a}, r) \in epi(f^c)$ if and only if $a \in \mathcal{C}\mathcal{A}_f$ with $a(\mathbf{x}) = c(\mathbf{a}^\top \mathbf{x}) - r$ and $c \in \mathcal{C}$. Additionally, it holds that $f^{\mathcal{CC}}(\mathbf{x}) = \sup\{a(\mathbf{x}) : a \in \mathcal{C}\mathcal{A}_f\}$ for every $\mathbf{x} \in \mathbb{R}^n$.*

Combining now Lemma 1.49 and Theorem 1.17 we immediately obtain for the sets $\mathcal{C}_i, i = 0, 1$ the following generalization of the Fenchel-Moreau theorem.

Theorem 1.21 *For any function $f : \mathbb{R}^n \to [-\infty, \infty]$ it follows that $f^{\mathcal{C}_0\mathcal{C}_0}(\mathbf{x}) = f_{ec}(\mathbf{x})$ and $f^{\mathcal{C}_1\mathcal{C}_1}(\mathbf{x}) = \overline{f_q}(\mathbf{x}) = f_{\overline{q}}(\mathbf{x})$ for every $\mathbf{x} \in \mathbb{R}^n$.*

Proof. By Lemma 1.49 we obtain $f^{\mathcal{C}_i\mathcal{C}_i}(\mathbf{x}) = \sup\{a(\mathbf{x}) : a \in \mathcal{C}_i\mathcal{A}_f\}, i = 0, 1$ and this shows by Theorem 1.17 the desired result. □

By Theorem 1.18, 1.19 and 1.21 we obtain the formulas

$$f^{\mathcal{C}_0\mathcal{C}_0}(\mathbf{x}) = f_{ec}(\mathbf{x}) = \sup_{\mathbf{a} \in \mathbb{R}^n} c_{\mathbf{a}}(\mathbf{a}^\top \mathbf{x})$$

and

$$f^{\mathcal{C}_1\mathcal{C}_1}(\mathbf{x}) = \overline{f_q}(\mathbf{x}) = f_{\overline{q}}(\mathbf{x}) = \sup_{\mathbf{a} \in \mathbb{R}^n} c_{\mathbf{a}}^{\diamondsuit}(\mathbf{a}^\top \mathbf{x}) \tag{1.90}$$

for every $\mathbf{x} \in \mathbb{R}^n$. Considering these formulas we now wonder whether it is possible to achieve the same result using a smaller set of extended real valued univariate functions.

Definition 1.31 *For any $r \in \mathbb{R}$ the function $c_r : \mathbb{R} \to [-\infty, \infty]$ is given by $c_r(t) = -\infty$ for $t < r$ and $c_r(t) = r$ for every $t \geq r$. The set $\mathcal{C}_r \subseteq \mathcal{C}_0$ consists now of all functions $c_r, r \in \mathbb{R}$, while the set $\overline{\mathcal{C}_r}$ consists of all functions $\overline{c_r}$, $r \in \mathbb{R}$ with $\overline{c_r}$ the l.s.c. hull of the function c_r.*

If $f : \mathbb{R}^n \to (-\infty, \infty]$ is an arbitrary function, then for $r \in \mathbb{R}$ and $\mathbf{a} \neq \mathbf{0}$ we obtain

$$f^{c_r}(\mathbf{a}) = \max\{-\infty, \sup\{r - f(\mathbf{y}) : \mathbf{a}^\top \mathbf{y} \geq r\}\} = r - c_{\mathbf{a}}(r) \tag{1.91}$$

with c_a defined in Definition 1.29. Moreover, for $\mathbf{a} = \mathbf{0}$ and $r \leq 0$, it follows that $f^{c_r}(\mathbf{0}) = \sup\{c_r(0) - f(\mathbf{y}) : \mathbf{y} \in \mathbb{R}^n\} = r - \inf\{f(\mathbf{y}) : \mathbf{y} \in \mathbb{R}^n\}$ and this shows

$$f^{c_r}(\mathbf{0}) = r - c_{\mathbf{0}}(r) = r - c_{\mathbf{0}}(0), r \leq 0. \tag{1.92}$$

Also for $r > 0$ it is easy to verify that $f^{c_r}(\mathbf{0}) = -\infty$ and so we have computed for every $r \in \mathbb{R}$ the c_r-conjugate function of the function f. To evaluate the $\overline{c_r}$-conjugate function of f we observe by Lemma 1.32 that $\overline{c_r}(t) = -\infty$ for every $t \leq r$ and $\overline{c_r}(t) = r$ for every $t > r$. Again considering $\mathbf{a} \neq \mathbf{0}$ it follows that

$$\begin{aligned} f^{\overline{c_r}}(\mathbf{a}) &= \max\{-\infty, \sup\{r - f(\mathbf{y}) : \mathbf{a}^\top \mathbf{y} > r\}\} \\ &= r - \inf\{f(\mathbf{y}) : \mathbf{a}^\top \mathbf{y} > r\}. \end{aligned} \tag{1.93}$$

Moreover, for $\mathbf{a} = 0$ and $r < 0$, we obtain that

$$\begin{aligned} f^{\overline{c_r}}(\mathbf{0}) &= \sup\{\overline{c_r}(0) - f(\mathbf{y}) : \mathbf{y} \in \mathbb{R}^n\} \\ &= r - \inf\{f(\mathbf{y}) : \mathbf{y} \in \mathbb{R}^n\} = r - c_{\mathbf{0}}(r), \end{aligned} \tag{1.94}$$

while for $r \geq 0$ it is easy to verify that $f^{\overline{c_r}}(\mathbf{0}) = -\infty$. Using the above computations we will first evaluate in the proof of Lemma 1.50 the bi-$\mathcal{C}_r$-conjugate function of a function $f : \mathbb{R}^n \to (-\infty, \infty]$, while in the proof of Lemma 1.51 the same computation will be carried out for a bi-$\overline{\mathcal{C}_r}$-conjugate function of the same function f.

Lemma 1.50 *For every* $\mathbf{x} \in \mathbb{R}^n$ *and* $f : \mathbb{R}^n \to (-\infty, \infty]$ *it follows for every* $\mathbf{x} \in \mathbb{R}^n$ *that*

$$f^{\mathcal{C}_r\mathcal{C}_r}(\mathbf{x}) = \sup_{\mathbf{a} \in \mathbb{R}^n} \inf\{f(\mathbf{y}) : \mathbf{a}^\top \mathbf{y} \geq \mathbf{a}^\top \mathbf{x}\} = f_{ec}(\mathbf{x}).$$

Proof. By relation (1.92) and $f^{c_r}(\mathbf{0}) = -\infty$ for every $r > 0$ we obtain using the convention $-\infty - (-\infty) = -\infty + \infty = -\infty$ that

$$\sup_{r \in \mathbb{R}}\{c_r(0) - f^{c_r}(\mathbf{0})\} = \sup_{r \leq 0} c_{\mathbf{0}}(0) = c_{\mathbf{0}}(0). \tag{1.95}$$

Also by relation (1.91) and $(-\infty) - (-\infty) = (-\infty) + \infty = -\infty$ it follows for every $\mathbf{x}$ that

$$\sup_{\mathbf{a} \neq \mathbf{0},\, r \in \mathbb{R}}\{c_r(\mathbf{a}^\top \mathbf{x}) - f^{c_r}(\mathbf{a})\} = \sup_{\mathbf{a} \neq \mathbf{0},\, r \leq \mathbf{a}^\top \mathbf{x},\, r \in \mathbb{R}} c_{\mathbf{a}}(r).$$

This shows, using $c_{\mathbf{a}}$ is nondecreasing for every $\mathbf{a} \neq \mathbf{0}$, that

$$\sup_{\mathbf{a} \neq \mathbf{0},\, r \in \mathbb{R}}\{c_r(\mathbf{a}^\top \mathbf{x}) - f^{c_r}(\mathbf{a})\} = \sup_{\mathbf{a} \neq \mathbf{0}} c_{\mathbf{a}}(\mathbf{a}^\top \mathbf{x}) \tag{1.96}$$

and so $f^{\mathcal{C}_r\mathcal{C}_r}(\mathbf{x}) = \sup_{\mathbf{a}\in\mathbb{R}^n} c_{\mathbf{a}}(\mathbf{a}^\top\mathbf{x})$ using relations (1.95) and (1.96). This shows the first equality and the second one is already listed in Theorem 1.18. □

The next result yields a similar result as Lemma 1.50 for a quasiconvex and l.s.c. function.

Lemma 1.51 *For every* $\mathbf{x} \in \mathbb{R}^n$ *and* $f : \mathbb{R}^n \to (-\infty, \infty]$ *it follows for every* $\mathbf{x} \in \mathbb{R}^n$ *that*

$$f^{\overline{\mathcal{C}}_r\overline{\mathcal{C}}_r}(\mathbf{x}) = \sup_{\mathbf{a}\in\mathbb{R}^n} \sup_{s<\mathbf{a}^\top\mathbf{x}} \inf\{f(\mathbf{y}) : \mathbf{a}^\top\mathbf{y} \geq s\} = \overline{f_q}(\mathbf{x}) = f_{\overline{q}}(\mathbf{x}).$$

Proof. By relation (1.93) and $f^{\overline{c_r}}(\mathbf{0}) = -\infty$ for every $r \geq 0$ we obtain using $-\infty - (-\infty) = -\infty + \infty = -\infty$ that

$$\sup_{r\in\mathbb{R}}\{\overline{c_r}(0) - f^{\overline{c_r}}(\mathbf{0})\} = \sup_{r<0} c_{\mathbf{0}}(r) = c_{\mathbf{0}}^{\Diamond}(0). \tag{1.97}$$

Also by relation (1.92) and $(-\infty)-(-\infty) = (-\infty)+\infty = -\infty$ it follows with $h(\mathbf{x}) := \sup_{\mathbf{a}\neq\mathbf{0},\ r\in\mathbb{R}}\{\overline{c_r}(\mathbf{a}^\top\mathbf{x}) - f^{\overline{c_r}}(\mathbf{a})\}$ that

$$h(\mathbf{x}) = \sup_{\mathbf{a}\neq\mathbf{0},\ r<\mathbf{a}^\top\mathbf{x},\ r\in\mathbb{R}} \inf\{f(\mathbf{y}) : \mathbf{a}^\top\mathbf{y} > r\}. \tag{1.98}$$

Since $\inf\{f(\mathbf{y}) : \mathbf{a}^\top\mathbf{y} > r\} \geq c_{\mathbf{a}}(r)$ for every $r \in \mathbb{R}$ and $\mathbf{a} \neq \mathbf{0}$ we obtain by relation (1.98) that

$$h(\mathbf{x}) \geq \sup_{\ \mathbf{a}\neq\mathbf{0},\ r<\mathbf{a}^\top\mathbf{x},\ r\in\mathbb{R}} c_{\mathbf{a}}(r) = \sup_{\mathbf{a}\neq\mathbf{0}} c_{\mathbf{a}}^{\Diamond}(\mathbf{a}^\top\mathbf{x}). \tag{1.99}$$

Applying now relations (1.90), (1.97) and (1.99) it holds for every $\mathbf{x} \in \mathbb{R}^n$ that

$$f^{\mathcal{C}_r\mathcal{C}_r}(\mathbf{x}) \geq \sup_{\mathbf{a}\in\mathbb{R}^n} c_{\mathbf{a}}^{\Diamond}(\mathbf{a}^\top\mathbf{x}) = \overline{f_q}(\mathbf{x}) = f^{\mathcal{C}_1\mathcal{C}_1}(\mathbf{x}). \tag{1.100}$$

Since $\overline{\mathcal{C}_r} \subseteq \mathcal{C}_1$ it follows that $f^{\mathcal{C}_1\mathcal{C}_1} \geq f^{\mathcal{C}_r\mathcal{C}_r}$ and this shows by relation (1.100) the desired result. □

In the last two lemmas we have shown that it is sufficient for any function f satisfying $f > -\infty$ to consider the class of $\mathcal{C}_r$-affine minorants and the class of $\overline{\mathcal{C}_r}$-affine minorants for approximating f_{ec}, respectively $\overline{f_q}$. This concludes the section on quasiconvex duality. In the next section we will discuss some important applications.

4. On applications of convex and quasiconvex analysis

In this section we will discuss different applications of the theory of convex and quasiconvex analysis. In Subsection 4.1 we consider applications to noncooperative game theory, while in Subsection 4.2 we discuss its applications to optimization problems and in particular to Lagrangian duality. Finally in Subsection 4.3 we will use the duality representation of evenly quasiconvex functions to show that every positively homogeneous evenly quasiconvex function satisfying $f(\mathbf{0}) = 0$ and $f > -\infty$ is actually the minimum of two positively homogeneous l.s.c. convex functions. This result was first verified by Crouzeix (cf. [15]) for a slightly smaller class of quasiconvex functions and serves as a very nice application of quasiconvex duality.

4.1 Minimax theorems and noncooperative game theory

To introduce the field of infinite antagonistic game theory (cf.[72]) we assume that the set of pure strategies of player 1 is given by some nonempty set $A \subseteq \mathbb{R}^n$, while the set of pure strategies of player 2 is given by $B \subseteq \mathbb{R}^m$. If player 1 chooses the pure strategy $\mathbf{a} \in A$ and player 2 chooses the pure strategy $\mathbf{b} \in B$, then player 2 has to pay to player 1 an amount $f(\mathbf{a}, \mathbf{b})$ with $f : A \times B \to [0, \infty]$ a given function. This function is called the payoff function and for simplicity this function is taken to be nonnegative. Since player 1 likes to gain as much profit as possible, but at the moment he does not know how to achieve this, he first decides to compute a lower bound on his profit. To compute this lower bound player 1 argues as follows : if he decides to choose action $\mathbf{a} \in A$, then it follows that he wins at least $\inf_{\mathbf{b} \in B} f(\mathbf{a}, \mathbf{b})$ irrespective of the action of player 2. Therefore a lower bound on the profit for player 1 is given by

$$r_* := \sup_{\mathbf{a} \in A} \inf_{\mathbf{b} \in B} f(\mathbf{a}, \mathbf{b}). \tag{1.101}$$

Similarly player 2 likes to minimize his losses but since he does not know how to achieve this he also decides to compute first an upper bound on his losses. To compute this upper bound player 2 argues as follows. If he decides to choose action $\mathbf{b}$ it follows that he loses at most $\sup_{\mathbf{a} \in A} f(\mathbf{a}, \mathbf{b})$ and this is independent of the action of player 1. Therefore an upper bound on his losses is given by

$$r^* := \inf_{\mathbf{b} \in B} \sup_{\mathbf{a} \in A} f(\mathbf{a}, \mathbf{b}). \tag{1.102}$$

Since the profit of player 1 is at least r_* and the losses of player 2 is at most r^* and the losses of player 2 are the profits of player 1 it follows directly that $r_* \leq r^*$. In general $r_* < r^*$, but under some properties on the action set and payoff function one can show that $r_* = r^*$. By the above inequality it follows immediately that $r_* = r^*$ for $r^* = -\infty$ and so we assume in the remainder of this section that $r^* > -\infty$. The equality $r_* = r^*$ is called a minimax result and if additionally inf and sup are attained an optimal strategy for both players can be easily derived. For player 1 it is possible to achieve at least a profit r_*, independent of the action of player 2, while for player 2 it is possible to achieve at most a loss r^* independent of the action of player 1. Since $r^* = r_* := v$ and both players have opposite interests, they will choose an action which achieves the value v and so player 1 will choose that action $\mathbf{a}_0 \in A$ satisfying

$$\inf_{\mathbf{b}\in B} f(\mathbf{a}_0, \mathbf{b}) = \max_{\mathbf{a}\in A} \inf_{\mathbf{b}\in B} f(\mathbf{a}, \mathbf{b}).$$

Moreover, player 2 will choose that strategy $\mathbf{b}_0 \in B$ satisfying

$$\sup_{\mathbf{a}\in A} f(\mathbf{a}, \mathbf{b}_0) = \min_{\mathbf{b}\in B} \sup_{\mathbf{a}\in A} f(\mathbf{a}, \mathbf{b}).$$

Since for $r_* = r^*$ and the additional assumption that the infimum and supremum are attained, it is clear how the optimal strategies should be chosen we will investigate in this subsection for which payoff functions and strategies the minimax result $r_* = r^*$ holds. Before discussing this, we give the following example for which this equality does not hold.

Example 1.14 Consider the continuous payoff function $f : [0,1] \times [0,1] \to [0,\infty)$ given by $f(a,b) = (a-b)^2$. For this function it holds for every $0 \leq a \leq 1$ that $\inf_{b\in[0,1]}(a-b)^2 = 0$ and so $r_* := \sup_{0\leq a\leq 1} \inf_{0\leq b\leq 1}(a-b)^2 = 0$. Moreover, it follows that $\sup_{0\leq a\leq 1}(a-b)^2 = (1-b)^2$ for every $0 \leq b < \frac{1}{2}$ and $\sup_{0\leq a\leq 1}(a-b)^2 = b^2$ for every $\frac{1}{2} \leq b \leq 1$. This shows $r^* := \inf_{0\leq b\leq 1} \sup_{0\leq a\leq 1}(a-b)^2 = \frac{1}{4}$ and so r_* does not equal r^*. For this example it is not obvious which strategies should be selected by the two players.

By extending the sets of the so-called pure strategies of each player it is possible to show under certain conditions that the extended game satisfies a minimax result. In the next definition we introduce the set of *mixed strategies.*

Definition 1.32 *For a nonempty set D of pure strategies and $\mathbf{d} \in D$ let $\epsilon_{\mathbf{d}}$ denote the one-point probability measure concentrated on the set $\{\mathbf{d}\}$ and denote by $\mathcal{P}_D$ the set of all probability measures on D with a finite support.*

Introducing the unit simplex $\Delta_k := \{\alpha : \sum_{i=1}^k \alpha_i = 1, \alpha_i \geq 0, 1 \leq i \leq k\}$ it follows by Definition 1.32 that λ belongs to the set $\mathcal{P}_D$ if and only if there exist some $k \in \mathbb{N}$ and set $\{\mathbf{d}_1, ..., \mathbf{d}_k\} \subseteq D$ consisting of different elements such that

$$\lambda = \sum_{i=1}^{k} \lambda_i \epsilon_{\mathbf{d}_i}, (\lambda_1, ..., \lambda_k) \in \Delta_k \text{ and } \lambda_i > 0.$$

Clearly the set $\mathcal{P}_D$ can seen as the convex hull of the set $\{\epsilon_{\mathbf{d}} : \mathbf{d} \in D\}$ and so it is convex. A game theoretic interpretation of a strategy $\lambda \in \mathcal{P}_D$ is now given by the following. If a player with pure strategy set D selects the mixed strategy $\lambda = \sum_{i=1}^{k} \lambda_i \epsilon_{\mathbf{d}_i} \in \mathcal{P}_D$, then with probability $\lambda_i, 1 \leq i \leq k$ this player will use the pure strategy $\mathbf{d}_i \in D$. By this interpretation it is clear that the set D of pure strategies can be identified with the set of one-point Borel probability measures $\{\epsilon_{\mathbf{d}} : \mathbf{d} \in D\}$. We now assume that player 1 uses the set $\mathcal{P}_A$ of mixed strategies and the same holds for player 2 using the set $\mathcal{P}_B$. This means that the payoff function f should be extended to a function $f_e : \mathcal{P}_A \times \mathcal{P}_B \to \mathbb{R}$ and this extension is given by

$$f_e(\lambda, \mu) := \sum_{i=1}^{k} \sum_{j=1}^{l} \lambda_i \mu_j f(\mathbf{a}_i, \mathbf{b}_j) \tag{1.103}$$

with $\lambda = \sum_{i=1}^{k} \lambda_i \epsilon_{\mathbf{a}_i} \in \mathcal{P}_A$ and $\mu = \sum_{j=1}^{l} \mu_j \epsilon_{\mathbf{b}_j} \in \mathcal{P}_B$. This extension represents the expected profit for player 1 or expected loss of player 2 if player 1 selects the mixed strategy $\lambda \in \mathcal{P}_A$ and player 2 selects the mixed strategy $\mu \in \mathcal{P}_B$. Without any conditions on the pure strategy sets A and B and the function f one can show the next result.

Lemma 1.52 *For any set A and B of pure strategies it follows that*

$$\inf_{\mu \in \mathcal{P}_B} \sup_{\lambda \in \mathcal{P}_A} f_e(\lambda, \mu) = \inf_{\mu \in \mathcal{P}_B} \sup_{\mathbf{a} \in A} f_e(\epsilon_{\mathbf{a}}, \mu)$$

and

$$\sup_{\lambda \in \mathcal{P}_A} \inf_{\mu \in \mathcal{P}_B} f_e(\lambda, \mu) = \sup_{\lambda \in \mathcal{P}_A} \inf_{\mathbf{b} \in B} f_e(\lambda, \epsilon_{\mathbf{b}}).$$

Proof. Since $\{\epsilon_{\mathbf{a}} : \mathbf{a} \in A\} \subseteq \mathcal{P}_A$ it follows that

$$\inf_{\mu \in \mathcal{P}_B} \sup_{\lambda \in \mathcal{P}_A} f_e(\lambda, \mu) \geq \inf_{\mu \in \mathcal{P}_B} \sup_{\mathbf{a} \in A} f_e(\epsilon_{\mathbf{a}}, \mu).$$

To verify the reverse inequality we observe for every mixed strategy $\mu \in \mathcal{P}_B, \lambda \in \mathcal{P}_A$ and relation (1.103) that $f_e(\lambda, \mu) \leq \sup_{\mathbf{a} \in A} f_e(\epsilon_{\mathbf{a}}, \mu)$. This implies

$$\sup_{\lambda \in \mathcal{P}_A} f_e(\lambda, \mu) \leq \sup_{\mathbf{a} \in A} f_e(\epsilon_{\mathbf{a}}, \mu)$$

and so the first formula is verified. The second formula can be shown by exactly the same argument. □

It is now possible to show that the extended game given by f_e and the mixed strategy sets $\mathcal{P}_A$ and $\mathcal{P}_B$ satisfies a minimax result under some topological conditions on the function f and the sets A and B of pure strategies. The next result was first given by Ville (cf. [70], [18], [72]) using a much more complicated proof. In the next alternative proof we only use the separation result for convex sets listed in Theorem 1.3 and the well-known result that a continuous function on a compact set is uniformly continuous (cf. [43]).

Theorem 1.22 *If the pure strategy sets $A \subseteq \mathbb{R}^n$ and $B \subseteq \mathbb{R}^m$ are compact and the function $f : A \times B \to \mathbb{R}$ is continuous, then it follows that*

$$\inf_{\mu \in \mathcal{P}_B} \sup_{\lambda \in \mathcal{P}_A} f_e(\lambda, \mu) = \sup_{\lambda \in \mathcal{P}_A} \inf_{\mu \in \mathcal{P}_B} f_e(\lambda, \mu).$$

Proof. It is easy to see that the inequality $\geq$ holds and so we only need to verify the reverse inequality. By Lemma 1.52 it is now sufficient to show that $\inf_{\mu \in \mathcal{P}_B} \sup_{\mathbf{a} \in A} f_e(\epsilon_{\mathbf{a}}, \mu) \leq \sup_{\lambda \in \mathcal{P}_A} \inf_{\mathbf{b} \in B} f_e(\lambda, \epsilon_{\mathbf{b}})$. By scaling we may assume that

$$\sup_{\lambda \in \mathcal{P}_A} \inf_{\mathbf{b} \in B} f_e(\lambda, \epsilon_{\mathbf{b}}) = 1 \tag{1.104}$$

and so need to show that

$$\inf_{\mu \in \mathcal{P}_B} \sup_{\mathbf{a} \in A} f_e(\epsilon_{\mathbf{a}}, \mu) \leq 1$$

Assume now by contradiction that there exists some $\gamma > 0$ satisfying

$$\sup_{\mathbf{a} \in A} f_e(\epsilon_{\mathbf{a}}, \mu) \geq 1 + \gamma \tag{1.105}$$

for every $\mu \in \mathcal{P}_B$. Since the function f is continuous on the compact set $A \times B$, it is well-known (cf. [64], [43]) that the function f is uniformly continuous on $A \times B$. Hence there exists some $\delta > 0$ such that for every $\mathbf{a}_1, \mathbf{a}_2 \in A$ satisfying $\|\mathbf{a}_1 - \mathbf{a}_2\| \leq \delta$ it follows that $\sup_{\mathbf{b} \in B} |f(\mathbf{a}_1, \mathbf{b}) - f(\mathbf{a}_2, \mathbf{b}| \leq \frac{\gamma}{2}$. This implies for every $\mathbf{a}_1, \mathbf{a}_2 \in A$ satisfying $\|\mathbf{a}_1 - \mathbf{a}_2\| \leq \delta$ that

$$\sup_{\mu \in \mathcal{P}_B} |f_e(\epsilon_{\mathbf{a}_1}, \mu) - f_e(\epsilon_{\mathbf{a}_2}, \mu)| \leq \frac{\gamma}{2}. \tag{1.106}$$

Since A is compact one can find a finite set $I \subseteq A$ satisfying $A \subseteq \cup_{\mathbf{a} \in I}(\mathbf{a} + \delta E)$ and this shows by relations (1.106) and (1.105) that

$$\max_{\mathbf{a} \in I} f_e(\epsilon_{\mathbf{a}}, \mu) \geq \sup_{\mathbf{a} \in A} f_e(\epsilon_{\mathbf{a}}, \mu) - \frac{\gamma}{2} \geq 1 + \frac{\gamma}{2} \tag{1.107}$$

for every $\mu \in \mathcal{P}_B$. Introducing the convex set V given by

$$V := co(\{(f_e(\epsilon_{\mathbf{a}}, \epsilon_{\mathbf{b}}))_{\mathbf{a} \in I}, \mathbf{b} \in B\}) \subseteq \mathbb{R}^{|I|}$$

it follows by the definition of V that $\mathbf{z}^\top = (z_1, ..., z_{|I|})$ belongs to V if and only if there exists some mixed strategy $\mu \in \mathcal{P}_B$ satisfying $\mathbf{z}^\top = (f_e(\epsilon_{\mathbf{a}}, \mu))_{\mathbf{a} \in I}$. This implies by relation (1.107) that

$$V \subseteq \{\mathbf{z} \in \mathbb{R}^{|I|} : \max_{1 \leq i \leq |I|} z_i \geq 1 + \frac{\gamma}{2}\}$$

and so the convex sets $\{\mathbf{z} \in \mathbb{R}^{|I|} : \max_{i \leq i \leq |I|} z_i < 1 + \frac{\gamma}{2}\}$ and V are disjoint. Applying now Theorem 1.3 one can find some mixed strategy $\lambda \in \mathcal{P}_I$ satisfying $\inf_{\mathbf{b} \in B} f_e(\lambda, \epsilon_{\mathbf{b}}) \geq 1 + \frac{\gamma}{2}$ and this contradicts relation (1.104). □

Actually the result in Theorem 1.22 holds under weaker topological conditions on the function f. However the proof of that result uses the Riesz representation theorem for the set of continuous functions on a compact Hausdorff space, the Banach-Alaoglu theorem and infinite dimensional separation (cf. [29]) and is beyond the scope of this chapter. The result listed in Theorem 1.22 is the most important result in infinite antagonistic game theory and fits within a chain of equivalent minimax theorems (cf. [28]). For one of these equivalent minimax results another alternative proof using also finite dimensional separation is given in [26]. Although not listed in [28], one result which also fits within this chain is the famous Sion's minimax theorem (cf. [66]) for quasiconcave-quasiconvex bifunctions.

Theorem 1.23 *If $A \subseteq \mathbb{R}^n$ is compact and convex, $B \subseteq \mathbb{R}^m$ is convex and the function $f : A \times B \to \mathbb{R}$ satisfies $\mathbf{a} \to f(\mathbf{a}, \mathbf{b})$ is quasiconcave and upper semicontinuous for every $\mathbf{b} \in B$ and $\mathbf{b} \to f(\mathbf{a}, \mathbf{b})$ is quasiconvex and lower semicontinuous for every $\mathbf{a} \in A$, then it follows that*

$$\max_{\mathbf{a} \in A} \inf_{\mathbf{b} \in B} f(\mathbf{a}, \mathbf{b}) = \inf_{b \in B} \max_{a \in A} f(\mathbf{a}, \mathbf{b}).$$

This result was proved using the Knaster-Kuratowski-Mazurkiewicz (KKM) lemma (cf. [74]). This lemma is the basis of fixed point theory and nonlinear functional analysis. It is also possible to give a more elementary proof of Sion's minimax theorem based on finite dimensional separation between convex sets. However, the most elementary proof of Sion's minimax theorem is given by an adaptation of the so-called level set method due to Joó (cf. [40], [41], [42]). This method first translates the minimax equality into an equivalent geometrical condition of a nonempty intersection of a collection of upper-level sets. Under the assumptions of Sion's minimax theorem it is now possible to verify this geometrical condition using compactness arguments and the well-known elementary topological result that every convex set is connected (cf. [36]). To start with our analysis we first introduce for every $\mathbf{a} \in A$ and $\mathbf{b} \in B$ the functions $f_{\mathbf{a}} : B \to \mathbb{R}$ and $f_{\mathbf{b}} : A \to \mathbb{R}$ defined by

$$f_{\mathbf{a}}(\mathbf{b}) = f_{\mathbf{b}}(\mathbf{a}) := f(\mathbf{a}, \mathbf{b}). \tag{1.108}$$

Also we introduce for every $r \in \mathbb{R}$ and $\mathbf{b} \in B$ the upper-level set $U(f_{\mathbf{b}}, r) \subseteq A$ given by

$$U(f_{\mathbf{b}}, r) := \{\mathbf{a} \in A : f_{\mathbf{b}}(\mathbf{a}) \geq r\}. \tag{1.109}$$

It is now easy to show the following result (cf. [41]).

Lemma 1.53 *It follows that* $r^* = r_*$ *if and only if* $\cap_{b \in B} U(f_{\mathbf{b}}, r) \neq \emptyset$ *for every* $r < r^*$.

Proof. If $r^* = r_* > -\infty$, then for every $r < r^*$ there exist by the definition of r_* some $\mathbf{a}_0 \in A$ satisfying $\inf_{\mathbf{b} \in B} f(\mathbf{a}_0, \mathbf{b}) > r$. This shows that $\mathbf{a}_0$ belongs to the intersection $\cap_{\mathbf{b} \in B} U(f_{\mathbf{b}}, r)$ and so $\cap_{\mathbf{b} \in B} U(f_{\mathbf{b}}, r)$ is nonempty. To verify the reverse implication it is sufficient to verify that $r_* \geq r^*$ or equivalently $r_* > r^* - \epsilon$ for every $\epsilon > 0$. Consider now $r := r^* - \epsilon$ for some $\epsilon > 0$. By our assumption it follows that the intersection $\cap_{\mathbf{b} \in B} U(f_{\mathbf{b}}, r)$ is nonempty and so there exists some $\mathbf{a}_0 \in A$ satisfying $\inf_{\mathbf{b} \in B} f(\mathbf{a}_0, \mathbf{b}) \geq r$. This implies that $r_* = \sup_{\mathbf{a} \in A} \inf_{\mathbf{b} \in B} f(\mathbf{a}, \mathbf{b}) \geq r$ and so the proof is completed. □

By Lemma 1.53 we need to show that $\cap_{\mathbf{b} \in B} U(f_{\mathbf{b}}, r) \neq \emptyset$ for every $r < r^*$. Before proving this result we consider an arbitrary finite set $\{\mathbf{b}_0, \mathbf{b}_1, ..., \mathbf{b}_k\} \subseteq B$ and introduce the affine mapping $p : [0, 1] \to B$, given by

$$p(\lambda) = \lambda \mathbf{b}_0 + (1 - \lambda)\mathbf{b}_1, \tag{1.110}$$

and the set valued mapping $\Phi_r : [0, 1] \to 2^A$, given by

$$\Phi_r(\lambda) = (\cap_{i=2}^k U(f_{\mathbf{b}_i}, r)) \cap U(f_{p(\lambda)}, r). \tag{1.111}$$

To verify the main result we need the following elementary lemma.

Lemma 1.54 *If the functions* $f_{\mathbf{a}} : B \to \mathbb{R}$ *are quasiconvex on the convex set* B *for every* $\mathbf{a} \in A$, *then it follows for every* $\lambda_0, \lambda_1 \in [0, 1]$ *and* $0 < \alpha < 1$ *that*

$$\Phi_r(\alpha\lambda_0 + (1 - \alpha)\lambda_1) \subseteq \Phi_r(\lambda_0) \cup \Phi_r(\lambda_1)$$

for every $r \in \mathbb{R}$.

Proof. If the vector $\mathbf{a}$ belongs to $\Phi_r(\alpha\lambda_0 + (1 - \alpha)\lambda_1)$, then by definition $\mathbf{a} \in \cap_{i=2}^k U(f_{\mathbf{b}_i}, r)$ and $f(\mathbf{a}, p(\alpha\lambda_0 + (1 - \alpha)\lambda_1)) \geq r$. This implies, using p is affine, that

$$f(\mathbf{a}, \alpha p(\lambda_0) + (1 - \alpha)p(\lambda_1)) \geq r \tag{1.112}$$

and by the quasiconvexity of the functions $f_{\mathbf{a}}$ we obtain by relation (1.112) that $\max\{f(\mathbf{a}, p(\lambda_0)), f(\mathbf{a}, p(\lambda_1))\} \geq r$. Hence it follows that $\mathbf{a}$ belongs to $\Phi_r(\lambda_0) \cup \Phi_r(\lambda_1)$ and the result is proved. □

In order to prove the next important lemma we denote by $\mathcal{F}(B)$ the set of all finite subsets of B.

Lemma 1.55 *If the functions $f_{\mathbf{b}} : A \to \mathbb{R}$ are quasiconcave and upper semicontinuous for every $\mathbf{b} \in B$ and the functions $f_{\mathbf{a}} : B \to \mathbb{R}$ are quasiconvex and lower semicontinuous for every $\mathbf{a} \in A$, then it follows for every J belonging to $\mathcal{F}(B)$ and $r < r^*$ that $\cap_{\mathbf{b}\in J} U(f_{\mathbf{b}}, r) \neq \emptyset$.*

Proof. If J is a subset of B consisting of one element the result clearly holds by the definition of r^* listed in relation (1.102). Suppose now for all sets J belonging to $\mathcal{F}(B)$ and consisting of at most k elements that

$$\cap_{b\in J} U(f_{\mathbf{b}}, r) \neq \emptyset \tag{1.113}$$

for every $r < r^*$. To prove the result for all sets J belonging to $\mathcal{F}(B)$ consisting of at most $k+1$ elements, we assume by contradiction that there exists some set $J = \{\mathbf{b}_0, ..., \mathbf{b}_k\} \subseteq B$ and some $r < r^*$ satisfying

$$\cap_{i=0}^{k} U(f_{\mathbf{b}_i}, r) = \emptyset. \tag{1.114}$$

Consider now for the points $\mathbf{b}_0$ and $\mathbf{b}_1$ the set valued mapping $\Phi_r : [0,1] \to 2^A$ given by relation (1.111). By our induction hypothesis listed in relation (1.113) and the assumption that the functions $f_{\mathbf{b}}, \mathbf{b} \in B$ are quasiconcave and upper semicontinuous we obtain that the sets $\Phi_r(\lambda)$ are nonempty, closed and convex for every $0 \leq \lambda \leq 1$. By relation (1.114) it follows that

$$\Phi_r(0) \cap \Phi_r(1) = \emptyset \tag{1.115}$$

and so the nonempty sets

$$S_i := \{0 \leq \lambda \leq 1 : \Phi_r(\lambda) \subseteq \Phi_r(i)\}, i = 0, 1$$

are disjoint and $S_0 \cup S_1 \subseteq [0,1]$. To show that $S_0 \cup S_1 = [0,1]$ consider for a given $0 \leq \lambda \leq 1$ the closed sets

$$A_i := \Phi_r(\lambda) \cap \Phi_r(i), i = 0, 1. \tag{1.116}$$

By Lemma 1.54 we obtain that $\Phi_r(\lambda) \subseteq \Phi_r(0) \cup \Phi_r(1)$ and so

$$A_0 \cup A_1 = \Phi_r(\lambda). \tag{1.117}$$

Also by relation (1.115) the sets A_0 and A_1 are disjoint and since $\Phi_r(\lambda)$ is convex and hence connected (cf. [36]) we obtain by relation (1.117) that either A_0 or A_1 is empty. This implies using again Lemma 1.54 that $\Phi_r(\lambda) \subseteq \Phi_r(0)$ or $\Phi_r(\lambda) \subseteq \Phi_r(1)$ and so $\lambda \in S_0 \cup S_1$. Hence we have shown that the sets S_0 and S_1 satisfy

$$S_0 \cap S_1 = \emptyset \text{ and } S_0 \cup S_1 = [0,1]. \tag{1.118}$$

We will now verify that the sets S_0 and S_1 are open in $[0,1]$ and to do so consider some $\lambda_0 \in S_0$ (a similar proof applies to S_1). Since $\Phi_r(\lambda_0)$ is nonempty for every $r < r^*$ it follows by the definition of $\Phi_r(\lambda_0)$ that

$$\sup_{\mathbf{a}\in A} \inf_{\mathbf{b}\in B_0} f(\mathbf{a},\mathbf{b}) \geq r^* > r$$

with $B_0 := \{\mathbf{b}_2, ..., \mathbf{b}_k, \lambda_0\mathbf{b}_0 + (1-\lambda_0)\mathbf{b}_1\}$. This means that there exists some $\mathbf{a}_0 \in \cap_{i=2}^k U(f_{\mathbf{b}_i}, r)$ satisfying

$$f(\mathbf{a}_0, \lambda_0\mathbf{b}_0 + (1-\lambda_0)\mathbf{b}_1) > r \tag{1.119}$$

and by lower semicontinuity of the function $f_{\mathbf{a}_0}$ and relation (1.119) there exist some $\epsilon > 0$ such that

$$f(\mathbf{a}_0, \lambda\mathbf{b}_0 + (1-\lambda)\mathbf{b}_1) > r$$

for every $\lambda \in \mathcal{N} := (\lambda_0 - \epsilon, \lambda_0 + \epsilon) \cap [0,1]$. Hence we obtain that $\mathbf{a}_0 \in \Phi_r(\lambda)$ for every $\lambda \in \mathcal{N}$ and since $\lambda_0 \in S_0$ this implies by relations (1.118) and (1.15) that $\Phi_r(\lambda) \subseteq \Phi_r(0)$ for every $\lambda \in \mathcal{N}$. Hence S_0 is an open set and since similarly S_1 is open we obtain by relation (1.118) and $[0,1]$ connected that either S_0 or S_1 is empty. This yields a contradiction with $S_i, i = 0,1$ nonempty and so relation (1.114) cannot hold. □

It is now possible to give a proof of Sion's minimax result.

Proof. (Sion's minimax theorem). Since A is compact and $f_{\mathbf{b}}$ is upper semicontinuous we obtain that the set $U(f_{\mathbf{b}}, r)$ is compact. By the finite intersection property for compact sets we obtain by Lemma 1.55 that $\cap_{\mathbf{b}\in B} U(f_{\mathbf{b}}, r) \neq \emptyset$ for every $r < r^*$ and this shows by Lemma 1.53 that $r^* = r_*$. Since A compact and $f_{\mathbf{b}}$ upper semicontinuous and $f_{\mathbf{a}}$ lower semicontinuous it follows by a standard argument that we may replace sup by max. □

Actually we can also apply Sion's minimax theorem to prove Theorem 1.22. Looking at the proof of Theorem 1.22 we observe in relation (1.107) that

$$\inf_{\mu\in\mathcal{P}_B} \max_{\mathbf{a}\in I} f_e(\epsilon_{\mathbf{a}}, \mu) \geq 1 + \frac{\gamma}{2}$$

with I belonging to $\mathcal{F}(A)$. This shows by Lemma 1.52 that

$$\inf_{\mu\in\mathcal{P}_B} \max_{\lambda\in\mathcal{P}_I} f_e(\lambda, \mu) \geq 1 + \frac{\gamma}{2}. \tag{1.120}$$

To the expression in relation (1.120) we may now apply Sion's minimax theorem and so we obtain

$$\max_{\lambda \in \mathcal{P}_I} \inf_{\mu \in \mathcal{P}_B} f_e(\lambda, \mu) \geq 1 + \frac{\gamma}{2}$$

and in a similar way we obtain a contradiction with relation (1.104). In the next subsection we will consider applications of convex analysis to optimization theory.

4.2 Optimization theory and duality

In this subsection we will show how the tools of convex analysis can be used within optimization theory. In particular we introduce the dual of an optimization problem and derive some important properties of this dual problem. To start with a general introduction to optimization theory let $f : \mathbb{R}^n \to [-\infty, \infty]$ be an arbitrary function and consider the so-called primal optimization problem given by

$$v(P) := \inf\{f(\mathbf{x}) : \mathbf{x} \in \mathbb{R}^n\}. \tag{P}$$

In this optimization problem the infimum need not be attained. Since f represents an extended real valued function the above optimization problem also covers optimization problems with restrictions. Associate now with the function f a function $F : \mathbb{R}^n \times \mathbb{R}^m \to [-\infty, \infty]$ satisfying $F(\mathbf{x}, \mathbf{0}) = f(\mathbf{x})$ for every $\mathbf{x}$ and consider the so-called *perturbation function* $p : \mathbb{R}^m \to [-\infty, \infty]$ given by

$$p(\mathbf{y}) := \inf\{F(\mathbf{x}, \mathbf{y}) : \mathbf{x} \in \mathbb{R}^n\}. \tag{1.121}$$

It is easy to verify (remember the strict epigraph and the effective domain of a function are listed in relation (1.43) and (1.45)!) that

$$\widetilde{epi}(p) = A(\widetilde{epi}(F)) \text{ and } dom(p) = A(dom(F)) \tag{1.122}$$

with $A : \mathbb{R}^{n+m} \to \mathbb{R}^m$ the projection of $\mathbb{R}^{n+m}$ onto $\mathbb{R}^m$ given by $A(\mathbf{x}, \mathbf{y}) = \mathbf{y}$. Also by the definition of the function F we obtain that $p(\mathbf{0}) = v(P)$. In the next definition we introduce the dual of the optimization problem (P) (cf. [63]).

Definition 1.33 *The so-called dual problem of optimization problem (P) is given by*

$$v(D) := \sup\{-p^*(\mathbf{a}) : \mathbf{a} \in \mathbb{R}^m\} \tag{D}$$

with p^* the conjugate function of p listed in Definition 1.25.

By Definitions 1.33 and 1.25 it follows that $v(D) = p^{**}(\mathbf{0})$ and since $p^{**}(\mathbf{0}) \leq p(\mathbf{0})$ the inequality $v(D) \leq v(P)$ always holds. We are now interested under which conditions on the perturbation function p it follows that $v(D) = v(P)$. If $v(P) = -\infty$, then the inequality $v(D) \leq v(P)$ implies $v(D) = v(P) = -\infty$ and every $\mathbf{a} \in \mathbb{R}^m$ is an optimal solution of the dual problem (D). Therefore we only need to consider $v(P) > -\infty$. Consider now the cases $v(P)$ is finite and $v(P) = \infty$. Observe the last case only happens if $\text{dom}(f)$ is empty. For $v(P)$ finite, one can now show the following result. This result is a direct consequence of Theorem 1.12 giving a dual characterization of a convex function (Fenchel-Moreau theorem) and Theorem 1.13.

Theorem 1.24 *If the function $p : \mathbb{R}^m \to [-\infty, \infty]$ is convex and $p(\mathbf{0})$ is finite, then it follows that*

$$v(P) = v(D) \Leftrightarrow \textit{the function } p \textit{ is l.s.c.at } \mathbf{0}.$$

Moreover, if $\mathbf{0}$ belongs to ri(dom(p)), then the dual problem has an optimal solution and $v(D) = v(P)$.

Proof. Since the function p is convex, l.s.c. at $\mathbf{0}$ and $p(\mathbf{0})$ finite it follows by relations (1.59) and (1.66) that $p_{\overline{c}}(\mathbf{0}) = \overline{p_{c}}(\mathbf{0}) = \overline{p}(\mathbf{0}) = p(\mathbf{0})$ is finite and this implies by Lemma 1.40 that $\mathcal{A}_p$ is nonempty. Therefore $\overline{p}(\mathbf{x}) > -\infty$ for every $\mathbf{x}$ and by the Fenchel-Moreau theorem (Theorem 1.12) we obtain $v(P) = p(\mathbf{0}) = \overline{p}(\mathbf{0}) = p^{**}(\mathbf{0}) = v(D)$. To prove the reverse implication we observe by Theorem 1.12 and $v(P) = v(D)$ is finite that $p(\mathbf{0}) = p^{**}(\mathbf{0}) = cl(p)(\mathbf{0})$ is finite. Hence it must follow by Definition 1.26 that $\overline{p}(\mathbf{0}) = p(\mathbf{0})$ and by relation (1.59) the function p is l.s.c. at $\mathbf{0}$. To show the second part it follows by Theorem 1.13 that $\partial p(\mathbf{0})$ is nonempty and by Lemma 1.43 it is now easy to verify that any $\mathbf{a}_0 \in \partial p(\mathbf{0})$ is an optimal solution of the dual problem. Moreover, by Lemma 1.36 we obtain that $p(\mathbf{0}) = \overline{p}(\mathbf{0})$ and we can apply the first part. □

Finally we consider the case $v(P) = \infty$. In general it does not hold even for p convex and l.s.c. in $\mathbf{0}$ that $v(P) = v(D)$. To show this we will discuss in Example 1.16 a linear programming problem satisfying $v(P) = \infty$ and $v(D) = -\infty$.

If $f : \mathbb{R}^n \to \mathbb{R}$ is some real valued function and $\mathbf{g} : \mathbb{R}^n \to \mathbb{R}^m$ a vector valued function represented by $\mathbf{g}(\mathbf{x}) := (g_1(\mathbf{x}), ..., g_m(\mathbf{x}))$, $g_i : \mathbb{R}^n \to \mathbb{R}$, then an important special case of optimization problem (P) is given by

$$\inf\{f(\mathbf{x}) : \mathbf{g}(\mathbf{x}) \in -K, \mathbf{x} \in D\} \qquad (P_1)$$

with $K \subseteq \mathbb{R}^m$ a nonempty convex cone and $D \subseteq \mathbb{R}^n$ some nonempty set. The above optimization problem includes some important classes of optimization problems listed in the following example.

Example 1.15

1 If $f(\mathbf{x}) = \mathbf{c}^\top \mathbf{x}$ and $\mathbf{g}(\mathbf{x}) = A\mathbf{x} - \mathbf{b}$ with A some $m \times n$ matrix, $K = \{\mathbf{0}\} \subseteq \mathbb{R}^m$ and $D = \mathbb{R}^n_+$, then optimization problem (P_1) reduces to the so-called *linear programming problem* (cf. [4], [54], [19])

$$\inf\{\mathbf{c}^\top \mathbf{x} : A\mathbf{x} = \mathbf{b}, \mathbf{x} \geq \mathbf{0}\}.$$

2 If $f(\mathbf{x}) = \mathbf{c}^\top \mathbf{x}$ and $\mathbf{g}(\mathbf{x}) = A\mathbf{x} - \mathbf{b}$ with A some $m \times n$ matrix, $K = \{\mathbf{0}\} \subseteq \mathbb{R}^m$ and $D \subseteq \mathbb{R}^n$ is some closed convex cone, then optimization problem (P_1) reduces to a so-called *conic convex programming problem* (cf. [53]), given by

$$\inf\{\mathbf{c}^\top \mathbf{x} : \mathbf{A}\mathbf{x} = \mathbf{b}, \mathbf{x} \in D\}.$$

3 If $m = n$ and $g(\mathbf{x}) = -\mathbf{x}$, then optimization problem (P_1) reduces to a so-called *generalized geometric programming problem* (cf. [57]), given by

$$\inf\{f(\mathbf{x}) : \mathbf{x} \in K \cap D\}.$$

4 If the nonempty convex cone $K \subseteq \mathbb{R}^m$ is given by $K = \mathbb{R}^p_+ \times \{\mathbf{0}\}$ with $\mathbf{0} \in \mathbb{R}^{m-p}, p \leq m$ and the set $D = \mathbb{R}^n$, then optimization problem (P_1) reduces to the classical *nonlinear programming problem* (cf. [3], [54], [19])

$$\inf\{f(\mathbf{x}) : g_i(\mathbf{x}) \leq 0, i = 1, ..., p, g_i(\mathbf{x}) = 0, p + 1 \leq i \leq n\}.$$

For optimization problem (P_1) the so-called *Lagrangian perturbation scheme* is used and this means that the function $F : \mathbb{R}^n \times \mathbb{R}^m \to [-\infty, \infty]$ is given by

$$F(\mathbf{x}, \mathbf{y}) = \begin{cases} f(\mathbf{x}) & \text{for } \mathbf{x} \in D \text{ and } \mathbf{g}(\mathbf{x}) \in -K + \mathbf{y} \\ \infty & \text{otherwise} \end{cases}.$$

For this specific choice of F we obtain by relation (1.121) that

$$p(\mathbf{y}) = \inf\{f(\mathbf{x}) : \mathbf{x} \in D, \mathbf{y} \in \mathbf{g}(\mathbf{x}) + K\}. \tag{1.123}$$

Using the representation of p, listed in relation (1.123), one can give a more detailed expression of the dual problem. Observe this dual problem is called the *Lagrangian dual problem.*

Lemma 1.56 *If the function $\theta : K^0 \to [-\infty, \infty]$ is given by $\theta(\mathbf{a}) = \inf\{f(\mathbf{x}) - \mathbf{a}^\top \mathbf{g}(\mathbf{x}) : \mathbf{x} \in D\}$, then the Lagrangian dual of optimization problem (P_1) equals*

$$v(LD) := \sup\{\theta(\mathbf{a}) : \mathbf{a} \in K^0\}. \tag{LD}$$

Proof. By the definition of the function p it follows for every $\mathbf{a} \in \mathbb{R}^n$ that

$$\begin{aligned} -p^*(\mathbf{a}) &= -\sup_{\mathbf{y}\in\mathbb{R}^m}\{\mathbf{a}^\top \mathbf{y} - \inf\{f(\mathbf{x}) : \mathbf{y} \in \mathbf{g}(\mathbf{x}) + K, \mathbf{x} \in D\}\} \\ &= -\sup_{\mathbf{y}\in\mathbb{R}^m} \sup\{\mathbf{a}^\top \mathbf{y} - f(\mathbf{x}) \colon \mathbf{y} \in \mathbf{g}(\mathbf{x}) + K, \mathbf{x} \in D\} \\ &= \inf\{f(\mathbf{x}) - \mathbf{a}^\top \mathbf{y} : \mathbf{y} \in \mathbf{g}(\mathbf{x}) + K, \mathbf{x} \in D\}. \end{aligned}$$

This shows

$$-p^*(\mathbf{a}) = \inf\{f(\mathbf{x}) - \mathbf{a}^\top(\mathbf{g}(\mathbf{x}) + \mathbf{k}) : \mathbf{k} \in K, \mathbf{x} \in D\}$$

and to simplify the above expression we first consider a vector $\mathbf{a}$ belonging to K^0. Since by definition $\mathbf{a}^\top \mathbf{k} \leq 0$ for every $\mathbf{k} \in K$ and $\mathbf{0} \in cl(K)$ this implies

$$-p^*(\mathbf{a}) = \inf\{f(\mathbf{x}) - \mathbf{a}^\top \mathbf{g}(\mathbf{x}) \colon \mathbf{x} \in D\} = \theta(\mathbf{a}).$$

Moreover, if the vector $\mathbf{a}$ does not belong to K^0 one can find some $\mathbf{k}_0 \subset K$ satisfying $\mathbf{a}^\top \mathbf{k}_0 > 0$. Since $\alpha \mathbf{k}_0 \in K$ for every $\alpha > 0$ and the set D is not empty this yields $-p^*(\mathbf{a}) = -\infty$ and the desired result is verified. □

By Lemmas 1.24 and 1.56 the following result about the Lagrangian dual problem is easy to derive.

Theorem 1.25 *If the primal problem is represented by* (P_1) *and the vector valued function* $\mathbf{h} : \mathbb{R}^n \to \mathbb{R}^{m+1}$ *is given by* $\mathbf{h}(\mathbf{x}) := (\mathbf{g}(\mathbf{x}), f(\mathbf{x}))$ *and satisfies* $\mathbf{h}(D) + (K \times (0, \infty))$ *is convex and* $\mathbf{0} \in ri(\mathbf{g}(D) + K)$*, then it follows that* $\infty > v(P_1) = v(LD)$ *and the Lagrangian dual problem (LD) has an optimal solution.*

Proof. Since by assumption $\mathbf{0}$ belongs to $ri(\mathbf{g}(D) + K) \subseteq \mathbf{g}(D) + K$ we obtain that the feasible region of the optimization problem (P_1) is not empty and this shows $v(P_1) < \infty$. For $v(P_1) = -\infty$ the result follows immediately and so we only consider $v(P_1)$ is finite. To apply Theorem 1.24 we first need to verify whether the function p is convex. It is easy to check that

$$\widetilde{epi}(F) = \{(\mathbf{x}, \mathbf{y}, r) \in \mathbb{R}^{n+m+1} : \mathbf{y} \in \mathbf{g}(\mathbf{x}) + K, \mathbf{x} \in D \text{ and } r > f(\mathbf{x})\}$$

and this implies by relation (1.122) that $\widetilde{epi}(p) = \mathbf{h}(D) + (K \times (0, \infty))$. By assumption this set is convex and hence by Lemma 1.24 the perturbation function p is convex. Also by relation (1.122) we obtain $ri(dom(p)) =$

$ri(\mathbf{g}(D) + K)$ and applying Lemma 1.56 and Theorem 1.24 the desired result follows. □

The condition $\mathbf{0} \in ri(\mathbf{g}(D) + K)$ is known in the literature as the *generalized Slater condition.* Observe, if f is a convex function and $\mathbf{g}$ is a so-called K-convex vector valued function (cf. [73], [6]), then it follows that $\widetilde{epi}(F)$ is a convex set and hence also $\mathbf{h}(D)+(K\times(0,\infty))$ is convex. Also it is possible to prove related results under slightly weaker conditions (cf. [25],[24]). As shown by the next lemma the Lagrangian dual (LD) of a conic convex programming problem is again a conic convex programming problem. Due to the recent developments in interior point methods this class of optimization problems became very important (cf. [53]).

Lemma 1.57 *If the primal problem* (P_1) *is a conic convex programming problem given by*

$$\inf\{\mathbf{c}^\top\mathbf{x} : A\mathbf{x} = \mathbf{b}, \mathbf{x} \in D\}$$

with $D \subseteq \mathbb{R}^n$ *some closed convex cone and there exists some* $\mathbf{x}_0 \in ri(D)$ *satisfying* $A\mathbf{x}_0 = \mathbf{b}$, *then it follows that*

$$\infty > v(P_1) = v(LD) = \inf\{\mathbf{a}^\top\mathbf{b} : A^\top\mathbf{a} - \mathbf{s} = \mathbf{c}, \mathbf{s} \in D^0\}$$

and the last dual conic convex optimization problem has an optimal solution.

Proof. By part 2 of Example 1.15 we know that a conic convex programming problem is a special case of optimization problem (P_1) with $K = \{\mathbf{0}\} \subseteq \mathbb{R}^m$, the vector valued function $\mathbf{h}$, listed in Theorem 1.25, given by $\mathbf{h}(\mathbf{x}) = (A\mathbf{x}-\mathbf{b}, \mathbf{c}^\top\mathbf{x})$ and $D \subseteq \mathbb{R}^n$ a closed convex cone. Clearly for this choice the set $\mathbf{h}(D) + (\{\mathbf{0}\} \times (0,\infty))$ is convex. Moreover, by Lemma 1.18 the generalized Slater condition reduces to

$$\mathbf{0} \in ri(\mathbf{g}(D) + \{\mathbf{0}\}) = ri(A(D) - \mathbf{b}) = A(ri(D)) - \mathbf{b}$$

and by our assumption this condition is satisfied. Therefore the above result is an immediate consequence of Theorem 1.25 once we have evaluated $\theta(\mathbf{a})$ for $\mathbf{a} \in K^0 = \mathbb{R}^m$. Observe now that

$$\begin{aligned}\theta(\mathbf{a}) &= \inf\{\mathbf{c}^\top\mathbf{x} - \mathbf{a}^\top(A\mathbf{x} - \mathbf{b}) : \mathbf{x} \in D\} \\ &= \mathbf{a}^\top\mathbf{b} + \inf\{(\mathbf{c} - A^\top\mathbf{a})^\top\mathbf{x} : \mathbf{x} \in D\}\end{aligned} \tag{1.124}$$

and since

$$\inf\{(\mathbf{c} - A^\top\mathbf{a})^\top\mathbf{x} : \mathbf{x} \in D\} = \begin{cases} 0 & \text{for } A^\top\mathbf{a} - \mathbf{c} \in D^0 \\ -\infty & \text{otherwise} \end{cases}$$

the desired result follows by Theorem 1.25. □

Using Lemma 1.57 with $D = \mathbb{R}^n_+$ it follows that the Lagrangian dual of the linear programming problem $\inf\{\mathbf{c}^\top \mathbf{x} : A\mathbf{x} = \mathbf{b}, \mathbf{x} \geq \mathbf{0}\}$ is given by $\sup\{\mathbf{b}^\top \mathbf{a} : A^\top \mathbf{a} \leq \mathbf{c}\}$ and so this dual problem reduces to the ordinary dual listed in many text books (cf. [4]). Since the set $D = \mathbb{R}^n_+$ is a polyhedral convex cone (cf. [63]), the generalized Slater condition in Lemma 1.57 can be replaced by the condition that the feasible region of the linear programming problem is nonempty. Actually it can be shown for every polyhedral convex cone D that the associated conic convex programming problem reduces to a linear programming problem and so it is only useful to consider conic convex programming problems with a nonpolyhedral convex cone D. It is also possible to extend the above duality results for conic convex programming problems to a larger class of problems than the one having a generalized Slater point and for more details on this the reader is referred to [67]. To conclude this section we consider the following example of a linear programming problem satisfying $v(P) = \infty$ and $v(D) = -\infty$.

Example 1.16 Consider the linear programming problem

$$\inf\{-x_1 - x_2 : x_1 - x_2 \geq 1, -x_1 + x_2 \geq 1, \mathbf{x} \in \mathbb{R}^2_+\}.$$

Clearly this optimization problem has an empty feasible region and so $v(P_1) = \infty$. Penalizing the restrictions $x_1 - x_2 - 1 \geq 0$ and $-x_1 + x_2 - 1 \geq 0$ using the nonpositive Lagrangian multipliers a_1 and a_2 we obtain that the Lagrangian function $\theta : \mathbb{R}^2_+ \to [-\infty, \infty)$ is given by

$$\theta(\mathbf{a}) = \inf\{x_1(\lambda_1 - \lambda_2 - 1) + x_2(\lambda_2 - \lambda_1 - 1) : \mathbf{x} \in \mathbb{R}^2_+\}$$

Observe now for every $\mathbf{a} \in \mathbb{R}^2_+$ that

$$a_1 - a_2 - 1 \geq 0 \Rightarrow a_2 - a_1 - 1 \leq -2$$

and

$$a_2 - a_1 - 1 \geq 0 \Rightarrow a_1 - a_2 - 1 \leq -2$$

and by this observation it follows that $\theta(\mathbf{a}) = -\infty$ for every $\mathbf{a} \in \mathbb{R}^2_+$ or equivalently $v(D) = -\infty$.

One can also use the same Lagrangian perturbation scheme and the dual representation of an evenly quasiconvex function and the corresponding c_r-conjugate function to introduce the so-called surrogate dual. Due to limited space we will not discuss the properties of such a dual but refer the reader to the literature cited in [27]. This concludes our discussion on duality and optimization problems. In the next subsection we will consider the structure of positively homogeneous evenly quasiconvex functions.

4.3 Positively homogeneous evenly quasiconvex functions and dual representations

In this subsection the dual representation of an evenly quasiconvex function is used to show a remarkable property of a positively homogeneous evenly quasiconvex function. In [11] a similar property is also derived for a positively homogeneous quasiconvex function. As such the results in [11] apply to a larger class of functions, but are slightly weaker. Also the proof technique used in [11] is more direct and based on the geometrical aspects of convexity, whereas the approach used in this chapter is a natural consequence of the dual representation of an evenly quasiconvex function discussed in Subsection 3.4. To start with the dual approach we consider a positively homogeneous evenly quasiconvex function $f : \mathbb{R}^n \to (-\infty, \infty]$ satisfying $\mathbf{0} \in dom(f)$. Since f is positively homogeneous and $f(\mathbf{x}) > -\infty$ for every $\mathbf{x}$ we obtain by Lemma 1.25 that $\mathbf{0} \in dom(f)$ if and only if $f(\mathbf{0}) = 0$. Considering for every $\mathbf{a} \in \mathbb{R}^n$ the function $c_{\mathbf{a}} : \mathbb{R} \to [-\infty, \infty)$,given by

$$c_{\mathbf{a}}(t) := \inf\{f(\mathbf{y}) : \mathbf{a}^\top \mathbf{y} \geq t\}, \tag{1.125}$$

(see also Definition 1.29) it is easy to verify the next result.

Lemma 1.58 *If $f : \mathbb{R}^n \to [-\infty, \infty]$ is positively homogeneous, then for every $\mathbf{a} \in \mathbb{R}^n$ it follows that the function $c_{\mathbf{a}} : \mathbb{R}^n \to [-\infty, \infty]$ is positively homogeneous and nondecreasing.*

Proof. For any nonzero vector $\mathbf{a}$ it is obvious by relation (1.125) that the function $c_{\mathbf{a}}$ is nondecreasing. Also by Lemma 1.25 we obtain for every $\alpha > 0$ and $t \in \mathbb{R}$ that

$$c_{\mathbf{a}}(\alpha t) = \inf\{f(\alpha \mathbf{y}) : \mathbf{a}^\top \mathbf{y} \geq t\} = \alpha c_{\mathbf{a}}(t)$$

and so the result is verified for every nonzero $\mathbf{a}$. Moreover, for $\mathbf{a} = \mathbf{0}$, we obtain for every $\alpha > 0$ and $t \leq 0$ that

$$c_{\mathbf{0}}(\alpha t) = \inf\{f(\mathbf{y}) : \mathbf{y} \in \mathbb{R}^n\} = \inf\{f(\alpha \mathbf{y}) : \mathbf{y} \in \mathbb{R}^n\} = \alpha c_{\mathbf{0}}(t),$$

while for $\alpha > 0$ and $t > 0$ it follows using the convention $\inf\{\varnothing\} = \infty$, that $c_{\mathbf{0}}(\alpha t) = \inf\{\varnothing\} = \infty = \alpha c_{\mathbf{0}}(t)$. Trivially the function $c_{\mathbf{0}}$ is nondecreasing and the proof is completed. □

To analyze the behaviour of a positively homogeneous evenly quasiconvex function f satisfying $f(\mathbf{x}) > -\infty$ for every $\mathbf{x}$ and $\mathbf{0} \in dom(f)$, we first decompose this function. Using a slightly different decomposition as done by Crouzeix (cf. [15],[11]) we introduce the nonnegative function $f_+ : \mathbb{R}^n \to [0, \infty]$, given by

$$f_+(\mathbf{x}) := \begin{cases} 0 & \text{if } \mathbf{x} \in \widetilde{L}(f, 0) \\ f(\mathbf{x}) & \text{otherwise} \end{cases} . \tag{1.126}$$

with the strict lower level set $\widetilde{L}(f, 0)$ of the function f of level 0 listed in relation (1.52). Using now $f(\mathbf{x}) > -\infty$ for every $\mathbf{x}$ and $\mathbf{0} \in dom(f)$ we immediately obtain that $f_+(\mathbf{0}) = f(\mathbf{0}) = 0$. Moreover, the function $f_- : \mathbb{R}^n \to (-\infty, \infty]$ is given by

$$f_-(\mathbf{x}) = \begin{cases} f(\mathbf{x}) & \text{if } \mathbf{x} \in cl(\widetilde{L}(f, 0)) \\ \infty & \text{otherwise} \end{cases} . \tag{1.127}$$

To analyze the function f_- it is only interesting to consider positively homogeneous evenly quasiconvex functions f satisfying $\widetilde{L}(f, 0)$ is nonempty. If this holds, we obtain by Lemma 1.25 that $\widetilde{L}(f, 0)$ is a nonempty convex cone and since $\mathbf{0} \in cl(\widetilde{L}(f, 0))$ it follows that $f_-(\mathbf{0}) = f(\mathbf{0}) = 0$. Also for every $r \in \mathbb{R}$ we obtain that

$$\widetilde{L}(f_-, r) = cl(\widetilde{L}(f, 0)) \cap \widetilde{L}(f, r) \tag{1.128}$$

and this yields for $r = 0$ that $\widetilde{L}(f_-, 0) = \widetilde{L}(f, 0)$. By relation (1.127) we therefore obtain for $\widetilde{L}(f, 0)$ is not empty that

$$dom(f_-) \subseteq cl(\widetilde{L}(f, 0)) = cl(\widetilde{L}(f_-, 0)). \tag{1.129}$$

Since trivially $f_-(\mathbf{x}) \geq f(\mathbf{x})$ for every $\mathbf{x}$ it is easy to verify considering the cases $f(\mathbf{x}) \geq 0$ and $f(\mathbf{x}) < 0$ that

$$f(\mathbf{x}) = \min\{f_+(\mathbf{x}), f_-(\mathbf{x})\} \tag{1.130}$$

for every $\mathbf{x} \in \mathbb{R}^n$. For the functions f_+ and f_- one can now show the following result.

Lemma 1.59 *If $f : \mathbb{R}^n \to [-\infty, \infty]$ is a positively homogeneous evenly quasiconvex function, then the functions f_+ and f_- are positively homogeneous and evenly quasiconvex.*

Proof. Since f is positively homogeneous and evenly quasiconvex (and hence quasiconvex) we obtain by Lemma 1.25 and 1.27 that $\widetilde{L}(f, 0)$ is a (possibly empty) convex cone. This implies again by Lemma 1.25 that f_+ is positively homogeneous. To show that f_+ is evenly quasiconvex we observe that $L(f_+, r) = L(f, r)$ for every $r > 0$. Also by the definition of f_+ we obtain

$$L(f_+, 0) = \widetilde{L}(f, 0) \cup \{\mathbf{x} : \mathbf{x} \notin \widetilde{L}(f, 0) \text{ and } f(\mathbf{x}) \leq 0\} = L(f, 0).$$

This shows, using the fact that f_+ is a nonnegative function and f is evenly quasiconvex, that also f_+ is evenly quasiconvex. To verify the same result for f_- we observe, since $cl(\widetilde{L}(f, 0))$ is also a (possibly empty) convex cone, that f_- is positively homogeneous. Moreover, for every $r \in \mathbb{R}$ we know by relation (1.128) that $L(f_-, r) = cl(\widetilde{L}(f, 0)) \cap L(f, r)$ and applying Lemma 1.21 and f is evenly quasiconvex it follows that f_- is evenly quasiconvex. □

We will now apply the dual representation of an evenly quasiconvex function and show the following result for a nonnegative positively homogeneous evenly quasiconvex function f with $\mathbf{0} \in dom(f)$. A related result is also discussed in [11]. Recall that a function is called sublinear, if it is positively homogeneous and convex.

Lemma 1.60 *If $f : \mathbb{R}^n \to [-\infty, \infty]$ is a nonnegative positively homogeneous evenly quasiconvex function with $\mathbf{0} \in dom(f)$, then f is a nonnegative l.s.c. sublinear function.*

Proof. By the dual representation of an evenly quasiconvex function (see Theorem 1.18) we obtain that

$$f(\mathbf{x}) = f_{ec}(\mathbf{x}) = \sup_{\mathbf{a} \in \mathbb{R}^n} c_{\mathbf{a}}(\mathbf{a}^\top \mathbf{x}). \tag{1.131}$$

Since $f \geq 0$ it follows by the definition of $c_{\mathbf{a}}$ that $c_{\mathbf{a}}$ is a nonnegative function for every $\mathbf{a} \in \mathbb{R}^n$. Moreover, using $f(\mathbf{0}) = 0$ and $0 \leq c_{\mathbf{a}}(0) \leq f(\mathbf{0})$ we obtain $c_{\mathbf{a}}(0) = 0$. Also for $\mathbf{x} \in \mathbb{R}^n$ and $\mathbf{a} \in \mathbb{R}^n$ satisfying $\mathbf{a}^\top \mathbf{x} \leq 0$ it follows by the monotonicity of $c_{\mathbf{a}}$ that $0 \leq c_{\mathbf{a}}(\mathbf{a}^\top \mathbf{x}) \leq c_{\mathbf{a}}(0) = 0$ and this implies $c_{\mathbf{a}}(\mathbf{a}^\top \mathbf{x}) = 0$ for every $\mathbf{a}^\top \mathbf{x} \leq 0$. Moreover, for $\mathbf{a}^\top \mathbf{x} > 0$ we obtain by Lemma 1.58 that $c_{\mathbf{a}}(\mathbf{a}^\top \mathbf{x}) = r_{\mathbf{a}} \mathbf{a}^\top \mathbf{x}$ with $r_{\mathbf{a}} := c_{\mathbf{a}}(1) \geq 0$ and combining both observations yields

$$c_{\mathbf{a}}(\mathbf{a}^\top \mathbf{x}) = \max\{r_{\mathbf{a}} \mathbf{a}^\top \mathbf{x}, 0\}$$

for every $\mathbf{a} \in \mathbb{R}^n$. Applying now relation (1.131) yields

$$f(\mathbf{x}) = \sup_{\mathbf{a} \in \mathbb{R}^n} \max\{r_{\mathbf{a}} \mathbf{a}^\top \mathbf{x}, 0\} = \max\{\sup_{\mathbf{a} \in \mathbb{R}^n} r_{\mathbf{a}} \mathbf{a}^\top \mathbf{x}, 0\} \tag{1.132}$$

and since $\mathbf{x} \to \sup_{\mathbf{a} \in \mathbb{R}^n} r_{\mathbf{a}} \mathbf{a}^\top \mathbf{x}$ is a l.s.c. sublinear function the desired result follows by relation (1.132). □

Since by relation (1.126) we obtain that $f_+(\mathbf{x}) = \max\{f(\mathbf{x}), 0\}$ and for f positively homogeneous and evenly quasiconvex the function f_+

is also positively homogeneous and evenly quasiconvex (Lemma 1.59) we may apply Lemma 1.60 and so we obtain the result that f_+ is a nonnegative l.s.c. sublinear function in case the function f is positively homogeneous, evenly quasiconvex, $\mathbf{0} \in dom(f)$ and $f(\mathbf{x}) > -\infty$ for every $\mathbf{x}$. Finally we will show the following result for a positively homogeneous evenly quasiconvex function f satisfying $\widetilde{L}(f,0)$ nonempty, $dom(f) \subseteq cl(\widetilde{L}(f,0))$ and $f(\mathbf{x}) > -\infty$ for every $\mathbf{x}$..

Lemma 1.61 *If $f : \mathbb{R}^n \to (-\infty, \infty]$ is a positively homogeneous evenly quasiconvex function with $\mathbf{0} \in dom(f) \subseteq cl(\widetilde{L}(f,0))$, then f is a nonpositive l.s.c. sublinear function.*

Proof. By the dual representation of an evenly quasiconvex function we obtain (see Theorem 1.18) that

$$f(\mathbf{x}) = \sup_{\mathbf{a} \in \mathbb{R}^n} c_{\mathbf{a}}(\mathbf{a}^\top \mathbf{x}). \tag{1.133}$$

If the vector $\mathbf{a}$ does not belong to the polar cone $(\widetilde{L}(f,0))^0$, then there exists some $\mathbf{x}_0$ satisfying $f(\mathbf{x}_0) < 0$ and $r := \mathbf{a}^\top \mathbf{x}_0 > 0$. By Lemma 1.58 this yields for every $t > 0$ that

$$c_{\mathbf{a}}(t) = c_{\mathbf{a}}(tr^{-1}r) = tr^{-1}c_{\mathbf{a}}(r) \leq tr^{-1}f(\mathbf{x}_0) < 0$$

and so $c_{\mathbf{a}}(\infty) := \lim_{t \uparrow \infty} c_{\mathbf{a}}(t) = -\infty$. Since the function $c_{\mathbf{a}}$ is nondecreasing, this shows that $c_{\mathbf{a}}(t) = -\infty$ for every $t \in \mathbb{R}$ and using the fact that $f(\mathbf{x}) > -\infty$ for every $\mathbf{x}$ and relation (1.133) we obtain

$$f(\mathbf{x}) = \sup\{c_{\mathbf{a}}(\mathbf{a}^\top \mathbf{x}) : \mathbf{a} \in (\widetilde{L}(f,0))^0\}. \tag{1.134}$$

If the vector $\mathbf{a}$ belongs to $(\widetilde{L}(f,0))^0$ and $\mathbf{a}^\top \mathbf{x} \geq t > 0$ for some $\mathbf{x} \in \mathbb{R}^n$, then clearly $\mathbf{x}$ does not belong to $cl(\widetilde{L}(f,0))$. Since $dom(f) \subseteq cl(\widetilde{L}(f,0))$ this implies that $f(\mathbf{x}) = \infty$ and so we have shown for every $\mathbf{a}$ belonging to $(\widetilde{L}(f,0))^0$ that

$$c_{\mathbf{a}}(t) = \infty \text{ for every } t > 0. \tag{1.135}$$

To analyze $c_{\mathbf{a}}(t)$ for $\mathbf{a} \in (\widetilde{L}(f,0))^0$ and $t \leq 0$ we first assume that there exists some $\mathbf{x}_0$ satisfying $f(\mathbf{x}_0) < 0$ and $\mathbf{a}^\top \mathbf{x}_0 = 0$. By Lemma 1.58 it holds that $\alpha c_{\mathbf{a}}(0) = c_{\mathbf{a}}(0)$ for every $\alpha > 0$ and since $c_{\mathbf{a}}(0) \leq f(\mathbf{x}_0) < 0$ we obtain that $c_{\mathbf{a}}(0) = -\infty$. Hence it follows that $c_{\mathbf{a}}(t) \leq c_{\mathbf{a}}(0) = -\infty$ for every $t \leq 0$ and we have shown for every $\mathbf{a} \in (\widetilde{L}(f,0))^0$, for which there exists some $\mathbf{x}_0 \in \widetilde{L}(f,0)$ satisfying $\mathbf{a}^\top \mathbf{x}_0 = 0$, that

$$c_{\mathbf{a}}(t) = -\infty \text{ for every } t \leq 0 \tag{1.136}$$

Using again the fact that $f(\mathbf{x}) > -\infty$ for every $\mathbf{x}$ and relations (1.134), (1.135) and (1.136) yields

$$f(\mathbf{x}) = \sup\{c_{\mathbf{a}}(\mathbf{a}^\top \mathbf{x}) : \mathbf{a} \in D\} \tag{1.137}$$

for every $\mathbf{x} \in \mathbb{R}^n$ with

$$D := \{\mathbf{a} \in (\widetilde{L}(f,0))^0 : \mathbf{a}^\top \mathbf{y} < 0 \text{ for every } \mathbf{y} \in \widetilde{L}(f,0)\}.$$

We will now analyze the behaviour of $\mathbf{x} \to c_{\mathbf{a}}(\mathbf{a}^\top \mathbf{x})$ for an arbitrary $\mathbf{a}$ belonging to D. If $\mathbf{a}^\top \mathbf{x} > 0$ it follows by relation (1.135) that $c_{\mathbf{a}}(\mathbf{a}^\top \mathbf{x}) = \infty$. Also, if $\mathbf{a}^\top \mathbf{x} = 0$, then for every $\mathbf{y}$ satisfying $\mathbf{a}^\top \mathbf{y} \geq \mathbf{a}^\top \mathbf{x} = 0$ we obtain, using $\mathbf{a} \in D$, that $f(\mathbf{y}) \geq 0$ and since $\mathbf{0} \in dom(f)$ this implies $c_{\mathbf{a}}(\mathbf{a}^\top \mathbf{x}) = 0$. Finally, for $\mathbf{a} \in D$ and $\mathbf{a}^\top \mathbf{x} < 0$ it follows by Lemma 1.58 that $c_{\mathbf{a}}(\mathbf{a}^\top \mathbf{x}) = q_{\mathbf{a}} \mathbf{a}^\top \mathbf{x}$ with $q_{\mathbf{a}} := -c_{\mathbf{a}}(-1)$ and since $\widetilde{L}(f,0)$ is nonempty we obtain by relation (1.137) that $0 < q_{\mathbf{a}} \leq \infty$. Hence we have shown for every $\mathbf{a} \in D$ that

$$c_{\mathbf{a}}(\mathbf{a}^\top \mathbf{x}) = \begin{matrix} q_{\mathbf{a}} \mathbf{a}^\top \mathbf{x} & \text{if } \mathbf{a}^\top \mathbf{x} < 0 \\ 0 & \text{if } \mathbf{a}^\top \mathbf{x} = 0 \\ \infty & \text{if } \mathbf{a}^\top \mathbf{x} < 0 \end{matrix} . \tag{1.138}$$

Again by relation (1.137) and $f(\mathbf{x}) > -\infty$ for every $\mathbf{x}$ we obtain that the set $D_0 := \{\mathbf{a} \in D : 0 < q_{\mathbf{a}} < \infty\}$ is nonempty and by relations (1.138) and (1.137) this shows

$$f(\mathbf{x}) = \sup\{c_{\mathbf{a}}(\mathbf{a}^\top \mathbf{x}) : \mathbf{a} \in D_0\}. \tag{1.139}$$

Since for $\mathbf{a} \in D_0$ it follows that $-\infty < c_{\mathbf{a}}(\mathbf{a}^\top \mathbf{x}) = q_{\mathbf{a}} \mathbf{a}^\top \mathbf{x}$ for $\mathbf{a}^\top \mathbf{x} \leq 0$ and ∞ otherwise, this is clearly a l.s.c. sublinear function and by relation (1.139) the desired result follows. □

Since by Lemma 1.59 and relation (1.127) the function f_- satisfies the conditions of Lemma 1.61 for f a positively homogeneous evenly quasiconvex function with $\mathbf{0} \in dom(f)$ and $f(\mathbf{x}) > -\infty$ for every $\mathbf{x}$ it follows that f_- is a nonpositive l.s.c. sublinear function. Using relation (1.130) and Lemma 1.59 up to 1.61 the following remarkable result follows immediately.

Theorem 1.26 *If $f : \mathbb{R}^n \to (-\infty, \infty]$ is a positively homogeneous evenly quasiconvex function and $\mathbf{0} \in dom(f)$, then f can be written as the minimum of a nonpositive l.s.c. sublinear function and a nonnegative l.s.c. sublinear function.*

Proof. If $\widetilde{L}(f,0)$ is empty then f is a nonnegative function and the result follows by Lemma 1.60. Moreover, if $\widetilde{L}(f,0)$ is nonempty, then by relation (1.130) it follows that $f = \min(f_+, f_-)$ and applying the observations after Lemma 1.60 and 1.61 yields the desired result. □

By Theorem 1.26 every positively homogeneous evenly quasiconvex function f satisfying $f(\mathbf{x}) > -\infty$ for every $\mathbf{x}$ and $\mathbf{0} \in dom(f)$ must be the minimum of two l.s.c. sublinear functions and so it is also l.s.c.. By relation (1.77) these l.s.c. sublinear functions can be written as support functions. This is a rather remarkable result, which does not hold in general for evenly quasiconvex functions. As an example we mention the evenly quasiconvex function sign(x) given by

$$sign(x) = -1 \text{ if } x < 0, \ sign(0) = 0 \text{ and } sign(x) = 1 \text{ if } x > 0$$

which is neither upper or lower semicontinuous at 0. To conclude this subsection we observe that Theorem 1.26 is an extension of the main result in Crouzeix (cf. [15]). For related results see also [14], [13], [12] and [11]. Introducing now the Dini upper directional derivative $\mathbf{d} \to f'_+(\mathbf{x}, \mathbf{d})$ given by

$$f'_+(\mathbf{x}, \mathbf{d}) := \limsup_{t \downarrow 0} t^{-1}(f(\mathbf{x} + t\mathbf{d}) - f(\mathbf{x}))$$

(cf. [30], [11]) it is possible to use the above so-called Crouzeix representation theorem for positively homogeneous quasiconvex functions to analyze the global behaviour of the function $\mathbf{d} \to f'_+(\mathbf{x}, \mathbf{d})$ for f quasiconvex (cf. [15], [44], [45], [33], [11]). This concludes our discussion of positively homogeneous evenly quasiconvex functions and dual representations. In the next section we mention some milestone papers and books within the long history of convex and quasiconvex analysis.

5. Some remarks on the history of convex and quasiconvex analysis

In this section[1] we will discuss the origin of the important notions used in convex and quasiconvex analysis. It seems that the field of convex geometry and convex bodies in two and three dimensional space was first studied systematically by H.Brunn (cf. [7], [8]) and Minkowski (cf. [51]). Brunn (cf. [9]) and Minkowski (cf. [52]) also proved the existence of support hyperplanes. Also at the end of 19*th* and the beginning of the 20*th* century Farkas showed in a series of papers (cf. [45], [61]) the alternative theorem for linear inequality systems and this result became known as Farkas lemma within linear programming. Although this result was listed with an incorrect proof in some of his earlier papers a

[1]The authors like to thank Prof. J. Kolumbán (Cluj) and Prof. S. Komlósi (Pecs) for pointing out some of the early developments.

correct proof of this result appeared in [20]. More fundamental ideas about the related field of necessary optimality conditions for nonlinear optimization subject to inequality constraints can be found in papers by Fourier, Cournot, Gauss, Ostrogradsky, and Hamel (cf. [61]). On the other hand, more early references related to the study of convex sets are listed in the reprinted version of the 1934 book of Bonnesen and Fenchel (cf. [5]), Fenchel (cf. [22]), Valentine (cf. [68]) and Varberg (cf. [62]). Also at the beginning of the 20th century convex functions were introduced by Jenssen (cf. [39]) and more than forty years later a thorough study of conjugate functions in $\mathbb{R}^n$ was initiated by Fenchel (cf. [21]). Although Mandelbrojt (cf. [48]) already introduced the conjugate function in $\mathbb{R}^n$ for $n = 1$ (cf. [69]), it was Fenchel, who first realized the importance of the conjugacy concept in convex analysis. Four years before the milestone paper of Fenchel, also the first book on convex functions written in French by Popoviciu (cf. [60]) was published. In the English scientific community the unpublished lecture notes by Fenchel (cf. [22]) were a long time the main source of references. This book served as the main inspiration for the classical book of Rockafellar (cf. [63]) as noted in its preface. Also in this preface it is mentioned that Prof. Tucker suggested the name convex analysis and this became the standard word for this field. The introduction of quasiconvex functions started later. Although in most of the literature de Finetti ([16]) is mentioned as being the first author introducing quasiconvex functions, these functions were already considered by von Neumann (cf. [71] and independently Popoviciu (cf. [59]). Actually von Neumann (cf. [71]) already proved in 1928 a minimax theorem on simplices for bifunctions which are quasiconcave in one variable and quasiconvex in the other variable. A generalization of this result was rediscovered by Sion (cf. [66]) 30 years later. For more details on the development of quasiconvex functions the reader is referred to [2]. To develop results for the surrogate dual concept developed by Glover (cf. [31]) an adhoc approach involving the $\overline{c_r}$-conjugate function was initiated by Greenberg and Pierskalla (cf. [32]). Their results were generalized and put into the proper framework of dual representations by Crouzeix in a series of milestone papers (cf. [12], [13], [15], [14]). In these papers Crouzeix focussed his attention on the dual representation of the l.s.c. hull of a quasiconvex function. Although Fenchel (cf. [23]) already introduced the concept of an evenly convex set the usefulness of this concept leading to a more symmetrical dual representation of an evenly quasiconvex function was discovered independently by Passy and Prisman (cf. [55]) and Martínez Legaz (cf. [50]). This concludes our short excursion, which is by no means complete, to the history of convex and quasiconvex analysis.

References

[1] Aubin, J.B., *Optima and Equilibria (An introduction to nonlinear analysis)*, Graduate Texts in Mathematics, v. 140, Springer Verlag, Berlin, 1993.

[2] Avriel, M., Diewert, W.E., Schaible, S. and I. Zang, *Generalized Concavity*, in Mathematical Concepts and Methods in Science and Engineering, v. 36, Plenum Press, New York, 1988.

[3] Bazaara, M.S., Sherali, H.D. and C.N. Shetty, *Nonlinear Programming (Theory and Applications)*, Wiley, New York, 1993.

[4] Bazaara, M.S., Jarvis, J.J. and H.D. Sherali, *Linear Programming and Network Flows (Second Edition)*, Wiley, New York, 1990.

[5] Bonnesen, T. and W. Fenchel, *Theorie der konvexen Körper*, Springer Verlag, Berlin, 1974.

[6] Breckner, W.W. and G.Kassay, *A systematization of convexity concepts for sets and functions*, Journal of Convex Analysis, 4, 1997, 1-19.

[7] Brunn, H., *Über Ovale und Eiflächen*, Inauguraldissertation, University of München, 1887.

[8] Brunn, H., *Über Kurven ohne Wendepunkte*, Habilitationsschrift, University of München, 1889.

[9] Brunn, H., *Zur Theorie der Eigebiete*, Arch. Math. Phys. Ser.3, 17, 1910, 289-300.

[10] Choquet, G., *Lectures on Analysis: Volume 2 Representation Theory*, in Mathematics Lecture Note Series, W.A. Benjamin, London, 1976.

[11] Crouzeix, J.-P., *Continuity and differentiability of quasiconvex functions*, Chapter 3 in this volume.

[12] Crouzeix, J.-P., *Contributions a l'étude des fonctions quasiconvexes*, Université de Clermond-Ferrand 2, France, 1977.

[13] Crouzeix, J.-P., *Conditions for convexity of quasiconvex functions*, Mathematics of Operations Research, 5, 1980, 120-125.

[14] Crouzeix, J.-P., *A duality framework in quasiconvex programming*, in Generalized Concavity in Optimization and Economics, Schaible, S. and W.T. Ziemba, eds., Academic Press, 1981, 207-225.

[15] Crouzeix, J.-P., *Continuity and differentiability properties of quasiconvex functions on* $\mathbb{R}^n$, in Generalized Concavity in Optimization and Economics, Schaible, S. and W.T. Ziemba, eds., Academic Press, 1981, 109-130.

[16] De Finetti, B., *Sulle stratificazioni convesse*, Ann. Mat. Pura Appl., [4] 30, 1949, 173-183.

[17] Dudley, R.M., *Real Analysis and Probability*, Wadsworth and Brooks/Cole, Pacific Grove, 1989.

[18] Edwards, R.E., *Functional Analysis: Theory and Applications*, Holt, Rinehart and Winston, Chicago, 1965.

[19] Faigle, U., Kern, W. and G. Still, *Algorithmic Principles of Mathematical Programming*, in Kluwer texts in Mathematical Sciences, v. 24, Kluwer Academic Publishers, Dordrecht, 2002.

[20] Farkas, J., *Uber die Theorie der einfache Ungleichungen*, Journal für die Reine und Angewandte Mathematik, 124, 1901, 1-27.

[21] Fenchel, W., *On conjugate convex functions*, Canadian Journal of Mathematics, 1, 1949, 73-77.

[22] Fenchel, W., *Convex Cones, Sets and Functions*, Lecture notes, Princeton University, 1951.

[23] Fenchel, W., *A remark on convex sets and polarity*, Communication Seminar on Mathematics, University of Lund supplementary volume, 1952, 22-89.

[24] Frenk, J.B.G and G. Kassay, *On classes of generalized convex functions, Gordan-Farkas type theorems and Lagrangian duality*, Journal of Optimization Theory and Applications, [2] 102, 1999, 315-343.

[25] Frenk, J.B.G. and G. Kassay, *Lagrangian duality and cone convexlike functions*, Econometric Institute, Erasmus University Rotterdam, Tech. Rep. 2000-27/A, 2000.

[26] Frenk, J.B.G. and G. Kassay, *Minimax results and finite dimensional separation*, Journal of Optimization Theory and Applications, [2] 113, 2002, 409-421.

[27] Frenk, J.B.G., Dias, D.M.L. and J. Gromicho, *Duality theory for convex and quasiconvex Functions and its application to optimization*, in Generalized Convexity: Proceedings Pecs, Hungary, 1992, Komlósi, S., Rapcsák, T. and S. Schaible, eds., Lecture notes in Economics and Mathematics, 405, 1994, 153-171.

[28] Frenk, J.B.G., Kassay, G. and J. Kolumbán, *On equivalent results in minimax theory*, European Journal of Operational Research 157, 2004, 46-58.

[29] Frenk, J.B.G., Protassov, V. and G. Kassay, *On Borel probability measures and noncooperative game theory*, Report Series Econometric Institute, Erasmus University Rotterdam, Tech. Rep. EI-2002-32, 2002.

[30] Giorgio, G. and S. Komlósi, *Dini derivatives in optimization-Part I*, Rivista di Matematica per le Scienze Economiche e Sociali, [1] 15, 1992, 3-30.

[31] Glover, F., *Surrogate constraints*, Operations Research, 16, 1967, 741-749.

[32] Greenberg, H.J. and W.P. Pierskalla, *Quasiconjugate functions and surrogate duality*, Cahiers du Centre d'étude de Recherche Operationelle, [4] 15, 1973, 437-448.

[33] Gromicho, J., *Quasiconvex Optimization and Location Theory*, in Applied Optimization, v. 9, Kluwer Academic Publishers, Dordrecht, 1998.

[34] Hiriart-Urruty, J.B. and C. Lemaréchal, *Convex Analysis and Minimization Algorithms I*, in Grundlehren der Mathematischen Wissenschaften, v. 305, Springer Verlag, Berlin, 1993.

[35] Hiriart-Urruty, J.B. and C. Lemaréchal, *Convex Analysis and Minimization Algorithms II*, in Grundlehren der Mathematischen Wissenschaften, v. 306, Springer Verlag, Berlin, 1993.

[36] Hocking, J.G and G.S. Young, *Topology*, Addison Wesley, Reading, Massachusetts, 1961.

[37] Holmes, R.B., *Geometric Functional Analysis and its Applications*, Springer, New York, 1975.

[38] Jahn, J., *Mathematical Vector Optimization in Partially Ordered Spaces*, Peter Lang, Fankfurt am Main, 1986.

[39] Jensen, J.L.W.V., *Sur les functions convexes et les inégalités entre des valeurs moyennes*, Acta Math., 30, 1906, 175-193.

[40] Joó, I., *A simple proof for von Neumann's minimax theorem*, Acta Sci. Math. Szeged, 42, 1980, 91-94.

[41] Joó, I., *Note on my paper: A simple proof for von Neumann's minimax theorem*, Acta Math. Hung., [3-4] 44, 1984, 171-176.

[42] Joó, I., *On some convexities*, Acta Math.Hung., [1-2] 54, 1989, 163-172.

[43] Kolmogorov, A.N. and S.V. Fomin, *Introductory Real Analysis*, in Dover books on Mathematics, Dover Publications, New York, 1975.

[44] Komlósi, S., *Quasiconvex first-order approximations and Kuhn-Tucker type optimality conditions*, European Journal of Operational Research, 65, 1993, 327-335.

[45] Komlósi, S., *Farkas theorems for positively homogeneous quasiconvex functions*, Journal of Statistics and Management Systems, 5, 2002, 107-123.

[46] Kreyszig, E., *Introductory Functional Analysis with Applications*, Wiley, New York, 1978.

[47] Lancaster, P. and M. Tismenetsky, *The Theory of Matrices (Second Edition with Applications)*, in Computer Science and Applied Mathematics, Academic Press, Orlando, 1985.

[48] Mandelbrojt, S., *Sur les fonctions convexes*, C.R. Acad. Sci. Paris, 209, 1939, 977-978.

[49] Martínez-Legaz, J.-E., *Generalized convex duality and its economic applications*, Chapter 6 in this volume.

[50] Martínez-Legaz, J.-E., *Un concepto generalizado deconjugacion, applicacion a las funciones quasiconvexas*, M.Sc. Thesis, Universidad de Barcelona, 1981.

[51] Minkowski, H., *Algemeine Lehrsätze über die convexen Polyeder*, Nachr. Ges. Wiss. Göttingen Math. Phys. K1, 1897, 198-219.

[52] Minkowski, H., *Geometrie der Zahlen*, Teubner, Leipzig, 1910.

[53] Nesterov, Y. and A. Nemirovski, *Interior Point Polynomial Algorithms in Convex Programming*, in SIAM Studies in Applied Mathematics, SIAM, Philadelphia, 1994.

[54] Nocedal, J. and S.J. Wright, *Numerical Optimization*, in Springer series in Operations Research, Springer Verlag, New York, 1999.

[55] Passy, U. and E.Z. Prisman, *Conjugacy in quasi-convex programming*, Mathematical Programming, 30, 1984, 121-146.

[56] Penot, J.-P. and M. Volle, *On quasi-convex duality*, Mathematics of Operations Research, [4] 15, 1990, 597-624.

[57] Peterson, E.L., *The fundamental relations between geometric programming duality, parametric programming duality and ordinary Lagrangian duality*, Annals of Operations Research, 105, 2001, 109-153.

[58] Ponstein, J., *Approaches to the Theory of Optimization*, Cambridge University Press, Cambridge, 1980.

[59] Popoviciu, T., *Deux remarques sur les fonctions convexes*, Bull. Sc. Acad. Roumaine, 20, 1938, 45-49.

[60] Popoviciu, T., *Les Fonctions Convexes*, Hermann, Paris, 1945.

[61] Prekopa, A., *On the development of optimization theory*, American Mathematical Monthly, [7] 87, 1980, 527-542.

[62] Roberts, A.V. and D.E. Varberg, *Convex Functions*, Academic Press, New York, 1973.

[63] Rockafellar, R.T., *Convex Analysis*, in Princeton Mathematical Series, v. 28, Princeton University Press, Princeton, New Jersey, 1972.

[64] Rudin, W., *Principles of Mathematical Analysis*, McGraw-Hill, New York, 1976.

[65] Rudin, W., *Functional Analysis*, McGraw-Hill, New York, 1991.

[66] Sion, M., *On general minimax theorems*, Pacific Journal of Mathematics, 8, 1958, 171-176.

[67] Sturm, J.F., *Primal-Dual Interior Point Approach to Semidefinite Programming*, PhD. Thesis, Econometric Institute, Erasmus University Rotterdam, 1997.

[68] Valentine, F.A., *Convex Sets*, Mc-Graw Hill, San Francisco, 1964.

[69] van Tiel, J., *Convex Analysis: an Introductory Text*, Wiley, New York, 1984.

[70] Ville, J., *Sur la théorie générale des jeux au intervient l'habilité des jouers*, [2] (5), in Traité du calcul des probabilités et de ses applications, E. Borel and others, eds., Gauthier-Villars Cie, Paris, 1938, 105-113.

[71] Von Neumann, J., *Zur Theorie der Gesellschaftsspiele*, Math. Ann, 100, 1928, 295-320.

[72] Vorob'ev, N.N., *Game Theory: Lectures for Economists and Systems Scientists*, in Applications of Mathematics, v. 7, Springer Verlag, New York, 1977.

[73] Wolkowicz, H., *Some applications of optimization in matrix theory*, Linear Algebra and Its Applications, 40, 1981, 101-118.

[74] Yuang, G.X.-Z., *KKM Theory and Applications in Nonlinear Analysis*, Marcel Dekker, New York, 1999.

Chapter 2

CRITERIA FOR GENERALIZED CONVEXITY AND GENERALIZED MONOTONICITY IN THE DIFFERENTIABLE CASE

Jean-Pierre Crouzeix
CUST and LIMOS
Université Blaise Pascal
63170 Aubière, France
jp.crouzeix@math.univ-bpclermont.fr

Abstract This chapter is devoted to first and second order characterizations of quasi/pseudo convexity of a function and first order characterizations of quasi/pseudo monotonicity of a single-valued map. Some applications are given.

Keywords: generalized convexity, generalized monotonicity, first and second order characterizations.

1. First order conditions for quasiconvexity

Given a convex subset C of a linear space E and $f : C \to \mathbb{R}$, we recall that f is said to be:
convex on C if for every $x, y \in C$ and $t \in (0,1)$ one has

$$f(x + t(y - x)) \leq f(x) + t(f(y) - f(x)),$$

strictly convex on C if for every $x, y \in C$, $x \neq y$ and $t \in (0,1)$

$$f(x + t(y - x)) < f(x) + t(f(y) - f(x)),$$

quasiconvex on C if for every $x, y \in C$ and $t \in (0,1)$

$$f(x + t(y - x)) \leq max[f(x), f(y)],$$

and *strictly quasiconvex* on C if for every $x, y \in C, x \neq y$ and $t \in (0,1)$

$$f(x + t(y - x)) < max[f(x), f(y)].$$

For $a \in C$ and $d \in E$, let us define

$$I_{a,d} = \{t \in \mathbb{R} : a + td \in C\},$$

and for $t \in I_{a,d}$

$$f_{a,d}(t) = f(a + td).$$

Then, f is convex (strictly convex, quasiconvex, strictly quasiconvex) on C if and only if $f_{a,d}$ is convex (strictly convex, quasiconvex, strictly quasiconvex) on $I_{a,d}$ for all $a \in C$ and $d \in E$. The proof is easy and left to the reader. Therefore, since first and second order conditions for convexity of a function of several variables are straightforwardly derived from the corresponding conditions for a function of one real variable, we turn one special attention to quasiconvex functions of one real variable.

Let $\theta : I \to \mathbb{R}$ be a differentiable function on the interval I of $\mathbb{R}$. Then, θ is quasiconvex on I if and only if there exists $t \in [-\infty, +\infty]$ so that θ is nonincreasing on $(-\infty, t] \cap I$ and nondecreasing on $[t, +\infty) \cap I$, one of the intervals $(-\infty, t] \cap I$ and $[t, +\infty) \cap I$ may be empty. Hence, we immediately deduce that each of the following four conditions

$$\begin{array}{rcl}
t_1, t_2 \in I \text{ and } \theta(t_1) < \theta(t_2) & \Rightarrow & \theta'(t_2)(t_1 - t_2) \leq 0, \\
t_1, t_2 \in I \text{ and } \theta(t_1) \leq \theta(t_2) & \Rightarrow & \theta'(t_2)(t_1 - t_2) \leq 0, \\
\left.\begin{array}{c} t_1, t_3 \in I, \ t_1 < t_2 < t_3, \\ \theta(t_1) < \theta(t_2) \text{ and } \theta'(t_2) = 0 \end{array}\right\} & \Rightarrow & \theta(t_2) \leq \theta(t_3), \\
t_1, t_2 \in I \text{ and } \theta'(t_1)(t_2 - t_1) > 0 & \Rightarrow & \theta'(t_2)(t_2 - t_1) \geq 0
\end{array}$$

is a sufficient and necessary condition for θ to be quasiconvex on I.

These characterizations are straightforwardly transcribed to general functions.

Proposition 2.1 *Assume that $f : C \to \mathbb{R}$ is differentiable on the convex subset C of E. Then each of the following four conditions*

$$x, y \in C \text{ and } f(y) < f(x) \implies \langle \nabla f(x), y - x \rangle \leq 0,$$

$$x, y \in C \text{ and } f(y) \leq f(x) \implies \langle \nabla f(x), y - x \rangle \leq 0,$$

$$\left.\begin{array}{c} x, x - h \in C, \\ f(x - h) < f(x) \\ \text{and } \langle \nabla f(x), h \rangle = 0 \end{array}\right\} \implies \left\{ \begin{array}{c} f(x) \leq f(x + th) \ \forall t > 0 \\ \text{such that } x + th \in C, \end{array}\right.$$

$$x, y \in C \text{ and } \langle \nabla f(x), y - x \rangle > 0 \implies \langle \nabla f(y), y - x \rangle \geq 0$$

is a sufficient and necessary condition for f to be quasiconvex on C.

Remark 2.1 Dealing with differentiability means that some topological structure has been defined on the space E. Several kinds of differentiability can be used. Here, only minimal conditions are required: it is sufficient that differentiability for f on E implies differentiability for all functions of one real variable $f_{a,d}$, $a \in C$, $d \in E$.

If f is convex and $\nabla f(x) = 0$ then f reaches its minimum at x. Unfortunately, this essential optimality condition is not preserved for quasiconvex functions as seen from the very simple example given by the function of one real variable $f(t) = t^3$. This motivates the introduction of the class of pseudoconvex functions: given a convex set C, a differentiable function $f : C \to \mathbb{R}$ is said to be:
pseudoconvex [39] on C if

$$x, y \in C \text{ and } f(y) < f(x) \Longrightarrow \langle \nabla f(x), y - x \rangle < 0,$$

and *strictly pseudoconvex* on C if

$$x, y \in C,\ x \neq y \text{ and } f(y) \leq f(x) \Longrightarrow \langle \nabla f(x), y - x \rangle < 0.$$

It is clear that a differentiable convex function is pseudoconvex. In view of Proposition 2.1, a differentiable pseudoconvex function is quasiconvex. It is easily seen that a differentiable strictly pseudoconvex function is strictly quasiconvex.

As for convexity and quasiconvexity, pseudoconvexity can be characterized via functions of one real variable. Indeed, it is clear that f is (strictly) pseudoconvex on C if and only if for all $a \in C$ and $d \in E$ the function of one real variable $f_{a,d}$ is (strictly) pseudoconvex on $I_{a,d}$. In connection with the last condition in Proposition 2.1, we have the following characterizations of pseudoconvexity.

Proposition 2.2 *Assume that f is differentiable on the convex set C. Then the three following conditions are equivalent:*
i) f is pseudoconvex on C,
ii) $x, y \in C$ and $\langle \nabla f(x), y - x \rangle > 0 \Longrightarrow \langle \nabla f(y), y - x \rangle > 0$,
iii) $x, y \in C$ and $\langle \nabla f(y), x - y \rangle \geq 0 \Longrightarrow \langle \nabla f(x), x - y \rangle \geq 0$.

Proof. It is clear that ii) and iii) are equivalent. Assume for contradiction that iii) holds but f is not pseudoconvex. Then there exist $x, y \in C$ such that $f(y) < f(x)$ and $\langle \nabla f(x), y - x \rangle \geq 0$. Then, in view of iii), $\langle \nabla f(x + t(y - x)), y - x \rangle \geq 0$ for all $t \in (0, 1)$. Hence $f(y) \geq f(x)$, a contradiction.

Next, assume that f is pseudoconvex and iii) does not hold. Then there exist $x, y \in C$ such that

$$\langle \nabla f(x), y - x \rangle \geq 0 \quad \text{and} \quad \langle \nabla f(y), x - y \rangle > 0.$$

Because f is pseudoconvex (and therefore quasiconvex too) the first inequality implies $f(y) \geq f(x)$ in contradiction with $f(x) > f(y)$ obtained from the second inequality. □

Proposition 2.3 *Assume that f is differentiable on the convex set C. Then the two following conditions are equivalent:*
i) f is strictly pseudoconvex on C,
ii) $x, y \in C$, $x \neq y$ and $\langle \nabla f(x), y - x \rangle \geq 0 \Longrightarrow \langle \nabla f(y), y - x \rangle > 0$.

Proof. Assume for contradiction that ii) holds but f is not strictly pseudoconvex. Then there exist $x, y \in C$, $x \neq y$ such that $f(y) \leq f(x)$ and $\langle \nabla f(x), y-x \rangle \geq 0$. Then, in view of ii), $\langle \nabla f(x+t(y-x)), y-x \rangle > 0$ for all $t \in (0, 1)$. Hence $f(y) > f(x)$, a contradiction.

Conversely, assume for contradiction that f is strictly pseudoconvex and ii) does not hold. Then there exist $x, y \in C$, $x \neq y$ such that $\langle \nabla f(x), y - x \rangle \geq 0$ and $\langle \nabla f(y), x - y \rangle \geq 0$. Hence, by strict pseudoconvexity, $f(y) > f(x)$ and $f(x) > f(y)$. □

The following proposition is a direct consequence of the definition.

Proposition 2.4 *Assume that f is differentiable and pseudoconvex on the convex set C. Assume that $x \in C$ and $\nabla f(x) = 0$. Then f has a global minimum on C at x.*

In fact, pseudoconvex functions on open convex sets are precisely those differentiable quasiconvex functions which reach their minimum at points where their gradients vanish. This is the object of the next theorem.

Theorem 2.1 *Assume that f is differentiable on the open convex set C.*
(i) If f is pseudoconvex on C then it is quasiconvex on C and has a global minimum at any $x \in C$ such that $\nabla f(x) = 0$.
(ii) If f is quasiconvex on C and has a local minimum at any $x \in C$ such that $\nabla f(x) = 0$, then it is pseudoconvex on C.

Proof. Only ii) needs to be proved. Assume, for contradiction, that the assumptions hold but f is not pseudoconvex. Then, $a \in C$ and $d \in E$ exist so that $a + d \in C$, $f(a + d) < f(a)$ and $\langle \nabla f(a), d \rangle \geq 0$. Since f is quasiconvex, $\langle \nabla f(a), d \rangle = 0$. Let us consider $\theta = f_{a,d}$. Then θ is quasiconvex on $[0,1]$ and $\theta'(0) = 0$. From $\theta(1) < \theta(0)$, we deduce that $\theta(t) \le \theta(0)$ for all $t \in [0,1]$. Set

$$\bar{t} = \max[\, t \in [0,1] : \theta(t) = \theta(0)].$$

Then $\bar{t} \in [0,1)$, $\theta(\bar{t}) = \theta(0)$ and $\theta(\bar{t}) > \theta(t)$ for all $t \in (\bar{t},1]$. If $\bar{t} > 0$, then $\theta(\bar{t}) \ge \theta(t)$ for all $t \in [0,\bar{t}]$ because θ is quasiconvex. Hence $\theta'(\bar{t}) = 0$ in case $\bar{t} > 0$ and in case $\bar{t} = 0$ as well.

Set $x = a + \bar{t}d$, then $\langle \nabla f(x), d \rangle = 0$ and $\nabla f(x) \neq 0$ since f has not a local minimum at x. Take k so that $\langle \nabla f(x), k \rangle > 0$ (for instance $k = \nabla f(x)$ when $E = \mathbb{R}^n$ or E is a Hilbert space). Since $a + d \in C$, C is open, f is continuous at $a+d$ and $f(a+d) < f(a) = f(x)$, then there is $r > 0$ such that

$$a + d + rk \in C \text{ and } f(a + d + rk) < f(x).$$

Next, since f is quasiconvex, $f(x + t(a + d + rk - x)) \le f(x)$ for all $t \in (0,1)$ which leads to the contradiction

$$0 \ge \langle \nabla f(x), a + d - x + rk \rangle = r\langle \nabla f(x), k \rangle > 0.$$

□

Remark 2.2 Notice that no continuity on ∇f is required. Also the argument of the proof involves only functions of two real variables. Hence, it is sufficient that the definition for the differentiability of f on E is such that it implies differentiability for the restrictions of f to the two-dimensional affine subspaces of E. Theorem 2.1 is due to Crouzeix-Ferland [9], a version for directionally differentiable functions has been given by Komlósi [34].

An immediate consequence of the above theorem is the following:

Corollary 2.1 *Assume that f is differentiable on the open convex set C and its gradient does not vanish on C. Then f is pseudoconvex on C if and only if it is quasiconvex on this set.*

2. Generalized monotonicity

In the same way that monotonicity is related to convexity, generalized monotonicity is related to generalized convexity. In this chapter, we are only concerned with gradients of functions, therefore we consider only generalized monotonicity for single-valued maps.

Let C be a convex subset of E and $F : C \to E'$, where E' denotes the dual space of E. The map F is said to be:
monotone on C if

$$x_1, x_2 \in C \Longrightarrow \langle F(x_1), x_2 - x_1 \rangle \leq \langle F(x_2), x_2 - x_1 \rangle,$$

pseudomonotone on C if

$$x_1, x_2 \in C, \ \langle F(x_1), x_2 - x_1 \rangle > 0 \Longrightarrow \langle F(x_2), x_2 - x_1 \rangle > 0,$$

or equivalently, if and only if

$$x_1, x_2 \in C, \ \langle F(x_1), x_2 - x_1 \rangle \geq 0 \Longrightarrow \langle F(x_2), x_2 - x_1 \rangle \geq 0,$$

quasimonotone on C if

$$x_1, x_2 \in C, \ \langle F(x_1), x_2 - x_1 \rangle > 0 \Longrightarrow \langle F(x_2), x_2 - x_1 \rangle \geq 0,$$

and *strictly pseudomonotone* on C if

$$x_1, x_2 \in C, x_1 \neq x_2, \langle F(x_1), x_2 - x_1 \rangle \geq 0 \Longrightarrow \langle F(x_2), x_2 - x_1 \rangle > 0.$$

Quasimonotonicity has been introduced by Hassouni ([23], 1983) and independently by Karamardian and Schaible ([30], 1990), pseudomonotonicity by Karamardian ([29], 1976). In fact, pseudomonotonicity, when applied to nonnegative maps on the positive orthant, is nothing else than the weak axiom of revealed preferences (Houthakker [26], 1950) for demand functions. A considerable literature is devoted to this axiom and its various avatars.

It is clear that a monotone map is pseudomonotone and a pseudomonotone map is quasimonotone. Assume that $f : C \to \mathbb{R}$ is differentiable on the convex set C. Then, besides the well known characterization of convex functions, we have, as direct consequences of Propositions 2.2, 2.3 and 2.1, the following characterizations of generalized convex functions.

Proposition 2.5 *f is convex (pseudoconvex, strictly pseudoconvex, quasiconvex) on C if and only if ∇f is monotone (pseudomonotone, strictly pseudomonotone, quasimonotone) on C.*

As for (generalized) convexity for functions, characterizations of (generalized) monotonicity for maps can be derived from the characterizations for maps of one real variable. As in the previous section, given $a \in C$ and $d \in E$ let us define

$$I_{a,d} = \{t \in \mathbb{R} \ : \ a + td \in C\}.$$

Next, for $t \in I_{a,d}$, let us define

$$F_{a,d}(t) = \langle F(a+td), d \rangle.$$

It is easily seen that F is monotone (pseudomonotone, strictly pseudomonotone, quasimonotone) on C if and only if for all $a \in C$ and $d \in E$, $F_{a,d}$ is monotone (pseudomonotone, strictly pseudomonotone, quasimonotone) on $I_{a,d}$. Furthermore, generalized monotonicity and generalized convexity are preserved under affine transformations as shown below. The proof is immediate and left to the reader.

Proposition 2.6 *Let X and Y be two linear spaces, $A : X \to Y$ be linear, $a \in Y$, C be a convex subset of X and D be a convex subset of Y such that $A(C) + a \subset D$.*
i) Assume that $f : D \to \mathbb{R}$ is quasiconvex (pseudoconvex) on D and let $g : C \to \mathbb{R}$ be defined by $g(x) = f(Ax + a)$. Then $g : C \to \mathbb{R}$ is quasiconvex (pseudoconvex) on C.
ii) Assume that $F : D \to Y'$ is quasimonotone (pseudomonotone) on D and let $G : C \to X'$ be defined by $G(x) = A^t F(Ax + a)$. Then $G : C \to X'$ is quasimonotone (pseudomonotone) on C.

The corresponding version of Theorem 2.1 is as follows:

Theorem 2.2 *Assume that F is continuous on the open convex set C. Then F is pseudomonotone on C if and only if F is quasimonotone on C and for all $a \in C$ such that $F(a) = 0$ and all $d \in E$ such that $a+d \in C$ there exists $t \in (0,1)$ so that $\langle F(a+td), d \rangle \geq 0$.*

Proof. The necessity follows from definitions. Let us prove the sufficiency by contradiction. Assume that F is not pseudomonotone, then $x, x+h \in C$ exist so that $\langle F(x), h \rangle \geq 0$ and $\langle F(x+h), h \rangle < 0$. Take

$$\widehat{t} = \sup_{t \in [0,1]} [t : \langle F(x+th), h \rangle \geq 0], \quad a = x + \widehat{t}h, \quad d = (1 - \widehat{t})h.$$

Then $\langle F(a), d \rangle = 0$, $0 \leq \widehat{t} < 1$ and $\langle F(a+td), d \rangle < 0$ for all $t \in (0,1]$. Hence $F(a) \neq 0$ in view of the assumption. Let k be such that $\langle F(a), k \rangle > 0$. By continuity, $r > 0$ exists so that $\langle F(a+d+rk), d+rk \rangle < 0$. But F is quasimonotone and therefore $\langle F(a), d + rk \rangle \leq 0$. It follows the contradiction

$$0 < r\langle F(a), k \rangle \leq 0.$$

□

The following immediate corollary concerns the special case where F does not vanish on C, it is a quasi transcription of Corollary 2.1.

Corollary 2.2 *Assume that F is continuous on the open convex set C and $F(x) \neq 0$ for all $x \in C$. Then F is pseudomonotone on C if and only if F is quasimonotone on C.*

Assume now that F is differentiable on C. Let us introduce the condition:

$$h \in E,\ \langle F(x), h\rangle = 0 \Longrightarrow \langle F'(x)h, h\rangle \geq 0 \qquad \text{(Sdp)}$$

Then (Sdp) gives a necessary condition for quasi/pseudomonotonicity as shown below.

Proposition 2.7 *Assume that $F : C \to E'$ is differentiable on the convex subset C of E. If F is quasimonotone on C, then (Sdp) holds for all $x \in int(C)$.*

Proof. Assume that (Sdp) does not hold for some $x \in int(C)$ and $h \in E$. Then, there is some $t > 0$ such that $x-th, x+th \in C$ and $\langle F(x+th), h\rangle < 0 < \langle F(x-th), h\rangle$. Hence F is not quasimonotone. □

This condition is necessary but not sufficient as seen from the following example: $f(t) = -t^4$ for $t \in \mathbb{R}$. Then, f is not quasiconvex and its derivative is not quasimonotone but (Sdp) holds for ∇f. A sufficient condition is as follows.

Proposition 2.8 *Assume that $F : C \to E'$ is differentiable on the convex subset C of E and the following condition holds for all $x \in C$*

$$h \in E,\ h \neq 0,\ \langle F(x), h\rangle = 0 \Longrightarrow \langle F'(x)h, h\rangle > 0 \qquad \textit{(dp)}$$

Then F is strictly pseudomonotone on C.

Proof. In a first step, we prove that

$$x, x+h \in C,\ \langle F(x), h\rangle \geq 0 \Longrightarrow \langle F(x+h), h\rangle \geq 0$$

If not, take

$$\bar{t} = \sup\,[\, t \in [0,1] \ :\ \langle F(x+th), h\rangle \geq 0\,].$$

Then, $\bar{t} \in [0,1)$, $\langle F(x+\bar{t}h), h\rangle = 0$ and $\langle F(x+th), h\rangle < 0$ for all $t \in (\bar{t}, 1]$ in contradiction with $\langle F'(x+\bar{t}h)h, h\rangle > 0$.

Next, if F is not strictly pseudomonotone, $a, b \in C$ exist so that $a \neq b$, $\langle F(a), b-a\rangle \geq 0$ and $\langle F(b), a-b\rangle \geq 0$. Hence, $\langle F(a+t(b-a)), b-a\rangle \geq 0$ and $\langle F(b+t(a-b)), a-b\rangle \geq 0$ for all $t \in [0,1]$. Condition (dp) contradicts $\langle F(a+t(b-a)), b-a\rangle = 0$ for all $t \in [0,1]$. □

Condition (dp) is sufficient but not necessary for strict pseudomonotonicity as shown by the function $f(t) = t^2$, $t \in \mathbb{R}$. Then, f is strictly pseudoconvex (and even strictly convex) but (dp) does not hold for the derivative at 0.

If we compare (Sdp) and (dp) respectively with the necessary condition for monotonicity and the sufficient condition for strict monotonicity:

$$x \in C,\, h \in E \Longrightarrow \langle F'(x)h, h\rangle \geq 0 \tag{p}$$

$$x \in C,\, h \in E,\, h \neq 0 \Longrightarrow \langle F'(x)h, h\rangle > 0 \tag{sp}$$

we see that (Sdp) and (dp) involve the (semi)-definiteness of the quadratic form $\langle F'(x)h, h\rangle$ over the linear subspace which is orthogonal to $F(x)$ instead of the whole space.

Before considering sufficient conditions for generalized monotonicity, we look at the semi-definiteness question in the next section.

3. Semi-definite positivity of a quadratic form over a linear subspace

In this section, given an $n \times n$ symmetric matrix A and an $n \times p$ matrix B, $1 \leq p \leq n-1$, we are interested in conditions equivalent to the condition

$$B^t x = 0 \Longrightarrow \langle x, Ax\rangle \geq 0 \tag{PSD}$$

or to the condition

$$x \neq 0,\;\; B^t x = 0 \Longrightarrow \langle x, Ax\rangle > 0 \tag{PD}$$

We start with some background in linear algebra.

The *inertia* of a $m \times m$ symmetric matrix M is the triple $In(M) = (\mu_+(M), \mu_-(M), \mu_0(M))$ where $\mu_+(M), \mu_-(M)$ and $\mu_0(M)$ denote respectively the numbers of positive, negative and null eigenvalues of M. Then $m = \mu_+(M) + \mu_-(M) + \mu_0(M)$. We make use of two basic results on inertia. The first of them is quoted in any good book of linear algebra.

Theorem 2.3 *(Lagrange-Sylvester law on inertia) Let M and Q be two $m \times m$ matrices, with M symmetric and Q nonsingular. Then*

$$In(M) = In(Q^t M Q).$$

The second result is less known. It concerns partitioned matrices and is a simple consequence of the previous one.

Theorem 2.4 *(Schur's complement) Let us consider the matrix*

$$M = \begin{pmatrix} A & B \\ B^t & C \end{pmatrix}$$

where A and C are symmetric and A is nonsingular. Then,
i) $\det(M) = \det(A) \times \det(C - B^t A^{-1} B)$,
ii) $In(M) = In(A) + In(C - B^t A^{-1} B)$.

Proof. Observe that

$$\begin{pmatrix} I & 0 \\ -B^t A^{-1} & I \end{pmatrix} \begin{pmatrix} A & B \\ B^t & C \end{pmatrix} \begin{pmatrix} I & -A^{-1}B \\ 0 & I \end{pmatrix} = \begin{pmatrix} A & 0 \\ 0 & C - B^t A^{-1} B \end{pmatrix}.$$

i) is a direct consequence of the rules on determinants and ii) follows from the Lagrange-Sylvester law. □

The matrix $C - B^t A^{-1} B$ is termed as the *Schur's complement* of M by A. Now, we introduce the *bordered* matrix associated with condition (PSD) and/or (PD)

$$M = \begin{pmatrix} A & B \\ B^t & 0 \end{pmatrix}.$$

The fundamental theorem is as follows

Theorem 2.5 *Assume that B has rank p. Then,*
i) $\mu_+(M) \geq p$ *and* $\mu_-(M) \geq p$,
ii) (PSD) *holds if and only if* $\mu_-(M) = p$,
iii) (PD) *holds if and only if* $\mu_+(M) = n$,
iv) (PD) *holds if and only if* $A + rBB^t$ *is positive definite for* $r > 0$ *large enough,*
v) if (PSD) *holds, then A has at most p negative eigenvalues.*

Proof. i) Since B has rank p, a nonsingular $n \times n$ matrix P exists so that $B^t P = (0, I)$ (I stands for the identity matrix, for simplicity, we do not precise its order, thus, in the following, I denotes the identity matrix of order $p, n-p$ or n according to the case). Then,

$$\begin{pmatrix} P^t & 0 \\ 0 & I \end{pmatrix} \begin{pmatrix} A & B \\ B^t & 0 \end{pmatrix} \begin{pmatrix} P & 0 \\ 0 & I \end{pmatrix} = \begin{pmatrix} \hat{A}_{11} & \hat{A}_{12} & 0 \\ \hat{A}_{21} & \hat{A}_{22} & I \\ 0 & I & 0 \end{pmatrix},$$

and the product of the three matrices

$$\begin{pmatrix} I & 0 & -\hat{A}_{12} \\ 0 & 0 & I \\ 0 & I & -\hat{A}_{22} \end{pmatrix} \begin{pmatrix} \hat{A}_{11} & \hat{A}_{12} & 0 \\ \hat{A}_{21} & \hat{A}_{22} & I \\ 0 & I & 0 \end{pmatrix} \begin{pmatrix} I & 0 & 0 \\ 0 & 0 & I \\ -\hat{A}_{21} & I & -\hat{A}_{22} \end{pmatrix}$$

is the matrix

$$\begin{pmatrix} \hat{A}_{11} & 0 & 0 \\ 0 & 0 & I \\ 0 & I & -\hat{A}_{22} \end{pmatrix}.$$

The Lagrange-Sylvester law on inertia implies

$$In(M) = In(\hat{A}_{11}) + In\begin{pmatrix} 0 & I \\ I & -\hat{A}_{22} \end{pmatrix} = In(\hat{A}_{11}) + (p, p, 0).$$

Hence, i) is obtained.

ii) and iii) Set $x = Py$, then (PSD) and (PD) are equivalent to say that $\hat{A}_{11}$ is positive semi-definite and positive definite respectively.

iv) The sufficient condition is obvious. Assume that (PD) holds, then $In(M) = (n, p, 0)$. By continuity, for $r > 0$ large enough

$$In\begin{pmatrix} A & B \\ B^t & -\frac{1}{r}I \end{pmatrix} = (n, p, 0).$$

It follows from Theorem 2.4 that

$$In\begin{pmatrix} A & B \\ B^t & -\frac{1}{r}I \end{pmatrix} = In(-\frac{1}{r}I) + In(A + rBB^t)$$

from what iv) is deduced.

v) follows from the fact that the number of negative eigenvalues of a leading submatrix (here A) cannot exceed that of the matrix (here M). □

A way of computing the inertia of a matrix consists in considering the sign changes in the sequence of the determinants of the leading submatrices. This is the object of the next theorem. For $R \subseteq \{1, 2, \cdots, n\}$, let us define

$$M_R = \begin{pmatrix} A_R & B_R \\ B_R^t & 0 \end{pmatrix},$$

where A_R is obtained from A by deleting rows and columns whose indices are not in R. Similarly, the matrix B_R is obtained from B by deleting rows whose indices are not in R. More specifically, if $R_k = \{k, k+1, \cdots, n\}$, where $k = 1, \cdots, n+1-p$, the matrices M_k, A_k and B_k correspond to the matrices associated to R_k. Notice that $card(R_k) = n + 1 - k$. Still, we consider the case where B has full rank, moreover we assume that the $p \times p$ matrix B_{n+1-p} (obtained in keeping the last p rows of B) is nonsingular.

Theorem 2.6 *Assume that B has rank p and B_{n+1-p} is nonsingular. Then,*

i) (PD) holds if and only if $(-1)^p \det(M_k) > 0$ for $k = 1, 2, \cdots, n+1-p$,

ii) (PSD) holds if and only if $(-1)^p \det(M_R) \geq 0$ for all $R \supseteq \{n+1-p, n+2-p, \cdots, n\}$.

Proof. Assume that (PSD) holds, then for all $R \supseteq \{n+1-p, n+2-p, \cdots, n\}$

$$x \neq 0,\ B_R^t x = 0 \Rightarrow \langle x, A_R x\rangle \geq 0.$$

If (PD) holds, the inequality is strict. Since B_R^t has rank p, it follows from Theorem 2.5 that $(-1)^p \det(M_R)$ is nonnegative or positive according to the case.

Let us prove i). For $k = n-p, n-p+1, \cdots, 2, 1$, the matrix M_k has the following form

$$M_k = \begin{pmatrix} a_{kk} & v_k^t \\ v_k & M_{k+1} \end{pmatrix}.$$

If M_{k+1} is nonsingular, we set $c_{kk} = a_{kk} - v_k^t M_{k+1}^{-1} v_k$. We start with $k = n-p$, we know that $In(M_{n+1-p}) = (p, p, 0)$ because B_{n+1-p} is nonsingular. Next, for $k = n-p, n-p+1, \cdots, 2, 1$, and as long as M_{k+1} is nonsingular, Theorem 2.4 implies

$$\det(M_k) = c_{kk} \det(M_{k+1})$$

and

$$In(M_k) = In(c_{kk}) + In(M_{k+1}) = In(\frac{\det(M_k)}{\det(M_{k+1})}) + In(M_{k+1}).$$

Proceed by induction.

Let us prove ii). The case where $\mu_0(M) = 0$ being already treated in i), we consider the case $\mu_0(M) \geq 1$. Since M_{n+1-p} is nonsingular and M has rank $n - \mu_0(M)$, then performing a suitable permutation on the first $n-p$ rows and columns of M, we have M_k nonsingular for $k = n-p, n-p-1, \cdots, \mu_0(M)+1$ and singular for $k \leq \mu_0(M)$. Proceed as for i), then $In(M) = (n - \mu_0(M), p, \mu_0(M))$. □

Next, we analyse more specially the case where $p = 1$. Then B is a column matrix that we represent by the vector b. Conditions (PSD) and (PD) become

$$\langle b, x\rangle = 0 \Longrightarrow \langle x, Ax\rangle \geq 0 \qquad (PSD_v)$$

and

$$x \neq 0,\ \langle b, x\rangle = 0 \Longrightarrow \langle x, Ax\rangle > 0. \qquad (PD_v)$$

Theorem 2.7 *Assume that $b \neq 0$ and $In(A) = (n_+, n_-, n_0)$. Denote by $A^\dagger$ the Moore-Penrose pseudoinverse matrix of A. Then,*

$$i) \quad (PSD_v) \iff \begin{cases} \text{either } n_- = 0, \\ \text{or } n_- = 1, \ b \in A(\mathbb{R}^n) \text{ and } b^t A^\dagger b \leq 0. \end{cases}$$

$$ii) \quad (PD_v) \iff \begin{cases} \text{either } n_+ = n, \\ \text{or } In(A) = (n-1, 0, 1) \text{ and } b \notin A(\mathbb{R}^n), \\ \text{or } In(A) = (n-1, 1, 0) \text{ and } b^t A^{-1} b < 0. \end{cases}$$

Proof. Let P be a $n \times n$ matrix and D be a nonsingular $(n-n_0)\times(n-n_0)$ diagonal matrix so that $PP^t = I$ and $P^tAP = \begin{pmatrix} D & 0 \\ 0 & 0 \end{pmatrix}$. Set $(c_1^t, c_2^t) = b^tP$ and $r = -c_2^t D^{-1} c_2 = -b^t A^\dagger b$. It follows from Theorem 2.4 that

$$In(M) = In(D^{-1}) + In \begin{pmatrix} 0 & c_2 \\ c_2^t & r \end{pmatrix}.$$

Hence,

$$In(M) = \begin{cases} (1+n_+, 1+n_-, -1+n_0) & \text{if } b \notin A(\mathbb{R}^n), \\ (1+n_+, n_-, n_0) & \text{if } b \in A(\mathbb{R}^n) \text{ and } b^tA^\dagger b < 0, \\ (n_+, 1+n_-, n_0) & \text{if } b \in A(\mathbb{R}^n) \text{ and } b^tA^\dagger b > 0, \\ (n_+, n_-, 1+n_0) & \text{if } b \in A(\mathbb{R}^n) \text{ and } b^tA^\dagger b = 0. \end{cases}$$

and the theorem follows. □

Historical comments: Positive and positive semi-definiteness of the restriction of a quadratic form to a linear subspace have been investigated since a long time. The necessary and sufficient condition iv) in Theorem 2.5 is known as the Finsler-Debreu Lemma (see references [22] and [18]). More recent references are Haynsworth [24], Cottle [6] and Chabrillac-Crouzeix [5].

4. Sufficient conditions for generalized monotonicity

Throughout this section, C is an open convex subset of $\mathbb{R}^n$ and $F : C \to \mathbb{R}$ is differentiable on C. We have already seen that the following condition:

$$x \in C, \ \langle F(x), h\rangle = 0 \Longrightarrow \langle F'(x)h, h\rangle \geq 0 \qquad \text{(Sdp)}$$

is necessary but not sufficient for F to be quasimonotone or pseudomonotone on C. With regard to sufficiency, we consider the following additional conditions:

$$\left.\begin{array}{c} x, x-h \in C \\ F(x)=0,\ F'(x)h=0, \\ \langle F(x-h), h\rangle > 0 \end{array}\right\} \Longrightarrow \left\{\begin{array}{c} \forall \bar{t}>0\ \exists t \in (0,\bar{t}] \text{ so that} \\ \langle F(x+th), h\rangle \geq 0 \end{array}\right. \tag{Bq}$$

and

$$\left.\begin{array}{c} x \in C,\ F(x)=0, \\ F'(x)h=0 \end{array}\right\} \Longrightarrow \left\{\begin{array}{c} \forall \bar{t}>0\ \exists t \in (0,\bar{t}] \text{ so that} \\ \langle F(x+th), h\rangle \geq 0 \end{array}\right. \tag{Bp}$$

In order to prepare the proof of the main result of this section, we begin with some very simple preliminary results.

Lemma 2.1 *Let $\gamma : [0,s] \to \mathbb{R}$ (with $s>0$) be differentiable, strictly positive on $(0,s]$ and such that $\gamma(0)=\gamma'(0)=0$. Assume in addition that there exists some $M>0$ such that $\gamma'(t) \leq M$ for all $t \in (0,s]$. Then there is a sequence $\{t_k\} \subset (0,s]$ converging to 0 such that for all k*

$$0 < \gamma(t_k) \leq 2t_k\gamma'(t_k).$$

Proof. Given $k>0$, let us define $m_k = \sup[\gamma'(t) : t \in [0, \frac{s}{k}]]$. Then $0 < m_k \leq M$ and there exists $t_k \in (0, \frac{s}{k}]$ such that $\gamma'(t_k) \geq \frac{m_k}{2}$. In view of the mean value theorem, there is $t \in (0, t_k)$ such that $0 < \gamma(t_k) = t_k\gamma'(t)$. But $0 < \gamma'(t) \leq m_k$. □

Lemma 2.2 *Given A a real $p \times p$ symmetric matrix, $a, b \in \mathbb{R}^p$, $\alpha, \beta \in \mathbb{R}$ with $\beta \neq 0$, then*

$$In \begin{pmatrix} A & a & b \\ a^t & \alpha & \beta \\ b^t & \beta & 0 \end{pmatrix} = (1,1,0) + In(A + \frac{\alpha}{\beta^2}bb^t - \frac{1}{\beta}ab^t - \frac{1}{\beta}ba^t).$$

Proof. Notice that

$$\begin{pmatrix} \alpha & \beta \\ \beta & 0 \end{pmatrix}^{-1} = -\frac{1}{\beta^2}\begin{pmatrix} 0 & -\beta \\ -\beta & \alpha \end{pmatrix}$$

and the inertia of this matrix is $(1,1,0)$. Then the lemma is a direct consequence of Theorem 2.4. □

Lemma 2.3 *Assume that F is differentiable on the open convex set C and (Sdp) holds but F is not quasimonotone on C. Then, there exist*

$\bar{x} \in C$, $h \in \mathbb{R}^n$ and two positive real numbers $\hat{t}$ and $\bar{t}$ such that $\langle F(\bar{x} - \hat{t}h), h\rangle > 0$ and

$$\langle F(\bar{x}), h\rangle = \langle F'(\bar{x})h, h\rangle = 0 \text{ and } \langle F(\bar{x} + th), h\rangle < 0 \;\; \forall t \in (0, \bar{t}] \quad (Ctr)$$

Proof. Because F is not quasimonotone, $a, a + h \in C$ exist so that $\langle F(a), h\rangle > 0$ and $\langle F(a + h), h\rangle < 0$. Take

$$\hat{t} = \sup_{t \in [0,1]} [t : \langle F(a + th), h\rangle \geq 0].$$

Then $0 < \hat{t} < 1$, $\langle F(a + \hat{t}h), h\rangle = 0$, $\langle F(a + th), h\rangle < 0$ for all $t \in (\hat{t}, 1]$ and $\langle F'(a+\hat{t}h)h, h\rangle \leq 0$. Then (Sdp) implies $\langle F'(a+\hat{t}h)h, h\rangle = 0$. Take $\bar{x} = a + \hat{t}h$ and $\bar{t} = 1 - \hat{t}$. □

Lemma 2.4 *Assume that F is differentiable on the open convex set C and (Sdp) holds but F is not pseudomonotone on C. Then, there exist $\bar{x} \in C$, $h \in \mathbb{R}^n$ and $\bar{t} > 0$ such that*

$$\langle F(\bar{x}), h\rangle = \langle F'(\bar{x})h, h\rangle = 0 \text{ and } \langle F(\bar{x} + th), h\rangle < 0 \, \forall t \in (0, \bar{t}] \quad (Ctr)$$

Proof. Because F is not pseudomonotone, there exist $a, a + h \in C$ so that $\langle F(a), h\rangle \geq 0$ and $\langle F(a + h), h\rangle < 0$. Define $\hat{t}, \bar{t}$ and $\bar{x}$ as in the proof of Lemma 2.3 (here $0 \leq \hat{t} < 1$). □

Theorem 2.8 *(Brighi, Crouzeix–Ferland, John) Assume that F is continuously differentiable on the open convex set C. Then:*
i) F is quasimonotone on C if and only if the two conditions (Sdp) and (Bq) hold.
ii) F is pseudomonotone on C if and only if the two conditions (Sdp) and (Bp) hold.

Proof. It is obvious that (Sdp) and (Bq) hold when F is quasimonotone and (Sdp) and (Bp) hold when F is pseudomonotone. Assume, for contradiction, that (Sdp) and (Bq) hold but F is not quasimonotone (or (Sdp) and (Bp) hold but F is not pseudomonotone). It results from Lemma 2.3 (Lemma 2.4) that (Ctr) holds for some $\bar{x}$. Without loss of generality, we assume that h is e_n, the n-th vector of the canonical basis of $\mathbb{R}^n$. For $x \in C$, let us consider the bordered matrices

$$\begin{pmatrix} F'(x) + [F']^t(x) & F(x) \\ F^t(x) & 0 \end{pmatrix} = \begin{pmatrix} A(x) & a(x) & b(x) \\ a^t(x) & \alpha(x) & \beta(x) \\ b^t(x) & \beta(x) & 0 \end{pmatrix}$$

where $A(x)$ is the $(n-1) \times (n-1)$ matrix corresponding to the $(n-1)$ first lines and rows of $F'(x) + [F']^t(x)$. Also, $(a^t(x)\ \alpha(x))$ corresponds to the last line of $F'(x) + [F']^t(x)$ and $(b^t(x)\ \beta(x))$ to $F^t(x)$. It results from (Ctr) that $\alpha(\bar{x}) = \beta(\bar{x}) = 0$ and $\beta(\bar{x} + te_n) < 0$ (and therefore $F(\bar{x} + te_n) \neq 0$) for all $t \in (0, \bar{t}]$. Next, define $x(t) = \bar{x} + te_n$ with $t \in (0, \bar{t}]$. For simplicity, we write $A(t)$ instead of $A(x(t))$ and so on. Notice that $\alpha(t)$ is nothing else than the derivative of $2\beta(t)$. Hence, in view of Lemma 2.1, there exists a sequence $\{t_k\}$ converging to 0 such that $0 < -\beta(t_k) \leq -t_k\alpha(t_k)$ for all k. It results from (Sdp) and Lemma 2.2 that the matrix

$$M(t) = A(t) + \frac{\alpha(t)}{\beta(t)^2} b(t)b(t)^t - \frac{1}{\beta(t)}[a(t)b^t(t) + b(t)a^t(t)]$$

is positive semi-definite. In particular, when $b(t) \neq 0$, it holds

$$0 \leq \langle A(t)\frac{b(t)}{\|b(t)\|}, \frac{b(t)}{\|b(t)\|}\rangle + \frac{\|b(t)\|}{\beta(t)}[\frac{\alpha(t)}{\beta(t)}\|b(t)\| - 2\langle a(t), \frac{b(t)}{\|b(t)\|}\rangle].$$

Notice that the quantities

$$\langle A(t)\frac{b(t)}{\|b(t)\|}, \frac{b(t)}{\|b(t)\|}\rangle \quad \text{and} \quad \langle a(t_k), \frac{b(t_k)}{\|b(t_k)\|}\rangle$$

stay bounded when $t \in (0, \bar{t}]$.

We consider the following cases:

1) $F(\bar{x}) \neq 0$. Then $b(0) \neq 0$ because $\beta(0) = 0$. By continuity of F, $b(t_k)$ converges to $b(0)$ when $k \to \infty$. It follows that the quantity

$$c_k = \frac{\|b(t_k)\|}{\beta(t_k)}[\frac{\alpha(t_k)}{\beta(t_k)}\|b(t_k)\| - 2\langle a(t_k), \frac{b(t_k)}{\|b(t_k)\|}\rangle]$$

converges to $-\infty$ and this is not possible since the matrix $M(t_k)$ is semi-definite positive. In that part, one needs only to require F' to be bounded, but not necessarily continuous, in a neighborhood of $\bar{x}$.

2) $F(\bar{x}) = 0$ and $F'(\bar{x})h \neq 0$. Then (Sdp) implies that the matrix $F'(\bar{x}) + [F']^t(\bar{x})$ is positive semi-definite positive. Hence $A(0)$ is positive semi-definite positive and $a(0) = 0$ because $\alpha(0) = 0$. Since F' is continuous, $a(t_k)$ converges to $a(0) = 0$ when $k \to \infty$. Next, let us consider the first order approximation of $F(\bar{x} + th)$. Since $F(\bar{x}) = 0$ and $\langle F'(\bar{x})e_n, e_n\rangle = 0$, we have

$$\begin{pmatrix} b(t) \\ \beta(t) \end{pmatrix} \simeq t \begin{pmatrix} b'(0) \\ 0 \end{pmatrix}.$$

It follows that $\|b(t)\| \simeq t\|F'(\bar{x})h\|$. Then, it is easily derived that $c_k \to -\infty$ when $k \to \infty$ which is not possible.

3) $F(\bar{x}) = 0$ and $F'(\bar{x})h = 0$. Then, there is a contradiction with Condition (Bq) in the quasimonotone case or with Condition (Bp) in the pseudomonotone case. □

It clearly follows that if F does not vanish on C, F is pseudomonotone and quasimonotone on C as soon as (Sdp) holds.

Historical comments: It was early observed ([42, 31]) that (Sdp) is a necessary condition for pseudo/quasimonotonicity (actually, the proof is quite immediate as seen in Section 2). If F is affine, then it is easily shown that (Sdp) is also sufficient (see for instance [13]). Sufficient conditions in the non-affine case are more difficult to get. It is shown in Hildenbrand [25] that (Sdp) is sufficient for pseudomonotonicity when F does not vanish on C (with a proof due to John). Independently, Crouzeix and Ferland [12] have shown that if (Sdp) holds with the additional condition

$$\left.\begin{array}{c} x \in C,\, F(x) = 0, \\ \langle F'(x)h, h\rangle = 0 \end{array}\right\} \Rightarrow \left\{\begin{array}{c} \forall \bar{t} > 0, \exists t \in (0,\bar{t}) \text{ so that} \\ \langle F(x+th), h\rangle > 0 \end{array}\right. \tag{Cfp}$$

then F is pseudomonotone on C. They also proved that if (Sdp) holds with

$$\left.\begin{array}{c} x \in C,\, x-h \in C, \\ F(x) = 0,\, \langle F'(x)h, h\rangle = 0, \\ \langle F(x-h), h\rangle > 0 \end{array}\right\} \Rightarrow \left\{\begin{array}{c} \forall \bar{t} > 0, \exists t \in (0,\bar{t}) \text{ so that} \\ \langle F(x+th), h\rangle \geq 0 \end{array}\right. \tag{Cfq}$$

then F is quasimonotone on C. The proof of Crouzeix-Ferland, based on the Cauchy-existence theorem for differential systems, is rather complicated. Recently, John [27] adapted the approach already used in Hildenbrand [25] to give an elegant proof of the Crouzeix-Ferland result for pseudomonotonicity. Very recently, Brighi [4] has shown that Condition (Cfp) can be relaxed in Condition (Bp). The proof of Theorem 2.8 given above is original and looks, at least for the author, much simpler than the former ones. Finally, we quote a generalization of the theorem to Lipschitzian maps by Dinh The Luc and Schaible [38] based on Crouzeix-Ferland's proof.

It is clear that Conditions (Bq) and (Bp) can be replaced in the theorem respectively by the following conditions:

$$\left.\begin{array}{c} x, x-h \in C, \\ F(x) = 0,\, F'(x)h = 0, \\ \langle F(x-h), h\rangle > 0 \end{array}\right\} \Rightarrow \left\{\begin{array}{c} \exists \bar{t} > 0 \text{ so that} \\ \langle F(x+th), h\rangle \geq 0\; \forall t \in (0,\bar{t}) \end{array}\right. \tag{Bqr}$$

and

$$\left.\begin{array}{c} x \in C, \\ F(x) = 0,\ F'(x)h = 0 \end{array}\right\} \Rightarrow \left\{ \begin{array}{c} \exists \bar{t} > 0 \text{ so that} \\ \langle F(x+th), h\rangle \geq 0\ \forall t \in (0, \bar{t}) \end{array}\right. \tag{Bpr}$$

Remark 2.3 For simplicity, the results have been established in a finite dimensional setting. As seen, the proof of the theorem proceeds by contradiction. The contradiction leads to condition (Ctr). Without loss of generality, it can be assumed in (Ctr) that $\bar{x} = 0$. Then the proof involves only the linear space generated by the vectors h and $F(\bar{x})$ when $F(\bar{x}) \neq 0$, or h and $F'(\bar{x})h$ when $F(\bar{x}) = 0$, a two-dimensional space in both cases. Thus, Theorem 2.8 is still valid in infinite dimensional settings.

5. Second order conditions for generalized convexity

Assume that $f :\longrightarrow \mathbb{R}$ is twice continuously differentiable on the convex set C. Then, by Proposition 2.5, f is quasiconvex (pseudoconvex) on C if and only if $F = \nabla f$ is quasimonotone (pseudomonotone) on C. Hence, Theorem 2.8 gives birth to necessary and sufficient condition for generalized convexity. Condition (Sdp) becomes

$$x \in C,\ \langle \nabla f(x), h\rangle = 0 \Longrightarrow \langle \nabla^2 f(x)h, h\rangle \geq 0 \tag{Sdp}$$

Next, let us have a look at conditions (Bp) and (Bq). The matrix $F'(x) = \nabla^2 f(x)$ is symmetric. If $\nabla f(x) = 0$, then condition (Sdp) implies that the matrix $\nabla^2 f(x)$ is positive semi-definite, hence $F'(x)h = 0$ is equivalent to $\langle \nabla^2 f(x)h, h\rangle = 0$ and conditions (Bp) and (Bq) become

$$\left.\begin{array}{c} x, x-h \in C, \nabla f(x) = 0, \\ \langle \nabla^2 f(x)h, h\rangle = 0, \\ \langle \nabla f(x-h), h\rangle > 0 \end{array}\right\} \Longrightarrow \left\{ \begin{array}{c} \forall \bar{t} > 0\ \exists t \in (0, \bar{t}] \text{ so that} \\ \langle \nabla f(x+th), h\rangle \geq 0 \end{array}\right. \tag{Bqf}$$

and

$$\left.\begin{array}{c} x \in C,\ \nabla f(x) = 0, \\ \langle \nabla^2 f(x)h, h\rangle = 0 \end{array}\right\} \Longrightarrow \left\{ \begin{array}{c} \forall \bar{t} > 0\ \exists t \in (0, \bar{t}] \text{ so that} \\ \langle \nabla f(x+th), h\rangle \geq 0 \end{array}\right. \tag{Bpf}$$

Let us consider the following conditions which involves the function f in place of its gradient:

$$\left.\begin{array}{c} x, x-h \in C, \nabla f(x) = 0, \\ \langle \nabla^2 f(x)h, h\rangle = 0, \\ f(x-h) < f(x) \end{array}\right\} \Longrightarrow \left\{ \begin{array}{c} \forall \bar{t} > 0\ \exists t \in (0, \bar{t}] \text{ so that} \\ f(x+th) \geq f(x) \end{array}\right. \tag{Cq}$$

and

$$x \in C,\ \nabla f(x) = 0 \Longrightarrow f \text{ has a local minimum at } x \qquad \text{(Cp)}$$

Theorem 2.9 *(Crouzeix, Crouzeix-Ferland) Assume that f is twice differentiable on the open convex set C and ∇f is locally Lipschitz on C. Then:*
i) f is quasiconvex on C if and only if the two conditions (Sdp) and (Cq) hold.
ii) f is pseudoconvex on C if and only if the two conditions (Sdp) and (Cp) hold.

Proof. It is clear that (Sdp) and (Cq) hold when f is quasiconvex and (Sdp) and (Cp) hold when f is pseudoconvex.

Assume that (Sdp) and (Cq) hold but f is not quasiconvex. Then there exist $a, a+d \in C$ such that $f(a) < f(a+d)$ and $\langle \nabla f(a+d), d\rangle < 0$. Take

$$\hat{t} = \sup[\, t \in [0,1] \ : \ \langle \nabla f(a+td), d\rangle \geq 0]$$

and $x = a + \hat{t}d$. Then $\hat{t} \in [0,1)$, $\langle \nabla f(x), d\rangle = 0$, $\langle \nabla f(x+td), d\rangle < 0$ and therefore $f(a) < f(a+d) < f(x+td) < f(x)$ for all $t \in (0, 1-\hat{t})$. It follows from (Sdp) that $\langle \nabla^2 f(x)d, d\rangle = 0$. Next, referring to the proof of Theorem 2.8, condition (Ctr) holds for x and d. Besides $F(x) = \nabla f(x) \neq 0$ because condition (Cq). Then, proceed as in case 1) of the proof of Theorem 2.8.

Assume that (Sdp) and (Cp) hold but f is not pseudoconvex. Then there exist $a, a+d \in C$ such that $f(a) > f(a+d)$ and $\langle \nabla f(a), d\rangle \geq 0$. Take

$$\hat{t} = \sup[\, t \in [0,1] \ : \ \langle \nabla f(a+td), d\rangle \geq 0]$$

and $x = a + \hat{t}d$. Then $\hat{t} \in [0,1)$, $\langle \nabla f(x), d\rangle = 0$, $\langle \nabla f(x+td), d\rangle < 0$ for all $t \in (0, 1-\hat{t})$. Proceed as above, here $F(x) = \nabla f(x) \neq 0$ follows from condition (Cp). □

Corollary 2.3 *Assume that f is twice differentiable on the open convex set C, ∇f is locally Lipschitz on C and ∇f does not vanish on C. Then f is quasiconvex on C (and pseudoconvex on C) if and only if (Sdp) holds.*

Historical comments: It was early seen that (Sdp) is a necessary condition for quasiconvexity (Arrow-Enthoven [1]). It is sufficient when f is quadratic as we shall see in the last section (Martos [40], Ferland [21] and Schaible [43]). For the non-quadratic case, using the implicit function theorem, Katzner [32] was able to prove that (Sdp) is also sufficient when the gradient stays strictly negative on the domain. Theorem 2.9 is due to

Crouzeix [8], see also Crouzeix-Ferland [9] and Diewert-Avriel-Zang [20]. As in Katzner, the proof in [8] is based on the implicit function theorem: when $\nabla f(\bar{x}) \neq 0$, it is shown that in a neighborhood of $\bar{x}$ the boundary of the level set $\{x : f(x) \leq f(\bar{x})\}$ is defined by a convex function. A very closed approach is followed by Komlosi [35, 36] where condition (Sdp) is expressed in terms of the so-called Quasi-Hessians. Also, Dinh The Luc [37], using a version of the implicit function theorem for Lipschitzian maps, has generalized the theorem to functions for which the gradient has a Lipschitz property.

Remark 2.4 In Theorem 2.9 as in Theorem 2.8, the assumption "C open" cannot be omitted, at least for pseudoconvexity and pseudomonotonicity. To see that, consider the function $f(x_1, x_2) = -x_1 x_2$, this function (its gradient) is pseudoconvex (pseudomonotone) on the positive orthant of $\mathbb{R}^n$ but not on its closure. Fortunately, for quasiconvexity and quasimonotonicity, things are better as seen below.

Proposition 2.9 *Assume that C is a convex subset of E with $int(C) \neq \emptyset$.*
i) Assume that $f : C \to \mathbb{R}$ is continuous on C and quasiconvex on $int(C)$. Then f is quasiconvex on C.
ii) Assume that $F : C \to E'$ is continuous on C and quasimonotone on $int(C)$. Then F is quasimonotone on C.

Proof. By assumption, there exists $x \in int(C)$.
i) Let $a, b \in C$, $\lambda \in (0,1)$ and $c = a + \lambda(b-a)$. For all $t \in (0,1)$ the points $a + t(x-a)$, $b + t(x-b)$ and $c + t(x-c)$ belong to the interior of C. Hence $f(c + t(x-c)) \leq \max[f(a + t(x-a)), f(b + t(x-b))]$. Set $t \to 0$. Then $f(c) \leq \max[f(a), f(b)]$.
ii) Let $a, b \in C$ and $\lambda \in (0,1)$ with $\langle F(a), b-a \rangle > 0$. For all $t \in (0,1)$ the points $a + t(x-a)$ and $b + t(x-b)$ belong to the interior of C. For t small enough, $\langle F(a + t(x-a)), b + t(x-b) - a - t(x-a) \rangle > 0$. Then $\langle F(b + t(x-b)), b + t(x-b) - a - t(x-a) \rangle > 0$. Set $t \to 0$, then we have $\langle F(b), b-a \rangle > 0$. □

6. Some applications

The following lemma will be used in several places of this section. Its proof is given in the chapter on continuity and differentiability of generalized convex functions in this volume.

Lemma 2.5 *(Newman-Crouzeix) Assume that $f : C \to \mathbb{R}$ is quasiconvex and positively homogeneous on the open convex set C.*

i) If $f(x) < 0$ for all $x \in C$, then f is convex on C.
ii) If $f(x) \geq 0$ for all $x \in C$, then f is convex on C.

6.1 Cobb-Douglas functions

Cobb-Douglas functions are of common use in economics. They are defined on the positive orthant of $\mathbb{R}^n$ by the relation $f(x) = \Pi x_i^{\alpha_i}$, $\alpha_i \neq 0$ for all i.

Proposition 2.10 *f is pseudoconvex on the positive orthant if and only if one of the following conditions holds*
i) $\alpha_i < 0$ for all i,
ii) $\alpha_i < 0$ for all i except one and $\Sigma\alpha_i \geq 0$.

Proof. Set $g(x) = \log(f(x))$. Then, f is pseudoconvex if and only if g is so. The proof directly follows from Theorems 2.9 and 2.7. □

The following result is well known. It can be derived from the previous one.

Proposition 2.11 *f is convex on the positive orthant if and only if one of the following conditions holds*
a) $\alpha_i < 0$ for all i,
b) $\alpha_i < 0$ for all i except one and $\Sigma\alpha_i \geq 1$.

Proof. A convex function is pseudoconvex. Hence either i) or ii) of the last proposition holds. If i) holds, then g is convex and therefore $f(x) = \exp(g(x))$ is convex. If ii) holds with $\Sigma\alpha_i < 1$, the function $t \to f(tx)$, $t > 0$, is not convex, whence f is not convex. If $\Sigma\alpha_i = 1$, then f is pseudoconvex and positively homogeneous of order 1, hence it is convex in view of Lemma 2.5. If $\alpha = \Sigma\alpha_i > 1$, then $f(x) = [\Pi x_i^{\frac{\alpha_i}{\alpha}}]^\alpha$ and therefore f is convex. □

6.2 Mereau-Paquet's type conditions

Let C be a convex set and $f : C \to \mathbb{R}$. f is said to be *convexifiable* or *convex transformable* on C if there exist a convex function $g : C \to \mathbb{R}$ and a *scaling function* $k : g(C) \to \mathbb{R}$ continuous, strictly increasing and such that $f(x) = k[g(x)]$ for all $x \in C$. A convexifiable function is quasiconvex; if in addition k is differentiable with $k'(t) > 0$ for all $t \in g(C)$ then f is pseudoconvex.

Assume that f is twice differentiable on the convex set C and let g be defined on C by $g(x) = \exp(rf(x))$ with $r > 0$. Then $\nabla g(x) = rg(x)\nabla f(x)$ and $\nabla^2 g(x) = rg(x)[\nabla f^2(x) + r\nabla f(x)\nabla^t f(x)]$ (where for simplicity we write $\nabla^t f(x)$ for $[\nabla f(x)]^t$). From this very simple observation we deduce the following results.

Proposition 2.12 *(Mereau-Paquet [41]) Assume that f is twice differentiable on the convex set C and for all $x \in C$ the matrix $\nabla f^2(x) + r\nabla f(x)\nabla^t f(x)$ is positive semi-definite. Then f is convexifiable on C, whence it is pseudoconvex on C.*

Proposition 2.13 *Assume that f is twice continuously differentiable on the convex set C and the following condition holds for some $\bar{x} \in int(C)$*

$$\langle \nabla f(\bar{x}), h\rangle = 0,\ h \neq 0 \Longrightarrow \langle \nabla f^2(\bar{x})h, h\rangle > 0$$

then f is convexifiable on a convex neighborhood of $\bar{x}$.

Proof. The proof of the first proposition is immediate. For the second proposition, combine item iv) of Theorem 2.5 with the continuity of $\nabla^2 f$.

□

In the same kind of idea, given a convex subset C of E and $F : C \to E'$, F is said *monotone transformable* on C if there exists $k : C \to (0, +\infty)$ such that the map G defined by $G(x) = k(x)F(x)$ is monotone on C. A monotone transformable map is pseudomonotone.

Assume that $F : C \to E'$ is differentiable on the convex set C and let G be the map defined on C by $G(x) = \exp(s(x))F(x)$ with $s : C \to \mathbb{R}$ differentiable on C. Then $G'(x) = \exp(s(x))[F'(x) + F(x)\nabla^t s(x)]$. From this observation, we deduce the following proposition.

Proposition 2.14 *Assume that F is continuously differentiable on the convex C and the following condition holds for some $\bar{x} \in int(C)$*

$$\langle F(\bar{x}), h\rangle = 0,\ h \neq 0 \Longrightarrow \langle F'(\bar{x})h, h\rangle > 0$$

then F is monotone transformable on a convex neighborhood of $\bar{x}$.

Proof. For $r > 0$ large enough, the matrix $F'(\bar{x}) + rF(\bar{x})(F(\bar{x}))^t$ is positive definite. Take $s(x) = r\langle F(\bar{x}), x - \bar{x}\rangle$. Then proceed by continuity.

□

6.3 Generalized convex quadratic functions

We start with a preliminary result.

Proposition 2.15 *Assume that the $n \times n$ symmetric matrix A ($n > 1$) has one and only one negative eigenvalue, then there exists a closed convex cone T such that*

$$T \cup (-T) = \{x : \langle Ax, x\rangle \le 0\}, \; int(T) \cup int(-T) = \{x : \langle Ax, x\rangle < 0\}$$

and

$$int(T) \cap int(-T) = \emptyset.$$

Proof. There exist a $n \times n$ diagonal matrix D and a $n \times n$ matrix P such that $A = P^t DP$, $P^t P = I$ and $d_1 < 0 \le d_2 \le \cdots \le d_n$ where d_i corresponds to the $i-$th diagonal entry of D. Then

$$\{x : \langle Ax, x\rangle \le 0\} = \{x : \langle DPx, Px\rangle \le 0\} = \{x = P^t y : \langle Dy, y\rangle \le 0\}.$$

Next,

$$\{y : \langle Dy, y\rangle \le 0\} = \{y : [\Sigma_{i=2,\cdots,n} d_i y_i^2]^{\frac{1}{2}} \le |y_1|[(-d_1)]^{\frac{1}{2}}\}.$$

Take

$$U = \{y : \langle Dy, y\rangle \le 0\} = \{y : [\Sigma_{i=2,\cdots,n} d_i y_i^2]^{\frac{1}{2}} \le y_1\sqrt{(-d_1)}\}$$

and $T = P^t U$. □

Now, we are concerned with the quadratic function

$$f(x) = \frac{1}{2}\langle Ax, x\rangle - \langle a, x\rangle.$$

where A is an $n \times n$ symmetric matrix and $a \in \mathbb{R}^n$. It is known that f is convex if and only if A is positive semi-definite. We look at the non positive semi-definite case.

Proposition 2.16 *Assume that A is a $n \times n$ symmetric matrix, $n > 1$, and A is not positive semi-definite. Then f is quasiconvex on the open convex set C if and only if A has one and only one negative eigenvalue, $a \in A(\mathbb{R}^n)$ and $C - A^\dagger a$ is contained either in $int(T)$ or in $int(-T)$ where T is the cone defined in Proposition 2.15.*

Proof. The necessary condition (Sdp) for quasiconvexity is in the present case

$$x \in C, h \in \mathbb{R}^n, \langle Ax - a, h\rangle = 0 \Longrightarrow \langle Ah, h\rangle \ge 0.$$

Apply Theorem 2.7. Then A has one negative eigenvalue and, for all $x \in C$, $Ax - a \in A(\mathbb{R}^n)$ and $\langle Ax - a, A^\dagger(Ax - a)\rangle \le 0$. It follows that $a \in A(\mathbb{R}^n)$ because C is open. Set $\bar{x} = A^\dagger a$. Then $\langle A(x - \bar{x}), (x - \bar{x})\rangle \le 0$

for all $x \in C$. Hence C is contained either in $int(T)$ or in $-int(T)$. Note that $\nabla f(x) = A(x - \bar{x}) \neq 0$ on $int(T)$ and on $-int(T)$. Hence the condition is sufficient as well. □

In particular, if the condition holds, f is pseudoconvex on $int(T)$ and $int(-T)$. By continuity, f is quasiconvex on T and $-T$. It is also easily seen that T and $-T$ are the maximal domains of quasiconvexity of f. Another interesting result is as follows.

Corollary 2.4 *Assume that A has one and only one negative eigenvalue and $a \in A(\mathbb{R}^n)$. Then f is convexifiable on T and $-T$. Actually, the function $g(x) = -[-\langle A(x - \bar{x}), (x - \bar{x})\rangle]^{\frac{1}{2}}$ is convex on T and $-T$.*

Proof. The function h defined by $h(u) = \langle Au, u\rangle$ is quasiconvex on T and $-T$. Hence $k(u) = -\sqrt{-h(u)}$ is also quasiconvex. Moreover, k is positively homogeneous and negative on the interiors of T and $-T$. Hence, in view of Lemma 2.5, it is convex on these interiors and by continuity on their closures. But, $g(x)$ is nothing else than $k(x - \bar{x})$. □

Historical comments: The main historical references on generalized convex quadratic functions are Martos [40], Ferland [21] and Schaible [43], see also [33].

6.4 Generalized monotone affine maps

We consider the map $F(x) = Ax - a$ where $n > 1$, $a \in \mathbb{R}^n$ and A is an $n \times n$ matrix which is not necessarily symmetric. We are concerned with the generalized monotonicity of F on the open convex set C. As for quadratic functions, we exclude the trivial case where A is positive semi-definite. The necessary condition

$$x \in C, \quad h \in \mathbb{R}^n, \ \langle Ax - a, h\rangle = 0 \Longrightarrow \langle Ah, h\rangle \geq 0$$

implies that the matrix $A + A^t$ has one and only one negative eigenvalue and, for all $x \in C$, $Ax - a \in (A + A^t)(\mathbb{R}^n)$ and $\langle Ax - a, (A + A^t)^{\dagger}(Ax - a)\rangle \leq 0$. It follows that a belongs to $(A + A^t)(\mathbb{R}^n)$ (and not necessarily to $A(\mathbb{R}^n)$). The analysis becomes a little more difficult than for functions. See references [13] and [17].

6.5 Generalized convexity over an affine space

Let C be a convex subset of $\mathbb{R}^n$, $f : C \to \mathbb{R}$, $a \in \mathbb{R}^p$ and A be a $p \times n$ matrix with rank p. We are interested in the (generalized) convexity of f on the set

$$D = \{x \in \mathbb{R}^n \ : \ x \in C \text{ and } Ax = a\}.$$

It is assumed that D is nonempty. Let $\bar{x}$ be fixed in D and B be a $n \times (n-p)$ matrix with rank $n-p$ such that $AB = 0$ (such a matrix exists). Then for all x such that $Ax = a$ there exists an unique $y \in \mathbb{R}^{n-p}$ so that $x = By + \bar{x}$. Furthermore, there is $\hat{D} \subset \mathbb{R}^{n-p}$ convex such that $D = B(\hat{D}) + \bar{x}$. Let g be defined on $\hat{D}$ by $g(y) = f(By + \bar{x})$. It is clear that f is convex (quasiconvex, pseudoconvex) on D if and only if g is so on $\hat{D}$. Notice that $\nabla g(y) = 0$ corresponds to $\nabla f(x) \in A^t(\mathbb{R}^n)$. The following theorem is a direct consequence of the second order conditions for (generalized) convexity.

Theorem 2.10 *Assume that C is open and f is twice continuously differentiable on D.*

i) f is convex on D if and only if the following condition holds

$$x \in D,\ Ah = 0 \Longrightarrow \langle \nabla^2 f(x)h, h\rangle \geq 0.$$

ii) Assume that $\nabla f(x) \notin A^t(\mathbb{R}^n)$ for all $x \in D$. Then f is pseudoconvex on D if and only if the following condition holds

$$x \in D,\ Ah = 0 \text{ and } \langle \nabla f(x), h\rangle = 0 \Longrightarrow \langle \nabla^2 f(x)h, h\rangle \geq 0.$$

iii) If the following condition holds

$$x \in D,\ Ah = 0, h \neq 0 \Longrightarrow \langle \nabla^2 f(x)h, h\rangle > 0$$

then f is strictly convex on D.

iv) If the following condition holds

$$x \in D,\ Ah = 0,\ h \neq 0 \text{ and } \langle \nabla f(x), h\rangle = 0 \Longrightarrow \langle \nabla^2 f(x)h, h\rangle > 0$$

then f is strictly convex on D.

As an application, given $\alpha > 0$, we consider the (generalized) convexity of the function

$$f(x,t) = \frac{t^\alpha}{\Pi x_i}$$

on the set $C = A \times (0, +\infty)$ where $A = \{x \in \mathbb{R}^n : x > 0, x_1 + \cdots + x_n = 1\}$. This function is convex when $\alpha \leq 0$. We consider the case $\alpha > 0$, such a function is of interest in interior point methods where it is termed as a *multiplicative potential function* and is used in the analysis of the convergence of Karmarkar's type methods.

Theorem 2.11 *f is convex on $A \times (0, +\infty)$ if and only if $\alpha \geq n$, quasiconvex and pseudoconvex if and only if $\alpha \geq n-1$. Moreover, f is strictly convex when $\alpha \geq n$ and strictly pseudoconvex when $\alpha \geq n-1$.*

Proof. For simplicity, we use the following notation: e is the vector of $\mathbb{R}^n$ and, for $x \in \mathbb{R}^n$, X is the $n \times n$ diagonal matrix defined by

$$e = \begin{pmatrix} 1 \\ 1 \\ \\ 1 \end{pmatrix}, \qquad X = \begin{pmatrix} x_1 & 0 & 0 \\ 0 & x_2 & 0 \\ & & \\ 0 & & x_n \end{pmatrix}.$$

Easy computations give

$$\frac{1}{f(x,t)} \nabla f(x,t) = \begin{pmatrix} X^{-1} & 0 \\ 0 & \frac{\alpha}{t} \end{pmatrix} \begin{pmatrix} e \\ 1 \end{pmatrix},$$

$$\frac{1}{f(x,t)} \nabla^2 f(x,t) = \begin{pmatrix} X^{-1} & 0 \\ 0 & \frac{\alpha}{t} \end{pmatrix} \begin{pmatrix} I + ee^t & -e \\ -e^t & 1 - \frac{1}{\alpha} \end{pmatrix} \begin{pmatrix} X^{-1} & 0 \\ 0 & \frac{\alpha}{t} \end{pmatrix}.$$

It is easily seen that

$$In(I + ee^t) = (n, 0, 0), \qquad (I + ee^t)^{-1} = I - \frac{1}{n+1} ee^t,$$

$$\inf[\, \|x\|^2 : x \in A\,] = \frac{1}{n} \quad \text{and} \quad \sup[\, \|x\|^2 : x \in A\,] = 1.$$

Convexity: f is convex on C if and only if for all $t > 0$ and $x \in A$ the matrix

$$\begin{pmatrix} \nabla^2 f(x,t) & & e \\ & & 0 \\ e & 0 & 0 \end{pmatrix}$$

has only one negative eigenvalue (it has at least one), i.e., because the Lagrange-Sylvester law on inertia, if and only if the following matrix M has also only one negative eigenvalue.

$$M = \begin{pmatrix} I + ee^t & -e & x \\ -e^t & 1 - \frac{1}{\alpha} & 0 \\ x^t & 0 & 0 \end{pmatrix}.$$

Apply Theorem 2.4. Then,

$$In(M) = (n, 0, 0) + In \begin{pmatrix} \frac{1}{n+1} - \frac{1}{\alpha} & \frac{1}{n+1} \\ \frac{1}{n+1} & \frac{1}{n+1} - \|x\|^2 \end{pmatrix}.$$

The 2×2 matrix has one negative eigenvalue. It has only one if and only if $1 \geq (n + 1 - \alpha)\|x\|^2$. This condition holds for all $x \in A$ if and only if $\alpha \geq n$. There is only one case where the second eigenvalue is zero, when $\alpha = n$ and $x_i = n^{-1}$ for all i. Then f is strictly convex.

Quasiconvexity: It is much simpler to deal with the function

$$g(x,t) = \alpha \log t - \Sigma \log x_i$$

Indeed, f is quasiconvex if and only if g is so. Here,

$$\nabla g(x,t) = \begin{pmatrix} -X^{-1}e \\ \frac{\alpha}{t} \end{pmatrix} \text{ and } \nabla^2 g(x,t) = \begin{pmatrix} X^{-2} & 0 \\ 0 & -\frac{\alpha}{t^2} \end{pmatrix}.$$

Next, g is quasiconvex if and only if for all $t > 0$ and all $x \in A$ the matrix

$$N = \begin{pmatrix} \nabla^2 g(x,t) & \nabla g(x,t) & \begin{matrix} e \\ 0 \end{matrix} \\ \nabla^t g(x,t) & 0 & 0 \\ e^t \;\; 0 & 0 & 0 \end{pmatrix}$$

has only two negative eigenvalues (it has at least 2). Hence, in view of the Lagrange-Sylvester law, if and only if the matrix

$$\widetilde{N} = \begin{pmatrix} I & 0 & -e & x \\ 0 & -\frac{1}{\alpha} & 1 & 0 \\ -e^t & 1 & 0 & 0 \\ x^t & 0 & 0 & 0 \end{pmatrix}$$

has two and only two negative eigenvalues. Next, in view of Theorem 2.4, $In(\widetilde{N}) = (n, 1, 0) + In(P)$ where

$$P = \begin{pmatrix} \alpha - n & 1 \\ 1 & -\|x\|^2 \end{pmatrix}$$

Hence f is quasiconvex (and pseudoconvex) if and only if P has at most one negative eigenvalue knowing that it has at least one. This condition is equivalent to

$$(n - \alpha)\|x\|^2 \leq 1,$$

it holds for all $x \in A$ if and only if $\alpha \geq n - 1$. The second eigenvalue is positive and therefore f is strictly pseudoconvex. □

6.6 Additive separability

Let us assume that, for $i = 1, 2, \cdots, p$, C_i is an open convex subset of $\mathbb{R}^{n_i}$, and $f_i : C_i \to \mathbb{R}$ is twice differentiable. Set $C = C_1 \times C_2 \times \cdots \times C_p$ and let $f : C \to \mathbb{R}$ be defined by $f(x) = f_1(x_1) + f_1(x_2) + \cdots + f_1(x_p)$ for all $x = (x_1, x_2, \cdots, x_p) \in C$. We are concerned with the generalized convexity of f on C. We exclude the degenerate cases where $p = 1$ or f_i is constant on C_i for some i.

It is clear that f is convex if and only if all f_i are convex. Suppose that f is quasiconvex on C, then it is clear that each function f_i is quasiconvex on C_i for all i. Suppose that one of them (say f_1) is not convex. Then there exists some $\bar{x_1} \in C_1$ for which $\nabla^2 f_1(\bar{x_1})$ is not positive semi-definite. Let us consider the matrix $\nabla^2 f(x)$ for $x = (\bar{x_1}, x_2, \cdots, x_p)$. In view of (Sdp) and Theorem 2.7, this matrix has at most one negative eigenvalue. It follows that for $i = 2, 3, \cdots, p$ the matrix $\nabla^2 f_i(x_i)$ is positive semi-definite at each $x_i \in C_i$, therefore f_i is convex. Thus, if f is quasiconvex on C, at most one of the functions f_i is not convex.

The conjunction of additive separability with generalized convexity has very strong implications which cannot be given here. We refer the reader to Debreu-Koopmans [19] and Crouzeix-Lindberg [10]. For maps, the conjunction of separability with generalized monotonicity has also strong implications, we refer to Crouzeix-Hassouni [14, 15].

References

[1] Arrow K.J. and Enthoven K.J., *Quasiconcave Programming*, Econometrica 29, 1961, 779–800.

[2] Avriel M., Diewert W.E., Schaible S. and Zang I., *Generalized Concavity*, Plenum Press, New York and London, 1988.

[3] Brighi L. and John R., *Characterizations of pseudomonotone maps and economic equilibrium*, Generalized convexity, generalized monotonicity, optimality conditions and duality in scalar and vector optimization. Journ. Stat. Manag. Syst. 5, 2002, 253–273.

[4] Brighi L., *A stronger criterion for the weaker axiom*, Technical report, Dept. di Scienze economiche, Università di Pisa, sept. 2000.

[5] Chabrillac Y. and Crouzeix J.-P., *Definiteness and semi-definiteness of quadratic forms revisited*, Linear Algebra and its Applications 63, 1984, 283–292.

[6] Cottle R.W., *Manifestations of the Schur complement*, Linear Algebra and its Applications 8, 1974, 189–211.

[7] Crouzeix J.-P., *Contribution à l'étude des fonctions quasiconvexes*, Thèse de Docteur es Sciences, Université de Clermont-Ferrand 2, 1977.

[8] Crouzeix J.-P., *A second order for quasiconvexity*, Mathematical Programming 18, 1980, 349–352.

[9] Crouzeix J.-P. and Ferland J.A., *Criteria for quasiconvexity and pseudoconvexity: relationships and comparisons*, Mathematical Programming 23, 1982, 193–205.

[10] Crouzeix J.-P. and Linberg P.O., *Additively decomposed quasiconvex functions*, Mathematical Programming 35, 1986, 42–57.

[11] Crouzeix J.-P., Ferland J.A., and Schaible S., *Generalized concavity on affine subspaces with an application to potential functions*, Mathematical Programming 56, 1992, 223–232.

[12] Crouzeix J.-P. and Ferland J.A., *Criteria for differentiable generalized monotone maps*, Mathematical Programming 75, 1996, 399–406.

[13] Crouzeix J.-P. and Schaible S., *Generalized monotone affine maps*, SIAM Journal on Matrix Analysis and Applications 17, 1996, 992–997.

[14] Crouzeix J.-P. and Hassouni A., *Quasimonotonicity of separable operators and monotonicity indices*, SIAM Journal on Optimization 3, 1994, 649–658.

[15] Crouzeix J.-P. and Hassouni A., *Generalized monotonicity of a separable product of operators: the multivalued case*, Set-Valued Analysis 3, 1995, 351–373.

[16] Crouzeix J.-P., Martinez-Legaz J.-E. and Volle M. editors, *Generalized convexity, generalized monotonicity*, Non convex optimization and its applications, Kluwer Academic Publishers, 1998.

[17] Crouzeix J.-P, Hassouni A., Lahlou A. and Schaible S., *Positive subdefinite matrices, generalized monotonicity and linear complementarity problems*, SIAM Journal on Matrix Analysis and Applications 22, 2001, 66–85.

[18] Debreu G., *Definite and semi-definite quadratic forms*, Econometrica 20, 1952, 295–300.

[19] Debreu G. and Koopmans T. C., *Additively decomposed quasiconvex functions*, Mathematical Programming 24, 1982, 1–38.

[20] Diewert W.E., Avriel M. and Zang I., *Nine kinds of quasiconcavity and concavity*, Journal of Economic Theory 25, 1981, 397–420.

[21] Ferland J.A., *Mathematical programming with quasiconvex objective functions*, Mathematical Programming 3, 1972, 296–301.

[22] Finsler P., *Über das Vorkommen definiter Formen und Scharen quadratischer Formen*, Commentarii Math. Helvet. 9, 1937, 188–192.

[23] Hassouni A., *Sous-differentiels des fonctions quasiconvexes*, Thèse de troisième cycle, Université Paul Sabatier, Toulouse, 1983.

[24] Haynsworth E.V., *Determination of the inertia of a partitioned hermitian matrix*, Linear Algebra and its Applications 1, 1968, 73–81.

[25] Hildenbrand W., *Market Demand: Theory and Empirical Evidence*, Princeton University Press, Princeton, NJ, 1994.

[26] Houthakker H.S., *Revealed preference and the utility function*, Economica, 1950, 159–174.

[27] John R., *A first order characterization of generalized monotonicity*, Mathematical Programming, 88, 2000, 147–155.

[28] Karamardian S., *The complementarity problem*, Mathematical Programming 2, 1972, 107–129.

[29] Karamardian S., *Complementarity problems over cones with monotone and pseudomonotone maps*, Journal of Optimization Theory and Applications 18, 1976, 445–454.

[30] Karamardian S. and Schaible S., *Seven kind of monotone maps*, Journal of Optimization Theory and Applications 66, 1990, 37–46.

[31] Karamardian S., Schaible S. and Crouzeix J.-P., *Characterizations of generalized monotone maps*, Journal of Optimization Theory and Applications 76, 1993, 399–413.

[32] Katzner D., *Static Demand Theory*, Mac Millan, New York, 1970.

[33] Komlósi S., *Generalized convexity of quadratic functions*, Izv. Vyssh. Uchebn. Zaved Math. 9, 1983, 38–43.

[34] Komlósi S., *Some properties of nondifferentiable pseudoconvex functions*, Mathematical Programming 26, 1983, 232–237.

[35] Komlósi S., *Second Order characterizations of pseudoconvex and strictly pseudoconvex functions in terms of quasi-Hessians*, Contribution to the theory of optimization, F. Forgo editor, University of Budapest, 1983, 19–46.

[36] Komlósi S., *On pseudoconvex functions*, Acta Sci. Math., 1993, 569–586.

[37] Luc D.T., *On generalized convex nonsmooth functions*, Bulletin of the Australian Mathematical Society 49, 1994, 139–149.

[38] Luc D.T. and Schaible S., *Generalized monotone nonsmooth maps*, Journal of Convex Analysis 3, 1996, 195–205.

[39] Mangasarian O.L., *Pseudoconvex functions*, SIAM Journal on Control A3, 1965, 281–290.

[40] Martos B., *Subdefinite matrices and quadratic forms*, SIAM Journal on Applied Mathematics 17, 1969, 1215–1223.

[41] Mereau P. and Paquet J.G., *Second order conditions for pseudo-convex functions*, SIAM Journal on Applied Mathematics 27, 1974, 131–137.

[42] Mitjushin L. G. and Polterovich W.M., *Criteria for monotonicity of demand functions* (in russian), Ekonomika i Matematicheskie Metody 14, 1978, 122–128.

[43] Schaible S., *Beiträge zur quasikonvexen Programmierung*, Doctoral Dissertation, Köln, Germany, 1971.

[44] Schaible S. and Ziemba W.T. editors, *Generalized Concavity in Optimization and Economics*, Academic Press, New York, 1981.

Chapter 3

CONTINUITY AND DIFFERENTIABILITY OF QUASICONVEX FUNCTIONS

Jean-Pierre Crouzeix
CUST and LIMOS
Université Blaise Pascal
63170 Aubière, France
jp.crouzeix@math.univ-bpclermont.fr

Abstract The convexity of the epigraph of a convex function induces important properties with respect to the continuity and differentiability of the function. Moreover, the function is locally Lipschitz in the interior of the domain of the function. If for a quasiconvex function, the convexity concerns the lower level sets and not the epigraph, some important properties on continuity and differentiability are still preserved. An important property of quasiconvex functions is that they are locally nondecreasing with respect to some positive cone.

Keywords: quasiconvexity, monotonic functions, continuity, differentiability, directional derivatives, Dini-differentiability, normal cones.

1. Introduction and notation

A convex function of one real variable admits right and left derivatives at any point in the interior of its domain, hence it is continuous at such a point. A convex function f defined on a normed linear space E is continuous at $x \in E$ if bounded in a neighbourhood of x. If $E = \mathbb{R}^n$ and x belongs to the interior of the domain of f, then f is bounded in a neighbourhood of x and thereby continuous at this point. Concerning differentiability, it follows from the observation on functions of one real variable that f admits a directional derivative $f'(x, d)$ with respect to any direction d at any point x in the interior of the domain. It is easily seen that the function $f'(x, d)$ is convex in d. The Fenchel-subdifferential

$\partial f(x)$ of f at x is defined from the directional derivatives by

$$\partial f(x) = \{x^\star \in E' : f'(x,d) \geq \langle x^\star, d\rangle \quad \forall d \in E\}$$

where E' denotes the topological dual of E. If that definition of $\partial f(x)$ appears analytical, its true essence is geometrical: the deep nature of convex analysis is geometrical, even if this nature is sometimes hidden by analytical considerations. We illustrate that below with the example of the subdifferential. Assume that f is convex and a belongs to the interior of its domain. Then, the point $(a, f(a))$ does not belong to the strict epigraph of f which is the set $\widetilde{epi} f = \{(y,\lambda) : f(y) < \lambda\}$, that epigraph is convex. Let us define the following set

$$N_e(a) = \{(u,\mu) : \langle u, y-a\rangle + \mu(\lambda - f(a)) \leq 0 \, \forall \, (y,\lambda) \in \widetilde{epi} f\} \quad (3.1)$$

Then, in view of separation theorems on convex sets and under suitable assumptions in the case where the dimension of E is not finite, $N_e(a)$ is a closed convex nonempty cone. The set $\partial f(a)$ is nothing else that the set obtained from $N_e(a)$ as follows

$$u \in \partial f(a) \iff (u,-1) \in N_e(a).$$

Convex analysis finds its main applications in optimization. A general formulation of a convex optimization problem is

$$\text{minimize } f(x) \text{ subject to } x \in C,$$

where C is convex and f is convex. A point $a \in C$ is an optimal solution if and only if $\widetilde{S}(a) \cap C = \emptyset$ where

$$\widetilde{S}(a) = \{x : f(x) < f(a)\}.$$

Both sets C and $\widetilde{S}(a)$ are convex. Hence, according to separation theorems and under the appropriate assumptions when the dimension of E is not finite, we are interested in vectors $a^\star \neq 0$ such that

$$\langle a^\star, x-a\rangle \leq 0 \leq \langle a^\star, z-a\rangle \;\; \forall x,z \text{ so that } z \in C, f(x) < f(a).$$

Such vectors belong to the closed convex cone

$$\widetilde{N}(a) = \{\, a^\star : \langle a^\star, x-a\rangle \leq 0 \text{ for all } x \text{ so that } f(x) < f(a)\}$$

which is nothing else that the polar cone of the cone $\widetilde{K}(a)$ generated by $\widetilde{S}(a)$. Namely,

$$\widetilde{K}(a) = \{d : a + td \in \widetilde{S}(a) \text{ for some } t > 0\}$$

and

$$\widetilde{N}(a) = \{\, a^\star \, : \, \langle a^\star, d\rangle \leq 0 \ \text{ for all } d \in \widetilde{K}(a)\}.$$

It is clear that $\widetilde{N}(a)$ is the projection on E of the set $N_e(a)$ defined in Equation (3.1). Hence, $\widetilde{N}(a)$ is the closed convex cone generated by $\partial f(a)$, the Fenchel-subdifferential of f at a. When the point a moves, the point $(a, f(a))$ moves on the boundary of the epigraph. Because the epigraph is convex, the map N_e has nice properties of continuity which afterwards implies nice properties of continuity for the maps ∂f and $\widetilde{N}$. These properties are the tools used to analyse sensitivity in convex programming.

The epigraph of a quasiconvex function is not convex and $N_e(a)$ cannot be used. However, convexity is present through the level sets $\widetilde{S}(a)$. The purpose of this chapter is to show how this convexity gives birth to some interesting properties concerning the continuity and the differentiability of f, how $\widetilde{N}(a)$ can be obtained from the directional derivatives of f at a and what can be said about the continuity of the point-to-set map $\widetilde{N}$.

Throughout thc chapter, we use the following notation:

E is a normed linear real space. Given $a \in E$ and $r > 0$, $B(a,r)$ denotes the open ball

$$B(a,r) = \{x \, : \, \|x - a\| < r\}.$$

Let $f : E \to [-\infty, +\infty]$ (as usual in convex analysis, we consider functions defined on the whole space; if we have a function $f : D \to \mathbb{R}$ with $D \subset E$, set $f(x) = +\infty$ for $x \notin D$). The epigraph $epi(f)$ (denoted also by $epi\, f$) and the strict epigraph $\widetilde{epi}(f)$ ($\widetilde{epi}\, f$) are respectively the following subsets of $E \times \mathbb{R}$

$$\begin{aligned} epi(f) &= \{(x,\lambda) \, : \, f(x) \leq \lambda\}, \\ \widetilde{epi}(f) &= \{(x,\lambda) \, : \, f(x) < \lambda\}. \end{aligned}$$

Given $\lambda \in \mathbb{R}$, we define

$$\begin{aligned} S_\lambda(f) &= \{x \, : \, f(x) \leq \lambda\}, \\ \text{and} \quad \widetilde{S}_\lambda(f) &= \{x \, : \, f(x) < \lambda\}. \end{aligned}$$

Clearly, for $\lambda < \mu$,

$$\widetilde{S}_\lambda(f) \subseteq S_\lambda(f) \subseteq \widetilde{S}_\mu(f) \subseteq S_\mu(f)$$

and

$$S_\lambda(f) = \cap_{\mu > \lambda} \widetilde{S}_\mu(f) = \cap_{\mu > \lambda} S_\mu(f).$$

We recall that $f : E \to [-\infty, +\infty]$ is said to be *quasiconvex* if

$$x, y \in E,\ 0 < t < 1 \Longrightarrow f(tx + (1-t)y) \leq max[\,f(x), f(y)\,],$$

and $f : D \to (-\infty, +\infty)$ is said to be *strictly quasiconvex* on the convex set D if

$$x, y \in D,\ x \neq y,\ 0 < t < 1 \Longrightarrow f(tx + (1-t)y) < max[\,f(x), f(y)\,].$$

$f : E \to [-\infty, +\infty]$ is said to be *lower semi-continuous* (*lsc* in short) at $a \in E$ if for all $\lambda < f(a)$ there is a neighbourhood V of a such that $\lambda < f(x)$ for all $x \in V$. The function f is said to be *upper semi-continuous* (*usc* in short) at $a \in E$ if $-f$ is lsc at a. It is clear that f is continuous at a if and only if it is both lsc and usc at a. Finally, f is continuous (lsc, usc) on E if continuous (lsc, usc) at any $a \in E$.

For the sake of completeness, we recall in the two following propositions the meaningful geometrical interpretations of lower semi-continuity and quasiconvexity.

Proposition 3.1 *The following statements are equivalent:*

- *f is lower semi-continuous on E,*
- *$epi(f)$ is closed,*
- *$S_\lambda(f)$ is closed for all $\lambda \in \mathbb{R}$.*

Proposition 3.2 *The following statements are equivalent:*

- *f is quasiconvex on E,*
- *$S_\lambda(f)$ is convex for all $\lambda \in \mathbb{R}$,*
- *$\widetilde{S_\lambda}(f)$ is convex for all $\lambda \in \mathbb{R}$.*

A set $C \subseteq E$ is said to be *evenly convex* (Fenchel [10]) if it is the intersection of open half spaces. It results from separation theorems that open and closed convex sets are evenly convex.

A function f is said to be *evenly quasiconvex* if all $S_\lambda(f)$ are evenly convex. A lower semi-continuous quasiconvex function is evenly quasiconvex since its level sets are convex and closed and thereby evenly convex. If f is an upper semi-continuous quasiconvex function then it is also evenly quasiconvex because its level sets $S_\lambda(f) = \cap_{\mu > \lambda} \widetilde{S_\mu}(f)$ are intersections of open convex sets.

Given $C \subset E$, $cl(C)$ and $int(C)$ denote the closure and the interior of C. The relative interior of C, denoted by $rint(C)$, is the interior of C with respect to the affine subspace generated by C. It is known that the relative interior of a nonempty convex set C of $\mathbb{R}^n$ is nonempty, furthermore C and $rint(C)$ have the same closure.

Sometimes, infinite values are not well praised when considering continuity and differentiability. It is easy to avoid this problem; indeed, given $g : E \to [-\infty, +\infty]$, let us consider the function f defined by $f(x) = \arctan g(x)$ for all $x \in E$. Then $f : E \to [-\frac{\pi}{2}, +\frac{\pi}{2}]$. It is easy to see that f is (evenly) quasiconvex if and only if g is (evenly) quasiconvex and f is lsc (usc) at a point a if and only if g is so. In this spirit, we shall say that g is differentiable at a if f is differentiable at this point. Thus, for continuity and differentiability questions in quasiconvex analysis, it is sufficient to consider functions which are finite on the whole space.

2. Quasiconvex regularizations of functions

We begin this section with the four following observations:

1) Given $f : E \to [-\infty, +\infty]$, we have

$$S_\lambda(f) = \cap_{\mu>\lambda}\widetilde{S}_\mu(f) = \cap_{\mu>\lambda}S_\mu(f).$$

2) Given $f, g : E \to [-\infty, +\infty]$, we have

$$f(x) \leq g(x)\ \forall\, x \in E \Longleftrightarrow \widetilde{epi}(g) \subset \widetilde{epi}(f) \Longleftrightarrow \widetilde{S_\lambda}(g) \subset \widetilde{S_\lambda}(f)\ \forall \lambda \in \mathbb{R},$$

$$f(x) \leq g(x)\ \forall\, x \in E \Longleftrightarrow epi(g) \subset epi(f) \Longleftrightarrow S_\lambda(g) \subset S_\lambda(f)\ \forall \lambda \in \mathbb{R}.$$

3) A function is thoroughly defined when its epigraph or its level sets are given. Indeed,

$$\begin{aligned} f(x) &= \inf\,[\lambda : (x,\lambda) \in epi\, f] &&= \inf\,[\lambda : (x,\lambda) \in \widetilde{epi}\, f] \\ &= \inf\,[\lambda : x \in S_\lambda(f)] &&= \inf\,[\lambda : x \in \widetilde{S_\lambda}(f)]. \end{aligned}$$

Given $T \subset E \times \mathbb{R}$ such that

$$(x,\lambda) \in T \text{ and } \mu > \lambda \Longrightarrow (x,\mu) \in T$$

(such a property holds for the epigraph and the strict epigraph of a function), a function g is defined by

$$g(x) = \inf[\lambda : (x,\lambda) \in T].$$

Then

$$epi(g) = \{(x,\lambda) : (x,\mu) \in T \text{ for all } \mu > \lambda\}.$$

Note that $epi(g) \supset T$ but $epi(g) \neq T$ in general. If $T = epi(f)$ then f and g coincide.

In the same manner, given a family of sets $\{T_\lambda\}_{\lambda \in \mathbb{R}}$ such that $T_\lambda \subset T_\mu$ for all $\lambda < \mu$, a function g is defined by

$$g(x) = \inf[\lambda : x \in T_\lambda].$$

Then

$$S_\lambda(g) = \cap_{\mu > \lambda} T_\mu.$$

We have $S_\lambda(g) \supset T_\lambda$ but $S_\lambda(g) \neq T_\lambda$ in general. If $T_\lambda = S_\lambda(f)$ for all λ, then f and g coincide.

4) Any intersection of closed (convex, closed convex, evenly convex) sets is a closed (convex, closed convex, evenly convex) set. Given $C \subset E$ we denote by $cl(C), co(C), \overline{co}(C)$ and $ec(C)$ the intersection of all the closed sets, the convex sets, the closed convex sets and the evenly convex sets, respectively, which contain C. These sets, named the closure (i.e. the closed hull), the convex hull, the closed convex hull, and the evenly convex hull are the smallest closed set, convex set, closed convex set and evenly convex set, respectively, which contains C.

Based on these four observations, we can construct regularisations of a function by considering the hulls of its epigraph or of its level sets. Thus, given a function $f : E \to [-\infty, +\infty]$, we define in connection with the epigraph

$$\begin{aligned} \bar{f}(x) &= \inf[\lambda : (x, \lambda) \in cl(epi\, f)], \\ f_c(x) &= \inf[\lambda : (x, \lambda) \in co(epi\, f)], \\ f_{\bar{c}}(x) &= \inf[\lambda : (x, \lambda) \in \overline{co}(epi\, f)]. \end{aligned}$$

It results that $\bar{f}, f_c$ and $f_{\bar{c}}$ are the greatest lsc function, the greatest convex function, the greatest lsc convex function which are majorized by f. It is clear that $epi\, \bar{f} = cl(epi\, f)$ and $epi\, f_{\bar{c}} = \overline{co}(epi\, f)$ but $epi\, f_c \neq co(epi\, f)$ in general. It is worth noticing that, in the three above relations, we can use the strict epigraph instead of the epigraph.

Before dealing with level sets, we recall that for all $\lambda < \mu$

$$\widetilde{S_\lambda}(f) \subset S_\lambda(f) \subset \widetilde{S_\mu}(f) \subset S_\mu(f).$$

Then, given f, we construct the functions

$$\begin{aligned} \bar{f}(x) &= \inf[\lambda : x \in cl(S_\lambda(f))] &&= \inf[\lambda : x \in cl(\widetilde{S_\lambda}(f))], \\ f_q(x) &= \inf[\lambda : x \in co(S_\lambda(f))] &&= \inf[\lambda : x \in co(\widetilde{S_\lambda}(f))], \\ f_{ec}(x) &= \inf[\lambda : x \in ec(S_\lambda(f))] &&= \inf[\lambda : x \in ec(\widetilde{S_\lambda}(f))], \\ f_{\bar{q}}(x) &= \inf[\lambda : x \in \overline{co}(S_\lambda(f))] &&= \inf[\lambda : x \in \overline{co}(\widetilde{S_\lambda}(f))]. \end{aligned}$$

The first line leads to alternative constructions of $\bar{f}$. With the last three lines, we obtain respectively the greatest quasiconvex function, the greatest evenly quasiconvex function and the greatest lsc quasiconvex function majorized by f. It is clear that for all $x \in E$

$$f_{\bar{c}}(x) \leq f_{\bar{q}}(x) \leq f_{ec}(x) \leq f_q(x) \leq f(x).$$

Also, we have $f_{\bar{q}} = \overline{f_{ec}} = \overline{f_q}$.

Proposition 3.3 *Let $f : E \to [-\infty, +\infty]$. Then*

$$f \text{ is lsc at } a \Longleftrightarrow \bar{f}(a) = f(a).$$

Proof. We know that $\bar{f} \leq f$. Assume that f is lsc at a and $\bar{f}(a) < f(a)$. Take λ be such that $\bar{f}(a) < \lambda < f(a)$. There is a neighbourhood V of a such that $V \cap S_\lambda(f) = \emptyset$. Hence $a \notin cl(S_\mu(f))$ for all $\mu \leq \lambda$, in contradiction with $\bar{f}(a) < \lambda$. Assume that $\bar{f}(a) = f(a)$. Since $\bar{f}$ is lsc at a, for all $\lambda < \bar{f}(a) = f(a)$, there is a neighbourhood V of a such that $\lambda < \bar{f}(x)$ for all $x \in V$. But $\bar{f}(x) \leq f(x)$ and therefore f is lsc at a. □

Since

$$S_\lambda(\bar{f}) = \cap_{\mu > \lambda} cl(S_\mu(f)) \text{ and } S_\lambda(f) = \cap_{\mu > \lambda} S_\mu(f),$$

it follows that for all $\lambda \in \mathbb{R}$

$$cl(S_\lambda(f)) \subset S_\lambda(\bar{f}).$$

The equality does not hold in general. However, for quasiconvex functions, we have the following result:

Proposition 3.4 *Assume that f is quasiconvex and $int(S_\lambda(f)) \neq \emptyset$. Then*

$$cl(S_\lambda(f)) = S_\lambda(\bar{f}).$$

Proof. Fix $a \in int(S_\lambda(f))$. Let $x \in S_\lambda(\bar{f})$, then $x \in cl(S_\mu(f))$ for all $\mu > \lambda$ and $x_t = x + t(a - x) \in int(S_\mu(f))$ for all $t \in (0, 1)$. It follows that $x_t \in S_\lambda(f)$ and therefore $x \in cl(S_\lambda(f))$. □

Furthermore, for quasiconvex functions, we also have the following result:

Proposition 3.5 *Assume that $f : E \to [-\infty, +\infty]$ is quasiconvex and $E = \mathbb{R}^n$. Then f is continuous at a if and only if $\bar{f}$ is continuous at a.*

Proof. Assume that f is continuous at a. By definition $\bar{f}$ is lsc, it is also usc at a since $\bar{f}(a) = f(a)$, $\bar{f} \leq f$ and f is usc at a. Next, assume that $\bar{f}$ is continuous at a. Let $\lambda > \bar{f}(a)$. There is an open neighbourhood

V of a such that $\bar{f}(x) < \lambda$ for all $x \in V$, i.e., $V \subset cl(S_\lambda(f))$. Since $E = \mathbb{R}^n$ and $S_\lambda(f)$ is convex, then $V \subset int(S_\lambda(f))$ because, in a finite dimensional space, a convex set and its closure have the same interior. It follows that $f(a) = \bar{f}(a)$ and f is usc at a. □

This result does not hold when the dimension of E is infinite, even for a convex function, as shown below.

Example 3.1 Let H be an hyperplane of E which is not closed (such H exists when the dimension of E is not finite). Take $f(x) = 0$ if $x \in H$ and $f(x) = +\infty$ otherwise. Then f is convex, $\bar{f}(x) = 0$ for all $x \in E$ and therefore $\bar{f}$ is continuous on E but f is not continuous at any $x \in E$.

Assume that f is quasiconvex. If f is lsc at a, then $f(a) = \bar{f}(a) = f_{\bar{q}}(a)$. If f is usc on the whole space, then f is evenly quasiconvex, hence $f(a) = f_{ec}(a)$ at any point. This equality does not hold when f is only usc at a (and even if f is usc in a neighbourhood of a) as shown by the following counter-example.

Example 3.2 $f(x, y) = 1$ if $(y > 0)$ or $(y = 0$ and $x \geq -1)$, $f(x, y) = 0$ otherwise. Then f is quasiconvex and is usc on a neighbourhood of $a = (0, 0)$ but $0 = f_{ec}(a) \neq f(a) = 1$.

Historical comments: If evenly convex sets have been introduced by Fenchel [10], the introduction of evenly quasiconvex functions is due to Passy and Prisman [15] and, independently, to Martinez-Legaz [13]. The description of the process for constructing regularized functions is given in Crouzeix [3, 6].

3. Quasiconvex functions are nondecreasing

Let $\theta : \mathbb{R} \to [-\infty, +\infty]$. Then θ is quasiconvex if and only if there exists $t \in [-\infty, +\infty]$ so that:

- either θ is nonincreasing on $(-\infty, t]$ and nondecreasing on $(t, +\infty)$,
- or θ is nonincreasing on $(-\infty, t)$ and nondecreasing on $[t, +\infty)$.

Thus, the simplest examples of quasiconvex functions are the nondecreasing functions of one real variable. It results that, unlike convex functions, quasiconvex functions are not continuous on the interior of their domain. A fortiori, directional derivatives are not necessarily defined. Still, nondecreasing functions of one real variable are almost everywhere continuous and differentiable, whence quasiconvex functions of one real variable are also almost everywhere continuous and differentiable on the interior of their domains.

When E is a linear space, quasiconvexity has again a connection with monotonicity. Let E be a normed linear space, $f : E \to [-\infty, +\infty]$ and K be a convex cone of E, f is said to be *nondecreasing with respect to* K if

$$x, y \in E,\;\; y - x \in K \Longrightarrow f(x) \leq f(y).$$

For $x \in E$, let us define

$$\widetilde{K}(x) = \{d \,:\, x + td \in \widetilde{S}(x) \text{ for some } t > 0\}.$$

Theorem 3.1 *Let $f : E \to \mathbb{R}$ be quasiconvex, $\lambda \in \mathbb{R}$ and $a \in E$ such that $int(S_\lambda(f)) \neq \emptyset$ and $a \notin cl(S_\lambda(f))$. Then there exists an open convex neighbourhood V of a and a nonempty open convex cone K so that*

$$x, y \in V,\; y - x \in K \Longrightarrow f(x) \leq f(y).$$

Futhermore, if f is strictly quasiconvex

$$x, y \in V,\; y - x \in K,\; x \neq y \Longrightarrow f(x) < f(y).$$

Moreover $-K \subseteq \widetilde{K}(x)$ for all $x \in V$.

Proof. Let $b \in int(S_\lambda(f))$, $r > 0$ and $R > 0$ be such that $B(b, r) \subseteq S_\lambda(f)$ and $S_\lambda(f) \cap B(a, R) = \emptyset$. Let some $\bar{t} > 0$. Set $c = a + \bar{t}(a - b)$ and

$$K \;=\; \{d \,:\, c - td \in B(b, r) \text{ for some } t > 0\}.$$

Then K is a nonempty open convex cone. Hence $y - K \subseteq c - K$ for all $y \in c - K$. Set $V = (c - K) \cap B(a, R)$. Assume that $x, y \in V$ with $y - x \in K$. Then there is $t > 1$ so that $z = y + t(x - y) \in B(b, r)$. Observe that $f(z) \leq \lambda < f(y)$, then the results follow from the (strict) quasiconvexity of f. □

In the particular case where $E = \mathbb{R}^n$, we derive the following result.

Corollary 3.1 *Let $f : \mathbb{R}^n \to (-\infty, +\infty)$ quasiconvex, $\lambda \in \mathbb{R}$ and $a \in \mathbb{R}^n$ such that $int(S_\lambda(f)) \neq \emptyset$ and $a \notin cl(S_\lambda(f))$. Then there exist an open convex neighbourhood V of a and $v_1, v_2, \cdots, v_n$, n linearly independent vectors such that*

$$x, x + \sum t_i v_i \in V,\; t_1, t_2, \cdots, t_n \geq 0 \Longrightarrow f(x) \leq f(x + \sum t_i v_i).$$

Furthermore, if f is strictly quasiconvex and $\sum t_i > 0$, then the inequality is strict.

Proof. Choose for vectors v_i n linearly independent vectors in K. □

A nondecreasing function is not necessarily quasiconvex as shown below.

Example 3.3 Let $f : \mathbb{R}^2 \to \mathbb{R}$ be defined by $f(x_1, x_2) = \min[x_1, x_2]$. Then f is nondecreasing with respect to the nonnegative orthant of $\mathbb{R}^2$ but is not quasiconvex.

It appears that the relationship between quasiconvex functions and nondecreasing functions looks like the relationship between convex functions and Lipschitz functions. Because the class of convex functions is contained in the class of quasiconvex functions, it is interesting to know if such a relation exists between the classes of Lipschitz and monotonic functions. The following result shows that, indeed, there is nearly such a relation: if f is Lipschitz, there is a linear function l such that $f + l$ is nondecreasing.

Proposition 3.6 *Assume that $f : \mathbb{R}^n \to \mathbb{R}$ is locally Lipschitz in a neighbourhood of a, i.e., there is $r > 0$ and $L > 0$ such that*

$$x_i, y_i \in [a_i - r, a_i + r] \text{ for all } i \Longrightarrow |f(y) - f(x)| \leq L \sum |y_i - x_i|.$$

Define a function g by $g(x) = f(x) + L \sum x_i$. Then

$$a_i - r \leq x_i \leq y_i \leq a_i + r, \quad i = 1, 2, \cdots, n \Longrightarrow g(x) \leq g(y).$$

Proof. The proof is immediate. □

It follows that a Lipschitz function enjoys the properties of monotonic functions.

4. Continuity

Given $f : E \to [-\infty, +\infty]$, $a \in E$ with $-\infty < f(a) < +\infty$ and a direction $d \in E$, we define the function of one real variable

$$f_{a,d}(t) = f(a + td).$$

It is know that f is convex (quasiconvex) on E if and only if, for any $d, a \in E$, $f_{a,d}$ is convex (quasiconvex) on $\mathbb{R}$. The first result concerns nondecreasing functions.

Proposition 3.7 *Assume that $f(a)$ is finite, U is a convex neighbourhood of a K is an open convex cone, $d \in K$ and f is nondecreasing with* *respect to K on U. Then f is lsc (usc) at a if and only if $f_{a,d}$ is lsc (usc) at* 0.

Proof. There exist $t_- < 0$ and $t_+ > 0$ such that

$$V = \{x : x - a - t_- d \in K \text{ and } a + t_+ d - x \in K\} \subset U,$$

then V is a neighbourhood of a. Assume that $f_{a,d}$ is lsc at 0. Let some $\lambda_- < f(a)$, t_- can be chosen close enough to 0 so that $\lambda_- < f(a+t_-d)$. and $\lambda_- < f(a + t_-d) \leq f(x)$ for all $x \in V$. If $f_{a,d}$ is usc at 0 and $\lambda_+ > f(a)$, t_+ is chosen close enough to 0 so that $\lambda_+ > f(a+t_+d)$, then $\lambda_+ > f(a + t_+d) \geq f(x)$ for all $x \in V$. □

Next, we combine that proposition with Theorem 3.1 in order to derive results for quasiconvex functions. Recall that, as far as continuity is considered, it is enough to deal with real-valued functions.

Corollary 3.2 *Assume that $f : E \to \mathbb{R}$ is quasiconvex, $a \in E$, $\lambda < f(a)$, $b \in int(S_\lambda(f))$ and $a \notin cl(S_\lambda(f))$. Set $d = a - b$. Then f is lsc (usc) at a as soon as $f_{a,d}$ is lsc (usc) at* 0.

Proof. Define K as done in the proof of Theorem 3.1 and apply the proposition. □

Proposition 3.8 *Assume that $f : E \to \mathbb{R}$ is quasiconvex, $a, b \in E$, $f(b) < f(a)$ and f is usc at b. Set $d = b - a$. Then f is lsc (usc) at a if and only if $f_{a,d}$ is lsc (usc) at* 0.

Proof. Take λ so that $f(b) < \lambda < f(a)$. Then there exists $r > 0$ such that the open ball $B(b, 2r)$ is contained in $S_\lambda(f)$. Notice that $2r \leq \|b - a\|$.

Assume that $f_{a,d}$ is lsc at 0 and show that f is lsc at a. Let μ so that $\lambda < \mu < f(a)$, there is $\bar{t} \in (0, \|b - a\| - r)$ so that $f(c = a + \bar{t}d) > \mu$. Set

$$V = \{x = c + s(c - y) : s > 0 \text{ and } \|b - y\| < r\}.$$

Then V is a neighbourhood of a and, because f is quasiconvex, $f(x) \geq f(c)$ for all $x \in V$.

Assume that $f_{a,d}$ is usc at 0 and show that f is usc at a. Let μ so that $\mu > f(a)$, there is $\bar{t} < 0$ so that $f(c = a + \bar{t}d) < \mu$. Set

$$V = \{x = c + s(y - c) : s \in (0, 1) \text{ and } \|b - y\| < r\}.$$

Then V is a neighbourhood of a and, because f is quasiconvex, $f(x) \leq f(c)$ for all $x \in V$. □

In these two corollaries, we have shown that the continuity of f holds if it holds along one specific direction. The following result can appear weaker since it involves continuity along all directions, but it deserves to be given in reason of its very simple formulation. Before, we introduce the following definition.

The function f is said to be *continuous along the lines* at a point a (or *hemi-continuous* or again *radially continuous*) at a if for each $d \in E$ the function $f_{a,d}$ is continuous at 0. Similar definitions apply to lower and upper semicontinuity.

Theorem 3.2 *[3, 6] Assume that f is quasiconvex on $\mathbb{R}^n$ and $f(a)$ is finite. Then f is lsc (usc) at a if and only if f is lsc (usc) along the lines at a.*

Proof. 1) Assume that, for every d, $f_{a,d}$ is lsc at 0 and let us prove that f is lsc at a. Let $\lambda < f(a)$ and let us prove that $f(x) > \lambda$ for all x in a neighbourhood V of a. If $S_\lambda(f) = \emptyset$, take $V = \mathbb{R}^n$. If not take b in $rint(S_\lambda(f))$, the relative interior of $S_\lambda(f)$. Set $d = b - a$, there is $\bar{t} \in (0,1)$ such that $f(a + \bar{t}(b-a)) > \lambda$. It results that $a \notin cl(S_\lambda(f))$. In view of separation theorems, there exist $x^\star$ and α such that for all $x \in S_\lambda(f)$

$$\langle x^\star, x\rangle \leq \alpha < \langle x^\star, a\rangle.$$

Take $V = \{y : \alpha < \langle x^\star, y\rangle\}$.

2) Next, assume that, for all d, $f_{a,d}$ is usc at 0 . Let $\lambda > f(a)$. Denote by e_i the i-th vector of the canonical basis of $\mathbb{R}^n$. There is $t_i > 0$ such that $f(a + te_i) < \lambda$ for all $t \in [-t_i, t_i]$. Take for V the convex hull of the $2n$ points $a \pm t_i e_i$. Then V is a neighborhood of a and $V \subseteq S_\lambda(f)$. □

Such a result does not hold if the dimension of E is not finite.

Example 3.4 Consider a function l which is linear but not continuous on E, such a function exists as soon as the dimension of E is infinite. This function is continuous along the lines at any point.

For other results on continuity in infinite dimensional spaces, we refer the reader to Hadjisavvas [11].

5. Differentiability: notation and first results

We start with the notation. Assume that $f(a)$ is finite and $h \in E$. Then the *upper* and the *lower Dini-derivative of f at a with respect to* *the direction* h are respectively defined by

$$f'_+(a,h) = \limsup_{t\to 0_+} \frac{f(a+th)-f(a)}{t},$$
$$f'_-(a,h) = \liminf_{t\to 0_+} \frac{f(a+th)-f(a)}{t}.$$

If $-\infty < f'_-(a,h) = f'_+(a,h) < \infty$ then *the directional derivative of* f *with respect to the direction* h exists and is defined by

$$f'(a,h) = f'_-(a,h) = f'_+(a,h).$$

If for all $h \in E$

$$f'(a,h) \text{ and } f'(a,-h) \text{ exist and } f'(a,h) + f'(a,-h) = 0$$

then f is said to be *differentiable at* a *along the lines* or again *weakly Gâteaux-differentiable* at a.

If there is a vector $c \in E'$ such that

$$f'(a,h) = \langle c,h\rangle \qquad \forall h \in E$$

then f is said to be *Gâteaux-differentiable* at a. Such a c is uniquely defined, it is called the *(Gâteaux-)gradient* of f at a and denoted by $\nabla f(a)$.

If f is Gâteaux-differentiable at a and

$$\frac{f(a+h)-f(a)-\langle\nabla f(a),h\rangle}{\|h\|} \to 0 \text{ when } h \to 0$$

then f is said to be *Fréchet-differentiable* at a.

Although Fréchet- and Gâteaux-differentiability do not coincide for an arbitrary function, they coincide when the function is monotone on $\mathbb{R}^n$.

Theorem 3.3 *(Chabrillac-Crouzeix [2]) Let* $f : \mathbb{R}^n \to [-\infty,\infty]$, K *be a nonempty open convex cone. Assume that* f *is nondecreasing with respect to* K *and* f *is Gâteaux-differentiable at* a. *Then* f *is Fréchet-differentiable at* a.

Proof. Assume that f is Gâteaux- but not Fréchet-differentiable at a. Then there exist $\epsilon > 0$ and a sequence $\{h_n\}_n$ converging to 0 such that for all n

$$7\epsilon < \frac{|f(a+h_n)-f(a)-\langle\nabla f(a),h_n\rangle|}{\|h_n\|}. \tag{3.2}$$

Set $d_n = \frac{1}{\|h_n\|}h_n$. Without loss of generality, we can assume that the whole sequence $\{d_n\}_n$ converges to some $\bar{d}$. Let $e \in K$. Then $\mu > 0$ exists so that

$$|\langle \nabla f(a), d - \bar{d} \rangle| < \epsilon \quad \forall\, d \in V = \{d \in E : d - d_-,\ d_+ - d \in K\} \quad (3.3)$$

where $d_- = \bar{d} - \mu e$ and $d_+ = \bar{d} + \mu e$. Then V is a neighbourhood of $\bar{d}$. For n large enough, $d_n \in V$ and therefore

$$f(a + \|h_n\| d_-) \leq f(a + h_n) \leq f(a + \|h_n\| d_+) \quad (3.4)$$

Since f is Gâteaux-differentiable at a, for n large enough

$$\frac{|f(a + \|h_n\| d_-) - f(a) - \langle \nabla f(a), \|h_n\| d_- \rangle|}{\|h_n\|} < \epsilon, \quad (3.5)$$

and

$$\frac{|f(a + \|h_n\| d_+) - f(a) - \langle \nabla f(a), \|h_n\| d_+ \rangle|}{\|h_n\|} < \epsilon. \quad (3.6)$$

The contradiction follows from Equations (3.2), (3.3), (3.4), (3.5) and (3.6). □

As an immediate corollary, we see that Gâteaux-differentiability and Fréchet-differentiability coincide for locally Lipschitz functions on $\mathbb{R}^n$. Also, Theorem 3.3 can be applied to quasiconvex functions.

Theorem 3.4 *Assume that $f : \mathbb{R}^n \to [-\infty, \infty]$ is quasiconvex. If f is Gâteaux-differentiable at a, then f is Fréchet-differentiable at a as well.*

Proof. Let $S = \{x : f(x) < f(a)\}$.

1) $S = \emptyset$. Then $f(x) \geq f(a)$ for all x and therefore $\nabla f(a) = 0$. Let $(e_1, e_2, \ldots, e_n)$ be the canonical basis of $\mathbb{R}^n$. Set $e_{i+n} = -e_i$ for $i = 1, \ldots, n$. For $h \in \mathbb{R}^n$ take $t_i = max[0, h_i]$ and $t_{i+n} = max[0, -h_i]$. Then

$$a + h = a + \sum_{i=1}^{i=2n} t_i e_i = \sum_{i=1}^{i=2n} (a + \frac{t_i}{\|h\|} \|h\| e_i)$$

where $\|h\| = \sum_{i=1}^{i=n} |h_i| = \sum_{i=1}^{i=2n} t_i$. Then, since f is quasiconvex

$$0 \leq \frac{f(a + th) - f(a)}{\|h\|} \leq \max_{i=1,\ldots,2n} \frac{f(a + \|h\| e_i) - f(a)}{\|h\|},$$

and the result follows.

2) $int(S) \neq \emptyset$. There is $\lambda \in \mathbb{R}$ such that $int(S_\lambda(f)) \neq \emptyset$ and $a \notin cl(S_\lambda(f))$. The result follows from Theorems 3.1 and 3.3.

3) We are left with the case where $int(S) = \emptyset$ but $S \neq \emptyset$. Of course, $\nabla f(a) = 0$. The proof is obtained in working on the affine set generated by S and combining the proofs in 1) and 2). □

A previous proof by Crouzeix [8] of that theorem does not make use of Theorem 3.3.

It is well known that a nondecreasing function of one real variable is almost everywhere differentiable. The result still holds in $\mathbb{R}^n$.

Theorem 3.5 *(Chabrillac-Crouzeix [2]) Let $f : \mathbb{R}^n \to [-\infty, \infty]$, K be an open nonempty convex cone. Assume that f is nondecreasing with respect to K. Then f is almost everywhere Fréchet-differentiable.*

The celebrated Rademacher's theorem on locally Lipschitz functions can be viewed as a corollary of this theorem. Another consequence is that quasiconvex functions on $\mathbb{R}^n$ are also almost everywhere Fréchet-differentiable, a previous and direct proof of this result was given in Crouzeix [7].

6. Directional derivatives

It is clear that, for a general function, the Dini-derivatives $f'_-(a,h)$ and $f'_+(a,h)$ are positively homogeneous of degree 1 with respect to the direction h. This is also the case for the directional derivatives $f'(a,h)$ when they are defined. The following proposition is a direct consequence of quasiconvexity.

Proposition 3.9 *Assume that $f : E \to [-\infty, \infty]$, f is quasiconvex and $f(a)$ is finite. Then $f'_+(a, .)$ is quasiconvex.*

Proof. Let $h, k \in E$, $\lambda, \mu \geq 0$ with $\lambda + \mu = 1$. Then, for all $t > 0$,

$$\frac{f(a + t\lambda h + t\mu k) - f(a)}{t} \leq \max[\frac{f(a + t\lambda h) - f(a)}{t}, \frac{f(a + t\mu k) - f(a)}{t}].$$

Pass to the limit when $t \to 0$, there is no problem because we consider the lim sup. □

There are counter-examples (Crouzeix [5]) where the function f is quasiconvex, but the lower Dini-derivative is not. The fact that, for a convex function, the directional derivative $f'(a,h)$ is convex and positively homogeneous in h has strong implications. Indeed, the indicator function of $\partial f(a)$, the Fenchel-subdifferential of f at a, is nothing else that the Fenchel-conjugate of the function $f'(a, .)$. Quasiconvex positively homogeneous functions also will play a fundamental role as seen below.

Theorem 3.6 *(Newman [14], Crouzeix [4]) Assume that $C \subseteq E$ is convex and $\theta : C \to (-\infty, +\infty]$ is quasiconvex and positively homogeneous of degree one.*

1 If $\theta(x) < 0$ for all $x \in C$, then θ is convex on C,

2 If $\theta(x) \geq 0$ for all $x \in C$, then θ is convex on C.

Proof. Several types of proofs can be given. The following one is based on the geometrical aspect of convexity. Consider in case 1)

$$S = S_{-1}(\theta) \times \{-1\} \subseteq E \times \mathbb{R},$$

and in case 2)

$$S = S_1(\theta) \times \{1\} \subseteq E \times \mathbb{R}.$$

Next, let us consider the cone T in $E \times \mathbb{R}$ generated by S. Both S and T are convex. In case 1) T corresponds to the epigraph of θ. In case 2) take $\widetilde{T} = T \cup (S_\infty \times \{0\})$ where S_∞ is the recession cone of $S_1(\theta)$. Then, $\widetilde{T}$ corresponds to the epigraph of θ as well. □

Now, assume that $\theta : E \to (-\infty, +\infty]$ is quasiconvex and positively homogeneous of degree 1. Define θ_- and θ_+ by

$$\begin{array}{llll} \text{if} \;\; \theta(x) < 0 & \text{then} \;\; \theta_-(x) = \theta(x) & \text{and} \;\; \theta_+(x) = 0, \\ \text{if} \;\; \theta(x) \geq 0 & \text{then} \;\; \theta_-(x) = +\infty & \text{and} \;\; \theta_+(x) = \theta(x). \end{array} \tag{3.7}$$

Then $\theta(x) = \min[\,\theta_-(x), \theta_+(x)\,]$. Hence θ is the minimum of two convex functions. This decomposition will be useful for the directional derivatives and the upper Dini-derivatives of quasiconvex functions, indeed they are quasiconvex in the direction.

Theorem 3.7 *(Crouzeix [8]) Assume that $f : E \to [-\infty, \infty]$ is quasiconvex, $f(a)$ is finite and f is weakly Gâteaux-differentiable at a. Then f is Gâteaux-differentiable at a as well.*

Proof. By assumption, $f'(a,h) + f'(a,-h) = 0$ for all h. If $f'(a,h) = 0$ for all h, take $\nabla f(a) = 0$. Next, assume that $f'(a,h) \neq 0$ for some h. Set $\theta(h) = f'(a,h)$, $S_- = \{h \in E : f'(a,h) < 0\}$, $S_0 = \{h \in E : f'(a,h) \leq 0\}$ and $S_+ = \{h \in E : f'(a,h) > 0\}$. Then, $S_+ = -S_-$. The sets S_-, S_0 and S_+ are convex. S_0 and S_+ are two complementary sets. Hence the closure of S_0 is a half-space. The functions θ_- and θ_+ are convex on S_- and S_+ respectively, furthermore $\theta_-(h) = -\theta_+(-h)$ for all $h \in S_- = -S_+$. Hence θ_- is linear on S_-. Finally, it is deduced that θ is linear on the whole space. □

Theorems 3.4 and 3.7 claim that for quasiconvex functions on $\mathbb{R}^n$, Fréchet-differentiability, Gâteaux-differentiability and weakly Gâteaux-differentiability coincide. For nondecreasing functions on $\mathbb{R}^n$, Gâteaux-differentiability and weakly Gâteaux-differentiability do not coincide in general.

7. More on the Dini-derivatives

For simplicity, throughout the sequel of the chapter, we shall use the following notation: given f quasiconvex on E and $a \in E$ with $f(a)$ finite, we define the sets

$$\begin{aligned} \widetilde{S}(a) &= \{x : f(x) < f(a)\}, \\ \widetilde{K}(a) &= \{d : a + td \in \widetilde{S}(a) \text{ for some } t > 0\}, \\ \widetilde{N}(a) &= \{a^\star : \langle a^\star, d\rangle \le 0 \quad \forall d \in \widetilde{K}(a)\}. \end{aligned}$$

If the first result looks rather technical, its consequences are quite meaningful.

Theorem 3.8 *Assume that f is quasiconvex, $f(a)$ is finite and $h \in int(\widetilde{K}(a))$. Let us define*

$$\rho(h) = \sup[\, r : B(h, r) \subset \widetilde{K}(a)].$$

i) If $f'_-(a,h) > -\infty$, then for all $d \in E$

$$f'_-(a,d) \ge \frac{\|d\|}{\rho(h)} f'_-(a,h).$$

ii) If $f'_+(a,h) > -\infty$, then for all $d \in E$

$$f'_+(a,d) \ge \frac{\|d\|}{\rho(h)} f'_+(a,h).$$

Proof. If $d \notin \widetilde{K}(a)$, then both $f'_-(a,d)$ and $f'_+(a,d)$ are nonnegative while both $f'_-(a,h)$ and $f'_+(a,h)$ are nonpositive; the inequalities hold. We are left with $d \in \widetilde{K}(a)$. Take

$$r \in (0, \rho(h)) \text{ and } k = h + \frac{r}{\|h-d\|}(h-d),$$

then $k \in int(\widetilde{K}(a))$ and

$$h = k + \lambda(d-k) \text{ where } \lambda = \frac{r}{r + \|h-d\|} \in (0,1).$$

Because $k \in \widetilde{K}(a)$, there exists $\bar{t} > 0$ so that

$$c = a + \bar{t}k \in \tilde{S}(a).$$

For all $t \in (0, \bar{t})$, let us define

$$m(t) = a + th \text{ and } n(t) = a + \xi(t)d,$$

where $\xi(t)$ is determined in such a way that $c, m(t)$ and $n(t)$ are on the same straight line. More precisely, we have

$$m(t) = c + \mu(t)(n(t) - c),$$

$$\text{where } \mu(t) = 1 - (1-\lambda)\frac{t}{\bar{t}} \quad \text{and} \quad \xi(t) = \frac{\lambda t}{\mu(t)}.$$

Since $\mu(t) \in (0,1)$ and f is quasiconvex, we have

$$\frac{f(m(t)) - f(a)}{t} \leq \max[\frac{f(c) - f(a)}{t}, \frac{f(n(t)) - f(a)}{t}].$$

The quantity $f(c) - f(a)$ is negative because $c \in \tilde{S}(a)$. Hence,

$$\frac{f(c) - f(a)}{t} \to -\infty \text{ when } t \to 0_+.$$

On the other hand,

$$\frac{\xi(t)}{t} \to \lambda \text{ when } t \to 0_+.$$

Because the concluding argument applies similarly to both lower and upper Dini-derivatives, we develop it below only for the lower Dini-derivative. Since $f'_-(a,h) > -\infty$, it follows that for $t > 0$ small enough

$$\frac{f(m(t)) - f(a)}{t} \leq \frac{f(n(t)) - f(a)}{t} = \frac{f(a + \xi(t)d) - f(a)}{\xi(t)} \frac{\lambda}{\mu(t)}.$$

Hence, $f'_-(a,h) \leq \lambda f'_-(a,d)$. Replace λ by its value and let $r \to \rho(h)$, then

$$f'_-(a,d) \geq (1 + \frac{\|h - d\|}{\rho(h)}) f'_-(a,h). \tag{3.8}$$

Recall that $\widetilde{K}(a)$ is a cone and the Dini-derivatives are positively homogeneous, thence for all $\theta > 0$,

$$\theta h \in int(\widetilde{K}(a)), \ \ \rho(\theta h) = \theta\rho(h) \ \text{ and } \ f'_-(a, \theta h) = \theta f'_-(a,h).$$

Therefore, replacing h by θh in (3.8), one obtains

$$f'_-(a,d) \geq (\theta + \frac{\|\theta h - d\|}{\rho(h)}) f'_-(a,h).$$

Finally, passing to the limit when $\theta \to 0_+$, we obtain

$$f'_-(a,d) \geq \frac{\|d\|}{\rho(h)} f'_-(a,h).$$

□

From this theorem, we immediately deduce the following result.

Corollary 3.3 *Assume that f is quasiconvex and $-\infty < f(a) < +\infty$.*

- *If $f'_-(a,h) = -\infty$ for some h, then $f'_-(a,d) = -\infty$ for all $d \in int(\widetilde{K}(a))$.*
- *If $f'_-(a,h) < 0$ for some h, then $f'_-(a,d) < 0$ for all $d \in int(\widetilde{K}(a))$.*
- *The same conclusions hold for the upper-Dini derivatives.*

Remark 3.1 When $E = \mathbb{R}^n$, the assumption "$h \in int(\widetilde{K}(a))$" can be replaced by "$h \in rint(\widetilde{K}(a))$" in the theorem and its corollary.

We recall that a function f which is differentiable at a is said to be *pseudoconvex at a* if

$$f(x) < f(a) \Longrightarrow \langle \nabla f(a), x - a \rangle < 0.$$

Dini-directional derivatives allow relaxations of this notion. According to the lower or the upper Dini derivative, f is said to be *D_--pseudoconvex at a* if

$$f(x) < f(a) \Longrightarrow f'_-(a, x-a) < 0,$$

and *D_+-pseudoconvex at a* if

$$f(x) < f(a) \Longrightarrow f'_+(a, x-a) < 0.$$

Another consequence of the theorem is given below.

Corollary 3.4 *Assume that f is quasiconvex, usc and $-\infty < f(a) < +\infty$. If there is some d such that $f'_-(a,d) < 0$ then f is D_--pseudoconvex at a. Similarly, if there is some d such that $f'_+(a,d) < 0$ then f is D_+-pseudoconvex at a.*

Proof. Assume that $f(x) < f(a)$. Since f is usc, $x \in int(\widetilde{S}(a))$ and therefore $h = x - a \in int(\widetilde{K}(a))$. Apply the theorem. □

We have said that, for a convex function f, the normal cone $\widetilde{N}(a)$ can be obtained from the Fenchel-subdifferential $\partial f(a)$ and therefore from the directional derivatives $f'(a,h)$. Such a result exists for quasiconvex functions. This is the object of the following theorem.

Theorem 3.9 *Assume that f is quasiconvex on E, $f(a)$ is finite and the interior of $\widetilde{S}(a)$ is not empty. If there exists d such that $f'_+(a,d) < 0$ then*

$$\widetilde{N}(a) = \{x^\star : \langle x^\star, h\rangle \leq 0 \ \forall h \text{ such that } f'_+(a,h) < 0\}.$$

The same result holds for the lower Dini-derivatives. If $E = \mathbb{R}^n$, the assumption $int(\widetilde{S}(a)) \neq \emptyset$ is not necessary.

Proof. Apply Corollary 3.3. □

Remark 3.2 This theorem says that if the quasiconvex function f is D_--pseudoconvex at a, then $\widetilde{N}(a)$ can be thoroughly recovered from the knowledge of the lower Dini-derivatives of f at a. The same conclusion holds for the upper Dini-derivatives.

We have seen the strong connection between quasiconvex functions and nondecreasing functions. The next result concerns nondecreasing functions. The proof is quite easy and left to the reader.

Proposition 3.10 *Assume that K is an open convex cone and f is nondecreasing with respect to K. Then $f'_-(a,.)$ and $f'_+(a,.)$ are nondecreasing with respect to K. Hence they are continuous on $K \cup -K$.*

If f is quasiconvex, Theorem 3.1 says that, under some mild assumptions, such a cone K exists and the proposition can be applied. However, a more general result exists. The next theorem involves $\widetilde{K}(a)$, since the cone K built in Theorem 3.1 is contained in $-\widetilde{K}(a)$, the following result leads to stronger results.

Theorem 3.10 *Assume that f is quasiconvex, $f(a)$ is finite and one of the following conditions holds:*
(1) $d \in E$ and $l \in int(\widetilde{K}(a))$,
(2) $d \in int(\widetilde{K}(a))$ and $l \in cl(\widetilde{K}(a))$.
Then

$$f'_+(a, d+l) \leq f'_+(a,d) \quad \text{and} \quad f'_-(a,d+l) \leq f'_-(a,d).$$

Moreover, if $d \in int(\widetilde{K}(a))$ and $l \in [cl(\widetilde{K}(a))] \cap [-cl(\widetilde{K}(a))]$, then

$$f'_+(a, d+l) = f'_+(a,d) \quad \text{and} \quad f'_-(a,d+l) = f'_-(a,d).$$

Proof. a) If $f'_+(a,d) = +\infty$ ($f'_-(a,d) = +\infty$), the property is trivial. If $f'_+(a,d) = -\infty$ ($f'_-(a,d) = -\infty$), then $d \in \widetilde{K}(a)$ in case (1) and therefore $d + l \in int(\widetilde{K}(a))$ in both cases. It results from Corollary 3.3 that $f'_+(a, d+l) = -\infty$ ($f'_-(a, d+l) = -\infty$).

b) We are left with the case $-\infty < f'_+(a,d) < \infty$ (or $-\infty < f'_-(a,d) < \infty$). In case (1), because $\widetilde{K}(a)$ is a convex cone and $l \in int(\widetilde{K}(a))$, there exists $\bar{\alpha} \geq 1$ such that

$$d + \alpha l \in int(\widetilde{K}(a)) \text{ for all } \alpha \geq \bar{\alpha}.$$

Such a $\bar{\alpha} \geq 1$ exists in case (2) as well. Let $\alpha > \bar{\alpha} \geq 1$, then there exists $\theta > 0$ such that $f(a + \theta d + \alpha\theta l) < f(a)$.

c) Firstly, we deal with the lower Dini-derivative. Let λ and μ such that $-\infty < \lambda < f'_-(a,d) < \mu < \infty$. By definition of $f'_-(a,d)$, there exists $\bar{t} > 0$ such that for all $t \in (0, \bar{t})$, there is $\xi(t) \in (0,t)$ such that

$$\frac{f(a + \theta d + \alpha\theta l) - f(a)}{\xi(t)} < \lambda < \frac{f(a + \xi(t)d) - f(a)}{\xi(t)} < \mu.$$

Since f is quasiconvex and

$$\begin{aligned} a + \frac{\alpha\theta\xi(t)}{\alpha\theta - \theta + \xi(t)}(d+l) \quad &= \quad \frac{\xi(t)}{\alpha\theta - \theta + \xi(t)}(a + \theta d + \alpha\theta l) \\ &\quad + \frac{\alpha\theta - \theta}{\alpha\theta - \theta + \xi(t)}(a + \xi(t)d) \end{aligned}$$

we have

$$f(a + \frac{\alpha\theta\xi(t)}{\alpha\theta - \theta + \xi(t)}(d+l)) - f(a) \leq f(a + \xi(t)d) - f(a).$$

Hence, dividing by $\xi(t)$ and passing to the limit when $t \to 0$, we obtain

$$f'_-(a, \frac{\alpha}{\alpha - 1}(d+l)) = \frac{\alpha}{\alpha - 1} f'_-(a, d+l) \leq \mu.$$

That inequality holds for all $\alpha > \bar{\alpha}$ and all $\mu > f'_-(a,d)$, thence

$$f'_-(a, d+l) \leq f'_-(a,d).$$

d) Next, we deal with the upper Dini-derivative. Firstly, we show that there exists $\bar{t} > 0$ so that

$$f(a + \theta d + \alpha\theta l) < f(a + td) \text{ for all } t \in (0, \bar{t}).$$

If not, there exists $\hat{t} > 0$ such that $f(a + \hat{t}d) < f(a + \theta d + \alpha\theta l)$. Since $f'_+(a,d) > -\infty$ there exists $\bar{t} \in (0, \hat{t})$ so that $f(a + \theta d + \alpha\theta l) < f(a + \bar{t}d)$. Since f is quasiconvex, $f(a + \theta d + \alpha\theta l) < f(a + td)$ for all $t \in (0, \bar{t})$.

Next, since f is quasiconvex and

$$a + \frac{\alpha\theta t}{\alpha\theta - \theta + t}(d+l) = \frac{t}{\alpha\theta - \theta + t}(a + \theta d + \alpha\theta l) + \frac{\alpha\theta - \theta}{\alpha\theta - \theta + t}(a + td),$$

one has

$$f(a + \frac{\alpha\theta t}{\alpha\theta - \theta + t}(d+l)) - f(a) \leq f(a+td) - f(a).$$

Hence, dividing by t and passing to the limit when $t \to 0$, one has

$$\frac{\alpha}{\alpha - 1} f'_+(a, d+l) = f'_+(a, \frac{\alpha}{\alpha-1}(d+l)) \leq f'_+(a,d).$$

Pass to the limit when $\alpha \to +\infty$.

e) Finally, assume that $d \in int(K(a))$ and $l \in [cl(\widetilde{K}(a)] \cap [-cl(\widetilde{K}(a))]$. Then, $d + l \in int(K(a))$ and $d = (d+l) - l$. Hence, in view of (2)

$$f'_+(a, d+l) \geq f'_+(a,d) \quad \text{and} \quad f'_-(a, d+l) \geq f'_-(a,d).$$

□

Remark 3.3 When $E = \mathbb{R}^n$, the theorem also holds under one of the following relaxed assumptions:

(3) $l \in rint(\widetilde{K}(a))$ and $d \in cl(\widetilde{K}(a))$;

(4) $d \in rint(\widetilde{K}(a))$ and $l \in cl(\widetilde{K}(a))$.

Indeed, in both cases, there exists $\bar{\alpha} \geq 1$ such that $d + \alpha l \in rint(\widetilde{K}(a))$ for all $\alpha > \bar{\alpha}$. The remaining of the proof is the same as in the theorem.

When the closure of $\widetilde{K}(a)$ is a half-space, the normal cone $\widetilde{N}(a)$ is reduced to one direction. The next theorem takes into account the implications of Theorem 3.10 in this very special case.

Theorem 3.11 *Assume that E is a Hilbert space, f is quasiconvex on E and $cl(\widetilde{K}(a))$ is a closed half-space. Take $x^\star \in \widetilde{N}(a)$ such that $\|x^\star\| = 1$. Then there exist $\alpha_-, \alpha_+, \beta_-$ and β_+ such that*

$$0 \leq \alpha_+ \leq \alpha_- \leq +\infty, \qquad 0 \leq \beta_- \leq \beta_+ \leq +\infty$$

and

i) if $\langle d, x^\star \rangle < 0$ then

$$f'_-(a,d) = \alpha_- \langle d, x^\star \rangle \text{ and } f'_+(a,d) = \alpha_+ \langle d, x^\star \rangle,$$

ii) if $\langle d, x^\star \rangle > 0$ then

$$f'_-(a,d) = \beta_- \langle d, x^\star \rangle \text{ and } f'_+(a,d) = \beta_+ \langle d, x^\star \rangle.$$

Moreover, if α_- and β_- are finite then

$$f'_-(a,d) = 0 \text{ for all } d \text{ such that } \langle d, x^\star \rangle = 0 \tag{3.9}$$

and if α_+ and β_+ are finite then

$$f'_+(a,d) = 0 \text{ for all } d \text{ such that } \langle d, x^\star \rangle = 0. \tag{3.10}$$

Proof. Notice that $int(\widetilde{K}(a)) = \{h : \langle h, x^\star \rangle < 0\}$. Set $y^\star = d - \langle d, x^\star \rangle x^\star$, then $\langle y^\star, x^\star \rangle = 0$. Furthermore, for all $\epsilon > 0$, $y^\star - \epsilon x^\star \in int(\widetilde{K}(a))$ and $y^\star + \epsilon x^\star \in -int(\widetilde{K}(a))$.

On the other hand

$$d = (\langle d, x^\star \rangle + \epsilon)x^\star - \epsilon x^\star + y^\star,$$

and

$$(\langle d, x^\star \rangle - \epsilon)x^\star = d - \epsilon x^\star - y^\star.$$

Hence, in view of Theorem 3.10,

$$f'_-(a, (\langle d, x^\star \rangle - \epsilon)x^\star) \leq f'_-(a,d) \leq f'_-(a, (\langle d, x^\star \rangle + \epsilon)x^\star).$$

Assume that $\langle d, x^\star \rangle > 0$. Pass to the limit when $\epsilon \to 0$, then

$$\langle d, x^\star \rangle f'_-(a, x^\star) \leq f'_-(a,d) \leq \langle d, x^\star \rangle f'_-(a, x^\star).$$

Similarly, assume that $\langle d, x^\star \rangle < 0$. Then,

$$-\langle d, x^\star \rangle f'_-(a, -x^\star) \leq f'_-(a,d) \leq -\langle d, x^\star \rangle f'_-(a, -x^\star).$$

Proceed in the same way for the upper Dini-derivative. Then, take $\alpha_- = -f'_-(a, -x^\star)$, $\alpha_+ = -f'_+(a, -x^\star)$, $\beta_- = f'_-(a, x^\star)$ and $\beta_+ = f'_+(a, x^\star)$.

Finally, assume that $\langle d, x^\star \rangle = 0$. Then for all $\epsilon > 0$

$$\epsilon f'_-(a, -x^\star) = f'_-(a, -\epsilon x^\star) \leq f'_-(a,d) \leq f'_-(a, \epsilon x^\star) = \epsilon f'_-(a, x^\star).$$

Hence $f'_-(a,d) = 0$ when both $f'_-(a, -x^\star)$ and $f'_-(a, x^\star)$ are finite. Proceed similarly for $f'_+(a,d)$. □

Remark 3.4 When (α_- and β_-) or (α_+ and β_+) are not finite the second part of the theorem does not hold as shown by the following counter-example: Take $E = \mathbb{R}^2$, $a = (0,0)$ and f defined by

$$f(x_1, x_2) = \begin{cases} -\pi/2 & \text{if } x_2 < 0, \\ \pi/2 & \text{if } x_2 > 0, \\ tan^{-1}(x_1) & \text{if } x_2 = 0. \end{cases}$$

Then,

$$\widetilde{K}(a) = \{(d_1, d_2) : d_2 < 0 \text{ or } (d_2 = 0 \text{ and } d_1 < 0)\}.$$

Here $cl(\widetilde{K}(a))$ is an halfspace and $x^\star = (0,1)$. Take $d = (1,0)$, then $\langle d, x^\star \rangle = 0$ but $f'_-(a,d) = f'_+(a,d) = 1$.

An important consequence of the last theorem is stated below.

Theorem 3.12 *Assume that f is quasiconvex on $\mathbb{R}^n$, $f(a)$ is finite and $d \in int(\widetilde{K}(a))$. Then f is differentiable at a if and only if $cl(\widetilde{K}(a))$ is an half-space and the function $\theta(t) = f(a+td)$ is differentiable at $t=0$.*

Proof. If f is differentiable at a, then $cl(\widetilde{K}(a))$ is an half-space. Conversely, the differentiability of θ at 0 is equivalent to the assertion that

$$-\infty < f'_-(a,d) = f'_+(a,d) = -f'_-(a,-d) = -f'_+(a,-d) < +\infty.$$

Apply Theorem 3.11, then $\alpha_- = \alpha_+ = \beta_- = \beta_+$ and then $f'(a,h) = k\langle x^\star, h\rangle$ for some $k \geq 0$. Thus f is Gâteaux-differentiable at a and therefore Fréchet-differentiable at this point since we are in $\mathbb{R}^n$. □

Generalized derivatives have been made popular with the Clarke approach. A good situation is when the function is locally Lipschitz. Unlike convex functions, quasiconvex functions are not locally Lipschitz on the interior of their domain: *Lipschitz properties do not belong to the essence of quasiconvexity.* Henceforth, our intimate opinion is that generalized derivatives for locally Lipschitz functions are quite inappropriate in quasiconvex analysis. However, because so many people have tried to adapt these generalized derivatives, we indicate that for quasiconvex functions, the Lipschitz condition holds if it holds in one direction. The result is as follows:

Theorem 3.13 *(Crouzeix [9]) Assume that f is quasiconvex on an open convex set C. Assume in addition that there exist an open convex cone K, $h \in K$ and a constant L such that*

- *$K \subseteq \widetilde{K}(x)$ for all $x \in C$,*
- *$|f(x+th) - f(x)| \leq L|t|$ for all t and x such that both x and $x+th$ lie in C.*

Then there exists a constant $\hat{L}$ such that

$$|f(x) - f(y)| \leq \hat{L}\|x - y\| \text{ for all } x, y \in C.$$

Proof. Without loss of generality, we assume that $f(x) > f(y)$. For all $t \in (0,1)$ define

$$\theta(t) = f(ty + (1-t)x) - tf(y) - (1-t)f(x).$$

The assumptions on K and $\widetilde{K}(x)$ imply that f is nondecreasing with respect to K, then Proposition 3.7 implies that f is continuous at any $z \in C$ and therefore θ is continuous on $[0,1]$. Choose $\bar{t} \in [0,1]$ so that

$$\theta(\bar{t}) = \max[\,\theta(t) : t \in [0,1]\,],$$

and take $a = \bar{t}y + (1-\bar{t})x$. Then

$$f'_-(a, y-x) + f(x) - f(y) = \theta'_-(\bar{t}, 1) \leq 0.$$

Next, define $\rho = \max[r : B(h,r) \subset K]$. By Theorem 3.8 and since $\rho \leq \rho(h)$

$$f'_-(a, y-x) \geq \frac{\|y-x\|}{\rho(h)} f'_-(a,h) \geq -\frac{L\|y-x\|}{\rho}.$$

Since

$$0 \geq f(y) - f(x) \geq f'_-(a, y-x),$$

the result follows with $\hat{L} = \frac{L}{\rho}$. □

Theorem 3.1 gives conditions for the existence of such a cone K.

8. Continuity of the normal cone

Recall that a point-to-set map $M : E \to F$ is said to be *closed at a point* a if for any sequence $\{(a_n, b_n)\}_n$ converging to (a,b) with $b_n \in M(a_n)$ we have $b \in M(a)$.

The map is said to be *closed on* E if closed at any point $a \in E$. It is known that a map is closed on E if and only if its graph $G(M) = \{(x,y) : y \in M(x)\}$ is a closed subset of $E \times F$.

The map M is said to be *USC at a point* a if for each open set $\Omega \supseteq M(a)$ there is a neighbourhood V of a such that $\Omega \supseteq M(x)$ for all $x \in V$. We shall use the following well known characterization.

Proposition 3.11 *Let $M : E \to K$ be a point-to-set map where E is a metric space and K is a compact set. Assume that, for all x in a neighbourhood of a, the set $M(x)$ is compact and nonempty and the map M is closed at x. Then M is USC at a.*

Proposition 3.12 *Let $f : E \to [-\infty, +\infty]$ be quasiconvex. Assume that f is finite and lsc at a. Then $\widetilde{N}$ is closed at a.*

Proof. Let $\{(a_n, b_n)\}_n$ be a sequence converging to (a,b) with $b_n \in \widetilde{N}(a_n)$. Let x be such that $f(x) < f(a)$. Then there is a neighbourhood V of a such that $f(x) < f(y)$ for all $y \in V$. For n large enough, $a_n \in V$ and therefore, because $b_n \in \widetilde{N}(a_n)$,

$$\langle b_n, x - a_n \rangle \leq 0.$$

Hence, passing to the limit

$$\langle b, x - a \rangle \leq 0.$$

Because the inequality holds for all x with $f(x) < f(a)$, it results that $b \in \widetilde{N}(a)$. □

The normal cone $\widetilde{N}$ is, of course, unbounded. Continuity with unbounded maps is sometimes difficult to handle. It is one of the reasons why the subdifferential is used in sensitivity in convex programming instead of the normal cone T at the epigraph.

We shall say that a point-to-set map $M : E \to F$ is *C-USC at a* (Borde-Crouzeix [1]) if there is a compact-valued map C such that C is USC at a and for all x in a neighbourhood of a

$$0 \notin C(x) \quad \text{and} \quad M(x) = \{\, y = \lambda d \,:\, \lambda \geq 0 \text{ and } d \in C(x)\,\}.$$

Theorem 3.14 *Let $f : \mathbb{R}^n \to (-\infty, +\infty)$ quasiconvex, $\lambda \in \mathbb{R}$ and $a \in \mathbb{R}^n$ such that $int(S_\lambda(f)) \neq \emptyset$ and $a \notin cl(S_\lambda(f))$. Then $\widetilde{N}$ is C-USC at a.*

Proof. By Theorem 3.1, there is K an open nonempty convex cone such that $K \subseteq \widetilde{K}(x)$ for x close to a. Let $\bar{d} \in K$ be fixed. We define a map C by

$$C(x) = \{\, x^\star \in \widetilde{N}(x) \,:\, \langle x^\star, \bar{d} \rangle = 1\}.$$

Then, for x close to a, $M(x)$ is generated by $C(x)$. Hence C is closed at x. Furthermore

$$C(x) \subseteq \bar{C} = \{\, x^\star \in K^o : \langle x^\star, \bar{d} \rangle = 1\}$$

where K^o is the polar cone of K. The set $\bar{C}$ is compact and the result follows. □

If a convex function is differentiable, then its gradient is continuous. Now, assume that we are faced with a function f which is differentiable and quasiconvex. Let a be such that $\nabla f(a) \neq 0$. Then

$$\widetilde{N}(a) = \{\lambda \nabla f(a) \,:\, \lambda \geq 0\}.$$

It results from Theorem 3.14 that the map

$$x \to \frac{1}{\|\nabla f(x)\|} \nabla f(x)$$

is continuous at any x where the gradient does not vanish. Recall that, for a convex function, the gradient itself is continuous, for a quasi convex function it is only the directiongiven by the gradient which is continuous.

Some people prefer to consider the normal cone

$$N(a) = \{ a^\star : \langle a^\star, x - a \rangle \leq 0 \text{ for all } x \text{ so that } f(x) \leq f(a) \}$$

instead of $\widetilde{N}(a)$. When f is pseudoconvex in one of the different senses we have given, the two cones coincide.

9. Are the generalized derivatives useful in quasiconvex analysis?

Assume that f is finite in a. A general formulation of a generalized derivative of f at a with respect to a direction d is as follows

$$\begin{array}{rcl} f^!(a,d) & = & \text{“some kind of limit”} \ \frac{1}{t} \left[f(x+th) - f(y) \right] \\ & & \text{when } t \to 0_+,\ x, y \to a \text{ and } h \to d. \end{array} \tag{3.11}$$

The different ways by which the arguments x, y converge to a, h converges to d, the different orders to take the limits with respect to the arguments, the different types of limit (sup, inf, ...) give birth to many combinations quite appropriate to exercice the skills of a mathematician. For instance, the upper (lower) Dini-derivative corresponds to the case where $x = y = a$, $h = d$ and the limit is the limit sup (limit inf).

Once a generalized derivative is defined, a generalized subdifferential is built as the closed convex set $\partial^! f(a)$ such that

$$\partial^! f(a) = \{ x^\star : \langle d, x^\star \rangle \leq f^!(a,d) \text{ for all } d \}.$$

Now, it is time to look at the uses of such a generalized derivative and such a subgradient. Because we deal with optimization problems, the first use concerns optimality conditions. We have seen that, in the quasiconvex setting, the normal cone $\widetilde{N}(a)$ plays a fundamental role. Hence, the first property to be asked to generalized derivatives and/or generalized subgradients is that they allow to recover the cone $\widetilde{N}(a)$. The second important use concerns sensitivity. For that, a standard approach consists to consider some continuity on the subdifferential, this continuity is provided in general by considering the convergences of the arguments x, y and h to a, a and d respectively in definition (3.11). *Is it really necessary to consider generalized derivatives and the associated subdifferential and not to consider directly the normal cone* $\widetilde{N}(a)$? We have seen, in the last section, that the normal cone has the wished continuity property.

Anyway, if we want to consider some generalized derivatives and their associated generalized subgradients, the simplest ones seem the best. Since, unlike in the convex case, directional derivatives cannot be considered, the best candidates are the Dini-derivatives. If f is pseudoconvex then Proposition 3.10 shows that $\widetilde{N}(a)$ can be recovered from the Dini-derivatives. For simplicity we assume that these Dini-derivatives are finite. Let us define

$$\partial^u f(a) = \{x^\star : f'_+(a,d) \geq \langle x^\star, d\rangle \text{ for all } d \in \widetilde{K}(a)\}.$$

and

$$\partial^l f(a) = \{x^\star : f'_-(a,d) \geq \langle x^\star, d\rangle \text{ for all } d \in \widetilde{K}(a)\}.$$

Then, $\widetilde{N}(a)$ is the cone generated by $\partial^u f(a)$ and/or $\partial^l f(a)$, hence it is thoroughly defined from the knowledge of these sets. Furthermore, because $f'_+(a,.)$ is quasiconvex, for all $h \in int(\widetilde{K}(a))$

$$f'_+(a,h) = \sup[\langle x^\star, h\rangle \ : \ x^\star \in \partial^u f(a)].$$

An exhaustive bibliography on generalized derivatives and generalized subgradients in quasiconvex programming is reference [16].

References

[1] Borde J. and Crouzeix J.-P., *Continuity of the normal cones to the level sets of quasi convex functions*, Journal of Optimization Theory and Applications, Vol. 66, No 3, 1990, 415–429.

[2] Chabrillac Y. and Crouzeix J.-P., *Continuity and differentiability properties of monotone real functions of several variables*, Mathematical Programming Study 30, 1987, 1–16.

[3] Crouzeix J.-P., *Contribution à l'étude des fonctions quasiconvexes*, Thèse de Docteur es Sciences, Université de Clermont-Ferrand 2, 1977.

[4] Crouzeix J.-P., *Conditions for convexity of quasiconvex functions*, Mathematics of Operations Research, Vol. 5, no 1, 1980, 120–125.

[5] Crouzeix J.-P., *Some differentiability properties of quasiconvex functions on* $\Re^n$, Optimization and Optimal Control, A. Auslender, W. Oettli and J. Stoer, editors, Springer-Verlag, Lecture Notes in Control and Information Sciences, Vol. 30, 1981, 9–20.

[6] Crouzeix J.-P., *Continuity and differentiability properties of quasiconvex functions*, Generalized Concavity in Optimization and Economics, S. Schaible and W.T. Ziemba editors, Academic Press, New-York, 1981, 109–130.

[7] Crouzeix J.-P., *A review of continuity and differentiability properties of quasiconvex functions on* $\Re^n$, Convex Analysis and Optimization, J.-P. Aubin and R. Vinter editors, Research Notes in Mathematics, Vol. 57, Pitman Advanced Publishing Programs, 1982, 18–34.

[8] Crouzeix J.-P., *About differentiability properties of quasiconvex functions*, Journal of Optimization Theory and Applications 36, 1982, 367–385.

[9] Crouzeix J.-P., *Some properties of Dini-derivatives of quasiconvex and pseudoconvex functions*, New Trends in Mathematical Programming, F. Giannessi, T. Rapcsák and S. Komlósi editors, Kluwer Academic Publishers, 1998, 41–57.

[10] Fenchel W., *Convex cones sets and functions*, Mimeographed lecture notes, Princeton University, 1951.

[11] Hadjisavvas N., *Continuity properties of quasiconvex functions in infinite-dimensional spaces*, Unpublished manuscript, University of the Aegean, Karlovassi, Greece, 1974.

[12] Laurent Jacques, *Les sous ensembles flous*, Grasset, Paris, 1981.

[13] Martinez-Legaz J.-E., *Un concepto generalizado de conjugacion, aplicacion a las funciones quasiconvexas*, Doctoral thesis, Universidad de Barcelona, 1981.

[14] Newman P., *Some properties of concave functions*, Journal of Economic Theory 1, 1969, 291–314.

[15] Passy U. and Prisman E. Z., *Conjugacy in quasiconvex Programming*, Mathematical Programming 30, 1984, 121–146.

[16] Penot J.-P., *Are generalized derivatives useful for generalized convex functions?*, Generalized Convexity, Generalized Monotonicity, Crouzeix J.-P., J.-E. Martinez-Legaz and M. Volle editors, Kluwer Academic Publishers, 1998, 3–59.

Chapter 4

GENERALIZED CONVEXITY AND OPTIMALITY CONDITIONS IN SCALAR AND VECTOR OPTIMIZATION

Alberto Cambini
Department of Statistics and Applied Mathematics
University of Pisa, Via Cosmo Ridolfi 10
56124 Pisa, ITALY
acambini@ec.unipi.it

Laura Martein
Department of Statistics and Applied Mathematics
University of Pisa, Via Cosmo Ridolfi 10
56124 Pisa, ITALY
lmartein@ec.unipi.it

Abstract In this chapter, the role of generalized convex functions in optimization is stressed. A particular attention is devoted to local-global properties, to optimality of stationary points and to sufficiency of first order necessary optimality conditions for scalar and vector problems. Despite of the numerous classes of generalized convex functions suggested in these last fifty years, we have limited ourselves to introduce and study those classes of scalar and vector functions which are more used in the literature.

Keywords: generalized convexity, vector optimization, optimality conditions.

1. Introduction

In classical scalar optimization theory, convexity plays a fundamental role since it guarantees the validity of important properties like as: a local minimizer is also a global minimizer, a stationary point is a global minimizer and the usual first order necessary optimality conditions are also sufficient for a point to be a global minimizer.

For many mathematical models used in decision sciences, economics, management sciences, stochastics, applied mathematics and engineering, the notion of convexity does not longer suffice. Various generalizations of convex functions have been introduced in literature. Many of such functions preserve one or more properties of convex functions and give rise to models which are more adaptable to real-world situations then convex models.

Starting from the pioneer work of Arrow-Enthoven [1], attempts have been made to weaken the convexity assumption and thus to explore the extent of optimality conditions' applicability. The results obtained in the scalar case have had a great influence on the area of vector optimization which has been widely developed in recent years with the aim to extend and generalize, in this field, such results (see for instance [15]), [42], [62].

In the scalar case various generalization of convexity have been suggested and their properties studied (see for instance [3], [86]). In this chapter we limit ourselves to consider only the classes which preserve the properties of convex functions related to optimality like as quasiconvex, semistrictly quasiconvex and pseudoconvex functions.

A particular attention is devoted to pseudolinear functions, that is functions which are both pseudoconvex and pseudoconcave, since they have also the nice property, which is very important from a computational point of view, that the set of all minimizers and the set of all maximizers are contained in the boundary of the feasible region; in particular, if such a region is a polyhedral set and if the minimum (maximum) value exists, it is attained at least at a vertex of the feasible set.

In Section 3 we have considered the class of the so-called invex functions since it is the wider class for which the Kuhn-Tucker conditions become sufficient. These functions can be characterized as the ones for which a stationary point is a minimum point and, like as the considered classes of generalized convex functions, they play an important role also in establishing constraint qualifications.

In vector optimization, the concept of minimum is usually translated by means of an ordering cone in the space of the objectives. For sake of simplicity in this chapter we refer to the Pareto cone, that is the non-negative orthant of the space of the objectives. This cone induces only a partial order and this is the main reason for which there are several ways to extend the notion of generalized convexity in vector optimization.

As in the scalar case, we have chosen to present some classes of vector generalized convex functions which preserve local-global properties and the sufficiency of the most important first order necessary vector optimality conditions. Furthermore, in Section 10, we have suggested a possible way to extend the concept of pseudolinearity for a vector function, while in Section 11 the notion of vector invexity is developed.

In this chapter, we have given the fundamental notions, ideas and properties of generalized convex scalar and vector functions. For sake of simplicity, we have considered differentiable functions even if recent papers are devoted to the nonsmooth case.

2. Generalized convex scalar functions and optimality conditions.

In this section, we will establish local-global properties and the sufficiency of the most important first order necessary optimality conditions. With this aim, we introduce some classes of real-valued generalized convex functions which contain properly the convex one.

Let $f : X \to \Re$ be a function defined on an open set X of $\Re^n$ and let S be a convex subset of X.

Definition 4.1 *The function f is quasiconvex on S if its lower-level sets $L(\alpha) = \{x : f(x) \leq \alpha, x \in S\}$ are convex sets for all real numbers α.*

As is known, an useful characterization of a quasiconvex function is the following:

$$f(x_2) \leq f(x_1) \Rightarrow f((1-\lambda)x_1 + \lambda x_2) \leq f(x_1) \tag{4.1}$$

for every $x_1, x_2 \in S$ and $0 \leq \lambda \leq 1$.

When f is a differentiable function, f is quasiconvex if and only if

$$x_1, x_2 \in S, f(x_2) \leq f(x_1) \Rightarrow (x_2 - x_1)^T \nabla f(x_1) \leq 0. \tag{4.2}$$

Definition 4.2 *The function f is semistrictly quasiconvex on S if*

$$f(x_2) < f(x_1) \Rightarrow f((1-\lambda)x_1 + \lambda x_2) < f(x_1) \tag{4.3}$$

for every $x_1, x_2 \in S$ and $0 < \lambda < 1$.

Definition 4.3 *The differentiable function f is pseudoconvex on S if*

$$x_1, x_2 \in S, f(x_2) < f(x_1) \Rightarrow (x_2 - x_1)^T \nabla f(x_1) < 0. \tag{4.4}$$

We recall that, in the differentiable case, a pseudoconvex function is a semistrictly quasiconvex function, that, in turn, is quasiconvex.

The essence of the difference between quasiconvexity and pseudoconvexity is stated in the following known theorem for which we will give a very simple proof.

Theorem 4.1 *Let f be a differentiable function on an open convex set $S \subset \Re^n$.*

If $\nabla f(x) \neq 0$ $\forall x \in S$ then f is pseudoconvex if and only if it is quasiconvex.

Proof. Taking into account that a pseudoconvex function is quasiconvex too, we must prove that a quasiconvex function is pseudoconvex.

Assume, to get a contradiction, that there exist $x_1, x_2 \in S$, with $f(x_2) < f(x_1)$ and $(x_2 - x_1)^T \nabla f(x_1) \geq 0$. Since f is quasiconvex, necessarily we have $(x_2 - x_1)^T \nabla f(x_1) = 0$. For the continuity of f, there exists $\epsilon > 0$ such that $y = x_2 + \epsilon \nabla f(x_1) \in S$ with $f(y) < f(x_1)$ (observe that $y \neq x_2$ since $\nabla f(x_1) \neq 0$). Consequently, for the quasiconvexity of f, we have $(y - x_1)^T \nabla f(x_1) \leq 0$. On the other hand $y - x_1 = (y - x_2) + (x_2 - x_1)$, so that

$$\begin{aligned}(y - x_1)^T \nabla f(x_1) &= (y - x_2)^T \nabla f(x_1) + (x_2 - x_1)^T \nabla f(x_1) \\ &= \epsilon \parallel \nabla f(x_1) \parallel^2 > 0\end{aligned}$$

and this is absurd. □

Let us note that a function is quasiconcave, semistrictly quasiconcave or pseudoconcave if and only if the function $-f$ is quasiconvex, semistrictly quasiconvex or pseudoconvex, so that all results that we are going to describe for generalized convex functions hold with obvious changes for the corresponding class of generalized concave functions. Like as the convex case, a semistrictly quasiconvex function f (in particular pseudoconvex) has the nice properties that the set of points at which f attains its global minimum over S is a convex set and that a local minimum is also global. This last property is lost for a quasiconvex function; with this regard it is sufficient to consider the quasiconvex function

$$f(x) = \begin{cases} -x^2 & x \in [-1, 0] \\ 0 & x \in]0, 2[\\ (x-2)^2 & x \in [2, 3] \end{cases} \tag{4.5}$$

and the point $x_0 = 1$, which is a local but not global minimum point.

Nevertheless, if x_0 is a strict local minimum point for a quasiconvex function, then it is also a strict global minimum point. The class of semistrictly quasiconvex functions is the wider class for which a local minimum is also global in the following sense:

if f is a continuous quasiconvex function, then f is semistrictly quasiconvex if and only if every local minimum is also global.

In order to analyze local and global optimality at a point x_0, we must compare $f(x_0)$ with $f(x)$, $x \in S$ and this suggests to consider generalized convexity at the point x_0. Following Mangasarian [65], we can also weaken the assumption of convexity of S, requiring that S is star-shaped at $x_0 \in S$, that is $x \in S$ implies the line-segment $[x_0, x]$ is contained in S.

The following definitions hold:

Definition 4.4 *Let f be a function defined on the star-shaped set S at x_0.*

i) f is quasiconvex at x_0 if

$$f(x) \leq f(x_0) \Rightarrow f((1-\lambda)x_0 + \lambda x) \leq f(x_0), \ \forall\, x \in S, \ \forall \lambda \in [0,1]. \quad (4.6)$$

ii) f is semistrictly quasiconvex at x_0 if

$$f(x) < f(x_0) \Rightarrow f((1-\lambda)x_0 + \lambda x) < f(x_0), \ \forall\, x \in S, \ \forall \lambda \in (0,1). \quad (4.7)$$

iii) f is pseudoconvex at x_0 if f is differentiable at x_0 and

$$f(x) < f(x_0) \Rightarrow (x - x_0)^T \nabla f(x_0) < 0 \ \forall\, x \in S. \quad (4.8)$$

It is well known that a point, which is minimum with respect to every feasible direction starting from it, is not necessarily a local minimum point. Consider for instance the function $f(x,y) = (y - x^4)(y - x^2)$ on the set $S = \Re^2_+$. The point $x_0 = (0,0)$ is minimum with respect to every feasible direction $d \in \Re^2_+$, $d \neq 0$, but it is not a local minimum, how it can be verified by considering the restriction of $f(x)$ to the curve $y = x^3, x \geq 0$. The following theorem points out that, requiring suitable generalized convexity assumptions, optimality along feasible directions at a point x_0 implies the optimality of x_0.

Theorem 4.2 *Let f be a function defined on the star-shaped set S at x_0.*

i) If f is quasiconvex at $x_0 \in S$ and x_0 is a strict local minimum point for every direction $d = x - x_0$, $x \in S, x \neq x_0$, then x_0 is a strict global minimum point of f on S.

ii) If f is semistrictly quasiconvex at $x_0 \in S$ and x_0 is a local minimum point for every direction $d = x - x_0$, $x \in S, x \neq x_0$, then x_0 is a global minimum point of f on S.

iii) If f is pseudoconvex at $x_0 \in S$ and x_0 is a local minimum point for every direction $d = x - x_0$, $x \in S, x \neq x_0$, then x_0 is a global minimum point of f on S.

Now we will stress the role of generalized convexity in establishing sufficient first order optimality conditions.

With this aim, in the following we will consider a real-valued differentiable function f defined on an open subset X of $\Re^n$. It is well known that a stationary point for f is not necessarily a minimum point for f; such a property holds requiring a suitable assumption of generalized convexity.

Theorem 4.3 *Consider a differentiable function f on a convex set $S \subset X$. If f is pseudoconvex at the stationary point $x_0 \in S$, then x_0 is a global minimum point for the function f on S.*

Theorem 4.3 does not hold if pseudoconvexity assumption is substituted with quasiconvexity or semistrictly quasiconvexity. In fact $f(x) = x^3$ is a strictly increasing function so that it is both semistrictly quasiconvex and quasiconvex, but the stationary point $x_0 = 0$ is not a minimum point for f.

Now we will see how generalized convexity assumptions are very important to guarantee the sufficiency of first order optimality conditions which are in general only necessary. When S is star-shaped at x_0, one of the most known conditions for x_0 to be a local minimum point is $d^T\nabla f(x_0) \geq 0$, $\forall d \in F(S, x_0)$ where $F(S, x_0)$ is the cone of feasible directions of S at x_0 given by $F(S, x_0) = \{d \in \Re^n : d = x - x_0, x \in S, x \neq x_0\}$. Such a condition is not sufficient even if strict inequality holds $\forall d \in F(S, x_0)$ as is shown in the following example.

Example 4.1 Consider the function $f(x, y) = y - x^2$ defined on the convex set $S = \{(x, y) : y \geq x^4\}$ and the point $x_0 = (0, 0) \in S$. We have $F(S, x_0) = \{(d_1, d_2) : d_2 > 0\}$ and $d^T\nabla f(x_0) = d_2 > 0$, $\forall d \in F(S, x_0)$, but x_0 is not a minimum point as it can be verified performing a restriction of the function on the curve $y = x^4$.

In Example 4.1 the closure $clF(S, x_0)$ of the cone $F(S, x_0)$ contains properly $F(S, x_0)$ and directions d belonging to $clF(S, x_0) \setminus F(S, x_0)$ are critical directions, that is $d^T\nabla f(x_0) = 0$. In order to be sure that x_0 is a minimum point, we must require the validity of the sufficient condition $d^T\nabla f(x_0) > 0$, $\forall d \in clF(S, x_0)$.

When the function is pseudoconvex the critical directions do not play any role as it is stated in the following theorem whose proof follows directly by the definition of pseudoconvexity.

Theorem 4.4 *Let $S \subset X$ be a star-shaped set at $x_0 \in S$ and let f be pseudoconvex at x_0. Then x_0 is a minimum point for f on S if and only if $(x - x_0)^T\nabla f(x_0) \geq 0$ $\forall x \in S$.*

Remark 4.1 Let us note that if x_0 is a stationary point, the condition $(x - x_0)^T \nabla f(x_0) \geq 0 \ \forall x \in S$ is always verified. Since quasiconvexity assumption in a stationary point does not guarantee the optimality of x_0, Theorem 4.4 does not hold for this class of functions.

On the other hand, if f is quasiconvex at x_0 and $\nabla f(x_0) \neq 0$, then f is pseudoconvex at x_0, so that we have the following corollary.

Corollary 4.1 *Let $S \subset X$ be a star-shaped set at $x_0 \in S$. If f is quasiconvex at x_0, $\nabla f(x_0) \neq 0$ and $(x - x_0)^T \nabla f(x_0) \geq 0 \ \forall x \in S$, then x_0 is a minimum point for f on S.*

Consider now the case where the feasible set S is described by means of constraint functions. More exactly consider the problem:

$$P: \quad min f(x), \ x \in S = \{x \in X \ : \ g_i(x) \leq 0, \ i = 1, ..., m\}$$

where $f, g_i : X \to \Re$ are differentiable functions defined on an open set $X \subset \Re^n$.

The most known first order necessary optimality conditions for a constrained problem are the following Kuhn-Tucker conditions:
let x_0 be a feasible point and set $I(x_0) = \{i \in \{1, ..., m\} : g_i(x_0) = 0\}$; if x_0 is a local minimum point for P and a constraint qualification holds, then there exist $\lambda_i \in \Re, i \in I(x_0)$, such that:

$$\nabla f(x_0) + \sum_{i \in I(x_0)} \lambda_i \nabla g_i(x_0) = 0 \tag{4.9}$$

$$\lambda_i \geq 0, \ i \in I(x_0) \tag{4.10}$$

The following example points out that (4.9) and (4.10) are not sufficient optimality conditions.

Example 4.2 Consider the problem

$$min \ (x + y)^3, \ (x, y) \in S = \{(x, y) \in \Re^2 : -y \leq 0\}.$$

It is easy to verify that for the point $(0, 0)$, conditions (4.9) and (4.10) hold with $\lambda = 0$, but $(0, 0)$ is not a local minimum point for the problem.

Let us now show that (4.9) and (4.10) are also sufficient when the objective and the constraint functions are certain generalized convex functions.

Theorem 4.5 *Let x_0 be a feasible point for problem P and assume that (x_0, λ) verifies the Kuhn-Tucker conditions (4.9) and (4.10). If f is* *pseudoconvex at x_0 and $g_i, i \in I(x_0)$, are quasiconvex at x_0, then x_0 is a global minimum point for problem P.*

Proof. Assume there exists a feasible point $\bar{x}$ such that $f(\bar{x}) < f(x_0)$. From the pseudoconvexity of f we have $(\bar{x} - x_0)^T \nabla f(x_0) < 0$ and from the quasiconvexity of $g_i, i \in I(x_0)$, we have $(\bar{x} - x_0)^T \nabla g_i(x_0) \leq 0$. Taking into account that $\lambda_i \geq 0$, it results

$$(\bar{x} - x_0)^T \nabla f(x_0) + \sum_{i \in I(x_0)} \lambda_i (\bar{x} - x_0)^T \nabla g_i(x_0) < 0$$

and this contradicts (4.9). □

Remark 4.2 In Example 4.2, the objective and the constraint functions are quasiconvex and also semistrictly quasiconvex on $\Re^2$ and this points out that Theorem 4.5 does not hold if f is quasiconvex or semistrictly quasiconvex.

Taking into account Remark 4.1, Theorem 4.5 holds if f is quasiconvex at x_0 and $\nabla f(x_0) \neq 0$. Such a condition is verified, for instance, in Consumer Theory, where it is assumed that the partial derivatives of the utility function U are positive and -U is a quasiconvex function.

3. Invex scalar functions

In [38] Hanson has introduced a new class of generalized convex functions (invex functions) with the aim to extend the validity of the sufficiency of the Kuhn-Tucker conditions. The term invex is due to Craven [27] and it steams for invariant convex.

Since the papers of Hanson and Craven, during the last twenty years, a great deal of contributions related to invex functions, especially with regard to optimization problems, have been made (see for instance [8, 28, 29, 40, 80, 93]).

Definition 4.5 *The differentiable real-valued function f defined on the set $X \subset \Re^n$ is invex if there exists a vector function $\eta(x, y)$ defined on $X \times X$ such that*

$$f(x) - f(y) \geq \eta^T(x, y) \nabla f(y) \ \ \forall x, y \in X \tag{4.11}$$

Obviously a differentiable convex function (on an open convex set X) is also invex (it is sufficient to choose $\eta(x, y) = (x - y)$).

A meaningful property characterizing invex functions is stated in the following theorem [8].

Theorem 4.6 *A differentiable function is invex (with respect to some η) if and only if every stationary point is a global minimum point.*

Proof. Let f be invex with respect to some $\eta(x,y)$. If x_0 is a stationary point of f, from (4.11) we have $f(x) - f(x_0) \geq \eta^T(x,x_0)\nabla f(x_0) = 0$ $\forall x \in X$, so that x_0 is a global minimum point. Now we will prove that (4.11) holds for the function $\eta(x,y)$ defined as

$$\eta(x,y) = \begin{cases} 0 & if \ \nabla f(y) = 0 \\ \frac{(f(x)-f(y))\nabla f(y)}{\|\nabla f(y)\|^2} & if \ \nabla f(y) \neq 0 \end{cases}$$

If y is a stationary point and also a global minimum for f, we have $f(x) - f(y) \geq 0 = \eta^T(x,y)\nabla f(y)$; otherwise $\eta^T(x,y)\nabla f(y) = f(x) - f(y)$, so that (4.11) holds. □

It follows immediately from Theorem 4.6 that every function without stationary points is invex.

The class of pseudoconvex functions is contained in the class of invex functions (it is sufficient to note that for a pseudoconvex function a stationary point is also a global minimum point), while there is not inclusion relationships between the class of quasiconvex functions and the class of invex functions. Indeed, the function $f(x) = x^3$ is quasiconvex but not invex, since $x = 0$ is a stationary point but it is not a minimum point; furthermore the following example shows that there exist invex functions which are not quasiconvex.

Example 4.3 Consider the function $f(x,y) = y(x^2-1)^2$ on the open convex set $X = \{(x,y) \in \Re^2 : y > 0\}$.

It is easy to verify that the stationary points $(1,y), (-1,y)$, $y > 0$ of f are global minimum points, so that f is invex. On the other hand, setting $A = (-1,1)$, $B = (\frac{1}{2},1)$, we have $f(A) = 0 < f(B) = \frac{9}{16}$ and $(A-B)^T\nabla f(B) = (-\frac{3}{2},0)(-\frac{3}{2},\frac{9}{16})^T = \frac{9}{4} > 0$, so that f is not quasiconvex.

Since a function which is not quasiconvex is also not pseudoconvex, the previous example shows also that the classes of pseudoconvex and convex functions are strictly contained in the one of invex functions.

Some nice properties of convex functions are lost in the invex case. In fact, unlike the convex or pseudoconvex case, the restriction of an invex function on a not open set, does not maintain the local-global property. With this regard, consider again the function $f(x,y) = y(x^2-1)^2$ on the closed set $S = \{(x,y) \in \Re^2 : x \geq -\frac{1}{2},\ y \geq 1\}$. The point $(-\frac{1}{2},1)$ is a local minimum for f on S but it is not global since $f(-\frac{1}{2},1) = \frac{9}{16} > f(1,1) = 0$.

For the same function defined on $X = \{(x,y) \in \Re^2 : y > 0\}$, the set of all minimum points is given by $\{(1,y) : y > 0\} \cup \{(-1,y) : y > 0\}$ which is a non convex set; as a consequence, for an invex function the set of all minimum points is not necessarily a convex set.

Following Hanson [38], we will prove the sufficiency of the Kuhn-Tucker conditions under suitable invex assumptions.

Theorem 4.7 *Let x_0 be a feasible point for problem P and assume that (x_0, λ) verifies the Kuhn-Tucker conditions (4.9) and (4.10). If $f, g_i, i \in I(x_0)$, are invex functions with respect to the same $\eta(x,y)$, then x_0 is a global minimum point for problem P.*

Proof. For any $x \in S$, we have

$$\begin{aligned} f(x) - f(x_0) &\geq \eta^T(x,x_0)\nabla f(x_0) = - \sum_{i \in I(x_0)} \lambda_i \eta^T(x,x_0)\nabla g_i(x_0) \\ &\geq - \sum_{i \in I(x_0)} \lambda_i (g_i(x) - g_i(x_0)) \geq 0. \end{aligned}$$

Consequently x_0 is a global minimum point. □

The proof of Theorem 4.7 points out that it is sufficient to require the invexity of the functions $f, g_i, i \in I(x_0)$, at $x_0 \in S$, that is (4.11) holds with $y = x_0$, $\forall x \in S$.

Invexity allowed also the weakening of convexity requirements in duality theory, since duality results involving Wolfe dual or alternative duals can be established [29, 37, 40, 67]. Furthermore, invex functions, as well as generalized convex functions, play some role in establishing constraint qualifications as it will be seen in the next section.

Since invexity requires differentiability assumption, in [8, 40] the following new class of functions, not necessarily differentiable, have been introduced.

Let f be a real valued function defined on a subset of $\Re^n$ and $\eta : \Re^n \times \Re^n \to \Re^n$. We say that a subset X of $\Re^n$ is η-invex if for every $x, y \in X$ the segment $[y, y + \eta(x,y)]$ is contained in X.

Definition 4.6 *Let f be a real valued function defined on a η-invex set X; f is pre-invex with respect to η if the following inequality holds:*

$$f(y + \lambda\eta(x,y)) \leq \lambda f(x) + (1-\lambda) f(y), \ \forall x, y \in X, \forall \lambda \in [0,1]. \quad (4.12)$$

A differentiable function satisfying (4.12) is also invex and this is the reason why functions verifying (4.12) are called pre-invex [93].

Like as convex functions, for a pre-invex function every local minimum is also global and nonnegative linear combinations of pre-invex functions with respect to the same η are pre-invex. Furthermore it is possible to establish saddle point and duality theorems following the classical approach used in the convex case [93].

4. Generalized convexity and constraint qualifications

As we have pointed out in Section 2, the Kuhn-Tucker conditions (4.9) and (4.10) are necessary optimality conditions for problem P when certain regularity conditions on the constraints are satisfied. The aim of this section is to stress the role of generalized convexity and invexity in establishing constraint qualifications.

For our purposes, it will be useful to define the following sets:

$$G^0 = \{d \in \Re^n : d^T \nabla g_i(x_0) < 0, i \in I(x_0)\}$$

$$G = \{d \in \Re^n : d^T \nabla g_i(x_0) \leq 0, i \in I(x_0)\}$$

$$G^p = \{d \in \Re^n : d^T \nabla g_i(x_0) \leq 0, i \in J, d^T \nabla g_i(x_0) < 0, i \in I(x_0) \backslash J\}$$

where $J = \{i \in I(x_0) : g_i(x) \textit{ is pseudoconcave at } x_0\}$.

Denoting with $\overline{co}T(S, x_0)$ the closure of the convex hull of the Bouligand tangent cone $T(S, x_0)$ to the feasible region S at $x_0 \in S$, the following inclusion relationships hold:

$$clG^0 \subseteq clG^p \subseteq \overline{co}T(S, x_0) \subseteq G \tag{4.13}$$

In order to guarantee the validity of the Kuhn-Tucker conditions, it must result $\overline{co}T(S, x_0) = G$ (Guignard constraint qualification), so that any condition which implies $\overline{co}T(S, x_0) = G$ becomes a constraint qualification.

In particular, condition $G^0 \neq \emptyset$ ($G^p \neq \emptyset$) is a constraint qualification; indeed it implies $clG^0 = G$ ($clG^p = G$), so that from (4.13) we have $\overline{co}T(S, x_0) = G$. We will prove that, under suitable assumption of generalized convexity on the constraint functions, it results $G^0 \neq \emptyset$ or $G^p \neq \emptyset$. The following theorem holds.

Theorem 4.8 *If one of the following conditions holds then Guignard constraint qualification is verified.*

i) The functions $g_i(x), i \in I(x_0)$, are pseudoconvex at x_0 and there exists $x^* \in S$ such that $g_i(x^*) < 0, i \in I(x_0)$ [65].

ii) The functions $g_i(x), i \in I(x_0)$, are quasiconvex at $x_0, \nabla g_i(x_0) \neq 0$ and there exists $x^* \in S$ such that $g_i(x^*) < 0, i \in I(x_0)$ [1].

iii) The functions $g_i(x), i \in I(x_0)$, *are pseudoconvex at* x_0 *and there is some vector* $x^* \in S$ *such that* $g_i(x^*) < 0, i \in I(x_0)\backslash J$ *[69].*

iv) The functions $g_i(x), i \in I(x_0)$, *are pseudoconcave at* x_0 *[65].*

v) The functions $g_i(x), i \in I(x_0)$, *are invex at* x_0 *with respect to the same* $\eta(x, x_0)$ *and there exists* $x^* \in S$ *such that* $g_i(x^*) < 0, i \in I(x_0)$ *[8].*

Proof. i) Since $g_i(x^*) < 0 = g_i(x_0)$, for the pseudoconvexity of g_i, we have $(x - x_0)^T \nabla g_i(x_0) < 0$, so that the direction $d = x - x_0 \in G^0$ and thus $G^0 \neq \emptyset$.

ii) It follows from i), taking into account that the quasiconvexity of $g_i(x)$ at x_0 and the assumption $\nabla g_i(x_0) \neq 0$ imply the pseudoconvexity of $g_i(x)$ at x_0.

iii) For $i \in J$, the function $g_i(x)$ is pseudoconvex and pseudoconcave so that (see section 6) $g_i(x^*) < g_i(x_0) = 0$ implies $(x - x_0)^T \nabla g_i(x_0) \leq 0$. On the other hand $g_i(x^*) < g_i(x_0) = 0$, $i \in I(x_0)\backslash J$ implies, for the pseudoconvexity of g_i, $(x - x_0)^T \nabla g_i(x_0) < 0$, so that $d = x - x_0 \in G^p$ and thus $G^p \neq \emptyset$.

iv) It is sufficient to note that $G^p = G$.

v) We have $g_i(x^*) < g_i(x_0) = 0$, $i \in I(x_0)$. For the invexity of g_i, $i \in I(x_0)$, it results $\eta(x^*, x_0)^T \nabla g_i(x_0) < 0, i \in I(x_0)$, so that $d = \eta(x^*, x_0) \in G^0$ and thus $G^0 \neq \emptyset$. □

At last we prove the necessity of the Kuhn-Tucker conditions requiring a generalized Karlin constraint qualification which involves invexity [37].

Theorem 4.9 *Let* x_0 *be an optimal solution for problem* P *where the functions* g_i, $i \in I(x_0)$, *are invex with respect to the same* η. *Assume that there exist no vector* $p \in \Re^s$, $p \geq 0$, $p \neq 0$, *such that* $\sum_{i \in I(x_0)} p_i g_i(x) \geq 0$, $\forall x \in X$. *Then conditions (4.9) and (4.10) hold.*

Proof. The optimality of x_0 implies the following F. John conditions: there exist $\lambda_0 \geq 0,\ \ \lambda_i \geq 0,\ \ (\lambda_0, \lambda_1, ..., \lambda_s) \neq 0,\ \ i \in I(x_0)$, such that

$$\lambda_0 \nabla f(x_0) + \sum_{i \in I(x_0)} \lambda_i \nabla g_i(x_0) = 0$$

Assume that $\lambda_0 = 0$ that is $\sum_{i \in I(x_0)} \lambda_i \nabla g_i(x_0) = 0$. For the invexity assumption, we have:

$$g_i(x) - g_i(x_0) \geq \eta^T(x, x_0) \nabla g_i(x_0) \ \ , \forall i \in I(x_0)$$

so that $\sum_{i \in I(x_0)} \lambda_i g_i(x) \geq \eta^T(x, x_0) \sum_{i \in I(x_0)} \lambda_i \nabla g_i(x_0) = 0$ and this contradicts the generalized Karlin constraint qualification. □

5. Maximum points and generalized convexity

In this section we will show that for a generalized convex function, a global maximum point, if one exists, is attained at the boundary of the feasible region S and, under suitable assumptions on S, it is an extreme point of S.

We will begin to prove that if the maximum value of the function is reached at a relative interior point of S, then f is constant on S.

We recall that the relative interior of a convex set $C \subset \Re^n$, denoted by riC, is defined as the interior which results when C is regarded as a subset of its affine hull $affC$. In other words,

$$riC = \{x \in affC : \exists \epsilon > 0, (x + \epsilon B) \cap affC \subset C\}$$

where B is the Euclidean unit ball in $\Re^n$.

Lemma 4.1 *Let f be a continuous and semistrictly quasiconvex function on a convex set S. If $x_0 \in riS$ is such that $f(x_0) = max_{x \in S} f(x)$, then f is constant on S.*

Proof. Assume that there exists $\bar{x} \in S$ such that $f(\bar{x}) < f(x_0)$. For a known property of convex sets [81], there exists $x^* \in S$ such that $x_0 \in]x^*, \bar{x}[$. Since f is a continuous function, without loss of generality, we can assume that $f(x^*) > f(\bar{x})$. The semistrict quasiconvexity of f implies $f(x) < f(x^*)$, $\forall x \in]x^*, \bar{x}[$ and this is absurd since $x_0 \in]x^*, \bar{x}[$. □

From Lemma 4.1, we have directly the following result.

Theorem 4.10 *Let f be a continuous and semistrictly quasiconvex function on a convex and closed set S. If f assumes maximum value on S, then it is reached at some boundary point.*

The previous theorem can be strengthened when the convex set S does not contain lines (such an assumption implies the existence of an extreme point [81]).

Theorem 4.11 *Let f be a continuous and semistrictly quasiconvex function on a convex and closed set S containing no lines. If f assumes maximum value on S, then it is reached at an extreme point.*

Proof. If f is constant, then the thesis is trivial. Let x_0 be such that $f(x_0) = max_{x \in S} f(x)$. From Theorem 4.10, x_0 belongs to the boundary of S. Let C be the minimal face of S containing x_0; if x_0 is not an extreme point, then $x_0 \in riC$. It follows from Lemma 4.1 that f is constant on C. On the other hand, C is a convex closed set containing no lines, so that C has at least one extreme point $\bar{x}$ which is also an extreme point of S [81]. Consequently $\bar{x}$ is a global maximum for f on S. □

Obviously the previous result holds for a pseudoconvex function, while a quasiconvex function can have a global maximum point which is not a boundary point. In fact the function

$$f(x) = \begin{cases} -x^2 + 2x & 0 \leq x \leq 1 \\ 1 & x > 1 \end{cases}$$

is nondecreasing, so that it is quasiconvex; on the other hand any point $x > 1$ where f assumes its maximum value is not a boundary point.

If we want to extend Theorem 4.11 to the class of quasiconvex functions, we must require additional assumptions on the convex set S.

Theorem 4.12 *Let f be a continuous and quasiconvex function on a convex and compact set S. Then there exists some extreme point at which f assumes its maximum value.*

Proof. From Weierstrass Theorem, there exists $\bar{x} \in S$ with $f(\bar{x}) = max_{x \in S} f(x)$. Since S is convex and compact, it is also the convex hull of its extreme points, so that there exists a finite number $x^1, ..., x^h$ of extreme points such that $\bar{x} = \sum_{i=1}^{h} \lambda_i x_i$, $\sum_{i=1}^{h} \lambda_i = 1$, $\lambda_i \geq 0$. From the quasiconvexity of f we have $f(\bar{x}) \leq max\{f(x^1), ..., f(x^h)\}$ and the thesis follows. □

Taking into account that a pseudoconvex function is also semistrictly quasiconvex, we have the following corollaries:

Corollary 4.2 *Let f be a pseudoconvex function on a convex and closed set S. If f assumes maximum value on S, then it is reached at some boundary point.*

Corollary 4.3 *Let f be a pseudoconvex function on a convex and closed set S containing no lines. If f assumes maximum value on S, then it is reached at an extreme point.*

From a computational point of view, Theorems 4.11 and 4.12 and Corollaries 4.2 and 4.3 are very important since they establish that we must investigate the boundary of the feasible set (in particular the extreme points if one exists) in order to find a global maximum of a generalized convex function. Nevertheless, for these classes of functions a local maximum is not necessarily global and the necessary optimality condition $(x - x_0)^T \nabla f(x_0) \leq 0, \ \forall x \in S$ is not sufficient for x_0 to be a local maximum point, so that the problem of maximize a quasiconvex or pseudoconvex function is a hard problem. This kind of difficulties vanishes if f is also quasiconcave or pseudoconcave. We will deep this aspect in the next section.

6. Quasilinear and pseudolinear scalar functions

A function defined on a convex subset S of $\Re^n$ is said to be quasilinear (pseudolinear) if it is both quasiconvex and quasiconcave (pseudoconvex and pseudoconcave).

The pseudolinear functions have some properties stated in [26, 54, 55] for which we propose simple proofs.

Theorem 4.13 *Let f be a function defined on an open convex set $S \subset \Re^n$.*

i) If f is pseudolinear and there exists $x_0 \in S$ such that $\nabla f(x_0) = 0$, then f is constant on S.

ii) f is pseudolinear if and only if

$$x, y \in S, \;\; f(x) = f(y) \Longleftrightarrow (y-x)^T \nabla f(x) = 0 \tag{4.14}$$

iii) Assume $\nabla f(x) \neq 0 \; \forall x \in S$. Then f is pseudolinear on S if and only if its normalized gradient mapping $x \to \frac{\nabla f(x)}{\|\nabla f(x)\|}$ is constant on each level set $f(x) =$constant.

Proof. i) It follows from Theorem 4.3, taking into account that the stationary point x_0 is also a global maximum, since f is pseudoconcave.

ii) Let f be pseudolinear. Since f is also quasilinear, then $f(x) = f(y)$ implies that f is constant on the line-segment $[x, y]$, so that the directional derivative $(y-x)^T \nabla f(x)$ is equal to zero.

Assume now $(y-x)^T \nabla f(x) = 0$. Since $f(y) < f(x)$ $(f(y) > f(x))$ implies $(y-x)^T \nabla f(x) < 0$ $((y-x)^T \nabla f(x) > 0)$, necessarily we have $f(x) = f(y)$.

Assume that (4.14) holds; we must prove that f is both pseudoconvex and pseudoconcave. If f is not pseudoconvex, there exist $x, y \in S$ with $f(y) < f(x)$ such that $(y-x)^T \nabla f(x) \geq 0$. Since $(y-x)^T \nabla f(x) = 0$ implies $f(x) = f(y)$, we must have $(y-x)^T \nabla f(x) > 0$, so that the direction $d = y - x$ is an increasing direction at x. The continuity of the function f implies the existence of $x^* = x + t^*(y-x), t^* \in]0,1[$ such that $f(x^*) = f(x)$. Consequently $(x^* - x)^T \nabla f(x) = t^*(y-x)^T \nabla f(x) > 0$ and this contradicts (4.14). It follows that f is pseudoconvex. In an analogous way it can be proven that f is pseudoconcave.

iii) Let f be pseudolinear with $\nabla f(x) \neq 0$ $\forall x \in S$. We must prove that

$$f(x) = f(y) \Rightarrow \frac{\nabla f(x)}{\| \nabla f(x) \|} = \frac{\nabla f(y)}{\| \nabla f(y) \|} \tag{4.15}$$

Set $\Gamma_1 = \{d \in \Re^n : d^T \nabla f(x) = 0\}$, $\Gamma_2 = \{d \in \Re^n : d^T \nabla f(y) = 0\}$.

We have $\Gamma_1 = \Gamma_2$. Indeed, if $d \in \Gamma_1$, from ii) it results $f(x + td) = f(x) = f(y)$ for every t such that $x + td \in S$. From ii), it follows $(y - x - td)^T \nabla f(y) = 0$ and $(y - x)^T \nabla f(y) = 0$ so that $d^T \nabla f(y) = 0$ and thus $d \in \Gamma_2$. In an analogous way we can prove that $\Gamma_2 \subset \Gamma_1$.

Since $\Gamma_1 = \Gamma_2$, it results $\frac{\nabla f(x)}{\|\nabla f(x)\|} = \pm \frac{\nabla f(y)}{\|\nabla f(y)\|}$. Set $u = \frac{\nabla f(y)}{\|\nabla f(y)\|}$ and assume that $\frac{\nabla f(x)}{\|\nabla f(x)\|} = -u$; for a suitable $t \in]0, \epsilon[$ the points $z_1 = x + tu$, $z_2 = y + tu$ are such that $f(z_1) < f(x)$, $f(z_2) > f(y)$. The continuity of f implies the existence of $\lambda \in]0, 1[$ such that $f(z) = f(x) = f(y)$ with $z = \lambda z_1 + (1 - \lambda) z_2$. From ii) we must have $(z - y)^T u = 0$; on the other hand $(z - y)^T u = (\lambda(x - y) + (1 - \lambda) tu)^T u = (1 - \lambda) t \| u \|^2 > 0$ so that from ii), $f(y) \neq f(z)$ and this is absurd.

Consequently we have $\frac{\nabla f(x)}{\|\nabla f(x)\|} = \frac{\nabla f(y)}{\|\nabla f(y)\|}$.

Assume now that (4.15) holds. Let $x, y \in S$ and set $\phi(t) = f(x + t(y - x)), t \in [0, 1]$. If $\phi'(t)$ is constant in sign, then $\phi(t)$ is quasilinear on the line segment $[0, 1]$. Otherwise, from elementary analysis, there exist $t_1, t_2 \in]0, 1[$ such that $\phi(t_1) = \phi(t_2)$ with $\phi'(t_1)\phi'(t_2) < 0$. We can assume, without loss of generality, that $t_1 < t_2$, $\phi'(t_1) > 0$, $\phi'(t_2) < 0$. Set $z_1 = x + t_1(y - x), z_2 = x + t_2(y - x)$. Since $f(z_1) = \phi(t_1) = \phi(t_2) = f(z_2)$, we have $\phi'(t_1) = (1 - t_1)(y - x)^T \nabla f(z_1) > 0$ and $0 > \phi'(t_2) = (1 - t_2)(y - x)^T \nabla f(z_2) = (1 - t_2)(y - x)^T \nabla f(z_1) \frac{\|\nabla f(z_2)\|}{\|\nabla f(z_1)\|} > 0$ and this is absurd.

It follows that the restriction of the function over every line-segment contained in S is quasilinear, so that f is quasilinear and also pseudolinear since $\nabla f(x) \neq 0$, $\forall x \in S$. □

Remark 4.3 Following the same lines of the proof given in ii) of the previuos theorem, it can be shown that (4.14) is equivalent to the following two statements:

$$x, y \in S, \ \ f(y) < f(x) \Longleftrightarrow (y - x)^T \nabla f(x) < 0 \tag{4.16}$$

$$x, y \in S, \ \ f(y) > f(x) \Longleftrightarrow (y - x)^T \nabla f(x) > 0 \tag{4.17}$$

which point out that pseudolinearity is equivalent to require that the logical implication in the definition of pseudoconvex (pseudoconcave) function can be reversed.

Let us note that i) and ii) of Theorem 4.13 do not hold if f is quasilinear, as it is easy to verify considering the function $f(x) = x^3$; furthermore, i) and ii) of Theorem 4.13 hold even if S is a relatively open convex set, while in iii) the assumption $intS \neq \emptyset$ cannot be weakened, as it is shown in the following example.

Example 4.4 Consider the function $f(x,y,z) = xy+xz+\frac{x+y+z}{x-y+z}$ defined on the relatively open convex set $D = \{(x,y,z) : x - y + z > 0\}$

Consider the convex set $S = \{(x,y,z) : x = 0, y = 0, z > 0\}$; obviously $intS = \emptyset$, while $riS \neq \emptyset$.
By simple calculation, it results $\nabla f(0,0,z) = (z, \frac{2}{z}, 0)$. Let $A = (0,0,1)$, $B = (0,0,2)$; we have $f(A) = f(B) = 1, \nabla f(A) = (1,2,0), \nabla f(B) = (2,1,0)$ so that $\frac{\nabla f(A)}{\|\nabla f(A)\|} \neq \frac{\nabla f(B)}{\|\nabla f(B)\|}$, while f is pseudolinear on S.

Condition iii) of Theorem 4.13 can be strengthened when the function f is defined on the whole space $\Re^n$, in the sense stated in the following theorem.

Theorem 4.14 *The non constant function f is pseudolinear on the whole space $\Re^n$ if and only if its normalized gradient mapping $x \to \frac{\nabla f(x)}{\|\nabla f(x)\|}$ is constant on $\Re^n$.*

Proof. $\Leftarrow$ It follows from iii) of Theorem 4.13.

$\Rightarrow$ Let f be pseudolinear on $\Re^n$ and assume that its normalized gradient mapping is not constant on $\Re^n$. Then there exist $x_1, x_2 \in \Re^n$ such that $\frac{\nabla f(x_1)}{\|\nabla f(x_1)\|} \neq \frac{\nabla f(x_2)}{\|\nabla f(x_2)\|}$. From iii) of Theorem 4.13, we have $f(x_1) \neq f(x_2)$. Set $\Gamma_1 = \{d \in \Re^n : d^T \nabla f(x_1) = 0\}$ and $\Gamma_2 = \{d \in \Re^n : d^T \nabla f(x_2) = 0\}$. Let us note that $x \in x_1 + \Gamma_1$ implies $(x - x_1) \in \Gamma_1$ so that $(x - x_1)^T \nabla f(x_1) = 0$ and, from ii) of Theorem 4.13, $f(x) = f(x_1), \forall x \in x_1 + \Gamma_1$. Analogously we have $f(x) = f(x_2), \forall x \in x_2 + \Gamma_2$. Since $\frac{\nabla f(x_1)}{\|\nabla f(x_1)\|} \neq \frac{\nabla f(x_2)}{\|\nabla f(x_2)\|}$, there exists $\bar{x} \in (x_1 + \Gamma_1) \cap (x_2 + \Gamma_2)$, so that $f(\bar{x}) = f(x_1), f(\bar{x}) = f(x_2)$ and this is absurd. □

From a geometrical point of view, the previous theorem states that the level sets of a non constant pseudolinear function, defined on the whole space $\Re^n$ are parallel hyperplanes; vice-versa if the level sets of a differentiable function, with no critical points, are hyperplanes, then the function is pseudolinear.

In any case, if the level sets of a function are hyperplanes, then the function is quasilinear, but the vice-versa is not true. In fact the function

$\phi(x, y) = f(x)$, where $f(x)$ is as in (4.5), is quasilinear and the level set $\{(x, y) : \phi(x, y) = 0\} = [0, 2] \times \Re$ is not a hyperplane.

When the non constant pseudolinear function is defined on a convex set $S \subset \Re^n$, from iii) of Theorem 4.13, the level sets are the intersection between S and hyperplanes which are not necessarily parallel (consider for instance the classic case of linear fractional functions).

The above considerations suggest a simple way to construct a pseudolinear function. Consider for instance the family of lines $y = \frac{kx+1}{\sqrt{k+1}}$. It is easy to verify that such lines are the level sets of the function

$$f(x, y) = \frac{-2x + y^2 + y\sqrt{y^2 - 4x + 4x^2}}{2x^2}$$

defined on $S = \{(x, y) : x > 1, y > 0\}$; since $\nabla f(x, y) \neq 0 \ \forall (x, y) \in S$, $f(x, y)$ is pseudolinear on S.

Another way to construct a pseudolinear function is to consider a composite function $\phi \circ f$ where f is pseudolinear and $\phi : \Re \to \Re$ is a differentiable function having a strictly positive (or negative) derivative.

With respect to an optimization problem having a pseudolinear objective function defined on a polyhedral set S, we have the nice property that when the maximum and the minimum value exist, they are reached at a vertex of S.

Setting $d_1, ..., d_n$ the edges starting from a vertex $x_0 \in S$, the necessary and sufficient optimality condition stated in Theorem 4.4 and the analogous one for a pseudoconcave function can be specified by means of the following theorem.

Theorem 4.15 *Let f be a pseudolinear function defined on a polyhedral set S. Then:*

i) A vertex $x_0 \in S$ is a minimum point for f on S if and only if $d_i^T \nabla f(x_0) \geq 0$, i=1,...,n.

ii) A vertex $x_0 \in S$ is a maximum point for f on S if and only if $d_i^T \nabla f(x_0) \leq 0$, i=1,...,n.

When S is a polyhedral compact set, the previous theorem can be extended to a quasilinear function.

Theorem 4.16 *Let f be a quasilinear function defined on a polyhedral compact set S. Then:*

i) A vertex $x_0 \in S$ is a minimum point for f on S if and only if $d_i^T \nabla f(x_0) \geq 0$, i=1,...,n.

ii) A vertex $x_0 \in S$ is a maximum point for f on S if and only if $d_i^T \nabla f(x_0) \leq 0$, i=1,...,n.

The optimality conditions stated in the previous theorem have suggested some simplex-like procedure for pseudolinear problems. These programs include linear programs and linear fractional programs which arise in many practical applications [30, 87, 89]. Algorithms for a linear fractional problem have been suggested by several authors [9, 25, 68]. Computational comparisons between algorithms for linear fractional programming are given in [33].

7. Generalized convex vector functions

Let X be an open set of the n-dimensional space $\Re^n$, $S \subset X$ a convex subset of X and F a vector function from X to $\Re^s$.

In what follows, we will consider in $\Re^s$ the partial order induced by the Paretian cone, even if most of the results, which are going to establish, hold when the partial order is induced by any closed convex cone. For convenience, we set

$$C = \Re^s_+ = \{z = (z_1, ..., z_s) \in \Re^s : z_i \geq 0, i = 1, ..., s\}$$

$$C^0 = C \setminus \{0\}; \;\; intC = \{z = (z_1, ..., z_s) \in \Re^s : z_i > 0, i = 1, ..., s\}$$

As is known, there are different ways in extending the definitions of generalized convex functions to the vector case; we will address to those classes which will allow to obtain several properties related to optimality, referring to bibliography for further deepenings (see for instance [15], [18], [62]).

As it happens in the scalar case, we can refer to vector generalized convexity or to vector generalized concavity. In this last case, it is sufficient to substitute in what follows $-C, -C^0, -intC$ with $C, C^0, intC$, respectively, in order to obtain the corresponding definitions and results.

Definition 4.7 *The function F is said to be C-convex (on S) if*

$$F(x_1 + \lambda(x_2 - x_1)) \in F(x_1) + \lambda(F(x_2) - F(x_1)) - C$$

$\forall \lambda \in [0, 1], \;\; \forall x_1, x_2 \in S$

It is easy to prove that F is C-convex if and only if any component of F is convex.

Definition 4.8 *The function F is said to be C-quasiconvex (on S) if*

$$x_1, x_2 \in S, \;\; F(x_2) \in F(x_1) - C \Rightarrow F(x_1 + \lambda(x_2 - x_1)) \in F(x_1) - C$$

$\forall \lambda \in [0, 1]$

When $s = 1$, Definition 4.8 reduces to (4.1). If any component of F is quasiconvex, then F is C-quasiconvex, but the converse is not true. For instance the function $F(x, y) = (x, -x, -x^2, y)$ is $\Re^4_+$-quasiconvex on $S = \Re^2$, but it is not componentwise quasiconvex.

In [62], Luc suggests another definition of a quasiconvex vector function which is equivalent to require componentwise quasiconvexity and which plays an important role in establishing the connectedness of the set of all efficient points. This class of functions is strictly contained in the class of C-quasiconvex ones [15].

Remark 4.4 When F is a differentiable function, C-quasiconvexity implies the following property:

$$x_1, x_2 \in S, F(x_2) \in F(x_1) - C \Rightarrow F'(x_1)(x_2 - x_1) \in -C$$

where $F'(x_1)$ denotes the Jacobian matrix of F evaluated at x_1.

Unlike the scalar case, the converse implication does not hold in general and this points out that the study of vector generalized convexity is more complicated than that of the scalar case; this remark motivates once again the variety of definitions which have been suggested in the literature for the vector case.

Assume now that F is a differentiable function and let $F'(u)$ be the Jacobian matrix of F evaluated at u. As for the quasiconvex case, there are different ways to extend scalar pseudoconvexity to the vector case. We introduce three classes of vector pseudoconvex functions which reduce, when $s = 1$, to the classical definition given by Mangasarian [65].

Definition 4.9 *The function F is said to be (C^0, C^0)-pseudoconvex (on S) if*

$$x_1, x_2 \in S, F(x_2) \in F(x_1) - C^0 \Rightarrow F'(x_1)(x_2 - x_1) \in -C^0$$

Definition 4.10 *The function F is said to be $(C^0, intC)$-pseudoconvex (on S) if*

$$x_1, x_2 \in S, F(x_2) \in F(x_1) - C^0 \Rightarrow F'(x_1)(x_2 - x_1) \in -intC$$

Definition 4.11 *The function F is said to be $(intC, intC)$-pseudoconvex (on S) if*

$$x_1, x_2 \in S, F(x_2) \in F(x_1) - intC \Rightarrow F'(x_1)(x_2 - x_1) \in -intC$$

If any component of F is pseudoconvex, then F is $(intC, intC)$- pseudoconvex and also (C^0, C^0)-pseudoconvex; if any component of F is strictly pseudoconvex (that is $F_i(x_2) \leq F_i(x_1) \Rightarrow (x_2 - x_1)^T \nabla F_i(x_1) < 0$ $\forall i$), then F is $(C^0, intC)$- pseudoconvex; if any component of F is quasiconvex and at least one is strictly pseudoconvex, then F is (C^0, C^0)-pseudoconvex. The converse of these statements are not true in general, as it is shown in the following example.

Example 4.5 The function $F(x, y) = (x, -x, -x^2, y)$ is $(intC, intC)$-pseudoconvex and (C^0, C^0)-pseudoconvex on $S = \Re^2$, with $C = \Re^4_+$, but it is not componentwise quasiconvex or pseudoconvex.

The function $F(x) = (x, -x, -x^2)$ is $(C^0, intC)$-pseudoconvex on $S = \Re$, with $C = \Re^3_+$, but its components are not strictly pseudoconvex.

The following theorem states the inclusion relationships among the introduced classes of functions.

Theorem 4.17 *Let F be a differentiable function.*

i) If F is C-convex (on S), then it is C-quasiconvex, $(intC, intC)$-pseudoconvex and (C^0, C^0)-pseudoconvex (on S).

ii) If F is $(C^0, intC)$-pseudoconvex (on S), then it is (C^0, C^0)-pseudoconvex and $(intC, intC)$-pseudoconvex (on S).

Proof. In order to prove that C-convexity implies (C^0, C^0)-pseudoconvexity and $(intC, intC)$-pseudoconvexity, it is sufficient to note that C-convexity implies $F'(x_1)(x_2 - x_1) \in F(x_2) - F(x_1) - C$ and that $C^0 + C = C^0$, $intC + C = intC$.

The other inclusion relationships follow directly from the definitions.

□

The following examples point out that there are not inclusion relationships between (C^0, C^0)-pseudoconvexity and $(intC, intC)$-pseudoconvexity and between C-convexity and $(C^0, intC)$-pseudoconvexity.

Example 4.6 Consider the function $F(x) = (-x^2, x^2(x-1)^3)$, $S = \{x \in \Re, x \geq 0\}$, $C = \Re^2_+$.

It is easy to prove that there do not exist $x_1, x_2 \in S$ such that $F(x_2) \in F(x_1) - intC$, so that F is $(intC, intC)$-pseudoconvex. On the other hand, setting $x_2 = 1, x_1 = 0$, we have $F(1) \in F(0) - C^0$, while $F'(0) \cdot 1 \notin -C^0$ and thus F is not (C^0, C^0)-pseudoconvex.

Consider now the function $F(x) = (-x, -x^2)$, $S = \{x \in \Re, x \geq 0\}$, $C = \Re^2_+$.

Setting $x_2 = 1, x_1 = 0$, we have $F(1) \in F(0) - intC$ and $F'(0) \cdot 1 \notin -intC$, so that F is not $(intC, intC)$-pseudoconvex, while simple calculations show that F is (C^0, C^0)-pseudoconvex.

Example 4.7 Consider the function $F(x) = (-1, x^2 - 4x)$, $S = \Re$, $C = \Re^2_+$.

F is C-convex since it is componentwise convex. Setting $x_2 = 1, x_1 = 0$, we have $F(1) \in F(0) - C^0$ and $F'(0) \cdot 1 \notin -intC$, so that F is not $(C^0, intC)$-pseudoconvex.

On the other hand, the function $F(x) = (-x, -x^2)$ is not C-convex on $S = \{x \in \Re, x > 0\}$, with $C = \Re^2_+$, but it is $(C^0, intC)$-pseudoconvex.

The following example points out that, unlike the scalar case, there is not inclusion relationship between quasiconvex and pseudoconvex vector functions.

Example 4.8 Consider the function $F(x, y) = (x^3, y^3)$, $S = \Re^2$, $C = \Re^2_+$. Since F is componentwise quasiconvex, it is also C-quasiconvex.

Setting $x_2 = (-1, -1), x_1 = (0, 0)$, we have $F(-1, -1) \in F(0, 0) - intC$, but $F'(0, 0)(x_2 - x_1) \notin -intC$, so that F is not $(intC, intC)$-pseudoconvex. Furthermore $F(-1, -1) \in F(0, 0) - C^0$, but $F'(0, 0)(x_2 - x_1) \notin -C^0$, and thus F is not (C^0, C^0)-pseudoconvex.

Consider now the function $F(x) = (x^2 - x, -x^2 + x)$, $S = \Re$, $C = \Re^2_+$. It is easy to verify that the relation $F(x_2) \in F(x_1) - C^0$ (in particular $F(x_2) \in F(x_1) - intC$) does not hold for every $x_2, x_1 \in \Re$, so that F is both (C^0, C^0)-pseudoconvex and $(intC, intC)$-pseudoconvex. Setting $x_2 = 1, x_1 = 0$, we have $F(x_2) = F(x_1)$, so that $F(x_2) \in F(x_1) - C$, but $F(x_1 + \lambda(x_2 - x_1)) = F(\lambda) \notin F(x_1) - C$, $\forall \lambda \in]0, 1[$ and thus F is not C-quasiconvex.

8. Efficiency

In vector optimization, because of the contradiction of the objective functions, it is not possible, in general, to find a point which is optimal for all the objectives simultaneously. For such a reason, a new concept of optimality was introduced by the economist Edgeworth in 1881. However this new concept is usually referred to the French-Italian economist Pareto who in 1896 developed it further.

Speaking roughly, a point is Pareto optimal if and only if it is possible to improve (in the sense of minimization) one of the objective functions only at the cost of making at least one of the remaining objective functions worse; a point is weakly Pareto optimal if and only if it is not possible to improve all the objective functions simultaneously.

For a formal definition, consider the following vector optimization problem:

$$P : minF(x), x \in S \subset X$$

where X be an open set of $\Re^n$, $F : X \to \Re^s$.

Definition 4.12 *A point $x_0 \in S$ is said to be :*
- weakly efficient or weakly Pareto optimal if

$$F(x) \notin F(x_0) - intC, \ \forall x \in S$$

- efficient or Pareto optimal if

$$F(x) \notin F(x_0) - C^0, \ \forall x \in S.$$

If the previous conditions are verified in $I \cap S$, where I is a suitable neighbourhood of x_0, then x_0 is said to be a local weak efficient point or a local efficient point, respectively.

In the scalar case (s=1) a (local) weak efficient point and an (local) efficient point reduce to the ordinary definition of a (local) minimum point.

Obviously (local) efficiency implies (local) weak efficiency.
With respect to the class of $(C^0, intC)$-pseudoconvex functions, a point is (local) weakly efficient if and only if it is (local) efficient.

Let us note that all the results which will be established for problem P hold for the problem $maxF(x)$ substituting $-C^0$ with C^0, -$intC$ with $intC$ and requiring generalized concavity instead of generalized convexity.

With regard to the existence of efficient points of problem P, we refer the interested reader to [12, 62, 84, 94].

Now we will stress the role played by vector generalized convexity in investigating relationships between local and global optima. As in the scalar case, from now on, we will consider generalized convexity at a point x_0 and we will require that the feasible set S is star-shaped at x_0; furthermore, referring to a pseudoconvex vector function, the differentiability of F at x_0 is assumed.

The following theorem shows that, under suitable assumption of generalized convexity, local efficiency implies efficiency.

Theorem 4.18 *i) If x_0 is a local efficient point for P and F is $(C^0, intC)$-pseudoconvex at x_0, then x_0 is an efficient point for P.*

ii) If x_0 is a local weak efficient point for P and F is (intC,intC)-pseudoconvex at x_0, then x_0 is a weak efficient point for P.

Proof. i) Assume that there exists $\bar{x} \in S$ such that $F(\bar{x}) \in F(x_0) - C^0$. Since F is $(C^0, intC)$-pseudoconvex at x_0, we have $F'(x_0)(\bar{x} - x_0) \in -intC$, that is $\lim_{t \to 0^+} \frac{F(x_0 + t(\bar{x} - x_0)) - F(x_0)}{t} \in -intC$ and this implies the existence of a suitable $\epsilon > 0$, such that $F(x_0 + t(\bar{x} - x_0)) \in F(x_0) - intC, \ \forall t \in (0, \epsilon)$. This contradicts the local efficiency of x_0.

ii) The proof is similar to the one given in i). □

In general a local efficient point for P is not an efficient point when F is $(intC, intC)$-pseudoconvex, as it is shown in the following example.

Example 4.9 Consider the function $F(x) = (F_1(x), F_2(x))$ where $F_1(x) = 0 \ \forall x \in \Re$ and

$$F_2(x) = \begin{cases} 0 & x \leq 1 \\ -(x-1)^2 & x > 1 \end{cases}$$

It is easy to prove that F is $(intC, intC)$-pseudoconvex at $x_0 = 0$ with $C = \Re^2_+$ and that x_0 is a local efficient point, but not efficient since $F(2) \in F(0) - C^0$.

Let us note that the function F is also C-quasiconvex since its components are quasiconvex, so that, also for this class of vector functions, local efficiency does not imply efficiency.

The class of (C^0, C^0)-pseudoconvex functions is not appropriate to guarantee that a local efficient point is also efficient as is pointed out in the following example.

Example 4.10 Consider the function $F(x) = (-x^3 + x^2, x^2 - 2x)$, $x \in S = \{x \in \Re, x \geq 0\}$; F is (C^0, C^0)-pseudoconvex at $x_0 = 0$, with $C = \Re^2_+$ and the feasible point $x_0 = 0$ is a local efficient point but it is not a (weak) efficient point since $F(\frac{3}{2}) \in F(0) - intC$.

As we have pointed out in the previous examples, (C^0, C^0)-pseudoconvexity and $(intC, intC)$-pseudoconvexity do not guarantee that a local efficient point is also efficient. An important subclass of these two classes of functions, for which such a property holds, is the componentwise convex one.

Theorem 4.19 *Let F be C-convex at x_0. If x_0 is a local efficient (weak efficient) point for P, then x_0 is efficient (weak efficient) for P.*

Proof. Assume that there exists $\bar{x} \in S$ such that $F(\bar{x}) \in F(x_0) - C^0$ $(F(\bar{x}) \in F(x_0) - intC)$. Since F is C-convex at x_0, it results $F(x_0 + \lambda(\bar{x} - x_0)) \in F(x_0) + \lambda(F(\bar{x}) - F(x_0)) - C$. Taking into account that $C^0 + C = C$ (intC+C=intC), we have $F(x_0 + \lambda(\bar{x} - x_0)) \in F(x_0) - C^0$ $(F(x_0 + \lambda(\bar{x} - x_0)) \in F(x_0) - intC)$, $\forall \lambda \in (0, 1)$ and this contradicts the local efficiency (weak efficiency) of x_0. □

If F is componentwise pseudoconvex at x_0, unlike the scalar case, local efficiency does not imply efficiency.

Consider for instance the function $F(x) = (-x^2 + x, x^2 - 2x)$, $S = [0,1]$, $C = \Re^2_+$. The first component $F_1(x)$ of F is pseudoconvex at $x_0 = 0$, since there does not exist $x \in (0,1]$ such that $F_1(x) < F_1(x_0)$. Since the second component of F is convex, then F is componentwise pseudoconvex at x_0. It is easy to verify that $x_0 = 0$ is a local efficient point, but not efficient since $F(1) = (0,-1) \in F(0) - C^0$.

The following theorem shows that requiring the pseudoconvexity of the components of F not only at a point but on the whole feasible region, local efficiency implies efficiency.

Theorem 4.20 *Consider problem P where F is componentwise pseudoconvex on the convex set S. If x_0 is a local efficient point for P, then x_0 is efficient for P.*

Proof. We recall that a pseudoconvex scalar function ϕ has the property that if $\phi(z) < \phi(z_0)$, then ϕ is decreasing at z_0, with respect to the direction $z - z_0$ and if $\phi(z) = \phi(z_0)$, then ϕ is decreasing at z_0 or z_0 is a minimum point with respect to the direction $z - z_0$. Assume now that there exists $\bar{x} \in S$ such that $F_i(\bar{x}) \leq F_i(x_0), i = 1, ..., s$, where at least one inequality is strict. For the mentioned property, it results that x_0 is not a local efficient point and this is a contradiction. □

Some other classes of generalized convex vector functions have been suggested in order to maintain local-global properties. For instance in [11], a class of functions is introduced verifying the following property

$$F(x_2) \in F(x_1) - C^0 \Rightarrow F(x_1 + \lambda(x_2 - x_1)) \in F(x_1) - C^0 \qquad (4.18)$$

$\forall x_1, x_2 \in S, \ \forall \lambda \in]0,1[$.

This class contains componentwise semistrictly quasiconvex functions, but, unlike the scalar case, an upper semicontinuous function verifying (4.18) is not necessarily C-quasiconvex. Consider in fact the continuous function

$$F(x) = \begin{cases} (x sin\frac{1}{x}, -x sin\frac{1}{x}) & x \neq 0 \\ (0,0) & x = 0 \end{cases}$$

and the point $x_1 = 0$.

Condition (4.18) is verified at x_1, with $C = \Re^2_+$, since $F(x_2) \notin F(0) - C^0, \ \forall x$. On the other hand F is not C-quasiconvex at x_1 since for $x_2 = \frac{1}{\pi}$, we have $F(x_2) \in F(0) - C$, but it is not true that $F(\lambda x_2) \in F(0) - C, \ \forall \lambda \in]0,1[$.

Another class of functions verifying local-global properties has been introduced in [64] with the following property:

F is componentwise quasiconvex and for all x_1, x_2 such that $F(x_2) \neq F(x_1), \ F(x_1 + \lambda(x_2 - x_1)) < max(F(x_2), F(x_1))$.

It is easy to prove that this last class is strictly contained in the previous one.

9. Optimality conditions in vector optimization

As in the scalar case, generalized convexity in vector optimization plays an important role in establishing sufficient optimality conditions. In order to develop a self-contained analysis, we will also present the proofs of the first order necessary optimality conditions, which are based on the following known separation theorems.

Theorem 4.21 *Let W be a linear subspace of $\Re^t$.*

i) $W \cap (-int\Re^t_+) = \emptyset$ if and only if

$$\exists \lambda \in \Re^t_+ \setminus \{0\} : \lambda^T w = 0, \ \ \forall w \in W.$$

ii) $W \cap (-\Re^t_+) = \{0\}$ if and only if

$$\exists \lambda \in int\Re^t_+ : \lambda^T w = 0, \ \ \forall w \in W.$$

Theorem 4.22 *Set $W = W_1 \times W_2$, where W_1 and W_2 are linear subspaces of $\Re^s$ and $\Re^m$ respectively. Then the following hold.*

i) $W \cap ((-int\Re^s_+) \times (-int\Re^m_+)) = \emptyset$ if and only if

$$\exists \lambda_1 \in \Re^s_+, \exists \lambda_2 \in \Re^m_+, (\lambda_1, \lambda_2) \neq 0 : \lambda_1^T w_1 + \lambda_2^T w_2 = 0, \ \ \forall (w_1, w_2) \in W.$$

ii) $W \cap ((-int\Re^s_+) \times (-\Re^m_+)) = \emptyset$ if and only if

$$\exists \lambda_1 \in \Re^s_+ \setminus \{0\}, \exists \lambda_2 \in \Re^m_+ : \lambda_1^T w_1 + \lambda_2^T w_2 = 0, \ \ \forall (w_1, w_2) \in W.$$

iii) $W \cap ((-\Re^s_+ \setminus \{0\}) \times (-\Re^m_+)) = \emptyset$ if and only if

$$\exists \lambda_1 \in int\Re^s_+, \exists \lambda_2 \in \Re^m_+ : \lambda_1^T w_1 + \lambda_2^T w_2 = 0, \ \ \forall (w_1, w_2) \in W.$$

Consider problem P where F is a differentiable function. The following theorem holds.

Theorem 4.23 *If x_0 is an interior local weak efficient point for P then*

$$\exists \alpha \in \Re^s_+ \setminus \{0\} \ \ such \ that \ \ \alpha^T F'(x_0) = 0 \tag{4.19}$$

Proof. Consider the line-segment $x = x_0 + td, t \in [0, \epsilon[, d \in \Re^n$. The local weak efficiency of x_0 implies $\frac{F(x_0+td)-F(x_0)}{t} \notin -intC$, so that $F'(x_0)d = \lim_{t\to 0^+} \frac{F(x_0+td)-F(x_0)}{t} \notin -intC$. Setting $W = \{F'(x_0)d, d \in \Re^n\}$, it results $W \cap (-intC) = \emptyset$; the thesis follows from i) of Theorem 4.21. $\square$

Remark 4.5 In the scalar case, condition (4.19) is equivalent to state that x_0 is a stationary point; for such a reason, we will refer to points verifying (4.19) as stationary points of a vector function.

The following example shows that (4.19) is not, in general, a sufficient optimality condition.

Example 4.11 Consider problem P where $F(x_1, x_2) = (-x_1^3 - x_2, x_2 - x_1^2)$, $S = \Re^2, C = \Re^2_+$ and the feasible point $x_0 = (0,0)$.
We have $F'(x_0) = \begin{bmatrix} 0 & -1 \\ 0 & 1 \end{bmatrix}$, so that $\alpha^T F'(x_0) = 0$ with $\alpha^T = (1,1)$. Consequently x_0 is a stationary point for F, but not a local weak efficient point, since $F(x_1, 0) = (-x_1^3, -x_1^2) \in F(x_0) - intC, \forall x_1 > 0$.

Theorem 4.24 *i) If x_0 is a stationary point for F and F is $(C^0, intC)$-pseudoconvex at $x_0 \in S$, then x_0 is an efficient point.*
ii) If x_0 is a stationary point for F and F is $(intC, intC)$-pseudoconvex at $x_0 \in S$, then x_0 is a weak efficient point.

Proof. Assume that x_0 is not an efficient point or a weak efficient point. Then there exists $\bar{x} \in S$ such that $F'(x_0)(\bar{x} - x_0) \in -intC$ and thus $\alpha^T F'(x_0)(\bar{x} - x_0) < 0$; this contradicts (4.19). Then i) and ii) hold. □

Taking into account that a C-convex function is also $(intC, intC)$-pseudoconvex function, we have the following corollary.

Corollary 4.4 *If x_0 is a stationary point for F and F is C-convex at $x_0 \in S$, then x_0 is a weak efficient point.*

Consider the C-convex function $F(x) = (x^2 - 2x, 0), x \in \Re$ and the cone $C = \Re^2_+$. The point $x_0 = 0$ is a stationary point for F and also a weak efficient point, but it is not efficient. From i) of Theorem 4.17, it follows that a stationary point is not in general an efficient point for the classes of $(intC, intC)$ and (C^0, C^0)-pseudoconvex functions. For this last class of functions, a stationary point is not in general a weak efficient point; for instance the function $F(x) = (-x, -x^2), x \in \Re$, $C = \Re^2_+$ is (C^0, C^0)-pseudoconvex at the stationary point $x_0 = 0$, but x_0 is not a weak efficient point.

Requiring in (4.19) the positivity of all multipliers, it is possible to state some other sufficient optimality conditions.

Theorem 4.25 *If F is (C^0, C^0)-pseudoconvex at $x_0 \in S$ and*

$$\exists \alpha \in int\Re^s_+ \text{ such that } \alpha^T F'(x_0) = 0 \tag{4.20}$$

then x_0 is an efficient point.

Proof. The proof is similar to the one given in Theorem 4.24. □

Taking into account Theorems 4.19 and 4.20, we have the following corollary.

Corollary 4.5 *i) If F is C-convex at x_0 and (4.20) holds, then x_0 is an efficient point.*

ii) If F is componentwise pseudoconvex on S and (4.20) holds, then x_0 is an efficient point.

Remark 4.6 Let us note that (4.20) is a sufficient condition for x_0 to be a local efficient point without any requirement of generalized convexity when $KerF'(x_0) = 0$ or, equivalently, when $rankF'(x_0) = n \leq s$ [11].

In section 10 we will prove, for a wide class of functions including the linear ones, that (4.20) is a necessary and sufficient condition for x_0 to be an efficient point.

Consider now the case where x_0 is not necessarily an interior point. A necessary condition for x_0 to be a weak efficient point for P is

$$F'(x_0)(x - x_0) \notin -intC, \ \ \forall x \in S \tag{4.21}$$

In the scalar case (4.21) reduces to $(x - x_0)^T \nabla F(x_0) \geq 0 \ \forall x \in S$ and this implies $(x - x_0)^T \nabla F(x_0) \geq 0 \ \forall x \in coS$, where coS denotes the convex hull of S. In the vector case, when S is star-shaped at x_0, but not convex, condition (4.21) cannot be extended to the elements of coS and this implies that (4.21) cannot be expressed by means of multipliers as it is shown in the following example.

Example 4.12 Consider the linear vector function $F(x_1, x_2) = (-x_1 + x_2, x_1 - 2x_2), (x_1, x_2) \in S = \{t(1,0), t \geq 0\} \cup \{k(0,1), k \geq 0\}$.

Setting $x_0 = (0,0)$, it results $F'(x_0)(t,0)^T = (-t,t)^T \notin -intC, \ \forall t > 0$ and $F'(x_0)(0,k)^T = (k, -2k)^T \notin -intC, \ \forall k > 0$.

We have $(4,3) \in coS$ and $F'(x_0)(4,3)^T = (-1,-2)^T \in -intC$, so that it is not possible to separate S and $-intC$.

In order to express (4.21) by means of multipliers, we must require the convexity of S. The following theorem holds.

Theorem 4.26 *Consider problem P where S is a convex set. If x_0 is a weak efficient point, then*

$$\exists \alpha \in \Re^s_+ \setminus \{0\} \ \ \text{such that} \ \ \alpha^T F'(x_0)(x - x_0) \geq 0, \ \ \forall x \in S \tag{4.22}$$

Proof. Consider the set $W = \{F'(x_0)(x - x_0), x \in S\}$; since $F'(x_0)$ is a linear function and S is a convex set, W is convex too. The necessary optimality condition (4.21) implies $W \cap (-intC) = \emptyset$, so that (4.22) follows from Theorem 4.21. □

Under suitable assumptions of generalized convexity, (4.22) becomes a sufficient optimality condition.

Theorem 4.27 *i) If (4.22) holds and F is $(C^0, intC)$-pseudoconvex at $x_0 \in S$ then x_0 is an efficient point.*

ii) If (4.22) holds and F is $(intC, intC)$-pseudoconvex at $x_0 \in S$, then x_0 is a weak efficient point.

iii) If (4.22) holds and F is C-convex at $x_0 \in S$, then x_0 is a weak efficient point.

Proof. The proofs are similar to the one given in Theorem 4.24. □

Consider now the case where the feasible region S is expressed by inequality constraints, that is the problem

$$P^* : min \;\; F(x), x \in S = \{x \in X : G(x) \in -V\}$$

where $F, G : X \subseteq \Re^n \to \Re^m$ are differentiable functions and $V = \Re^m_+$.

For sake of simplicity, corresponding to a feasible point x_0, we will assume, without loss of generality, that x_0 is binding to all the constraints, that is $G(x_0) = 0$.

The following theorem states a first order necessary optimality condition which can be considered the natural extension in vector optimization of the classical Fritz John condition.

Theorem 4.28 *If x_0 is a weak efficient point for P^* then*

$$\exists \;\alpha \in \Re^s_+, \exists \;\beta \in \Re^m_+, (\alpha, \beta) \neq 0 : \alpha^T F'(x_0) + \beta^T G'(x_0) = 0 \qquad (4.23)$$

Proof. Let us note that $G'(x_0)d \in -intV$ implies that d is a feasible direction of S at $x_0 \in S$. Since x_0 is a weak efficient point, $F'(x_0)d \notin -intC \; \forall d : \;\; G'(x_0)d \in -intV$ or, equivalently, $(F'(x_0)d, G'(x_0)d) \notin (-intC) \times (-intV) \; \forall d \in \Re^n$. Setting $W_1 = \{F'(x_0)d, \;\; d \in \Re^n\}$, $W_2 = \{G'(x_0)d, \;\; d \in \Re^n\}$, we have $W_1 \times W_2 \cap (-int\Re^s_+) \times (-int\Re^m_+) = \emptyset$. The thesis follows from i) of Theorem 4.22. □

As in the scalar case, it can happen that $\alpha = 0$ in (4.23); if $\alpha \neq 0$, we will refer to (4.23) as the Kuhn-Tucker conditions in vector optimization.

Under suitable assumptions of generalized convexity, the Kuhn-Tucker conditions become sufficient optimality conditions.

Theorem 4.29 *i) If (4.23) holds with $\alpha \neq 0$, F is $(C^0, intC)$ -pseudoconvex at x_0 and G is C-quasiconvex at x_0, then x_0 is an efficient point for P^*.*

ii) If (4.23) holds with $\alpha \neq 0$, F is $(intC, intC)$-pseudoconvex at x_0 and G is C-quasiconvex at x_0, then x_0 is a weak efficient point for P^*.

iii) If (4.23) holds with $\alpha \in int\Re^s_+$, F is (C^0, C^0)-pseudoconvex at x_0 and G is C-quasiconvex at x_0, then x_0 is a local efficient point for P^*.

iv) If (4.23) holds with $\alpha \in int\Re^s_+$, F is C-convex at x_0 and G is C-quasiconvex at x_0, then x_0 is an efficient point for P^*.

Corollary 4.6 *i) If (4.23) holds with $\alpha \neq 0$, F is componentwise strictly pseudoconvex at x_0 and G is C-quasiconvex at x_0, then x_0 is an efficient point for P^*.*

ii) If (4.23) holds with $\alpha \neq 0$, F is componentwise pseudoconvex at x_0 and G is C-quasiconvex at x_0, then x_0 is a weak efficient point for P^*.

iii) If (4.23) holds with $\alpha \in int\Re^s_+$, F is componentwise pseudoconvex on S and G is C-quasiconvex at x_0, then x_0 is an efficient point for P^*.

10. Pseudolinearity in vector optimization

Conditions (4.16) and (4.17) suggest to define pseudolinearity with respect to the Paretian cone requiring that the logical implication in the definitions of (C^0, C^0)-pseudoconvexity and (C^0, C^0)-pseudoconcavity can be reversed.

More exactly, let $X \subset \Re^n$ be an open set, $F : X \to \Re^s$ a differentiable function and $S \subset X$ a convex set.

Definition 4.13 *The function F is (C^0, C^0)-pseudolinear (on S) if the following two statements hold:*

$$x_1, x_2 \in S,\ \ F(x_2) \in F(x_1) - C^0 \Leftrightarrow F'(x_1)(x_2 - x_1) \in -C^0 \qquad (4.24)$$

$$x_1, x_2 \in S,\ \ F(x_2) \in F(x_1) + C^0 \Leftrightarrow F'(x_1)(x_2 - x_1) \in C^0 \qquad (4.25)$$

Remark 4.7 If F is componentwise pseudolinear, then F is (C^0, C^0)-pseudolinear; the converse is not true, as it can be easily verified considering the class of functions $F : \Re \to \Re^2$, $F(x) = (F_1(x), F_2(x))$, where F_1, F_2 are such that $F_1'(x) > 0\ \forall x \in \Re, F_2'(x) \geq 0\ \forall x \in \Re$ and there exist $\bar{x}, x^* \in \Re$ such that $F_2'(\bar{x}) > 0, F_2'(x^*) = 0$.

(C^0, C^0)-pseudolinearity implies that a local efficient point is efficient too, even if such a property does not hold for (C^0, C^0)-pseudoconvex functions (see Example 4.10).

Theorem 4.30 *Let F be (C^0, C^0)-pseudolinear (on S). If $x_0 \in S$ is a local efficient point for P, then x_0 is efficient for P.*

Proof. Assume that there exists $\bar{x} \in S$ such that $F(\bar{x}) \in F(x_0) - C^0$. Then $F'(x_0)(\bar{x} - x_0) \in -C^0$. Consider the line-segment $]x_0, \bar{x}] = \{x = x_0 + t(\bar{x} - x_0), t \in]0,1]\}$. It results $F'(x_0)(x - x_0) = tF'(x_0)(\bar{x} - x_0) \in -C^0$, so that for the pseudolinearity of F, $F(x) \in F(x_0) - C^0 \;\; \forall x \in]x_0, \bar{x}]$ and this contradicts the local efficiency of x_0. □

Remark 4.8 If we are interested in maximizing a vector pseudolinear function, it is sufficient to consider efficiency with respect to the cone C instead of -C; so that, taking into account condition (4.25), the previous result and the ones that we are going to establish hold for a maximum vector problem too.

Another important property of vector pseudolinear functions is that the sufficient optimality condition (4.20) becomes also necessary.

Theorem 4.31 *Let F be (C^0, C^0)-pseudolinear (on S) and let $x_0 \in S$ be an interior point of S. Then x_0 is an efficient point for P if and only if*

$$\exists \alpha \in int\Re^s_+ \;\; such \;\; that \;\; \alpha^T F'(x_0) = 0 \tag{4.26}$$

Proof. The pseudolinearity of F and the efficiency of x_0 imply $F'(x_0)(x - x_0) \notin -C^0$, $\forall x \neq x_0$. (4.26) follows from ii) of Theorem 4.21. The converse statement follows from Theorem 4.25. □

Corollary 4.7 *Let F be a componentwise pseudolinear or an affine function (on S). Then $x_0 \in S$ is an interior efficient point if and only if*

$$\exists \alpha \in int\Re^s_+ \;\; such \;\; that \;\; \alpha^T F'(x_0) = 0.$$

When all the components F_i of F are pseudolinear on the whole space $\Re^n$, the existence of a stationary point implies that any point of $\Re^n$ is efficient as it is shown in the following theorem.

Theorem 4.32 *Let F be componentwise pseudolinear on $\Re^n$. If there exists an efficient point for F, then any point of $\Re^n$ is efficient.*

Proof. Without loss of generality, we can suppose that $\nabla F_i(x) \neq 0$, $\forall i = 1, ..., s$, $\forall x \in \Re^n$; indeed if there exists z such that $\nabla F_j(z) = 0$, then F_j is a constant function and the efficiency of a point x_0 with respect to F is equivalent to the efficiency of x_0 with respect to the function $(F_1, ..., F_{j-1}, F_{j+1}, ..., F_s)$.

Let x_0 be an efficient point for F; then from Corollary 4.7, there exists $\alpha^T = (\alpha_1, ..., \alpha_s) \in int\Re^s_+$ such that $\sum_{i=1}^{s} \alpha_i \nabla F_i(x_0) = 0$. From Theorem 4.14, we have $\nabla F_i(x_0) = \beta_i \nabla F_i(x)$ with $\beta_i = \frac{\|\nabla F_i(x_0)\|}{\|\nabla F_i(x)\|} > 0 \ \forall i$, so that $\sum_{i=1}^{s} \alpha_i \beta_i \nabla F_i(x) = 0 \ \forall x \in \Re^n$ and thus, taking into account Corollary 4.7, $x \in \Re^n$ is an efficient point. □

Let us note that if F is componentwise pseudolinear on a convex subset of $\Re^n$, the property stated in Theorem 4.32 may not hold even if there exist stationary points, as is shown in the following example.

Example 4.13 Consider the function

$$F(x_1, x_2) = (\frac{x_1 + x_2 - 1}{x_2 + 1}, \frac{-x_1 + x_2 - 3}{-x_2 + 3}),$$

$(x_1, x_2) \in S = \{(x_1, x_2) : 0 \leq x_1 \leq 2, 0 \leq x_2 \leq 1\}$.

F is (C^0, C^0)-pseudolinear on S since its components are pseudolinear on S. Consider the interior point $x_0 = (\frac{5}{4}, \frac{1}{2})$. It results $F'(x_0) = \begin{bmatrix} \frac{2}{3} & \frac{1}{3} \\ -\frac{2}{5} & -\frac{1}{5} \end{bmatrix}$, so that, setting $\alpha^T = (3, 5)$, we have $\alpha^T F'(x_0) = 0$, that is x_0 is a stationary point for F and, consequently, x_0 is an efficient point either with respect to C or with respect to -C, as it can be verified applying the definitions. On the other hand $x_0 = (0, 0) \in S$ is not an efficient point, so that Theorem 4.32 does not hold if we substitute $\Re^n$ with a subset S. Let us note that the line-segment $[A, B]$ with $A = (\frac{3}{2}, 0)$, $B = (1, 1)$, is the set of all efficient points of F.

As we have pointed out in Section 7, all the given definitions of vector pseudoconvex functions collapse, in the scalar case, to the ordinary definition of pseudoconvexity given by Mangasarian. This suggests the possibility to extend pseudolinearity in multiobjective programming as follows:

- F is $(intC, intC)$-pseudolinear if

$$x_1, x_2 \in S, F(x_2) \in F(x_1) - intC \Leftrightarrow F'(x_1)(x_2 - x_1) \in -intC$$

$$x_1, x_2 \in S, F(x_2) \in F(x_1) + intC \Leftrightarrow F'(x_1)(x_2 - x_1) \in intC$$

- F is $(C^0, intC)$-pseudolinear if

$$x_1, x_2 \in S, F(x_2) \in F(x_1) - C^0 \Leftrightarrow F'(x_1)(x_2 - x_1) \in -intC$$

$$x_1, x_2 \in S, F(x_2) \in F(x_1) + C^0 \Leftrightarrow F'(x_1)(x_2 - x_1) \in intC$$

In order to outline that the class of (C^0, C^0)-pseudolinear functions is more appropriate to deal with efficiency, let us note that a componentwise pseudolinear function is not in general $(C^0, intC)$-pseudolinear and that for a $(intC, intC)$-pseudolinear function an interior efficient point cannot be characterized as a stationary point with positive multipliers, as is shown in the following example which points out also that the classes of (C^0, C^0) and $(intC, intC)$-pseudolinear functions are not comparable as it happens for vector pseudoconvexity.

Example 4.14 Consider the function

$$F(x_1, x_2) = (x_1, -x_1, x_1 + x_2^3, -x_2), (x_1, x_2) \in \Re^2,$$

$C = \Re^4_+$. It can be verified that F is $(intC, intC)$-pseudolinear, but F is not (C^0, C^0)-pseudolinear even if F is (C^0, C^0)-pseudoconvex (pseudoconcave). Obviously $x_0 = (0, 0)$ is an efficient point for F, but $\alpha^T F'(x_0) = 0$, with $\alpha^T = (\alpha_1, \alpha_2, \alpha_3, \alpha_4)$, holds if and only if $\alpha_4 = 0, \alpha_1 - \alpha_2 + \alpha_3 = 0$, so that it is not possible to have the positivity of all multipliers.

In order to point out that there is not inclusion relationships between the two classes of functions, consider $F(x) = (x^3 + x, x^3), x \in \Re, C = \Re^2_+$. Such a function is (C^0, C^0)-pseudolinear but it is not $(intC, intC)$-pseudolinear.

Consider now the case where the feasible region is expressed by means of constraint functions, that is problem P^*. The following theorem states a necessary and sufficient optimality condition of the Kuhn-Tucker type.

Theorem 4.33 *Consider problem P^*, where F is (C^0, C^0)-pseudolinear, G is (V^0, V^0)-pseudolinear and such that*

$$G'(y)(x - y) = 0 \Leftrightarrow G(x) = G(y) \tag{4.27}$$

Then a feasible point x_0 is efficient point for P^* if and only if

$$\exists\ \alpha \in int\Re^s_+, \exists\ \beta \in \Re^m_+ : \alpha^T F'(x_0) + \beta^T G'(x_0) = 0 \tag{4.28}$$

Proof. **Necessity**. Consider the linear subspace

$$W = \{(F'(x_0)(x - x_0), G'(x_0)(x - x_0)), x \in \Re^n\}.$$

Now we will prove that $W \cap (-\Re^s_+ \setminus \{0\}) \times (-\Re^m_+)) = \emptyset$.
Assume that there exists $\bar{x} \in \Re^n$ such that

$$F'(x_0)(\bar{x} - x_0) \in -C^0 \tag{4.29}$$

$$G'(x_0)(\bar{x} - x_0) \in -V \tag{4.30}$$

Let us note that (4.29) and (4.30) hold for any point of the intersection between the convex set X and the line-segment $[x_0, \bar{x}]$, so that we can assume without loss of generality that $\bar{x} \in X$.

The pseudolinearity of G and property (4.27) imply the feasibility of $\bar{x}$. On the other hand, (4.29) contradicts the efficiency of x_0, so that $W \cap ((-C^0) \times (-V)) = \emptyset$. From iii) of Theorem 4.22, we have (4.28).

Sufficiency. It follows from iii) of Theorem 4.29 and from Theorem 4.30. □

In the previous theorem, we have assumed that the constraint vector function G belongs to the subclass of (V^0, V^0)-pseudolinear functions verifying the property (4.27); this subclass contains the componentwise pseudolinear functions and the inclusion is proper since the function $F(x) = (x, x^3)$, $x \in \Re$ is $(\Re^2_+ \setminus 0, \Re^2_+ \setminus 0)$-pseudolinear and verifies 4.27 but it is not componentwise pseudolinear.

As a direct consequence of the previous theorem, we obtain the following result given by Chew-Choo [26].

Corollary 4.8 *Consider problem P^*, where the objective function and the constraints are componentwise pseudolinear. A point x_0 is an efficient solution of problem P^* if and only if*

$$\exists\ \alpha \in int\Re^s_+, \exists\ \beta \in \Re^m_+ : \alpha^T F'(x_0) + \beta^T G'(x_0) = 0$$

In [26] it is also shown the strict relation among the efficiency of the multiobjective pseudolinear problem P^*, the efficiency of the linearized problem and the optimality of the scalarized problem associated to P^*, as it is stated in the following theorem.

Theorem 4.34 *Let x_0 be a feasible point of the problem P^*, where the objective function and the constraints are componentwise pseudolinear. Then the following statements are equivalent:*

i) x_0 is an efficient solution of P^*;

ii) x_0 is an efficient solution of the problem:

$$min\ \ F'(x_0)x, \ \ x \in S;$$

iii) there exist positive multipliers $\lambda_1, ..., \lambda_s$ such that x_0 minimizes the linear function $\sum_{i=1}^{s} \lambda_i \nabla F_i(x_0)x$, $x \in S$.

11. Invexity in vector optimization

The notion of scalar invexity can be extended to the vector case by means of the following definition.

Definition 4.14 *Let $F : X \to \Re^s$ be a differentiable vector-valued function defined on an open set $X \subset \Re^n$. F is C-invex on $S \subset X$ if there exists a vector function $\eta : S \times S \to \Re^n$ such that*

$$F(x_1) - F(x_2) - F'(x_2)\eta(x_1, x_2) \in C, \ \forall x_1, x_2 \in S \tag{4.31}$$

It is easy to verify that the class of C-invex functions, with respect to the same function η, is closed under addition and multiplication by a positive scalar and furthermore that C-invexity is equivalent to the invexity (with respect to the same η) of each component of F.

In [29], conditions are obtained, necessary or sufficient, for F to be K-invex with respect to some η, where K is a convex cone.

Obviously, C-convex functions are C-invex since (4.31) holds with $\eta(x_1, x_2) = (x_1 - x_2)$.

The properties of C-convex functions related to stationary points and to the sufficiency of the Kuhn-Tucker conditions can be extended to the class of C-invex functions as expressed in the following theorems.

Theorem 4.35 *Let F be C-invex on S.*

i) If x_0 is a stationary point for F, that is

$$\exists \alpha \in \Re^s_+ \setminus \{0\} \ \textit{such that} \ \alpha^T F'(x_0) = 0 \tag{4.32}$$

then x_0 is a weak efficient point.

ii) If (4.32) holds with $\alpha \in int\Re^s_+$, then x_0 is an efficient point.

Proof. i) If x_0 is not a weak efficient point, then there exists $x \in S$ such that $F(x) - F(x_0) \in -intC$; from the invexity assumption, we have $F'(x_0)\eta(x, x_0) \in -intC$ and thus $\alpha^T F'(x_0)\eta(x, x_0) < 0$ which contradicts (4.32).

ii) The nonefficiency of x_0 and the invexity of F imply the existence of $x \in S$ with $F'(x_0)\eta(x, x_0) \in -C^0$. Taking into account that $\alpha \in int\Re^s_+$, we have once again $\alpha^T F'(x_0)\eta(x, x_0) < 0$ and this is absurd. □

Theorem 4.36 *Consider problem P^* where F is C-invex and G is V-invex with respect to the same η.*

i) If

$$\exists \ \alpha \in \Re^s_+ \setminus \{0\}, \exists \ \beta \in \Re^m_+ : \alpha^T F'(x_0) + \beta^T G'(x_0) = 0 \tag{4.33}$$

then x_0 is a weak efficient point.

ii) If (4.33) holds with $\alpha \in int\Re^s_+$, then x_0 is an efficient point.

Proof. i) If x_0 is not a weak efficient point, then there exists $x \in S$ such that $F(x) - F(x_0) \in -intC$, $G(x) - G(x_0) \in -V$ (we have assumed, without loss of generality, $G(x_0) = 0$). The C-invexity of F and the V-invexity of G imply $F'(x_0)\eta(x, x_0) \in -intC$, $G'(x_0)\eta(x, x_0) \in -V$ so that $\alpha^T F'(x_0)\eta(x, x_0) + \beta^T G'(x_0)\eta(x, x_0) < 0$ and this contradicts (4.33).

ii) The proof is similar to i). □

A scalar function is invex if and only if every stationary point is a global minimum point. This property is lost for a vector invex function as it is shown in the following example.

Example 4.15 Consider the function $F : \Re \to \Re^2$, $F(x) = (x^3, -x^3)$, $C = \Re^2_+$. The function F is not C-invex since, choosing $x_1 = 1, x_2 = 0$, we have $F(x_1) - F(x_2) = (1, -1)$, $F'(x_2) = (0, 0)$, so that $F(x_1) - F(x_2) \notin F'(x_2)\eta(x_1, x_2) - C$ whatever the function η be.
On the other hand any $x \in \Re$ is a stationary point and also an efficient point.

In order to characterize the class of vector functions for which a stationary point is a weak efficient point, in [73] a class of vector pseudoinvex function is introduced.

Definition 4.15 *Let $F : S \subset \Re^n \to \Re^s$ be a differentiable function on the open set S. Then F is C-pseudoinvex on S if there exists a vector function $\eta : S \times S \to \Re^n$ such that*

$$x_1, x_2 \in S \;\; F(x_1) - F(x_2) \in -intC \Rightarrow F'(x_2)\eta(x_1, x_2) \in -intC \quad (4.34)$$

The following theorem points out that the class of C-invex functions is contained in the class of C-pseudoinvex functions; the inclusion is strict since the function in the previous example is C-pseudoinvex but not C-invex.

Theorem 4.37 *Let $F : S \subset \Re^n \to \Re^s$ be a differentiable function defined on the open set S. If F is C-invex with respect to some η, then F is C-pseudoinvex with respect to the same η.*

Proof. Let $x_1, x_2 \in S$ such that $F(x_1) - F(x_2) \in -intC$. Since F is C-invex, we have $F'(x_2)\eta(x_1, x_2) \in F(x_1) - F(x_2) - C$. Taking into account that $intC + C = intC$, it results $F'(x_2)\eta(x_1, x_2) \in -intC$ and thus F is C-pseudoinvex. □

Unlike the vector case, for scalar functions the concept of invexity and pseudoinvexity coincide.

A differentiable scalar function f is pseudoinvex on the open set $S \subset \Re^n$ if there exists $\eta : S \times S \to \Re^n$ such that

$$x_1, x_2 \in S, \;\; f(x_1) - f(x_2) < 0 \Rightarrow \eta(x_1, x_2)^T \nabla f(x_2) < 0.$$

Obviously invexity implies pseudoinvexity. On the other hand a stationary point $x_2 \in S$ is also a global minimum for a pseudoinvex function; indeed if there exists $x_1 \in S$ such that $f(x_1) < f(x_2)$, then we have $\eta(x_1, x_2)^T \nabla f(x_2) < 0$ and this contradicts the assumption $\nabla f(x_2) = 0$. From Theorem 4.6 f is invex, so that scalar invexity is equivalent to scalar pseudoinvexity.

The following theorem gives a characterization of C-pseudoinvex functions which reduces to Theorem 4.6 in the scalar case.

Theorem 4.38 *A vector function F is C-pseudoinvex on S if and only if every stationary point of F is a weak efficient point.*

Proof. **Necessity.** The proof is similar to i) of Theorem 4.35.
Sufficiency. Let $x_1, x_2 \in S$ be such that $F(x_1) - F(x_2) \in -intC$. Then x_2 is not a weak efficient point and so it is not a stationary point. Set $W = \{F'(x_2)d, d \in \Re^n\}$; from i) of Theorem 4.21 there exists $d \in \Re^n$ such that $F'(x_2)d \in -intC$. The thesis follows setting $\eta(x_1, x_2) = d$. □

Let us note that Theorem 4.36 can be extended to the case where F is C-pseudoinvex; further developments are given in [73].

Invexity has been studied also in the nondifferentiable case in order to obtain saddle point and duality results in multiobjective programming. We refer the interested reader for instance to [36, 49, 76, 80, 82, 93].

References

[1] Arrow K. J. and Enthoven A. C., *Quasi-concave programming*, Econometrica, 29, 1961, 779-800.

[2] Arrow K. J., Hurwicz L. and Uzawa H., *Constraint qualifications in maximization problems*, Nav.Res.Logist., 8, 1961, 175-191.

[3] Avriel M., Diewert W. E., Schaible S. and Zang I., *Generalized concavity*, Plenum Press, New York, 1988.

[4] Bazaraa M.S and Shetty C.M., *Foundations of optimization*, Lect. Notes Econ. Math. Syst., 122, Springer-Verlag, 1976.

[5] Bector C. R. and Singh C., *B-vex functions*, J. Optimization Theory Appl., 71, 1991, 237-253.

[6] Bector C. R., Suneja S. K. and Lalitha C. S., *Generalized B-vex functions and generalized B-vex programming*, J. Optimization Theory Appl., 76, 1993, 561-576.

[7] Ben-Israel A., Ben Tal A. and Charnes A., *Necessary and sufficient conditions for a Pareto optimum in convex programming*, Econometrica, 45, 1977, 811-820.

[8] Ben-Israel A. and Mond B., *What is invexity?*, J. Aust. Math. Soc.,Ser.B, 28, 1986, 1-19.

[9] Cambini A. and Martein L., *A modified version of Martos's algorithm for the linear fractional problem*, Methods of Operations Research, 53, 1986, 33-44.

[10] Cambini A. and Martein L., *Some optimality conditions in vector optimization*, J.Inf. Optimization Sci., 10, 1989, 141-151.

[11] Cambini A. and Martein L., *An approach to optimality conditions in vector and scalar optimization*, in Mathematical modelling in Economics (Diewert W. E. et. al. Eds.), Springer Verlag, Berlin, 1993, 345-358.

[12] Cambini A. and Martein L., *On the existence of efficient points*, Optimization, 28, 1994, 283-290.

[13] Cambini A. and Martein L., *Generalized concavity and optimality conditions in vector and scalar optimization*, "Generalized convexity" (Komlosi S. et al.Eds.), Lect. Notes Econom. Math. Syst., 405, Springer-Verlag, 1994, 337-357.

[14] Cambini A., Martein L. and Cambini R., *Some optimality conditions in multiobjective programming*, Multicriteria Analysis (Climaco J., Ed.), Springer Verlag, 1997, 168-178.

[15] Cambini, A. and Martein L., *Generalized concavity in multiobjective programming*, Generalized convexity, Generalized monotonicity (Crouzeix J. P. et al. Eds.), Nonconvex Optim. Appl., 27, Kluwer Academic Publishers, Dordrecht-Boston-London, 1998, 453-467.

[16] Cambini A. and Martein L., *On the pseudoconvexity of a quadratic fractional function*, Optimization 51, 2002, 677-687.

[17] Cambini A. and Martein L., *Pseudolinearity in scalar and vector optimization*, Rendiconti del Circolo Matematico di Palermo, Serie II, Suppl. 70, 2002, 135-153.

[18] Cambini R., *Some new classes of generalized concave vector valued functions*, Optimization, 36, 1996, 11-24.

[19] Cambini R., *Composition theorems for generalized concave vector valued functions*, J. Inf. Optimization Sci., 19, 133-150.

[20] Cambini R., *Generalized concavity and optimality conditions in vector optimization*, Operations Research and its Applications (Du D. Z. et al. Eds.), World Publishing Corporation, 1996, 172-180.

[21] Cambini R., *Generalized concavity for bicriteria functions*, Generalized convexity, Generalized monotonicity (Crouzeix J. P. et al. Eds.), Nonconvex Optim. Appl., 27, Kluwer Academic Publishers, Dordrecht-Boston-London, 1998, 439-451.

[22] Cambini R. and Martein L., *First and second order characterizations of a class of pseudoconcave vector functions*, Generalized convexity, Generalized monotonicity (Hadjisavvas N. et al. Eds.), Lect. Notes Econom. Math. Syst., Springer, 502, 2001.

[23] Castagnoli E. and Mazzoleni P. , *Towards a unified type of concavity*, J. Inf. Optimization Sci., 10, 1989, 225-240.

[24] Chankong V. and Haimes Y.Y., *Multiobjective decision making: theory and methodology*, North-Holland, 1983.

[25] Charnes A. and Cooper W. W., *Programming with linear fractional functionals*, Nav. Res. Logist., 9, 1962, 181-186.

[26] Chew K.L. and Choo E. U., *Pseudolinearity and efficiency* , Math. Prog., 28, 1984, 226-239.

[27] Craven B. D., *Duality for generalized convex fractional programs*, Generalized Concavity in Optimization and Economics (Schaible S. and Ziemba W. T. Eds), Academic Press, 1981, 473-489.

[28] Craven B. D., *Invex functions and constrained local minima*, Bull. Austral. Math. Soc., 24, 1981, 357-366.

[29] Craven B. D. and Glover B.M., *Invex functions and duality*, J. Aust. Math. Soc., Ser. A, 39, 1985, 1-20.

[30] Craven B. D., *Fractional Programming*, Sigma Ser. Appl. Math. 4, Heldermann Verlag, 1988.

[31] Egudo. R.R. and Mond B, *Duality with generalized convexity*, J. Aust. Math. Soc., Ser. B, 28, 1986, 10-21.

[32] Egudo R.R. and Hanson M.A., *Multiobjective duality with invexity*, J. Math. Anal. Appl., 126, 1987, 469-477.

[33] Ellero A. and Moretti Tomasin E., *A computational comparison between algorithms for linear fractional programming*, J. Inf. Optimization Sci., 13, 1992, 343-362.

[34] Giorgi G. and Guerraggio A., *Various types of nonsmooth invexity*, J. Inf. Optimization Sci., 17, 1994, 137-150.

[35] Giorgi G. and Guerraggio A., *Constraints qualifications in the invex case*, J. Inf. Optimization Sci., 19, 1998, 373-384.

[36] Giorgi G. and Guerraggio A., *The notion of invexity in vector optimization: smooth and nonsmooth case*, Generalized convexity, Generalized monotonicity (Crouzeix J. P. et al. Eds.), Nonconvex Optim.

Appl., 27, Kluwer Academic Publishers, Dordrecht-Boston-London, 1998, 389-405.

[37] Giorgi G. and Molho E., *Generalized invexity: relationships with generalized convexity and application to the optimality and duality conditions*, Generalized Concavity for economic applications (Mazzoleni P. Ed.), Tecnoprint, Bologna, 1992, 53-70.

[38] Hanson M.A., *On sufficiency of the Kuhn-Tucker conditions*, J. Math. Anal. Appl., 80, 1981, 545-550.

[39] Hanson M.A. and Mond B., *Further generalizations of convexity in mathematical programming*, J. Inf. Optimization Sci., 3, 1982, 25-32.

[40] Hanson M.A. and Mond B., *Convex transformable programming problems and invexity*, J. Inf. Optimization Sci., 8, 1987, 201-207.

[41] Hanson M.A. and Rueda N. G., *A sufficient condition for invexity*, J. Math. Anal. Appl., 138, 1989, 193-198.

[42] Jahn J., *Mathematical vector optimization in partially ordered linear space*, Verlag Peter Lang, 1986.

[43] Jahn J. and Sach E., *Generalized quasiconvex mappings and vector optimization*, SIAM J. Control Optimization, 24, 1986, 306-322.

[44] Jahn J., *Introduction to the theory of nonlinear optimization*, Springer Verlag, Berlin, 1996.

[45] Jeyakumar V., *Strong and weak invexity in mathematical programming*, Collection: Methods of Operations Research, 55, 1980, 109-125.

[46] Jeyakumar V. and Yang X. Q., *On characterizing the solution sets of pseudolinear programs*, J. Optimization Theory Appl., 87, 1995, 747-755.

[47] Kaul R.N. and Kaur S., *Generalizations of convex and related functions*, Eur. J. Oper. Res., 9, 1982, 369-377.

[48] Kaul R.N. and Kaur S., *Optimality criteria in nonlinear programming involving nonconvex functions*, J. Math. Anal. Appl., 105, 1985, 104-112.

[49] Kaul R.N., Suneja S. K. and Lalitha C. S., *Generalized nonsmooth invexity*, J. Inf. Optimization Sci., 15, 1994, 1-17.

[50] Khanh P. Q., *On necessary and sufficient conditions in vector optimization* , J. Optimization Theory Appl., 63, 1989, 391-413.

[51] Khanh P. Q., *Invex-convexlike functions and duality*, J. Optimization Theory Appl., 87, 1995, 141-165.

[52] Khanh P. Q., *Sufficient optimality conditions and duality in vector optimization with invex-convexlike functions*, J. Optimization Theory Appl., 87, 1995, 359-378.

[53] Kim D.S. and Lee G.M., *Optimality conditions and duality theorems for multiobjective invex programs*, J. Inf. Optimization Sci., 12, 1991, 235-242.

[54] Komlosi S., *First and second-order characterizations of pseudolinear functions*, Eur. J. Oper. Res., 67, 1993, 278-286.

[55] Kortanek K. O. and Evans J.P., *Pseudoconcave programming and Lagrange regularity*, Oper. Res., 15, 1967, 882-892.

[56] Kuhn H. W. and Tucker A. W., *Nonlinear programming*, Proceedings of the Second Berkeley Symposium in Mathematical Statistics and Probability (Neyman J. eds.), University of California Press, Berkeley, 1951, 481-492.

[57] Lee G.M., *Optimality conditions in multiobjective optimization problems*, J. Inf. Optimization Sci., 13, 1992, 107-111.

[58] Lee G.M., *Nonsmooth invexity in multiple programming*, J. Inf. Optimization Sci., 15, 1994, 127-136.

[59] Li S. X., *Quasiconcavity and nondominated solutions in multiobjective programming*, J. Optimization Theory Appl., 88, 1996, 197-208.

[60] Li X. F., Dong J. L. and Liu Q. H., *Lipschitz B-vex functions and nonsmooth programming*, J. Optimization Theory Appl., 93, 1997, 557-574.

[61] Luc D. T., *Connectedness of the efficient point set in quasiconcave vector maximization*, J. Math. Anal. Appl., 122, 1987, 346-354.

[62] Luc D.T., *Theory of vector optimization*, Springer-Verlag, Berlin, 1989.

[63] Luc D. T., *On three concepts of quasiconvexity in vector optimization*, Acta Mat. Vietnam., 15, 1990, 3-9.

[64] Luc D. T. and Schaible S., *Efficiency and generalized concavity*, J. Optimization Theory Appl., 94, 1997, 147-153.

[65] Mangasarian O.L., *Nonlinear programming*, McGraw-Hill, New York, 1969.

[66] Martein L., *Stationary points and necessary conditions in vector extremum problems*, J. Inf. Optimization Sci., 10, 1989, 105-128.

[67] Martin D.H., *The essence of invexity*, J. Optimization Theory Appl., 47, 1985, 65-76.

[68] Martos B., *Hyperbolic programming*, Nav. Res. Logist., 11, 1960, 135-155.

[69] Martos B., *Nonlinear programming Theory and Methods*, North-Holland, Amsterdam, 1975.

[70] Miettinen K., *Nonlinear multiobjective optimization*, Kluwer Academic Publishers, Dordrecht-Boston-London, 1999.

[71] Mititelu S., *Generalized invexities and global minimum properties*, Balkan Journal of Geometry and its Applications, 2, 1997, 61-72.

[72] Mond B. and Weir T., *Generalized concavity and Duality*, Generalized Concavity in Optimization and Economics (Schaible S. and Ziemba W. T. Eds), Academic Press, 1981, 263-279.

[73] Osuna-Gomez R., Rufian-Lizana A. and Ruiz-Canales P., *Invex functions and generalized convexity in multiobjective programming*, J. Optimization Theory Appl., 98, 1998, 651-661.

[74] Pini R., *Invexity and generalized convexity*, Optimization, 22, 1991, 513-525.

[75] Preda V., *On duality with generalized convexity*, Bull. U.M.I., 1991, 291-305.

[76] Preda V., Stancu-Minasian I. M. and Batatorescu A., *Optimality and duality in nonlinear programming involving semilocally preinvex and related functions*, J. Inf. Optimization Sci., 17, 1996, 585-596.

[77] Preda V. and Stancu-Minasian I. M., *Duality in multiple objective programming involving semilocally preinvex and related functions*, Glasnik Matematicki, 32, 1997, 153-165.

[78] Rapcsak T., *On pseudolinear functions*, Eur. J. Oper. Res., 50, 1991, 353-360.

[79] Reiland T. W., *Generalized invexity for nonsmooth vector-valued mappings*, Numer. Funct. Anal. and Optimization, 10, 1989, 1191-1202.

[80] Reiland T. W., *Nonsmooth invexity*, Bull. Austral. Math. Soc., 42, 1990, 437-446.

[81] Rockafellar R. T., *Convex Analysis*, Princeton University Press, 1969.

[82] Rueda N. G. and Hanson M.A., *Optimalty criteria in mathematical programming involving generalized invexity*, J. Math. Anal. Appl., 130, 1988, 375-385.

[83] Ruiz-Canales P. and Rufian-Lizana A., *A characterization of weakly efficient points*, Math. Progr. 68, 1995, 205-212.

[84] Sawaragi,Y., Nakayama,H. and Tanino,T., *Theory of multiobjective optimization*, Academic Press, New York, 1985.

[85] Singh C., *Optimality conditions in multiobjective differentiable programming*, J. Optimization Theory Appl., 53, 1987, 115-123.

[86] Schaible S. and Ziemba W.T. (Eds.), *Generalized Concavity in Optimization and Economics*, Academic Press, New York, 1981.

[87] Schaible S., *Fractional programming*, Handbook of Global Optimization (Horst R. and Pardalos Eds.), Kluwer Academic Publishers, Dordrecht-Boston-London, 1995, 495-608.

[88] Stadler W., *A survey of multicriteria optimization or the vector maximum problem, Part I: 1776-1960*, J. Optimization Theory Appl., 29, 1979, 1-52.

[89] Stancu-Minasian I. M., *Fractional Programming. Theory, Methods and Applications*, Kluwer Academic Publishers, Dordrecht-Boston-London, 1997.

[90] Suneja S. K., Singh C. and Bector C. R., *Generalization of preinvex and B-vex functions*, J. Optimization Theory Appl., 76, 1993, 577-587.

[91] Tanaka T., *Note on generalized convex functions*, J. Optimization Theory Appl., 66, 1990, 345-349.

[92] Tanaka T., *Generalized quasiconvexities, cone-saddle points, and minimax theorem for vector-valued functions*, J. Optimization Theory Appl., 81, 1994, 355-377.

[93] Weir T. and Mond B., *Pre-invex functions in multiple objective optimization*, J. Math. Anal. Appl., 136, 1988, 29-38.

[94] Yu P. L., *Cone convexity, cone extreme points, and nondominated solutions in decision problems with multiobjectives*, J. Optimization Theory Appl., 14, 1974, 319-377.

[95] Yu P.L., *Multiple-criteria decision making*, Plenum Press, New York, 1985.

Chapter 5

GENERALIZED CONVEXITY IN VECTOR OPTIMIZATION

Dinh The Luc
Department of Mathematics
University of Avignon
84000 Avignon, France
and
Hanoi Institute of Mathematics,
18 Hoang Quoc Viet Road
Hanoi, Vietnam
dtluc@univ-avignon.fr

Abstract In this chapter we introduce the notion of convexity and generalized convexity including invexity for vector valued functions. Some characterizations of these functions are provided. Then we study vector problems involving generalized convex functions. The major aspects of this study concern the existence of efficient solutions, optimality conditions using contingent derivatives and approximate Jacobians, scalarization for convex and quasiconvex problems, and topological properties of efficient solution sets of generalized convex problems.

Keywords: Generalized convex vector functions, efficient solutions, vector optimization problems.

1. Introduction and Preliminaries

The last decades of the passed century witnessed a tremendous development of vector optimization and of generalized convexity, because these two branches of mathematics, on one hand, provide a fertile ground where the usefulness of important theoretical results of mathematics can be confirmed and certain improvements can be suggested, and on the other hand, these branches themselves have many applications in applied sciences such as mechanics, economics, industry, finance etc. Vector optimization problems with generalized convex data are of particular interest because they offer exciting premises for theoretical research and challenging computational methods that are beyond the reach of scalar optimization. In this chapter we shall concentrate on the study of vector optimization problems involving generalized convex data.

The chapter is divided into 6 sections. In this first section we introduce the notion of *partial order* in a topological vector space by means of convex cones. Correct cones and cone-complete sets are of particular interest in the study of efficient solutions and are given in the two next subsections. Approximate Jacobians and contingent derivatives are the two models of generalized derivatives we have chosen to use in our analysis, and they are briefly presented in the subsections 1.3 and 1.4. In section 2 we investigate properties of generalized convex vector functions, their relationship with scalar counterparts. Invex vector functions are a part of this generalization. Section 3 introduces fundamental concepts of vector optimization: efficient points, efficient solutions, and presents conditions for the existence of efficient points. In section 4, we discuss optimality conditions of a vector optimization problem by using the classical derivative, the approximate Jacobians and the contingent derivatives. Whenever the problem is convex or invex, sufficient conditions are obtained. Section 5 treats the scalarization method for solving vector problems with convex structure. The final section gives most important topological properties of efficient point sets for problems with convex and quasiconvex data.

Notations in this chapter: E, X, Y, Z, W are real Hausdorff topological vector spaces if otherwise not mentioned; $L(X, Y)$ is the space of continuous linear maps from X to Y, $L(X, R)$ is usually denoted by X' which is the topological dual of X. Given $A \subseteq E$, the notations $A^c, riA, intA, clA, coA, \overline{co}A$ stand for the complement, the relative interior, the interior, the closure, the convex hull, the closed convex hull of A respectively. When E is a normed space, $B(a, \varepsilon)$ denotes the closed ball centered at $a \in E$ with radius $\varepsilon > 0$.

1.1 Partial orders and convex cones

Let E be a nonempty set and let $B \subseteq E \times E$ be a binary relation on E. We say that B is a partial order on E if it is reflexive, i.e. $(x, x) \in B$ for every $x \in E$, and transitive, i.e.$(x, y), (y, z) \in B$ imply $(x, z) \in B$. We shall write $x \geq y$ instead of $(x, y) \in B$.
Throughout this chapter we shall deal with a special case where E is a topological vector space and is equipped with a partial order B which is

linear in the sense that $(x, y) \in B$ implies $(x + z, y + z), (tx, ty) \in B$ for all $z \in E$ and $t > 0$.
Recall that a nonempty set $C \subseteq E$ is said to be a cone if $tx \in C$ for each $x \in C$ and $t \geq 0$. The linear part of a cone C is denoted by

$$\ell(C) := C \cap (-C),$$

and when $\ell(C) = \{0\}$, the cone is said to be pointed. A set $C_0 \subseteq E$ is called a base of C if its closure does not meet the origin and $C = \text{cone}(C_0)$, where $\text{cone}(C_0)$ is the cone generated by the set C_0, i.e.

$$\text{cone}(C_0) = \{tx : x \in C_0, t \in R, t \geq 0\}.$$

The *positive polar cone* of C is given by

$$C' := \{\xi \in E' : \langle \xi, x \rangle \geq 0, \text{ for all } \ x \in C\}$$

and the *strictly positive polar cone* of C is given by

$$C^+ := \{\xi \in E' : \langle \xi, x \rangle > 0, \text{ for all } \ x \in C, x \neq 0\}.$$

The next result, which is immediate from the definition, characterizes linear partial orders by means of convex cones.

Proposition 5.1 *If $(\geq)$ is a linear partial order in E, then the set*

$$C := \{x \in E : x \geq 0\}$$

is a convex cone in E. Conversely, if C is a convex cone in E, then the relation $(\geq_C)$, defined by $x \geq_C y$ if and only if $x - y \in C$, is a linear partial order in E.

Sometimes we simply write $x \geq y$ instead of $x \geq_C y$, and $x >_C y$ if $x - y \in C \setminus \ell(C)$. When *int*$C$ is nonempty, we also write $x \gg_C y$ instead of $x - y \in intC$. Below are some examples of linear partial orders.

Example 5.1 1. The usual order: Let $E = R^n$ and $C = R^n_+$ (the nonnegative orthant). Then $(\geq_C)$ is the usual componentwise order, i.e. for $x = (x_1, \cdots, x_n), y = (y_1, \cdots, y_n)$ one has $x \geq_C y$ if and only if $x_i \geq y_i$, $i = 1, \cdots, n$.

2. The lexicographic order: Let C be the cone in R^n consisting of all vectors, whose first nonzero component is positive. Then $(\geq_C)$ is a linear partial order. Actually this order is complete in the sense that ány two elements of R^n are comparable (either $x \geq_C y$ or $y \geq_C x$).

3. The ubiquitous order: Let ℓ_0 denote the space of sequences whose terms are all zero except for a finite number. This is a normed space if

we equip it with the max-norm. Let C be a cone consisting of sequences whose last nonzero component is positive. Then the order generated by C is a linear partial order in ℓ_0. The cone C is called ubiquitous because of the following property : for each $x \in \ell_0$, there exists $y \in C$ such that $[y, x) \subseteq C$. Here $[y, x) = \{ty + (1-t)x : 0 < t \leq 1\}$.

Definition 5.1 *We say that the cone C is correct if $clC + C \setminus \ell(C) \subseteq C$ or equivalently $clC + C \setminus \ell(C) \subseteq C \setminus \ell(C)$.*

Note that the cone R^n_+ is pointed and correct, while the lexicographic cone and the ubiquitous cone are pointed and not correct. This kind of cones is very useful in establishing the existence of optimal solutions for vector problems. Let us mention some cases where C is correct.

Proposition 5.2 *Each of the following conditions is sufficient for a convex cone C to be correct*

(i) C is closed;

(ii) $C \setminus \ell(C)$ is open;

(iii) C consists of the origin and an intersection of half-spaces that are either open or closed.

1.2 Order complete sets

Let C be a convex cone in E. Let $\{x_\alpha\}_{\alpha \in \Gamma}$ be a net in E. It is said to be decreasing if $x_\alpha >_C x_\gamma$ for $\alpha < \gamma$.

Definition 5.2 *A set $A \subseteq E$ is said to be $C-$complete (resp. strongly $C-$complete) if it has no covering of the form*

$$\{(x_\alpha - \text{ cl } C)^c : \alpha \in \Gamma\} \quad (\text{resp. } \{(x_\alpha - C)^c : \alpha \in \Gamma\})$$

where $\{x_\alpha\}_{\alpha \in \Gamma}$ is a decreasing net in A.

We note that every strongly $C-$complete set is $C-$complete. The converse is not always true. When C is a closed cone the two concepts coincide.

Recall that a subset $A \subseteq E$ is said to be $C-$compact if any cover of A of the form $\{U_i + C : i \in I, U_i \text{ are open}\}$ admits a finite subcover (see [42]), and $C-$semicompact if any cover of A of the form $\{(x_i - \text{cl}C)^c : i \in I, x_i \in A\}$ admits a finite subcover (see [19]). It is clear that every compact set is $C-$compact, and every $C-$compact set is $C-$semicompact. The converse is not true in general. Below are some criteria for $C-$completeness. The first criterion and the second part of the second one can be found in [42], the second criterion can be derived from [66].

Proposition 5.3 *The following assertions hold:*

(i) Every $C-$semicompact set is $C-$complete. In particular every weakly compact set in a locally convex space is $C-$complete.

(ii) Every compact set is strongly C-complete if C has Sterna-Karwat's property: for every linear subspace $L \subseteq E$, the set $C \cap L$ is a linear subspace if and only if $\ell(C \cap L)$ is a linear subspace. In particular, when E is finite dimensional, every compact set is strongly $C-$complete whatever C be.

1.3 Approximate Jacobians

Let f be a continuous map from R^n to R^m. A closed set of $(m \times n)$-matrices $\partial f(x) \subseteq L(R^n, R^m)$ is said to be an *approximate Jacobian* of f at x if for every $u \in R^n$ and $v \in R^m$, one has

$$(vf)^+(x,u) \leq \sup_{M \in \partial f(x)} \langle v, M(u) \rangle,$$

where vf is the real function $\sum_{i=1}^m v_i f_i$, here $v_1, ..., v_m$ are components of v and $f_1, ..., f_m$ are components of f, and $(vf)^+(x,u)$ is the upper Dini directional derivative of the function vf at x in the direction u, that is

$$(vf)^+(x,u) := \limsup_{t \downarrow 0} \frac{(vf)(x+tu) - (vf)(x)}{t}.$$

If for every $x \in R^n$, $\partial f(x)$ is an approximate Jacobian of f at x, then the set-valued map $\partial f : R^n \rightrightarrows L(R^n, R^m)$ is called an approximate Jacobian map of f. When $m = 1$, approximate Jacobian is also called approximate differential.

It follows from the definition that if a function is Gâteaux differentiable at a point, then its Gâteaux derivative is an approximate Jacobian and any other approximate Jacobian contains this derivative in its convex hull. Conversely, if a function admits a singleton approximate Jacobian, then the function is Gâteaux differentiable and its Gâteaux derivative coincides with that singleton element. It is important to notice that several known generalized derivatives of vector functions such as Clarke's generalized Jacobians, Warga's unbounded derivate containers (see [70]), Ioffe's matrix prederivatives (see [32]) etc. are examples of approximate Jacobians. Moreover, as seen in the next example of [36], a locally Lipschitz function may admit an approximate Jacobian whose convex hull is strictly smaller than Clarke's generalized Jacobian. Define $f : R^2 \to R$ by

$$f(x, y) = |x| - |y|.$$

Then it can easily be verified that

$$\partial_1 f(0) = \{(1,1), (-1,-1)\} \text{ and } \partial_2 f(0) = \{(1,-1), (-1,1)\}$$

are approximate Jacobians of f at 0; whereas Clarke's generalized subdifferential is given by

$$\partial_{Cl} f(0) = co(\{(1,1), (-1,1), (1,-1), (-1,-1)\}).$$

Clearly, this example shows that certain results such as mean value conditions and necessary optimality conditions that are expressed in terms of $\partial f(x)$ may provide sharp conditions even for locally Lipschitz functions.

Approximate Jacobians were introduced and studied in [36]. Further developments and applications of this concept were given in [36], [37], [38], [39]. We refer the interested reader to these papers for more details on approximate Jacobian. For our purposes we shall need the following elementary calculus rules of approximate Jacobians which were already established in [36], [38]:

a) Suppose that $f : R^n \to R$ is continuous. If f admits an approximate Jacobian $\partial f(x)$ at x and attains its minimum at x, then $0 \in \overline{co}(\partial f(x))$.

b) Suppose thatf_1 and $f_2 : R^n \to R^m$ are continuous. If $\partial f_1(x)$ and $\partial f_2(x)$ are approximate Jacobians of f_1 and f_2 respectively at x, then $\mathrm{cl}\{\partial f_1(x) + \partial f_2(x)\}$ is an approximate Jacobian of $f_1 + f_2$ at x.

c) *Mean value theorem*: Let f be a continuous map from R^n to R^m. Let $a, b \in R^n$ and let $\partial f(x)$ be an approximate Jacobian of f at $x \in [a, b]$. Then $f(b) - f(a) \in \overline{co}(\partial f[a, b](b - a))$.

Some more terminologies are in order. Let $A \subset R^n$ be a nonempty set. The recession cone of A, which is denoted by A_∞, consists of all limits $\lim_{i \to \infty} t_i a_i$ where $a_i \in A$ and $\{t_i\}$ is a sequence of positive numbers converging to 0. It is important to notice that when A is closed and convex, $A + A_\infty = A$. Elements of $(\partial f(x))_\infty$ are called recession approximate Jacobian matrices.

1.4 Contingent derivatives

Let $F : X \rightrightarrows Y$ be a set-valued map. The graph of F which is denoted by $Gr(F)$, is defined by

$$Gr(F) := \{(x,y) \in X \times Y : x \in X, y \in F(x)\}$$

and the domain of F which is denoted by $dom(F)$, is defined by

$$dom(F) := \{x \in X : F(x) \neq \emptyset\}.$$

We say that F is upper semi-continuous at x_0 if for every neighborhood $V \subset Y$ of $F(x_0)$, there is a neighborhood $U \subset X$ of x_0 such that $F(U) \subset V$, and it is compact at $x_0 \in dom(F)$ if every net $\{(x_\alpha, y_\alpha)\} \subset Gr(F)$ possesses a convergent subnet with the limit belonging to the graph of F as soon as $\{x_\alpha\}$ converges to x_0. The monograph [3] gives a detailed study of set-valued maps.

The *upper and lower contingent derivatives* of F at $(x_0, y_0) \in Gr(F)$, which are denoted by $D_{upp}F(x_0, y_0)$ and $D_{low}F(x_0, y_0)$ respectively, are set-valued maps from X to Y defined by

$$\begin{aligned} D_{upp}F(x_0,y_0)(u) &= \limsup_{(t,u') \to (0^+,u)} \frac{F(x_0 + tu') - y_0}{t} \\ D_{low}F(x_0,y_0)(u) &= \liminf_{(t,u') \to (0^+,u)} \frac{F(x_0 + tu') - y_0}{t}, \end{aligned}$$

for every $u \in X$, where limsup and liminf above stand for the upper limit and the lower limit in the sense of Kuratowski-Painlevé. Thus, $v \in D_{upp}F(x_0, y_0)(u)$ if and only if there exist a net $\{u_\alpha\} \subseteq X$ converging to u, a net $\{t_\alpha\}$ of positive numbers converging to 0 and $y_\alpha \in F(x_0 + t_\alpha u_\alpha)$ such that

$$v = \lim \frac{y_\alpha - y_0}{t_\alpha};$$

and $w \in D_{low}F(x_0, y_0)(u)$ if and only if for every net $\{u_\alpha\} \subset X$ converging to u and for every net $\{t_\alpha\}$ of positive numbers converging to 0, there exists $y_\alpha \in F(x_0 + t_\alpha u_\alpha)$ such that

$$w = \lim \frac{y_\alpha - y_0}{t_\alpha}.$$

We observe that whenever F is a single-valued map, the value set $D_{low}F(x_0, y_0)(u)$ is either empty or a singleton, while the value set $D_{upp}F(x_0, y_0)(u)$ is not necessarily a singleton. Since in this case y_0 is uniquely defined, namely $\{y_0\} = F(x_0)$, one writes $D_{low}F(x_0)$ and $D_{upp}F(x_0)$ instead of $D_{low}F(x_0, y_0)$ and of $D_{upp}F(x_0, y_0)$ respectively.

When X and Y are normed spaces and F is a single-valued Frechet differentiable map around x_0, the Frechet derivative of F at x_0 coincides with both the upper and the lower contingent derivatives of F at $(x_0, F(x_0))$.

To expose calculus rules for contingent derivatives we need the following differentiability. We say that F is *upper semidifferentiable* at $(x, y) \in Gr(F)$ if for every net $\{(x_\alpha, y_\alpha)\} \subseteq Gr(F) \setminus \{(x, y)\}$ converging to (x, y), there is $t_\beta > 0$ such that the net $\{t_\beta(x_\beta - x, y_\beta - y)\}$ converges to a nonzero vector in $X \times Y$, where $\{(x_\beta, y_\beta)\}$ is a subnet of the net $\{(x_\alpha, y_\alpha)\}$. Note that this is always the case when X and Y are finite dimensional. In the literature $D_{upp}F(x_0, y_0)$ is called the contingent derivative of F at (x_0, y_0), while $D_{low}F(x_0, y_0)$ is called the Dini lower derivative of F at (x_0, y_0) (see [42], [60]). Here are some useful properties of contingent derivatives, that will be needed in the sequel (see [51] for the proof).

Proposition 5.4 *Let $F, F_1, F_2 : X \rightrightarrows Y$, $G : X \rightrightarrows Z$ and $H : X \rightrightarrows Z$ be set-valued maps. Then the following assertions hold for every $u \in X$,*

(i) $D_{low}(F, G)(x_0, y_0, z_0)(u) = (D_{low}F(x_0, y_0), D_{low}G(x_0, z_0))(u)$, *where* $x_0 \in dom(F) \cap dom(G), y_0 \in F(x_0)$ *and* $z_0 \in G(y_0)$;

(ii) $(D_{low}F(x_0, y_0), D_{upp}G(x_0, z_0))(u) \cup (D_{upp}F(x_0, y_0), D_{low}G(x_0, z_0))(u) \subseteq D_{upp}(F, G)(x_0, y_0, z_0)(u) \subseteq (D_{upp}F(x_0, y_0), D_{upp}G(x_0, z_0))(u)$, *where* $x_0 \in dom(F) \cap dom(G), y_0 \in F(x_0)$ *and* $z_0 \in G(y_0)$;

(iii) $D_{upp}(F_1 + F_2)(x_0, y_0)(u) \subseteq \cup_{y_1 \in F_1(x_0), y_2 \in F_2(x_0), y_1+y_2=y_0} (D_{upp}F_1(x_0, y_1)(u) + D_{upp}F_2(x_0, y_2)(u))$ *provided* F_1 *and* F_2 *are upper semi-continuous at* x_0, *one of them is compact at* x_0 *and one of them, say* F_1, *is upper semidifferentiable at* (x_0, y_1) *with* $D_{upp}F_1(x_0, y_1)(0) = \{0\}$ *for* $y_1 \in F_1(x_0)$;

(iv) $D_{upp}(H \circ F)(x_0, z_0)(u) \subseteq D_{upp}H(y_0, z_0)[D_{upp}F(x_0, y_0)(u)]$ *provided* H *is upper semi-continuous,* F *is compact at* x_0 *and upper semidifferentiable at* (x_0, y_0) *with* $D_{upp}F(x_0, y_0)(0) = \{0\}$, *where* $y_0 \in F(x_0)$ *and* $z_0 \in H(y_0)$;

(v) $D_{low}(F_1 + F_2)(x_0, y_0)(u) \supseteq D_{low}F_1(x_0, y_1)(u) + D_{low}F_2(x_0, y_2)(u)$, *where* $y_1 \in F_1(x_0), y_2 \in F_2(x_0)$ *and* $y_1 + y_2 = y_0$;
$D_{low}(H \circ F)(x_0, z_0)(u) \supseteq D_{low}H(y_0, z_0)[D_{low}F(x_0, y_0)(u)]$, *where* $y_0 \in F(x_0)$ *and* $z_0 \in H(y_0)$.

These extensions and some other modified versions of contingent derivatives are studied in [3], [15], [42] and references given therein.

2. Generalized Convex Vector Functions

2.1 Convex vector functions

The class of convex vector functions came to attention of researchers around seventies of the last century by the works of Valadier [69], Zowe [73] among others. These functions have many points in common with convex scalar functions. However, due to the incompleteness of orders in vector spaces, analysis over convex vector functions is much more complicated.

Let $A \subseteq X$ be a nonempty convex set, and let Y be partially ordered by a convex cone C_Y.

Definition 5.3 *Let $f : A \to Y$. We say that f is convex (respectively strictly convex) if for each $x_1, x_2 \in A, x_1 \neq x_2$ and $t \in (0,1)$ one has*

$$f(tx_1 + (1-t)x_2) \leq_{C_Y} tf(x_1) + (1-t)f(x_2)$$

(resp. $f(tx_1 + (1-t)x_2) \ll_{C_Y} tf(x_1) + (1-t)f(x_2)$).

Recall that the epigraph and the graph of f are defined by

$$\begin{aligned} Epi(f) :&= \{(x,y) \in X \times Y : x \in A, y \geq_{C_Y} f(x)\} \\ Gr(f) :&= \{(x,y) \in X \times Y : y = f(x), x \in A\}. \end{aligned}$$

For $y \in Y$, the lower level set of f at y is defined by

$$\mathrm{Lev}_f(y) := \{x \in A : f(x) \leq_{C_Y} y\}.$$

Immediate elementary properties of convex functions is given in the next proposition.

Proposition 5.5 *The following assertions hold.*

(i) If f is convex, then for each $y \in Y$, the lower level set $Lev_f(y)$ is convex.

(ii) f is convex if and only if $Epi(f)$ is a convex set;

(iii) f is convex (respectively strictly convex) if and only if the functions $\xi \circ f$ are convex (resp. strictly convex) for every $\xi \in C_Y' \setminus \{0\}$.

We turn now to the operations with convex functions. Let Z be partially ordered by a convex cone C_Z. Let $g : Y \to Z$ be a function. We say that g is non-decreasing if $y_1 \geq_{C_Y} y_2$ implies $g(y_1) \geq_{C_Z} g(y_2)$, and increasing if

$$y_1 \gg_{C_Y} y_2 \text{ implies } g(y_1) \gg_{C_Z} g(y_2).$$

Proposition 5.6 *Let $f_1, f_2 : A \to Y$ be two convex functions. The following assertions hold:*

(i) $f_1 + f_2$ and λf_1 are convex for each $\lambda \geq 0$, and they are strictly convex provided f_1 is strictly convex and $\lambda > 0$;

(ii) $g \circ f_1$ is convex if g is non-decreasing and convex; $g \circ f_1$ is strictly convex if f is strictly convex and g is increasing convex.

Observe that another generalization of the concept of strictly convex functions and increasing scalar functions is to use the order $(>)$ instead of $(\gg)$. We leave this kind of generalization to the interested reader. It follows from Proposition 5.5 the following continuity property of convex vector functions.

Corollary 5.1 *Assume that X and Y are of finite dimension, and the closure of the cone C is pointed. Then the convex function $f : A \to Y$ is locally Lipschitz on the relative interior of A.*

As in the scalar case, one may define the subdifferential of a convex function f at $x_0 \in A$ as follows

$$\partial f(x_0) := \{M \in L(X, Y) : f(x) - f(x_0) \geq_{C_Y} M(x - x_0), \text{ for all } x \in A\}.$$

The following result (see [73]) shows the existence of the subdifferential defined above.

Theorem 5.1 *Assume that X is a separable reflexive Banach space, f is convex, continuous at $x_0 \in intA$, C_Y is closed, convex and pointed, and with respect to the Mackey topology on Y' the interior of the cone C_Y' is nonempty. Then $\partial f(x_0)$ is a nonempty convex set.*

In the remaining of this section, X and Y are assumed to be finite dimensional. By assuming $\text{int}A \neq \emptyset$, in view of Corollary 5.1, f is locally Lipschitz near $x_0 \in \text{int}A$. Then, by the famous theorem of Rademacher f is differentiable almost everywhere in a neighborhood of x_0. Hence Clarke's generalized Jacobian $Jf(x_0)$ can be defined as the convex hull of the set of all limits of Jacobians $f'(x_i)$ with $x_i \to x_0$ and f is differentiable at x_i. The following rules of subdifferentials for convex vector functions have been proven in [54].

Proposition 5.7 *Assume that C_Y is pointed, convex and closed. The following assertions hold.*

(i) For every $x \in A, \partial f(x)$ is a closed convex set; and $\partial f(x) \neq \emptyset$ for $x \in riA$. In particular, if $x \in intA$, then $\partial f(x)$ is nonempty convex compact.

(ii) *f is differentiable at $x \in intA$ if and only if $\partial f(x)$ is a singleton.*

(iii) *The set-valued map $x \mapsto \partial f(x)$ is closed at any point $x \in A$ at which f is continuous.*

(iv) *$Jf(x) \subseteq \partial f(x)$ for all $x \in intA$.*

(v) *$\xi Jf(x) = \xi \partial f(x) = \partial(\xi f)(x)$ for all $x \in intA$ and $\xi \in C'$.*

(vi) *Let f and g be two convex functions from $A \subseteq X$ to Y. Then for $x \in intA$,*

$$\partial(f+g)(x) \supseteq \partial f(x) + \partial g(y).$$

Equality holds if C is generated by linearly independent vectors.

It should be noticed that in general, the inclusion of (iv) of the above proposition is strict. In order to characterize convex vector functions via subdifferential, we recall the notion of monotone maps.
Let $F : A \rightrightarrows L(X,Y)$ be a set-valued map. We say that F is *monotone* if for every $x_1, x_2 \in A, M_1 \in F(x_1), M_2 \in F(x_2)$ one has

$$(M_1 - M_2)(x_1 - x_2) \geq_{C_Y} 0.$$

It is evident that F is monotone if and only if the set-valued maps $\xi \circ F : A \rightrightarrows X'$, $\xi \in C_Y^{'}$, are monotone (with respect to the usual order of real numbers). When F is monotone, it is said to be maximal if there is no monotone map from A to $L(X,Y)$ the graph of which contains the graph of F as a proper subset.

Proposition 5.8 *The following assertions hold:*

(i) *If $f : A \to Y$ is convex, where A is open, then ∂f is a nonvoid-valued maximal monotone map from A to $L(X,Y)$;*

(ii) *If $\partial(x) \neq \emptyset$ for each $x \in A$ where A is a convex set, then f is convex;*

(iii) *Assuming f locally Lipschitz from an open convex subset $A \subseteq X$ to Y, then f is convex on A if and only if Jf is monotone.*

For the proof of this proposition see [54] and for more details about convex vector functions see [24], [42], [54], [68], [73].

2.2 Quasiconvex and pseudoconvex vector functions

As in the previous subsection $f : A \subseteq X \to Y$ is a function, A is a convex subset of X. For $y_1, y_2 \in Y$ denote

$$[y_1, y_2]^+ := \{y \in Y : y \geq_{C_Y} y_1, y \geq_{C_Y} y_2\},$$

and $a \leq_{C_Y} [y_1, y_2]^+$ means $a \leq_{C_Y} y$ for every $y \in [y_1, y_2]^+$.

Definition 5.4 *We say that f is*

(i) quasiconvex (on A) if for each $x_1, x_2 \in A$, for each $t \in (0,1)$ one has

$$f(tx_1 + (1-t)x_2) \leq_{C_Y} [f(x_1), f(x_2)]^+$$

(ii) semistrictly quasiconvex if it is quasiconvex and

$$f(tx_1 + (1-t)x_2) \ll_{C_Y} [f(x_1), f(x_2)]^+$$

whenever $f(x_1) \neq f(x_2)$;

(iii) strictly quasiconvex if it is quasiconvex and

$$f(tx_1 + (1-t)x_2) \ll_{C_Y} [f(x_1), f(x_2)]^+$$

whenever $x_1 \neq x_2$;

(iv) pseudoconvex if $f(x_2) >_{C_Y} f(x_1)$, one can find $y \gg_{C_Y} 0$ and $\gamma \in (0,1)$ such that

$$f(tx_1 + (1-t)x_2) \leq_{C_Y} f(x_2) - ty \text{ for all } 0 < t < \gamma.$$

(v) strictly pseudoconvex if $f(x_2) \geq_{C_Y} f(x_1)$ and $x_1 \neq x_2$, one can find $y \gg_{C_Y} 0$ and $\gamma \in (0,1)$ such that

$$f(tx_1 + (1-t)x_2) \leq_{C_Y} f(x_2) - ty \text{ for all } 0 < t < \gamma.$$

Above is one among many possibilities of defining generalized convex vector functions which are a direct extension of generalized convex scalar functions. For instance, for quasiconvexity one may require (see [10], [43])

(i') $f(x_1) \leq_{C_Y} f(x_2)$ implies $f(tx_1 + (1-t)x_2) \leq_{C_Y} f(x_2)$;

(ii') $f(tx_1 + (1-t)x_2) \leq_{C_Y} sf(x) + (1-s)f(x_2)$ for some $s \in (0,1)$.

One may also go further in generalizing (i), (ii) and (iii) by using separately order relations induced by $C_Y, C_Y \setminus \{0\}$ and $\text{int}C_Y$, (see [10], and also [12], [13], [14] for other generalizations). The following implications are immediate: (v)$\Rightarrow$ (iv); (iii) $\Rightarrow$ (ii) $\Rightarrow$ (i). Moreover, strict convexity implies strict quasiconvexity, convexity implies quasiconvexity. In the literature some authors use the terminology "strictly quasiconvex functions" instead of "semistrictly quasiconvex functions" and "strongly quasiconvex functions" instead of "strictly quasiconvex functions" (see [2]). Below are some properties of generalized convex vector functions that can be obtained without any difficulties. Recall that a set $B \subseteq Y$ is said to be strictly convex if for each $a, b \in B$ and $t \in (0,1)$ one has $ta + (1-t)b \in \text{ri}B$.

Proposition 5.9 *Assume that f is continuous. The following assertions hold:*

(i) f is quasiconvex if and only if its level sets are convex;

(ii) If f is strictly quasiconvex, then its level sets are strictly convex;

(iii) If f is directionally differentiable at x_2, then it is pseudoconvex if and only if

$$f'(x_2; x_1 - x_2) \ll_{C_Y} 0 \text{ when } f(x_1) <_{C_Y} f(x_2),$$

and it is strictly pseudoconvex if and only if

$$f'(x_2; x_1 - x_2) \ll_{C_Y} 0 \text{ when } f(x_1) \leq_{C_Y} f(x_2) \text{ and } x_1 \neq x_2.$$

We note that the characterization of (iii) has been used in [10] as a definition of pseudoconvexity. The converse of (ii) is not always true. Moreover some of the assertions of the above proposition remain valid without the continuity of f.

When Y is a finite dimensional Euclidean space R^n and C_Y is the nonnegative orthant R^m_+, generalized convex vector functions are characterized by its components.

Proposition 5.10 *Let $f = (f_1, ..., f_m) : A \to R^m$ and let R^m be partially ordered by the nonnegative orthant. Then the following assertions hold:*

(i) f is quasiconvex (resp. strictly quasiconvex) if and only if the components $f_1, ..., f_n$ have the same property;

(ii) If f is semistrictly quasiconvex, then the components $f_1, ..., f_n$ are semistrictly quasiconvex;

(iii) If the components $f_1, ..., f_n$ are strictly pseudoconvex, then f is strictly pseudoconvex.

The following scalar characterization of quasiconvex vector functions in Banach spaces which generalizes (ii) of Proposition 5.10, has recently been established in [7] (see [42] for the sufficient condition and for the particular case when Y is a finite dimensional Banach lattice, and [11] for the case in which C_Y is a polyhedral cone).

Proposition 5.11 *Assume that $Y = C_Y - C_Y$ and that C_Y' is the weak* closed convex hull of extremal directions of C_Y'. Then f is quasiconvex if and only if for every extremal direction ξ of the cone C_Y' the composite function $\xi \circ f$ is a scalar quasiconvex function.*

Recall that for a function $\phi : A \to R$, where A is an open set, the directional upper derivative (resp. lower derivative) of ϕ at $x \in A$, in direction $u \in X$ is defined by

$$\phi^+(x; u) = \limsup_{t \downarrow 0} \frac{\phi(x + tu) - \phi(x)}{t}$$

$$(\text{resp. } \phi^-(x; u) = \liminf_{t \downarrow 0} \frac{\phi(x + tu) - \phi(x)}{t}).$$

We define the directional upper derivative of $f = (f_1, ..., f_m)$ at $x \in A$ in direction $u \in X$ as

$$f^+(x; u) = (f_1^+(x; u), ..., f_m^+(x; u))$$

and the directional lower derivative of f at x in direction $u \in X$ as

$$f^-(x; u) = (f_1^-(x; u), ..., f_m^-(x; u)).$$

Note that R^m, equipped with the partial order R_+^m is a Banach lattice, hence for every $a, b \in R^m$, $\min\{a, b\}$ and $\max\{a, b\}$ exist. By extending the concept of generalized monotonicity for vector bifunctions (see [49]), we can characterize generalized convexity by means of directional derivatives. As examples, let us give some extensions.

Definition 5.5 *Let $F : A \times X \to R^m$. We say that F is*

(i) quasimonotone if for each $x_1, x_2 \in A$ one has

$$\min\{F(x_1, x_2 - x_1), F(x_2, x_1 - x_2)\} \leq 0;$$

(ii) strictly quasimonotone if it is quasimonotone and if $x_1, x_2 \in A$, $x_1 \neq x_2$, *there is* $x_3 \in (x_1, x_2)$ *such that*

$$\max\{F(x_3, x_2 - x_1), F(x_3, x_1 - x_2)\} \gg 0;$$

(iii) strictly pseudomonotone if for each $x_1, x_2 \in A$ *with* $x_1 \neq x_2$, *one has*

$$\min\{F(x_1, x_2 - x_1), F(x_2, x_1 - x_2)\} \ll 0.$$

The following implications hold:(iii) $\Rightarrow$ (ii) $\Rightarrow$ (i). These definitions were given for the case when $m = 1$ in [49] (see also [47], [53]). By using the criteria for scalar functions of [49], one can derive the following result.

Theorem 5.2 *Let* A *be an open set in* X *and let* $f = (f_1, ..., f_m) : A \to R^m$ *be lower semicontinuous so that* $f^+(x; u)$ *is finite at every* $x \in A$ *in every direction* $u \in X$. *Then the following assertions hold:*

(i) f *is quasiconvex if and only if* f^+ *is quasimonotone;*

(ii) If f^+ *is strictly quasimonotone (resp. strictly pseudomonotone), then* f *is strictly quasiconvex (resp. strictly pseudomonotone).*

It is worth noticing that the foregoing result is true when f^+ is substituted by f^- for quasiconvexity and strict quasiconvexity, while this fails for strict pseudoconvexity in general. Similar results can also be obtained in terms of generalized derivatives of vector functions (see [39], [49], [68]).

2.3 Invex vector functions

Let $f : X \to Y$ be a function. Let $F : X \rightrightarrows Y$ be a set-valued map and let $\eta : X \to X$ be a function.

Definition 5.6 *We say that* f *is* $F-$*invex at* x_0 *with respect to* η *if for every* $x \in X$, *one has*

$$f(x) - f(x_0) \in F(\eta(x)) + C.$$

We shall simply say that f is $D_{upp}-$invex if $F = D_{upp}f(x_0)$, and $D_{low}-$invex if $F = D_{low}f(x_0)$. The concept of invexity was initially introduced by Hanson [29] and Craven [20] for Fréchet differentiable functions. Nonsmooth invex functions and set-valued invex maps were given in [51], [64] and some others. The three propositions below have been given in [51] for a more general case in which the function f is set-valued.

Proposition 5.12 *Assume that $f : X \to Y$ is $D_{low}-$invex at x_0 with respect to η and $g : X \to Z$ is $D_{upp}-$invex (resp. $D_{low}-$invex) at x_0 with respect to η. Then (f, g) is $D_{upp}-$invex (resp. $D_{low}-$invex) at x_0 with respect to η.*

The relationship between convexity and invexity is seen in the next result.

Proposition 5.13 *Assume that f is convex. Then it is $D_{upp}-$invex (and $D_{low}-$invex when f is continuous) at every $x_0 \in X$ with respect to $\eta(x) = x - x_0$. Conversely, if f is $F-$invex at every $x_0 \in X$ with respect to $\eta(x) = x - x_0$ and if F is convex in the sense*

$$\lambda F(x_1) + (1 - \lambda)F(x_2) \subseteq F(\lambda x_1 + (1 - \lambda)x_2) + C$$

for each $x_1, x_2 \in X, \lambda \in (0, 1)$ with $F(0) \subseteq C$, then f is convex.

Proposition 5.14 *Let $f_1, f_2 : X \to Y$ be $F-$invex at x_0 with respect to η and let $L : Y \to Z$ be a linear nondecreasing map. Then*

(i) $f_1 + f_2$ is $(2F)-$invex with respect to $\eta; tf_1$ is $(tF)-$ invex with respect to η for every $t \geq 0$;

(ii) $L \circ f$, is $L \circ F-$invex with respect to η.

Invex functions form a class of functions in which necessary optimality conditions are also sufficient conditions and exact duality theory can be developed.

3. Vector Optimization Problems

3.1 Efficient points

Definition 5.7 *Let $A \subseteq Y$ be a nonempty set and let Y be partially ordered by a convex cone C_Y. We say that a point $a \in A$ is*

(i) an ideal point of A if $x \geq a$ for every $x \in A$. The set of all ideal points of A is denoted by IMin(A) or IMin$(A|C_Y)$;

(ii) an efficient point of A if whenever $a \geq x$ for some $x \in A$ one has $x \geq a$. The set of all efficient points of A is denoted by Min(A) or Min$(A|C_Y)$.

Sometimes one is interested also in the set of efficient points with respect to the ordering generated by the cone $\{0\} \cup \text{int}C_Y$ if $\text{int}C \neq \emptyset$. This is the set of weakly efficient points and denoted by WMin(A) or WMin$(A|C_Y)$. If there exists a convex cone $K \neq Y$ with $\text{int}K \supseteq C_Y \setminus \ell(C_Y)$, such that

$a \in \text{Min}(A|K)$, then we call it properly efficient. The set of all properly efficient points of A is denoted by $\text{PrMin}(A)$ or $\text{PrMin}(A|C_Y)$.
When Y is a locally convex space, if for each neighborhood V of the origin, there is a smaller neighborhood U such that $(U - C_Y) \cap \text{cone}(A - a) \subseteq V$, then a is said to be a superefficient point of A. The set of all superefficient points of A is denoted by $\text{SuMin}(A)$ or $\text{SuMin}(A|C_Y)$.
The definition of proper efficient points is due to Henig [30]. There are some other definitions of proper efficient points. They coincide with the above one when A is a convex set in a finite dimensional space (see [25], [27], [48] for more details). The definition of superefficient points is due to Borwein and Zhuang [9]. The results of this subsection are standard and can be found in [42].

Example 5.2 1. Let $A = \{(x, y) \in R^2 : x^2 + y^2 \leq 1 \quad \text{or} \quad x \geq 0, |y| \leq 1\} \subseteq R^2$ and let $C_Y = R^2_+$. Then

$$\begin{array}{rl} \text{IMin } A & = \emptyset \\ \text{PrMin } A & = \{(x, y) \in R^2 : \quad x^2 + y^2 = 1\ ,\ x < 0,\ y < 0\} \\ \text{Min } A & = \{(x, y) \in R^2 : \quad x^2 + y^2 = 1,\ x \leq 0,\ y \leq 0\} \\ \text{WMin } A & = \text{Min } A \cup \{(x, -1) : x \geq 0\}\ . \end{array}$$

2. For $E = \ell_0$ (Example 5.1(3)), C the ubiquitous cone, the unit ball has no efficient points.

Below is an equivalent definition of efficient points.

Proposition 5.15 *Let $A \subseteq Y$. Then*

(i) $a \in IMin(A)$ if and only if $a \in A$ and $A \subseteq a + C_Y$;

(ii) $a \in Min(A)$ if and only if $a \in A$ and $A \cap (a - C_Y) \subseteq a + \ell(C_Y)$. In other words $a \in Min(A)$ if and only if $a \in A$ and there is no $y \in A$ with $a > y$;

(iii) $a \in WMin(A)$ if and only if $a \in A$ and $A \cap (a-\ \text{int}\ C_Y) = \emptyset$. In other words, $a \in WMin(A)$ if and only if $a \in A$ and there is no $y \in A$ with $a \gg y$.

The relationship between the different concepts of efficiency is seen in the next result. We suppose always that $C_Y \neq Y$.

Proposition 5.16 *For every nonempty set $A \subseteq E$ one has*

(i) $PrMin(A) \subseteq Min(A) \subseteq WMin(A)$;

(ii) *SuMin*$(A) \subseteq$ *Min*(A), *and when* E *is a normed space and* C_Y *has a convex closed base,* *SuMin*$(A) \subseteq$ *PrMin*(A)*;*

(iii) *If* *IMin*$(A) \neq \emptyset$, *then* *IMin*$(A) =$ *Min*(A) *and this set is a singleton whenever* C_Y *is pointed.*

If the space Y is equipped with two orders, then the relationship between efficiencies with respect to these orders is expressed by the next proposition.

Proposition 5.17 *Assume that* K *is a pointed convex cone with* $C_Y \subseteq K$*. Then we have*

(i) *IMin*$(A|K) =$ *IMin*$(A|C_Y)$ *provided* *IMin*$(A|C_Y)$ *is nonempty ;*

(ii) *Min*$(A|K) \subseteq$ *Min*$(A|C_Y)$ *(similarly for WMin and PrMin).*

Note that the above result is no longer true if K is not pointed except for the particular case where K is a closed half-space. We shall denote by $A_x := A \cap (x - C_Y)$ for $x \in Y$ and call it a *section* of A at x.

Proposition 5.18 *Let* $x \in Y$ *with* $A_x \neq \emptyset$*. The following assertions hold*

(i) *IMin*$(A_x) \subseteq$ *IMin*(A) *if* *IMin*$(A) \neq \emptyset$ *;*

(ii) *Min*$(A_x) \subseteq$ *Min*(A) *(similarly for WMin).*

Remark that the inclusion PrMin$(A_x) \subseteq$ PrMin A is not true in general except for very specific cases.

3.2 Existence criteria

Theorem 5.3 *Let* A *be a nonempty set in* Y*. The following assertions hold*

(i) *Min*$(A) \neq \emptyset$ *if and only if there is* $x \in Y$ *such that* A_x *is nonempty and strongly* C_Y*-complete;*

(ii) *When* C *is correct,* *Min*$(A) \neq \emptyset$ *if and only if there is* $x \in Y$ *such that* A_x *is nonempty and* C_Y*-complete.*

Proof. (Brief proof of the first assertion). The necessity is obvious. For the sufficiency, suppose to the contrary that for some $x \in E$, the section A_x is nonempty and strongly C_Y-complete, but Min$(A) = \emptyset$. Denote by $\mathcal{P}$ the set of all decreasing nets in A_x and introduce a partial order on $\mathcal{P}$ by inclusion, i.e. for $a, b \in \mathcal{P}$ one writes $a \geq b$ if and only if $b \subseteq a$ as sets. We observe that $\mathcal{P}$ is nonempty because $\mathrm{Min}(A) = \emptyset$ and the above introduced order is a partial order on $\mathcal{P}$. Now we prove that $\mathcal{P}$ satisfies the hypothesis of Zorn's lemma : every chain $D = \{a_\lambda : \lambda \in \Lambda\} \subseteq \mathcal{P}$ has an upper bound. Indeed, denote by $\mathcal{B}$ the family of all finite subsets of Λ. For each $B \in \mathcal{B}$ we set $a_B := \bigcup_{\lambda \in B} a_\lambda$. It is evident that $a_B \in \mathcal{P}$. Now we put $a_0 = \cup\{a_B : B \in \mathcal{B}\}$. Let I_0 be the index set consisting of all elements of a a_0 with $\alpha > \beta$ if $\beta >_{C_Y} \alpha$ being considered as elements of a_0. In other words the index set order is defined by the cone $(-C_Y \setminus \ell(C_Y)) \cup \{0\}$. Then I_0 is a directed index set because for $\alpha, \beta \in I_0$ there exist $B_1, B_2 \in \mathcal{B}$ such that $\alpha \in a_{B_1}$ and $\beta \in a_{B_2}$. Taking $B = B_1 \cup B_2$ we see that $\alpha, \beta \in a_B$. Since a_B is a decreasing net, there is $\gamma \in a_B$ such that $\alpha >_{C_Y} \gamma$ and $\beta >_{C_Y} \gamma$. Then $\gamma \in I_0$ with $\gamma > \alpha$ and $\gamma > \beta$. Moreover, it is evident that $a_0 \geq a$ for all $a \in D$. Hence a_0 is an upper bound of X. Now we apply Zorn's lemma to obtain a maximal element a_*, say $a_* = \{x_i\}_{i \in I} \in \mathcal{P}$. Then the family $\{(x_i - C_Y)^c : i \in I\}$ is a covering of A which is impossible because A_x is strongly C_Y-complete. □

Corollary 5.2 *If A is a nonempty compact set in a finite dimensional space, then Min$(A) \neq \emptyset$ whatever the cone C_Y be. If A is a nonempty compact set in an infinite dimensional space and the cone C_Y is closed, then Min$(A) \neq \emptyset$.*

Note that in an infinite dimensional space a compact set may have no efficient points if the cone C_Y is not correct. In fact, let Y be ℓ_0 and let C_Y be the ubiquitous cone. Let $x_0 = (1, 0, 0, \cdots)$, $x_n = (1, -\frac{1}{2^n}, \cdots, -\frac{1}{2^n}, 0, \cdots 0)$ and $A = \{x_i : i = 0, 1, 2, \cdots\}$. It is evident that $\lim_{n\to\infty} x_n = x_0$. Hence A is a compact set. Despite of this, $\mathrm{Min}(A) = \emptyset$ because $x_0 > x_1 > x_2 \cdots$

The existence of superefficient points is given in the next result of [28]. See also [33], [63] and [72] for more on existence criteria.

Theorem 5.4 *If Y is a locally convex space, C_Y has a closed bounded base and A is C_Y-compact, then A possesses superefficient points.*

3.3 Vector optimization problems

Let $A \subseteq X$ be a nonempty set, let Y be partially ordered by a convex cone C_Y. The vector optimization problem associated with A and f, which is denoted by (VP), is written as

$$\begin{array}{ll} \mathrm{Min} & f(x) \\ \text{subject to} & x \in A. \end{array}$$

This problem consists of finding an element $x_0 \in A$, called an efficient solution such that

$$f(x_0) \in \mathrm{Min}(f(A)|C_Y).$$

The set of all efficient solutions of (VP) is a denoted by $S(A, f)$. By replacing IMin, PrMin and WMin instead of Min in the above definition, we obtain the set of ideal solutions $\mathrm{IS}(A, f)$, the set of properly efficient solutions $\mathrm{PrS}(A, f)$ and the set of weakly efficient solutions $\mathrm{WS}(A, f)$ respectively.
The set A is sometimes given in the form of constraints

$$g(x) \leq_{C_Z} 0 \tag{5.1}$$
$$h(x) = 0 \tag{5.2}$$

where $g : X \to Z, Z$ is partially ordered by a convex cone $C_Z; h : X \to W$. These two constraints can be combined in one by equipping the product space $Z \times W$ with a partial order generated by the convex cone $K := C \times \{0\}$ and by the constraint

$$(g, h) \leq_K 0.$$

Note that the cone K has no interior points even when C has, therefore those results which require a nonempty interior of the cone, cannot apply. In such a situation the separated form of constraints 5.1 and 5.2 works better.
The following inclusions are immediate:

$$\mathrm{PrS}(A, f) \subseteq \mathrm{S}(A, f) \subseteq \mathrm{WS}(A, f).$$

Furthermore, if $\mathrm{IS}(A, f)$ is nonempty, then $\mathrm{IS}(A, f) = \mathrm{S}(X, f)$. Some criteria for the existence of efficient solutions are given below, which can be deduced from Theorem 5.3 and Corollary 5.2.

Proposition 5.19 *The following assertions hold*

(i) If A is compact, C_Y is correct and f is continuous, then (VP) has efficient solutions;

(ii) If A is compact, f is continuous and Y is finite dimensional, then (VP) has efficient solutions.

The notion of local solutions of (VP) can be defined by considering a neighborhood of a point instead of the whole set A. Namely, $x_0 \in A$ is said to be a local efficient solution of (VP) if there is a neighborhood U of x_0 in X such that $f(x_0) \in \mathrm{Min}(f(U \cap A)|C_Y)$. The other kinds of local solutions are defined similarly.

3.4 Local-global properties

Let us consider the problem (VP) of the previous subsection. We say that f has a *local-global minimum property* if every local efficient solution of (VP) is a global efficient solution. This property is an important feature in numerical computing because most existing less costly algorithms allow to compute local solutions only. The next result was established in [52].

Proposition 5.20 *Assume that A is convex. Each of the following conditions is sufficient for the local-global property of (VP):*

(i) f is convex;

(ii) f verifies $f(\lambda x_1+(1-\lambda)x_2) <_{C_Y} [f(x_1), f(x_2)]^+$ whenever $f(x_1) >_{C_Y} f(x_2)$, and $0 < \lambda < 1$;

(iii) f is semistrictly quasiconvex;

(iv) f is pseudoconvex.

We note that a similar result can be formulated for weakly efficient solutions. In fact, by considering the cone C_Y^0 =int$C_Y \cup \{0\}$ instead of C_Y, we have $\text{Min}(f(X)|C_Y^0)$ =$\text{WMin}(f(X)|C_Y)$. Observe that (ii) does not guarantee the local-global property for weakly efficient solutions as seen by the next example.

Example 5.3 Let $f = (f_1, f_2) : [-1, 1] \to R^2$, where

$$f_1(x) = \begin{cases} x & \text{for } x \in [-1,0] \\ 0 & \text{for } x \in (0,1], \end{cases}$$
$$f_2(x) = x.$$

Then f verifies (ii). Despite of this, the point $x = 1$ is a local weakly efficient solution, but it is not a global weakly efficient solution.

4. Optimality Conditions

The readers are already acquainted with several conditions for efficient solutions of a vector problem in the previous chapter by Cambini and Martein. In this section we provide a few complementary results, which enlighten the role of generalized convexity in vector optimization. We choose three kinds of derivatives to express the optimality conditions: the classical derivatives, the contingent derivatives and the approximate Jacobians. Classical derivatives provide an easy way to understand the meaning of multipliers (see [34]). Approximate Jacobians give a general approach and sometimes sharp conditions when treating problems with nonsmooth continuous data. Contingent derivatives are useful in expanding the analysis to set-valued optimization problems (see [15], [42]).

4.1 Differentiable problems

Let us consider the following vector problem (VP):

$$\begin{aligned} \text{Min} \quad & f(x) \\ \text{subject to} \quad & g(x) \leq_{C_Z} 0, \\ & h(x) = 0, \end{aligned}$$

where f, g and h are functions from X to Y, Z and W respectively with X, Y, Z and W Banach spaces. We assume that Y and Z are partially ordered by convex pointed cones C_Y and C_Z having nonempty interiors. In this section we shall derive a necessary condition for local weakly efficient solutions. Two classic results of analysis will be needed:

1) Mean Value Theorem (MVT): If f is Gâteaux differentiable on X, then for each $a, b \in X$ one has

$$||f(b) - f(a)|| \leq \sup\{||f'(c)|| \cdot ||b - a|| : c \in [a, b]\} .$$

2) Open Mapping Theorem (Lyusternik's Theorem): If h is Fréchet differentiable with h' continuous at x_0 and if $h'(x_0)$ is surjective, then the tangent cone to the set $M := \{x \in X : h(x) = 0\}$ at $x_0 \in M$ defined by

$$T_M(x_0 := \{v \in X : v = \lim_{i \to \infty} t_i(x_i - x_0),\ t_i > 0, \quad x_i \longrightarrow x_0,\ x_i \in M\}$$

coincides with Ker $h'(x_0)$.

Theorem 5.5 *Assume that f, g and h are Fréchet differentiable with f' and g' bounded and h' continuous in a neighborhood of x_0. If x_0 is a local weakly efficient solution of (VP), then there exist multipliers $(\xi, \theta, \gamma) \in (C_Y, C_Z, \{0\})' \setminus \{0\}$ such that*

$$\xi f'(x_0) + \theta g'(x_0) + \gamma h'(x_0) = 0 , \qquad \theta g(x_0) = 0 .$$

Proof. (Sketch) If $h'(x_0)$ is not surjective, then there exists a nonzero functional $\gamma \in W' \setminus \{0\}$ such that

$$\prec \gamma, h'(x_0)(u) \succ = 0 \quad \text{for all } u \in X .$$

This implies $\gamma h'(x_0) = 0$. Now setting $\xi = 0$ and $\theta = 0$ we obtain multipliers (ξ, θ, γ) as requested. When $h'(x_0)$ is surjective, by using the (MVT) and the Lyusternik's theorem one may prove that

$$(f'(x_0), g'(x_0), h'(x_0)(X) \cap (-\text{int } C_Y, -g(x_0) - \text{int } C_Z, \{0\}) = \emptyset \, .$$

Then one separates these convex sets by a linear functional $(\xi, \theta, \gamma) \in (Y, Z, W)' \setminus \{0\}$:

$$\xi f'(x_0)(u) + \theta[g'(x_0)(v) + g(x_0)] + \gamma h'(x_0)(w) \geq \langle \xi, -c \rangle + \langle \theta, -k \rangle$$

for all $u \in Y, \; v \in Z, \; w \in W, \; c \in C_Y, \; k \in C_Z$. It follows from the above inequality that $\xi \in C_Y^{'}$, $\theta \in X_Z^{'}$, $\gamma \in W'$ and $\theta g(x_0) \geq 0$. Remember that $g(x_0) \in -C_Z$, hence $\theta g(x_0) = 0$. Moreover, one has

$$\xi f'(x_0)(u) + \theta g'(x_0)(v) + \gamma h'(x_0)(w) \geq 0$$

for all $u \in Y$, $v \in Z$, $w \in W$ which implies

$$\xi f'(x_0) + \theta g'(x_0) + \gamma h'(x_0) = 0$$

as required. □

When (VP) is a convex problem, that is f and g are convex functions and h is linear (therefore we may omit h), the necessary condition is also sufficient.

Theorem 5.6 *Assume that f and g are convex and there exist multipliers $(\xi, \theta) \in (C_Y, C_Z)' \setminus [0\}$ such that*

$$0 = \xi f'(x_0) + \theta g'(x_0), \;\; \textit{and} \;\; \theta g(x_0) = 0 \, .$$

Then x_0 is an efficient (resp. weakly efficient) solution of (VP) if $\xi \in$ int C_Y^+ (resp. $\xi \in C_Y^{'} \setminus \{0\}$).

Proof. An "ab absurdo" argument will achieve the proof. □

4.2 Conditions using approximate Jacobian

Let us consider the constrained problem (CP_2),

$$\begin{array}{rl} \text{Min} & f(x) \\ \text{subject to} & x \in A \quad , g(x) \leq_{C_Z} 0, h(x) = 0. \end{array}$$

We assume that X, Y, Z and W are finite dimensional and define $H := (f, g, h)$. It is a continuous function from A to $Y \times Z \times W$. The product space $Y \times Z \times W$ is equipped with the euclidean norm. The space $L(X, Y \times Z \times W)$ is equipped with the norm of linear operators, i.e. for an $M \in L(X, Y \times Z \times W)$:

$$\|M\| = \max_{x \in X, \|x\| \leq 1} \|M(x)\|.$$

The closed unit ball of this space is denoted by B. We also denote by T the set of all vectors $\lambda \in (C_Y, C_Z, \{0\})'$ with $\|\lambda\| = 1$. The following lemma will be needed (see [50]).

Lemma 5.1 *Let $\omega_0 \in Y \times Z \times W$ be a nonzero vector with*

$$\max_{\lambda \in T} \langle \lambda, \omega_0 \rangle > 0.$$

Then there exists a unique point $\lambda_0 \in T$ such that

$$\langle \lambda_0, \omega_0 \rangle = \max_{\lambda \in T} \langle \lambda, \omega_0 \rangle.$$

Moreover, for every $\epsilon > 0$, there is some $\delta > 0$ such that

$$\max_{\lambda \in T} \langle \lambda, \omega \rangle = \max_{\lambda \in T, \|\lambda - \lambda_0\| \leq \epsilon} \langle \lambda, \omega \rangle$$

for all ω with $\|\omega - \omega_0\| \leq \delta$.

Recall that a set-valued map $F : X \rightrightarrows Y$ is said to be upper semicontinuous at x_0 if for every $\epsilon > 0$, there is some $\delta > 0$ such that $F(x_0 + \delta B_X) \subset F(x_0) + \epsilon B_Y$, where B_X and B_Y denote the closed unit balls in X and Y respectively.

Theorem 5.7 *Assume that ∂H is an approximate Jacobian map of H which is upper semicontinuous at x_0. If x_0 is a local weakly efficient solution of (CP$_2$), then there is a vector $\lambda_0 = (\xi_0, \theta_0, \gamma_0) \in T$ such that*

$$0 \in \lambda_0(\overline{co}\partial H(x_0) \cup co[(\partial H(x_0))_\infty \setminus \{0\}]),$$

$$\theta_0 g(x_0) = 0.$$

Proof. (Sketch) Let us choose a vector $e \in intC$ so that

$$\max_{\xi \in C', \|\xi\| \leq 1} \langle \xi, e \rangle = 1.$$

For each $\epsilon > 0$, define functions $H_\epsilon : R^n \to R^m \times R^k \times R^l$ and $P_\epsilon : R^n \to R$ as follows

$$H_\epsilon(x) := (f(x) - f(x_0) + \epsilon e, g(x), h(x)),$$
$$P_\epsilon(x) := \max_{\lambda \in T}\langle \lambda, H_\epsilon(x)\rangle.$$

It is clear that these functions are continuous. Let $U \subset R^n$ be a neighborhood that exists by the definition of the local weakly efficient solution x_0. Then $P_\epsilon(x) > 0$ for all $x \in U$. Furthermore, since $P_\epsilon(x_0) = \epsilon < \inf P_\epsilon + \epsilon$, by Ekeland's variational principle (see [17]), there is x_ϵ such that $\|x_0 - x_\epsilon\| < \sqrt{\epsilon}$, and

$$P_\epsilon(x_\epsilon) < P_\epsilon(x) + \sqrt{\epsilon}\|x - x_\epsilon\| \text{ for all } x \neq x_\epsilon.$$

In particular, the net $\{x_\epsilon\}$ converges to x_0 as ϵ tends to 0, and x_ϵ provides a minimum of the function

$$Q_\epsilon(x) := P_\epsilon(x) + \sqrt{\epsilon}\|x - x_\epsilon\|.$$

If $\partial Q_\epsilon(x_\epsilon)$ is an approximate Jacobian of Q_ϵ at x_ϵ, then $0 \in \overline{co}\partial Q_\epsilon(x_\epsilon)$. It can be seen that for each integer $r \geq 1$, there is some $\epsilon(r) > 0$ such that for every $\epsilon \in (0, \epsilon(r)]$ the set

$$L_\epsilon := \{\lambda(M + \frac{1}{r}N) : \lambda \in T, \|\lambda - \lambda_\epsilon\| \leq \epsilon, M \in \partial H(x_0), N \in B\},$$

is an approximate Jacobian of P_ϵ at x_ϵ. Hence for each $r \geq 1$, there is $\epsilon(r) > 0$ such that for $0 < \epsilon \leq \epsilon(r)$, the set

$$\partial Q_\epsilon(x_\epsilon) := L_\epsilon + \sqrt{\epsilon}B_n$$

is an approximate Jacobian of Q_ϵ at x_ϵ. By choosing $\epsilon(r) \downarrow 0$ as $r \to \infty$ and taking into account the fact that B, B_n and T are all compacts, we derive the existence of vectors

$$\xi_r \in co\{\lambda M : \lambda \in T, \|\lambda - \lambda_\epsilon(r)\| \leq \epsilon(r), M \in \partial H(x_0)\}$$

such that $\lim_{r\to\infty} \xi_r = 0$. We apply Caratheodory's theorem to express the vectors ξ_r as

$$\xi_r = \sum_{j=1}^{n+1} a_{rj}\lambda_{rj}M_{rj},$$

where $\sum_{j=1}^{n+1} a_{rj} = 1$, $a_{rj} \geq 0$, $\lambda_{rj} \in T$ with $\|\lambda_{rj} - \lambda_{\epsilon(r)}\| \leq \epsilon(r)$, and $M_{rj} \in \partial H(x_0)$, $j = 1, ..., n+1$. Since T is compact, without loss of generality, we may assume that the sequence $\{\lambda_{\epsilon(r)}\}$ converges to some $\lambda_0 \in T$. Then

$$\lim_{r\to\infty} \lambda_{rj} = \lambda_0, \text{for all } \ j = 1, ..., n+1.$$

Moreover, we also may assume that the sequences $\{a_{rj}\}_r$ converge to $a_{0j}, j = 1, ..., n+1$, and that

$$\xi_r = \sum_{j \in I_1} a_{rj}\lambda_{rj}M_{rj} + \sum_{j \in I_2} a_{rj}\lambda_{rj}M_{rj} + \sum_{j \in I_3} a_{rj}\lambda_{rj}M_{rj}$$

where the above sums have the following properties:
1) for each $j \in I_1$, the sequence $\{M_{rj}\}_r$ is bounded and converges to some $M_{0j} \in \partial H(x_0)$;
2) for each $j \in I_2$, the sequence $\{M_{rj}\}_r$ is unbounded, but the sequence $\{a_{rj}M_{rj}\}_r$ is bounded and converges to some M_{*j};
3) for each $j \in I_3$, the sequence $\{a_{rj}M_{rj}\}_r$ is unbounded and there is some $j_0 \in I_3$ such that the sequences $\{a_{rj}M_{rj}/\|a_{rj_0}M_{rj_0}\|\}_r$ converge to some $M_{\infty j}, j \in I_3$.
If I_3 is nonempty, then we have $M_{\infty j} \in [\partial H(x_0)]_\infty$ and $M_{\infty j_0} \neq 0$. Hence $0 \in \lambda_0 co([\partial H(x_0)]_\infty \setminus \{0\})$. If I_3 is empty, then for $j \in I_2$, one has $a_{0j} = 0$, which implies that $\sum_{j \in I_1} a_{0j} = 1$ and $M_{*j} \in [\partial H(x_0)]_\infty$. Thus,

$$0 = \lim_{r \to \infty} \xi_r = \lambda_0(\sum_{i \in I_1} a_{0j}M_{0j} + \sum_{j \in I_2} M_{*j}) \in \lambda_0 \overline{co}\partial H(x_0).$$

The complementary slackness $\theta_0 g(x_0) = 0$ is obtained by observing that if $g_i(x_0) < 0$, then the vector λ_ϵ must have the corresponding component $\theta_{\epsilon i} = 0$ and by passing it to limit when ϵ tends to 0. □

We end up this subsection by noticing that when the data of the problem are locally Lipschitz, one can use Clarke's generalized Jacobian as an approximate Jacobian. In this case the recession part of the multiplier rule disappears and the above theorem gives the multiplier rule by Craven [20]. When the problem is convex, the necessary condition is sufficient too.

4.3 Conditions using contingent derivatives

To easily understand the analysis, let us begin with the unconstrained problem (VP)

$$\begin{array}{rl} \text{Min} & f(x) \\ \text{subject to} & x \in A \end{array} ,$$

where A is an open subset of X. The following optimality conditions are quite straightforward (see [42] and [51]).

Theorem 5.8 *If $x_0 \in A$ is a local weakly efficient solution of (VP), then for every $u \in X$ one has*

$$D_{upp}f(x_0)(u) \cap -intC_Y = \emptyset.$$

Conversely, if f is $D_{upp}-$invex (respectively $D_{low}-$invex) at $x_0 \in A$ and for each $u \in X$, the above relation holds (resp. $D_{low}f(x_0)(u) \cap -intC_Y = \emptyset$), then x_0 is a weakly efficient solution of (VP).

We turn to the problem with inequality constraints of the form (CP_1)

$$\begin{array}{rl} \text{Min} & f(x) \\ \text{subject to} & x \in A \quad, g(x) \leq_{C_Z} 0. \end{array}$$

where $g : A \to Z$. We assume that $\text{int}C_Z \neq \emptyset$. The proof of the following theorem can be found in [51].

Theorem 5.9 *Let x_0 be a local weakly efficient solution of (CP_1). Then for every $u \in domD_{upp}(f,g)(x_0)$ and for every $(v,z) \in D_{upp}(f,g)(x_0)(u)$, there exist a nonzero multiplier $(\xi,\zeta) \in (C_Y,C_Z)'$ such that*

$$\xi(v) + \zeta(z + g(x_0)) \geq 0.$$

The multiplier (ξ,ζ) can be chosen independent of (v,z) provided that the set $D_{upp}(f,g)$ $(x_0)(u) + (C_Y,C_Z)$ is convex, and independent of (u,v,z) provided that the set $D_{upp}(f,\ g)(x_0)(X) + (C_Y,C_Z)$ is convex. In the latter case one has $\zeta g(x_0) = 0$ and the vector ξ can be guaranteed nonzero under an additional hypothesis, called constraint qualification, that

$$D_{upp}(f,g)(x_0)(X) \cap (Y, -intC_Z) \neq \emptyset.$$

The problem with mixed constraints takes the form (CP_2)

$$\begin{array}{rl} \text{Min} & f(x) \\ \text{subject to} & x \in A \quad, g(x) \leq_{C_Z} 0, h(x) = 0, \end{array}$$

where $g : A \to Z$ and $h : A \to W$. Below F denotes a set-valued map from X to $Y \times Z \times W$, and $L^+(Z,Y)$ the space of linear nondecreasing operators from Z to Y. We have the following sufficient condition under an invexity hypothesis (see [51]).

Theorem 5.10 *Assume that the following conditions hold*

(i) $x_0 \in A$ is a feasible solution of (CP_2);

(ii) The function (f,g,h) is $F-$invex at x_0 with respect to some function η;

(iii) For every $u \in domF$ and for every $(y,z,w) \in F(u)$ there exist linear operators $p \in L^+(Z,Y)$ and $q \in L(W,Y)$ such that

$$y + p(z + g(x_0)) + q(w) \notin -intC_Y.$$

Then x_0 is a weakly efficient solution of (CP_2).

Observe that condition (iii) can be replaced by the following more familiar one:
(iii') For every $u \in domF$ and for every $(y,z,w) \in F(u)$, there exist a multiplier $(\xi,\zeta,\nu) \in (C_Y, C_Z, \{0\})'$ with $\xi \neq 0$ such that

$$\xi(y) + \zeta(z + g(x_0)) + \nu(w) \geq 0.$$

A particular case is obtained when f, g and h are differentiable and F is given by (f', g', h').

Corollary 5.3 *Let x_0 be a feasible solution of (CP_2). Assume that*

(i) f, g, h are differentiable at x_0 and for every $x \in X$ one has

$$(f,g,h)(x) - (f,g,h)(x_0) - (f',g',h')(x_0)(x - x_0) \in (C_Y, C_Z, \{0\})$$

(ii) There is $(\xi,\zeta,\nu) \in (C_Y, C_Z, \{0\})'$ with $\xi \neq 0, \zeta g(x_0) = 0$ such that

$$\xi f(x_0) + \zeta g'(x_0) + \nu h'(x_0) = 0.$$

Then x_0 is a weakly efficient solution of (CP_2).

Note that when (CP_2) is a convex problem, condition (ii) is satisfied and we obtain the same result as Theorem 5.6.

As we have seen in the above theorems, the convexity of the sets $D_{upp}(f,g)\,(x_0)(X)+(C_Y, C_Z)$ presents a sufficient condition for the existence of a multiplier at a local weakly efficient solution. This set actually is the projection of the epigraph of the set-valued map $D_{upp}(f,g)(x_0)$ on the product space $Y \times Z$. When the map $D_{upp}(f,g)(x_0)$ is single-valued, the convexity of the epigraph is equivalent to the convexity of this map. In Subsection 1.4, contingent derivatives are given in a general setting for set-valued maps. They are very convenient for the study of those problems whose objective function and constraint functions are set-valued. This, in fact, was done in [51] in which necessary conditions for locally weakly efficient solutions of (CP_2) are also derived by means of contingent lower derivatives.

5. Scalarization of vector problems

One of the most important methods for solving problem (VP) is scalarization which consists of converting problem (VP) into a scalar problem whose optimal solutions are weakly efficient or efficient for problem (VP). Usually, one finds a scalarizing function $g : R^n \to R$ which preserves order relations of Y, and solves the scalar problem of the form, which is denoted by (P):

$$\begin{aligned} \max \quad & g \circ f(x) \\ \text{subject to} \quad & x \in A. \end{aligned}$$

The scalarizing function g depends on the structure of the vector problem. In this section we shall describe this function for particular cases in which convexity and compactness structures are imposed.

5.1 General case

In this section we say that $g : B \subseteq R^n \to R$ is increasing (with respect to the cone C_Y) if $a >_{C_Y} b$ implies $g(a) > g(b)$. An increasing function with respect to the cone C_Y^0 (remember that $C_Y^0 = \{0\} \cup \text{int} C_Y$) is called weakly increasing. The following standard result expresses the relationship between the solution set $S(g \circ f, X)$ of problem (P) and that of (VP) (see [42]).

Proposition 5.21 *The following assertions are true:*

(i) If g is increasing (respectively, weakly increasing), then every optimal solution of (P) is an efficient solution (respectively, weakly efficient solution) of (VP);

(ii) Conversely, for every weakly efficient solution x of (VP), there exists a continuous weakly increasing function g such that x is an optimal solution of (P).

(iii) There exists a continuous weakly increasing function g such that the weakly efficient solution set of (VP) coincides with the solution set of (P).

(iv) If Y is a normed space, C_Y has a compact base and if both $f(A)$ and $Min(f(A)|C_Y)$ are compact, then there is a continuous, increasing function g such that the efficient solution set of (VP) coincides with the solution set of (P).

The so-called *smallest weakly increasing function* $h_{e,a}$ which is defined as below is quite useful (see [42] for this definition, and [40] for early use).

Let $a \in Y$ and $e \in \text{int } C_Y$. Set

$$h_{e,a}(y) = \min\{t \in R : y \in a + te - C_Y\}.$$

This function has the property that if h is a weakly increasing function on Y, then the level set of h at $h(a)$ must contain the level set of $h_{e,a}$ at 0. By taking $a = f(x)$ and any $e \in \text{int}C_Y$ one obtains $h_{a,e}$ as a scalarizing function that verifies (ii) of the above proposition.

5.2 Convex case

In this subsection we consider a convex problem i.e. problem (VP) with A being a convex set and f a convex function.

Proposition 5.22 *Assume that (VP) is a convex problem. Then*

(i) A point $x \in A$ is a properly efficient solution of (VP) if and only if it is an optimal solution of (P) with $g \in C_Y^+$;

(ii) A point $x \in A$ is a weakly efficient solution of (VP) if only if it is an optimal solution of (P) with $g \in C_Y' \setminus \{0\}$.

It should be noticed that an efficient solution of a convex vector problem, which is not proper, can be a solution of no scalar problem (P) with $g \in C_Y^+$. Thus we have the following inclusions for a convex problem

$$\begin{aligned} PrS(f, A) &= \cup_{g \in C_Y^+} S(g \circ f, A) \subseteq S(f, A) \subseteq WS(f, A) \\ &= \cup_{g \in C_Y' \setminus \{0\}} S(g \circ f, A). \end{aligned}$$

Under some additional hypotheses on C_Y, A and f (for instance when C_Y is closed and has a bounded convex base, A is compact and f is continuous) one may have

$$S(f, A) \subseteq cl[PrS(f, A)].$$

The equality

$$S(f, A) = WS(f, A)$$

can be realized if f is strictly convex. Another important feature of the scalarizing method is that for $g \in C_Y'$, problem (P) is convex, which allows us to apply convex optimization techniques to solve the vector problem.

5.3 Quasiconvex case

Now assume that (VP) is a quasiconvex problem, i.e f is a quasiconvex function and A is a convex set. If Y is finite dimensional, say R^m, and C_Y is closed, then for each extreme vector g of C'_Y, problem (P) is a quasiconvex problem, whose optimal solution are solutions of (VP), with $g \in C'_Y$. This shows that there is no hope of getting weakly efficient solutions of (VP) via scalarizing functions from C'_Y. The smallest weakly increasing functions are quite useful in this situation (see [40] and [42]).

Proposition 5.23 *Let $e \in intC_Y$. Then $x \in A$ is a weakly efficient solution of (VP) if and only if x is an optimal solution of (P) with $g = h_{e,f(x)}$. Moreover, for a fixed $a \in Y, x \in A$ with $f(x) \gg a$ is a weakly efficient solution of (VP) if and only if x is an optimal solution of (P) with $g = g_{e,a}$ where $e = f(x) - a$. An optimal solution of (P) is efficient if and only if it is a unique optimal solution. The problem (P) mentioned above is quasiconvex if (VP) is quasiconvex.*

5.4 Density properties

The scalarization results for convex problems of Subsection 5.2 suggest an idea of finding efficient points of a convex set A by solving a family of scalar problems of the form

$$\min \langle \xi, x \rangle$$

$$\text{s.t. } x \in A$$

where $\xi \in C_Y^+$. This in fact was proven by Arrow, Barankin and Blackwell as early as in 1953 (see [1]).

Theorem 5.11 *Let $Y = R^m, C_Y = R^m_+$ and let A be a convex compact set in R^m. Then the set $PoMin(A) := \{x \in A : x \text{ minimizes } < \xi, . > \text{ over } A \text{ for some } \xi \in R^m_+, \xi \gg 0\}$ is dense in $Min(A)$.*

This density result is then generalized by several authors to more general spaces (see [22], [26], [35], [61]) for more general ordering cones). One of the most interesting generalizations is due to Jahn [35] and then improved by Petschke [61], and Gong [26] as follows.

Theorem 5.12 *Assume that Y is a Banach space, C_Y is a convex closed pointed cone and $A \subseteq Y$ is a weakly compact convex set. Then $PoMin(A)$ is dense in $Min(A)$ under each of the following conditions*

(i) C_Y has a bounded convex base;

(ii) *0 is a point of continuity of C_Y in the sense that for each $\varepsilon > 0$, the weak closure of $C_Y \setminus B(0,\varepsilon)$ does not contain 0 (here $B(0,\varepsilon)$ is the closed ball centered at 0 with radius ε);*

(iii) *Every efficient point of A is denting in the sense for each $\varepsilon > 0, x_0 \notin \overline{co}A \setminus (x_0 + B(0,\varepsilon))$.*

Note that conditions (i) and (ii) are equivalent. When A is not convex, the set PoMin$A$ is very poor. For instance with $Y = R^2, C_Y = R^2_+$ and

$$A = \{(x_1,x_2) \in R^2 : (x_1+1)^2 + (x_2+1)^2 = 1, x_1 \leq 0, x_2 \leq 0\}$$

We have Min$(A) = A$, while PoMin$(A) = \{(-1,0),(0,-1)\}$. In such a situation, one looks for density of properly efficient points or superefficient points. The next result is due to Ha ([28]), some similar results have been obtained in [9], [56] etc. Recall that a closed convex cone C_Y in a normed space is said to be normal if for every $x_1, x_2 \in C_Y$, one has $\|x_1\| \leq \|x_1 + x_2\|$.

Theorem 5.13 *Assume that E is a Banach space, C_Y is a normal cone with a convex base, and A is C_Y-complete. Then $PrMin(A)$ is dense in Min(A). If in addition C_Y has a bounded base, then SuMin(A) is dense in Min(A).*

Proof. (Sketch) Let $a \in Min(A)$ which can be assumed to be 0. Thus, $A \cap (-C_Y) = \{0\}$. Define a sequence of delating cones by $C_n = $ cone$(cl(C_0 + B(0,\frac{1}{n})))$, where C_0 is a closed convex base of C_Y. There exists $n_i \geq 1$ such that

$$A_i := A \cap (-C_{n_i}) \subseteq B(0, \frac{1}{2^i}).$$

Since A is C-compact, one sees that A_i is C_{n_i}-compact. Furthermore, as C_n is a closed normal cone with a closed convex base (because C_Y is), $C_n^+ \neq \emptyset$. Consequently, $PoMin(A_i|C_{n_i}) \neq \emptyset$ and there exists $a_i \in Min(A_i|C_{n_i})$ (which contains $PoMin(A_i|C_{n_i})$) for $i \geq 1$. It is clear that $a_i \in Min(A|C_{n_i}) \subseteq PrMin(A|C_Y)$ and $a_i \to 0$. By this $PrMin(A|C_Y)$ is dense in $Min(A|C_Y)$. When C_Y has a bounded convex base, so does C_{n_i}, by which $a_i \in SuMin(A|C_Y)$ and the result follows. □

6. Structure of efficient solution sets

One of the key and most challenging topics of vector optimization theory is to investigate the structure of efficient point sets. Among the topological properties of these sets, the connectedness and the contractibility are of interest as they provide a possibility of continuous moving from one optimal solution to another along optimal alternatives only, and they guarantee the stability of numerical algorithms under limiting processes. In this chapter we shall concentrate on the above properties for problems that have some convexity structure.

6.1 Closeness and compactness properties

Consider the vector problem (VP)

$$\begin{array}{rl} \text{Min} & f(x) \\ \text{subject to} & x \in A, \end{array}$$

where $A \subseteq X$ is a nonempty set.

Proposition 5.24 *If $f(A)$ is closed, then $WMin(f(A)|C)$ is closed. If A is closed and f is continuous, then $WS(A, f)$ is closed. In particular, if $f(A)$ is compact, then $WMin(f(A)|C)$ is a compact set; and if A is compact, f is continuous, then $WS(f, A)$ is a compact set.*

Note that the conclusion of the proposition remains true if the continuity of f is replaced by a weaker condition, the C-continuity: for each $x \in A$ and each neighborhood V of $f(x)$, there is a neighborhood U of x such that $f(U \cap A) \subseteq V + C$. Note also that in general, $S(f, A)$ and $Min(f(A)|C)$ are not closed even when A is (convex) compact and f is continuous.

6.2 Convex problems

One of the most interesting properties of the efficient point set of a convex set is the contractibility. Let us recall that a set $B \subseteq Y$ is said to be *contractible* if there is a continuous function $H : B \times [0, 1] \to B$ such that

$$H(x, 1) = x \text{ and } H(x, 0) = x_0 \text{ for all } x \in B \text{ and some } x_0 \in B.$$

The following result was essentially given in [45].

Theorem 5.14 *Assume that Y is a uniformly convex Banach space, C_Y is a closed convex cone with a bounded convex base and B is a weakly compact set such that $B + C_Y$ is convex and has a nonempty relative interior. Then $Min(B|C_Y)$ is a contractible set and $WMin(B|C_Y)$ is arcwise connected.*

Proof. (Sketch) Let B_0 be a bounded convex base of C_Y. There exists $\xi \in intC_Y$ such that $\langle \xi, b\rangle > \varepsilon$ for every $b \in B_0$ and for $\varepsilon > 0$ sufficiently small. Let $B_1 := \{x \in C_y : \langle \xi, x\rangle = 1\}$. Then B_1 is a bounded convex base of C_Y. For k sufficiently large, $B_1 \subseteq intB(0,k)$, where $B(0,k)$ is the ball centered at 0 with radius k. Let

$$D := \{x \in Y : \langle \xi, x\rangle = t, \|x\| \leq k(1+\sqrt{t}), t \geq 0\}$$

and

$$g(x) = int\{t \in R : x \in te - D\}.$$

One can see that g is continuous, increasing and strictly quasiconvex. Moreover, it attains its minimum on B at a unique point. Define a set-valued map $G : Y \rightrightarrows Y$ by $G(x) = (x - C_Y) \cap (B + C_Y)$. This map is closed and lower semicontinuous at every point of $ri(B + C_Y) \cup Min(B|C_Y)$. Define a function $h : B + C_Y \to Min(B|C_Y)$ by $h(x) = \{y \in G(x) : y$ minimized g on $G\{x\}$. It can be seen that h is well defined and continuous. Now a continuous function $H : Min(B|C_Y) \times [0,1] \to Min(B|C_Y)$ can be constructed by $H(x,t) = h(tx + (1-t)a)$ for $x \in Min(B|C_Y)$ and $t \in [0,1]$, where a is some point chosen from $ri(B + C_Y)$. This function shows the contractibility of $Min(B|C_Y)$. □

Some weaker results can be obtained under weaker conditions (see [42], [57], [58]).

Theorem 5.15 *Assume that Y is a normed space, C_Y is a closed cone with a convex base and B is a compact (respectively weakly compact) with $B + C_Y$ convex. Then $Min(B|C_Y)$ is connected (respectively weakly connected).*

Proof. (Sketch) Under the hypothesis of the theorem, $Min(B|C_Y)$ is sandwiched between the weakly connected set $\cup_{\xi \in C_Y^+}\{y \in B : y$ minimizes $\langle \xi, .\rangle$ on $B\}$ and its (weak) closure, hence is (weakly) connected.

□

Theorem 5.16 *Let Y be a real Hausdorff topological vector space, C_Y a closed convex cone with a convex base, and B a compact and convex set. Then $Min(B|C_Y)$ is arcwise connected.*

In a two dimensional space, the structure of the efficient set of a convex closed set is very simple.

Theorem 5.17 *Assume that $B \subset R^2$ is a closed and convex set, and $C \subset R^2$ is a convex cone with nonempty interior and $\overline{C}$ is pointed. Then $Min(B|C)$ is homeomorphic to an interval if it is nonempty.*

As a particular case of this theorem, if $B+C$ is convex, closed and if there is some bounded set A such that $B+C = A+C$, then $Min(B|C)$ is compact. Consequently, $Min(B|C)$ is homeomorphic to a simplex. This observation is a corrected version of Theorem 3.1 of [41] in which $B+C \subset A+C$ was used instead of $B+C = A+C$.

Now we turn to the convex problem (VP):

$$\begin{array}{rl} \text{Min} & f(x) \\ \text{subject to} & x \in A, \end{array}$$

where $A \subseteq X$ is a nonempty set. The following result is immediate from Theorem 5.14.

Corollary 5.4 *Assume that Y is a normed space, C_Y is a closed cone with a convex base, A is convex and compact, and f is convex with $Epi(f)$ closed with respect to the weak topology of Y. Then $S(f, A)$ is connected and $Min(f(A)|C_Y)$ is weakly connected.*

6.3 Quasiconvex problems

Consider the vector problem (VP) as in the previous section:

$$\begin{array}{rl} \text{Min} & f(x) \\ \text{subject to} & x \in A \quad , \end{array}$$

where A is a nonempty convex set, Y is partially ordered by a closed, convex and pointed cone C_Y, having a nonempty interior. The two theorems below can be found in [42].

Theorem 5.18 *Assume that f is a continuous, quasiconvex function and A is a convex closed set. The following assertions hold.*

(i) $WS(f, A)$ is nonempty, closed and connected provided that for every $a \in Y$ the set $WS(f, lev_f(a))$ is nonempty compact whenever the level set $lev_f(a)$ is nonempty;

(ii) $WMin(f(A)|C_Y)$ is nonempty, closed and connected provided that for each $a \in Y$ the set $WMin(f(A) \cap (a - C_Y)|C_Y)$ is nonempty compact whenever $f(A) \cap (a - C_Y) \neq \emptyset$.

Theorem 5.19 *Assume that f is $C-$continuous, strictly quasiconvex and A is closed convex. The following assertions hold*

(i) $S(f, A)$ is nonempty, closed and connected provided that for each $a \in E$, $S(lev_f(a), f)$ is compact;

(ii) *$Min(f(A)|C_Y)$ is nonempty, closed and connected provided that f is continuous and that for each $a \in E$, the set$Min([f(A)]_a|C_Y)$ is compact. In this case, $Min(f(A)|C_Y)$ and $S(A,f)$ are homeomorphic and they are a retract of A and of $f(A)$ respectively. In particular, they are contractible.*

Observe that the sets $S(f,A)$ and $Min(f(A)|C)$ are no longer connected if f is continuous and quasiconvex. When X and Y are finite dimensional and C_Y is the nonnegative orthant, some connectedness and contractibility properties are still available by weakening the strict quasiconvexity to the semistrict quasiconvexity, as shown by the next results of [5].

Theorem 5.20 *Assume that $A \subseteq R^n$ is a compact set and $f = (f_1, ..., f_m)$ with $f_1, ..., f_m$ semistrictly quasiconvex. Then $S(f,A)$ is contractible.*

The connectedness of $S(f,A)$ under the same hypothesis above was established by Sun (see [67]). The following interesting result has been proved by Schaible [65] for $m = 2$, by Daniilidis, Hadjisavvas and Schaible [23] (see also [18] and [6] for an improvement), for $m = 3$, and finally by Benoist [4] for an arbitrary m. A particular case where $f_1, ..., f_m$ are linear fractional has been proven by Choo and Atkins [16] for $m \leq 3$, and by Yen and Phuong [71] for m arbitrary.

Theorem 5.21 *Assume that $A \subseteq R^n$ is convex, compact and $f = (f_1, ..., f_m)$ with $f_i, i = 1, ..., m$ semistrictly quasiconvex. Then $S(f,A)$ and $Min(f(A)|R^n_+)$ are connected.*

Proof. (Main ideas) We say that $B \subseteq R^m$ is sequentially strictly quasiconvex if for each $a, b \in B$ there is a sequence $\{a^k\} \subseteq B$ converging to a such that

$$a_i^k < max(a_i, b_i) \text{ if } a_i \neq b_i \text{ and } a_i^k \leq a_i \text{ if } a_i = b_i, i = 1, ..., m.$$

If B is sequentially strictly quasiconvex, and if for every $a \in R^m$, the section $B \cap (a - R^m_+)$ is compact, then $Min(B|R^M_+)$ is connected. When $f_1, ..., f_m$ are semistrictly quasiconvex, the set $f(A)$ is continuously strictly quasiconvex in the sense that for each $a, b \in f(A)$, there is a continuous function $\gamma : [0,1] \to A$ such that $\gamma(0) = a, \gamma(1) = b$ and $\gamma(t) < max(a_i, b_i)$ when $a_i \neq b_i, \gamma_i(t) \leq a_i$ when $a_i = b_i$ for every $t \in (0,1), i = 1, ..., m$. Hence $f(A)$ is sequentially strictly quasiconvex and the result follows. □

References

[1] Arrow, K., Barankin, E. and Blackwell, D., *Admissible points of convex sets*, in Contributions to the Theory of Games, eds. Kuhn H. W. and Tucker F. H. , Princeton University Press, Princeton, NJ., 1953, pp. 87-92.

[2] Avriel, M., Diewert, W.E., Schaible, S. and Zang, I., *Generalized Concavity*, Plenum Publishing Corporation, New York, 1988.

[3] Aubin, J.P. and Frankowska, H., *Set-Valued Analysis*, Birkhauser, Boston, 1990.

[4] Benoist, J., *Connectedness of the efficient set for strictly quasiconcave sets*, J. Optimization Theory and Applications, 96 (1998), pp. 627-654.

[5] Benoist, J., *Contractibility of the efficient set in strictly quasiconcave vector maximization*, J. Optimization Theory and Applications, 110 (2001), pp. 325-336.

[6] Benoist, J. and Popovici, N., *Contractibility of the efficient frontier of three-dimensional simply-shaded sets*, J. Optimization Theory and Applications, 111 (2001), pp. 81-116.

[7] Benoist, J., Borwein, J.M., and Popovici, N., *A characterization of quasiconvex vector-valued functions*, Proceedings of the American Mathematical Society 131 (2003), pp. 1109-1113.

[8] Borwein, J.M., *The geometry of Pareto efficiency over cones*, Mathematische Operationsforschung und Statistik, Serie Optimization, 11 (1980), pp. 235-248.

[9] Borwein, J.M., and Zhuang, D., *Supper efficiency in vector optimization*, Transactions of the American Mathematical Society, 339 (1993), pp. 105-122.

[10] Cambini, A., and Martein, L., *Generalized concavity and optimality conditions in vector and scalar optimization*, in "Generalized Convexity", eds. Komlosi S., Rapcsak T. and Schaible S., Lecture Notes on Economics and Mathematical Systems 405, Springer-Verlag, Berlin, 1994, pp. 337-357.

[11] Cambini, A., Luc, D.T. and Martein, L., *Order preserving transformations and applications*, J. Optimization Theory and Applications 118 (2003), pp. 275-293.

[12] Cambini, R., *Some new classes of generalized convex vector valued functions*, Optimization 36 (1996), pp. 11-24.

[13] Cambini, R., and Komlosi, S., *On the scalarization of pseudoconcavity and pseudomonotonicity concepts for vector valued functions*,

in "Generalized Convexity, Generalized Monotonicity " (Proceedings of the Fifth Symposium on Generalized Convexity, Luminy, June 1996), eds. Crouzeix J.P., Martinez-Legas J.E. and Volle M., Kluwer Academic Publishers, London, 1998, pp. 277-290.

[14] Cambini, R., and Komlosi, S., *On polar generalized monotonicity in vector optimization*, Optimization 47 (2000), pp. 111-121.

[15] Chen, G.Y. and Jahn, J., *Optimality conditions for set-valued optimization problems*, Mathematical Methods of Operations Research, 48 (1998), pp. 187-200.

[16] Choo, E.U. and Atkins, D.R., *Connectedness in multiple linear fractional programming*, Management Science, 29 (1983), pp. 250-255.

[17] Clarke, F.H., *Optimization and Nonsmooth Analysis*, John Wiley and Sons, New York 1983.

[18] Choo, E.U., Schaible, S. and Chew, S., *Connectedness of the efficient set in three-objective quasiconcave programming*, Cahiers du Centre d'Etudes de Recherche Operationnelle 27 (1985), pp. 213-220.

[19] Corley, H.W., *An existence result for maximization with respect to cones*, J. Optimization Theorey and Applications 31 (1980), 277-281.

[20] Craven, B.D., *Nonsmooth multiobjective programming,* Numerical Functional Analysis and Optimization, 10 (1989), pp. 49-64.

[21] Craven, B.D., *Nonsmooth multiobjective programming,* Numerical Functional Analysis and Optimization, 10 (1989), pp. 49-64.

[22] Daniilidis, A., *Arrow-Barankin-Blackwell theorems and related results in cone duality: a survey*, Optimization (Namur, 1998), Lecture Notes in Economics and Mathematical Systems, 481, Springer Verlag, Berlin, 2000, pp. 119-131.

[23] Daniilidis, A., Hadjisavvas, N. and Schaible, S., *Connectedness of the efficient set for three-objective quasiconcave maximization Problems,* J. Optimization Theory and Applications, 93 (1997), pp. 517-524.

[24] Durousseau, C., *Contribution à l'analyse convexe et à l'optimization vectorielle*, PhD. Thesis, Faculté des Sciences, Université de Limoges, 1999.

[25] Giannessi, F., Maestroeni, G. and Pellegrini, L., *On the theory of vector optimization and variational inequalities. Image space analysis and separation,* in Vector Variational Inequalities and Vector Equilibria, Ed. F. Giannessi, Kluwer Academic Publishers, London, pp. 153 - 216, 2000.

[26] Gong, X.H., *Connectedness of the efficient solution set of a convex vector optimization problem in a normed space*, Nonlinear Analysis 23 (1994), pp. 1105-1114.

[27] Guerraggio, A., Molho, E. and Zaffaroni, A., *On the notion of proper efficiency in vector optimization*, J. Optimization Theory and Applications, 82 (1994), pp. 1-21.

[28] Ha, T.X.D., *Existence and density results for proper efficiency in cone compact sets*, Optimization, 111 (2001), pp. 173-194.

[29] Hanson, M.A., *On sufficiency of the Kuhn-Tucker conditions*, J. Mathematical Analysis and Applications, 80 (1981), pp. 545-550.

[30] Henig, M.I., *Proper efficiency with respect to cones*, J. Optimization Theory and Applications, 36 (1982), pp. 387-407.

[31] Huy, N.Q., Phuong, T.D., and Yen, N.D., *On the contractibility of the efficient and weakly efficient sets in* R^2, in "Equilibrium Problems and Variational Models", Maugeri A. (Ed.), Kluwer, 2001.

[32] Ioffe, A.D., *Nonsmooth analysis: Differential calculus of non-differentiable mappings,* Transactions of the American Mathematical Society, 266 (1981), pp. 1-56.

[33] Isac, G., *Pareto optimization in infinite dimensional spaces: The importance of nuclear cones*, J. Math. Anal. Appl., 182 (1994), 393-404.

[34] Jahn, J., *Mathematical Vector Optimization in Partially Ordered Linear Spaces*, Peter Lang, Frankfurt, 1986.

[35] Jahn, J., *A generalization of a theorem of Arrow, Barankin and Blackwell*, SIAM J. Control and Optimization, 26 (1988), pp. 999-1005.

[36] Jeyakumar, V. and Luc, D.T., *Approximate Jacobian matrices for nonsmooth continuous maps and* C^1*-optimization,* SIAM Journal on Control and Optimization, 36 (1998), pp.1815-1832.

[37] Jeyakumar, V. and Luc, D.T., *Nonsmooth calculus, minimality and monotonicity of convexificators,* J. Optimization Theory and Applications, 101 (1999), pp. 599-621.

[38] Jeyakumar, V. and Luc, D.T., *Open mapping theorem using unbounded generalized Jacobians,* Nonlinear Analysis, 50 (2002), pp. 647-663.

[39] Jeyakumar, V., Luc, D.T. and Schaible, S., *Characterizations of generalized monotone nonsmooth continuous maps using approximate Jacobians,* J. Convex Analysis, 5 (1998), pp. 119-132.

[40] Luc, D.T., *On scalarizing method in vector optimization*, in Large-Scale Modelling and Interactive Decision Analysis, Proceedings, Eisenach 1985, Eds. Fandel, Grauer, Kurzahanski and Wiersbicki, LNEMS 273 (1986), pp. 46-51.

[41] Luc, D.T., *Structure of the efficient point set*, Proc. Amer. Math. Soc. 95 (1985), pp. 433-440.

[42] Luc, D.T., *Theory of Vector Optimization*, Lecture Notes on Economics and Mathematical Systems 319, Springer-Verlag,Germany, 1989.

[43] Luc, D.T., *On three concepts of quasiconvexity in vector optimization,* Acta Math.Vietnamica, 15 (1990), pp. 3-9.

[44] Luc, D.T., Continuity properties of cone-convex functions, Acta Math.Hungar.,55 (1990), pp. 57-61.

[45] Luc, D.T., *Contractibility of efficient point setrs in normed spaces,* J. Nonlinear Analysis and Applications 15 (1990), pp.3-9.; and Corrigendum, idem., 38 (1999), pp. 547.

[46] Luc, D.T., *Contingent derivative of set-valued maps and applications to vector optimization,* Mathematical Programming, 50 (1991), pp. 99-111.

[47] Luc, D.T., *On generalized convex nonsmooth functions,* Bulletin of the Austral. Math.Soc., 49 (1994), pp. 139-149.

[48] Luc, D.T., *On the properly efficient points of nonconvex sets,* European Journal of Operational Research, 86 (1995), pp. 332-336.

[49] Luc, D.T., *Generalized monotone maps and bifunctions,* Acta Math.Vietnamica, 21 (1996), pp. 213-253.

[50] Luc, D.T., *A multiplier rule for multiobjective programming problems with continuous data*, SIAM J. Optimization, 13 (2002), pp. 168-178.

[51] Luc, D.T. and Malivert, C., *Invex optimization problems,* Bulletin of the Australian Mathematical Society, 46 (1992), pp.47-66.

[52] Luc, D.T., and Schaible, S., *Efficiency and generalized concavity*, J. Optimization Theory and Applications, 94 (1997), pp. 147-153.

[53] Luc, D.T., and Swaminathan, S., *A characterization of convex functions,* J.Nonlinear Analysis: TMA, 20 (1993), pp. 697-701.

[54] Luc, D.T., Tan, N.X., and Tinh, P.N., *Convex vector functions and their subdifferential*, Acta Mathematica Vietnamica, 23 (1998), pp. 107-127.

[55] Luc, D.T., and Volle, M., *Level sets, infimal convolution and level addition,* J. Optimization Theory and Application, 94 (1997), pp. 695-714.

[56] Makarov, E.K. and Rachkovski, N.N., *Density theorems for generalized Henig proper efficiency*, J. Optimization Theory and Applications, 91 (1996), pp. 419-437.

[57] Makarov, E.K. and Rachkovski, N.N., *Efficient sets of convex compacta are arcwise connected*, J. Optimization Theory and Applications, 110 (2001), pp. 159-172.

[58] Makarov, E.K., Rachkovski, N.N. and SONG, W., *Arcwise connectedness of closed point sets*, J. Mathematical Analysis and Applications, 247 (2000), pp. 377-383.

[59] Peleg, B., *Topological properties of the efficient point set*, Proceedings Amer. Math. Soc, 35 (1972), pp. 531-536.

[60] Penot, J.-P., *Differentiability of relations and differential stability of perturbed optimization problems*, SIAM J. Control and Optimization, 22 (1984), pp. 529-551.

[61] Petschke, M., *On a theorem of Arrow, Barankin and Blackwell*, SIAM J. Control and Optimization, 28 (1990), pp. 395-401.

[62] Popovici, N., *Contribution à l' optimization vectorielle*, PhD. Thesis, Faculté des Sciences, Université de Limoges, 1995.

[63] Postolica, V., *New existence results for efficient points in locally convex spaces ordered by supernormal cones*, J. Global Optimization, 3 (1993), pp. 233-243.

[64] Sach, P.H., Yen, N.D. and Craven, B.D., *Generalized invexity and duality theories with multifunctions*, Numerical Functional Analysis and Optimization, 15 (1994), pp. 131-153.

[65] Schaible, S., *Bicriteria quasiconvex programs,* Cahiers du Centre d'Etudes de Recherche Opérationnelle, 25 (1983), pp. 93-101.

[66] Sterna-Karwat, A., *On existence of cone-maximal points in real topological linear spaces*, Israel J.Math. 54 (1986), pp. 33-41.

[67] Sun, E.J., *On the connectedness of the efficient set for strictly quasiconvex vector minimization problem*, J. Optimization Theory and Applications, 89 (1996), pp. 475-581.

[68] Tinh, P.N., Luc, D.T., and Tan, N.X., *Subdifferential characterization of quasiconvex and convex vector functions*, Vietnam J. Mathematics, 26 (1998), pp. 53-69.

[69] Valadier, M., *Sous-différentiabilité de fonctions convexes à valeurs dans un espace vectoriel ordonné,* Math. Scand., 30 (1972), pp. 65-74.

[70] Warga, J., *Fat homeomorphisms and unbounded derivate containers*, J. Mathematical Analysis and Applications, 81 (1981), pp. 545-560.

[71] Yen, N.D. and Phuong, T.D., *Connectedness and stability of the solution set in linear fractional vector optimization inequalities*, In: Vector Variational Inequalities and Vector Equilibria. Mathematical Theorems, Giannessi F., Ed., Kluwer Academic Publishers, Dordrecht, 2000, pp. 479-489.

[72] Zalinescu, C., and Sonntag, Y., *Comparison of existence results for efficient points*, J. Optimization Theory and Applications, 105 (2000), pp. 161-188.

[73] Zowe, J., *Subdifferentiability of convex functions with values in an ordered vector space*, Math. Scandi., 34 (1974), pp. 69-83.

Chapter 6

GENERALIZED CONVEX DUALITY AND ITS ECONOMIC APPLICATIONS*

Juan Enrique Martínez-Legaz
CODE and Departament d'Economia i d'Història Econòmica
Universitat Autònoma de Barcelona
08193 Bellaterra
Spain
JuanEnrique.Martinez@uab.es

Abstract This article presents an approach to generalized convex duality theory based on Fenchel-Moreau conjugations; in particular, it discusses quasiconvex conjugation and duality in detail. It also describes the related topic of microeconomics duality and analyzes the monotonicity of demand functions.

Keywords: Generalized convex duality, quasiconvex optimization, demand function, utility function.

1. Introduction

A central topic in optimization is convex duality theory. In its modern approach, mainly due to Rockafellar [121], [122], given a (primal) convex optimization problem one embeds it into a family of perturbed optimization problems and then, relative to these perturbations, one associates to it a so-called dual problem. The deep relations existing between the

*Research supported by the Ministerio de Ciencia y Tecnología (Spain), Project BEC2002-00642, and by the Comissionat per Universitats i Recerca de la Generalitat de Catalunya, Grant 2001SGR-00162. This work has been completed during a visit to the Laboratoire de Mathématiques Appliquées of the University of Pau (November 2001). I am grateful to Rosalind Elster, Jean-Paul Penot, Ivan Singer and the editors for their careful reading of earlier drafts of this manuscript and their helpful remarks and to John K.-H. Quah for his useful comments on Section 7.

primal and the dual are helpful for analyzing the properties of the original problem and, in particular, for obtaining optimality conditions; they are also used to devise numerical algorithms. In the case of problems arising in applications to other sciences, particularly in economics, the dual problems usually have nice interpretations that shed new light into the nature of the associated primal problems and yield a new perspective for analyzing them.

Convex duality is based on the theory of convex conjugation. In its finite dimensional version, which we adopt in this article for simplicity[1], to each extended real-valued function $f : \mathbb{R}^n \to \overline{\mathbb{R}} = [-\infty, +\infty]$ one associates another function $f^* : \mathbb{R}^n \to \overline{\mathbb{R}}$, called its conjugate, defined by $f^*(x^*) = \sup \{\langle x, x^* \rangle - f(x)\}$; here $\langle \cdot, \cdot \rangle$ stands for the scalar product in $\mathbb{R}^n$. Being the pointwise supremum of the collection

$$\{\langle x, \cdot \rangle - f(x)\}_{x \in f^{-1}(\mathbb{R})}$$

of affine functions, f^* is convex, lower semicontinuous (l.s.c.) and proper (in the sense that it does not take the value $-\infty$ except if it is identically $-\infty$)[2]. So is, in particular, f^{**}, the second conjugate of f, which, on the other hand, is a minorant of f (an easy consequence of the definition of conjugate function); it actually follows from the separation theorem for closed convex sets that f^{**} is the largest l.s.c. proper convex minorant of f. Thus a function f is convex, proper and l.s.c if and only if it coincides with its second conjugate.

Another essential tool in duality theory is the notion of subgradient. One says that $x^* \in \mathbb{R}^n$ is a subgradient of $f : \mathbb{R}^n \to \overline{\mathbb{R}}$ at $x_0 \in f^{-1}(\mathbb{R})$ if

$$f(x) - f(x_0) \geq \langle x - x_0, x^* \rangle \qquad (x \in \mathbb{R}^n).$$

The set of all subgradients of f at x_0, denoted $\partial f(x_0)$, is called the subdifferential of f at x_0. For every $f : \mathbb{R}^n \to \overline{\mathbb{R}}, x_0 \in f^{-1}(\mathbb{R})$ and $x^* \in \mathbb{R}^n$, the following properties hold:

$$x^* \in \partial f(x_0) \text{ if and only if } f(x_0) + f^*(x^*) = \langle x_0, x^* \rangle,$$
$$x^* \in \partial f^{**}(x_0) \text{ if and only if } x_0 \in \partial f^*(x^*),$$
$$\partial f(x_0) \neq \emptyset \text{ implies } f^{**}(x_0) = f(x_0),$$
$$f^{**}(x_0) = f(x_0) \text{ implies } \partial f^{**}(x_0) = \partial f(x_0).$$

[1]However convex and quasiconvex conjugation and duality theories extend easily to the framework of locally convex real topological vector spaces.

[2]This notion of properness differs from the usual one in that it is satisfied by the constant functions $+\infty$ and $-\infty$. Under this slightly modified definition, some statements become simpler.

One can easily verify that the subdifferential mapping $x \rightrightarrows \partial f(x)$ is cyclically monotone, that is, one has

$$\left.\begin{array}{r}\langle x_1 - x_0, x_0^*\rangle + \langle x_2 - x_1, x_1^*\rangle + \cdots + \langle x_0 - x_m, x_m^*\rangle \le 0 \\ \text{for any set of pairs } (x_i, x_i^*),\ i = 0, 1, ..., m \ (m \text{ arbitrary}) \\ \text{such that } x_i^* \in \partial f(x_i).\end{array}\right\} \quad (6.1)$$

This property is characteristic of "submappings" of subdifferential operators [121, Thm. 24.8]: A multivalued mapping ρ from $\mathbb{R}^n$ to $\mathbb{R}^n$ is cyclically monotone, i.e. (6.1) holds with ∂f replaced with ρ, if and only if there exists a l.s.c. proper convex function $f : \mathbb{R}^n \to \mathbb{R} \cup \{+\infty\}$ such that $\rho(x) \subseteq \partial f(x)$ for every $x \in \mathbb{R}^n$.

To apply convex conjugation theory to duality in optimization, one considers a family of problems

$$(\mathcal{P}_u) \qquad \text{minimize } \varphi(x, u),$$

$\varphi : \mathbb{R}^n \times \mathbb{R}^m \to \overline{\mathbb{R}}$ being an objective function and $u \in \mathbb{R}^m$ denoting a given parameter vector; the minimization variable is thus $x \in \mathbb{R}^n$. Problem $(\mathcal{P}_u)$ is regarded as a perturbation of an (unperturbed) primal problem, defined as

$$(\mathcal{P}) \qquad \text{minimize } \varphi(x, 0).$$

The perturbation function associated to the family of perturbed optimization problems is $p : \mathbb{R}^m \to \overline{\mathbb{R}}$, defined by $p(u) = \inf_{x \in \mathbb{R}^n} \varphi(x, u)$; it thus assigns to each perturbation parameter $u \in \mathbb{R}^m$ the optimal value of the corresponding problem. One associates to $(\mathcal{P})$ a dual problem, relative to the family of perturbed problems $(\mathcal{P}_u)$:

$$(\mathcal{D}) \qquad \text{maximize } -\varphi^*(0, u^*).$$

One can easily check that the objective function of this dual problem satisfies $-\varphi^*(0, u^*) = -p^*(u^*)$, and hence the optimal value is $p^{**}(0)$, therefore it is less than or equal to the optimal value $p(0)$ of the primal problem; moreover, the set of optimal dual solutions is $\partial p^{**}(0)$. If the perturbation function p is convex (which happens, in particular, if φ is convex), proper and l.s.c. then both optimal values are the same and the optimal solution set is simply $\partial p(0)$. In fact, it turns out that for the optimal values to be the same a necessary and sufficient condition is the coincidence at the origin of the perturbation function with its largest l.s.c. proper convex minorant. Notice that this duality theory is fully symmetric when applied to problems with a l.s.c. proper convex perturbed objective function φ. Indeed, in this case, by embedding the dual problem $(\mathcal{D})$ into the family of perturbed problems

$$(\mathcal{D}_{x^*}) \qquad \text{maximize } -\varphi^*(x^*, u^*),$$

or, equivalently,

$$(\mathcal{D}_{x^*}) \qquad \text{minimize } \psi(u^*, x^*),$$

with $\psi : \mathbb{R}^m \times \mathbb{R}^n \to \overline{\mathbb{R}}$ defined by $\psi(u^*, x^*) = \varphi^*(x^*, u^*)$, one gets, as a dual problem to $(\mathcal{D})$,

$$\text{maximize } -\psi^*(0, x),$$

which is clearly equivalent to the primal problem $(\mathcal{P})$ (as $\psi^*(u, x) = \varphi(x, u)$ for every $(u, x) \in \mathbb{R}^m \times \mathbb{R}^n$). In fact, what are in symmetric duality are the families $(\mathcal{P}_u)$ and $(\mathcal{D}_{x^*})$ of perturbed problems rather than the unperturbed problems $(\mathcal{P})$ and $(\mathcal{D})$ alone.

The above duality theory specializes very nicely in the case of inequality constrained minimization problems of the form

$$\begin{array}{ll} \text{minimize} & f(x) \\ \text{subject to} & g(x) \le 0, \end{array}$$

with $f : \mathbb{R}^n \to \overline{\mathbb{R}}$ and $g : \mathbb{R}^n \to \mathbb{R}^m$; the inequality $\le$ in $\mathbb{R}^m$ is to be understood in the componentwise sense. The classical way to embed this problem into a family of perturbed ones is by introducing so-called vertical perturbations, namely, one considers the perturbed objective function $\varphi : \mathbb{R}^n \times \mathbb{R}^m \to \overline{\mathbb{R}}$ given by

$$\varphi(x, u) = \begin{cases} f(x) & \text{if } g(x) + u \le 0 \\ +\infty & \text{otherwise} \end{cases}.$$

This function is convex whenever f and the component functions of g are convex. Clearly, minimizing $\varphi(x, 0)$ is equivalent to the original problem. A straightforward computation of $\varphi^*(0, u^*)$ shows that, in this case, the dual problem $(\mathcal{D})$ reduces to

$$\begin{array}{ll} \text{maximize} & \inf_{x \in \mathbb{R}^n} L(x, u^*) \\ \text{subject to} & u^* \ge 0, \end{array}$$

$L : \mathbb{R}^n \times \mathbb{R}^m \to \overline{\mathbb{R}}$ being the Lagrangian function defined by $L(x, u^*) = f(x) + \langle g(x), u^* \rangle$. In this way, the classical Lagrangian duality theory becomes a particular case of the perturbational duality theory we have briefly described.

It is worth mentioning that, historically, the first dual problems discovered in optimization theory were defined without using any perturbation; the perturbational approach to duality proposed by Rockafellar later on provided us with a unifying scheme for all those duals. In a series of papers, [143], [146], [147], Singer has shown that the converse way also works, that is, some unperturbational dual problems induce the perturbational dual problems.

Another example of the general duality scheme described above is Fenchel duality for the unconstrained minimization problem

$$\text{minimize } f(x) - g(x), \tag{6.2}$$

with $f : \mathbb{R}^n \to \mathbb{R} \cup \{+\infty\}$ and $g : \mathbb{R}^n \to \mathbb{R} \cup \{-\infty\}$. By embedding this problem into the family of perturbed minimization problems with objective function $\varphi : \mathbb{R}^n \times \mathbb{R}^n \to \overline{\mathbb{R}}$ defined by $\varphi(x,u) = f(x+u) - g(x)$, one arrives at the dual problem

$$\text{maximize } g_*(u^*) - f^*(u^*),$$

$g_* : \mathbb{R}^n \to \overline{\mathbb{R}}$ denoting the concave conjugate of g, given by $g_*(u^*) = -(-g)^*(-u^*) = \inf\{\langle x, u^*\rangle - g(x)\}$.

Of a completely different nature, though also based on convex conjugation, is Toland-Singer duality theory [159], [137]. It also deals with problem (6.2), but assuming that $f, g : \mathbb{R}^n \to \mathbb{R} \cup \{+\infty\}$ are l.s.c. proper convex functions and using the convention $(+\infty) - (+\infty) = +\infty$. Thus the problem consists in minimizing a d.c. (difference of convex) function, which is in general a nonconvex problem. Since $g = g^{**}$, one has

$$\begin{aligned}
\inf_{x\in\mathbb{R}^n} \{f(x) - g(x)\} &= \inf_{x\in\mathbb{R}^n} \{f(x) - g^{**}(x)\} \\
&= \inf_{x\in\mathbb{R}^n} \left\{ f(x) - \sup_{x^*\in\mathbb{R}^n} \{\langle x, x^*\rangle - g^*(x^*)\} \right\} \\
&= \inf_{x,x^*\in\mathbb{R}^n} \{f(x) - \langle x, x^*\rangle + g^*(x^*)\} \\
&= \inf_{x^*\in\mathbb{R}^n} \left\{ g^*(x^*) + \inf_{x\in\mathbb{R}^n} \{f(x) - \langle x, x^*\rangle\} \right\} \\
&= \inf_{x^*\in\mathbb{R}^n} \{g^*(x^*) - f^*(x^*)\}.
\end{aligned}$$

Notice that the preceding proof of Toland-Singer theorem does not require any property of f; however there is no loss of generality in assuming that f is a l.s.c. proper convex function, since the dual problem $\inf_{x^*\in\mathbb{R}^n} \{g^*(x^*) - f^*(x^*)\}$ depends on f only through f^{**}.

Duality theory in d.c. optimization is further developed in [164], [152] and [165]. Problems with d.c. objective and constraint functions are discussed in [57], [58] and [93]. In this last paper, Lagrangian and Toland-Singer duality theories are unified, as they arise as special cases of the general framework developed there. In [94], d.c. duality on compact sets is studied; the compactness assumption is used there as a substitute for the restrictive constraint qualifications required in [93].

Convex duality theory admits an extension to the generalized convex case; it will be presented in Section 3. It is based on the so-called Fenchel-Moreau generalized conjugation scheme, which will be outlined in the next section. A special type of generalized conjugation operators are the so-called level sets conjugations, useful in quasiconvex analysis; they will be described in Section 4. Their applications to duality theory in quasiconvex optimization will be the object of Section 5. The rest of the chapter is devoted to economic applications. Section 6 deals with duality between direct and indirect utility functions in consumer theory. The related topic of monotonicity of demand functions will be treated in Section 7, in which a new result characterizing utility functions inducing monotone demands will be presented. In the last section we will be concerned with the extension of consumer duality theory to the case when consumer's preferences are not represented by utility functions.

The subject of generalized convex duality is covered in several monographs [148], [105], [124], where the reader can find some further developments as well as other topics not covered in this article.

2. Generalized Convex Conjugation

Convex conjugation theory was extended by Moreau [102] to an abstract framework, which we are going to present in this section.

Let X and Y be arbitrary sets and $c : X \times Y \to \overline{\mathbb{R}}$ be a function, which will be called the coupling function. For any $f : X \to \overline{\mathbb{R}}$, we define its $c-$conjugate $f^c : Y \to \overline{\mathbb{R}}$ by

$$f^c(y) = \sup_{x \in X} \{c(x,y) - f(x)\};$$

here and in the sequel we use the conventions $+\infty + (-\infty) = -\infty + (+\infty) = +\infty - (+\infty) = -\infty - (-\infty) = -\infty$, except in the statement of Theorem 6.1 and formula (6.4), where $+\infty + (-\infty) = +\infty - (+\infty) = -\infty - (-\infty) = +\infty$. Similarly, the $c'-$conjugate of $g : Y \to \overline{\mathbb{R}}$ is the function $g^{c'} : X \to \overline{\mathbb{R}}$ defined by

$$g^{c'}(x) = \sup_{y \in Y} \{c(x,y) - g(y)\};$$

notice that this notation is consistent with considering the coupling function $c' : Y \times X \to \overline{\mathbb{R}}$ given by $c'(y,x) = c(x,y)$.

Functions of the form $x \in X \to c(x,y) - \beta \in \overline{\mathbb{R}}$, with $y \in Y$ and $\beta \in \overline{\mathbb{R}}$, are called $c-$elementary; in the same way, $c'-$elementary functions are those of the form $y \in Y \to c(x,y) - \beta \in \overline{\mathbb{R}}$, for given $x \in X$ and $\beta \in \overline{\mathbb{R}}$. We denote by Φ_c and $\Phi_{c'}$ the sets of $c-$elementary functions and $c'-$elementary functions, respectively.

Let Φ be a set of extended real-valued functions on X. According to the duality theory introduced by Dolecki and Kurcyusz [32] (see also the pioneering paper [56], in which a more abstract duality theory was first developed), a function $f : X \to \overline{\mathbb{R}}$ is called $\Phi-$convex if it is the pointwise supremum of a subset of Φ. Clearly, the class of $\Phi-$convex functions is closed under pointwise supremum. Hence, every function $f : X \to \overline{\mathbb{R}}$ has a largest $\Phi-$convex minorant, which is called its $\Phi-$convex hull. Notice that these notions make also sense in the more general case when Φ is a set of functions from X into a complete lattice A and $f : X \to A$.

The proofs of the following propositions are easy.

Proposition 6.1 *Let $f : X \to \overline{\mathbb{R}}$, $g : Y \to \overline{\mathbb{R}}$, $x \in X$ and $y \in Y$. Then*
(i) $f^c(y) \geq c(x,y) - f(x)$, $g^{c'}(x) \geq c(x,y) - g(y)$,
(ii) $f^{cc'c} = f^c$, $g^{c'cc'} = g^{c'}$,
(iii) f^c *and* $g^{c'}$ *are* $\Phi_{c'}-$*convex and* Φ_c-*convex, respectively.*

Proposition 6.2 *The Φ_c-convex hull of $f : X \to \overline{\mathbb{R}}$ coincides with $f^{cc'}$. In the same way, the $\Phi_{c'}-$convex hull of $g : Y \to \overline{\mathbb{R}}$ coincides with $g^{c'c}$.*

Corollary 6.1 *A function $f : X \to \overline{\mathbb{R}}$ is Φ_c-convex if and only if it coincides with its second $c-$conjugate $f^{cc'}$. In the same way, a function $g : Y \to \overline{\mathbb{R}}$ is $\Phi_{c'}-$convex if and only if it coincides with its second $c'-$conjugate $g^{c'c}$.*

In view of the preceding proposition, we shall say that $f : X \to \overline{\mathbb{R}}$ is Φ_c-convex at $x_0 \in X$ ($g : Y \to \overline{\mathbb{R}}$ is $\Phi_{c'}-$convex at $y_0 \in Y$) if $f^{cc'}(x_0) = f(x_0)$ (resp., $g^{c'c}(y_0) = g(y_0)$). Consequently, Φ_c-convexity (or $\Phi_{c'}-$convexity) of a function is equivalent to the corresponding property at every point.

We next give some useful examples of generalized conjugation operators:

Example 6.1 [102, p. 125] Let X and Y be a dual pair of vector spaces and $\langle \cdot, \cdot \rangle : X \times Y \to \mathbb{R}$ denote the duality pairing. Define $c : X \times Y \to \overline{\mathbb{R}}$ by $c(x,y) = \log \langle x, y \rangle$, with the convention $\log t = -\infty$ for $t \leq 0$. Then $f : X \to \overline{\mathbb{R}}$ is Φ_c-convex if and only if e^f (with the convention $e^{-\infty} = 0$) is sublinear (i.e. subadditive and positively homogeneous) or, equivalently, the support function of a set $B \subseteq Y$ that contains the origin. The function e^{f^c} is the support function of $B^0 = \{x \in X : \langle x, y \rangle \leq 0 \ \forall y \in B\}$.

Example 6.2 [102, p. 126] Let $X = Y = [0, +\infty]$, and define $c : X \times Y \to \overline{\mathbb{R}}$ by $c(x, y) = xy$, with the convention $\alpha(+\infty) = (+\infty)\alpha = +\infty$ for every $\alpha \in [0, +\infty]$. Then $f : X \to \overline{\mathbb{R}}$ is Φ_c–convex if and only if it is convex and nondecreasing. The conjugate function f^c is the so-called Young transform of f.

Example 6.3 [6, Thm 2] Let X be a topological space, Y an arbitrary set and $c : X \times Y \to \mathbb{R}$ be of needle type on X, i.e. such that for every $(x_0, y, \eta) \in X \times Y \times \mathbb{R}$ and every neighborhood N of x_0 there exist $y' \in Y$ and a neighborhood $N' \subseteq N$ of x_0 such that

$$c(x, y') - c(x_0, y') \leq c(x, y) + \eta \qquad (x \in X \backslash N')$$

and

$$c(x, y') - c(x_0, y') \leq 0 \qquad (x \in X).$$

Then $f : X \to \mathbb{R} \cup \{+\infty\}$ is Φ_c–convex if and only if it is l.s.c. and has a finite-valued c–elementary minorant.

If X is a Hilbert space and $Y = \mathbb{R}_+ \times X$ then the function $c : X \times Y \to \mathbb{R}$ defined by $c(x, (\rho, y)) = -\rho \|x - y\|^2$ is of needle type on X.

Example 6.4 [71, Prop. 2.2] Let $X = Y = \mathbb{R}^n$, $0 < \alpha \leq 1$ and $N > 0$. Define $c : X \times Y \to \mathbb{R}$ by $c(x, y) = -N \|x - y\|^\alpha$. Then $f : X \to \mathbb{R}$ is Φ_c–convex if and only if it is α–Hölder continuous with constant N.

Example 6.5 [80, Thm. 5.4] Let X be a normed space with dual X^*, $0 < \alpha \leq 1$ and $Y = \mathcal{B}^*(0, N) \times \mathbb{R}$, with $\mathcal{B}^*(0, N)$ denoting the closed ball in X^* with radius $N > 0$. Define $c : X \times Y \to \mathbb{R}$ by $c(x, (\omega, k)) = \min\{-(k - \omega(x))^\alpha, 0\} + k$, with the convention $t^\alpha = -\infty$ if $t < 0$ and $\alpha \neq 1$. Then $f : X \to \mathbb{R}$ is Φ_c–convex if and only if it is quasiconvex[3] and α–Hölder continuous with constant N^α.

Example 6.6 [85, Thm. 5.3] Let $X = \{0, 1\}^n$, $Y = C_1 \times C_2 \times \cdots \times C_n$, where $C_i \subseteq \mathbb{R}$ are unbounded from above and from below, and

[3]See Section 4 for the definition of quasiconvexity and for another conjugation scheme for quasiconvex functions that does not require the very restrictive Hölder continuity condition of this example.

$c : X \times Y \to \mathbb{R}$ be the restriction of the scalar product. Then $f : X \to \overline{\mathbb{R}}$ is Φ_c-convex if and only if it does not take the value $-\infty$ unless it is identically $-\infty$. This generalized conjugation operator (in the case $Y = \mathbb{R}^n$) has been used in [76] to propose a dual representation of cooperative games.

Example 6.7 [92, Thm. 2.1] Let $X = \mathbb{R}^n$ and

$$Y = \bigcup_{k=0}^{n} \left(\mathcal{L}\left(\mathbb{R}^n, \mathbb{R}^k\right) \times \mathbb{R}^k\right) \times \mathbb{R}^n,$$

with $\mathcal{L}\left(\mathbb{R}^n, \mathbb{R}^k\right)$ denoting the set of all linear mappings $u : \mathbb{R}^n \to \mathbb{R}^k$. Define $c : X \times Y \to \overline{\mathbb{R}}$ by

$$c\left(x, \left(u, z, x^*\right)\right) = \begin{cases} -\infty & \text{if } u\left(x\right) <_L z \\ \langle x, x^* \rangle & \text{if } u\left(x\right) = z \\ +\infty & \text{if } u\left(x\right) >_L z \end{cases};$$

$<_L$ and $>_L$ stand here for "lexicographically less than" and "lexicographically greater than", respectively. Then $f : X \to \overline{\mathbb{R}}$ is Φ_c-convex if and only if it is convex. This example shows that the notion of $\Phi-$convexity is a true generalization of convexity (not only of lower semicontinuous proper convexity)!

Example 6.8 [128, Thm. 2.1] Let $X = Y = \mathbb{R}^n_+$, and define $c : X \times Y \to \mathbb{R}$ by

$$c\left(x, y\right) = \begin{cases} \min_{i \text{ s.t. } y_i > 0} x_i y_i & \text{if } y \neq 0 \\ 0 & \text{if } y = 0 \end{cases}.$$

Then $f : X \to \overline{\mathbb{R}}$ is Φ_c-convex if and only if it is nondecreasing and its restriction to any open ray emanating from the origin is convex. These functions are called ICAR (increasing and convex-along-rays). Examples of ICAR functions are all nondecreasing positively homogeneous functions of degree $k \geq 1$ (in particular, all Cobb-Douglas functions $f(x) = x_1^{\delta_1} x_2^{\delta_2} \cdot \ldots \cdot x_n^{\delta_n}$, $\delta_i > 0$, with $\sum_{i=1}^n \delta_i \geq 1$, which are of utmost importance in economic modeling) and all polynomials with nonnegative coefficients (in particular, all quadratic forms $f(x) = \langle x, Ax \rangle$ with nonnegative matrix A) [128, Ex. 2.1 and 2.2]. Based on this generalized convexity property of ICAR functions, the so-called cutting angle method (a generalization of the well known cutting plane method of convex programming) has been proposed and successfully implemented for solving a very broad class of nonconvex global optimization problems (see [3], [125] and [7]).

Example 6.9 [78, Lemma A.1(a)] Let $X = Y = \mathbb{R}^n_{++} \cup \{0\}$, and define $c : X \times Y \to \mathbb{R}$ by $c(x, y) = -\max_{i=1,\ldots,n} x_i y_i$. Then $f : X \to \overline{\mathbb{R}}$ is Φ_c-convex if and only if it is nonincreasing and its restriction to any ray emanating from the origin is convex and continuous.

The possibility of extending Fenchel duality (see the Introduction) to the unconstrained minimization of the sum of two generalized convex functions in terms of conjugates is discussed in [75].

As for Toland-Singer duality, extensions to the generalized conjugation framework have been developed by Singer [142], [144] and Volle [163]. According to [72, Thm. 3.1], one has:

Theorem 6.1 *For any $h : X \to \overline{\mathbb{R}}$, the following statements are equivalent:*

(i) h is Φ_c-convex.

(ii) $\inf_{x\in X} \{f(x) - h(x)\} = \inf_{x\in X}\{f(x) - h^{cc'}(x)\} \qquad \left(f \in \overline{\mathbb{R}}^X\right).$

(iii) $\inf_{x\in X} \{f(x) - h(x)\} = \inf_{y\in Y} \{h^c(y) - f^c(y)\} \qquad \left(f \in \overline{\mathbb{R}}^X\right).$

(iv) $\inf_{x\in X} \{f(x) - h(x)\} = \inf_{x\in X}\{f^{cc'}(x) - h^{cc'}(x)\} \qquad \left(f \in \overline{\mathbb{R}}^X\right).$

(v) $\inf_{x\in X} \{f(x) - h(x)\} = \inf_{x\in X}\{f^{cc'}(x) - h(x)\} \qquad \left(f \in \overline{\mathbb{R}}^X\right).$

For some related results involving generalized conjugation for differences of functions (under suitable assumptions on the coupling function), see [72, Thm. 4.1]. Applications to global optimality conditions for unconstrained minimization are discussed in [41]. For an abstract extension of d.c. duality theory, we refer to [90].

Following [6], we say that $f : X \to \overline{\mathbb{R}}$ is $c-$subdifferentiable at $x_0 \in X$ if $f(x_0) \in \mathbb{R}$ and there exists $y_0 \in Y$ such that $c(x_0, y_0) \in \mathbb{R}$ and

$$f(x) - f(x_0) \geq c(x, y_0) - c(x_0, y_0) \qquad (x \in X).$$

One then says that y_0 is a $c-$subgradient of f at x_0. The set of all $c-$subgradients of f at x_0, denoted $\partial_c f(x_0)$, is called the $c-$subdifferential of f at x_0. We set $\partial_c f(x_0) = \emptyset$ if $f(x_0) \notin \mathbb{R}$. The following properties hold:

Proposition 6.3 *Let $f : X \to \overline{\mathbb{R}}$, $x_0 \in X$ and $y_0 \in Y$. If $c(x_0, y_0) \in \mathbb{R}$ then*

$$y_0 \in \partial_c f(x_0) \textit{ if and only if } f(x_0) + f^c(y_0) = c(x_0, y_0),$$
$$y_0 \in \partial_c f^{cc'}(x_0) \textit{ if and only if } x_0 \in \partial_{c'} f^c(y_0),$$
$$\partial_c f(x_0) \neq \emptyset \textit{ implies that } f \textit{ is } \Phi_c - \textit{convex at } x_0,$$
$$f \textit{ is } \Phi_c - \textit{convex at } x_0 \textit{ implies } \partial_c f^{cc'}(x_0) = \partial_c f(x_0).$$

We omit the obvious corresponding notions and properties for functions $g : Y \to \overline{\mathbb{R}}$.

From Proposition 6.3, it follows that, for a Φ_c-convex function $f : X \to \overline{\mathbb{R}}$, the inverse to the $c-$subdifferential operator $\partial_c f$ is $\partial_{c'} f^c$, that is, for any $x_0 \in X$.and $y_0 \in Y$ one has $y_0 \in \partial_c f(x_0)$ if and only if $x_0 \in \partial_{c'} f^c(y_0)$.

The equivalence between submappings of convex subdifferential operators and the cyclic monotonicity property [121, Thm. 24.8] can be extended to the nonconvex case [44, Thm. 2.7]:

Definition 6.1 *Let $c : X \times Y \to \mathbb{R}$. A multivalued mapping ρ from X to Y is $c-$cyclically monotone if the expression*

$$(c(x_1, y_0) - c(x_0, y_0)) + (c(x_2, y_1) - c(x_1, y_1)) + \cdots + (c(x_0, y_m) - c(x_m, y_m))$$

is nonpositive for any set of pairs $(x_i, y_i), i = 0, 1, ..., m$ (m arbitrary) such that $y_i \in \rho(x_i)$.

Theorem 6.2 *A multivalued mapping ρ from X to Y is $c-$cyclically monotone if and only if there exists a Φ_c-convex function $f : X \to \mathbb{R} \cup \{+\infty\}$ such that $\rho(x) \subseteq \partial_c f(x)$ for every $x \in X$.*

In classical convex analysis, one has the strongest result that maximal cyclically monotone mappings are exactly the subdifferentials of l.s.c. convex functions, and two such functions having the same subdifferential mapping coincide up to an additive constant. However, these results do not extend to our general setting, as the following example shows:

Example 6.10 [60, p. 15] Let $X = Y = \{1, 2\}$ and define $c : X \times Y \to \mathbb{R}$ by $c(i, j) = \begin{cases} 1 & \text{if } i = j \\ 0 & \text{if } i \neq j \end{cases}$. Consider the functions $f_1, f_2, f_3 : X \to \mathbb{R}$ given by $f_1(1) = f_1(2) = 1$, $f_2(1) = 1$, $f_2(2) = 0$, $f_3(1) = 1$ and $f_3(2) = \frac{3}{2}$. One can easily check that, for $x \in \{1, 2\}$, one has

$$f_1(x) = \max\{c(x, 1), c(x, 2)\},$$
$$f_2(x) = \max\{c(x, 1), c(x, 2) - 1\},$$
$$f_3(x) = \max\left\{c(x, 1), c(x, 2) + \frac{1}{2}\right\},$$

whence all three functions are Φ_c-convex. The subdifferential mappings are given by

$$\partial_c f_1(x) = \{x\} \qquad (x \in X),$$
$$\partial_c f_2(1) = \{1\}, \ \partial_c f_2(2) = Y,$$
$$\partial_c f_3(x) = \{x\} \qquad (x \in X).$$

Thus $\partial_c f_1$ and $\partial_c f_3$ are not maximal $c-$cyclically monotone, since their graphs are strictly contained in that of $\partial_c f_2$, which is $c-$cyclically monotone. Moreover, $\partial_c f_1 = \partial_c f_3$ but $f_1 - f_3$ is not constant. Notice that $\partial_c f_2$ is maximal $c-$cyclically monotone, because the only mapping that strictly dominates $\partial_c f_2$ has all of $X \times Y$ as its graph and so is not $c-$cyclically monotone.

From an axiomatic point of view, conjugation operators $f \in \overline{\mathbb{R}}^X \mapsto f^c \in \overline{\mathbb{R}}^Y$ were characterized by Singer [140, Thm. 3.1]:

Theorem 6.3 *For a mapping $\Delta : \overline{\mathbb{R}}^X \to \overline{\mathbb{R}}^Y$, the following statements are equivalent:*

(i) There exists a coupling function $c : X \times Y \to \overline{\mathbb{R}}$ such that

$$\Delta(f) = f^c \qquad (f \in \overline{\mathbb{R}}^X).$$

(ii) One has

$$\Delta(\inf_{i\in I} f_i) = \sup_{i\in I} \Delta(f_i) \qquad (\{f_i\}_{i\in I} \subseteq \overline{\mathbb{R}}^X) \tag{6.3}$$

and

$$\Delta(f + d) = \Delta(f) - d \qquad (f \in \overline{\mathbb{R}}^X,\ d \in \mathbb{R}).$$

Moreover, in this case c is uniquely determined by Δ, namely, one has

$$c(x, y) = \Delta(\delta_{\{x\}})(y) \qquad (x \in X,\ y \in Y),$$

$\delta_{\{x\}}$ denoting the indicator function[4] of $\{x\}$.

Other generalized conjugation schemes are developed in [36], [61], [27], [28], [29], [38] and [37]. For generalized conjugation theory with functions taking values in ordered groups, we refer to [161] and [91]. An application of generalized conjugation to solving infimal convolution and deconvolution equations is given in [97]. The relationship between generalized conjugation and the theory of lower semicontinuous linear mappings for dioid-valued functions in idempotent analysis is explored in [151].

Operators satisfying (6.3) are called dualities. This notion makes also sense for operators acting on functions that take their values in arbitrary complete lattices [148, p. 419]:

[4] We recall that the indicator function $\delta_C : X \to \overline{\mathbb{R}}$ of $C \subseteq X$ is defined by

$$\delta_C(x) = \begin{cases} 0 & \text{if } x \in C \\ +\infty & \text{if } x \notin C \end{cases}.$$

Definition 6.2 *Let A and B be complete lattices. A mapping $\Delta : A^X \to B^Y$ is a duality if*

$$\Delta(\inf_{i\in I} f_i) = \sup_{i\in I} \Delta(f_i) \qquad (\{f_i\}_{i\in I} \subseteq A^X).$$

For this broader class of mappings, the following representation (of which Theorem 6.3 is a corollary) was given in [85, Thm. 3.1].

Theorem 6.4 *Let A and B be complete lattices. A mapping $\Delta : A^X \to B^Y$ is a duality if and only if there exists a function $G : X \times Y \times A \to B$ such that*

$$G(x, y, \inf_{i\in I} a_i) = \sup_{i\in I} G(x, y, a_i) \qquad (x \in X,\ \ y \in Y,\ \ \{a_i\}_{i\in I} \subseteq A)$$

and

$$\Delta(f)(y) = \sup_{x\in X} G\left(x, y, f\left(x\right)\right) \qquad (f \in A^X,\ \ y \in Y).$$

Moreover, in this case G is uniquely determined by Δ, namely, one has

$$G(x, y, a) = \Delta(\delta_{\{x\}} + a)(y) \qquad (x \in X,\ \ y \in Y,\ \ a \in A). \tag{6.4}$$

When $\Delta : \overline{\mathbb{R}}^X \to \overline{\mathbb{R}}^Y$ is a conjugation operator with coupling function $c : X \times Y \to \overline{\mathbb{R}}$, the function $G : X \times Y \times \overline{\mathbb{R}} \to \overline{\mathbb{R}}$ associated to it in the sense of the preceding theorem is given by $G(x, y, r) = c(x, y) - r$.

The dual operator of a duality is defined as follows:

Definition 6.3 *The dual operator of a duality $\Delta : A^X \to B^Y$ is the mapping $\Delta' : B^Y \to A^X$ given by*

$$\Delta'\left(g\right) = \inf\left\{f \in A^X : \Delta(f) \le g\right\}.$$

In the preceding definition, as in the sequel, the infimum in A^X is to be interpreted in the pointwise sense.

It is well known that the dual Δ' of a duality Δ is a duality, too [148, Thm. 5.3]; moreover, for any $f : X \to A$ and $g : Y \to B$ one has [148, Cor. 5.3]:

$$\Delta(f) \le g \qquad \Leftrightarrow \qquad \Delta'\left(g\right) \le f.$$

From this equivalence, it easily follows that any duality Δ coincides with its second dual [148, Thm. 5.3], i.e. $\Delta'' = \Delta$. The relationship between the representations of Δ and Δ' (in the sense of Theorem 6.4) was described in [85, Thm. 3.5]:

Theorem 6.5 *Let $\Delta : A^X \to B^Y$ be a duality , with dual $\Delta' : B^Y \to A^X$, and $G : X \times Y \times A \to B$, $G' : Y \times X \times B \to A$ be the mappings corresponding to them by Theorem 6.4. Then*

$$G'(y,x,b) = \min\{a \in A : G(x,y,a) \leq b\} \qquad (y \in Y,\ x \in X,\ b \in B).$$

In particular, the dual operator $\Delta' : \overline{\mathbb{R}}^Y \to \overline{\mathbb{R}}^X$ of a conjugation operator $\Delta : \overline{\mathbb{R}}^X \to \overline{\mathbb{R}}^Y$ with coupling function $c : X \times Y \to \overline{\mathbb{R}}$ is the conjugation operator associated with the coupling function $c' : Y \times X \to \overline{\mathbb{R}}$.

Composing a duality $\Delta : A^X \to B^Y$ with its dual $\Delta' : B^Y \to A^X$ yields a hull operator $\Delta'\Delta : A^X \to A^X$ (see, e.g., [148, Cor. 5.5]). The functions that are closed under this operator were identified in [85, Thm. 3.6]:

Theorem 6.6 *Under the assumptions of Theorem 6.5, for every function $f : X \to A$ one has*

$$\Delta'\Delta(f) = \sup\left\{G'(y,\cdot,b) : y \in Y,\ b \in B,\ G'(y,\cdot,b) \leq f\right\}.$$

Hence, $\Delta'\Delta(f) = f$ if and only if f is $\Phi^{G'}$−convex, with

$$\Phi^{G'} = \left\{G'(y,\cdot,b) : y \in Y,\ b \in B\right\}.$$

Example 6.11 [87, Thm. 5.2] Let X be a (real) locally convex space, with dual X^*, and denote by $\langle\cdot,\cdot\rangle : X \times X^* \to \mathbf{R}$ the canonical bilinear pairing. Define $\Delta : \overline{\mathbb{R}}_+^X \to \overline{\mathbb{R}}_+^{X^*}$ by

$$\Delta(f)(x^*) = \sup_{x \in X} \frac{\max\{\langle x,x^*\rangle,0\}}{f(x)},$$

with the convention $\frac{0}{0} = 0$. Then $f : X \to \overline{\mathbb{R}}_+$ satisfies $\Delta'\Delta(f) = f$ if and only if it is convex, positively homogeneous and l.s.c..

One can easily check that the dual operator $\Delta' : \overline{\mathbb{R}}_+^{X^*} \to \overline{\mathbb{R}}_+^X$ is given by

$$\Delta'(g)(x) = \sup_{x^* \in X^*} \frac{\max\{\langle x,x^*\rangle,0\}}{g(x^*)}.$$

Example 6.12 [91, Thm. 7.2] Let X, X^* and $\langle\cdot,\cdot\rangle$ be as in the preceding example. Define $\Delta : \overline{\mathbb{R}}_+^X \to \overline{\mathbb{R}}_+^{X^*}$ by

$$\Delta(f)(x^*) = \sup_{x \in X} \frac{|\langle x,x^*\rangle|}{f(x)},$$

with the convention $\frac{0}{0} = 0$. Then $f : X \to \overline{\mathbb{R}}_+$ satisfies $\Delta'\Delta(f) = f$ if and only if it is a l.s.c. extended semi-norm.

One can easily check that the dual operator $\Delta' : \overline{\mathbb{R}}_+^{X^*} \to \overline{\mathbb{R}}_+^X$ is given by

$$\Delta'(g)(x) = \sup_{x^* \in X^*} \frac{|\langle x, x^* \rangle|}{g(x^*)}.$$

Example 6.13 [81] Let $X = Y = \mathbb{R}_+^n$.Define $\Delta : \overline{\mathbb{R}}_+^X \to \overline{\mathbb{R}}_+^Y$ by

$$\Delta(f)(x^*) = -\inf \{f(x) : f(x) + \langle x, x^* \rangle < 0\}.$$

Then $f : X \to \overline{\mathbb{R}}_+$ satisfies $\Delta'\Delta(f) = f$ if and only if it is quasiconvex, nonincreasing, l.s.c. and co-radiant (i.e. it satisfies $u(\beta x) \geq \beta u(x)$ for $\beta \geq 1$ or, equivalently, $u(\alpha x) \leq \alpha u(x)$ for $\alpha \in (0, 1]$).

One can easily check that the dual operator $\Delta' : \overline{\mathbb{R}}_+^Y \to \overline{\mathbb{R}}_+^X$ is given by

$$\Delta'(g)(x) = -\inf_{x^* \in \mathbb{R}_+^n} \max\{\langle x, x^* \rangle, g(x^*)\};$$

a function $g : X \to \overline{\mathbb{R}}_+$ satisfies $\Delta\Delta'(g) = g$ if and only if it is quasiconvex, nonincreasing and l.s.c.. Hence, the mapping $f \longmapsto \Delta(f)$ is a bijection, with inverse $g \longmapsto \Delta'(g)$, from the set of l.s.c. nonincreasing co-radiant quasiconvex functions $f : \mathbb{R}_+^n \to \overline{\mathbb{R}}_+$ onto the set of l.s.c. nonincreasing quasiconvex functions $g : \mathbb{R}_+^n \to \overline{\mathbb{R}}_+$.

In the special case when $\Delta : \overline{\mathbb{R}}^X \to \overline{\mathbb{R}}^Y$ is a conjugation operator with coupling function $c : X \times Y \to \overline{\mathbb{R}}$, the hull operator $\Delta'\Delta : \overline{\mathbb{R}}^X \to \overline{\mathbb{R}}^X$ assigns to every function $f : X \to \overline{\mathbb{R}}$ its second $c-$conjugate $f^{cc'}$.

The usefulness of dualities in generalized convexity theory is made evident in [85, Thm. 3.7 and Remark 3.3], where it is proved that for every set Φ of extended real-valued functions one can construct a duality Δ such that the functions closed under $\Delta'\Delta$ are precisely the $\Phi-$convex functions. Some more results on dualities can be found in [85]. In [86], [87], [89], [90] and [45], dualities associated to certain operations are studied. Subdifferentials with respect to dualities are introduced in [88]. Some generalizations of the notion of duality have been proposed by Penot [110].

3. Generalized Convex Duality

Generalizing the convex duality theory we have presented in the Introduction, in this section we consider two arbitrary sets X and U and a family of optimization problems

$$(\mathcal{P}_u) \qquad \text{minimize } \varphi(x,u),$$

$\varphi : X \times U \to \overline{\mathbb{R}}$ being an objective function and $u \in U$ denoting a parameter; the minimization variable is thus $x \in X$. This family is regarded as consisting of perturbations of an (unperturbed) primal problem, defined as the one corresponding to a distinguished element $u_0 \in U$:

$$(\mathcal{P}) \qquad \text{minimize } \varphi(x,u_0).$$

The set U is thus interpreted as a perturbation space. The associated perturbation function is $p : U \to \overline{\mathbb{R}}$, defined by $p(u) = \inf_{x\in X} \varphi(x,u)$. We further consider a "dual" set V to the perturbation space and a coupling function $c : U \times V \to \overline{\mathbb{R}}$. One then associates to $(\mathcal{P})$ the dual problem

$$(\mathcal{D}) \qquad \text{maximize } c(u_0,v) - p^c(v).$$

The following result is immediate (the second part of its statement follows from Proposition 6.3):

Proposition 6.4 *The optimal value of the dual problem $(\mathcal{D})$ is $p^{cc'}(u_0)$. If it is finite, the optimal solution set is $\partial_c p^{cc'}(u_0)$.*

Since the optimal value of the primal problem is obviously $p(u_0)$, the following results hold:

Theorem 6.7 *The optimal value of the dual problem $(\mathcal{D})$ is not greater than the optimal value of $(\mathcal{P})$. They coincide if and only if the perturbation function p is Φ_c-convex at u_0. In this case, if the optimal value is finite then the optimal solution set to $(\mathcal{D})$ is $\partial_c p(u_0)$.*

Corollary 6.2 *If $x \in X$ and $v \in V$ satisfy $\varphi(x,u_0) = c(u_0,v) - p^c(v)$ then they are optimal solutions to $(\mathcal{P})$ and $(\mathcal{D})$, respectively.*

One can introduce a Lagrangian function in connection with this duality theory. One defines the $c-$Lagrangian $L : X \times V \to \overline{\mathbb{R}}$ of problem $(\mathcal{P})$, relative to the family of perturbed problems $(\mathcal{P}_u)$, by

$$L(x,v) = c(u_0,v) - \varphi_x^c(v),$$

$\varphi_x : U \to \overline{\mathbb{R}}$ denoting the partial mapping $\varphi_x(u) = \varphi(x,u)$. If φ_x is Φ_c-convex at u_0 for every $x \in X$, the supremum of the $c-$Lagrangian with respect to its second argument coincides with the objective function of $(\mathcal{P})$:

$$\sup_{v\in V} L(x,v) = \sup_{v\in V} \{c(u_0,v) - \varphi_x^c(v)\} = \varphi_x^{cc}(u_0) = \varphi_x(u_0) = \varphi(x,u_0).$$

It follows that the optimal value of $(\mathcal{P})$ is $\inf_{x\in X}\sup_{v\in V} L(x,v)$. Similarly, if c does not take the value $+\infty$, the infimum of the c–Lagrangian with respect to its first argument coincides with the objective function of the dual problem $(\mathcal{D})$:

$$\begin{aligned}\inf_{x\in X} L(x,v) &= \inf_{x\in X}\{c(u_0,v)-\varphi_x^c(v)\}\\ &= \inf_{x\in X}\left\{c(u_0,v)-\sup_{u\in U}\{c(u,v)-\varphi_x(u)\}\right\}\\ &= c(u_0,v)-\sup_{u\in U}\left\{c(u,v)-\inf_{x\in X}\varphi(x,u)\right\}\\ &= c(u_0,v)-\sup_{u\in U}\{c(u,v)-p(u)\} = c(u_0,v)-p^c(v).\end{aligned}$$

Therefore in this case the dual optimal value is $\sup_{v\in V}\inf_{x\in X} L(x,v)$. Thus, under all these conditions, the optimal values of $(\mathcal{P})$ and $(\mathcal{D})$ coincide if and only if $\sup_{v\in V}\inf_{x\in X} L(x,v) = \inf_{x\in X}\sup_{v\in V} L(x,v)$. In this case the set of saddlepoints of L coincides with the Cartesian product of the optimal solution sets.

This c–Lagrangian function is a generalization of the classical one of convex optimization, which corresponds to the special case when $U=V=\mathbb{R}^m$ and c is the usual scalar product; in particular, for vertically perturbed inequality constrained optimization problems one gets the standard Lagrangian mentioned in the Introduction (after restricting the dual vectors to the nonnegative orthant, which is possible because only nonnegative values of the dual variable are actually relevant).

From a practical point of view, the most useful example of generalized convex duality is the one described next:

Example 6.14 [61, Ex. 1"] For the inequality constrained optimization problem

$$(\mathcal{P})\qquad \begin{array}{ll}\text{minimize} & f(x)\\ \text{subject to} & g(x)\leq 0,\end{array}$$

with $f:\Omega\subseteq\mathbb{R}^n\to\overline{\mathbb{R}}$ and $g:\Omega\subseteq\mathbb{R}^n\to\mathbb{R}^m$, consider the vertically perturbed objective function $\varphi:\mathbb{R}^n\times\mathbb{R}^m\to\overline{\mathbb{R}}$ defined by

$$\varphi(x,u)=\begin{cases} f(x) & \text{if } x\in\Omega \text{ and } g(x)+u\leq 0\\ +\infty & \text{otherwise}\end{cases}$$

and the coupling function $c:\mathbb{R}^m\times(\mathbb{R}_+\times\mathbb{R}^m)\to\mathbb{R}$ given by $c(u,(\rho,y)) = -\rho\|u-y\|^2$ (cf. Example 6.3). A straightforward computation shows that the c–Lagrangian function $L:\mathbb{R}^n\times\mathbb{R}_+\times\mathbb{R}^m\to\overline{\mathbb{R}}$ satisfies, for every $(x, \rho, y) \in \mathbb{R}^n \times \mathbb{R}_+ \times \mathbb{R}^m$,

$$L(x,\rho,y) = \begin{cases} f(x) + \rho \sum_{i=1}^m \left(2y_i \max\{g_i(x), -y_i\} + (\max\{g_i(x), -y_i\})^2\right) & \text{if } x \in \Omega \\ +\infty & \text{if } x \notin \Omega \end{cases}.$$

This $c-$Lagrangian function coincides, up to a very simple change of variables, with the augmented Lagrangian introduced by Rockafellar [123]. Since c is of needle type on $\mathbb{R}^m$, for every $x \in X$ the function φ_x is Φ_c-convex and hence $\sup_{(\rho,y) \in \mathbb{R}_+ \times \mathbb{R}^m} L(x, \rho, y) = \varphi(x, 0)$. Moreover, since c is finite-valued the dual problem is

$(\mathcal{D})$ maximize $\inf_{x \in \mathbb{R}^n} L(x, \rho, y)$.

By Theorem 6.7, the optimal values of $(\mathcal{P})$ and $(\mathcal{D})$ coincide if and only if the perturbation function is l.s.c. at the origin and has a finite-valued $c-$elementary minorant. Notice that a sufficient condition for the existence of a $c-$elementary minorant of the perturbation function is the objective function f to be bounded from below. A necessary and sufficient condition for the existence of a dual optimal solution is given in [123, Thm. 5].

Other works on nonconvex duality based on generalized conjugation theory are [154], [155] and [104]. Some other approaches to generalized convex duality for single, multiobjective or optimal control problems are presented, e.g., in [33], [103], [170], [53], [108], [100] and [54]. Applications of duality theory to generalized convex fractional programming problems, based on geometric programming [117], are given in [134] and [135]. The literature on the applications of generalized convexity to duality theory during the decade 1985-1995 is surveyed in [118].

4. Quasiconvex Conjugation

Let X be an arbitrary set. We recall that the (lower) level sets of a function $f : X \to \overline{\mathbb{R}}$ are

$$S_\lambda(f) = \{x \in X : f(x) \leq \lambda\} \qquad (\lambda \in \mathbb{R}).$$

A function $f : \mathbb{R}^n \to \overline{\mathbb{R}}$ is called quasiconvex when its level sets are convex or, equivalently, when it satisfies

$$f((1-\alpha)x + \alpha y) \leq \max\{f(x), f(y)\} \quad (x, y \in \mathbb{R}^n,\ \alpha \in [0,1]); \quad (6.5)$$

it is called quasiconcave if $-f$ is quasiconvex. The functions that are both quasiconvex and quasiconcave are said to be quasiaffine. If (6.5) holds with strict inequality whenever $x \neq y$ and $\alpha \in (0,1)$ then f and $-f$

are called strictly quasiconvex and strictly quasiconcave, respectively. Notice that all these notions, like that of a convex function, are of a purely algebraic nature; indeed they make sense for functions defined on an arbitrary real vector space instead of $\mathbb{R}^n$.

For duality purposes, the most suitable quasiconvex functions are those whose level sets are not just convex, but evenly convex. We recall [40] that a subset of $\mathbb{R}^n$ is called evenly convex if it is an intersection of open halfspaces. As a consequence of the Hahn-Banach theorem [34, Cor. 1.4], every open or closed convex set is evenly convex (note that any closed halfspace is an intersection of open halfspaces). It follows from the definition that the class of evenly convex sets is closed under intersection. A function $f : \mathbb{R}^n \to \overline{\mathbb{R}}$ is said to be evenly quasiconvex [106] (or normal quasiconvex [66]) if all of its level sets are evenly convex, and evenly quasiaffine if it is quasiconcave and evenly quasiconvex[5]. Obviously, every l.s.c. quasiconvex function is evenly quasiconvex, and it is easy to prove that upper semicontinuous (u.s.c.) quasiconvex functions are evenly quasiconvex, too. Evenly quasiconvex functions are closed under pointwise supremum, given that the level sets of the supremum of a family of functions are intersections of level sets of the members of the family. Therefore every function has a largest evenly quasiconvex minorant, which is called its evenly quasiconvex hull. Clearly, the evenly quasiconvex hull of any function lies between its l.s.c. quasiconvex and quasiconvex hulls. One says that a function is evenly quasiconvex at a point if it coincides with its evenly quasiconvex hull at that point. A characterization of evenly quasiconvex functions, which is not expressed in terms of separation of level sets, is given in [26].

In the same way that the essence of classical convex conjugation is the fact that l.s.c. proper convex functions are upper envelopes of affine functions, it will follow from the quasiconvex conjugation theory in this section that the evenly quasiaffine functions are supremal generators of the class of evenly quasiconvex functions. Evenly quasiaffine functions have a simple structure, as shown by the next theorem [70, Thm. 2.36] (an earlier version of the equivalence *(i)*$\Longleftrightarrow$*(iii)* for u.s.c. functions was given in [158, Thm. 1]):

Theorem 6.8 *For any function $f : \mathbb{R}^n \to \overline{\mathbb{R}}$, the following statements are equivalent:*

(i) f is evenly quasiaffine.

[5] It easily follows from the characterization below of evenly quasiconvex functions that an evenly quasiaffine function is evenly quasiconcave (that is, $-f$ is evenly quasiconvex), too.

(ii) Every nonempty level set of f is either an (open or closed) halfspace or the whole space.

(iii) There exists $x^* \in \mathbb{R}^n$ and a nondecreasing function $h : \mathbb{R} \to \overline{\mathbb{R}}$ such that $f = h \circ \langle \cdot, x^* \rangle$.

Except in the trivial case of constant functions, the decomposition given in *(iii)* above is unique up to a multiplicative constant, i.e. if $f = h_1 \circ \langle \cdot, x_1^* \rangle = h_2 \circ \langle \cdot, x_2^* \rangle$, with $x_1^*, x_2^* \in \mathbb{R}^n$ and nondecreasing $h_1, h_2 : \mathbb{R} \to \overline{\mathbb{R}}$, then one has $x_2^* = c x_1^*$ and $h_2 = h_1 \left(c^{-1} \cdot \right)$ for some positive real number c [114, Prop. 2.4].

To apply the generalized convex conjugation theory that we have developed in Section 2 to the analysis of quasiconvex functions, in this section we first present level sets conjugations, a specialization of that theory that is useful for the dual description of functions whose level sets are generalized convex in a certain sense. This will in particular yield a conjugation theory for quasiconvex functions. Level set conjugations have been studied in [65], [139], [162], [142], [163], [109], [110] and [111].

Let Y be another arbitrary set and $c : X \times Y \to \overline{\mathbb{R}}$ be the opposite of the indicator function of $G \subseteq X \times Y$, i.e.

$$c(x,y) = \begin{cases} 0 & \text{if } (x,y) \in G \\ -\infty & \text{otherwise} \end{cases} . \tag{6.6}$$

The set G is the graph of the multivalued mapping $F : X \rightrightarrows Y$ given by $F(x) = \{y \in Y : (x,y) \in G\}$; the inverse of F is $F^{-1} : Y \rightrightarrows X$, defined by $F^{-1}(y) = \{x \in X : y \in F(x)\} = \{x \in X : (x,y) \in G\}$. According to the definitions in Section 2, the $c-$ and $c'-$conjugates of $f : X \to \overline{\mathbb{R}}$ and $g : Y \to \overline{\mathbb{R}}$, respectively, are $f^c : Y \to \overline{\mathbb{R}}$ and $g^{c'} : X \to \overline{\mathbb{R}}$, given by

$$f^c(y) = - \inf_{x \in F^{-1}(y)} f(x), \tag{6.7}$$

$$g^{c'}(x) = - \inf_{y \in F(x)} g(y).$$

Thus for the second $c-$conjugate of f one has

$$f^{cc'}(x_0) = \sup_{y \in F(x_0)} \inf_{x \in F^{-1}(y)} f(x) \qquad (x_0 \in X).$$

The $c-$elementary functions are those taking a constant value on $F^{-1}(y)$, for some $y \in Y$, and the value $-\infty$ on $X \backslash F^{-1}(y)$.

As for the $c-$subgradients of $f : X \to \overline{\mathbb{R}}$, for any $x_0 \in f^{-1}(\mathbb{R})$ and $y_0 \in Y$ one has:

$$y_0 \in \partial_c f(x_0) \quad \text{if and only if} \quad y_0 \in F(x_0) \text{ and } f(x_0) = \inf_{x \in F^{-1}(y_0)} f(x).$$

The following results [70, Thm. 4.1 and Cor. 4.2] describe local and global Φ_c–convexity:

Theorem 6.9 *Let $f : X \to \overline{\mathbb{R}}$ and $x_0 \in X$. Then f is Φ_c–convex at x_0 if and only if for every $\lambda < f(x_0)$ there exists $y_\lambda \in F(x_0)$ such that $S_\lambda(f) \cap F^{-1}(y_\lambda) \neq \emptyset$.*

Corollary 6.3 *A function $f : X \to \overline{\mathbb{R}}$. is Φ_c–convex if and only if each level set of f is an intersection of sets of the form $X \backslash F^{-1}(y)$ with $y \in Y$.*

Volle [164, Thm. 3] obtained an analogue of Toland-Singer duality theorem for level set conjugations. According to [72, Thm. 5.1], one has (compare Theorem 6.1):

Theorem 6.10 *For any $h : X \to \overline{\mathbb{R}}$, the following statements are equivalent:*

(i) h is Φ_c–convex.

(ii) $\inf_{x\in X} \max\{f(x), -h(x)\} = \inf_{x\in X} \max\left\{f(x), -h^{cc'}(x)\right\} \quad \left(f \in \overline{\mathbb{R}}^X\right).$

(iii) $\inf_{x\in X} \max\{f(x), -h(x)\} = \inf_{y\in Y} \max\{h^c(y), -f^c(y)\} \quad \left(f \in \overline{\mathbb{R}}^X\right).$

(iv) $\inf_{x\in X} \max\{f(x), -h(x)\} = \inf_{x\in X} \max\left\{f^{cc'}(x), -h^{cc'}(x)\right\} \quad \left(f \in \overline{\mathbb{R}}^X\right).$

(v) $\inf_{x\in X} \max\{f(x), -h(x)\} = \inf_{x\in X} \max\left\{f^{cc'}(x), -h(x)\right\} \quad \left(f \in \overline{\mathbb{R}}^X\right).$

For a related formula on the c–conjugate of $\max\{f, -h\}$ (under a suitable assumption on G), see [72, Thm. 5.2]. For the case when f and h are convex see [168, Thm. 2.1].

Now we set $X = \mathbb{R}^n$, $Y = \mathbb{R}^n \times \mathbb{R}$ and

$$G = \{(x, x^*, t) \in \mathbb{R}^n \times \mathbb{R}^n \times \mathbb{R} : \langle x, x^*\rangle \geq t\},$$

in order to specialize the preceding scheme to quasiconvex conjugation[6]. The conjugation formulas are then

$$f^c(x^*, t) = -\inf\{f(x) : \langle x, x^*\rangle \geq t\} \qquad (f : \mathbb{R}^n \to \overline{\mathbb{R}},\ x^* \in \mathbb{R}^n,\ t \in \mathbb{R}) \tag{6.8}$$

and

$$g^{c'}(x) = -\inf\{g(x^*, t)) : \langle x, x^*\rangle \geq t\} \qquad (g : \mathbb{R}^n \times \mathbb{R} \to \overline{\mathbb{R}},\ x \in \mathbb{R}^n).$$

---

[6]Alternatively, one can consider $G = \{(x, x^*, t)\ /\ \langle x, x^*\rangle > t\}$ (see [70] for details). One then obtains another quasiconvex conjugation scheme, which is suitable for l.s.c. quasiconvex functions; however the approach used in this section yields a simpler theory and is applicable to the broader class of evenly quasiconvex functions, as will be shown below.

Thus the second c−conjugate of $f : \mathbb{R}^n \to \mathbb{R}$ satisfies

$$f^{cc'}(x_0) = \sup_{x^* \in \mathbb{R}^n} \inf \{f(x) : \langle x, x^* \rangle \geq \langle x_0, x^* \rangle\} \qquad (x_0 \in \mathbb{R}^n). \qquad (6.9)$$

Since in this setting one has $F^{-1}(x^*, t) = \{x \in \mathbb{R}^n : \langle x, x^* \rangle \geq t\}$ for every $(x^*, t) \in Y$, the c−elementary functions are those that take a constant value on a closed halfspace (or the empty set or the whole space) and the value $-\infty$ on its complement. Given that $X \backslash F^{-1}(x^*, t) = \{x \in \mathbb{R}^n : \langle x, x^* \rangle < t\}$, in view of Corollary 6.3, one has:

Theorem 6.11 *A function $f : \mathbb{R}^n \to \overline{\mathbb{R}}$ is Φ_c−convex if and only if it is evenly quasiconvex.*

Corollary 6.4 *The second c−conjugate $f^{cc'}$ of any function $f : \mathbb{R}^n \to \overline{\mathbb{R}}$ coincides with the evenly quasiconvex hull of f.*

Since the c−elementary functions are evenly quasiaffine, from the preceding theorem one immediately gets:

Corollary 6.5 *Every evenly quasiconvex function is the pointwise supremum of a collection of evenly quasiaffine functions.*

In fact, given that $f^{cc'} = f$ if f is evenly quasiconvex, (6.9) yields an explicit family of evenly quasiconvex functions whose supremum is f, namely one has

$$f = \sup_{x^* \in \mathbb{R}^n} \varphi_{x^*}, \text{ with } \varphi_{x^*} = \inf \{f(x) : \langle x, x^* \rangle \geq \langle \cdot, x^* \rangle\}.$$

Notice that a (finite) real-valued version of Corollary 6.5 would be false, as shown, e.g., by the one variable evenly quasiconvex function $f(x) = 0$ if $x \leq 0$, $\ln x$ if $x > 0$, which cannot be expressed as a supremum of real-valued evenly quasiaffine functions; indeed, this function has no real-valued evenly quasiaffine minorant since, by Theorem 6.8, a one variable function is evenly quasiconvex if and only if it is monotonic.

The next result [70, Prop. 4.3] shows that the c−subdifferential of a function at a point is closely related to its quasi-subdifferential in the sense of Greenberg and Pierskalla [49]:

Theorem 6.12 *Let $f : \mathbb{R}^n \to \overline{\mathbb{R}}$, $x_0 \in \mathbb{R}^n$ be such that $f(x_0) \in \mathbb{R}$, and denote by*

$$\partial^{GP} f(x_0) = \{x^* \in \mathbb{R}^n : \langle x, x^* \rangle < \langle x_0, x^* \rangle \;\; \forall\, x \in \mathbb{R}^n \text{ s. t. } f(x) < f(x_0)\}$$

the quasi-subdifferential of f *at* x_0. *Then*

$$\partial_c f(x_0) =$$
$$\left\{(x^*,t) \in \mathbb{R}^n \times \mathbb{R} : x^* \in \partial^{GP} f(x_0),\ t \leq \langle x_0, x^* \rangle,\ \inf\{f(x) : \langle x, x^* \rangle \geq t\} = f(x_0)\right\}.$$

In fact, since the existence of $t \leq \langle x_0, x^* \rangle$ such that

$$\inf \{f(x) : \langle x, x^* \rangle \geq t\} = f(x_0)$$

occurs (if and) only if $x^* \in \partial^{GP} f(x_0)$, an alternative simpler, though less explicit, description of $\partial_c f(x_0)$ is

$$\partial_c f(x_0) = \{(x^*, t) \in \mathbb{R}^n \times \mathbb{R} : t \leq \langle x_0, x^* \rangle,\ \inf \{f(x) : \langle x, x^* \rangle \geq t\} = f(x_0)\}.$$

From Theorem 6.12, it easily follows that $\partial_c f(x_0)$ is an evenly convex cone. The quasi-subdifferential can be obtained from the $c-$ subdifferential by means of the formula presented next [62, p. 10]:

Corollary 6.6 *Let* f *and* x_0 *be as in Theorem 6.12. Then*

$$\partial^{GP} f(x_0) = \{x^* \in \mathbb{R}^n : (x^*, \langle x_0, x^* \rangle) \in \partial_c f(x_0)\}.$$

Corollary 6.7 *Let* f *and* x_0 *be as in Theorem 6.12. Then* $\partial^{GP} f(x_0)$ *coincides with the projection of* $\partial_c f(x_0)$ *onto* $\mathbb{R}^n$.

A more detailed presentation of quasiconvex conjugation theory, which uses a similar approach as the one in this section, can be found in [70]. For some other approaches and results, we refer to [17], [19], [4], [138], [66], [68], [106] (applied to quasiconvex optimization duality in [107]), [162], [142], [112], [145], [113], [153], [35], [42], [129], [160], [127], [150]... In particular, in [67], [69] and [70] a conjugation scheme based on a lexicographic separation theorem for convex sets is presented, which is exact for quasiconvex functions in the sense that second conjugates coincide with quasiconvex (rather than evenly quasiconvex) hulls. The following characterization of quasiconvex functions [70, Cor. 3.6] (on arbitrary vector spaces) is closely related to that conjugation scheme:

Theorem 6.13 *Let* X *be a real vector space. A function* $f : X \to \overline{\mathbb{R}}$ *is quasiconvex if and only if there is a family* $\{\rho_i\}_{i \in I}$ *of total order relations on* X *that are compatible with its linear structure*[7] *and a corresponding*

[7]We recall that an order relation ρ on a real vector space X is said to be compatible with the linear structure of X if

$$x_1\ \rho\ y_1\ ,\ x_2\ \rho\ y_2 \implies x_1 + x_2\ \rho\ y_1 + y_2$$

and

$$x\ \rho\ y\ ,\ \lambda \geq 0 \implies \lambda\, x\ \rho\ \lambda y.$$

family $\{f_i\}_{i\in I}$ *of isotonic functions* $f : (X, \rho_i) \to (\overline{\mathbb{R}}, \leq)$ *such that*

$$f(x) = \max_{i\in I} f_i(x) \qquad (x \in X).$$

Since the only total order relations on $\mathbb{R}$ that are compatible with its linear structure are the standard orderings $\geq$ and $\leq$, the preceding theorem generalizes the well-known fact that a function of one real variable is quasiconvex if and only if it is the pointwise maximum of a nondecreasing function and a nonincreasing function.

Applications of quasiconvex conjugacy to the study of Hamilton-Jacobi equations are developed in [8], [9], [11], [126], [166], [10], [169], [1] and [116]; more abstract generalized convex conjugation concepts have also been considered for their analysis in [110] and [115].

5. Quasiconvex Duality

As in Section 3, we here consider two arbitrary sets X and U, a function $\varphi : X \times U \to \overline{\mathbb{R}}$, the family of perturbed optimization problems

$$(\mathcal{P}_u) \qquad \text{minimize } \varphi(x, u)$$

and the associated perturbation function $p : U \to \overline{\mathbb{R}}$, defined by $p(u) = \inf_{x\in X} \varphi(x, u)$. The unperturbed, primal problem is

$$(\mathcal{P}) \qquad \text{minimize } \varphi(x, u_0),$$

for some fixed $u_0 \in U$. To apply the level sets conjugation scheme described in the preceding section, we further consider another set V, "dual" to U, and $G \subseteq U \times V$; G induces the multivalued mappings $F : U \rightrightarrows V$ and $F^{-1} : V \rightrightarrows U$, given by $F(u) = \{v \in V : (u, v) \in G\}$ and $F^{-1}(v) = \{u \in U : v \in F(u)\} = \{u \in U : (u, v) \in G\}$, respectively, and the coupling function $c : U \times V \to \overline{\mathbb{R}}$ given by

$$c(u, v) = \begin{cases} 0 & \text{if } (u, v) \in G \\ -\infty & \text{otherwise} \end{cases}.$$

Then the dual problem to $(\mathcal{P})$ relative to this coupling function is

$$(\mathcal{D}) \qquad \text{maximize } c(u_0, v) - p^c(v).$$

Its objective function takes the value $-\infty$ at any $v \in V\backslash F(u_0)$ and, for $v \in F(u_0)$, one has
$c(u_0, v) - p^c(v) = -p^c(v) = \inf_{u\in F^{-1}(v)} p(u) = \inf_{x\in X,\, u\in F^{-1}(v)} \varphi(x, u)$.
Thus $(\mathcal{D})$ can be equivalently written as

$$(\mathcal{D}) \qquad \begin{array}{c} \text{maximize } \inf_{x\in X,\ u\in F^{-1}(v)} \varphi(x,u) \\ \text{subject to } v \in F(u_0). \end{array}$$

The c–Lagrangian $L : X \times V \to \overline{\mathbb{R}}$ can be easily computed:

$$L(x,v) = c(u_0,v) - \varphi_x^c(v) = c(u_0,v) + \inf_{u\in F^{-1}(v)} \varphi(x,v)$$
$$= \begin{cases} \inf_{u\in F^{-1}(v)} \varphi(x,v) & \text{if } v \in F(u_0) \\ -\infty & \text{otherwise} \end{cases}.$$

Let us now consider the case when $U = \mathbb{R}^m$, $V = \mathbb{R}^m \times \mathbb{R}$, $G = \{(u,u^*,t) \in \mathbb{R}^m \times \mathbb{R}^m \times \mathbb{R} : \langle u,u^*\rangle \geq t\}$ and $u_0 = 0$. We then obtain the following formulation for the dual problem

$$(\mathcal{D}) \qquad \begin{array}{c} \text{maximize } \inf_{x\in X,\ \langle u,u^*\rangle\geq t} \varphi(x,u) \\ \text{subject to } t \leq 0, \end{array}$$

which, by setting $t = 0$, reduces to the unconstrained problem, in the variable u^* alone,

$$(\mathcal{D}) \qquad \text{maximize} \quad \inf_{x\in X,\ \langle u,u^*\rangle\geq 0} \varphi(x,u).$$

This dual problem was introduced by Crouzeix [18], who was the first to propose, as Rockafellar did in the convex case, a unifying perturbational approach to duality theory in quasiconvex optimization [17], [19]. The c–Lagrangian $L : \mathbb{X} \times \mathbb{R}^m \times \mathbb{R} \to \overline{\mathbb{R}}$ is given in this case by

$$L(x,u^*,t) = \begin{cases} \inf\{\varphi(x,u) : \langle u,u^*\rangle \geq t\} & \text{if } t \leq 0 \\ -\infty & \text{otherwise} \end{cases}.$$

As particular cases of the results of Section 3, we can state

Proposition 6.5 *The optimal value of the dual problem $(\mathcal{D})$ coincides with the value of the evenly quasiconvex hull of the perturbation function p at 0.*

Theorem 6.14 *The optimal value of $(\mathcal{D})$ is not greater than the optimal value of $(\mathcal{P})$. They coincide if and only if the perturbation function p is evenly quasiconvex at 0.*

Corollary 6.8 *Let $x_0 \in \mathbb{X}$ be an optimal solution to $(\mathcal{P})$. Then u^* is an optimal solution to $(\mathcal{D})$ if and only if $(x_0, 0) \in X \times \mathbb{R}^m$ minimizes $\varphi(x,u)$ subject to the constraint $\langle u,u^*\rangle \geq 0$.*

In the case of a vertically perturbed inequality constrained minimization problem, that is, when φ has the form

$$\varphi(x,u) = \begin{cases} f(x) & \text{if } x \in \Omega,\ g(x)+u \leq 0 \\ +\infty & \text{otherwise} \end{cases},$$

with $f : \Omega \subseteq \mathbb{R}^n \to \overline{\mathbb{R}}$ and $g : \Omega \to \mathbb{R}^m$, the dual objective function satisfies

$$\begin{aligned} \inf_{x\in X,\ \langle u,u^*\rangle \geq 0} \varphi(x,u) &= \inf\{f(x) : g(x)+u \leq 0,\ \ \langle u,u^*\rangle \geq 0\} \\ &= \begin{cases} \inf\{f(x) : \langle g(x),u^*\rangle \leq 0\} & \text{if } u^* \geq 0 \\ \inf f(\Omega) & \text{if } u^* \ngeq 0 \end{cases}. \end{aligned}$$

Thus one obtains, as the dual problem to

$$(\mathcal{P}) \qquad \begin{array}{ll} \text{minimize} & f(x) \\ \text{subject to} & g(x) \leq 0, \end{array}$$

the so-called surrogate dual problem

$$(\mathcal{D}) \qquad \begin{array}{ll} \text{maximize} & \inf\{f(x) : \langle g(x),u^*\rangle \leq 0\} \\ \text{subject to} & u^* \geq 0; \end{array}$$

it is associated to the $c-$Lagrangian $L : \mathbb{R}^n \times \mathbb{R}^m \times \mathbb{R} \to \overline{\mathbb{R}}$ given by

$$L(x,u^*,t) = \begin{cases} \inf\{f(x) : \langle g(x),u^*\rangle + t \leq 0\} & \text{if } u^* \geq 0 \text{ and } t \leq 0 \\ \inf f(\Omega) & \text{if } u^* \ngeq 0 \text{ and } t \leq 0 \\ -\infty & \text{if } t > 0 \end{cases}.$$

For the perturbation function $p : \mathbb{R}^m \to \overline{\mathbb{R}}$, defined in this case by $p(u) = \inf\{f(x) : g(x)+u \leq 0\}$, to be quasiconvex it suffices that Ω be convex, f be quasiconvex and the component functions of g be convex. A strong duality theorem holds under an additional mild topological assumption on f and a Slater constraint qualification [64, Thm. 3]:

Theorem 6.15 *If f is quasiconvex and u.s.c. along lines (i.e. for every $x_1, x_2 \in \Omega$, $f((1-\lambda)x_1 + \lambda x_2)$ is an u.s.c. function of $\lambda \in [0,1]$), the component functions of g are convex and there is an $x \in \Omega$ such that $g(x) < 0$ (componentwise) then*

$$\inf\{f(x) : g(x) \leq 0\} = \max_{u^* \geq 0} \inf\{f(x) : \langle g(x),u^*\rangle \leq 0\}.$$

Hence, $u^* \geq 0$ is an optimal solution to $(\mathcal{D})$ if and only if the optimal value of $(\mathcal{P})$ coincides with that of the surrogate problem

$$(\mathcal{S}_{u^*}) \qquad \textit{minimize} \qquad f(x)$$
$$\textit{subject to} \quad \langle g(x), u^* \rangle \leq 0.$$

In this case, any optimal solution $x_0 \in \Omega$ *to* $(\mathcal{P})$ *is also an optimal solution to* $(\mathcal{S}_{u^*})$.

Other works on surrogate duality in quasiconvex optimization and integer programming are [64], [46], [47], [48] and [141]; for vector optimization problems, surrogate duality was introduced and studied in [84].

The earliest approach to quasiconvex duality that uses generalized conjugate functions in a similar way as convex conjugates are classically employed in convex duality theory is due to Crouzeix; an extensive study can be found in [19]. A detailed treatment of quasiconvex duality theory from the viewpoint of generalized conjugation is presented in [70] and [114]. Duality for optimization problems involving quasiconvex functions is also studied in [130], [157], [109], [167], [42], [2], [149] and [111]; in particular, applications to generalized fractional programming are discussed in [21], [23] and [69].

6. Duality in Consumer Theory

Consider an economy in which n different type of commodities are available, so that the set of commodity bundles is the nonnegative orthant $\mathbb{R}^n_+$. It is usually assumed that the preferences of a consumer in the commodity space are represented by a so-called utility function[8] $u : \mathbb{R}^n_+ \to \overline{\mathbb{R}}$; even though, traditionally, utility functions are assumed to take only finite values, it is convenient for some of the developments in this section to consider extended real-valued utility functions. The problem a consumer faces consists in spending his income $M > 0$ in an optimal way subject to a budget constraint; thus, if the exogenously given prices of the goods are represented by the price vector $p \in \mathbb{R}^n_+$, the problem to solve is

$$(\mathcal{P}) \qquad \text{maximize} \qquad u(x)$$
$$\text{subject to} \quad \langle x, p \rangle \leq M.$$

[8] A preference relation on $\mathbb{R}^n_+$ is said to be represented by a utility function $u : \mathbb{R}^n_+ \to \overline{\mathbb{R}}$ if, for any $x, y \in \mathbb{R}^n_+$, x is preferred or indifferent to y if and only if $u(x) \geq u(y)$. A preference relation represented by a utility function must be a total preorder (see Section 8 for the definitions of these terms). We refer to [13] for an extensive discussion on the problem of representing preference relations by utility functions.

The function $\widetilde{v}$ that associates to (p, M) the optimal value of this problem,

$$\widetilde{v}(p, M) = \sup \{u(x) : \langle x, p \rangle \leq M\},$$

is called the indirect utility function associated with u. Since $\widetilde{v}$ is positively homogeneous of degree zero, there is no loss of generality by assuming that $M = 1$; in this way one obtains the (normalized) indirect utility function $v : \mathbb{R}^n_+ \to \overline{\mathbb{R}}$, defined by

$$v(p) = \widetilde{v}(p, 1) = \sup \{u(x) : \langle x, p \rangle \leq 1\} \qquad \left(p \in \mathbb{R}^n_+\right). \tag{6.10}$$

One obviously has $\widetilde{v}(p, M) = v(M^{-1}p)$ for every $p \in \mathbb{R}^n_+$ and $M > 0$. The function v can be interpreted as a utility function representing the preferences that the consumer has on price vectors; indeed, the consumer is supposed to prefer $p \in \mathbb{R}^n_+$ to $q \in \mathbb{R}^n_+$ if $v(p) > v(q)$, as he can get a larger utility by purchasing goods under the prices represented by p than under those represented by q.

From the definition of the indirect utility function v, it immediately follows that it is a nonincreasing function; on the other hand, a straightforward computation shows that its level sets satisfy

$$S_\lambda(v) = \bigcap_{x : u(x) > \lambda} \left\{p \in \mathbb{R}^n_+ : \langle x, p \rangle > 1\right\} \qquad (\lambda \in \mathbb{R}).$$

Thus these level sets are intersections of collections of open halfspaces, and hence they are evenly convex. It follows that v is evenly quasiconvex. Notice that no special properties of the utility function u are required for v to be nonincreasing and evenly quasiconvex. Indirect utility functions (arising from arbitrary utility functions) are almost characterized by these two conditions; just an additional minor condition on the behavior on the boundary $bd\ \mathbb{R}^n_+$ is required to get a complete characterization [73, Thm. 2.2]:

Theorem 6.16 *Let $v : \mathbb{R}^n_+ \to \overline{\mathbb{R}}$. There exists a utility function $u : \mathbb{R}^n_+ \to \overline{\mathbb{R}}$ having v as its associated indirect utility function if and only if v is nonincreasing, evenly quasiconvex and satisfies*

$$v(p) \leq \lim_{\alpha \to 1^-} \overline{v}(\alpha p) \qquad (p \in bd\ \mathbb{R}^n_+), \tag{6.11}$$

$\overline{v}$ denoting the l.s.c. hull of v.

In this case, one can take u nondecreasing, evenly quasiconcave and satisfying

$$u(x) \geq \lim_{\alpha \to 1^-} \underline{u}(\alpha x) \qquad (x \in bd\ \mathbb{R}^n_+). \tag{6.12}$$

Under these conditions, u is unique, namely, u is the pointwise largest utility function inducing v; furthermore, it satisfies

$$u(x) = \inf \{v(p) : \langle x, p \rangle \leq 1\} \qquad (x \in \mathbb{R}^n_+). \tag{6.13}$$

Condition (6.11) is actually satisfied at any $p \in \mathbb{R}^n_+$, not just on the boundary, for any indirect utility function v. In fact, any nonincreasing function satisfies it on the interior. This condition is weaker than lower semicontinuity at p; for $p = 0$, both conditions are equivalent.

Comparing expressions (6.10) and (6.7), we observe that the transformation assigning to a utility function its corresponding indirect utility function is, up to a sign change, of the level set conjugation type. Indeed, for c of (6.6), with $G = \{(x,p) \in \mathbb{R}^n_+ \times \mathbb{R}^n_+ : \langle x,p \rangle \leq 1\}$, one has $v = (-u)^c$. Therefore the results in this section on the duality between direct and indirect utility functions can be interpreted in the framework of level set conjugations. It thus turns out that this is, in a sense, the most appropriate conjugation theory for nonincreasing quasiconvex functions on $\mathbb{R}^n_+$. Unlike the one presented in Section 4 for quasiconvex functions on the whole space, it does not require introducing an extra parameter; the conjugate of a function on $\mathbb{R}^n_+$ is also defined on $\mathbb{R}^n_+$. It follows from Theorem 6.16 that the conjugation operator $f \longmapsto f^c$ is an involution from the set of nonincreasing evenly quasiconvex functions satisfying (6.11) onto itself; it therefore follows that this conjugation scheme is fully symmetric. Another related symmetric approach to quasiconvex conjugacy, which does not require the introduction of an extra parameter either, was proposed by Thach [156] (see [77, Thm. 2.1] for a simple characterization of the functions that coincide with their second conjugates).

According to Theorem 6.16, any nonincreasing evenly quasiconvex function $v : \mathbb{R}^n_+ \to \overline{\mathbb{R}}$ satisfying (6.11) is the indirect utility function associated with a unique nondecreasing evenly quasiconcave function $u : \mathbb{R}^n_+ \to \overline{\mathbb{R}}$ satisfying (6.12).

For an extension of Theorem 6.16 to utility functions taking values in arbitrary complete chains, we refer to [82, Thm. 1]. The usefulness of such an extension lies in that it allows one to consider preference orders that do not admit real-valued utility representations; notice that any preference order can be represented by a utility function taking values in a sufficiently large complete chain.

As an immediate consequence of the preceding theorem, one gets [73, Cor. 2.3]:

Corollary 6.9 *For every* $v : \mathbb{R}^n_+ \to \overline{\mathbb{R}}$*, the function* $v_0 : \mathbb{R}^n_+ \to \overline{\mathbb{R}}$ *defined by*

$$v_0(p) = \sup \{u(x) : \langle x, p \rangle \leq 1\},$$

with u *given by (6.13), is the pointwise largest nonincreasing evenly quasiconvex minorant of* v *that satisfies*

$$v_0(p) \leq \lim_{\alpha \to 1^-} \overline{v_0}(\alpha p) \qquad (p \in bd\ \mathbb{R}^n_+).$$

In view of the already observed symmetry between the definition of v and formula (6.13), Theorem 6.16 and Corollary 6.9 have the following dual versions [73, Thm. 2.4 and Cor. 2.5]:

Theorem 6.17 *Let* $u : \mathbb{R}^n_+ \to \overline{\mathbb{R}}$*. There exists a function* $v : \mathbb{R}^n_+ \to \overline{\mathbb{R}}$ *such that (6.13) holds if and only if* u *is nondecreasing, evenly quasiconcave and satisfies (6.12).*

In this case, one can take v *nonincreasing, evenly quasiconvex and satisfying (6.11). Under these conditions,* v *is unique, namely,* v *is the pointwise smallest function such that (6.13) holds; furthermore, it is the indirect utility function associated with* u*.*

Corollary 6.10 *For every* $u : \mathbb{R}^n_+ \to \overline{\mathbb{R}}$*, the function* $u^0 : \mathbb{R}^n_+ \to \overline{\mathbb{R}}$ *defined by*

$$u^0(x) = \inf \{v(p) : \langle x, p \rangle \leq 1\}, \tag{6.14}$$

v *being the indirect utility function associated with* u*, is the pointwise smallest nondecreasing evenly quasiconcave majorant of* u *that satisfies*

$$u^0(x) \geq \lim_{\alpha \to 1^-} \underline{u^0}(\alpha x) \qquad (x \in bd\ \mathbb{R}^n_+).$$

It follows from Theorem 6.17 that every nondecreasing quasiconcave utility function $u : \mathbb{R}^n_+ \to \overline{\mathbb{R}}$ satisfying (6.12) can be recovered from its associated indirect utility function $v : \mathbb{R}^n_+ \to \overline{\mathbb{R}}$ by (6.13).

All the results presented so far on the duality between direct and indirect utility functions have been stated for extended real-valued functions; however, one can easily check that they remain valid after replacing $u : \mathbb{R}^n_+ \to \overline{\mathbb{R}}$ and $v : \mathbb{R}^n_+ \to \overline{\mathbb{R}}$ by $u : \mathbb{R}^n_+ \to [a, b]$ and $v : \mathbb{R}^n_+ \to [a, b]$, respectively, in their statements. The case of nonnecessarily bounded, but finite-valued, utility functions is dealt with in the next theorem [73, Thm. 2.6]:

Theorem 6.18 *A nondecreasing evenly quasiconcave function* $u : \mathbb{R}^n_+ \to \overline{\mathbb{R}}$ *satisfying (6.12) is finite-valued if and only if its associated indirect utility function* $v : \mathbb{R}^n_+ \to \overline{\mathbb{R}}$ *is bounded from below and finite-valued on the interior of* $\mathbb{R}^n_+$*.*

To compare the preceding theorem with the situation in the bounded case, notice that, since any indirect utility function is nonincreasing, it is bounded from above if and only if it is finite-valued at the origin.

The dual version of Theorem 6.18 is [73, Thm. 2.7]:

Theorem 6.19 *The indirect utility function $v : \mathbb{R}^n_+ \to \overline{\mathbb{R}}$ induced by a nondecreasing evenly quasiconcave function $u : \mathbb{R}^n_+ \to \overline{\mathbb{R}}$ satisfying (6.12) is finite-valued if and only if u is bounded from above and finite-valued on the interior of $\mathbb{R}^n_+$.*

As an immediate consequence of theorems 6.18 and 6.19, one gets [73, Cor. 2.8]

Corollary 6.11 *Let $u : \mathbb{R}^n_+ \to \overline{\mathbb{R}}$ be a nondecreasing evenly quasiconcave function satisfying (6.12) and let $v : \mathbb{R}^n_+ \to \overline{\mathbb{R}}$ be its associated indirect utility function. The following statements are equivalent:*

(i) u and v are finite-valued.

(ii) u is bounded.

(iii) v is bounded.

Some axiomatic characterizations of the mapping assigning to every utility function its associated indirect utility function are given in [74].

Quasiconcave utility functions admit another dual representation, namely, by the so-called expenditure functions. From a purely mathematical point of view, an expenditure function is nothing else than the support function of the upper level sets of the utility function. Let $u : \mathbb{R}^n_+ \to \mathbb{R}$ be a utility function (from now on, we shall restrict ourselves to the real-valued case). One defines the associated expenditure function $e_u : \mathbb{R}^n_+ \times \mathbb{R} \to \mathbb{R}_+ \cup \{+\infty\}$ by

$$e_u(p, \lambda) = \inf \{\langle x, p\rangle : u(x) \geq \lambda\} \qquad \left((p, \lambda) \in \mathbb{R}^n_+ \times \mathbb{R}\right).$$

It shows the amount of money that a consumer with preferences represented by u needs to spend under the prices p to achieve a utility level λ at least.

Expenditure functions induced by arbitrary utility functions are characterized in [83, Thm. 2.2]:

Theorem 6.20 *A function $e : \mathbb{R}^n_+ \times \mathbb{R} \to \mathbb{R}_+ \cup \{+\infty\}$ is the expenditure function e_u for some utility function $u : \mathbb{R}^n_+ \to \mathbb{R}$ if and only if the following conditions hold:*

(i) For each $\lambda \in \mathbb{R}$, either $e(\cdot, \lambda)$ is finite-valued, concave, linearly homogeneous and u.s.c., or it is identically equal to $+\infty$.

(ii) For each $p \in \mathbb{R}^n_+$, $e(p, \cdot)$ is nondecreasing.

(iii) $\bigcup_{\lambda \in \mathbb{R}} \partial e(\cdot, \lambda)(0) = \mathbb{R}^n_+$, *with* $\partial e(\cdot, \lambda)(0)$ *denoting the superdifferential of the concave function* $e(\cdot, \lambda)$ *at the origin, i.e.*

$$\partial e(\cdot, \lambda)(0) = \left\{ x \in \mathbb{R}^n_+ : \langle x, p \rangle \geq e(p, \lambda) \qquad \forall\, p \in \mathbb{R}^n_+ \right\}.$$

The duality between expenditure and utility functions, under the weakest possible assumptions, is described next [83, Thm. 2.5]:

Theorem 6.21 *The mapping* $u \longmapsto e_u$ *is a bijection from the set of u.s.c. nondecreasing quasiconcave functions* $u : \mathbb{R}^n_+ \to \mathbb{R}$ *onto the set of functions* $e : \mathbb{R}^n_+ \times \mathbb{R} \to \mathbb{R}_+ \cup \{+\infty\}$ *that satisfy (i)-(iii) (of Theorem 6.20),*

(iv) $\bigcap_{\lambda \in \mathbb{R}} \partial e(\cdot, \lambda)(0) = \emptyset$

and

(v) $\bigcap_{\mu < \lambda} \partial e(\cdot, \mu)(0) = \partial e(\cdot, \lambda)(0) \qquad (\lambda \in \mathbb{R})$.

Furthermore, the inverse mapping is $e \longmapsto u_e$, *with* $u_e : \mathbb{R}^n_+ \to \mathbb{R}$ *given by*

$$u_e(x) = \sup \{ \lambda \in \mathbb{R} : x \in \partial e(\cdot, \lambda)(0) \}.$$

According to the preceding theorem, a utility function $u : \mathbb{R}^n_+ \to \mathbb{R}$ can be recovered from its associated expenditure function e_u (that is, one has $u = u_{e_u}$) if and only if it is quasiconcave, nondecreasing and u.s.c.. On the other hand, for a function $e : \mathbb{R}^n_+ \times \mathbb{R} \to \mathbb{R}_+ \cup \{+\infty\}$ satisfying *(i)-(v)*, u_e is the pointwise largest utility function whose associated expenditure function e_{u_e} is e; moreover, it is the only one that is quasiconcave, nondecreasing and u.s.c.

A dynamic analogue of the duality in consumer theory is provided in [15]. Some applications of duality theory to the study of rationality of choice are given in the fundamental paper [120]. More material on economics duality can be found in [136], [12], [30], [5], [16] and [39]. The survey paper [31] provides an extensive review of the literature on duality in microeconomics up to 1982. More recent surveys are [55] and [14]. Among the papers dealing with the relations between the properties of direct and indirect utility functions that have appeared after that survey paper was published, let us mention [22], where a symmetric duality in the continuously differentiable case is developed, and [43], which relates continuity properties of direct utility functions to those of the corresponding indirect ones.

7. Monotonicity of Demand Functions

The mapping assigning to each pair (p, M) the (possibly empty) solution set $\widetilde{X}(p, M)$ to the problem $(\mathcal{P})$ stated in the preceding section is called the Walrasian demand correspondence. Since $\widetilde{X}$, as $\widetilde{v}$, is positively homogeneous of degree zero, as in the preceding section there is no loss of generality by assuming that $M = 1$. One can then consider the (normalized) demand correspondence $p \in \mathbb{R}^n_{++} \rightrightarrows X(p) = \widetilde{X}(p, 1)$; clearly, $\widetilde{X}(p, M) = X(M^{-1}p)$. Notice that we only consider strictly positive prices $p \in \mathbb{R}^n_{++}$; in fact, under the natural assumption that the utility function is strictly increasing in each variable, $X(p)$ would be empty for $p \in \mathbb{R}^n_+ \backslash \mathbb{R}^n_{++}$ since there would be "an infinite demand" for a good with zero price.

If preferences are locally nonsatiated, the demand correspondence is cyclically quasimonotone (the notion of cyclically quasimonotone operator was introduced and studied in [59], [24] and [25]):

Theorem 6.22 *If the utility function $u : \mathbb{R}^n_+ \to \overline{\mathbb{R}}$ has no local maximum then the demand function $X : \mathbb{R}^n_{++} \rightrightarrows \mathbb{R}^n_+$ is cyclically quasimonotone (in the decreasing sense), i.e. if $p^i \in \mathbb{R}^n_{++}$, $x^i \in X(p^i)$ $(i = 1, ..., k)$ then*

$$\min_{i=1,\dots,k} \left\langle x^i - x^{i+1}, p^i \right\rangle \le 0, \qquad \textit{with } x^{k+1} = x^1.$$

The proof of the preceding theorem is quite simple: If we had $\min_{i=1,\dots,k} \left\langle x^i - x^{i+1}, p^i \right\rangle > 0$, from the inequalities $\left\langle x^{i+1}, p^i \right\rangle < \left\langle x^i, p^i \right\rangle$ $(i = 1, ..., k)$ and the local nonsatiation assumption it would follow that $u\left(x^{i+1}\right) < u\left(x^i\right)$ for $i = 1, ..., k$, which is impossible.

Quasimonotonicity is the property resulting from imposing only for $k = 2$ the condition in the definition of cyclic quasimonotonicity; thus, X is quasimonotone (in the decreasing sense) if

$$p, q \in \mathbb{R}^n_{++},\ x \in X(p),\ y \in Y(q) \implies \min\left\{\langle x - y, p\rangle, \langle y - x, q\rangle\right\} \le 0.$$

Hence every cyclically quasimonotone mapping is quasimonotone. As observed in [59], quasimonotonicity (cyclic quasimonotonicity) of a demand function (i.e., a single-valued demand correspondence) is a consequence of the related weak (strong, resp.) axiom of revealed preference introduced by Samuelson [131], [132] (Houthakker [51], resp.).

Quasimonotonicity is weaker than monotonicity. One says that X is monotone (in the decreasing sense) if

$$p, q \in \mathbb{R}^n_{++},\ x \in X(p),\ y \in Y(q) \qquad \implies \qquad \langle x - y, p\rangle + \langle y - x, q\rangle \le 0. \tag{6.15}$$

Obviously, if the preceding sum is nonpositive then at least one of its terms must be nonpositive, so that a monotone mapping is quasimonotone, too.

In general, a demand correspondence X need not be monotone, as shown by the following example. Let $u : \mathbb{R}^2_+ \to \mathbb{R}$ be the utility function defined by

$$u(x_1, x_2) = \begin{cases} x_1 + x_2 - 1 & \text{if } x_1 + x_2 < 1 \\ x_1 & \text{if } x_1 + x_2 \geq 1 \end{cases}.$$

One can easily check that u is quasiconcave, nondecreasing, u.s.c. and has no local maximum. This utility function can be interpreted as follows. Suppose x_1 and x_2 denote liters of wine and water, respectively, and the consumer needs to drink at least one liter (of anything) to stay healthy, so that he derives a negative utility $x_1 + x_2 - 1$, the drinking deficit, from drinking a total amount $x_1 + x_2 < 1$; in case he drinks enough, $x_1 + x_2 \geq 1$, he derives a nonnegative utility equal to the amount of wine he consumes (our consumer is assumed to enjoy only wine and not water). For $p_1 \geq 1$ and $1 \geq p_2 > 0$, the optimal solution of

$$\begin{array}{ll} \text{maximize} & u(x_1, x_2) \\ \text{subject to} & p_1 x_1 + p_2 x_2 \leq 1 \end{array}$$

is $x_1 = \frac{1-p_2}{p_1-p_2}$, $x_2 = \frac{p_1-1}{p_1-p_2}$. Clearly, x_2 is increasing in p_2, so that the demand function is not monotone in our sense. The interpretation of this fact is clear; if wine is too expensive and water is cheap enough, a rise of the price of water prevents the consumer to drink as much wine as he was drinking before, so that his consumption of water must increase in order to meet the one liter drinking requirement. Goods, like water in the example, for which this kind of situation occurs are called inferior, or Giffen, goods.

Under mild assumptions, the demand correspondence is actually a function (i.e. it is single-valued) and the values it takes can be easily obtained from the indirect utility function. This is the case, for instance, when the utility function u is u.s.c. (which ensures the nonemptiness of X) and its associated indirect utility function v is continuously differentiable on $\mathbb{R}^n_{++}$ with a nonzero gradient (conditions on u ensuring the differentiability of v and a symmetric duality for the continuously differentiable case have been obtained by Crouzeix in [22]). Indeed, let $\overline{p} \in \mathbb{R}^n_{++}$ and $\overline{x} \in X(\overline{p})$. Then one has $\langle \overline{x}, \overline{p} \rangle \leq 1$ and $v(\overline{p}) = u(\overline{x})$, whence, as $v(p) \geq u(x)$ for every pair $(x, p) \in \mathbb{R}^n_+ \times \mathbb{R}^n_+$ such that $\langle x, p \rangle \leq 1$, it follows that $\overline{p}$ is an optimal solution to the (dual) problem

$$\begin{array}{ll} \text{minimize} & v(p) \\ \text{subject to} & \langle \overline{x}, p \rangle \leq 1. \end{array}$$

Thus, by the Kuhn-Tucker theorem there is a real number $\lambda \geq 0$ such that

$$\nabla v(\overline{p}) + \lambda \overline{x} = 0 \qquad \text{and} \qquad \lambda \left(\langle \overline{x}, \overline{p} \rangle - 1 \right) = 0.$$

From these equalities one gets $\langle \nabla v(\overline{p}), \overline{p}\rangle + \lambda \langle \overline{x}, \overline{p}\rangle = 0$ and $\lambda \langle \overline{x}, \overline{p}\rangle = \lambda$, and hence $\lambda = -\langle \nabla v(\overline{p}), \overline{p}\rangle$, so that one arrives at

$$\nabla v(\overline{p}) - \langle \nabla v(\overline{p}), \overline{p}\rangle \overline{x} = 0.$$

To solve this equation for $\overline{x}$ one just needs $\nabla v(\overline{p})$ to be different from zero (in which case, since v is nonincreasing, $\langle \nabla v(\overline{p}), \overline{p}\rangle$ is negative). We have thus proved (see, e.g., [96, Prop. 3.G.4]):

Theorem 6.23 *(Roy's Identity) If the utility function u is u.s.c. and its associated indirect utility function v is continuously differentiable at $\overline{p} \in \mathbb{R}^n_{++}$, with $\nabla v(\overline{p}) \neq 0$, then*

$$X(\overline{p}) = \left\{ \frac{1}{\langle \nabla v(\overline{p}), \overline{p}\rangle} \nabla v(\overline{p}) \right\}.$$

Sufficient conditions for monotonicity of demand were given a long time ago by Mitjushin and Polterovich [101] under the assumption that the utility function is concave. This assumption is somewhat artificial, since concavity of the utility function is not an intrinsic property of the consumer's preferences; it depends on the specific utility representation. Notice that, e.g., composing a concave utility function with an increasing function one gets a new utility function representing the same preferences as the initial one; however, concavity is not necessarily preserved by this operation. This fact is in sharp contrast with the corresponding situation regarding quasiconcavity: If a utility function is quasiconcave, any other utility representation of the same preferences is necessarily quasiconcave. Conditions for a preference relation to admit a concave utility representation can be found in [95, Section 2.6].

Theorem 6.24 *If the utility function $u : \mathbb{R}^n_+ \to \mathbb{R}$ is concave, C^2, has a componentwise strictly positive gradient on $\mathbb{R}^n_{++}$, induces a demand function $\varphi : \mathbb{R}^n_{++} \to \mathbb{R}^n_+$ (i.e. a single-valued demand correspondence $p \in \mathbb{R}^n_{++} \rightrightarrows X(p) = \{\varphi(p)\}$), with φ of class C^1, and satisfies*

$$-\frac{\langle x, \nabla^2 u(x) x\rangle}{\langle x, \nabla u(x)\rangle} < 4 \qquad (x \in \mathbb{R}^n_{++}) \tag{6.16}$$

then φ is strictly monotone, i.e. it satisfies (6.15) as a strict inequality whenever $p \neq q$.

At this point it is interesting to observe the symmetry between the roles played by u and v in connection with the demand correspondence. From the very definitions of X and v it follows that

$$X(p) = \left\{ x \in \mathbb{R}^n_+ : u(x) = v(p) \right\} \qquad \left(p \in \mathbb{R}^n_{++}\right).$$

Therefore, for the inverse demand correspondence X^{-1} one has

$$X^{-1}(x) = \{p \in \mathbb{R}^n_{++} : -v(p) = -u(x)\} \qquad (x \in \mathbb{R}^n_+);$$

the minus signs in the description of this set are included in order to make evident the analogy between the expressions for $X(p)$ and $X^{-1}(x)$. Indeed, according to (6.13), under the standard assumptions on u one has

$$-u(x) = \sup\{-v(p) : \langle x, p\rangle \leq 1\} \qquad (x \in \mathbb{R}^n_+),$$

which shows that $-u$ can be regarded as "the indirect utility function associated with $-v$". Therefore, as the monotonicity of X is obviously equivalent to that of X^{-1}, every result relating properties of u to the monotonicity of X admits a dual version in terms of v, which can be obtained by replacing u with $-v$ in the original statement (after taking care of some little technical details that are needed to deal with the lack of symmetry due to the fact that X and X^{-1} have slightly different domains). In particular, a dual version of Theorem 6.24 is given in [119, Thm. 2.2].

Condition (6.16) is in fact related to generalized convexity properties, as shown by the next proposition [79]:

Proposition 6.6 *For a C^2 nondecreasing quasiconcave utility function $u : \mathbb{R}^n_+ \to \mathbb{R}$ with no stationary points*[9]*, the following statements are equivalent:*

(i) $-\frac{\langle x, \nabla^2 u(x) x\rangle}{\langle x, \nabla u(x)\rangle} \leq 4 \qquad (x \in \mathbb{R}^n_{++})$.

(ii) The function $(x_1, ..., x_n) \in \mathbb{R}^n_{++} \longmapsto u(x_1^{-\frac{1}{3}}, ..., x_n^{-\frac{1}{3}})$ *is convex-along-rays.*

(iii) The restriction $v : \mathbb{R}^n_{++} \to \mathbb{R} \cup \{+\infty\}$ *of the indirect utility function to the positive orthant has a representation of the type*

$$v(p) = \max_{(y,c)\in U} \{c - (\langle y, p\rangle)^3\} \qquad (p \in \mathbb{R}^n_{++}),$$

with $U \subseteq (\mathbb{R}^n_{++} \cup \{0\}) \times \mathbb{R}$.

A simple modification of the proof of Theorem 6.24 (as presented, e.g., in [50, Appendix 4]) yields the next result [79], which states a necessary and sufficient condition for the monotonicity of demand functions. Notice that it only requires strict quasiconcavity of the utility function (in contrast with Theorem 6.24, which assumes concavity).

[9] A stationary point of a differentiable function is a point at which the gradient of the function vanishes.

Theorem 6.25 *Let $\varphi : \mathbb{R}^n_{++} \to \mathbb{R}^n_+$ be a C^1 demand function induced by a strictly quasiconcave utility function $u : \mathbb{R}^n_+ \to \mathbb{R}$, which is C^2 on $\mathbb{R}^n_{++}$ and has a componentwise strictly positive gradient on $\mathbb{R}^n_{++} \cup \varphi\left(\mathbb{R}^n_{++}\right)$. Then φ is monotone if and only if*

$$-\frac{\left\langle x, \nabla^2 u(x) x\right\rangle}{\langle x, \nabla u(x)\rangle} \leq 4 - \frac{\langle x, \nabla u(x)\rangle}{\left\langle \nabla u(x), \left(\nabla^2 u(x)\right)^{-1} \nabla u(x)\right\rangle} \tag{6.17}$$

$$\forall\, x \in \mathbb{R}^n_{++} \textit{ such that } \nabla^2 u(x) \textit{ is nonsingular}$$

and

$$-\frac{\left\langle x, \nabla^2 u(x) x\right\rangle}{\langle x, \nabla u(x)\rangle} \leq 4 \qquad \forall\, x \in \mathbb{R}^n_{++} \textit{ such that } \nabla^2 u(x) \textit{ is singular.}$$

The strict version of inequality (6.17) was first considered in [99, formula (4)]; in the same paper it was proved that it is invariant under monotone transformations of the utility function [99, Annexe II]. Thus it only depends on the consumer's preferences rather than on any particular utility representation.

An earlier characterization of monotone demand functions for consumers with concave utility functions $u : \mathbb{R}^n_+ \to \mathbb{R}$ was given by Kannai [52, Thm. 2.1] in terms of differential geometric properties of the indifference surfaces $u^{-1}(\lambda)$, $\lambda \in \mathbb{R}$.

The following theorem [79] gives a sufficient condition for the monotonicity of demand correspondences, without requiring any differentiability assumption:

Theorem 6.26 *Let $u : \mathbb{R}^n_+ \to \mathbb{R}$ be a utility function and let $v : \mathbb{R}^n_+ \to \mathbb{R} \cup \{+\infty\}$ be its associated indirect utility function. If the set $\left\{(p, x) \in \mathbb{R}^n_{++} \times \mathbb{R}^n_+ : u(x) - v(p) \geq 0\right\}$ is convex (in particular, if the function $\psi : \mathbb{R}^n_{++} \times \mathbb{R}^n_+ \to \mathbb{R} \cup \{+\infty\}$ defined by $\psi(p, x) = u(x) - v(p)$ is quasiconcave) and u has no maximum then the demand correspondence X is monotone.*

Corollary 6.12 *Let $u : \mathbb{R}^n_+ \to \mathbb{R}$ be a utility function and let $v : \mathbb{R}^n_+ \to \mathbb{R} \cup \{+\infty\}$ be its associated indirect utility function. If u is concave and has no maximum and v is convex then the demand correspondence X is monotone.*

Corollary 6.12 is essentially due to Milleron [99]. It suggests to investigate conditions under which an indirect utility function is convex. In [79], the following result is proved:

Theorem 6.27 *If $u : \mathbb{R}^n_+ \to \mathbb{R}$ is nondecreasing and the function $x \in \mathbb{R}^n_{++} \longmapsto u(x^{-1})$ is convex-along-rays then the restriction $v : \mathbb{R}^n_{++} \to \mathbb{R} \cup \{+\infty\}$ of the indirect utility function to the positive orthant is convex. Conversely, if $v : \mathbb{R}^n_{++} \to \mathbb{R}$ is bounded, nonincreasing and convex then there is a nondecreasing quasiconcave utility function $u : \mathbb{R}^n_+ \to \mathbb{R}$ such that $x \in \mathbb{R}^n_{++} \longmapsto u(x^{-1})$ is convex-along-rays and whose associated indirect utility function extends v.*

The first part of the preceding theorem refines statement *ii)* in [20, Thm. 11], which uses the additional assumption that u can be recovered from v by (6.13). This theorem can also be found in [119, Prop. 2.4(i)] under the extra hypothesis that u is continuous and quasiconcave. The following corollary is immediate [119, Prop. 2.4(ii)]:

Corollary 6.13 *Let $u : \mathbb{R}^n_+ \to \mathbb{R}$ be a C^2 nondecreasing quasiconcave utility function. The restriction $v : \mathbb{R}^n_{++} \to \mathbb{R}$ of its associated indirect utility function to the positive orthant is convex if and only if*

$$< x, \nabla^2 u(x) x > +2 < \nabla u(x), x > \geq 0 \qquad (x \in \mathbb{R}^n_{++}). \quad (6.18)$$

Condition (6.18) is stronger than Mitjushin - Polterovich inequality (6.16), as one has

$$< \nabla u(x), x >> 0 \qquad (x \in \mathbb{R}^n_{++})$$

for any nondecreasing utility function u with no stationary points. Thus, combining the preceding corollary with Mitjushin - Polterovich's result, one obtains again Corollary 6.12 (under differentiability assumptions, which are actually superfluous).

8. Consumer Theory without Utility

The theory developed in the preceding sections admits a partial extension to the case in which consumer's preferences are not necessarily represented by a utility function. The relevance of such an extension stems from the fact that preference orders, rather than utility representations, are the primitive objects of consumer theory. Since not all preference orders can be represented by a utility function, by analyzing preferences directly one gets more general results. Besides, a utility function representing a given preference order is not unique, and some utility representations may satisfy the regularity conditions that make the full duality relations possible while some others may not. This shows that duality theory based on utility functions is not intrinsic to the economic model, an undesirable fact that disappears by focusing directly on preference orders. Two pioneering contributions to duality theory for preference orders are [98] and [63].

There is still another reason why the preference orders approach to duality theory is convenient: An indirect utility function does not always represent accurately the preferences of a consumer with respect to prices, since the definition (6.10) does not distinguish whether the supremum is attained or not. If, for an indirect utility function $v : \mathbb{R}^n_+ \to \mathbb{R}$ and price vectors $p_1, p_2 \in \mathbb{R}^n_+$, one has $v(p_1) = v(p_2)$ then p_1 and p_2 are considered to be indifferent, but it might be the case that the supremum in the definition of $v(p_1)$ is attained while that in the definition of $v(p_2)$ is not; then the consumer should prefer p_1 to p_2, since he can achieve a utility $v(p_1) = v(p_2)$ under p_1, but not under p_2. In fact, in this situation one would have $v'(p_1) > v'(p_2)$ for the indirect utility function v' associated to some other utility representation of the same preference order, which shows that comparing prices through indirect utility functions might be inconsistent and misleading. Of course this inconsistency is absent under the common assumption that maximal elements exist whenever the utility function is bounded on the budget set, which is the case, e.g., when it is continuous and strictly increasing in each component.

Let us assume that the preferences of a consumer on commodity bundles are represented by a preorder $\succsim$, that is, a reflexive and transitive binary relation on $\mathbb{R}^n_+$. One says that $\succsim$ is total if, for every $x_1, x_2 \in \mathbb{R}^n_+$, at least one of $x_1 \succsim x_2$ and $x_2 \succsim x_1$ holds true. The assertion $x_1 \succsim x_2$ is to be interpreted as meaning that x_1 is at least as good as x_2 in the consumer's preference ranking. When both $x_1 \succsim x_2$ and $x_2 \succsim x_1$ hold true, one says that x_1 and x_2 are indifferent and write $x_1 \sim x_2$. The indifference relation $\sim$ is obviously an equivalence relation. We write $x_1 \succ x_2$ when $x_1 \succsim x_2$ and $x_1$and x_2 are not indifferent; the relation $\succ$ is called the strict preorder associated to $\succsim$.

The indirect preorder $\succsim^i$ induced by $\succsim$ is defined as follows: $p_1 \succsim^i p_2$ if and only if for every x_2 in the budget set $B(p_2) = \left\{x \in \mathbb{R}^n_+ : \langle x, p_2\rangle \leq 1\right\}$ there exists $x_1 \in B(p_1)$ such that $x_1 \succsim x_2$. The meaning of this definition is clear: The consumer prefers the vector price p_1 to p_2 if he can get under p_1 commodity bundles that are at least as good as those that are available to him under p_2. One can easily check that $\succsim^i$ is a total preorder, too. The associated indifference relation and strict preorder will be denoted by $\backsim^i$ and $\succ^i$, respectively.

To be able to recover preference orders on goods from indirect preorders, one has first to define the direct preorder $\succsim^{*d}$ induced by a total preorder $\succsim^*$ on prices $p \in \mathbb{R}^n_+$. This is done as follows: For $x_1, x_2 \in \mathbb{R}^n_+$, one writes $x_1 \succsim^{*d} x_2$ if and only if for every p_1 in the inverted budget set $B^{-1}(x_1) = \left\{p \in \mathbb{R}^n_+ : \langle x_1, p\rangle \leq 1\right\}$ there exists $p_2 \in B^{-1}(x_2)$ such that

$p_1 \precsim^* p_2$. This definition is in fact an adaptation to the total preorders setting of formula (6.13), which defines the utility function on goods induced by a given utility function on price vectors. Since $\precsim^*$ is a total preorder, $\precsim^{*d}$ is a total preorder, too.

Thus, given a total preorder $\precsim$ on goods one can construct its associated total preorder $\precsim^i$ on prices, which in turn induces a new total preorder $\precsim^{id}$ on goods. The next theorem [82, Thm. 2] characterizes the case when one recovers in this way the original total preorder:

Theorem 6.28 *Let $\precsim$ be a total preorder on $\mathbb{R}^n_+$. The following statements are equivalent:*

(i) $\precsim$ coincides with $\precsim^{id}$.

(ii) $\precsim$ has the following properties:

(a) $\precsim$ is nondecreasing[10]*.*

(b) For every $x_1 \in \mathbb{R}^n_+$, the upper contour set $\{x \in \mathbb{R}^n_+ : x \precsim x_1\}$ is evenly convex.

(c) For every $x_1 \in \mathbb{R}^n_+$, if $\alpha > 1$ and x_2 belongs to the closure of $\{x \in \mathbb{R}^n_+ : x \precsim x_1\}$, then $\alpha x_2 \precsim x_1$.

(d) For every $x_1, x_2 \in \mathbb{R}^n_+$, if $x_1 \sim x_2$ and x_1 is a $\precsim$ – maximal element of $B(p_1)$ for some $p_1 \in \mathbb{R}^n_+$ then x_2 is a $\precsim$ – maximal element of $B(p_2)$ for some $p_2 \in \mathbb{R}^n_+$.

Moreover, if conditions (a)-(d) hold, then $\precsim^i$ has the following properties:

(a') $\precsim^i$ is nonincreasing[11]*.*

(b') For every $p_1 \in \mathbb{R}^n_+$, the lower contour set $\{p \in \mathbb{R}^n_+ : p_1 \precsim^i p\}$ is evenly convex.

(c') For every $p_1 \in \mathbb{R}^n_+$, if $\alpha > 1$ and p_2 belongs to the closure of $\{p \in \mathbb{R}^n_+ : p_1 \precsim^i p\}$, then $p_1 \precsim^i \alpha p_2$.

(d') For every $p_1, p_2 \in \mathbb{R}^n_+$, if $p_1 \backsim^i p_2$ and p_1 is a $\precsim^i$ – minimal element of $B^{-1}(x_1)$ for some $x_1 \in \mathbb{R}^n_+$ then p_2 is a $\precsim^i$ – minimal element of $B^{-1}(x_2)$ for some $x_2 \in \mathbb{R}^n_+$.

The duality mapping $\precsim \mapsto \precsim^i$ is a bijection, with inverse $\preceq^* \mapsto \preceq^{*d}$, from the set of all total preorders $\preceq$ on $\mathbb{R}^n_+$ with properties (a)-(d) onto the set of all total preorders $\precsim^i$ on $\mathbb{R}^n_+$ with properties (a')-(d').

It is important to notice that conditions *(a')-(c')* are satisfied by the indirect preorder $\precsim^i$ induced by an arbitrary preorder $\precsim$, that is, conditions *(a)-(d)* are not required to this effect. An example of a total

[10] One says that $\precsim$ is nondecreasing if it is an extension of the componentwise ordering $\geq$, that is, if one has $x_1 \precsim x_2$ whenever $x_1 \geq x_2$.

[11] One says that $\precsim^i$ is nondecreasing if it is an extension of the reverse componentwise ordering $\leq$, that is, if one has $p_1 \precsim p_2$ whenever $p_1 \leq p_2$.

preorder $\precsim$ whose induced indirect preorder fails to possess property *(d')* is the one on $\mathbb{R}^2_+$ represented by the utility function $u : \mathbb{R}^2_+ \longrightarrow \mathbb{R}$ defined by

$$u(y,z) = \begin{cases} -\frac{1}{y+1} & \text{if } z \leq 1 \\ 1 & \text{if } z > 1 \end{cases} .$$

Its associated indirect utility function $v : \mathbb{R}^2_+ \longrightarrow \mathbb{R}$ satisfies

$$v(q,r) = \begin{cases} 1 & \text{if } r < 1 \\ 0 & \text{if } r \geq 1 \text{ and } q = 0 \\ -\frac{q}{q+1} & \text{if } r \geq 1 \text{ and } q > 0 \end{cases} .$$

One can easily check that if $v(q,r) = 0$ then no $\succeq$ – maximal element exists in the corresponding budget set $B(q,r)$; on the contrary, if $v(q,r) \neq 0$ then $B(q,r)$ has at least a maximal element. Hence v is a utility representation for the indirect preorder $\precsim^i$. Consider now the price vectors $(0,1)$ and $(0,2)$, which are clearly indifferent under $\precsim^i$. While $(0,1)$ is a $\precsim^i$ – minimal element of $B^{-1}(1,1)$, one can verify that no $(y,z) \in \mathbb{R}^2_+$ exists such that $(0,2)$ is a $\precsim^i$ – minimal element of $B^{-1}(y,z)$. Indeed, if $(y,z) \in \mathbb{R}^2_+$, with $y \neq 0$, is such that $(0,2) \in B^{-1}(y,z)$ then $(0,2) \succ^i \left(\frac{2-3z}{2y}, \frac{3}{2}\right) \in B^{-1}(y,z)$; if $(0,z) \in \mathbb{R}^2_+$ is such that $(0,2) \in B^{-1}(0,z)$ then $(0,2) \succ^i \left(\frac{3}{2}, \frac{3}{2}\right) \in B^{-1}(0,z)$. We have thus seen that condition *(d')* does not hold. Therefore, in view of Theorem 6.28, $\precsim^i$ and $\precsim^{idi}$ must be different. To check that this is the case, let $u^0 : \mathbb{R}^2_+ \longrightarrow \mathbb{R}$ be defined by (6.14). A straightforward computation shows that

$$u^0(y,z) = \begin{cases} -1 & \text{if } z \leq 1 \text{ and } y = 0 \\ \frac{z-1}{y-z+1} & \text{if } z \leq 1 \text{ and } y > 0 \\ 1 & \text{if } z > 1 \end{cases}$$

and that the infimum in the right hand side of (6.14) is not attained if and only if $u^0(x) = -1$. Therefore u^0 is a utility representation of $\precsim^{id}$. For every $(y,z) \in B(0,2)$ one has $u^0(y,z) < 0 = u^0(1,1)$, whence, as $(1,1) \in B(0,1)$, it follows that $(0,1) \succ^{idi} (0,2)$. Since $(0,1) \backsim^i (0,2)$, we have an evidence that $\precsim^i$ and $\precsim^{idi}$ are different.

An obvious manipulation of the preceding example would yield a total preorder $\precsim$ satisfying conditions *(a)-(c)* but not *(d)*. Conditions *(a)-(c)* are actually satisfied by any total preorder $\precsim$ that can be written in the form $\precsim^{*d}$ for some total preorder $\precsim^*$; however $\precsim^*$ must be different from $\precsim^i$ unless condition *(d)* also holds. The following result [82, Prop. 3] shows the discrepancy between $\precsim$ and $\precsim^{id}$ when $\precsim$ satisfies *(a)-(c)* but not *(d)*.

Proposition 6.7 *Let $\succsim$ be a total preorder on $\mathbb{R}^n_+$ satisfying properties (a)-(c) of Theorem 6.28. Then $\succsim^{id}$ is the total preorder whose strict preorder $\succ^{id}$ is defined as follows:*

$x_1 \succ^{id} x_2$ if and only if

either $x_1 \succ x_2$

or $x_1 \backsim x_2$, x_1 is not a $\succsim$ −maximal element in $B(p)$ for any $p \in \mathbb{R}^n_+$ and x_2 is a $\succsim$ − maximal element of $B(p)$ for some $p \in \mathbb{R}^n_+$.

The preceding proposition tells us that $\succ^{id}$ is an extension of $\succ$ (or, equivalently, $\succsim$ is an extension of $\succsim^{id}$). In other words, the only difference between $\succsim^{id}$ and $\succsim$, if $\succsim$ satisfies *(a)-(c)* but not *(d)*, is that some pairs that are indifferent under $\succsim$ are not indifferent under $\succsim^{id}$. In a sense, $\succsim^{id}$ can be regarded as a regularized version of $\succsim$. Indeed, in spite of the fact that, for an arbitrary total preorder $\succsim$, $\succsim^{idi}$ does not necessarily coincide with $\succsim^i$ (as shown by the example above), one has [82, Thm. 4]:

Theorem 6.29 *For every total preorder $\succsim$ on $\mathbb{R}^n_+$, $\succsim^{idid}$ coincides with $\succsim^{id}$.*

For total preorders such that all budget sets corresponding to strictly positive prices have maximal elements, one has the following duality result [74, Thm. 5], which, in contrast with the general case, also provides a characterization of the associated indirect preorders:

Theorem 6.30 *Let $\succsim$ be a total preorder on $\mathbb{R}^n_+$ such that for every $p \in \mathbb{R}^n_{++}$ the set $B(p)$ has a $\succsim$ −maximal element. Then $\succsim^i$ has the following properties:*

(a) $\succsim^i$ is nonincreasing.

(b) For every $p_1 \in \mathbb{R}^n_+$, the strict lower contour set $\{p \in \mathbb{R}^n_+ : p_1 \succ^i p\}$ is evenly convex.

(c) For every $p_1 \in \mathbb{R}^n_+$, if p_2 belongs to the closure of $\{p \in \mathbb{R}^n_+ : p_1 \succ^i p\}$ and $\alpha > 1$ then $p_1 \succsim \alpha p_2$.

Conversely, if $\succsim^*$ is a total preorder on $\mathbb{R}^n_+$ such that (a)-(c) hold with $\succsim^i$ replaced by $\succsim^*$ then, for every $p \in \mathbb{R}^n_{++}$, the set $B(p)$ has a $\succsim^{*d}$ −maximal element and $\succsim^{*di}$ coincides with $\succsim^*$.

Therefore, the mapping $\succsim \longmapsto \succsim^i$ is a bijection from the set of total preorders $\succsim$ on $\mathbb{R}^n_+$ that satisfy conditions (a)-(c) of Theorem 6.28 and are such that, for every $p \in \mathbb{R}^n_{++}$, the set $B(p)$ has a $\succsim$ −maximal element onto the set of all total preorders $\succsim^*$ on $\mathbb{R}^n_+$ for which (a)-(c) hold with $\succsim^i$ replaced by $\succsim^*$.

The class of total preorders that, besides being in perfect duality with their associated indirect preorders, have the additional property that all budget sets corresponding to strictly positive prices have maximal elements admits a very easy characterization [74, Thm. 6]:

Theorem 6.31 *Let $\succsim$ be a total preorder on $\mathbb{R}^n_+$ that coincides with $\succsim^{id}$. The following statements are equivalent:*

(i) For every $p_1 \in \mathbb{R}^n_{++}$ the set $B(p_1)$ has a $\succsim$ −maximal element.

(ii) For every $p_1 \in \mathbb{R}^n_+$, the strict lower contour set $\{p \in \mathbb{R}^n_+ : p_1 \succ^i p\}$ is evenly convex.

(iii) For every $p_1 \in \mathbb{R}^n_{++}$, the strict lower contour set $\{p \in \mathbb{R}^n_+ : p_1 \succ^i p\}$ is evenly convex.

An analogue of the concept of expenditure function also exists in the context of consumer's preferences represented by preorders instead of utility functions. For a (partial) preorder $\succsim$ on $\mathbb{R}^n_+$, the associated expenditure function $e_{\succsim} : \mathbb{R}^n_+ \times \mathbb{R}^n_+ \to \mathbb{R}_+$ is defined by

$$e_{\succsim}(p,x) = \inf \{\langle x', p\rangle : x' \succsim x\} \qquad (p \in \mathbb{R}^n_+, \; x \in \mathbb{R}^n_+) \,.$$

It shows the amount of money that a consumer with preferences represented by $\succsim$ needs to spend under the prices p to purchase a commodity bundle at least as good as x.

The duality relationship between expenditure functions and preorders is described in the next theorem, whose statement involves the following *Hull Cancellation Property:*

(HCP) For all $x_1, x_2 \in \mathbb{R}^n_+$, $\overline{co}\left(S^{x_1}_{\succsim} + \mathbb{R}^n_+\right) \subseteq \overline{co}\left(S^{x_2}_{\succsim} + \mathbb{R}^n_+\right)$ only if $S^{x_1}_{\succsim} \subseteq S^{x_2}_{\succsim}$, with $\overline{co}$ denoting closed convex hull.

One can easily observe that, for a total preorder, the Hull Cancellation Property can be equivalently expressed in terms of equalities instead of inclusions. However, this is not the case, in general, for partial preorders [83, Thm. 3.3]:

Theorem 6.32 *The mapping $\succsim \;\longmapsto\; e_{\succsim}$ is a bijection from the set of all preorders $\succsim$ on $\mathbb{R}^n_+$ whose upper contour sets $S^x_{\succsim} = \{x' \in \mathbb{R}^n_+ : x' \succsim x\}$ satisfy (HCP) onto the set of functions $e : \mathbb{R}^n_+ \times \mathbb{R}^n_+ \to \mathbb{R}_+$ that satisfy the following properties:*

(a) For every $x \in \mathbb{R}^n_+$, the mapping $e(\cdot, x)$ is concave, positively homogeneous and upper semicontinuous.

(b) For every $x \in \mathbb{R}^n_+$, the closed convex hull of the set $\{x' \in \mathbb{R}^n_+ : e(\cdot, x') \geq e(\cdot, x)\} + \mathbb{R}^n_+$ coincides with $\partial e(\cdot, x)(0)$, the superdifferential of the concave function $e(\cdot, x)$ at the origin (see Theorem 6.20).

The inverse mapping is $e \longmapsto \succsim_e$, with $\succsim_e$ denoting the preorder on $\mathbb{R}^n_+$ defined by $x_1 \succsim_e x_2$ if and only if $e(\cdot, x_1) \geq e(\cdot, x_2)$ (pointwise).

According to the preceding theorem, for a preorder $\succsim$ whose upper contour sets satisfy the Hull Cancellation Property, $\succsim_{e_{\succsim}}$ and $\succsim$ coincide. In fact, one can easily prove that the converse statement is also true. The class of preorders satisfying the Hull Cancellation Property includes all nondecreasing preorders whose upper contour sets are closed and convex, since for any such preorder $\succsim$ and any $x \in \mathbb{R}^n_+$ one has $\overline{co}\left(S^x_{\succsim} + \mathbb{R}^n_+\right) = S^x_{\succsim}$. For this subclass of preorders, the following duality theorem [83, Thm. 3.5] holds:

Theorem 6.33 *The mapping $\succsim \;\longmapsto\; e_{\succsim}$ is a bijection from the set of all nondecreasing preorders $\succsim$ on $\mathbb{R}^n_+$ whose upper contour sets are closed and convex onto the set of functions $e : \mathbb{R}^n_+ \times \mathbb{R}^n_+ \to \mathbb{R}_+$ that satisfy the following properties:*

(a) For every $x \in \mathbb{R}^n_+$, the mapping $e(\cdot, x)$ is concave, positively homogeneous and upper semicontinuous.

(b')For every $x \in \mathbb{R}^n_+$, $\{x' \in \mathbb{R}^n_+ : e(\cdot, x') \geq e(\cdot, x)\} = \partial e(\cdot, x)(0)$.

The inverse mapping $e \longmapsto \succsim_e$ is given by: $x_1 \succsim_e x_2$ if and only if $x_1 \in \partial e(\cdot, x_2)(0)$.

Some classical results of the standard utility framework, like the Slutsky equation, can be generalized to the preference setting; see [98, formula (5)] and [83, Thm. 4.13]. Demand correspondences can also be introduced in this context. Given a total preorder $\succsim$ on the commodity space $\mathbb{R}^n_+$, the associated demand correspondence X assigns to each price vector $p \in \mathbb{R}^n_{++}$ the set $X(p) = \{x \in B(p) : x \succsim y \quad \forall\, y \in B(p)\}$. The last results [79] in this section state sufficient conditions for the monotonicity of X. We recall that a total preorder $\succsim$ on $\mathbb{R}^n_+$ is said to be locally nonsatiated if no relatively open subset of $\mathbb{R}^n_+$ has a $\succsim$ − maximal element. The next theorem generalizes Theorem 6.26.

Theorem 6.34 *Let $\succsim$ be a locally nonsatiated total preorder in $\mathbb{R}^n_+$ and let X be its associated demand correspondence. If the set $\{(x,p) \in \mathbb{R}^n_+ \times \mathbb{R}^n_{++} : x \succsim y \quad \forall\, y \in B(p)\}$ is convex then X is monotone.*

From this theorem the following result follows:

Proposition 6.8 *Let $\succsim$ be a nondecreasing total preorder in $\mathbb{R}^n_+$, X be its associated demand correspondence, and assume that $\succsim$ satisfies the following condition:*

$$\left.\begin{array}{r} x_1 \succsim \lambda y \\ x_2 \succsim \mu y \\ \lambda > 0,\ \mu > 0 \end{array}\right\} \Longrightarrow \tfrac{1}{2}\left(x_1 + x_2\right) \succsim 2\tfrac{\lambda\mu}{\lambda+\mu} y.$$

Then X is monotone.

In the case when the upper contour sets are closed, they must also be closed if the condition in the preceding proposition holds (take $\lambda = \mu = 1$). Notice that this sufficient condition is satisfied when $\succsim$ is nondecreasing and its graph, $graph(\succsim) = \{(x, y) \in \mathbb{R}^n_+ \times \mathbb{R}^n_+ : x \succsim y\}$, is a convex set; indeed, if $graph(\succsim)$ is convex then one has

$$\left.\begin{array}{r} x_1 \succsim \lambda y \\ x_2 \succsim \mu y \\ \lambda > 0,\ \mu > 0 \end{array}\right\} \Longrightarrow \tfrac{1}{2}(x_1 + x_2) \succsim \tfrac{\lambda+\mu}{2} y$$

and, since $\frac{\lambda+\mu}{2} \geq 2\frac{\lambda\mu}{\lambda+\mu}$ for every positive λ and μ, if $\succsim$ is nondecreasing then one also has $\frac{\lambda+\mu}{2} y \succsim 2\frac{\lambda\mu}{\lambda+\mu} y$ for every $y \in \mathbb{R}^n_+$.

To conclude, let us mention the recent paper [133], in which the possibility of reconstructing demand correspondences from indirect preferences is studied.

Refcrences

[1] O. Alvarez, E.N.Barron and H. Ishii, *Hopf-Lax formulas for semicontinuous data*, Indiana Univ. Math. J. **48**, pp. 993-1035 (1999).

[2] Ch. Amara and M. Volle, *Compléments de dualité quasi convexe*, Ann. Sci. Math. Québec **23**, pp. 119-137 (1999).

[3] M. Andramonov; A.M. Rubinov and B.M. Glover, *Cutting angle methods in global optimization*, Appl. Math. Lett. **12**, pp. 95-100 (1999).

[4] M. Atteia and A. Elqortobi, *Quasi-convex duality*, in Optimization and Optimal Control, A. Auslender, W. Oettli and J. Stoer, eds., Springer-Verlag, Berlin, pp. 3-8 (1981).

[5] M. Avriel, W.T. Diewert, S. Schaible and I.Zang, *Generalized concavity*, Plenum Publishing Corporation (1988).

[6] E.J. Balder, *An extension of duality-stability relations to non-convex optimization problems*, SIAM J. Control Optimization **15**, pp. 329-343 (1977).

[7] A.M. Bagirov and A.M.Rubinov, *Global minimization of increasing positively homogeneous functions over the unit simplex*, Ann. Oper. Res. **98,** pp. 171-187 (2000).

[8] E.N. Barron, R. Jensen and W. Liu, *Hopf-Lax formula for* $u_t + H(u, Du) = 0$, J. Differ. Eq. **126**, pp. 48-61 (1996).

[9] E.N. Barron, R. Jensen and W. Liu, *Hopf-Lax formula for* $u_t + H(u, Du) = 0$ *II*, Comm. P.D.E. **22**, pp. 1141-1160 (1997).

[10] E.N. Barron, R. Jensen and W. Liu, *Applications of Hopf-Lax formula for* $u_t + H(u, Du) = 0$, SIAM J. Math. Anal. **29**, pp. 1022-1039 (1998).

[11] E.N. Barron and W. Liu, *Calculus of variations in* L^∞, Applied Math. Opt. **35**, pp. 237-243 (1997).

[12] C. Blackorby, D. Primont and R.R. Russell, *Duality, Separability, and Functional Structure,* North-Holland Publishing Company (1978).

[13] D.S. Bridges, G.B. Mehta, *Representations of preferences orderings*, Springer-Verlag (1995).

[14] M. Browning, *Dual approaches to utility*, in Handbook of Utility Theory. Volume 1. Principles, S. Barberà, P. Hammond and Chr. Seidl, eds., Kluwer Academic Publishers, Dordrecht, pp. 122-144 (1998).

[15] R.J. Cooper and K.R. McLaren, *Atemporal, temporal and intertemporal duality in consumer theory*, Internat. Econom. Rev. **21**, pp. 599–609 (1980).

[16] R. Cornes, *Duality and Modern Economics,* Cambridge University Press (1992).

[17] J.-P. Crouzeix, *Polaires quasi-convexes et dualité*, C.R. Acad. Sci. Paris série A **279**, pp. 955-958 (1974).

[18] J.-P. Crouzeix, *Conjugacy in quasiconvex analysis*, in Convex Analysis and Its Applications, A. Auslender, ed., Springer-Verlag, Berlin, pp. 66-99 (1977).

[19] J.-P. Crouzeix, *Contribution à l'étude des fonctions quasi-convexes*, Thèse d'Etat, Univ. de Clermont II (1977).

[20] J.-P. Crouzeix, *Continuity and differentiability properties of quasiconvex functions on* $\mathbb{R}^n$, in Generalized Concavity in Optimization and Economics, S. Schaible and W.T. Ziemba, eds., Academic Press, New York, pp. 109-130 (1981).

[21] J.-P. Crouzeix, *A duality framework in quasiconvex programming*, in Generalized Concavity in Optimization and Economics, S. Schaible and W.T. Ziemba, eds., Academic Press, New York, pp. 207-225 (1981).

[22] J.-P. Crouzeix, *Duality between direct and indirect utility functions. Differentiability properties*, J. Math. Econ. **12**, pp. 149-165 (1983).

[23] J.-P. Crouzeix, J.A. Ferland and S. Schaible, *Duality in generalized linear fractional programming*, Math. Programming **27**, pp. 342-354 (1983).

[24] A. Daniilidis and N. Hadjisavvas, *On the subdifferentials of quasiconvex and pseudoconvex functions and cyclic monotonicity*, J. Math. Anal. Appl. **237**, pp. 30-42 (1999).

[25] A. Daniilidis and N. Hadjisavvas, *On generalized cyclically monotone operators and proper quasimonotonicity*, Optimization **47**, pp. 123-135 (2000).

[26] A. Daniilidis and J.-E. Martínez-Legaz, *Characterizations of evenly convex sets and evenly quasiconvex functions*, J. Math. Anal. Appl. **273**, pp. 58-66 (2002).

[27] R. Deumlich and K.-H. Elster, *Φ-conjugation and nonconvex optimization. A survey. Part I*, Math. Operationsforsch. Stat. Ser. Optim. **14**, pp. 125-149 (1980).

[28] R. Deumlich and K.-H. Elster, *Φ-conjugation and nonconvex optimization. A survey. Part II*, Math. Operationsforsch. Stat. Ser. Optim. **15**, pp. 499-515 (1980).

[29] R. Deumlich and K.-H. Elster, *Φ-conjugation and nonconvex optimization. A survey. Part III*, Optimization **16**, pp. 789-803 (1980).

[30] W.E. Diewert, *Generalized concavity and economics,* in Generalized Concavity in Optimization and Economics, S. Schaible and W.T. Ziemba, eds., Academic Press, New York, pp. 511-541 (1981).

[31] W.E. Diewert, *Duality approaches to microeconomics theory*, in Handbook of Mathematical Economics vol. 2, K.J. Arrow and M.D. Intriligator, eds., North-Holland Publishing Company, Amsterdam, pp. 535-599 (1982).

[32] S. Dolecki and S. Kurcyusz, *On Φ-convexity in extremal problems*, SIAM J. Control Optimization **16**, pp. 277-300 (1978).

[33] R.R. Egudo, *Efficiency and generalized convex duality for multiobjective programs*, J. Math. Anal. Appl. **138**, pp. 84–94 (1989).

[34] I. Ekeland and R. Temam, *Convex Analysis and Variational Problems*, North-Holland Publishing Company (1976).

[35] A. Elqortobi, *Conjugaison quasi-convexe des fonctionnelles positives*, Ann. Sci. Math. Québec **17**, pp. 155-167 (1993).

[36] K.-H. Elster and R. Nehse, *Zur Theorie der Polarfunktionale*, Math. Oper. Stat. **5**, pp. 3-21 (1974).

[37] K.-H. Elster and A. Wolf, *Recent results on generalized conjugate functions,* in Trends in Mathematical Optimization, K.-H. Hoffmann, J.-B. Hiriart-Urruty, C. Lemarechal and J. Zowe, eds., Birkhauser Verlag, Basel, pp. 67-78 (1988).

[38] J.J.M. Evers and H. van Maaren, *Duality principles in mathematics and their relations to conjugate functions*, Nieuw Arch. Wiskd. IV. Ser. **3**, pp. 23-68 (1985).

[39] R. Färe and D. Primont, *Multi-Output Production and Duality: theory and applications*, Kluwer-Nijhoff Publishing (1995).

[40] W. Fenchel, *A remark on convex sets and polarity*, Meddel. Lunds Univ. Mat. Sem. Suppl.-band M. Riesz, pp. 82-89 (1952).

[41] F. Flores-Bazán and J.E. Martínez-Legaz, *Simplified global optimality conditions in generalized conjugation theory*, in Generalized Convexity, Generalized Monotonicity, J.-P. Crouzeix, J.-E. Martínez-Legaz and M. Volle, eds., Kluwer Academic Publishers, Dordrecht, pp. 305-329 (1998).

[42] J.G.B. Frenk, D.M.L. Dias and J. Gromicho, *Duality theory for convex/quasiconvex functions and its application to optimization*, in Generalized Convexity, S. Komlosi, T. Rapcsák and S. Schaible, eds., Springer-Verlag, Berlin, pp. 153-170 (1994).

[43] S. Fuchs-Seliger, *A note on duality in consumer theory*, Econom. Theory **13**, pp. 239–246 (1999).

[44] W. Gangbo and R.J. McCann, *The geometry of optimal transportation*, Acta Math. **177**, pp. 113-161 (1996).

[45] J. Getán, J.-E. Martínez-Legaz and I. Singer, *$(*, s)$-dualities*, J. Math. Sci. (N. Y.) **115**, pp. 2506-2541 (2003).

[46] F. Glover, *A multiphase-dual algorithm for the zero-one integer programming problem*, Oper. Res. **13**, pp. 879-919 (1965).

[47] F. Glover, *Surrogate constraints*, Oper. Res. **16**, pp. 741-749 (1968).

[48] H.J. Greenberg and W.P. Pierskalla, *Surrogate mathematical programming*, Oper. Res. **18**, pp. 924-939 (1970).

[49] H.J. Greenberg and W.P. Pierskalla, *Quasiconjugate function and surrogate duality*, Cahiers du Centre d'Etude de Recherche Oper. **15**, pp. 437-448 (1973).

[50] W. Hildenbrand, *Market Demand: Theory and Empirical Evidence*, Princeton University Press (1994).

[51] H.S. Houthakker, *Revealed preference and the utility function*, Economica N.S. **17**, pp. 159-170 (1950).

[52] Y. Kannai, *A characterization of monotone individual demand functions*, J. Math. Econ. **18**, pp. 87-94 (1989).

[53] D.S. Kim, G.M. Lee and H. Kuk, *Duality for multiobjective fractional variational problems with generalized invexity*, Nihonkai Math. J. **9**, pp. 17–25 (1998).

[54] D.S. Kim and S. Schaible, *Optimality and duality for invex nonsmooth multiobjective programming problems,* Working Paper 02-06, A. Gary Anderson Graduate School of Management Univ. of California Riverside (2002).

[55] V. Krishna and H. Sonnenschein, *Duality in consumer theory,* in Preferences, Uncertainty and Optimality, J. Chipman, D. McFadden and M.K. Richter, eds., Westview Press, pp. 44-55 (1990).

[56] S.S. Kutateladze and A.M. Rubinov, *Minkowski duality and its applications,* Russian Math. Surveys **27**, pp. 137-192 (1972).

[57] B. Lemaire and M. Volle, *Duality in DC programming,* in Generalized Convexity, Generalized Monotonicity, J.-P. Crouzeix, J.-E. Martínez-Legaz and M. Volle, eds., Kluwer Academic Publishers, Dordrecht, pp. 331-345 (1998).

[58] B. Lemaire and M. Volle, *A general duality scheme for nonconvex minimization problems with a strict inequality constraint,* J. Global Opt. **13**, pp. 317-327 (1998).

[59] V. Levin, *Quasi-convex functions and quasi-monotone operators,* J. Convex Anal. **2**, pp. 167-172 (1995).

[60] V. Levin, *Abstract cyclical monotonicity and Monge solutions for the general Monge-Kantorovich problem,* Set-Valued Analysis **7**, pp. 7-32 (1999).

[61] P.O. Lindberg, *A generalization of Fenchel conjugation giving generalized lagrangians and symmetric nonconvex duality,* in Survey of Mathematical Programming 1, A. Prékopa, ed., Akad. Kiado and North-Holland Publishing Company, Budapest and Amsterdam, pp. 249-268 (1979).

[62] P.O. Lindberg, *On quasiconvex duality,* Preprint, Dept. Math. Royal Inst. Technology Stockholm (1981).

[63] J.T. Little, *Indirect preferences,* J. Econom. Theory **20**, pp. 182-193 (1979).

[64] D.G. Luenberger, *Quasi-convex programming,* SIAM J. Appl. Math. **16**, pp. 1090-1095 (1968).

[65] J.-E. Martínez-Legaz, *Conjugación asociada a un grafo,* Acta Salmant. Cienc. **46**, pp. 837-839 (1982).

[66] J.-E. Martínez-Legaz, *A generalized concept of conjugation methods,* in Optimization, Theory and Algorithms, J.-B. Hiriart-Urruty, W. Oettli and J. Stoer, eds., Marcel Dekker, New York, pp. 45-49 (1983).

[67] J.-E. Martínez-Legaz, *Exact quasiconvex conjugation,* Z. Oper. Res. Ser. A **27**, pp. 257-266 (1983).

[68] J.-E. Martínez-Legaz, *Lexicographical order, inequality systems and optimization*, in System Modelling and Optimization, P. Thoft-Christensen, ed., Springer-Verlag, Berlin, pp. 203–212 (1984).

[69] J.-E. Martínez-Legaz, *Some new results on exact quasiconvex duality*, Methods Oper. Res. **49**, pp. 47-62 (1985).

[70] J.-E. Martínez-Legaz, *Quasiconvex duality theory by generalized conjugation methods*, Optimization **19**, pp. 603-652 (1988).

[71] J.-E. Martínez-Legaz, *On lower subdifferentiable functions*, in Trends in Mathematical Optimization, K.-H. Hoffmann, J.-B. Hiriart-Urruty, C. Lemarechal and J. Zowe, eds., Birkhauser Verlag, Basel, pp. 197-232 (1988).

[72] J.-E. Martínez-Legaz, *Generalized conjugation and related topics*, in Generalized convexity and fractional programming with economic applications, A. Cambini, E. Castagnoli, L. Martein, P. Mazzoleni and S. Schaible, eds., Springer-Verlag, Berlin, pp. 168-197 (1990).

[73] J.-E. Martínez-Legaz, *Duality between direct and indirect utility functions under minimal hypothesis*, J. Math. Econ. **20**, pp. 199-209 (1991).

[74] J.-E. Martínez-Legaz, *Axiomatic characterizations of the duality correspondence in consumer theory*. J. Math. Econ. **22**, pp. 509-522 (1993).

[75] J.-E. Martínez-Legaz, *Fenchel duality and related properties in generalized conjugation theory*, Southeast Asian Bull. Math. **19**, pp. 99-106 (1995).

[76] J.-E. Martínez-Legaz, *Dual representation of cooperative games based on Fenchel-Moreau conjugation*, Optimization **36**, pp. 291-319 (1996).

[77] J.-E. Martínez-Legaz, *Characterization of R-evenly quasiconvex functions*, J. Optim. Th. Appl. **95**, pp. 717-722 (1997).

[78] J.-E. Martínez-Legaz and J.K.-H. Quah, *A contribution to duality theory, applied to the measurement of risk aversion*, submitted.

[79] J.-E. Martínez-Legaz and J.K.-H. Quah, *Monotonic demand: direct and indirect approaches*, in preparation.

[80] J.-E. Martínez-Legaz and S. Romano-Rodríguez, *α−lower subdifferentiable functions*, SIAM J. Optim. **4**, pp. 800-825 (1993).

[81] J.-E. Martínez-Legaz, A.M. Rubinov and S. Schaible, *Increasing quasiconcave co-radiant functions with applications in mathematical economics*, Math. Methods Oper. Res. (to appear).

[82] J.-E. Martínez-Legaz and M.S. Santos, *Duality between direct and indirect preferences*, Econom. Theory **3**, 335-351 (1993).

[83] J.-E. Martínez-Legaz, and M.S. Santos, *On expenditure functions*, J. Math. Econ. **25**, pp. 143-163 (1996).

[84] J.-E. Martínez-Legaz and I. Singer, *Surrogate Duality for Vector Optimization*, Numerical Functional Analysis and Optimization **9**, pp. 547-568 (1987).

[85] J.-E. Martínez-Legaz and I. Singer, *Dualities between complete lattices*, Optimization **21**, pp. 481-508 (1990).

[86] J.-E. Martínez-Legaz and I. Singer, $\vee$*-dualities and* $\perp$*-dualities*, Optimization **22**, pp. 483-511 (1991).

[87] J.-E. Martínez-Legaz and I. Singer, $*$*-dualities*, Optimization **30**, pp. 295-315 (1994).

[88] J.-E. Martínez-Legaz and I. Singer, *Subdifferentials with respect to dualities*, ZOR-Math. Methods Oper. Res. **42**, pp. 109-125 (1995).

[89] J.-E. Martínez-Legaz and I. Singer, *Dualities associated to binary operations on* $\overline{\mathbb{R}}$, J. Convex Anal. **2**, pp. 185-209 (1995).

[90] J.-E. Martínez-Legaz and I. Singer, *An extension of D.C. duality theory, with an appendix on* $*$*-subdifferentials*, Optimization **42**, pp. 9-37 (1997).

[91] J.-E. Martínez-Legaz, and I. Singer, *On conjugations for functions with values in extensions of ordered groups*, Positivity **1**, pp. 193-218 (1997).

[92] J.-E. Martínez-Legaz, and I. Singer, *On* Φ*-convexity of convex functions*, Linear Algebra Appl. **278**, pp. 163-181 (1998).

[93] J.-E. Martínez-Legaz, and M. Volle, *Duality in D. C. programming: The case of several D. C. constraints,* J. Math. Anal. Appl. **237**, pp. 657-671 (1999).

[94] J.-E. Martínez-Legaz, and M. Volle, *Duality for D. C. optimization problems over compact sets*, in Optimization Theory; Recent Developments from Matrahaza, F. Giannessi, P. Pardalos and T. Rapcsák, eds., Kluwer Academic Publishers, Dordrecht, pp. 139-146 (2001).

[95] A. Mas-Colell, *The Theory of General Economic Equilibrium. A Differentiable Approach*, Cambridge University Press (1989).

[96] A. Mas-Colell, M.D. Whinston and J.R. Green, *Microeconomic Theory,* Oxford University Press (1995).

[97] M.-L. Mazure and M. Volle, *Équations inf-convolutives et conjugaison de Moreau-Fenchel,* Ann. Fac. Sci. Toulouse, V. Sér., Math. 12, pp. 103-126 (1991).

[98] L. McKenzie, *Demand theory without a utility index*, Review of Economic Studies **24**, p. 185-189 (1956-57).

[99] J.C. Milleron, *Unicité et stabilité de l'équilibre en économie de distribution*, Séminaire d'Econometrie Roy-Malinvaud (1974).

[100] S.K. Mishra, *Second order symmetric duality in mathematical programming with F-convexity*, European J. Oper. Res. **127**, pp. 507–518 (2000).

[101] L.G. Mitjushin and V.M. Polterovich, *Monotonicity criterion for demand function*, Ekonom. Mat. Metody **14**, pp. 122-128 (1978).

[102] J.-J. Moreau, *Inf-convolution, sous-additivité, convexité des fonctions numériques*, J. Math. Pures et Appl. **49**, pp. 109-154 (1970).

[103] R.N. Mukherjee, *Generalized convex duality for multiobjective fractional programs*, J. Math. Anal. Appl. **162**, pp. 309-316 (1991).

[104] W. Oettli and D. Schläger, *Conjugate functions for convex and nonconvex duality*, J. Global Optim. **13**, pp. 337–347 (1998).

[105] D. Pallaschke and S. Rolewicz, *Foundations of Mathematical Optimization. Convex Analysis without Linearity*, Kluwer Academic Publishers (1998).

[106] U. Passy and E.Z. Prisman, *Conjugacy in quasi-convex programming*, Math. Programming **30**, pp. 121-146 (1984).

[107] U. Passy and E.Z. Prisman, *A convexlike duality scheme for quasiconvex programs*, Math. Programming **32**, pp. 278-300 (1985).

[108] R.B. Patel, *Duality and sufficiency for fractional control problems with invexity*, Nonlinear Anal. **35**, pp. 541-548 (1999).

[109] J.P. Penot, *Duality for radiant and shady programs*, Acta Math. Vietnam. **22**, pp. 541-566 (1997).

[110] J.P. Penot, *What is quasiconvex analysis?,* Optimization **47**, pp. 35-110 (2000).

[111] J.P. Penot, *Duality for anticonvex programs* J. Global Optim. **19**, pp. 163–182 (2001).

[112] J.P. Penot and M. Volle, *Dualité de Fenchel et quasi-convexité*, C.R. Acad. Sc. Paris série I **304**, pp. 269-272 (1987).

[113] J.P. Penot and M. Volle, *Another duality scheme for quasiconvex problems*, in Trends in Mathematical Optimization, K.-H. Hoffmann, J.-B. Hiriart-Urruty, C. Lemarechal and J. Zowe, eds., Birkhauser Verlag, Basel, pp. 259-275 (1988).

[114] J.P. Penot and M. Volle, *On quasi-convex duality*, Math. Operat. Research **15**, pp. 597-625 (1990).

[115] J.P. Penot and M. Volle, *Explicit solutions to Hamilton-Jacobi equations under mild continuity and convexity assumptions*, J. Nonlinear Convex Anal. **1**, pp. 177-199 (2000).

[116] J.P. Penot and M. Volle, *Convexity and generalized convexity methods for the study of Hamilton-Jacobi equations*, in Generalized Convexity and Generalized Monotonicity, N. Hadjisavvas, J.-E. Martínez-Legaz and J.P. Penot, eds., Springer-Verlag, Berlin, pp. 294-316 (2001).

[117] E.L. Peterson, *Geometric Programming*, SIAM Review 18, pp. 1-52 (1976).

[118] R. Pini and C. Singh, *A survey of recent advances in generalized convexity with applications to duality theory and optimality conditions (1985-1995)*, Optimization **39**, pp. 311-360 (1997).

[119] J.K.-H. Quah, *The monotonicity of individual and market demand*, Econometrica **68**, pp. 911-930 (2000).

[120] M.K. Richter, *Duality and rationality*, J. Econom. Theory **20**, pp. 131-181 (1979).

[121] R.T. Rockafellar, *Convex Analysis*, Princeton University Press (1970).

[122] R.T. Rockafellar, *Conjugate duality and optimization*, SIAM (1974).

[123] R.T. Rockafellar, *Augmented Lagrange multiplier functions and duality in nonconvex programming*, SIAM J. Control Optimization **12**, pp. 268-285 (1974).

[124] A.M. Rubinov, *Abstract convexity and global optimization*, Kluwer Academic Publishers (2000).

[125] A.M. Rubinov and M. Andramonov, *Lipschitz programming via increasing convex-along-rays functions*, Optim. Methods Softw. **10**, pp. 763-781 (1999).

[126] A.M. Rubinov and B.M. Glover, *On generalized quasiconvex conjugation*, in Recent Developments in Optimization Theory and Nonlinear Analysis, Y. Censor and S. Reich, eds., American Mathematical Society, Providence, pp. 199–216 (1997).

[127] A.M. Rubinov and B.M. Glover, *Quasiconvexity via two steps functions*, in Generalized Convexity, Generalized Monotonicity, J.-P. Crouzeix, J.-E. Martínez-Legaz and M. Volle, eds., Kluwer Academic Publishers, Dordrecht, pp. 159-183 (1998).

[128] A.M. Rubinov and B.M. Glover, *Increasing convex-along-rays functions with applications to global optimization*, J. Optim. Th. Appl. **102**, pp. 615-642 (1999).

[129] A.M. Rubinov and B. Simsek, *Conjugate quasiconvex nonnegative functions*, Optimization **35**, pp. 1-22 (1995).

[130] A.M. Rubinov and B. Simsek, *Dual problems of quasiconvex maximization*, Bull. Aust. Math. Soc. **51**, pp. 139-144 (1995).

[131] P.A. Samuelson, *A note on the pure theory of consumer's behavior*, Economica N.S. **5**, pp. 61-71 (1938).

[132] P.A. Samuelson, *A note on the pure theory of consumer's behavior: an addendum*, Economica N.S. **5**, pp. 353-354 (1938).

[133] P. Scapparone, *The indirect preference relation,* Riv. Mat. Sci. Econom. Commerc. **48**, pp. 331-351 (2001).

[134] C.H. Scott and T.R.Jefferson, *Conjugate duality in generalized fractional programming,* J. Optim. Th. Appl. 60, pp. 475-483 (1989).

[135] C.H. Scott and T.R.Jefferson, *Duality for a nonconvex sum of ratios*, J. Optim. Th. Appl. 98, pp. 151-159 (1998).

[136] Shephard, R.W., *Theory of cost and production functions*, Princeton University Press (1970).

[137] I. Singer, *A Fenchel-Rockafellar type duality theorem for maximization*, Bull. Aust. Math. Soc. **20**, pp. 193-198 (1979).

[138] I. Singer, *Pseudo-conjugate functionals and pseudo-duality*, in Mathematical Methods in Operations Research, A. Dontchev, ed., Bulgarian Academy of Sciences, Sofia, pp. 115-134 (1981).

[139] I. Singer, *Generalized convexity, functional hulls and applications to conjugate duality in optimization*, in Selected Topics in Operations Research and Mathematical Economics, G. Hammer and D. Pallaschke, eds., Springer-Verlag, Berlin, pp. 49-79 (1984).

[140] I. Singer, *Conjugation operators*, in Selected Topics in Operations Research and Mathematical Economics, G. Hammer and D. Pallaschke, eds., Springer-Verlag, Berlin, pp. 80-97 (1984).

[141] I. Singer, *Surrogate dual problems and surrogate Lagrangians*, J. Math. Anal. Appl. **98**, pp. 31-71 (1984).

[142] I. Singer, *Some relations between dualities, polarities, coupling functions and conjugations*, J. Math. Anal. Appl. **115**, pp. 1-22 (1986).

[143] I. Singer, *A general theory of dual optimization problems*, J. Math. Anal. Appl. **116**, pp. 77-130 (1986).

[144] I. Singer, *Generalizations of convex supremization duality*, in Nonlinear and Convex Analysis, B.-L. Lin and S. Simons, eds., Marcel Dekker, New York, pp. 253-270 (1987).

[145] I. Singer, *Infimal generators and dualities between complete lattices*, Ann. Mat. Pura Appl. IV. Ser. **148**, pp. 289-352 (1987).

[146] I. Singer, *A general theory of dual optimization problems. II. On the perturbational dual problem corresponding to an unperturbational dual problem*, Z. Oper. Res. 33 (1989), 241-258.

[147] I. Singer, *Some further relations between unperturbational and perturbational dual optimization problems*, Optimization **22**, pp. 317-339 (1991).

[148] I. Singer, *Abstract Convex Analysis*, Wiley (1997).

[149] I. Singer, *Duality in quasi-convex supremization and reverse convex infimization via abstract convex analysis, and applications to approximation*, Optimization **45**, pp. 255–307 (1999).

[150] I. Singer, *Dual representations of hulls for functions satisfying* $f(0) = \inf f(X\setminus\{0\})$, Optimization **45**, pp. 309-341 (1999).

[151] I. Singer, *Some relations between linear mappings and conjugations in idempotent analysis*, J. Math. Sci. (N. Y.) **115**, pp. 2610-2630 (2003).

[152] P.D. Tao and S. El Bernoussi, *Duality in D.C. (difference of convex functions). Optimization. Subgradient methods*, in Trends in Mathematical Optimization, K.-H. Hoffmann, J.-B. Hiriart-Urruty, C. Lemarechal and J. Zowe, eds., Birkhauser Verlag, Basel, pp. 277-293 (1988).

[153] P.T. Thach, *Quasiconjugates of functions, duality relationships between quasiconvex minimization under a reverse convex constraint and quasiconvex maximization under a convex constraint and applications*, J. Math. Anal. Appl. **159**, pp. 299-322 (1991).

[154] P.T. Thach, *Global optimality criterion and a duality with a zero gap in nonconvex optimization*, SIAM J. Math. Anal. **24**, pp. 1537-1556 (1993).

[155] P.T. Thach, *A nonconvex duality with zero gap and applications*, SIAM J. Optim. **4**, pp. 44-64 (1994).

[156] P.T. Thach, *Diewert-Crouzeix conjugation for general quasiconvex duality and applications*, J. Optim. Th. Appl. **86**, pp. 719-743 (1995).

[157] P.T. Thach and M. Kojima, *A generalized convexity and variational inequality for quasi-convex minimization*, SIAM J. Optim. **6**, pp. 212–226 (1996).

[158] W.A. Thompson and D.W. Parke, *Some properties of generalized concave functions*, Oper. Res. **21**, pp. 305-313 (1973).

[159] J.F. Toland, *Duality in nonconvex optimization*, J. Math. Anal. Appl. **66**, pp. 399-415 (1978).

[160] S. Traoré and M. Volle, *Quasi-convex conjugation and Mosco convergence*, Richerche di Mat. **44**, pp. 369-388 (1995).

[161] M. Volle, *Regularizations in abelian complete ordered groups*, J. Math. Anal. Appl. **84**, pp. 418-430 (1981).

[162] M. Volle, *Conjugaison par tranches*, Ann. Mat. Pura Appl. **139**, pp. 279-312 (1985).

[163] M. Volle, *Contributions à la dualité en optimisation et à l'épiconvergence*, Thesis, Univ. de Pau (1986).

[164] M. Volle, *Conjugaison par tranches et dualité de Toland*, Optimization **18**, pp. 633-642 (1987).

[165] M. Volle, *Concave duality: application to problems dealing with difference of functions*, Math. Programming 41, pp. 261-278 (1988).

[166] M. Volle, *Conditions initiales quasiconvexes dans les équations de Hamilton-Jacobi*, C.R. Acad. Sci. Paris série I **325**, pp. 167-170 (1997).

[167] M. Volle, *Quasiconvex duality for the max of two functions*, in Recent Advances in Optimization, P. Gritzmann, R. Horst, E. Sachs and R. Tichatschke, eds., Springer-Verlag, Berlin, pp. 365-379 (1997).

[168] M. Volle, *A formula on the conjugate of the max of a convex function and a concave function*, J. Math. Anal. Appl. 220, 313–321 (1998).

[169] M. Volle, *Duality for the level sum of quasiconvex functions and applications*, ESAIM: Control, Optimisation and Calculus of Variations **3**, pp. 329-343 (1998), available at http://www.edpsciences.org/articles/cocv/abs/1998/01/cocvEng-Vol3.15.html.

[170] X.M. Yang, *Generalized convex duality for multiobjective fractional programs*, Opsearch **31**, pp. 155–163 (1994).

Chapter 7

ABSTRACT CONVEXITY

Alexander Rubinov*
School of Information Technology and Mathematical Science
University of Ballarat
3353 Victoria, Australia
amr@ballarat.edu.au

Joydeep Dutta†
Department of Mathematics, Indian Institute of Technology
Kanpur, 208016, India
jdutta@iitk.ac.in

Abstract In this paper we study the emerging area of abstract convexity. The theory of abstract convex functions and sets arises out of the properties of convex functions related to their global nature. One of the main applications of abstract convexity is global optimization. Apart from discussing the various fundamental facts about abstract convexity we also study quasiconvex functions in the light of abstract convexity. We further describe the surprising applications of the ideas of abstract convexity to the study of Hadamard type inequalities for quasiconvex functions.

Keywords: abstract convexity, subdifferential, conjugation, quasiconvexity, Hadamard inequality, global optimization.

*The work of the first author was carried out by the Grant No. SAB1999-0184 of the Spanish Ministry of Education and Culture.
†The work of the second author was carried out by the Grant No. SB99-B0771103B of the Spanish Ministry of Education and Culture.

1. Introduction

It is a well known fact that convex analysis is one of the cornerstones of modern Operations Research, including optimization and its application to various fields. The notion of subdifferential of a convex function at a given point of its domain plays a central role in convex analysis. The subdifferential plays two different roles. One of them is local: the subdifferential $\partial f(x)$ furnishes a local approximation to a convex function f in a neighborhood of a given point x. The other is global: $\partial f(x)$ is a tool for constructing supporting hyperplanes to the epigraph of the convex function f. An affine function h is a global support of f at x, if h is a minorant ($h(y) \leq f(y)$ for all $y \in \operatorname{dom} f$) and $h(x) = f(x)$. Indeed, if $l \in \partial f(x)$ then the affine function $h(y) = l(y) - c$, where $c = l(x) - f(x)$, is a global support of f at x.

Generalizations of the first trail of the subdifferential lead to *nonsmooth analysis* while the second trail leads to *abstract convexity*. Thus abstract convexity in a sense is the generalization of the global properties of convex functions. The existence of global affine supports is guaranteed by the separation theorem for convex sets. This idea also leads to the following fundamental result of convex analysis: each lower semicontinuous convex function f is the upper envelope (pointwise supremum) of all affine functions majorized by it. The main motivation for the development of *abstract convexity* lies in the fact that the envelope representation is very convenient, even when we consider the upper envelope of sets of non-affine functions.

Thus one is motivated to examine the main global notions of convex analysis in a non-convex setting. Two kinds of objects are studied in the framework of abstract convexity: abstract convex functions and abstract convex sets. It is assumed that a class of not necessary linear or affine *elementary* functions is given. Abstract convex functions can be represented via the envelope representation of sets of elementary functions. Abstract convex sets bring in the notion of non-linear separation. A point not belonging to an abstract convex set can be separated from it by an elementary function.

Various classes of functions and sets can be studied from the point of view of abstract convexity. Since this volume deals with generalized convexity, special attention in this paper is devoted for the study of quasiconvex functions from the point of view of abstract convexity.

Abstract convexity allows one to consider a new interesting area of applications of quasiconvexity, namely Hadamard type inequalities. This is one of the surprising applications of abstract convexity.

The structure of the paper is as follows. In Section 2 we present main definitions related to abstract convexity. In Section 3 we examine Minkowski duality and Fenchel-Moreau conjugation and some links between the distance and the plus-Minkowski gauge. Examples of abstract convex functions and sets based on various sets of elementary functions are discussed in Section 4. In particular, we study supremal generators of sets of lower semicontinuous functions in this section. Supremal generators of sets of quasiconvex functions are examined in Section 5. Some applications of abstract convexity can be found in Section 6. In the final Section 7 we discuss applications to Hadamard type inequalities for quasiconvex functions.

2. Main definitions

Order relations will play a very important role in this paper. In the sequel we assume, unless the opposite is explicitly mentioned, that all sets of functions under consideration will be equipped with the point wise order relation $\geq$. In other words, if Y is a set of extended real-valued functions defined on a set X and $f_1, f_2 \in Y$, then $f_1 \geq f_2 \iff f_1(x) \geq f_2(x)$ for all $x \in X$. If $U \subset Y$, then $\sup U$ is the pointwise supremum (the upper envelope) of the set U and $\inf U$ is the pointwise infimum (the lower envelope) of U. By definition $(\sup U)(x) = \sup\{f(x) : f \in U\}$, $(\inf U)(x) = \inf\{f(x) : f \in U\}$.

Let H be a set of extended real-valued functions defined on a set X. Consider a function $f : X \to \bar{R} := [-\infty, +\infty]$. The set

$$\text{supp}\,(f, H) = \{h \in H : h \leq f\} \tag{7.1}$$

is called the *support set* of the function f with respect to H. The support set accumulates the global information about the function f in terms of the set H. The function

$$\text{co}_H f = \sup\{h : h \in \text{supp}\,(f, H)\}, \tag{7.2}$$

is called the *H-convex hull* of a function f. Clearly $\text{co}_H f \leq f$. A function f defined on X is called *abstract convex* with respect to H (or H-convex) if $f = \text{co}_H f$, that is

$$f(x) = \sup\{h(x) : h \in H, h \leq f\}, \qquad x \in X. \tag{7.3}$$

Remark 7.1 For application it is interesting to consider only sets H of *sufficiently simple* functions. The set H will be called in the sequel the set of *elementary functions.*

Assume now that the set H consists of finite functions. If the support set $\text{supp}\,(f, H)$ is nonempty, then $\text{co}_H f(x) > -\infty$ for all $x \in X$. If supp (f, H) is empty then $\mathrm{co}_H f \equiv -\infty$. Let $R_{+\infty} = R \cup \{+\infty\}$. Denote by F_X the union of the set of all functions $f : X \to R_{+\infty}$ and the function identically equal to $-\infty$. Then each H-convex function belongs to F_X.

Let H be a set of finite functions defined on X. A set $V \subset X$ is called *abstract convex* with respect to H (or (X, H)-convex) if each point $x \notin V$ can be separated from V by a function from H, that is, there exists $h \in H$ such that $h(x) > \sup\{h(v) : v \in V\}$. Similarly a set $U \subset H$ is called *abstract convex* with respect to X (or (H, X)-convex) if for each $h' \notin U$ there exists $x \in X$ such that $h'(x) > \sup\{h(x) : h \in U\}$. It is easy to check that U is (H, X)-convex if and only if there exists a function f defined on X such that $U = \mathrm{supp}\,(f, H)$.

The intersection of all (H, X)-convex sets containing a set $U \subset H$ is called the *abstract convex hull* or (H, X)-convex hull of the set U. We shall denote the (H, X)-convex hull of U by $\mathrm{co}_{H,X} U$, or, if it does not cause misunderstanding, by $\mathrm{co}_H U$.

Example 7.1 Let X be a locally convex Hausdorff topological vector space. Consider the conjugate space $L = X^*$ as a set of elementary functions. Then a function $f : X \to R_{+\infty}$ is abstract convex with respect to X^* if and only if this function is lower semicontinuous and sublinear. (This is a version of Hahn-Banach theorem.) Consider now the set H of all continuous affine functions as a set of elementary functions. Then a function $f : X \to R_{+\infty}$ is abstract convex with respect to H if and only if f is a lower semicontinuous convex function. Let $U \subset X$. Then the following assertions are equivalent:

1) U is (X, L)-convex;

2) U is (X, H)-convex;

3) U is closed and convex.

A notion of abstract convex set allows one to define abstract quasiconvex functions. A function $f : X \to \bar{R}$ is called *abstract quasiconvex* with respect to a set of elementary functions H (or H-quasiconvex) if its level sets $S(c) = \{x \in X : f(x) \le c\}$ are (X, H)-convex for all $c \in \bar{R}$.

Remark 7.2 Abstract concave functions, that is, the lower envelopes of subsets of elementary functions, abstract concave sets and abstract quasiconcave functions may equally be studied in this fashion. For the sake of definiteness we shall discuss here mainly the abstract convex situation.

Let H be a set of finite functions defined on a set X and $f : X \to R_{+\infty}$ be an H-convex function. Then

$$f(x) = \sup\{h(x) : h \in \operatorname{supp}(f, H)\} \tag{7.4}$$

for all $x \in X$. The set

$$\partial^*_H f(y) = \{h \in \operatorname{supp}(f, H) : h(y) = f(y)\}$$

is called the *support set of f at a point $y \in X$*. Clearly $\partial^*_H f(y)$ is not empty if and only if the supremum in (7.4) with x replaced by y is attained.

We now give the definition of abstract subdifferential. Let L be a set of finite functions $l : X \to R$ and H_L be the set of functions h of the form $h = l - c\mathbf{1}$ with $l \in L$ and $c \in R$. Here $\mathbf{1}$ is the function defined by $\mathbf{1}(x) = 1$ for all $x \in X$. Let $f : X \to R_{+\infty}$ be an H_L-convex function. The *L-subdifferential* $\partial_L f(y)$ of f at a point y is the following set:

$$\partial_L f(y) = \{l \in L : l(x) - l(y) \le f(x) - f(y)\} \tag{7.5}$$

There is a simple link between the L-subdifferential $\partial_L f(y)$ and the support set $\partial^*_{H_L} f(y)$ with respect to H_L.

Proposition 7.1 *Let f be an H_L-convex function, $y \in X$ and $l \in L$. Let $c = l(y) - f(y)$ and $h_{l,c} = l - c\mathbf{1}$. Then $l \in \partial_L f(y)$ if and only if $h_{l,c} \in \partial^*_{H_L} f(y)$.*

The proof is straightforward.

3. Minkowski duality and Fenchel-Moreau conjugation

Abstract convexity provides a convenient framework for the study of various types of dual objects related to abstract convex functions and sets. Various types of conjugacies, dualities and polarities have been defined and studied. A detailed presentation and historical survey of corresponding definitions and results can be found in the book [32]. The reader can find a survey of corresponding results in the paper [17] in this Handbook. Some discussions on this topic (from the point of view of quasiconvexity) can be found also in [20]. In this paper we consider only Minkowski Duality and Fenchel-Moreau conjugation and relations between them.

Let H be a set of finite functions defined on a set X. Consider the set $P(H, X) \subset F_X$ of all H-convex functions and the set $S(H, X)$ of all (H, X)-convex sets. We assume that $P(H, X)$ is equipped with the pointwise order relation and $S(H,X)$ is equipped with the order relation by inclusion.

Definition 7.1 *The mapping* $\varphi : P(H,X) \to S(H,X)$ *defined by* $\varphi(f) = \operatorname{supp}(f,H)$ *is called the Minkowski duality.*

Remark 7.3 The notion of Minkowski duality was introduced by S.S. Kutateladze and A.M. Rubinov [13, 14]. The concept of Minkowski duality in more general setting can be found in [24].

We now present the main result related to Minkowski duality. Its proof, based on the well-known Moore scheme [2], can be found in [13, 14, 24].

Proposition 7.2 *The ordered sets* $P(H,X)$ *and* $S(H,X)$ *are complete lattices and the Minkowski duality* φ *is an isomorphism between these lattices.*

For arbitrary family $(f_t)_{t\in T}$ of H-convex functions we have

$$(\overset{*}{\sup_t} f_t)(x) = \sup_t f_t(x), \qquad (\overset{*}{\inf_t} f_t)(x) = (\mathrm{co}_H(\inf_t f_t))(x),$$

where $\sup_t^* f_t$ and $\inf_t^* f_t$ are boundaries in the lattice $P(H,X)$ while $\sup_t f_t$ and $\inf_t f_t$ are pointwise boundaries. For an arbitrary family $(U_t)_{t\in T}$ of (H,X)-convex sets we have $\sup_t U_t = \mathrm{co}_H(\bigcup U_t)$, where co_H denotes the (H,X)-convex hull of a set and $\inf_t U_t = \bigcap_t U_t$, where sup and inf are boundaries in the lattice $S(H,X)$.

We now consider the Fenchel-Moreau conjugation. Let L be a set of elementary functions defined on a set X. For a given function $f \in F_X$, the function $f_L^* : L \to \bar{R}$ defined by

$$f^*(l) = \sup_{x\in X}(l(x) - f(x)), \qquad (l \in L)$$

is called the Fenchel-Moreau L-conjugate to f.

Let $l \in L, c \in R$. The function $h_{l,c} = l - c\mathbf{1}$ is referred to as the abstract affine function, corresponding to (l,c). The set

$$H_L = \{h_{l,c} : l \in L, c \in R\}$$

is called the set of abstract affine functions corresponding to L. (We have already used this set under the definition of the abstract subdifferential.) If L possesses the following property $l \in L \implies l - c\mathbf{1} \in L$, then $L = H_L$.

The Fenchel-Moreau conjugate function possesses some good properties, if the set L is the so-called set of abstract linear functions. We now give a corresponding definition.

A set L of finite functions l defined on X is called a set of *abstract linear* functions if $l \in L \implies l - c\mathbf{1} \notin L$ for all $c \neq 0$. (Note that we do not define an individual abstract linear function, but only the set of all such functions.) It follows directly from the definition of a set of abstract linear functions that the mapping $(l, c) \mapsto h_{l,c}$ is one-to-one correspondence, so we can identify a pair (l, c) with the function $h_{l,c}$. Keeping in mind this identification, we show that the support set $\text{supp}\,(f, H_L)$ of a function $f \in F_X$ can be identified with the epigraph $\text{epi}\, f_L^* := \{(l, c) : c \geq f^*(l)\}$ of the conjugate function f_L^* .

Proposition 7.3 *Let* $f \in F_X$. *Then* $\text{supp}\,(f, H_L) = \text{epi}\, f_L^*$.

Proof. We have

$$\begin{aligned} (l, c) \in \text{epi}\, f_L^* &\iff c \geq f_L^*(l) \iff c \geq l(x) - f(x) \ (\forall\, x \in X) \\ &\iff f(x) \geq l(x) - c \ (\forall\, x \in X) \\ &\iff (l, c) \in \text{supp}\,(f, H_L). \end{aligned}$$

□

Let φ be the Minkowski duality between the sets $P(H_L, X)$ and $S(H_L, X)$. Then $\text{supp}\,(f, H_L) = \varphi(f)$, hence for each $f \in P(H_L, X)$ the following equality holds

$$\varphi(f) = \text{epi}\, f_L^*, \tag{7.6}$$

which expresses links between Minkowski duality and Fenchel-Moreau conjugation.

It is convenient to study the Fenchel-Moreau conjugation in terms of coupling functions. Let X and L be a pair of sets with a coupling function $\langle\cdot|\cdot\rangle : L \times X \to R$. We can consider each element $l \in L$ as a function defined on X by $x \mapsto \langle l|x\rangle$ and each element $x \in X$ as a function defined on L by $l \mapsto \langle l|x\rangle$. So we can consider L as a set of functions defined on X, and X as a set of functions defined on L.

Let $f : X \to R_{+\infty}$ and $\text{dom}\, f := \{x \in X : f(x) < +\infty\}$ be nonempty. Clearly $f_L^*(l) = \sup\{\langle l|x\rangle - f(x) : x \in \text{dom}\, f\}$. Since for each $x \in \text{dom} f(x)$ the function $l \mapsto \langle l|x\rangle - f(x)$ is an abstract affine function defined on X, it follows that f_L^* is abstract convex with respect to the set H_X of X-affine functions. Let $f^{**} = (f_L^*)_X^*$ be the second conjugate of the function f. Then f^{**} is L-convex. The following theorem is one of the main results of abstract convex analysis.

Theorem 7.1 *(Fenchel- Moreau) Let L and X be a pair of sets equipped with coupling function* $\langle\cdot|\cdot\rangle : L \times X \to R$. *Let* $f \in F_X$. *Then* $f^{**} = co_{H_L} f$.

Corollary 7.1 $f = f^{**}$ if and only if f is H_L-convex.

The Fenchel-Moreau conjugation can be defined also with respect to a set of functions, mapping into the extended real line $\bar{R}$ (see [32] and [17]). The abstract subdifferentials can be defined in this setting with the help of this conjugation (for more details see [32, 17]).

4. Examples of abstract convex functions and sets. Supremal generators

For a meaningful application of abstract convexity we first need to find sufficiently simple classes of elementary functions with respect to which the functions/sets involved in the problem at hand are abstract convex. From this point of view it is interesting and important to have a description of abstract convex functions and sets for some given sets of elementary functions. We now present some examples.

4.1 Multiplicative min-type functions

Consider the cone R^n_+ of all n-vectors with nonnegative coordinates and the cone R^n_{++} of all n-vectors with positive coordinates. We assume that both of them are equipped with coordinate-wise order relation. Let K be either R^n_+ or R^n_{++}. A function $f : K \to \bar{R}$ is called increasing if $x \geq y$ implies $f(x) \geq f(y)$. A conic structure of the set K allows one to define properties along-rays. We say that a certain property (P) of a function $f : K \to R$ holds along-rays if this property is valid for the restriction of f to the each ray $R_x = \{\lambda x : \lambda \geq 0\}$ with $x \in K \setminus \{0\}$ starting from zero and passing through x. In other words, the property (P) holds along-rays if for each $x \in K,\ x \neq 0$ the function of one variable f_x defined by

$$f_x(\alpha) = f(\alpha x), \qquad (\alpha \geq 0) \tag{7.7}$$

satisfies (P). We now give some examples of functions with properties along-rays:

1) lower semicontinuous-along-rays functions f: the function f_x is lower semicontinuous for all $x \in K \setminus \{0\}$;

2) convex-along-rays functions f: the function f_x is convex for all $x \in K \setminus \{0\}$;

3) positively homogeneous functions. A function p is called positively homogeneous (of degree k) if $p(\lambda x) = \lambda^k p(x)$ for all $\lambda \geq 0$ (in particular if $k = 1$ then function p is linear-along-rays);

We shall usually study functions with properties along-rays, which in addition, possess some "good and not along-rays" property. As a rule we shall consider increasing functions with properties along-rays. A combination of monotonicity with properties along-rays is very useful. For example, the following result holds ([25]):

Proposition 7.4 *Let $f : R^n_+ \to R_{+\infty}$ be an increasing and lower semicontinuous along-rays function. Then f is lower semicontinuous.*

In particular we consider in the sequel IPH (increasing and positively homogeneous of degree one) function and ICAR (increasing and convex-along-rays) functions defined on K, where either $K = R^n_+$ or $K = R^n_{++}$. The order relation allows one to define *normal* subsets of the cone K. A set $U \subset K$ is called normal, if $(x \in U, x' \leq x, x' \in K) \implies x' \in U$. It is easy to check that a function $f : K \to \bar{R}$ is increasing if and only if its level sets $S(c) = \{x \in K : f(x) \leq c\}$ are normal for all c.

First we consider $K = R^n_{++}$. Consider a coupling function $\langle \cdot, \cdot \rangle$ defined on $R^n_{++} \times R^n_{++}$ by

$$\langle l, x \rangle = \min_{i \in I} l_i x_i, \tag{7.8}$$

where $I = \{1, \ldots, n\}$. This coupling function generates functions

$$x \mapsto \langle l, x \rangle \tag{7.9}$$

with $l \in R^n_{++}$. We denote function (7.9) by the same symbol l as the vector, which generates it. Function $l(x) = \min_{i \in I} l_i x_i$ is called a multiplicative min-type function. Consider the set L of all multiplicative min-type functions as the set of abstract linear functions. Note that we can identify L and R^n_{++} as topological ordered sets, however we cannot identify them as convex cones. The following results hold (see [24] and references therein).

Theorem 7.2 *A function $p : R^n_{++} \to R_{+\infty}$ is L-convex if and only if this function is IPH.*

Theorem 7.3 *A set $U \subset R^n_{++}$ is abstract convex (or, more precisely (R^n_{++}, L)-convex) if and only if this set is normal and closed (in the topological space R^n_{++}).*

Remark 7.4 It follows from Theorem 7.3 that a function f is L-quasiconvex if and only if this function is increasing and lower semicontinuous.

To describe the L-subdifferential we need the following notation. If $l \in R^n_{++}$ and $c \in R$ then

$$\frac{c}{l} = \left(\frac{c}{l_1}, \ldots, \frac{c}{l_n} \right). \tag{7.10}$$

Theorem 7.4 *Let p be an IPH function. Assume that p is proper, that is $p \not\equiv 0$, $p \not\equiv +\infty$. Then $\partial p_L(x)$ is not empty for all $x \in R^n_{++}$ and*

$$\partial p_L(x) = \left\{ l \in L : l \geq \frac{p(x)}{x}, p\left(\frac{1}{l}\right) = 1 \right\}.$$

Remark 7.5 It follows from Theorem 7.4 that $p(x)/x \in \partial p_L(x)$ for all $x \in R^n_{++}$.

Consider now the set H_L of abstract affine functions with respect to the set of abstract linear functions L. By definition, $H_L = \{h_{l,c} : l \in L, c \in R\}$, where $h_{l,c}(x) = \langle l, x\rangle - c$ for all $x \in R^n_{++}$.

Clearly the sets, which are abstract convex with respect to H_L, coincide with L-convex subsets, hence with normal closed subsets of R^n_{++}. Hence a function f is H_L-quasiconvex if and only if this function is increasing and lower semicontinuous. We now describe H_L-convex functions.

Theorem 7.5 *A function $f : R^n_{++} \to R_{+\infty}$ is H_L-convex if and only if this function is ICAR.*

Let f be an ICAR function. Then for each $x \in R^n_{++}$ the function f_x defined on $[0, +\infty)$ by $f_x(t) = f(tx)$ is convex, hence its subdifferential (in the sense of convex analysis) $\partial f_x(1)$ at the point $t = 1$ is nonempty. Let

$$A_x(f) = \left\{ \frac{u}{x} : u \in \partial f_x(1) \right\}. \tag{7.11}$$

Theorem 7.6 *Let $f : R^n_{++} \to R_{+\infty}$ be an ICAR function and $x \in \operatorname{int}\operatorname{dom} f$, where $\operatorname{dom} f = \{x : f(x) < +\infty\}$. Then $\partial_L f(x) \supset A_x(f)$, hence $\partial_L f(x)$ is nonempty. If f is strictly increasing at the point x, that is $y \leq x, y \neq x \implies f(y) < f(x)$, then $\partial_L f(x) = A_x(f)$.*

Consider now the cone $K = R^n_+$. The coupling function (7.8) is not suitable for R^n_+, since $\min l_i x_i = 0$ for all $x \in R^n_+$ if at least one coordinate of l is equal to zero. We substitute (7.8) for the coupling function

$$\langle l, x\rangle = \min_{i \in I(l)} l_i x_i \text{ where } I(l) = \{i \in I : l_i > 0\}. \tag{7.12}$$

Note that the restriction of the function (7.12) to R^n_{++} coincides with (7.8), however in contrast with (7.8) the function (7.12) is not symmetric. The description of abstract convex sets, abstract convex functions and their L-subdifferentials can be generalized for the case under consideration, see [24, 28] and references therein for details.

4.2 Additive min-type functions

Consider a space R^n, equipped with the coordinate-wise order relation. A mapping $f : R^n \to R^m$ is called topical if f is increasing ($x \geq y \implies f(x) \geq f(y)$), where $\geq$ means coordinate-wise order relation) and $f(x + \lambda\mathbf{1}) = f(x) + \lambda\mathbf{1}$ for all $x \in R^n$ and $\lambda \in R$. Here $\mathbf{1}$ is the vector of corresponding dimension , whose all coordinates is equal to one. Topical mappings are intensively studied and they have many applications in various parts of applied mathematics (see [10]). Note that a mapping f is topical if and only if its coordinate functions are topical. A set $U \subset R^n$ is called downward if $(x \in U, y \leq x) \implies y \in U$. If $m = 1$ it is natural to study topical functions, mapping into the extended real line $\bar{R}$. However it is not very difficult to show ([27]) that each topical function mapping into $\bar{R}$ is either finite or identically $+\infty$ or $-\infty$.

Topical functions $f : R^n \to \bar{R}$ and downward sets can be studied in the framework of abstract convexity (see [15, 27]). Consider a coupling function defined on $R^n \times R^n$ by

$$\langle w, x\rangle^+ = \min_{i \in I}(w_i + x_i) \tag{7.13}$$

where $I = \{1, \dots, n\}$. For each $w \in R^n$ this coupling function generates the function defined on R^n, which we denote by the same symbol w. By definition $w(x) = \langle w, x\rangle^+$. Let W be the set of all functions $w(\cdot)$ with $w \in W$. Note that the function w_c, where $w_c(x) = w(x) + c$, belongs to W if $w \in W$, so we cannot consider W as a set of abstract linear functions. A function $w \in W$ is called additive min-type function. The following results hold:

Theorem 7.7 *[27] A function $f : R^n \to R$ is W-convex if and only if f is topical.*

Theorem 7.8 *[15] A set $U \subset R^n$ is abstract convex (more precisely, (R^n, W)-convex) if and only if U is downward and closed.*

Direct proofs of Theorem 7.7 and Theorem 7.8 can be found in [27] and [15], respectively. We now present a construction, which allows one to easily derive these proofs from known results for IPH functions. A detailed presentation of this construction can be found in [15, 27]. We shall use the following notation:

$$e^x = (e^{x^1}, \dots, e^{x^n}), \qquad (x = (x_1, \dots, x_n) \in R^n)$$

$$\ln y = (\ln y_1, \dots, \ln y_n), \qquad (y = (y_1, \dots, y_n) \in R^n_{++}).$$

The mapping E defined by $E(x) = e^x$ is an isomorphism between ordered sets R^n and R^n_{++} with the inverse mapping L, where $L(y) = \ln y$. It is easy to see $U \subset R^n_{++}$ is normal if and only $L(U) \subset R^n$ is downward and f is topical if and only if the function p defined by $p(y) = e^{f(\ln y)}$ is IPH. Using this isomorphism we can obtain Theorem 7.7 and Theorem 7.8 as a simple corollary of Theorem 7.2 and Theorem 7.3, respectively. This isomorphism and Theorem 7.4 also allows one to describe easily the support set at a point for topical functions. Namely the following result holds [27]:

Theorem 7.9 *Let f be a topical function and $y \in R^n$. Then*

$$\partial^*_W f(y) = \{w \in R^n : w \geq f(y)\mathbf{1} - y, f(-w) = 0\}.$$

Note that a function $f : R^n \to \bar{R}$ is increasing if and only if its level sets $S(c) = \{x : f(x) \leq c\}$ are downward for all $c \in \bar{R}$. Hence, due to Theorem 7.8 a function $f : R^n \to \bar{R}$ is W-quasiconvex if and only if f is increasing and lower semicontinuous.

As it turned out, the Fenchel-Moreau conjugation has a very simple form in the case under consideration.

Theorem 7.10 *[27] For a function $f : R^n \to \bar{R}$ the following are equivalent:*

1) f is a topical function;

2) we have $f^*(w) = -f(-w)$. (Here f^* is the Fenchel-Moreau conjugate with respect to the set W.)

Proof. 1) $\implies$ 2). First we show that $f^*(w) \geq -f(-w)$ for each function f and each $w \in R^n$. Indeed,

$$f^*(w) = \sup_{x \in R^n} (\langle x, w\rangle^+ - f(x)) \geq \langle -w, w\rangle^+ - f(-w) = -f(-w).$$

We now show that the opposite inequality holds if f is a topical function. It follows from the definition of the coupling function $\langle \cdot, \cdot\rangle^+$ that $x + w \geq \langle x, w\rangle^+\mathbf{1}$, which implies that $x \geq -w + \langle x, w\rangle^+\mathbf{1}$. Since f is topical we have

$$f(x) \geq f(-w + \langle x, w\rangle^+\mathbf{1}) = f(-w) + \langle x, w\rangle^+, \qquad x, w \in R^n.$$

Hence

$$f^*(w) = \sup_{x \in R^n} (\langle x, w\rangle^+ - f(x)) \leq -f(-w).$$

2) $\implies$ 1) Let f be an arbitrary function and $x \in R^n$. Then the function h_x, where $h_x(w) = \langle x, w\rangle^+ - f(x)$ is topical. Applying the equality $f^*(w) = \sup_{x \in R^n} h_x(w)$ we can easily check that f^* is a topical function. Indeed f^* is increasing as the upper envelope of a family of increasing functions. Then

$$f^*(w + \lambda\mathbf{1}) = \sup_{x \in R^n} h_x(w + \lambda\mathbf{1}) = \sup_{x \in R^n} (h_x(w) + \lambda) = f^*(w) + \lambda.$$

Thus, f^* is topical. Due to 2), we conclude that the function $w \mapsto -f(-w)$ is topical. It follows directly from this assertion that f is also topical. □

Corollary 7.2 *For an IPH function p define the polar function p° by*

$$p^\circ(l) = \sup_{x \in R^n_{++}} \frac{\langle l, x\rangle}{p(x)}.$$

Applying Theorem 7.10 and the isomorphism E described above we can deduce the following result (see [24] and references therein):

$$p^\circ(l) = \frac{1}{p\left(\frac{1}{l}\right)},$$

where $1/l$ is defined by (7.10).

Remark 7.6 Let k be a positive integer and let $\mathcal{L}_k$ be the set of functions $\ell : R^n \to R$ of the form $\ell(x) = \min_{i=1,\dots,k}[l_i, x]$, where $l_1, \dots, l_k \in R^n$ and $[l, x]$ stands for the inner product of vectors l and x. Abstract convexity with respect to sets $\mathcal{L}_n$ and $\mathcal{L}_{n+1}$ has been studied in [24]. In particular, it was shown in [24] that a set $U \subset R^n$, which contains the origin, is $\mathcal{L}_n$-abstract convex if and only if this set is closed and radiant (in another terminology, star-shaped with respect to zero). The latter means that $(x \in U, 0 < \lambda < 1) \implies \lambda x \in U$.

4.3 Supremal generators

Let F be a set of functions defined on a set X. A set $H \subset F$ is called a *supremal generator* of F if each $f \in F$ is abstract convex with respect to H. Some examples of supremal generators can be found in previous subsections. For instance, the set L of multiplicative min-type functions is a supremal generator of the set of all IPH functions (see Subsection 4.1) and the set of additive min-type functions is a supremal generator of the set of all topical functions (see Subsection 4.2). If H is a supremal generator of F and if the set H_1 is such that $H \subset H_1 \subset F$ then H_1 is a supremal generator of F as well. There exist very small supremal generators of very large sets, in particular of the set of all lower semicontinuous functions defined on a compact set X. To show it we need the following assertion.

Lemma 7.1 *Let H be a set of continuous functions defined on a metric space X with the metric d. Assume that*

1) H is a conic set (that is, $(h \in H, \lambda > 0) \implies \lambda h \in H$);

2) H contains negative constant functions;

3) for each $z \in X$ and $\delta > 0$ there exists a function $h \in H$ such that $h(z) = 1, \quad h(x) \leq 1$ if $d(x,z) < \delta$ and $h(x) \leq 0$ if $d(x,z) \geq \delta$.

Then H is a supremal generator of the set of all lower semicontinuous functions defined on X and bounded from below.

Proof. Let f be a lower semicontinuous function defined on X and bounded from below. Without loss of generality assume that f is finite. There exists a constant c such that the function $f_c(x) = f(x) - c \ (x \in X)$ is positive. Let $z \in X$ and $g(x) = f_c(x)/f_c(z)$. The function g is lower semicontinuous and $g(z) = 1$, so for each $\varepsilon > 0$ there exists $\delta > 0$ such that $g(x) > 1 - \varepsilon$ if $d(x,z) < \delta$. Applying the property 3) we can find a function $h' \in H$ such that $h'(z) = 1$, $h'(x) \leq 1$ if $d(x,z) < \delta$ and $h'(x) < 0$ if $d(x,z) \geq \delta$. Let $\bar{h}(x) = (1-\varepsilon)h'(x)$. Then $\bar{h}(x) \leq g(x)$ for all $x \in X$. Indeed, $\bar{h}(x) \leq 1 - \varepsilon < g(x)$ if $d(x,z) \leq \delta$ and $\bar{h}(x) \leq 0 < g(x)$ if $d(x,z) \geq \delta$. We have also $\bar{h}(z) = 1 - \varepsilon$. Since ε is an arbitrary positive number, it follows that $g(z) = \sup\{h(z) : h \in H, h \leq g\}$. Since H is a conic set, we have $f_c(x) = f_c(z)g(x) = \sup\{h(x) : h \in H, h \leq f_c\}$. Hence $f(z) = f_c(z) + c = \sup\{h(z) : h \in H, h \leq f\}$. □

Remark 7.7 If X is a compact space then a set H with properties 1)-3) is a supremal generator of the set of all lower semicontinuous functions.

Remark 7.8 A more general result, based on the notion of the *support to a Urysohn peak* can be found in [13].

We now give a simple example of a supremal generator.

Proposition 7.5 *Let Y be a Hilbert space with the inner product $[\cdot,\cdot]$ and let $X \subset Y$. For each triplet $\alpha = (a, l, c)$, where $a \geq 0, l \in Y, c \in R$ consider the function h_α defined on X by*

$$h_\alpha(x) = -a\|x\|^2 + [l, x] - c, \tag{7.14}$$

Let H be the set of all functions of the form (7.14). Then H is a supremal generator of the set of all lower semicontinuous, bounded from below functions defined on X.

Proof. Clearly, H is a conic set, which contains negative constants. For each $z \in X$ and $a > 0$ the function

$$\bar{h}(x) = -a\|z - x\|^2 + 1 = -a\|x\|^2 + [2az, x] - (a\|x\|^2 - 1)$$

belongs to H. Note that $\bar{h}(x) \leq \bar{h}(z) = 1$ for all $x \in X$ such that $d(x, z) \geq \delta$ and all fairly large numbers a. Let $h(x) = (1/\|z\|^2)\bar{h}(x)$. Then $h(x) \leq 1 = h(z)$ for all $x \in X$. If $\delta > 0$ then $h(x) < 0$ for all fairly large numbers a. Thus the desired result follows from Lemma 7.1. □

Remark 7.9 Assume that Y is a n-dimensional Euclidean space. Then the set H from Proposition 7.5 is a half space of the $(n+2)$-dimensional space, so this set is a "very thin" subset of the set of all lower semicontinuous functions. It can be shown [13] that there exists convex cones, which are spanned by $n + 2$ functions, which are supremal generators of the set of all lower semicontinuous functions defined on the compact set $X \subset R^n$. On the other hand, if the set of all lower semicontinuous functions defined on a compact space X contains a convex cone, which is spanned by $n+2$ functions, then X can be topologically embedded to R^n. (See [13] and also [24] for details.)

Various versions of Proposition 7.5 have been proposed in order to take into account some functions unbounded from below (see, for example [3, 5, 24]). We now present (without proof) one of the most general results in this direction ([30]). Let X be a normed space, k be a positive number and let $\mathcal{P}^k$ be the set of all lower semicontinuous functions $f : X \to R_{+\infty}$, which are bounded from below on each bounded set and such that

$$\liminf_{\|x\|\to+\infty} \frac{f(x)}{\|x\|^k} > -\infty.$$

Consider a continuous sublinear function p defined on X such that $\inf_{\|x\|=1} p(x) > 0$. Let H^k be the set of all functions h of the form $h(x) = -ap^k(x - z) - c$ with $z \in X$, $c \in R$ and $a \geq 0$.

Theorem 7.11 *The set H^k is a supremal generator of the set $\mathcal{P}^k$.*

In particular, if X is a Hilbert space then the set H^2 of all functions h of the form $h(x) = -a\|x\|^2 + [l, x] - c$ with $l \in X$, $c \in R$ and $a \geq 0$ is a supremal generator of the set $\mathcal{P}^2$.

5. Supremal generators of sets of quasiconvex functions

Abstract convexity is a convenient tool in the study of quasiconvex functions. There are numerous papers in this direction and we are not able to give a comprehensive survey of this research. So we restrict ourselves only to some results directly related to abstract convexity of quasiconvex functions. We also do not quote numerous results pertaining to different types of dual objects for quasiconvex functions. A detailed survey of many of these results can be found in the book [32] and papers [17] and [20]. In this section we mainly present results due to J.E. Martinez Legaz [16], J.P. Penot and M. Volle [21, 22], I. Singer [32] and M. Volle [33], which are related to abstract convexity of quasiconvex functions with respect to a certain subset of quasiaffine functions. In contrast to original papers, we give direct proofs of the corresponding results, which are not based on conjugation. We do not present a historical survey related to these results. A detailed historical survey can be found in [32], see also [17, 20].

Recall the notion of a quasiconvex function.

Definition 7.2 *Let X be a locally convex Hausdorff topological vector space. Let $f : X \to \bar{R}$. Then f is said to be quasiconvex if for all $x, y \in X$ and $\lambda \in [0,1]$*

$$f(\lambda x + (1-\lambda)y) \leq \max(f(x), f(y)). \tag{7.15}$$

Quasiconvex functions can also be characterized via their level subsets. A function $f : X \to R$ is quasiconvex if and only if its level sets $S(c) = \{x \in X : f(x) \leq c\}$, $c \in R$ is convex or equivalently if its strict level sets $T_f(c) = \{x \in X : f(x) < c\}$, $c \in R$ are convex. In fact this was the starting point for the definition of quasiconvex functions.

A function $f : X \to \bar{R}$ is called *quasiconcave* if $-f$ is quasiconvex. Note that a function h defined on a convex set is affine if and only if this function is simultaneously convex and concave. This observation leads to the following definition.

Definition 7.3 *Let X be a locally convex Hausdorff topological vector space. A function $f : X \to \bar{R}$ is called quasiaffine if it is both quasiconvex and quasiconcave.*

It is well known (see Example 7.1) that the set of all affine functions is a supremal generator to the set of all lower semicontinuous convex functions mapping into $R_{+\infty}$. As it turned out, the set of quasiaffine functions plays the similar role with respect to the set of quasiconvex functions. First we need to characterize quasiaffine functions. In what follows we shall always denote by H the class of all increasing functions $h : R \to \bar{R}$ and by H_+ we shall denote the class of all increasing functions $h : R \to \bar{R}$, which are continuous from the left (i.e lower semicontinuous). A function $f : X \to \bar{R}$ is said to be *proper* if $f(x) > -\infty$ for all $x \in X$ and there exists at least a $x \in X$ such that $f(x) < +\infty$.

Theorem 7.12 *Let $f : X \to \bar{R}$ be a proper function. Let X^* denote the topological dual of X. Then the following statements are equivalent*

(i)There exists $h \in H_+$ and $\omega \in X^*$ such that $h \circ \omega = f$.

(ii) f is proper lower semicontinuous and quasiaffine.

(iii) Every non-empty level set of f is either a closed half-space or the whole space.

Proof. (i) $\Longrightarrow$ (ii) Note that the linearity of ω implies its convexity and concavity. Let us assume that (i) holds . Then since h is increasing, the linearity of ω implies both quasiconvexity and quasiconcavity of f. Hence f is quasiaffine. Moreover h is lower-semicontinuous and ω is continuous and hence f is proper and lower semicontinuous.

(ii) $\Longrightarrow$ (iii). Since f is quasiconvex and lower semicontinuous all its level sets $S(c)$ are convex and closed. Again $X \setminus S(c)$ is convex since f is also quasiconcave. This set is also open. If $X \setminus S(c)$ is nonempty then by applying the standard separation theorem we can assert that there exists $0 \neq \omega \in X^*$ and $k \in R$ such that $\omega(x) \leq k$, for all $x \in S(c)$ and $\omega(x) > k$ for all $x \in X \setminus S(c)$. This shows that $S(c) = \{x \in X : \omega(x) \leq k\}$. This proves the assertion.

(iii) $\Longrightarrow$ (i) Assume first that each of the level sets are empty or the whole space X. In this case f must be constant and (i) holds with any $\omega \in X^*$ if we take h as the constant with the same value as f. Assume now that $S(c_0) = \{x \in X : \omega(x) \leq k_0\}$ for some c_0, $k_0 \in R$ and $\omega \in X^*$. Then, if $c < c_0$, then either $S(c) = \emptyset$, that is $\{x : \omega(x) \leq k_c\}$, with $k_c = -\infty$, or $S(c)$ is a closed half-space on which ω is bounded above by k_0. From this argument it is clear that $S(c) = \{x \in X : \omega(x) \leq k_c\}$, where k_c must satisfy $k_c \leq k_0$. Similarly if $k_c > k_0$ then $S(c) = \{x \in X : \omega(x) \leq k_c\}$ for some $k_c \geq k_0$ (possibly $k_c = +\infty$). Let $h : R \to \bar{R}$ be defined by $h(t) = \inf\{c : t \leq k_c\}$. Clearly h is increasing and it is easy to check that h is left continuous. Finally for any $x \in X$ we have

$$f(x) = \inf\{c : f(x) \leq c\} = \inf\{c : \omega(x) \leq k_c\} = h(\omega(x)).$$

Hence $f = h \circ \omega$. □

Lemma 7.2 *The supremum of any family of lower semicontinuous quasiaffine functions is a lower semicontinuous quasiconvex function.*

Proof. Consider any arbitrary family of lower semicontinuous quasiaffine functions on X, given as $\{h_\gamma : \gamma \in \Gamma\}$ where Γ is an arbitrary index set. Let $f(x) = \sup_{\gamma\in\Gamma} h_\gamma(x)$. This shows that $S(c) = \bigcap_{\gamma\in\Gamma} S_\gamma(c)$, where $S(c)$, $c \in R$ denotes a lower level set of f and $S_\gamma(c)$ denotes a lower level set of h_γ , $\gamma \in \Gamma$. If there exists $\gamma \in \Gamma$, $S_\gamma(c) = \emptyset$ then $S(c) = X$ by convention. If for all γ, $S_\gamma(c) \neq \emptyset$ then from Theorem 7.12 we have that for all γ, $S_\gamma(c)$ is a closed half space or the whole space. Hence $S(c)$ is either the whole space or a closed convex set. Again if there exists $\gamma, \delta \in \Gamma$ such that $S_\gamma(c) = \emptyset$ and $S_\delta(c) \neq \emptyset$ then $S(c) = \emptyset$. Since $c \in R$ is arbitrary we have that $S(c)$ is a closed convex set for every c. This shows that f is quasiconvex and lower semicontinuous. □

We are now in a position to prove one of the most significant results of this section.

Theorem 7.13 *The set of all lower semicontinuous quasiaffine functions forms a supremal generator of the set of all lower semicontinuous quasiconvex functions.*

Proof. We need to show that given any proper lower semicontinuous quasiconvex function $f : X \to \bar{R}$ it can be supremally generated by a family of proper lower semicontinuous quasiaffine function. In fact we have to demonstrate that

$$f(x) = \sup\{g : \exists h \in H_+, \omega \in X^*, g = h \circ \omega \leq f\} \tag{7.16}$$

where H_+ is as before the family of increasing extended real functions on the real line which are left continuous. To show this we need to show that for $y \in X$ and $r \in R$ with $r < f(y)$, there exist $h \in H_+$ and $\omega \in X^*$ such that $h \circ \omega \leq f$ and $r < h(\omega(y))$. Since $y \notin S(r)$ we have that there exist $\omega \in X^* \setminus \{0\}$ and $k \in R$ such that $\omega(y) > k$ and $\omega(x) < k$ for all $x \in S(r)$. Let us define the function $h : R \to \bar{R}$ as follows

$$h(t) = \inf\{f(x) : \omega(x) \geq t\}. \tag{7.17}$$

It is clear that $h \in H$ and $h \circ w \leq f$. Again let us observe that $\bar{h} \circ \omega \leq f$ where $\bar{h}$ denotes the lower semicontinuous hull of h. Now we put $g = \bar{h} \circ \omega$. From the definition of the lower semicontinuous hull we can write that

$$\bar{h}(\omega(y)) = \sup\{h(p) : p < \omega(y)\}. \tag{7.18}$$

Hence $g(y) = \bar{h}(\omega(y)) \geq h(k)$. Now from the definition of h and noting the fact that $w(x) \geq k \implies f(x) > r$, it is clear that $h(k) \geq r$. This proves the assertion. □

We shall now turn to another important subclass of quasiconvex functions and shall also describe a supremal generator for them. Before that we need the notion of an evenly convex set. A set $S \subseteq X$ is called evenly convex if for each $y \notin S$, there exists $v \in X^*$ such that $[v, y] > [v, x]$ for all $x \in S$. It is clear that open and closed convex sets are evenly convex.

Definition 7.4 *A proper function $f : X \to \bar{R}$ is called evenly quasiconvex if every level set of f is evenly convex.*

The class of evenly quasiconvex functions is very broad. In fact every lower semicontinuous and every upper semicontinuous quasiconvex function is evenly quasiconvex.

Definition 7.5 *A function $f : X \to R$ is called evenly quasiaffine if it is evenly quasiconvex and also quasiaffine.*

This definition was introduced by Martinez-Legaz [16] and Penot and Volle [21]. Recall that we denote by H the set of all increasing functions $h : R \to \bar{R}$.

Theorem 7.14 *For any $f : X \to R$ the following assertions are equivalent:*

(i) There exists $h \in H$ and $\omega \in X^*$ such that $f = h \circ \omega$.

(ii) f is evenly quasiaffine.

(iii) Every non empty level set of f is either the whole space X or a closed half space or an open half space of X.

We now state the following result:

Theorem 7.15 *Let G denote the set of all evenly quasiaffine functions $g : X \to \bar{R}$. A function $f : X \to \bar{R}$ is evenly quasiconvex if and only if there exists $M \subseteq G$ such that*

$$f(x) = \sup_{g \in M} g(x). \tag{7.19}$$

A proof of Theorem 7.15 can be found for example in [21].

We now mention an important property of the set A of all affine functions defined on a locally convex Hausdorff topological vector space.

This set is a supremal generator of the set Γ of all lower semicontinuous convex functions, which is minimal in the following sense: for each $a \in A$ and $\varepsilon > 0$ the set $A' := A \setminus \{a' \in A : \|a - a'\| < \varepsilon\}$ is not a supremal generator of Γ. (If $a(x) = l(x) - c$ with $l \in X^*$ and $c \in R$, then $\|a\| = \|l\| + |c|$.) It is enough to show that a does not coincide with the upper envelope of its support set supp$\,(a, A')$. Indeed, if

$$a'(x) := l'(x) + c' \leq a(x) := l(x) + c, \qquad x \in X$$

then $c' \leq c$ and $l = l'$. Assume then $a'(0) = c' > a(0) - \varepsilon = c + \varepsilon$, then $|c - c'| < \varepsilon$, which is impossible for $a' \in A'$.

Such a property does not hold for the set of quasiaffine functions. There exist very thin subsets of the set of all lower semicontinuous (evenly) quasiaffine functions, which are supremal generators of the set of lower semicontinuous (evenly, respectively) quasiconvex functions. To describe such sets we can use supremal generators of the set H_+ (H, respectively). First, we introduce the following notation. Let K be a set of functions $k : R \to \bar{R}$. We shall denote by $K \circ X^*$ the set $\{k \circ \omega : k \in K, \omega \in X^*\}$. Consider now a supremal generator K_+ of the set H_+. Then for each $h \in H_+$ we have $h(y) = \sup_{k \in K_+, k \leq h} k(y)$. Since each lower semicontinuous quasiaffine function g can be represented in the form $g(x) = h(\omega(x))$ with $h \in H_+$ and $\omega \in X^*$, we have

$$g(x) = h(\omega(x)) = \sup\{k(\omega(x)) : k \in K_+, k \leq h\},$$

so $K_+ \circ X^*$ is a supremal generator of the set $H_+ \circ X^*$ of all lower semicontinuous quasiaffine functions. Since, in turn, $H_+ \circ X^*$ is a supremal generator of the set Q_+ of all lower semicontinuous quasiconvex functions, it follows that $K_+ \circ X^*$ is a supremal generator of Q_+. The same argument shows that $K \circ X^*$ is a supremal generator of the set Q of all evenly quasiconvex functions if K is a supremal generator of the set H. The simplest and very small supremal generator of the set H_+ is formed by the functions of the form

$$h(y) = \begin{cases} c & y > d \\ -\infty & y \leq d, \end{cases}$$

where $c \in \bar{R}, d \in R$.

The set of the so-called subaffine functions can serve as a small supremal generator of Q_+. Let us consider $t \in \bar{R}$ and for each t we define a function $h_t : R \to \bar{R}$ as follows

$$h_t(y) = \min\{y, t\}, \quad y \in R. \tag{7.20}$$

Again observe that for a given t the function h_t is increasing. The following function

$$h_t(p(x)) = \min\{p(x), t\}, \tag{7.21}$$

where $p : X \to R$ is an affine function, is called a *subaffine* function. Observe that every subaffine function is also quasiaffine. All subaffine functions are continuous. It can be shown that the subaffine functions form a supremal generator of the class of all lower semicontinuous quasiconvex functions. We now present an example of a conjugation scheme, which is closely related to the discussed supremal generation. This scheme has relations to the lower subdifferential of Plastria (see Martinez-Legaz [16] for more details).

Consider $f : X \to \bar{R}$ and let $f^c : X^* \times R \to \bar{R}$ denote the generalized conjugate of f given by

$$f^c(\omega, k) = \sup_{x \in X}\{\min\{\omega(x), k\} - f(x)\}. \tag{7.22}$$

From the discussions above it is clear that f^c is a lower semicontinuous quasiconvex function on $X^* \times R$.

A conjugate can also be defined by defining the conjugate function on a restricted set $B^*(0, N) \times R$, where $B^*(0, N)$ is a ball of radius N centered at the origin in X^* and hence the subaffine function that is used in (7.21) will be now restricted by $\omega \in B^*(0, N)$. Denote by ω_N the restriction of a linear function ω to the ball $B^*(0, N)$. We have

$$f_N^c(\omega, k) = \sup_{x \in X}\{\min\{\omega_N(x), k\} - f(x)\}. \tag{7.23}$$

With ω restricted to $B^*(0, N)$ we now just have subaffine functions which are Lipschitz with rank N. Let us denote the set of subaffine functions of the form $\min\{\omega(x), k\} + d$ where $d \in R$ is a constant and $\omega \in B^*(0, N)$, as $H(N)$. Let $Q(N)$ be the set of all $H(N)$-convex functions. We now present the following result (see, Martinez-Legaz [16]).

Theorem 7.16 *Let $f : X \to \bar{R}$. Then the following statements are equivalent*

i) $f \in Q(N)$.

ii) Either $f(X) \subseteq R$ and f is quasiconvex and Lipschitz with rank N, or f identically equal to $+\infty$ or $-\infty$.

For some applications it is interesting to consider only quasiconvex functions mapping into $R_{+\infty}$ or even bounded from below. Let H'(H'_+, respectively) be the set of all increasing (lower semicontinuous increasing, respectively) functions $h : R \to R_{+\infty}$ bounded from below. It is easy to see that the set $H' \circ X^*$ is a supremal generator of the set Q' of all evenly quasiconvex functions $f : X \to R_{+\infty}$ bounded from below and the set $H'_+ \circ X^*$ is a supremal generator of the set Q'_+ of all lower semicontinuous quasiconvex functions $f : X \to R_{+\infty}$ bounded from below. If K' is a supremal generator of H' then $K' \circ X^*$ is a supremal generator of Q'. If K'_+ is a supremal generator of H'_+ then $K'_+ \circ X^*$ is a supremal generator of Q'_+. Simple supremal generators of H' and H'_+ are formed by the so-called two step functions, that is either functions of the form

$$l(y) = \begin{cases} c : & y \geq d \\ c' : & y < d \end{cases} \tag{7.24}$$

or functions of the form

$$l(y) = \begin{cases} c : & y > d \\ c' : & y \leq d \end{cases} \tag{7.25}$$

where $c, c' \in R$ with $c \geq c'$ and $d \in R$. It is easy to see that the set K' of all functions of the form (7.24) is a supremal generator of H' and the set K'_+ of all functions of the form (7.25) is a supremal generator of H'_+.

It is interesting to consider supremal generators of some subsets of the set Q' and Q'_+. One of such subsets is the set of all nonnegative lower semicontinuous (evenly, respectively) quasiconvex functions q such that $q(0) = 0$. We now present the following result (see, for example, [24] and references therein).

Theorem 7.17 *1) Let Q^0 be the class of all evenly quasiconvex functions $q : X \to R_{+\infty}$ such that $0 = q(0) = \inf\{q(x) : x \in X\}$. Then the class L^0 of all two steps functions of the form*

$$l_{v,c}(x) = \begin{cases} c & : v(x) \geq 1 \\ 0 & : v(x) < 1 \end{cases}$$

with $c \geq 0$ and $v \in X^*$, is a supremal generator of Q^0.
2) Let Q^0_+ be the class of all lower semicontinuous quasiconvex functions $q : X \to R_{+\infty}$ such that $q(0) = \inf\{q(x) : x \in X\}$. Then the class L^0_+ of all two steps functions of the form

$$l_{v,c}(x) = \begin{cases} c & : v(x) > 1 \\ 0 & : v(x) \leq 1 \end{cases}$$

with $c \geq 0$ and $v \in X^*$ is a supremal generator of Q^0_+.

As it was mentioned in [31], the set L^0_+ is a minimal conic set, which is a supremal generator of Q^0_+. The above form of representation of the supremal generators for quasiconvex functions will be important in the study of Hadamard type inequalities for quasiconvex functions. In particular we shall use Theorem 7.17 in the simplest case, when $X = R$. Let us note that in this case each quasiconvex function is evenly quasiconvex. The following result (see, for example, [18]) can be easily proved.

Theorem 7.18 *Let H_1 denote the class of two step functions h given as*

$$l_{v,c}(x) = \begin{cases} c & : v(x) \geq d \\ 0 & : v(x) < d \end{cases}$$

with $v \in \{-1, 1\}$, $c \geq 0$, $d \in R$. Then H_1 is a supremal generator of the class of all non-negative quasiconvex functions $q : R \to R_{+\infty}$.

6. Some applications of abstract convexity

Abstract convexity has found many applications in various areas of mathematics. In particular, abstract convexity forms a natural framework in the study of various form of duality in not necessary linear setting. This question is discussed in details in the book [32] and in papers [17] and [20]. At the same time abstract convexity can be applied in the study of many concrete problems. We now present some known examples of such applications.

6.1 Boundary values of convex functions defined on a ball

Consider the unit ball B of the Hilbert space Y. Let $f : B \to R_{+\infty}$ be a lower semicontinuous convex function and g be the restriction of f to the unit sphere S. Then g is a lower semicontinuous function and moreover, g is bounded from below. Indeed, since f is lower semicontinuous and convex, there exists an affine continuous function a defined on B, such that $f(x) \geq a(x)$ for all $x \in B$. In particular $g(x) \geq \inf_{x \in S} a(x) = -\|a\| > -\infty$. We now show that the reverse assertion holds.

Proposition 7.6 *Let g be a lower semicontinuous bounded from below function defined on S. Then there exist a lower semicontinuous convex function $f : B \to R_{+\infty}$, such that $f(x) = g(x)$ for all $x \in S$.*

Proof. Let H be the set of all functions h of the form

$$h(x) = -a\|x\|^2 + [l, x] - c \tag{7.26}$$

with $l \in Y$, $a > 0$ and $c \in R$. For a function h defined by (7.26) consider the affine function k_h, where $k_h(x) = [l, x] - a - c$. Clearly

$$k_h(x) = h(x) \qquad (x \in S). \tag{7.27}$$

Consider the support set supp (g, H) of the function g with respect to H. Let $U = \{k_h : h \in \text{supp}\,(g, H)\}$ and $f(x) = \sup\{k_h(x) : k_h \in U\}$. Clearly f is a lower semicontinuous convex function. Applying Proposition 7.5 and (7.27) we conclude that for all $x \in S$:

$$g(x) = \sup\{h(x) : h \in \text{supp}\,(g, H)\} = \sup\{k_h(x) : k_h \in U\} = f(x).$$

□

6.2 The principle of preservation of inequalities

The following simple, however very useful result, was established in [13, 14],

Theorem 7.19 (Principle of preservation of inequalities) *Let L be a supremal generator of a set of functions Y. Let ψ be an increasing functional defined on Y, that is $(f, g \in Y, f \geq g) \implies \psi(f) \geq \psi(g)$. Let further, $u \in X$. Then*

$$(l(u) \leq \psi(l) \textit{ for all } l \in L) \implies (f(u) \leq \psi(f) \textit{ for all } f \in Y).$$

Proof. Let $f \in Y$. Then

$$f(x) = \sup\{l(x) : l \in \text{supp}\,(f, L)\} = \sup\{l : l \in \text{supp}\,(f, L)\}(x) \quad (7.28)$$

For $l' \in \text{supp}\,(f, L)$ we have $l' \leq \sup\{l : \text{supp}\,(f, L)\} = f$. Since ψ is increasing we have $\psi(l') \leq \psi(f)$. Hence $\sup\{\psi(l') : l' \in \text{supp}\,(f, L)\} \leq \psi(f)$. Assume that $l(u) \leq \psi(l)$ for al $l \in L$. Applying (7.28) with $x = u$, we conclude that

$$f(u) = \sup\{l(u) : l \in \text{supp}\,(f, L)\} \leq \sup\{\psi(l) : l \in \text{supp}\,(f, L)\} \leq \psi(f).$$

□

Applications of the Principle of Preservation of Inequalities to inequalities of Hadamard type will be studied in the next section. We now consider applications of different type.

Proposition 7.7 *Consider the space $C(X)$ of all continuous functions defined on the compact topological space X equipped with the pointwise order relation. Let $L \subset C(X)$ be a supremal generator of $C(X)$. Let $\psi : C(X) \to C(X)$ be an increasing operator such that*

$$\psi(f) \leq -\psi(-f), \qquad (f \in C(X). \quad (7.29)$$

If $\psi(l) \geq l$ for all $l \in L$ then ψ coincides with the identity.

Proof. Let $u \in X$ and $\psi_u(f) = \psi(f)(u)$ for all $f \in C(X)$. Then ψ_u is an increasing function defined on $C(X)$ and $\psi_u(l) \geq l(u)$ for all $l \in L$. Let $f \in C(X)$. It follows from the Principle of Preservation of Inequalities that $\psi_u(f) \geq f(u)$. Applying this principle to $-f$, we get $\psi_u(-f) \geq -f(u)$, hence due to (7.29) we obtain the opposite inequality: $\psi_u(f) \leq -(\psi_u)(-f) \leq f(u)$. Since u is an arbitrary point of X, we have $\psi(f) = f$. □

Remark 7.10 The inequality (7.29) holds for superadditive operators, that is operators ψ such that $\psi(f+g) \geq \psi(f) + \psi(g)$ and $\psi(0) = 0$. In particular, (7.29) is valid for linear operators. If ψ is linear operator then Proposition 7.7 is a special case of the well-known Korovkin theorem. (See, for example [13, 14] for corresponding discussion, where some versions of Proposition 7.7 are presented and also a convergence of a sequence of linear operators to identity is discussed.)

Remark 7.11 Proposition 7.7 is of special interest since there exist very small supremal generators of $C(Q)$ (see Subsection 4.3). In particular the following assertion follows from Proposition 7.14 and Proposition 7.7: Let $[a,b]$ be the segment on the real line and

$$f_1(x) = 1, \quad f_2(x) = x, \quad f_3(x) = x^2, \qquad (x \in [a,b]).$$

Let $\psi : C([a,b] \to C([a,b])$ be a linear operator such that $\psi(f) \geq 0$ for $f \geq 0$ and $\psi(f_i) = f_i$ $(i = 1,2,3)$. Then ψ is the identity.

6.3 Numerical methods of global optimization

Let us recall a well-known cutting plane method (see for example [12]) in convex optimization. Consider a problem of convex minimization

$$f(x) \to \min \text{ subject to } x \in X, \tag{7.30}$$

where $X \subset R^n$ is a convex compact set and f is a convex finite function defined on an open set containing X. We shall use the support set of a convex function f at a point. Let L be the set of affine functions defined on R^n. Due to Proposition 7.1 the support set $\partial_L^* f(x)$ of a function f at a point x with respect to L consists of affine functions l of the form $l(x) = [v,x] - c$, where $v \in \partial f(x)$ and $c = f(x) - [v,x]$.

Cutting plane method

Step 0. Let $k := 0$. Choose an arbitrary point $x^0 \in X$.

Step 1. Let $l_k \in \partial_L^* f(x_k)$. Consider
$h_k(x) = \max_{i=0,\dots,k} l_i(x) = \max(h_{k-1}, l_k(x))$.

Step 2. Solve the problem

$$h_k(x) \to \min \quad \text{subject to} \quad x \in X. \tag{7.31}$$

Step 3. Let y^* be a solution of (7.31). Let $k := k+1$, $x^k = y^*$ and go to step 1.

Only abstract convexity of convex functions with respect to the set of affine functions has been used here. So we can easily generalize the cutting plane method for almost any abstract convex situation.

Let L be a set of elementary functions defined on set X and let f be an L-convex function. Assume that the support set $\partial_L^* f(x) = \{l \in \operatorname{supp}(f, L) : l(x) = f(x)\}$ of a function f with respect to L is not empty for all $x \in X$. Consider the problem of global optimization:

$$f(x) \to \min \quad \text{subject to} \quad x \in X.$$

Generalized cutting plane method

Step 0. Let $k := 0$. Choose an arbitrary initial point $x^0 \in X$.

Step 1. Calculate $l_k \in \partial_L^* f(x^k)$. Let

$$h_k(x) = \max_{i=0,\dots,k} l_i(x) \text{ for all } x \in X$$

Step 2. Find a global optimum of the problem

$$h_k(x) \to \min \text{ subject to } x \in X. \tag{7.32}$$

Step 3. Let y^* be a solution of the problem (7.32). Let $k := k + 1, x^k = y^*$ and go to step 1.

Generalized cutting plane method was studied by many authors. (See [24], where some works related to this method have been mentioned.) The convergence of this method can be proved under very mild assumptions. We present here a particular case of the result obtained by D. Pallaschke and S. Rolewicz [18].

Theorem 7.20 *([18]) Let X be a compact set, L consist of continuous functions and f be a continuous function, abstract convex with respect to L. Then each limit point of the sequence (x^k), which is produced by the generalized cutting plane algorithm, is a global minimizer of function f over X.*

There are two major difficulties in the numerical implementation of the generalized cutting plane algorithm.

1) to find $l^k \in \partial_L f(x_k)$

2) to solve problem (7.32).

For special classes L of elementary functions these difficulties can be overcome. In particular, if L consists of multiplicative min-type functions, both problems mentioned above can be solved. In such a case generalized cutting plane method is called the *cutting angle method* (since a level set $S(c) = \{x \in R^n_{++} : l(x) \leq c\}$ of a multiplicative min-type function l is the complement to right angle). A presentation of the cutting angle method can be found in [24].

7. Inequalities of Hadamard type for quasiconvex functions

The main field of application of quasiconvexity is optimization problems with the quasiconvex objective function. These problems are very important for economical applications (see, for example [1]). Another interesting field of applications of quasiconvexity is the so-called Hamilton-Jacobi equations (see for example [20] and references therein). We now show that abstract convexity (in particular, Principle of Preservation of inequalities) allows one to find interesting applications of quasiconvexity in the theory of inequalities. In particular, we shall study Hadamard type inequalities and their generalizations with quasiconvex functions involved.

Given a convex function $f : [a, b] \to R$, where $[a, b]$ is a closed interval in R, the famous Hadamard inequality tells us that the following holds,

$$f\left(\frac{a+b}{2}\right) \leq \frac{1}{b-a}\int_a^b f(x)dx \leq \frac{1}{2}(f(a) + f(b)). \tag{7.33}$$

In this article we attempt to study the generalizations of the left side of the inequality for some classes of quasiconvex functions defined on R^n. We shall also introduce a class of generalized convex functions called P-functions in our study and will also develop the Hadamard inequality for such a class of functions in context of the real line.

Remark 7.12 The term "Hadamard-Hermite"-type inequality (see [8]) more precisely reflect the history, however we shall use for the sake of simplicity the term Hadamard type inequality.

We will first present a well-known generalization of the inequality (7.33) for convex functions defined on compact subsets of R^n. (See for example [23], where functions defined on a compact subset of a locally convex Hausdorff topological space have been considered.)

Before stating the Hadamard type inequality for convex functions in higher dimensions we shall state certain formal notions required in the theorem. Consider $Z \subset R^n$ which is a convex and compact set. Let μ be a Borel nonnegative normed measure on the set Z, that is, μ is a σ-additive function defined on the Borel σ-algebra of subsets of Z, such that $\mu(E) \geq 0$ for all Borel sets $E \subset Z$ and $\mu(Z) = 1$.

Theorem 7.21 *Let Z be a convex and compact subset of R^n. Let us define $\bar{z} = \int_Z x d\mu$. Then $\bar{z} \in Z$ and*

$$f(\bar{z}) \leq \int_Z f(x) d\mu \tag{7.34}$$

for each lower semicontinuous convex function defined on Z.

Proof. Let us note that $\int_Z x d\mu$ is the limit of sums of the form $\sigma_k = \sum^k x_k \mu(E_k)$ such that $\max_k \mu(E_k) \to 0$ as $k \to \infty$, where E_k is a family of disjoint subsets such that $\cup_k E_k = Z$ and $x^k \in E_k$. Since Z is convex and $\mu(E_k) \geq 0$ for all k and $\sum_k \mu(E_k) = 1$ we have that $\sigma_k \in Z$ and since Z is compact we have $\bar{z} \in Z$. Let $\bar{z}_i$ be the i-th coordinate of $\bar{z}$. Clearly $\bar{z}_i = \int_Z x_i d\mu$ for all $i = 1, 2, \ldots, n$, where x_i is the i-th coordinate of x. Let us consider any $l \in R^n$. Then

$$[l, \bar{z}] = [l, \int_Z x d\mu] = \int_Z [l, x] d\mu.$$

Now consider any affine function h defined on Z. Then there exists $l \in R^n$ such that $h(y) = [l, y] - c \ (y \in Z)$, where $c \in R$ is some constant. Hence we have

$$h(\bar{z}) = [l, \bar{z}] - c = \int_Z [l, x] d\mu - c \int_Z d\mu = \int_Z h(x) d\mu.$$

Now by applying the Principle of Preservation of Inequalities we establish the theorem. □

We will now define the notion of a P- function.

Definition 7.6 *Let S be a convex subset of a locally convex Hausdorff topological vector space X and $f : S \to R_{+\infty}$ be a non-negative function on S. Then f is said to be a P-function if for all $x, y \in S$ and $\lambda \in [0,1]$ we have*

$$f(\lambda x + (1-\lambda)y) \leq f(x) + f(y). \tag{7.35}$$

Though we have defined the notion of a P-function in the setting of a locally convex Hausdorff topological vector space the original definition of P- function (see [6]) was provided on a segment of the real line. In [19, 24] certain interesting properties of P- functions is noted for case defined over the real line. We now present these properties in our general setting. Denote the class of P-functions defined on a convex set S by $P(S)$.

Proposition 7.8 *The following properties hold.*

1) $P(S)$ is a convex cone, that is for $f_1, f_2 \in P(S) \implies f_1 + f_2 \in P(S)$ and $\lambda \geq 0, f \in P(S) \implies \lambda f \in P(S)$.

2) $P(S)$ is a complete upper semi-lattice , that is for an arbitrary index set A, with $f_\alpha \in P(S)$ for all $\alpha \in A$ and $f(x) = \sup_{\alpha \in A} f_\alpha(x)$, then $f \in P(S)$.

3) $P(S)$ is closed under pointwise convergence.

We would now like to assert that the class of $P(S)$ is indeed very broad and the following proposition justifies this assertion.

Proposition 7.9 *Let $h : S \subset X \to R$ be a non-negative bounded function. Then there exists a number $c > 0$ such that the function $h + c\mathbf{1} \in P(S)$. Here $\mathbf{1}(x) = 1$, for all x.*

Proof. Let us consider $x, y \in S$. Let $A := \sup\{(h(z) - h(x) - h(y) : x, y, z \in S\}$ and let $c = \max(A, 0)$. So writing $z = \lambda x + (1-\lambda)y$, where $\lambda \in [0,1]$ we have

$$h(\lambda x + (1-\lambda)y) \leq h(x) + h(y) + c, \qquad x, y \in S,\ \lambda \in [0,1]. \tag{7.36}$$

Let us put $f(x) = h(x) + c$. Then

$$f(\lambda x + (1-\lambda)y) = h(\lambda x + (1-\lambda)y) + c \leq (h(x) + c) + (h(y) + c). \tag{7.37}$$

This shows that

$$f(\lambda x + (1-\lambda)y) \leq f(x) + f(y) \tag{7.38}$$

This clearly shows that $f \in P(S)$. □

We shall now study P-functions for the case where $X = R$ and S a segment in R. We will be more interested in the case where $S = R$ and we shall in that case denote $P(S)$ by P. Our next task is to describe a small supremal generator for the set P, (see [19]). Let T be the set of all collections $t = \{u, c_1, c_2\}$, with , $u, c_1, c_2 \in R$ and $c_1 \geq 0, c_2 \geq 0$. For any such triplet t we define h_t as the following function:

$$h_t(x) = \begin{cases} c_1 & : x < u \\ c_1 + c_2 & : x = u \\ c_2 & : x > u. \end{cases}$$

It is clear that $h_t \in P$. Let us denote by H_T the set of all functions h_t with $t \in T$. It has been proved (see [19, 24] that H_T forms the supremal generator of the class P.

Remark 7.13 Note that each function h_t ($t = \{u, c_1, c_2\} \in T$) can be represented as the sum of two quasiconvex functions h_t^1 and h_t^2, where

$$h_t^1 = \begin{cases} c_1 & x \leq u \\ 0 & x > u \end{cases}, \qquad h_t^2 = \begin{cases} 0 & x < u \\ c_2 & x \geq u \end{cases},$$

It easily follows from these observation that P is the least set, which contains the set of all nonnegative quasiconvex functions defined on R, and is a convex cone and complete upper semi-lattice. See [19, 24] for details.

Before actually presenting the result in which we deduce the Hadamard type inequality for the P-functions we would like to state a few words regarding the nature of the inequality. For P - functions and quasiconvex functions we want to establish an inequality of the form

$$f(u) \leq \gamma \int_{[a,b]} f d\mu. \tag{7.39}$$

We can refer to γ as the modulus of the inequality. The above inequality is called *asymptotically sharp* if for every $\epsilon > 0$ there exits f such that the above inequality is violated when the modulus is changed to $\gamma - \epsilon$. The inequality is called *sharp* if equality holds at least for one f. We will now state certain preliminaries required to establish the inequality.

Consider the σ-algebra Σ of Borel subsets of the segment $(0, 1)$ and a non-negative measure μ on Σ such that $\mu((0, 1)) = 1$. Let P' be the set of all Borel measurable P-functions defined on $(0, 1)$. We denote by h'_t the restriction of the function h_t to the segment $(0, 1)$. We also denote by H'_T the class of all functions h'_t. It is clear that each h'_t is Borel measurable and it can be shown that H'_T is indeed the supremal generator of P'.

Consider the functional

$$J(f) = \int_0^1 f d\mu \tag{7.40}$$

where $\int_0^1$ stands for $\int_{(0,1)}$. For a fixed $y \in (0, 1)$ consider the following functions

$$e_y^1(x) = \begin{cases} 1 & : x \leq y \\ 0 & : x > y \end{cases}$$

$$e_y^2(x) = \begin{cases} 0 & : x \leq y \\ 1 & : x > y \end{cases}$$

Let $g_1(y) = \int_0^1 e_y^1 d\mu = \mu((0, y])$ and $g_2(y) = \int_0^1 e_y^2 d\mu = \mu((y, 1))$. Clearly g_1 is increasing while g_2 is decreasing and further $g_1(y)+g_2(y) = \mu((0,1)) = 1$. A measure μ on a σ-algebra Σ is called *atomless* if $\mu(\{x\}) = 0$ for each $x \in (0, 1)$. We have the following lemma

Lemma 7.3 *Let μ be an atomless Borel measure with $\mu((0,1)) = 1$ and let $h'_t \in H'_T$ with $t = \{y, c_1, c_2\}$ with $c_1 + c_2 > 0$ and $y \in (0,1)$. Then $J(h'_t) = c_1 g_1(y) + c_2 g_2(y)$.*

Proof. It is easy to see that since μ is atomless we have

$$J(h'_t) = \int_0^y c_1 d\mu + \int_y^1 c_2 d\mu = c_1\mu((0, y)) + c_2\mu((y, 1)). \tag{7.41}$$

Again as μ is atomless we have

$$J(h'_t) = c_1\mu((0, y]) + c_2\mu((y, 1)) = c_1 g_1(y) + c_2 g_2(y). \tag{7.42}$$

Hence the result follows. □

Now for any $u \in (0, 1)$ define

$$\gamma_u = \min_{c_1 \geq 0, c_2 \geq 0} \frac{c_1 g_1(u) + c_2 g_2(u)}{c_1 + c_2}. \tag{7.43}$$

It is now easy to check that $\gamma_u = \min(g_1(u), g_2(u))$.

Theorem 7.22 *Let μ be an atomless measure and $u \in (0,1)$. Then*

$$f(u) \leq \frac{1}{\min(g_1(u), g_2(u))} \int_0^1 f d\mu \tag{7.44}$$

for all $f \in P'$.

Proof. Let us as before define $J(f) = \int_0^1 f d\mu$, where $f \in P'$. It is clear that J is an increasing function. We will first check that the inequality (7.44) holds for all $h'_t \in H'_T$. Then the Principle of Preservation of Inequalities at once will guarantee the result. We will consider the distinct cases h'_t with $t = \{u, c_1, c_2\}$ and h'_t with $t = \{y, c_1, c_2\}$ with $y \neq u$.
In the first case let us observe that $h'_t(u) = c_1 + c_2$. Hence

$$h'_t(u) = c_1 + c_2 \leq \frac{c_1 g_1(u) + c_2 g_2(u)}{\gamma_u}. \tag{7.45}$$

This clearly shows that by applying Lemma we have

$$h'_t(u) \leq \frac{1}{\gamma_u} J(h'_t) = \frac{1}{\gamma_u} \int_0^1 h'_t d\mu. \tag{7.46}$$

Now plugging the expression for γ_u we establish the result in the first case.
Now we consider the case where $y \neq u$. Let us calculate h'_t. Now from the definition of h'_t we have either $h'_t(u) = \min(c_1, c_2)$ or $h'_t(u) = \max(c_1, c_2)$. In the first case noting that $g_1(y) \geq 0$, $g_2(y) \geq 0$ and $g_1(y) + g_2(y) = 1$ we have

$$h'_t(u) = \min(c_1, c_2) \leq c_1 g_1(y) + c_2 g_2(y) \leq J(h'_t). \tag{7.47}$$

Since $g_1(u) \leq 1$ and $g_2(u) \leq 1$ it is clear that

$$H'_t(u) \leq J(h'_t) \leq \frac{1}{\min(g_1(u), g_2(u))} J(h'_t). \tag{7.48}$$

Let us now assume that $h'_t(u) = \max(c_1, c_2)$. If $c_1 \geq c_2$ and hence $h'_t(u) = c_1$ and thus from the definition of the function h'_t it is clear that $y > u$. Since g_1 is an increasing function we have

$$g_1(u) h'_t(u) \leq g_1(y) c_1 \leq c_1 g_1(y) + c_2 g_2(y) = J(h'_t). \tag{7.49}$$

Thus

$$h'_t(u) \leq \frac{1}{g_1(u)} J(h'_t) \leq \frac{1}{\min(g_1(u), g_2(u))} J(h'_t).$$

When $c_2 \geq c_1$ we can prove the assertion along similar lines. As we have already mentioned before the Principle of Preservation of Inequalities at once proves that the inequality is indeed true for all $f \in P'$. □

Remark 7.14 Let us remark that the Hadamard type inequality proved in the theorem above is indeed a sharp inequality.

If we consider μ to be the Lebesgue measure then $\int_0^1 f d\mu = \int_0^1 f(x)dx$. Then $g_1(y) = y$ and $g_2(y) = 1 - y$. Thus from the above theorem we have

$$f(u) \leq \frac{1}{\min(u, 1-u)} \int_0^1 f(x)dx \tag{7.50}$$

In particular let us note that

$$f\left(\frac{1}{2}\right) \leq 2 \int_0^1 f(x)dx \tag{7.51}$$

This result was established by Dragomir, Pecaric and Persson [6]. Let us observe that if we instead of the interval $(0, 1)$ considered some arbitrary interval (a, b) with $a, b \in R$ and let μ remain to be the Lebesgue measure then it can be shown that for $u \in (a, b)$ we have

$$f(u) \leq \frac{1}{\min(u-a, b-u)} \int_a^b f(x)dx \tag{7.52}$$

Let us also mention that the Hadamard type inequality can be proved in a similar manner for a non-negative quasiconvex function without requiring that μ be atomless. We can find a supremal generator consisting on two-steps functions, such that one step is equal to zero (see Theorem 7.18). Let us also note that the classical Hadamard type inequality is used to estimate the function value at the center of a symmetrical set. The version of Hadamard type inequality with respect to an arbitrary interior point was first established by Pearce and Rubinov [19] which is in fact the theorem presented above.

We shall now present a simple and very elegant proof to derive a Hadamard type inequality at an arbitrary point for a quasiconvex function defined on a closed and bounded interval of the real line and bounded from below. The fundamental notion of supremal generators is very beautifully used in the proof. The proof is due to Professor J. E. Martinez-Legaz who very kindly agreed to our request to reproduce it here.

Theorem 7.23 *Let $f : [0,1] \to R$ be a quasiconvex function which is bounded from below by m. Let $u \in (0,1)$. Then*

$$f(u) \leq \frac{1}{\min(u, 1-u)} \left[\int_0^1 f(x)dx - m \right] + m.$$

Proof. Without loss of generality assume that $m = 0$. (If $m \neq 0$ we can consider the quasiconvex nonnegative function $f(x) - m$ instead of f.) Let us consider that $c = f(u)$ where $u \in (0,1)$. Consider the two step function $g : [0,1] \to R$ given as

$$g(x) = \begin{cases} 0 & : x \in [0, u) \\ c & : x \in [u, 1]. \end{cases}$$

If $c > f(x_0)$ for some $x_0 \in (0, u)$ then by quasiconvexity we have $f \geq g$. Further let us consider the following two step function $h : [0,1] \to R$ given as

$$h(x) = \begin{cases} c & : x \in [0, u] \\ 0 & : x \in (u, 1]. \end{cases}$$

If $c \leq f(x_0)$, for all $x_0 \in (0, u)$ then it is clear that $f \geq h$. Hence either

$$\int_0^1 f(x)dx \geq \int_0^1 g(x)dx = c(1-u)$$

or

$$\int_0^1 f(x)dx \geq \int_0^1 h(x)dx = cu$$

This shows that

$$\int_0^1 f(x)dx \quad \geq \quad \min(cu, c(1-u)) = c \min(u, 1-u).$$

This clearly shows that

$$f(u) = c \quad \leq \quad \frac{1}{\min\{u, 1-u\}} \int_0^1 f(x)dx.$$

□

Remark 7.15 Observe that if the function f in the above theorem is non-negative then $m = 0$ and the result reduces to that of the previous theorem. A further more important fact is that if $m > 0$ then the result in the above theorem provides a stronger estimate (or rather a tighter bound) than the usual Hadamard type inequality for non-negative quasiconvex functions. Another important point to observe in the above proof is underlying use of the notion of supremal generators. In the above proof either g or h are in fact belong to the support set of f with respect to the supremal generator of the set of quasiconvex functions, which consists of two step functions. The above theorem has a simple approach since it does not have to depend on other stand alone results like the Principle of Preservation of Inequalities. The above proof also illustrates that the supremal generators are one of the most elegant and powerful ways to view a quasiconvex function.

Remark 7.16 A different approach for Hadamard-type inequalities for quasiconvex functions was suggested by Hadjisavvas [11].

We will now turn our attention to prove a Hadamard type inequality for quasiconvex functions in higher dimensions. For a special class of quasiconvex functions defined on R^2 we will give an exact expression for calculating the modulus of the Hadamard-type inequality. The results that will now be presented are due to a recent work by Rubinov and Dutta [26]. We will in fact consider only the class of non-negative evenly quasiconvex functions which is large enough since it contains both non-negative upper and lower semicontinuous quasiconvex functions. As one will observe, we indeed prove the Hadamard type inequality for certain important subclasses of the class of non-negative evenly quasiconvex functions.

Denote by Q^+ the set of all nonnegative evenly quasiconvex functions $f : R^n \to R_{+\infty} \equiv R \cup \{+\infty\}$. Let $Q_0^+ = \{f \in Q^+ : f(0) = 0\}$. For each vector $v \in R^n$ and each number $c \geq 0$ consider the function $l^{v,c}$ defined by

$$l^{v,c}(x) = \begin{cases} c & \text{if } [v, x] \geq 1 \\ 0 & \text{if } [v, x] < 1. \end{cases} \tag{7.53}$$

Let $L_0 = \{l^{v,c} : v \in R^n, c \in R_+\}$. It is easy to check that $L_0 \subset Q_0^+$. The following result holds (see, for example, [24]).

Proposition 7.10 *L_0 is a supremal generator of Q_0^+.*

Let $X \subset R^n$ be a closed convex set. Denote by $Q^+(X)$ the set of all nonnegative evenly quasiconvex functions $f : X \to R_{+\infty}$ such that the minimum $\min\{f(x) : x \in X\}$ is attained and less than $+\infty$. If X is compact then each non-negative lower semicontinuous quasiconvex function f defined on X belongs to $Q^+(X)$. We now define a set $L(X)$ of two-step functions defined on X. A function l belongs to $L(X)$ if and only if there exists a vector $v \in R^n$, a point $x_0 \in X$ and nonnegative numbers c and d such that $c \geq d$ and for all $x \in X$ we have:

$$l(x) = \begin{cases} c & \text{if } [v, x - x_0] \geq 1 \\ d & \text{if } [v, x - x_0] < 1. \end{cases} \tag{7.54}$$

Clearly $L(X) \subset Q^+(X)$.

Proposition 7.11 *[26]The set* $L(X)$ *is a supremal generator of the set* $Q^+(X)$.

Let

$$\psi(f) = \int_X f(x)d\mu, \qquad (f \in Q^+(X)) \tag{7.55}$$

where μ is a Borel measure defined on X. Let $u \in X$. We wish to establish the asymptotically sharp inequality of the form

$$f(u) \leq \gamma \int_X f(x)d\mu,$$

which is valid for all $f \in Q^+(X)$. For this purpose we need to calculate the constant γ. In our case it is sufficient to calculate γ in the following form:

$$\gamma = \sup_{l \in L(X)} \frac{l(u)}{\int_X l(x)d\mu}. \tag{7.56}$$

Let $u \in X$. In order to calculate γ, we need the following sets A_u^+, A_u^-:

$$A_u^+ = \{(v, x_0) \in R^n \times X : [v, u - x_0] \geq 1\} \tag{7.57}$$

and

$$A_u^- = \{(v, x_0) \in R^n \times X : [v, u - x_0] < 1\}. \tag{7.58}$$

For the given u and each $v \in R^n$ and $x_0 \in X$ consider the sets

$$X_{v,x_0}^+ = \{x \in X : [v, x - x_0] \geq 1\}, \qquad X_{v,x_0}^- = \{x \in X : [v, x - x_0] < 1\} \tag{7.59}$$

Clearly

$$A_u^+ \cap A_u^- = \emptyset \quad \text{and} \quad A_u^+ \cup A_u^- = R^n \times X$$

and

$$X_{v,x_0}^+ \cap X_{v,x_0}^- = \emptyset \quad \text{and} \quad X_{v,x_0}^+ \cup X_{v,x_0}^- = X.$$

If $(v, x_0) \in A_u^+$ then $X_{v,x_0}^+ \neq \emptyset$; indeed, $u \in X_{v,x_0}^+$. The same argument shows that $X_{v,x_0}^- \neq \emptyset$ for $(v, x_0) \in A_u^-$.

Proofs of the following results can be found in [26]

Theorem 7.24 *Let γ defined by (7.56). Then*

$$\gamma = \sup_{(v,x_0)\in A_u^+} \frac{1}{\mu(X_{v,x_0}^+)}. \tag{7.60}$$

Corollary 7.3 Let $u \in X$ and γ is defined by (7.60). Then

$$f(u) \leq \gamma \int_X f(x)d\mu \tag{7.61}$$

for each $f \in Q^+(X)$ and the constant γ is asymptotically sharp.

Corollary 7.4 If $\inf_{(v,x_0)\in A_u^+} \mu(X_{v,x_0}^+)$ is attained, then the constant γ in the inequality (7.61) is sharp.

We shall now discuss the special case of non-negative evenly quasiconvex functions which vanish at the point $x = 0$ and thus forms a subclass of all non-negative evenly quasiconvex functions which attains a minimum over the set $X \subset R^n$. Obviously we assume that $0 \in X$. We will present the results without a proof. For the proofs the reader is referred to Rubinov and Dutta [26].

Let $X \subset R^n$ be a closed convex set such that $x \in X_0$. Let $Q_0^+(X)$ be the set of all evenly quasiconvex nonnegative functions defined on X and such that $f(0) = 0$ and let $L_0(X)$ be the set of restrictions to X of functions from L_0. In other words, $l \in L_0(X)$ if there exists $v \in R^n$ and a number $c > 0$ such that $l(x) = l^{v,c}(x)$, $(x \in X)$, where $l^{v,c}$ is defined by (7.53). It follows from Proposition 7.10 that $L_0(X)$ is a supremal generator of $Q_0(X)$.

Then we consider the inequality of the form

$$f(u) \leq \gamma_h \int_X h(f(x))d\mu, \qquad f \in Q_0(X).$$

Here $h : R_+ \to R_+$ is an increasing function such that

$$\lambda_h := \sup_{c>0}(c/h(c)) < +\infty.$$

We show that the asymptotically sharp estimate γ_h in this inequality is the product $\lambda_h \cdot \gamma_*$ where γ_* is the asymptotically sharp constant in the inequality $f(u) \leq \gamma_* \int_X f(x)d\mu$, $f \in Q_0(X)$.

Consider an increasing function $h : R_+ \to R_+$ such that $h(x) > 0$ for all $x > 0$ and

$$\lambda_h = \sup_{c>0} \frac{c}{h(c)} < +\infty. \tag{7.62}$$

and let

$$\psi_h(f) = \int_X h(f(x))d\mu, \qquad f \in Q_0(X), \tag{7.63}$$

where μ is a finite Borel measure on X. Clearly ψ_h is an increasing functional defined on $Q_0(X)$. Let $u \in X$ and

$$\gamma_h = \sup_{l \in L_0(X)} \frac{l(u)}{\int_X h(l(x))d\mu}. \tag{7.64}$$

Hence in this case the modulus of the Hadamard inequality is denoted by γ_h. Consider the sets

$$B_u = \{v \in R^n : [v, u] \geq 1\}$$

and

$$X_v = \{x \in X : [v, x] \geq 1\}, \quad (v \in B_v).$$

Note that B_u is not empty for all $u \neq 0$ and X_v is not empty for all $u \neq 0$, since $u \in X_v$. Let

$$\gamma_* = \sup_{v \in B_u} \frac{1}{\mu(X_v)}. \tag{7.65}$$

Theorem 7.25 *We have $\gamma_h \leq \lambda_h \cdot \gamma_*$. If either $h(0) = 0$ or $\lambda_h := \sup_{c>0}(h(c)/c) = \lim_{c \to +\infty}(h(c)/c)$, then $\gamma_h = \lambda_h \cdot \gamma_*$.*

For our study we need h to be the identity function and hence $\lambda_h = 1$. Also observe that all the conditions of Theorem 7.25 are satisfied. This shows that $\gamma_h = \gamma_*$. Hence for the non-negative evenly quasiconvex functions vanishing at $x = 0$ we indeed have to calculate γ_* by formula (7.65).

Corollary 7.5 Let $u \in X$ and let λ_h and γ_* be defined by (7.64) and (7.65), respectively and let $\gamma_h = \lambda_h \cdot \gamma_*$. Then

$$f(u) \leq \gamma_h \int_X f(x)d\mu \tag{7.66}$$

for each $f \in Q_0^+(X)$ and the constant $\gamma_h = \lambda_h \gamma_*$ is asymptotically sharp.

Corollary 7.6 If $\inf_{(v,x_0) \in B_u} \mu(X_v)$ is attained and $\sup_{c>0}(c/h(c))$ is attained then the constant γ_h in the inequality (7.66 is sharp.

We are now going to mention a theorem in which we provide an explicit calculation for γ_* in the two dimensional case.

Theorem 7.26 *Consider* $X = [-1,1] \times [-1,1]$ *in the plane* R^2. *For any point* $(u_1, u_2) \in int X \setminus \{0\}$ *we have*

$$\gamma_* = \frac{1}{2(1-|u_1|)(1-|u_2|)}.$$

We end this section with the following result.

Proposition 7.12 *Consider* $n = 1$ *and* $X = [-1, +1]$, *and* $u \in int X$ *then*

$$\gamma = \gamma_* = \frac{1}{\min(1+u, 1-u)}, \quad u \neq 0$$
$$\gamma = 1, \gamma_* = 0, \quad , u = 0.$$

References

[1] Avriel, M., Diewert, W.E., Schaible, S. and I. Zang, *Generalized concavity*, Plenum Press, New York 1988.

[2] Birkhoff, G., *Lattice Theory*, third ed., AMS Colloquium Publications, vol. 25, AMS, 1967

[3] Balder E.J., *An extension of duality-stability relations to non-convex optimization problems*, SIAM J. Contr. Optimization, 15, 1977, 329-343.

[4] Daniilidis, A. and N. Hadjisavvas, *Characterization of nonsmooth semistrictly quasiconvex and strictly quasiconvex functions*, J. Optim. Theory Appl. 102, 1999, 525-536.

[5] Dolecki S. and S. Kursyusz, *On Φ-convexity in extremal problems*, SIAM J. Contr. Optimization, 16, 1978, 277-300.

[6] Dragomir S.S., Pečarić J.E. and L. E. Persson, *Some inequalities of Hadamard type*, Soochow J. Math. 21, 1995, 335-341.

[7] Dragomir S.S and C. E. M. Pearce, *Quasi-convex functions and Hadamard's inequality*, Bull. Austral. Math. Soc., 57, 1998, 377-385.

[8] Dragomir S.S. and C. E. M. Pearce, *Selected Topics on Hermite-Hadamard Inequalities and Applications*, http://melba.vut.edu/ rgmia/reports.html

[9] Penot, J.P., *Are generalized derivatives useful for generalized convex functions?*, in Generalized Convexity, Generalized Monotonicity, J.-P. Crouzeix, J.-E. Martínez-Legaz and M. Volle, eds., Kluwer Academic Publishers, Dordrecht, 1998, 3-59.

[10] Gunawardena J., *An introduction to idempotency*, in Idempotency, J. Gunawardena, ed., Cambridge Univ. Press, Cambridge,1998, 1-49.

[11] Hadjisavvas N., *Hadamard-type inequalities for quasiconvex functions*, J. Inequal. Pure Appl. Math. (electronic) Vol 4, Issue 1, Article 13, 2003.

[12] Hiriart-Urruty J.-B. and C. Lemarechal, *Convex Analysis and Minimization algorithms*, II, Springer-Verlag, Berlin-Heidelberg, 1993.

[13] Kutateladze S.S. and A.M. Rubinov *Minkowski duality and its applications*, Russian Mathem. Surveys, 27, 1972, 137-191.

[14] Kutateladze S.S and A. M. Rubinov A.M. *Minkowski Duality and its Applications*, Nauka, Novosibirsk, 1976. (In Russian.)

[15] Martinez-Legaz J.E., Rubinov A.M. and I. Singer, *Downward sets and their separation and approximation properties*, Journal of Global Optimization, 23, 2002, 11-137.

[16] Martinez Legaz J.E., *Quasiconvex duality theory by generalized conjugation methods*, Optimization, 19, 1988, 603-652.

[17] J.-E. Martinez Legaz, this volume

[18] Pallaschke D. and S. Rolewicz, *Foundations of Mathematical Optimization (Convex Analysis without Linearity)* Kluwer Academic Publishers, Dordrecht, 1997.

[19] Pearce C.E.M. and A.M. Rubinov, *P-functions, quasiconvex functions and Hadamard-type inequalities*, J. Mathem. Analysis and Applications, 240, 1999, 92-104.

[20] Penot J.-P., *What is quasiconvex analysis?* Optimization, 47, 2000, 35-110.

[21] Penot J.P. and M. Volle, *On quasiconvex duality*, Math. Oper. Research, 15, 1990, 597-625.

[22] Penot J.P. and M. Volle, *Another duality scheme for quasiconvex problems*, In Trends in Mathematical Optimization, K.-H. Hoffmann et al., eds, Birkhauser, Basel, 1988, 259-275.

[23] Phelps R., *Lectures on Choquet's Theorem*, D. Van Nostrand Company, Inc., Princeton, 1966.

[24] Rubinov A.M., *Abstract Convexity and Global Optimization*, Kluwer Academic Publishers, Dordrecht, 2000.

[25] Rubinov A.M., *Distance to the solution set of an inequality with an increasing function*, in Equilibrium Problems and Variational Models, F. Giannessi, A. Maugeri and P. Daniele, eds., Kluwer Academic Publishers, 2003, 417-431.

[26] Rubinov A.M and J. Dutta, *Hadamard type inequality for quasiconvex functions in higher dimensions*, Journal of Mathematical Analysis and Applications, 270, 2002, 80-91.

[27] Rubinov A.M. and I. Singer, *Topical and sub-topical functions, downward sets and abstract convexity*, Optimization, 50, 2001, 301-357.

[28] Rubinov A.M. and I. Singer, *Best approximation by normal and conormal sets* J. Approx. Theory, 107, 2000, 212-243.

[29] Rubinov A.M. and I. Singer, *Distance to the intersection of normal sets and applications*, Numerical Functional Analysis and Optimization, 21, 2000, 521-536.

[30] Rubinov A.M. and A. Shveidel, *On some classes of abstract convex functions*, in Optimization and Control with Applications, Kluwer Academic Publishers, L.Q. Qi, K.L. Teo and X.Q. Yang, eds., 2004.

[31] Rubinov A.M. and B. Simsek, *Conjugate quasiconvex nonnegative functions*, Optimization, 35, 1995, 1 - 22.

[32] Singer I., *Abstract convex analysis*, Wiley-Interscience, New York, 1997.

[33] Vollc M.,*Conjugation par transhes*, Annali. Math. Pura Appl., 139, 1985, 279-312.

Chapter 8

FRACTIONAL PROGRAMMING

Johannes B.G. Frenk
Econometric Institute,
Erasmus University,
3000 DR Rotterdam, The Netherlands
frenk@few.eur.nl

Siegfried Schaible
A.G. Anderson Graduate School of Management
University of California,
Riverside, CA 92521-0203, USA
siegfried.schaible@ucr.edu

Abstract Single-ratio and multi-ratio fractional programs in applications are often generalized convex programs. We begin with a survey of applications of single-ratio fractional programs, min-max fractional programs and sum-of-ratios fractional programs. Given the limited advances for the latter class of problems, we focus on an analysis of min-max fractional programs. A parametric approach is employed to develop both theoretical and algorithmic results.

Keywords: Single-ratio fractional programs, min-max fractional programs, sum-of-ratios fractional programs, parametric approach.

1. Introduction.

In various applications of nonlinear programming a ratio of two functions is to be maximized or minimized. In other applications the objective function involves more than one ratio of functions. Ratio optimization problems are commonly called fractional programs. One of the earliest fractional programs (though not called so) is an equilibrium model for an expanding economy introduced by von Neumann (cf. [74])

in 1937. The model determines the growthrate of an economy as the maximum of the smallest of several output-input ratios. At a time when linear programming hardly existed, the author already proposed a duality theory for this nonconvex program. However, apart from a few isolated papers like von Neumann's, a systematic study of fractional programming began much later.

In 1962 Charnes and Cooper (cf. [18]) published their classical paper in which they show that a linear fractional program with one ratio can be reduced to a linear program using a nonlinear variable transformation. Separately, Martos [49] in 1964 (from his Ph.D. dissertation in Hungarian in 1960) showed that linear fractional programs can be solved with an adjacent vertex-following procedure, the same way as linear programs are solved with the simplex method. He recognized that generalized convexity properties (pseudolinearity) of linear ratios enables such a technique which is successfully used in linear programming.

The study of fractional programs with only one ratio has largely dominated the literature in this field until about 1980. Many of the results known then are presented in the first monograph on fractional programming (cf. [65]) which the second author published in 1978. Since then two other monographs solely devoted to fractional programming appeared, one in 1988 authored by Craven (cf. [21]) and one in 1997 by Stancu-Minasian (cf. [71]). An overview of solution methods for single-ratio and multi-ratio fractional location problems appeared in the monograph by Barros (cf. [5]).

Fractional programs with one or more ratios have often been studied in the broader context of generalized convex programming (cf. [4]). Ratios of convex and concave functions as well as composites of such ratios are not convex in general, even in the case of linear ratios. But often they are generalized convex in some sense. From the beginning, fractional programming has benefited from advances in generalized convexity, and vice versa (cf. [50]).

Fractional programming also overlaps with global optimization. Several types of ratio optimization problems have local, nonglobal optima. An extensive survey of fractional programming with one or more ratios appeared in the Handbook of Global Optimization [61]. The survey also contains the largest bibliography on fractional programming with one or multiple ratios so far. It has almost twelve-hundred entries. For a separate, rich bibliography [71] may be consulted.

Very recently two surveys have appeared updating some of the developments reviewed in [61]. The single-ratio and min-max case is treated in [59] and the sum-of-ratios case in [60].

2. Classification of Fractional Programs.

To start with single-ratio fractional programs, let $B \subseteq \mathbb{R}^n$ be a nonempty closed set and $f, g : \mathbb{R}^n \to [-\infty, \infty]$ be extended real-valued functions which are finite-valued on B. Assuming $g(x) > 0$ for every $x \in B$, consider the nonlinear program

$$\inf_{x \in B} \frac{f(x)}{g(x)}. \tag{P_1}$$

The problem (P_1) is called a *single-ratio fractional program.* In most applications the nonempty feasible region B has more structure and is given by

$$B = \{x \in C : h_k(x) \leq 0, k = 1, ..., l\} \tag{8.1}$$

with $C \subseteq \mathbb{R}^n$ and $h_k : \mathbb{R}^n \to \mathbb{R}$, $1 \leq k \leq l$ some set of real-valued continuous functions. So far, the functions in the numerator and denominator were not specified. If f, g and $h_k, 1 \leq k \leq l$ are affine functions (linear plus a constant) and $C = \mathbb{R}^n_+$ denotes the nonnegative orthant of $\mathbb{R}^n$, then the optimization problem (P_1) is called a *single-ratio linear fractional program.* Moreover, we call (P_1) a *single-ratio quadratic fractional program* if $C = \mathbb{R}^n_+$, the functions f and g are quadratic and the functions $h_k, 1 \leq k \leq l$ are affine. The minimization problem (P_1) is called a *single-ratio convex fractional program* if C is a convex set, $h_k, 1 \leq k \leq l$ and f are convex functions and g is a positive concave function on B. In addition it is assumed that f is nonnegative on B if g is not affine. In case of a maximization problem the single-ratio fractional program is called a *single-ratio concave fractional program* if f is concave and g is convex. Under these restrictive convexity\concavity assumptions the minimization problem (P_1) is in general a nonconvex problem.

In some applications more than one ratio appears in the objective function. One form of such an optimization problem is the nonlinear programming problem

$$\inf_{x \in B} \sup_{1 \leq i \leq m} \frac{f_i(x)}{g_i(x)} \tag{P_2}$$

with extended real-valued functions $f_i, g_i : \mathbb{R}^n \to [-\infty, \infty], 1 \leq i \leq m$ which are finite-valued on B with $g_i(x) > 0$ for every $1 \leq i \leq m$ and $x \in B$. The problem (P_2) is often called a *generalized fractional program.* As for single-ratio fractional programs we can specify the functions and make a distinction between multi-ratio linear fractional programs and multi-ratio convex fractional programs. If one g_i is not affine, we need to assume that all functions f_i are nonnegative. Clearly both problems (P_1) and (P_2) are special cases of the following problem.

Let $A \subseteq \mathbb{R}^m$ and $B \subseteq \mathbb{R}^n$ be nonempty closed sets and $f : \mathbb{R}^{m+n} \to [-\infty, \infty]$ be a finite-valued function on $A \times B$. In case $g : \mathbb{R}^{m+n} \to [-\infty, \infty]$ is a finite-valued positive function on $A \times B$, consider the min-max nonlinear programming problem

$$\inf_{x \in B} \sup_{y \in A} \frac{f(y,x)}{g(y,x)}. \tag{P}$$

Problem (P) is called a *(primal) min-max fractional program.* In order to unify the theory for single-ratio and multi-ratio fractional programs, we consider in Section 6 the so-called parametric approach applied to problem (P) and derive from this approach duality results and algorithmic procedures for problem (P). This yields immediately duality results and algorithmic procedures for problems (P_1) and (P_2).

Another multi-ratio fractional program we encounter in applications is the so-called *sum-of-ratios fractional program* given by

$$\inf_{x \in B} \sum_{i=1}^{m} \frac{f_i(x)}{g_i(x)} \tag{P_3}$$

with $g_i(x) > 0$ for every $x \in B$ and $1 \leq i \leq m$. It is a more challenging problem than (P_2) as recent studies have shown. We also encounter in applications the so-called *multi-objective fractional program*

$$\inf_{x \in B} \left(\frac{f_1(x)}{g_1(x)}, \ldots, \frac{f_m(x)}{g_m(x)}\right) \tag{P_4}$$

which is related to (P_2) and (P_3).

In Sections 3 and 4 we will review applications of fractional programs (P_1) and (P_2), respectively. Section 5 focuses on applications of the fractional program (P_3). In addition we review here some of the solution procedures for this rather challenging problem. Finally in Section 6 we return to problems (P_1) and (P_2). In a joint treatment of both involving the more general problem (P) a parametric approach is used for the analysis and development of solution procedures of (P).

3. Applications of Single-Ratio Fractional Programs (P_1).

Single-ratio fractional programs (P_1) arise in management decision making as well as outside of it. They also occur sometimes indirectly in modelling where initially no ratio is involved. The purpose of the following overview is to demonstrate the diversity of problems which can be cast in the form of a single-ratio fractional program. A more comprehensive coverage which also includes additional references for the models below is contained in [61]. For other surveys of applications of a single-ratio fractional program see [21, 59, 62, 64, 65, 71].

Economic Applications.
The efficiency of a system is sometimes characterized by a ratio of technical and/or economical terms. Maximizing the efficiency then leads to a fractional program. Some applications are given below.

- *Maximization of Productivity.*
  Gilmore and Gomory [37] discuss a stock cutting problem in the paper industry for which under the given circumstances it is more appropriate to minimize the ratio of wasted and used amount of raw material rather than just minimizing the amount of wasted material. This stock cutting problem is formulated as a linear fractional program. In a case study, Hoskins and Blom [43] use fractional programming to optimize the allocation of warehouse personnel. The objective is to minimize the ratio of labor cost to the volume entering and leaving the warehouse.

- *Maximization of Return on Investment.*
  In some resource allocation problems the ratio profit/capital or profit/revenue is to be maximized. A related objective is return per cost maximization. Resource allocation problems with this objective are discussed in more detail by Mjelde in [53]. In these models the term 'cost' may either be related to actual expenditure or may stand, for example, for the amount of pollution or the probability of disaster in nuclear energy production. Depending on the nature of the functions describing return, profit, cost or capital, different types of fractional programs are encountered. For example, if the price per unit depends linearly on the output and cost and capital are affine functions, then maximization of the return on investment gives rise to a concave quadratic fractional program (assuming linear constraints). In location analysis maximizing the profitability index (rate of return) is in certain situations preferred to maximizing the net present value, according to [5] and [6] and the cited references.

- *Maximization of Return/Risk.*
  Some portfolio selection problems give rise to a concave nonquadratic fractional program of the form (8.3) below which expresses the maximization of the ratio of expected return and risk. For related concave and nonconcave fractional programs arising in financial planning see [61]. Markov decision processes may also lead to the

maximization of the ratio of mean and standard deviation. A very recent application of fractional programming in portfolio theory is given in [48]. The authors argue that the ratio of two variances gives sophisticated forecasting models with significant predictive power.

- *Minimization of Cost/Time.*
  In several routing problems a cycle in a network is to be determined which minimizes the cost-to-time ratio or maximizes the profit-to-time ratio. Some of these models are combinatorial fractional programs (cf. [56]). Also the average cost objective used within the theory of stochastic regenerative processes (cf. [2]) leads to the minimization of cost per unit time. A particular example occurring within this framework is the determination of the optimal ordering policy of the classical periodic and continuous review single item inventory control models (cf. [12, 13, 34]). Other examples of this framework are maintenance and replacement models. Here the ratio of the expected cost for inspection, maintenance and replacement and the expected time between two inspections is to be minimized (cf. [7, 35]).

- *Maximization of Output/Input.*
  Charnes and Cooper use a linear fractional program as a model to evaluate the efficiency of decision making units (Data Envelopment Analysis (DEA)). Given a collection of decision making units, the efficiency of each unit is obtained from the maximization of a ratio of weighted outputs and inputs subject to the condition that similar ratios for every decision making unit are less than or equal to unity. The variable weights are then the efficiency of each member relative to that of the others. For an extensive recent treatment of DEA see [17]. In the management literature there has been an increasing interest in optimizing relative terms such as relative profit. No longer are these terms merely used to monitor past economic behavior. Instead the optimization of rates is receiving more attention in decision making processes for future projects (cf. [5, 42]). We mention here a case study in which the effectiveness of medical institutions in the area of trauma and burned management was analyzed with help of linear fractional programming (cf. [24]).

Non-Economic Applications.
In information theory the capacity of a communication channel can be defined as the maximal transmission rate over all probabilities. This is a

concave nonquadratic fractional program. Also the eigenvalue problem in numerical mathematics can be reduced to the maximization of the Rayleigh quotient, and hence gives rise to a quadratic fractional program which is generally not concave. An example of a fractional program in physics is given by Falk (cf. [25]). He maximizes the signal-to-noise ratio of an optical filter which is a concave quadratic fractional program.

Indirect Applications.

There are a number of management science problems that indirectly give rise to a concave fractional program. We begin with a recent study which shows that the sensitivity analysis of general decision systems leads to linear fractional programs (cf. [52]). The developed software was used in the appraisal of Hungarian hotels. A concave quadratic fractional program arises in location theory as the dual of a Euclidean multifacility min-max problem. In large scale mathematical programming, decomposition methods reduce the given linear program to a sequence of smaller problems. In some of these methods the subproblems are linear fractional programs. The ratio originates in the minimum-ratio rule of the simplex method.

Fractional programs are also met indirectly in stochastic programming, as first shown by Charnes and Cooper [19] and by Bereanu [14]. This will be illustrated by two models below (cf. [65, 71]).

Consider the following stochastic mathematical program

$$\max\{a^{\mathsf{T}}x : x \in B\} \tag{8.2}$$

where the coefficient vector a is a random vector with a multivariate normal distribution and B is a (deterministic) convex feasible region. It is assumed that the decision maker replaces the above optimization problem by the *maximum probability model*

$$\max\{P(a^{\mathsf{T}}x \geq k) : x \in B\},$$

i.e., he wants to maximize the probability that the random variable $a^{\mathsf{T}}x$ attains at least a value equal to a prescribed level k. Then the optimization problem listed in (8.2) reduces to

$$\max\{\frac{e^{\mathsf{T}}x - k}{\sqrt{x^{\mathsf{T}}Vx}} : x \in B\} \tag{8.3}$$

where e is the mean vector of the random vector a and V its variance-covariance matrix. Hence the maximum probability model of the concave program (8.2) gives rise to a fractional program. If in problem (8.2) the linear objective function is replaced by other types of nonlinear functions, then the maximum probability model leads to various other fractional programs as demonstrated in [65] and [71].

Consider a second stochastic program

$$\max\{f_0(x) + \theta f_1(x) : x \in B\} \tag{8.4}$$

where f_0, f_1 are concave functions on the convex feasible region B, $f_1 > 0$ and θ is a random variable with a continuous cumulative distribution function. Then the maximum probability model for (8.4) gives rise to the fractional program

$$\max\{\frac{f_0(x) - k}{f_1(x)} : x \in B\}. \tag{8.5}$$

For a linear program (8.4) the deterministic equivalent (8.5) becomes a linear fractional program. If f_0 is concave and f_1 linear, then (8.5) is still a concave fractional program. However, if f_1 is also a (nonlinear) concave function, then (8.5) is no longer a concave fractional program. Obviously a quadratic program (8.4) reduces to a quadratic fractional program. For more details on (8.4) and (8.5) see [65, 71].

Stochastic programs (8.2) and (8.4) are met in a wide variety of planning problems. Whenever the maximum probability model is used as a deterministic equivalent, such decision problems lead to a fractional program of one type or another. Hence fractional programs are encountered indirectly in many different applications of mathematical programming, although initially the objective function is not a ratio.

4. Applications of Min-Max Fractional Programs (P_2).

In mathematical economics the multi-ratio fractional program (P_2) arises when the growthrate of an expanding economy is defined as follows (cf. [74]):

$$\text{growthrate} = \max_x \left(\min_{1 \le i \le m} \frac{\text{output}_i(x)}{\text{input}_i(x)} \right) \tag{8.6}$$

where x denotes a feasible production plan of the economy.

In management science simultaneous maximization of rates such as those discussed in the previous section can also lead to a multi-ratio fractional program. This is the case if either in a worst-case approach the model

$$\min_{1 \le i \le m} \frac{f_i(x)}{g_i(x)} \to \sup \tag{8.7}$$

is used or with the help of prescribed ratio goals r_i the model

$$\max_{1 \leq i \leq m} |\frac{f_i(x)}{g_i(x)} - r_i| \to \inf \tag{8.8}$$

is employed. Examples of the second approach are found in financial planning with different fractional goals or in the allocation of funds under equity considerations. Financial planning with fractional goals is discussed in [38]. Furthermore, multi-facility location-queueing problems giving rise to (P_2) are introduced in [5].

A third area of application of min-max fractional programs is numerical mathematics (cf. [41]). Given the values F_i of a function $F(t)$ in finitely many points t_i of an interval for which an approximating ratio of two polynomials $N(t, x_1)$ and $D(t, x_2)$ with coefficient vectors x_1, x_2 is sought. If the best approximation is defined in the sense of the L_∞-norm, then the following problem is to be solved:

$$\max_i |\frac{N(t_i, x_1)}{D(t_i, x_2)} - F_i| \to \inf \tag{8.9}$$

with variables x_1, x_2.

At the end of this section on applications of (P_2) we point out that in case of infinitely many ratios (P_2) is related to a fractional semi-infinite program (cf. [41]). Several applications in engineering give rise to such a problem when a lower bound for the smallest eigenvalue of an elliptical differential operator is to be determined (cf. [40]).

For further applications of (P_2) we refer to the very recent survey [59].

5. Sum-of-Ratios Fractional Programs (P_3).

Problem (P_3) arises naturally in decision making when several rates are to be optimized simultaneously and a compromise is sought which optimizes a weighted sum of these rates. In light of the applications of single-ratio fractional programming numerators and denominators may be representing output, input, profit, cost, capital, risk or time, for example. A multitude of applications of the sum-of-ratios problem can be envisioned in this way. Included is the case where some of the ratios are not proper quotients. This describes situations where a compromise is sought between absolute and relative terms like profit and return on investment (profit/capital) or return and return/risk, for example.

Almogy and Levin (cf. [1]) analyze a multistage stochastic shipping problem. A deterministic equivalent of this stochastic problem is formulated which turns out to be a sum-of-ratios problem.

Rao (cf. [57]) discusses various models in cluster analysis. The problem of optimal partitioning of a given set of entities into a number of mutually exclusive and exhaustive groups (clusters) gives rise to various mathematical programming problems depending on which optimality criterion is used. If the objective is to minimize the sum of the squared distances within groups, then a minimum of a sum of ratios is to be determined.

The minimization of the mean response time in queueing-location problems gives rise to (P_3) as well, as shown by Drezner et al. (cf. [23]); see also [75].

Furthermore we mention an inventory model analyzed in [63] which is designed to determine simultaneously optimal lot sizes and an optimal storage allocation in a warehouse. The total cost to be minimized is the sum of fixed cost per unit, storage cost per unit and material handling cost per unit.

In [46] Konno and Inori formulate a bond portfolio optimization problem as a sum-of-ratios problem.

More recently other applications of the sum-of-ratios problem have been identified. Mathis and Mathis [51] formulate a hospital fee optimization problem in this way. The model is used by hospital administrators in the State of Texas to decide on relative increases of charges for different medical procedures in various departments.

According to [20] a number of geometric optimization problems give rise to the sum-of-ratios problem. These often occur in layered manufacturing, for instance in material layout and cloth manufacturing. Quite in contrast to other applications of the sum-of-ratios problem mentioned before, the number of variables is very small (one, two or three), but the number of ratios is large; often there are hundreds or even thousands of ratios involved.

Our current understanding of the structural properties of the sum-of-ratios problem is rather limited. In [36] Freund and Jarre showed that this problem is essentially NP-hard, even in the case of one concave ratio and a concave function. Hence (P_3) is a global optimization problem in contrast to (P_1) and (P_2).

Given the small theoretical basis, it is not surprising that algorithmic advances have been rather limited too. However in recent years some progress has been made. Some of the proposed algorithms have been computationally tested. Typically execution times grow very rapidly in the number of ratios. At this time problems up to about ten ratios can be handled. We refer to the algorithms by Konno and Fukaishi (cf. [45]) (see also [44]) and by Kuno (cf. [47]). The former is superior to several earlier methods (cf. [45]) while the latter is seemingly faster than the former. Clearly a more thorough testing of various proposed algorithms is needed before further conclusions can be drawn. Also some of the applications call for methods which can handle a large number of ratios; e.g., fifty (cf. [1]). Currently such methods are not available.

For a special class of sum-of-ratios problems with up to about one thousand ratios, but only very few variables an algorithm is given in [20]. This method by Chen et al. is superior to the other algorithms on the particular class of problems in manufacturing These are geometric optimization problems arising in layered manufacturing. In contrast to general-purpose algorithms for (P_3), the method in [20] is rather robust with regard to the number of ratios.

Focus of the remainder of this review of fractional programming will be the min-max fractional program (P). It includes as special cases (P_1) and (P_2). For a very recent survey of applications, theoretical results and solution methods for (P_1) and (P_2) since [61] was published we refer to [59]. A corresponding survey for (P_3) since [61] appeared is given in [60]. For a survey of some recent developments for multi-objective fractional programs (P_4) we refer to [33].

6. Analysis of Min-Max Fractional Programs.

In this section we will analyze min-max fractional programs by means of a parametric approach. Although other approaches are also available, this one makes it possible to derive duality results for the (primal) min-max fractional program (P) and at the same time to construct an algorithm which solves problem (P). As already mentioned in Section 2, let $B \subseteq \mathbb{R}^n$ and $A \subseteq \mathbb{R}^m$ be some nonempty closed sets and $f : \mathbb{R}^{m+n} \to [-\infty, \infty]$ a finite-valued function on $A \times B$. Moreover, consider the function $g : \mathbb{R}^{m+n} \to [-\infty, \infty]$ which is a finite-valued positive function on $A \times B$. For the related functions $g_{\inf} : \mathbb{R}^n \to [-\infty, \infty]$ and $g_{\sup} : \mathbb{R}^n \to [-\infty, \infty]$ given by

$$g_{\inf}(x) := \inf_{y \in A} g(y, x) \text{ and } g_{\sup}(x) := \sup_{y \in A} g(y, x)$$

we assume, unless stated otherwise, that the following condition holds.

Condition 8.1 *For every $x \in B$ we have $0 < g_{\inf}(x) \leq g_{\sup}(x) < \infty$.*

For every $x \in B$ we now consider the single-ratio fractional program

$$\lambda_*(x) := \sup_{y \in A} \frac{f(y, x)}{g(y, x)}. \qquad (P^x)$$

This optimization problem is well-defined and its objective function value satisfies $-\infty < \lambda_*(x) \leq \infty$. A more complicated optimization problem is given by the already introduced (primal) min-max fractional program

$$\lambda_* := \inf_{x \in B} \sup_{y \in A} \frac{f(y,x)}{g(y,x)}. \qquad (P)$$

Clearly $-\infty \leq \lambda_* = \inf_{x \in B} \lambda_*(x) \leq \infty$. It is not assumed beforehand that the optimization problems (P) and (P^x) have an optimal solution. Therefore we cannot replace sup by max or inf by min. The simpler optimization problem (P^x) is introduced since it will be part of the so-called primal Dinkelbach-type approach discussed in subsection 6.2 to solve the (primal) min-max fractional program (P).

Another optimization problem is to consider for every $y \in A$ the single-ratio fractional program

$$\mu_*(y) := \inf_{x \in B} \frac{f(y,x)}{g(y,x)}. \qquad (D^y)$$

Also this problem is well-defined and it satisfies $-\infty \leq \mu_*(y) < \infty$. Clearly for every $y \in A$ we obtain $\mu_*(y) \leq \lambda_*$. Similarly as for the (primal) min-max fractional program we introduce the more complicated optimization problem

$$\mu_* := \sup_{y \in A} \inf_{x \in B} \frac{f(y,x)}{g(y,x)}. \qquad (D)$$

This problem is called a *(dual) max-min fractional program.* Clearly its optimal objective function value μ_* satisfies $\mu_* \leq \lambda_*$. Like for the (primal) min-max fractional program we introduce the functions $\underline{g}_{\text{inf}} : \mathbb{R}^m \to [-\infty, \infty]$ and $\overline{g}_{\text{sup}} : \mathbb{R}^m \to [-\infty, \infty]$ given by

$$\underline{g}_{\text{inf}}(y) := \inf_{x \in B} g(y,x) \text{ and } \overline{g}_{\text{sup}}(y) := \sup_{x \in B} g(y,x).$$

Analyzing the so-called dual Dinkelbach-type approach to solve problem (D), we need the following counterpart of Condition 8.1.

Condition 8.2 *For every $y \in A$ we have $0 < \underline{g}_{\text{inf}}(y) \leq \overline{g}_{\text{sup}}(y) < \infty$.*

The simpler optimization problem (D^y) is introduced since it will be part of the dual Dinkelbach-type approach discussed in subsection 6.4 to solve the (dual) max-min fractional program (D). If we consider a single-ratio fractional program, A consists of one element and the optimization problems (P) and (D) are identical. For a classical multi-ratio fractional program (generalized fractional program) A is a finite set consisting of more than one element; hence optimization problems (P) and (D) are different from each other. If programs (P) and (D) are different and additionally $\mu_* = \lambda_*$, both the primal and dual Dinkelbach-type approach can be used to solve optimization problem (P). As already observed before, many results (cf. [4, 5, 21]) were derived for generalized fractional programs. In this section we will consider the more general (primal) min-max and (dual) max-min fractional program and derive similar structural properties for this problem as it was done for the more specialized primal and dual generalized fractional program before.

We selected these more general optimization problems not often considered in the fractional programming literature since one can use similar parametric techniques as for generalized fractional programs and at the same time unify the existing theory for single-ratio and multi-ratio fractional programs. Using the parametric approach one can reduce the max-min and min-max fractional program to so-called (semi-infinite) max-min and min-max programs. Unfortunately, solving these semi-infinite optimization problems efficiently on a computer is very difficult. For an extensive discussion of some of the used procedures the reader should consult [55]. However for special cases there is still room for improvement, and this seems to be a new area of research (cf. [15]). In the theoretical analysis of max-min and min-max fractional programs it will turn out that convexity plays a major role, not only in establishing the equality $\lambda_* = \mu_*$ (a so-called strong duality result), but also in the rate of convergence analysis for the primal and dual Dinkelbach-type parametric approach. Due to symmetry arguments similar type of convergence results hold for these two algorithms.

In case we analyze the primal Dinkelbach-type approach, not all the results are valid under Condition 8.1, and we sometimes need the following stronger condition.

Condition 8.3 *The set $A \subseteq \mathbb{R}^m$ is compact, the function g is positive on $A \times B$ and for every $x \in B$ the functions $y \to f(y,x)$ and $y \to g(y,x)$ are finite-valued and continuous on some open set $U \subseteq \mathbb{R}^m$ containing A.*

If Condition 8.3 holds, then it follows from Corollary 1.2 of [3] that

$$0 < g_{\inf}(x) \leq g_{\sup}(x) < \infty$$

for every $x \in B$, and so this condition implies Condition 8.1. Moreover, the single-ratio fractional program (P^x) has an optimal solution and $\lambda_*(x)$ is finite for every $x \in B$.

In case we also analyze the dual Dinkelbach-type approach, not all results are valid under Condition 8.2, and so we sometimes need the following counterpart of Condition 8.3.

Condition 8.4 *The set $B \subseteq \mathbb{R}^n$ is compact, the function g is positive on $A \times B$ and for every $y \in A$ the functions $x \to f(y,x)$ and $x \to g(y,x)$ are finite-valued and continuous on some open set $V \subseteq \mathbb{R}^n$ containing B.*

Again, if Condition 8.4 holds, it follows from Corollary 1.2 of [3] that

$$0 < \underline{g}_{\inf}(y) \leq \overline{g}_{\sup}(y) < \infty$$

for every $y \in A$, and so this condition implies Condition 8.2. Moreover, the single-ratio fractional program (D^y) has an optimal solution, and $\mu_*(y)$ is finite for every $y \in A$.

Before analyzing in the next subsection the parametric approach applied to (P), we will derive an alternative representation of a generalized fractional program. This alternative representation satisfies automatically Condition 8.3. For a generalized fractional program the set A is given by $\{1, ..., m\}, m < \infty$, and the functions f and g are replaced by the functions $f_i : B \to \mathbb{R}, i \in A$ and $g_i : B \to \mathbb{R}, i \in A$. This means

$$\sup_{y \in A} \frac{f(y,x)}{g(y,x)} = \max_{1 \leq i \leq m} \frac{f_i(x)}{g_i(x)} = \lambda_*(x).$$

In this case the optimization problem (P^x) can be solved trivially.

To obtain a different representation of a generalized fractional program, we introduce the unit simplex

$$\Delta_m := \{y \in \mathbb{R}^m : \sum_{i=1}^{m} y_i = 1, y_i \geq 0, 1 \leq i \leq m\}.$$

If the vector b belongs to $\mathbb{R}^m_{++}$, the strictly positive orthant of $\mathbb{R}^m$, it is well-known (cf. [4]) that the function $h : \Delta_m \to \mathbb{R}$ given by $h(y) := (y^\top b)^{-1} y^\top a$ is quasiconvex on Δ_m for every $a \in \mathbb{R}^m$. By Condition 8.1 it follows for $g : \mathbb{R}^n \to \mathbb{R}^m$ given by $g(x) := (g_1(x), ..., g_m(x))^\top$ that $g(x) \in \mathbb{R}^m_{++}$ for every $x \in B$. Then for $f : \mathbb{R}^n \to \mathbb{R}^m$ given by $f(x) = (f_1(x), .., f_m(x))^\top$ we have

$$\max_{i \in A} \frac{f_i(x)}{g_i(x)} = \max_{y \in \Delta_m} \frac{y^\top f(x)}{y^\top g(x)} \tag{8.10}$$

for every $x \in B$. Applying relation (8.10) yields

$$\inf_{x \in B} \max_{1 \leq i \leq m} \frac{f_i(x)}{g_i(x)} = \inf_{x \in B} \max_{y \in \Delta_m} \frac{y^\top f(x)}{y^\top g(x)}. \tag{8.11}$$

With this we have found another representation of a generalized fractional program. Using this representation, the corresponding (dual) generalized fractional program is given by

$$\sup_{y \in \Delta_m} \inf_{x \in B} \frac{y^\top f(x)}{y^\top g(x)}.$$

In subsection 6.3 we will give sufficient conditions to guarantee that the primal and dual optimal objective function values coincide. However before discussing this, we will first consider in the next subsection the so-called primal parametric approach for solving the (primal) min-max fractional program (P).

6.1 The Primal Parametric Approach.

To analyze the properties of the (primal) min-max fractional program (P) and at the same time construct some generic algorithm to solve this problem we introduce the function $p : \mathbb{R} \times A \times B \to \mathbb{R}$ given by

$$p(\lambda, y, x) := f(y, x) - \lambda g(y, x)$$

and consider for every $(\lambda, x) \in \mathbb{R} \times B$ the optimization problem

$$p_1(\lambda, x) := \sup_{y \in A} p(\lambda, y, x). \qquad (P_\lambda^x)$$

For every $x \in B$ the function $p_{1,x} : \mathbb{R} \to (-\infty, \infty]$ is now given by

$$p_{1,x}(\lambda) := p_1(\lambda, x). \qquad (8.12)$$

Since $g > 0$ on $A \times B$ and $p_{1,x}$ is the supremum of affine functions, it is obvious that $p_{1,x}$ is a decreasing lower semicontinuous convex function. Its so-called effective domain $dom(p_{1,x})$ is defined by (cf. [58])

$$dom(p_{1,x}) := \{\lambda \in \mathbb{R} : p_{1,x}(\lambda) < \infty\} \subseteq \mathbb{R}.$$

By the finiteness of p on $\mathbb{R} \times A \times B$ it is obvious that for every $x \in B$ $dom(p_{1,x}) = \{\lambda \in \mathbb{R} : p_{1,x}(\lambda) \text{ finite}\}$. A more difficult optimization problem than (P_λ^x) is the parametric min-max optimization problem

$$p_2(\lambda) := \inf_{x \in B} p_1(\lambda, x). \qquad (P_\lambda)$$

For this function it holds that $-\infty \leq p_2(\lambda) \leq \infty$ for every $\lambda \in \mathbb{R}$. For the function p_2 the so-called effective domain $dom(p_2)$ is given by

$$dom(p_2) := \{\lambda \in \mathbb{R} : p_2(\lambda) < \infty\} \subseteq \mathbb{R}.$$

By the definition of the functions p_2 and $p_{1,x}$ it is easy to verify that

$$dom(p_2) = \cup_{x \in B} dom(p_{1,x})$$

In the next result we identify for $\lambda_* < \infty$ and $\lambda_*(x) < \infty$ the effective domains of the functions p_2 and $p_{1,x}$.

Lemma 8.1 *Assume Condition 8.1 holds. Then* $\lambda_* < \infty$ *if and only if* $dom(p_2) = \mathbb{R}$, *and* $\lambda_*(x)$ *is finite if and only* $dom(p_{1,x}) = \mathbb{R}$.

Proof. Assume $\lambda_* < \infty$. Suppose by contradiction that there exists some $\lambda \in \mathbb{R}$ satisfying $p_2(\lambda) = \infty$. This implies for every $x \in B$ that $p_1(\lambda, x) = \infty$. Hence for a given $x \in B$ one can find some sequence $\{y_i : i \in \mathbb{N}\} \subseteq A$ satisfying

$$i \leq (\frac{f(y_i, x)}{g(y_i, x)} - \lambda)g(y_i, x) \leq (\frac{f(y_i, x)}{g(y_i, x)} - \lambda)g_{\text{sup}}(x). \tag{8.13}$$

Since $g_{\text{sup}}(x) < \infty$ and λ is finite, we obtain by relation (8.13) that $\lambda_*(x) = \infty$ for every $x \in B$ yielding $\lambda_* = \infty$ which contradicts our assumption.

Conversely, if $dom(p_2) = \mathbb{R}$, then clearly $0 \in dom(p_2)$, and so there exists some $x_0 \in B$ satisfying $\sup_{y \in A} f(y, x_0) < \infty$. Due to $g_{\text{inf}}(x_0) > 0$ it is easy to see that $\lambda_*(x_0) < \infty$, and so $\lambda_* < \infty$ which completes the proof of the first part. By identifying B with $\{x\}$, the second part follows immediately from the first part. □

Using similar algebraic manipulations as in [22] applied to a generalized fractional program one can show the following important result for the optimal value function p_2 of a parametric min-max problem (P_λ). The validity of the so-called parametric approach to solve problem (P) is based on this result.

Theorem 8.1 *Assume Condition 8.1 holds and* $\lambda_* < \infty$. *Then* $\lambda_* < \lambda < \infty$ *if and only if* $p_2(\lambda) < 0$. *Moreover, if* $\lambda_*(x) < \infty$, *then* $\lambda_*(x) < \lambda < \infty$ *if and only if* $p_1(\lambda, x) < 0$.

Proof. If $\lambda_* < \infty$ and $\lambda > \lambda_* = \inf_{x \in B} \lambda_*(x)$, then there exist some $x_0 \in B$ and $\epsilon > 0$ satisfying

$$\lambda > \lambda_*(x_0) + \epsilon \geq \frac{f(y, x_0)}{g(y, x_0)} + \epsilon$$

for every $y \in A$. Since $g_{\text{inf}}(x_0) > 0$, this yields

$$f(y, x_0) - \lambda g(y, x_0) \leq -\epsilon g(y, x_0) \leq -\epsilon g_{\text{inf}}(x_0)$$

for every $y \in A$. It follows that

$$p_2(\lambda) \leq p_1(\lambda, x_0) \leq -\epsilon g_{\text{inf}}(x_0) < 0.$$

Conversely, if $p_2(\lambda) < 0$, then there exist some $\epsilon > 0$ and $x_0 \in B$ satisfying $p_1(\lambda, x_0) \leq -\epsilon$. This implies $f(y, x_0) - \lambda g(y, x_0) \leq -\epsilon$ for every $y \in A$, and we obtain for every $y \in A$ that

$$\frac{f(y,x_0)}{g(y,x_0)} \le \lambda - \frac{\epsilon}{g(y,x_0)} \le \lambda - \frac{\epsilon}{g_{\sup}(x_0)}. \tag{8.14}$$

Since $g_{\sup}(x_0) < \infty$, it follows from relation (8.14) that $\lambda_* \le \lambda_*(x_0) < \lambda$, and the proof of the first part is completed. By identifying B with $\{x\}$ the second part follows from the first part. □

A useful implication of Theorem 8.1 is given by the following result.

Lemma 8.2 *Assume Condition 8.1 holds and $\lambda_*(x) < \infty$ for some $x \in B$. Then $p_1(\lambda_*(x), x) = 0$.*

Proof. By the definition of $\lambda_*(x)$ we obtain $f(y,x) - \lambda_*(x)g(y,x) \le 0$ for every $y \in A$. This implies $p_1(\lambda_*(x), x) \le 0$. From Theorem 8.1 it follows that $p_1(\lambda_*(x), x) \ge 0$, and this shows the desired result. □

If Condition 8.1 holds and $\lambda_* < \infty$, we obtain from Theorem 8.1 and Lemma 8.1 that $p_2(\lambda_*) \ge 0$, and $p_2(\lambda)$ is finite for every $\lambda \le \lambda_*$. In case we only assume that g is positive on $A \times B$ it is easy to verify that $p_2(\lambda) \le 0$ for every $\lambda > \lambda_*$, and $p_2(\lambda) < 0$ implies $\lambda > \lambda_*$. However as shown by the following single-ratio fractional program satisfying Condition 8.1 and $\lambda_* = 1$, it may happen that $p_2(\lambda) = -\infty$ for every $\lambda > \lambda_*$ and $p_2(\lambda_*) \ne 0$ (cf. [22]).

Example 8.1 For $A = \{1\}$, $f_1(x) = x + 1$, $g_1(x) = x$ and $B = \{x \in \mathbb{R} : x \ge 1\}$ it follows that the optimization problem (P) reduces to $inf_{x \in B} \frac{x+1}{x}$, and so $\lambda_* = 1$. Also $0 < g_{\inf}(x) = g_{\sup}(x) = x < \infty$ for every $x \in B$ and $p_2(\lambda_*) = \inf_{x \in B}\{x + 1 - x\} = 1$. Moreover, the optimal solution set of the optimization problem (P_{λ_*}) equals B, and $p_2(\lambda) = -\infty$ for every $\lambda > 1$.

To derive some other properties of the so-called parametric approach we need to investigate in detail the functions p_2 and $p_{1,x}$. We first observe that the positivity of the function g on $A \times B$ implies that the functions p_2 and $p_{1,x}, x \in B$, are decreasing. In the next result it is shown that the decreasing function p_2 is upper semicontinuous.

Theorem 8.2 *Assume Condition 8.1 holds. Then the function $p_2 : \mathbb{R} \to [-\infty, \infty]$ is upper semicontinuous.*

Proof. To prove that the function p_2 is upper semicontinuous, let $\alpha \in \mathbb{R}$ and consider the upper level set $U(p_2, \alpha) := \{\lambda \in \mathbb{R} : p_2(\lambda) \ge \alpha\}$. If $U(p_2, \alpha) = \varnothing$, then this set is closed. So we assume that $U(p_2, \alpha) \ne$

$\emptyset$. To show that this set is closed consider some accumulation point $\lambda_\infty \in \mathbb{R}$ of the set $U(p_2, \alpha)$. Hence there exists some sequence $\{\lambda_i : i \in \mathbb{N}\} \subseteq U(p_2, \alpha)$ satisfying $\lim_{i\uparrow\infty} \lambda_i = \lambda_\infty$. If for some $i \in \mathbb{N}$ it holds that $\lambda_i \geq \lambda_\infty$, then by the monotonicity of the function p_2 we obtain $p_2(\lambda_\infty) \geq p_2(\lambda_i) \geq \alpha$, and so $\lambda_\infty \in U(p_2, \alpha)$. Therefore we may assume without loss of generality that $\lambda_i < \lambda_\infty$ for every $i \in \mathbb{N}$. Observe now for every $x \in B$ and $i \in \mathbb{N}$ that

$$p_1(\lambda_\infty, x) \geq p(\lambda_i, y, x) + (\lambda_i - \lambda_\infty) g(y, x)$$

for every $y \in A$. This implies using $\lambda_i < \lambda_\infty$ and $g > 0$ that

$$p_1(\lambda_\infty, x) \geq p(\lambda_i, y, x) + (\lambda_i - \lambda_\infty) g_{\text{sup}}(x)$$

for every $y \in A$, and hence

$$p_1(\lambda_\infty, x) \geq p_1(\lambda_i, x) + (\lambda_i - \lambda_\infty) g_{\text{sup}}(x). \tag{8.15}$$

Since $\lambda_i \in U(p_2, \alpha)$, we obtain for every $x \in B$ that $p_1(\lambda_i, x) \geq \alpha$. By relation (8.15), $\lim_{i\uparrow\infty} \lambda_i = \lambda_\infty$ and $0 < g_{\text{sup}}(x) < \infty$ this yields for every $x \in B$ that $p_1(\lambda_\infty, x) \geq \alpha$. Hence $p_2(\lambda_\infty) \geq \alpha$, and so $\lambda_\infty \in U(p_2, \alpha)$. Applying Theorem 1.7 of [29] yields that p_2 is upper semicontinuous. □

By Theorem 8.2 and Lemma 1.30 of [29] we obtain

$$\lim_{s\uparrow\lambda} p_2(s) = \limsup_{s\uparrow\lambda} p_2(s) \leq p_2(\lambda).$$

Since for every $s < \lambda$ we know that $p_2(s) \geq p_2(\lambda)$, this yields

$$\lim_{s\uparrow\lambda} p_2(s) = p_2(\lambda).$$

Again by the monotonicity of p_2 it follows that $\lim_{s\downarrow\lambda} p_2(s)$ exists. But this limit might not be equal to $p_2(\lambda)$. Therefore the function p_2 is left-continuous with right-hand limits.

An important consequence of Theorem 8.2 is given by the next result. To show this result we first introduce a so-called set-valued mapping $S : X \to 2^Y$ (cf. [3]) with 2^Y denoting the set of all subsets of the nonempty set $Y \subseteq \mathbb{R}^m$ and X a nonempty closed subset of $\mathbb{R}^n$. If $S : X \to 2^Y$ is a set-valued mapping, it is always assumed that $S(x) \subseteq Y$ is nonempty for every $x \in X$. The graph of a set-valued mapping $S : X \to 2^Y$ is given by

$$\text{graph}(S) = \{(x, y) \in X \times Y : y \in S(x)\}.$$

An important subclass of set-valued mappings is introduced in the next definition (cf. [11]).

Definition 8.1 *The set-valued mapping $S : X \to 2^Y$ where X is a closed set is called closed if its graph is a closed set.*

By the definition of a closed set it is immediately clear that the set-valued mapping $S : X \to 2^Y$ is closed if and only if for any sequence $\{x_k : k \in \mathbb{N}\} \subseteq X$ and $y_k \in S(x_k)$, $k \in \mathbb{N}$ it follows that

$$\lim_{k\uparrow\infty} x_k = x_\infty \text{ and } \lim_{k\uparrow\infty} y_k = y_\infty \Rightarrow y_\infty \in S(x_\infty).$$

Examples of set-valued mappings occurring in min-max optimization are the set-valued mappings $S_{p_1} : \mathbb{R} \times B \to 2^A$ and $S_{p_2} : \mathbb{R} \to 2^B$ given by

$$S_{p_1}(\lambda, x) := \{y \in A : p_1(\lambda, x) = p(\lambda, y, x)\} \tag{8.16}$$

and

$$S_{p_2}(\lambda) := \{x \in B : p_2(\lambda) = p_1(\lambda, x)\}. \tag{8.17}$$

The set $S_{p_1}(\lambda, x)$ represents the set of optimal solutions of the optimization problem (P_λ^x), while the set $S_{p_2}(\lambda)$ denotes the set of optimal solutions in B of the optimization problem (P_λ). Also we consider the set-valued mapping $S_p : \mathbb{R} \to 2^{A\times B}$ given by

$$S_p(\lambda) := \{(y, x) \in A \times B : p_2(\lambda) = p_1(\lambda, x) = p(\lambda, y, x)\}. \tag{8.18}$$

This set represents the set of optimal solutions of the optimization problem (P_λ). For the above set-valued mappings one can show the following result. It is always assumed in the next result that the sets $S_{p_1}(\lambda, x), S_{p_2}(\lambda)$ and $S_p(\lambda)$ are nonempty on their domain.

Lemma 8.3 *Assume Condition 8.1 holds and the functions f and g are finite-valued and continuous on some open set $W \subseteq \mathbb{R}^{m+n}$ containing $A \times B$. Then the set-valued mappings S_{p_1}, S_{p_2} and S_p are closed.*

Proof. We first show that the set-valued mapping S_{p_1} is closed. To start with this, consider some sequence $\{(\lambda_k, y_k, x_k) : y_k \in S_{p_1}(\lambda_k, x_k)\}_{k\in\mathbb{N}}$ satisfying $\lim_{k\uparrow\infty} \lambda_k = \lambda_\infty \in \mathbb{R}$, $\lim_{k\uparrow\infty} x_k = x_\infty$ and $\lim_{k\uparrow\infty} y_k = y_\infty$. Since A and B are closed sets, this yields $x_\infty \in B$ and $y_\infty \in A$ and by the definition of p_1 we obtain

$$p(\lambda_\infty, y_\infty, x_\infty) \leq p_1(\lambda_\infty, x_\infty). \tag{8.19}$$

Since the function p is continuous on $\mathbb{R} \times A \times B$, it is easy to verify using Theorem 1.7 of [29] that the function p_1 is lower semicontinuous on $\mathbb{R} \times B$. Using this together with Lemma 1.30 of [29] and $p_1(\lambda_k, x_k) = p(\lambda_k, y_k, x_k)$ we obtain

$$p(\lambda_\infty, y_\infty, x_\infty) = \liminf_{k\uparrow\infty} p_1(\lambda_k, x_k) \geq p_1(\lambda_\infty, x_\infty).$$

Then by relation (8.19) it follows that $y_\infty \in S_{p_1}(\lambda_\infty, x_\infty)$. This shows that the set S_{p_1} is closed. To prove that the set-valued mapping S_{p_2} is closed we consider some sequence $\{(\lambda_k, x_k) : x_k \in S_{p_2}(x_k)\}_{k\in\mathbb{N}}$ satisfying $\lim_{k\uparrow\infty} \lambda_k = \lambda_\infty \in \mathbb{R}$ and $\lim_{k\uparrow\infty} x_k = x_\infty$. By Theorem 8.2 and Lemma 1.30 of [29] we obtain

$$p_2(\lambda_\infty) \geq \limsup_{k\uparrow\infty} p_2(\lambda_k). \tag{8.20}$$

Since p_1 is lower semicontinuous on $\mathbb{R} \times B$, it follows that

$$\limsup_{k\uparrow\infty} p_2(\lambda_k) \geq \liminf_{k\uparrow\infty} p_1(\lambda_k, x_k) \geq p_1(\lambda_\infty, x_\infty).$$

Hence by relation (8.20) we obtain

$$p_2(\lambda_\infty) \geq p_1(\lambda_\infty, x_\infty).$$

Using $x_\infty \in B$ this shows that $x_\infty \in S_{p_2}(\lambda_\infty)$. Hence we have verified that S_{p_2} is closed.

Finally, to show that S_p is closed, consider a sequence $\{(\lambda_k, y_k, x_k) : (y_k, x_k) \in S_p(\lambda_k)\}_{k\in\mathbb{N}}$ satisfying $\lim_{k\uparrow\infty} \lambda_k = \lambda_\infty \in \mathbb{R}$, $\lim_{k\uparrow\infty} x_k = x_\infty$ and $\lim_{k\uparrow\infty} y_k = y_\infty$. Since $y_k \in S_{p_1}(\lambda_k, x_k)$, it follows that $y_\infty \in S_{p_1}(\lambda_\infty, x_\infty)$ using the fact that S_{p_1} is closed. This shows $p(\lambda_\infty, y_\infty, x_\infty) = p_1(\lambda_\infty, x_\infty)$. Moreover, since $x_k \in S_{p_2}(\lambda_k)$, we obtain $x_\infty \in S_{p_2}(\lambda_\infty)$ using the fact that S_{p_2} is closed. Hence $p_1(\lambda_\infty, x_\infty) = p_2(\lambda_\infty)$. Therefore (y_∞, x_∞) is an optimal solution of the min-max fractional program (P). This completes the proof. □

We will now consider for every $x \in B$ the decreasing convex function $p_{1,x} : \mathbb{R} \to \mathbb{R}$, introduced in relation (8.12). In the next result it is shown for $\lambda_*(x)$ finite that this function is Lipschitz continuous with Lipschitz constant $g_{\sup}(x)$.

Lemma 8.4 *Assume Condition 8.1 holds and $\lambda_*(x)$ is finite for $x \in B$. Then the function $p_{1,x} : \mathbb{R} \to (-\infty, \infty)$ is strictly decreasing and Lipschitz continuous with Lipschitz constant $g_{\sup}(x)$ and this function satisfies $\lim_{\lambda\uparrow\infty} p_{1,x}(\lambda) = -\infty$ and $\lim_{\lambda\downarrow-\infty} p_{1,x}(\lambda) = \infty$.*

Proof. If $\lambda_*(x)$ is finite for some $x \in B$, then we know by Lemma 8.1 that $p_{1,x}(\lambda)$ is finite for every $\lambda \in \mathbb{R}$. Selecting some $\mu \in \mathbb{R}$, using $g_{\sup}(x) < \infty$ and the fact that $p_{1,x}(\mu)$ is finite, it is easy to verify that

$$|p_{1,x}(\lambda) - p_{1,x}(\mu)| \leq g_{\sup}(x)|\lambda - \mu| \tag{8.21}$$

for every $\lambda \in \mathbb{R}$. Hence $p_{1,x}$ is a Lipschitz continuous convex function with Lipschitz constant $g_{\sup}(x) < \infty$. Also it is easy to verify using $g_{\inf}(x) > 0$ that

$$p_{1,x}(\lambda) - p_{1,x}(\mu) \geq (\mu - \lambda)g_{\inf}(x) \tag{8.22}$$

for every $\lambda < \mu$. This shows that $p_{1,x}$ is strictly decreasing on $\mathbb{R}$. Again by relation (8.22) we obtain for a given μ and $\lambda \downarrow -\infty$ that $\lim_{\lambda\downarrow-\infty} p_{1,x}(\lambda) = \infty$ and for a given λ and $\mu \uparrow \infty$ that $\lim_{\mu\uparrow\infty} p_{1,x}(\mu) = -\infty$. □

If $\lambda_*(x)$ is finite, it follows from Lemma 8.4 and Theorem 1.13 of [29] that the finite-valued convex function $p_{1,x}$ has a nonempty subgradient set $\partial p_{1,x}(\lambda)$ for every $\lambda \in \mathbb{R}$. Hence for every $a \in \partial p_{1,x}(\lambda)$ and $\mu, \lambda \in \mathbb{R}$ the subgradient inequality

$$p_{1,x}(\mu) \geq p_{1,x}(\lambda) + a(\mu - \lambda)$$

holds. Applying relation (8.21) and the fact that $p_{1,x}$ is strictly decreasing we obtain

$$g_{\text{sup}}(x) \geq p_{1,x}(\lambda - 1) - p_{1,x}(\lambda) \geq -a \tag{8.23}$$

for every $a \in \partial p_{1,x}(\lambda)$. Furthermore, applying relation (8.22) yields

$$-g_{\text{inf}}(x) \geq p_{1,x}(\lambda + 1) - p_{1,x}(\lambda) \geq a \tag{8.24}$$

for every $a \in \partial p_{1,x}(\lambda)$. Hence by relations (8.23) and (8.24) it follows that

$$\partial p_{1,x}(\lambda) \subseteq [-g_{\text{sup}}(x), -g_{\text{inf}}(x)]. \tag{8.25}$$

To give a more detailed representation of the subgradient set $\partial p_{1,x}(\mu)$ it is convenient to assume that the set $S_{p_1}(\lambda, x)$ introduced in relation (8.16) is nonempty. As already observed, this set represents the set of optimal solutions of the parametric problem (P^x_λ). It is easy to see that $-g(y, x) \in \partial p_{1,x}(\lambda)$ for every $y \in S_{p_1}(\lambda, x)$. Since $\partial p_{1,x}(\lambda)$ is a closed convex set, this implies

$$[-\sup_{y\in S_{p_1}(\lambda,x)} g(y, x), -\inf_{y\in S_{p_1}(\lambda,x)} g(y, x)] \subseteq \partial p_{1,x}(\lambda). \tag{8.26}$$

Although it is possible for a finite $\lambda_*(x)$ to give a complete representation of the subgradient set $\partial p_{1,x}(\lambda)$ for every $\lambda \in \mathbb{R}$, we only consider the following important subcase.

Lemma 8.5 *Assume Condition 8.3 holds. Then it follows for every $x \in B$ that $\lambda_*(x)$ is finite, $S_{p_1}(\lambda, x)$ is a nonempty compact set for every $(\lambda, x) \in \mathbb{R} \times B$ and*

$$\partial p_{1,x}(\lambda) = [-\max_{y\in S_{p_1}(\lambda,x)} g(y, x), -\min_{y\in S_{p_1}(\lambda,x)} g(y, x)].$$

Also for every $a_\lambda \in \partial p_{1,x}(\lambda)$ and $a_\mu \in \partial p_{1,x}(\mu)$ and $\lambda > \mu$ it holds that $0 > a_\lambda \geq a_\mu$.

Proof. Since the functions $y \to f(y,x)$ and $y \to g(y,x)$ are continuous, $g > 0$ on $A \times B$ and A is compact, we obtain that $\lambda_*(x)$ is finite. By the same argument it also follows that $S_{p_1}(\lambda, x)$ is nonempty for every $\lambda \in \mathbb{R}$. Also by the continuity of the function $y \to f(y,x) - \lambda g(y,x)$ the set $S_{p_1}(\lambda, x) \subseteq A$ is closed and hence compact. Using now the proof of Lemma 3.2 in [8] and the fact that $S_{p_1}(\lambda, x)$ is a compact set yields the desired representation of the subgradient set $\partial p_{1,x}(\lambda)$. To show the last part we observe by the subgradient inequality that $p_{1,x}(\mu) \geq p_{1,x}(\lambda) + a_\lambda(\mu - \lambda)$. Moreover, applying the same argument it follows that $p_{1,x}(\lambda) \geq p_{1,x}(\mu) + a_\mu(\lambda - \mu)$. Adding these two inequalities yields

$$p_{1,x}(\mu) + p_{1,x}(\lambda) \geq p_{1,x}(\lambda) + p_{1,x}(\mu) + (a_\mu - a_\lambda)(\lambda - \mu),$$

and since $\lambda > \mu$, it follows that $a_\mu - a_\lambda \leq 0$. □

Looking at the proof of the last inequality it is only needed that the subgradient sets $\partial p_{1,x}(\lambda)$ and $\partial p_{1,x}(\mu)$ are nonempty. In view of Lemma 8.1 this is true if $\lambda_*(x)$ is finite and Condition 8.1 holds. By relation (8.11) the above conditions are clearly satisfied for a generalized fractional program.

In the next lemma we show the following important improvement of Lemma 8.1 and Lemma 8.2.

Lemma 8.6 *Assume Condition 8.1 holds. Then the set* $\{\lambda \in \mathbb{R} : p_1(\lambda, x) = 0\}$ *is nonempty if and only if* $\lambda_*(x) < \infty$. *Moreover, if this set is nonempty, it only contains the finite value* $\lambda_*(x)$.

Proof. If the set $\{\lambda \in \mathbb{R} : p_1(\lambda, x) = 0\}$ is nonempty, then it follows for any λ belonging to this set that $f(y,x) \leq \lambda g(y,x)$ for every $y \in A$. This shows by the positivity of g on $A \times B$ that $\lambda_*(x) \leq \lambda < \infty$. Also by Lemma 8.2 we obtain for $\lambda_*(x)$ finite that $p_1(\lambda_*(x), x) = 0$. This proves the first part of the above result. To prove the second part, we observe that by Lemma 8.4 the function $p_{1,x}$ is strictly decreasing. This completes the proof. □

Up to now we did not assume that there exists some $x \in B$ satisfying $\lambda_* = \lambda_*(x) < \infty$, i.e., that the min-max fractional program (P) has an optimal solution in B. In the next theorem we show the implications of this assumption. To do so, consider the (possibly empty) set $D_0 \subseteq \mathbb{R}$ given by

$$D_0 := \{\lambda \in \mathbb{R} : p_2(\lambda) = 0 \text{ and } S_{p_2}(\lambda) \text{ is nonempty}\}. \tag{8.27}$$

It is now possible to prove the following theorem.

Theorem 8.3 *If Condition 8.1 holds, then* $\lambda_* = \lambda_*(x_0) < \infty$ *for some* $x_0 \in B$ *if and only if* $D_0 = \{\lambda_*\}$. *Moreover, if* $\lambda_* = \lambda_*(x_0) < \infty$ *for some* $x_0 \in B$, *then*

$$S_{p_2}(\lambda_*) = \{x \in B : \lambda_* = \lambda_*(x)\}.$$

Proof. By Lemma 8.2 it follows for $\lambda_* = \lambda_*(x) < \infty$ that $p_1(\lambda_*, x) = 0$. Since $p_1(\lambda_*, x) \geq p_2(\lambda_*) \geq 0$, this shows that

$$0 = p_1(\lambda_*, x) = p_2(\lambda_*). \tag{8.28}$$

Using relation (8.28) with x replaced by x_0 it follows for $\lambda_* = \lambda_*(x_0) < \infty$ that λ_* belongs to D_0. Hence we still need to show that D_0 only contains λ_*. Consider therefore an arbitrary λ belonging to D_0. By the definition of D_0 in relation (8.27) one can find some $x_0 \in B$ satisfying $0 = p_2(\lambda) = p_1(\lambda, x_0)$, and this implies by Lemma 8.6 that $\lambda_*(x_0) = \lambda$. Since $p_2(\lambda) = 0$, it follows by Theorem 8.1 that $\lambda \leq \lambda_*$, and this shows that $\lambda_*(x_0) = \lambda \leq \lambda_* \leq \lambda_*(x_0)$. Hence $\lambda = \lambda_*$, and we have verified that D_0 only contains λ_*.

To prove the converse we obtain for $\lambda_* \in D_0$ that $0 = p_2(\lambda_*) = p_1(\lambda_*, x_0)$ for some $x_0 \in B$. Applying Lemma 8.6 yields $\lambda_*(x_0)$ is finite and $\lambda_* = \lambda_*(x_0)$ which proves the "only if" implication. To verify the second part it follows by relation (8.28) that x belongs to $S_{p_2}(\lambda_*)$ for every $x \in B$ satisfying $\lambda_* = \lambda_*(x)$, and so

$$\{x \in B : \lambda_* = \lambda_*(x)\} \subseteq S_{p_2}(\lambda_*).$$

To prove the reverse inclusion, let $x \in S_{p_2}(\lambda_*)$. Since $\lambda_* = \lambda_*(x_0) < \infty$ for some $x_0 \in B$, it follows from relation (8.28) with x replaced by x_0 that $0 = p_2(\lambda_*)$. Since $x \in S_{p_2}(\lambda_*)$, this implies $p_1(\lambda_*, x) = p_2(\lambda_*) = 0$. Applying now Lemma 8.6 yields $\lambda_* = \lambda_*(x)$. □

If we introduce the (possibly empty) set $D_1 \subseteq \mathbb{R}$ given by

$$D_1 := \{\lambda \in \mathbb{R} : p_2(\lambda) = 0 \text{ and } (P_\lambda) \text{ has an optimal solution}\},$$

then without Condition 8.1 one can show, using similar techniques as before, the following result. Note the vector (y, x) is an optimal solution of the (primal) min-max problem (P) if and only if $(y, x) \in A \times B$ and $\lambda_* = \lambda_*(x) = f(y, x)(g(y, x))^{-1}$.

Theorem 8.4 *The (primal) min-max fractional program* (P) *has an optimal solution if and only if* $D_1 = \{\lambda_*\}$. *Moreover, if* (P) *has an optimal solution, then the set* $S_p(\lambda_*)$ *listed in relation (8.18) is nonempty and*

$$S_p(\lambda_*) = \{(y, x) \in A \times B : \lambda_* = \lambda_*(x) = \frac{f(y, x)}{g(y, x)}\}.$$

For the moment this concludes our discussion of some of the theoretical properties related to the parametric approach. In the next subsection we will consider the (primal) Dinkelbach-type algorithm and use the previously derived properties to show its convergence.

6.2 The Primal Dinkelbach-Type Algorithm.

In this section we will introduce the so-called primal Dinkelbach-type algorithm to solve the (primal) min-max fractional program (P). A similar approach for a slightly different min-max fractional program satisfying some compactness assumptions on the feasible sets A and B was considered by Ţigan (cf. [72, 73]). Contrary to [72] the feasible set A in this section does not depend on y. Due to this our assumptions are less restrictive. Using Lemma 8.1 and the fact that the (primal) Dinkelbach-type algorithm is based on solving a sequence of parametric optimization problems (P_λ) for $\lambda \geq \lambda_*$ it is natural to assume that the (primal) min-max fractional program (P) satisfies the next condition.

Condition 8.5

- Condition 8.1 holds and $\lambda_*(x)$ is finite for every $x \in B$.
- If λ_* is finite, then for every $\lambda \geq \lambda_*$ the set $S_{p_2}(\lambda)$ is nonempty while for $\lambda_* = -\infty$ the set $S_{p_2}(\lambda)$ is nonempty for every $\lambda \in \mathbb{R}$.

Contrary to the analysis in [22] for generalized fractional programs we do not assume that the min-max fractional program (P) has an optimal solution. Also for generalized fractional programs the first part of Condition 8.5 is automatically satisfied. If Condition 8.5 holds, then one can execute the following so-called primal Dinkelbach-type algorithm. The geometrical interpretation of this algorithm is as follows. By Theorem 8.3 we need to find the zero point λ_* of the optimal value function p_2. Starting at a given point $\lambda > \lambda_*$ it follows by Theorem 8.1 that $p_2(\lambda) < 0$. Since the function p_2 is nonconvex and it is too ambitious to compute in one step its zero point λ_*, we replace this function by the easier convex function $p_{1,x}(.)$ with x belonging to $S_{p_2}(\lambda)$. We know by the definition of $p_{1,x}$ and $S_{p_2}(\lambda)$ that $p_2(\lambda) = p_{1,x}(\lambda)$ and $p_{1,x}(.) \geq p_2(.)$. For the function $p_{1,x}(.)$ it is easy to compute its zero point. By Lemma 8.2 this is given by $\lambda_*(x)$. We now replace the original point λ in the parametric problem (P_λ) by the smaller value $\lambda_*(x) \geq \lambda_*$ and repeat the procedure.

Primal Dinkelbach-type algorithm.

1 Select $x_0 \in B$ and $k := 1$ and compute

$$\lambda_k := \lambda_*(x_0).$$

2 Determine $x_k \in S_{p_2}(\lambda_k)$. If $p_1(\lambda_k, x_k) \geq 0$ stop and return λ_k and x_k. Otherwise compute

$$\lambda_{k+1} := \lambda_*(x_k),$$

let $k := k + 1$ and go to step 1.

To determine $\lambda_*(x)$ in step 1 and 2 one has to solve a single-ratio fractional program. If A is a finite set, then this is easy. Also in order to select $x_k \in S_{p_2}(\lambda_k)$, one has to solve for A finite a finite min-max problem. Algorithms for such a problem can be found in part 2 of [55]. In case A is not finite, one needs to solve a much more difficult problem, a semi-infinite min-max problem (cf. [27, 55]). Therefore to apply the above generic primal Dinkelbach-type algorithm in practice one needs to have an efficient algorithm to determine an element of the set $S_{p_2}(\lambda_k)$, and this is in most cases the bottleneck. In general one cannot expect that an efficient and fast algorithm exists. But for special cases this might be the case. Including the construction of approximate solutions of the problem (P_{λ_k}) by using smooth approximations of the max operator, thus speeding up the computations and at the same time bounding the errors (cf. [16]) seems to be an important topic for future research.

By Lemma 8.6 it is sufficient to find in step 2 of the primal Dinkelbach-type algorithm the solution of the equation $p_1(\lambda, x_k) = 0$. As already observed, we can give an easy geometrical interpretation of the above algorithm (cf. [5, 16]). The next result shows that the sequence λ_k generated by the primal Dinkelbach-type algorithm is strictly decreasing.

Lemma 8.7 *If Condition 8.5 holds, then the sequence λ_k generated by the primal Dinkelbach-type algorithm is strictly decreasing and satisfies $\lambda_k \geq \lambda_* \geq -\infty$ for every $k \in \mathbb{N}$.*

Proof. If the algorithm stops at $k = 1$, then by the stopping rule we know that $p_2(\lambda_1) \geq 0$. This implies by Theorem 8.1 for $\lambda_1 = \lambda_*(x_0)$ that $\lambda_*(x_0) \leq \lambda_*$ which shows that $\lambda_*(x_0) = \lambda_*$. If the algorithm does not stop at the first step, then $p_2(\lambda_1) < 0$. Since $S_{p_2}(\lambda_1)$ is nonempty, the algorithm finds some $x_1 \in S_{p_2}(\lambda_1)$. Hence

$$0 > p_2(\lambda_1) = p_1(\lambda_1, x_1) = \sup_{y \in A} p(\lambda_1, y, x_1). \tag{8.29}$$

Thus for every $y \in A$ we obtain $f(y, x_1) - \lambda_1 g(y, x_1) < 0$, and so

$$\frac{f(y, x_1)}{g(y, x_1)} < \lambda_1$$

for every $y \in A$. This shows $\lambda_2 \leq \lambda_1$. To verify that $\lambda_*(x_1) = \lambda_2 < \lambda_1$ we assume by contradiction that $\lambda_*(x_1) = \lambda_1$. Since $x_1 \in S_{p_2}(\lambda_1)$, this yields by relation (8.29) and Lemma 8.6 that

$$0 > p_2(\lambda_1) = p_1(\lambda_1, x_1) = p_1(\lambda_*(x_1), x_1) = 0,$$

and we obtain a contradiction. Therefore $\lambda_2 < \lambda_1$, and by the definition of λ_2 it is obvious that $\lambda_2 \geq \lambda_*$. Applying now the same argument iteratively shows the desired result. □

By Lemma 8.7 it follows that the sequence λ_k generated by the primal Dinkelbach-type algorithm converges to some limit $w \geq -\infty$. In case the generated sequence is finite, it is easy to show the following result.

Lemma 8.8 *If Condition 8.5 holds and the primal Dinkelbach-type algorithm stops at* λ_i, *then* $\lambda_* = \lambda_i = \lambda_{i+1}$ *and* $p_2(\lambda_i) = 0$.

Proof. Since Condition 8.5 holds, we obtain $\lambda_* < \infty$. Also by the stopping rule of the Dinkelbach-type algorithm it follows that $p_2(\lambda_i) \geq 0$. This implies by Theorem 8.1 that $\lambda_i \leq \lambda_*$. Since always $\lambda_i \geq \lambda_*$, we obtain $\lambda_i = \lambda_*$. To show that $\lambda_{i+1} = \lambda_i$ with $\lambda_i := \lambda_*(x_{i-1})$ and $p_2(\lambda_i) = 0$, we observe by Lemma 8.6 and by using $x_i \in S_{p_2}(\lambda_i)$ that

$$0 \leq p_2(\lambda_i) = p_1(\lambda_i, x_i) \leq p_1(\lambda_i, x_{i-1}) = 0. \tag{8.30}$$

Hence it follows that $p_2(\lambda_i) = p_1(\lambda_i, x_i) = 0$. Applying again Lemma 8.6 we obtain $\lambda_{i+1} := \lambda_*(x_i) = \lambda_i$ which completes the proof. □

In the remainder of this subsection we only consider the case that the primal Dinkelbach-type algorithm generates an infinite sequence $\lambda_k, k \in \mathbb{N}$. By Lemma 8.7 it follows that $\lim_{k\uparrow\infty} \lambda_k = w \geq -\infty$ exists. Imposing some additional condition it will be shown in Lemma 8.9 that this limit equals λ_*. To simplify the notation in the following lemmas we introduce for the sequence $\{(\lambda_k, x_k) \in \mathbb{R} \times B : x_k \in S_{p_2}(\lambda_k)\}$ generated by the primal Dinkelbach-type algorithm the sequence $\{a_k : k \in \mathbb{N}\}$ with

$$a_k \in \partial p_{1,x_k}(\lambda_{k+1}) \tag{8.31}$$

and for λ_* finite the sequence $\{b_k : k \in \mathbb{N}\}$ with

$$b_k \in \partial p_{1,x_k}(\lambda_*). \tag{8.32}$$

By the observation after Lemma 8.4 these subgradient sets are nonempty. It is now possible to derive the next result.

Lemma 8.9 *If Condition 8.5 holds and there exists a subsequence* $\{a_{n_k} : k \in \mathbb{N}\}$ *satisfying* $\sum_{k=1}^{\infty} a_{n_k}^{-1} = -\infty$, *then* $\lim_{k\uparrow\infty} \lambda_k = \lambda_*$. *Moreover, for* λ_* *finite it follows that* $\lim_{k\uparrow\infty} p_2(\lambda_k) = 0 \leq p_2(\lambda_*)$.

Proof. By Lemma 8.7 the sequence $\{\lambda_k : k \in \mathbb{N}\}$ is strictly decreasing, and so $\lim_{k\uparrow\infty} \lambda_k := w \geq -\infty$ exists. If $w = -\infty$, we obtain using $\lambda_k \geq \lambda_*$ for every $k \in \mathbb{N}$ that $-\infty = w \geq \lambda_*$, and so for $w = -\infty$ the result is proved. Therefore assume that w is finite. Since $p_2(\lambda_k) = p_1(\lambda_k, x_k) < 0$ and the function p_2 and the sequence $\{\lambda_k : k \in \mathbb{N}\}$ are decreasing, it follows that the sequence $\{p_1(\lambda_k, x_k) : k \in \mathbb{N}\}$ is increasing and $-\infty < \alpha := \lim_{k\uparrow\infty} p_1(\lambda_k, x_k) \leq 0$ exists. If we assume that $\alpha < 0$, then one can find some $\epsilon > 0$ satisfying $p_1(\lambda_k, x_k) \leq -\epsilon$ for every $k \in \mathbb{N}$. By Lemma 8.2 we also know that $p_1(\lambda_{k+1}, x_k) = 0$. Applying the subgradient inequality to the convex function p_{1,x_k} we obtain for every $k \in \mathbb{N}$ that

$$a_k(\lambda_k - \lambda_{k+1}) \leq p_1(\lambda_k, x_k) - p_1(\lambda_{k+1}, x_k) = p_1(\lambda_k, x_k) \leq -\epsilon$$

with $a_k \in \partial p_{1,x_k}(\lambda_{k+1})$. Since by relation (8.25) it follows that $-\infty < a_k < 0$, the above inequality shows $\lambda_k - \lambda_{k+1} \geq -\epsilon a_k^{-1}$. This yields by our assumption and w finite that

$$\lambda_1 - w = \sum_{k=1}^{\infty} (\lambda_k - \lambda_{k+1}) \geq -\epsilon \sum_{k=1}^{\infty} a_k^{-1} \geq -\epsilon \sum_{k=1}^{\infty} a_{n_k}^{-1} = \infty,$$

and so $w = -\infty$. This contradicts that w is finite, and we have shown that $\lim_{k\uparrow\infty} p_2(\lambda_k) = 0$. Applying now Theorem 8.2 and Lemma 1.30 of [29] yields $p_2(w) \geq \limsup_{k\uparrow\infty} p_2(\lambda_k) = 0$. Then by Theorem 8.1 it follows that $w \leq \lambda_*$. Since by Lemma 8.7 it is obvious that $w = \lim_{k\uparrow\infty} \lambda_k \geq \lambda_*$, we obtain $w = \lambda_*$ completing the proof. □

By relation (8.25) it follows that

$$0 > a_k \geq -g_{\sup}(x_k)$$

for every $a_k \in \partial p_{1,x_k}(\lambda_{k+1})$, and so one can apply Lemma 8.9 in case $\sum_{k=1}^{\infty} g_{\sup}(x_{n_k})^{-1} = \infty$. To achieve a rate of convergence result for the sequence λ_k generated by the primal Dinkelbach-type algorithm, we need to assume in the proof that $p_2(\lambda_*) = 0$. To apply our procedure we always impose that $S_{p_2}(\lambda_*)$ is nonempty for λ_* finite. Then it follows by Theorem 8.3 that $p_2(\lambda_*) = 0$ if and only if the min-max fractional program (P) has an optimal solution in B or equivalently there exists some

$x_0 \in B$ satisfying $\lambda_* = \lambda_*(x_0)$. However, if the condition of Lemma 8.9 holds, we conjecture for λ_* finite that the min-max fractional program (P) might not have an optimal solution in B, and so $p_2(\lambda_*)$ is not equal to zero. Using a stronger condition than in Lemma 8.9, we show in the next lemma for finite λ_* that the sequence $\{p_2(\lambda_k) : k \in \mathbb{N}\}$ generated by the primal Dinkelbach-type algorithm satisfies $\lim_{k\uparrow\infty} p_2(\lambda_k) = p_2(\lambda_*) = 0$. This sufficient condition implies the existence of an optimal solution of the (primal) min-max fractional program (P) in B.

Lemma 8.10 *If Condition 8.5 holds, λ_* is finite and there exists a subsequence $\{b_{n_k} : k \in \mathbb{N}\}$ satisfying* $\inf_{k\in\mathbb{N}} b_{n_k} > -\infty$, *then* $\lim_{k\uparrow\infty} \lambda_k = \lambda_*$ *and* $\lim_{k\uparrow\infty} p_2(\lambda_k) = 0 = p_2(\lambda_*)$.

Proof. By the convexity of the function p_{1,x_k} and the subgradient inequality we obtain for every $k \in \mathbb{N}$ that

$$0 \geq p_2(\lambda_k) \geq p_1(\lambda_*, x_k) + b_k(\lambda_k - \lambda_*) \geq p_2(\lambda_*) + b_k(\lambda_k - \lambda_*) \quad (8.33)$$

with $b_k \in \partial p_{1,x_k}(\lambda_*)$. Since $\lambda_{k+1} > \lambda_*$, it follows by our assumption and the monotonicity of the subgradient sets as shown in Lemma 8.5 that one can find some finite M satisfying $M \leq b_{n_k} \leq a_{n_k} < 0$ for every $k \in \mathbb{N}$ and every sequence $\{a_{n_k} : k \in \mathbb{N} \text{ and } a_{n_k} \in \partial p_{1,x_k}(\lambda_{k+1})\}$. This shows

$$0 > M^{-1} \geq b_{n_k}^{-1} \geq a_{n_k}^{-1} \quad (8.34)$$

for every $k \in \mathbb{N}$, and so $\sum_{k=1}^{\infty} a_{n_k}^{-1} = -\infty$. Hence by Lemma 8.9 we obtain $\lim_{k\uparrow\infty} \lambda_k = \lambda_*$. Using relations (8.34) and (8.33) yields $0 \geq \lim_{k\uparrow\infty} p_2(\lambda_{n_k}) \geq p_2(\lambda_*)$. Since by Theorem 8.1 we know that $p_2(\lambda_*) \geq 0$, the proof is completed. □

By relation (8.25) it follows in case $\sup_{k\in\mathbb{N}} g_{\sup}(x_k) < \infty$ that the condition of Lemma 8.10 is satisfied. A similar condition is also given in [22] for a generalized fractional program. In the next lemma we consider the generated sequence $\{x_k : x_k \in S_{p_2}(\lambda_k)\}_{k\in\mathbb{N}}$ and show for B compact and some additional topological properties on the functions f and g that this sequence contains a converging subsequence.

Lemma 8.11 *If Condition 8.5 holds, the functions f and g are finite-valued and continuous on some open set $W \subseteq \mathbb{R}^{m+n}$ containing $A \times B$, the set B is compact and there exists a subsequence $\{a_{n_k} : k \in \mathbb{N}\}$ satisfying $\sum_{k=1}^{\infty} a_{n_k}^{-1} = -\infty$, then the sequence $\{x_k : x_k \in S_{p_2}(\lambda_k)\}_{k\in\mathbb{N}}$ has a converging subsequence and every limit point x_∞ of the sequence $\{x_k : k \in \mathbb{N}\}$ satisfies $\lambda_* = \lambda_*(x_\infty)$ with λ_* finite. Additionally, if there exist a unique $x_* \in B$ satisfying $\lambda_* = \lambda_*(x_*)$, then* $\lim_{k\uparrow\infty} x_k =$

x_*. Moreover, for $A \times B$ compact, the generated sequence $\{(y_k, x_k) : (y_k, x_k) \in S_p(\lambda_k)\}_{k\in\mathbb{N}}$ has a converging subsequence and every limit point of the sequence $\{(y_k, x_k) : k \in \mathbb{N}\}$ is an optimal solution of problem (P). If the optimization problem (P) has a unique optimal solution (y_*, x_*), then $\lim_{k\uparrow\infty} x_k = x_*$ and $\lim_{k\uparrow\infty} y_k = y_*$.

Proof. To verify that λ_* is finite we obtain by Condition 8.5 and f, g continuous that the finite-valued function $x \to \lambda_*(x)$ is lower semicontinuous. By the compactness of B this implies, using Corollary 1.2 of [3], that there exists some $x \in B$ satisfying $\lambda_* = \lambda_*(x)$, and so λ_* is finite. Again by the compactness of B it is also obvious that the sequence $\{x_k : k \in \mathbb{N}\}$ contains a convergent subsequence. To show that every limit point x_∞ of the sequence $x_k, k \in \mathbb{N}$ satisfies $\lambda_* = \lambda_*(x_\infty)$ we observe by Lemma 8.9 that $\lim_{k\uparrow\infty} \lambda_k = \lambda_*$. This implies by Lemma 8.3 that $x_\infty \in S_{p2}(\lambda_*)$. Using now Theorem 8.3 we obtain $\lambda_* = \lambda_*(x_\infty)$. If there exists a unique $x_* \in B$ satisfying $\lambda_* = \lambda_*(x_*)$, then again by Theorem 8.3 we obtain $S_{p_2}(\lambda_*) = \{x_*\}$. Since every converging subsequence of the sequence $x_k, k \in \mathbb{N}$ converges to an element of $S_{p_2}(\lambda_*)$, it follows that every convergent subsequence converges to the element x_*. By contradiction and B compact we obtain $\lim_{k\uparrow\infty} x_k = x_*$, and the proof of the first part is completed. If $A \times B$ is compact, then by the continuity of the function g we obtain

$$\sup_{(x,y)\in A\times B} g(y, x) < \infty.$$

Again by the observation after Lemma 8.10 we obtain $\lambda_k \downarrow \lambda_*$. By Lemma 8.3 the set-valued mapping S_p is closed and using a similar proof as for the first part one can show the last part. □

If we consider a generalized fractional program, then clearly A is compact, and if additionally the conditions of Lemma 8.11 hold, then the second part of this lemma applies. Unfortunately it is not clear to the authors whether in the first part of this lemma the condition $\sum_{k=1}^{\infty} a_{n_k}^{-1} = -\infty$ can be omitted.

We now want to investigate how fast the sequence λ_k converges to λ_*. Before discussing this in detail, we list for λ_* finite the following inequality for the sequence $\{\lambda_k : k \in \mathbb{N}\}$ generated by the primal Dinkelbach-type algorithm. A similar inequality can also be derived for the dual Dinkelbach-type algorithm to be discussed in subsection 6.4.

Theorem 8.5 *If Condition 8.5 holds and there exists some $x \in B$ satisfying $\lambda_* = \lambda_*(x)$, then it follows for every $c_k \in \partial p_{1,x}(\lambda_k)$ and*

$a_k \in \partial p_{1,x_k}(\lambda_{k+1})$ that

$$0 \leq \frac{\lambda_{k+1} - \lambda_*}{\lambda_k - \lambda_*} \leq (1 - c_k a_k^{-1}).$$

Proof. Since $\lambda_* = \lambda_*(x)$ for some $x \in B$, we obtain by Lemma 8.6 that $p_1(\lambda_*, x) = p_1(\lambda_*(x), x) = 0$. Applying now the subgradient inequality to the function $p_{1,x}$ at the point λ_* it follows for $c_k \in \partial p_{1,x}(\lambda_k)$ that

$$-p_1(\lambda_k, x) = p_1(\lambda_*, x) - p_1(\lambda_k, x) \geq c_k(\lambda_* - \lambda_k).$$

Hence

$$p_2(\lambda_k) \leq p_1(\lambda_k, x) \leq c_k(\lambda_k - \lambda_*). \tag{8.35}$$

Moreover, for every $x_k \in S_{p_2}(\lambda_k)$ and $\lambda_{k+1} = \lambda_*(x_k)$ we obtain again by Lemma 8.6 that $p_1(\lambda_{k+1}, x_k) = 0$. Applying now the subgradient inequality to the function p_{1,x_k} at the point λ_{k+1} yields for $a_k \in \partial p_{1,x_k}(\lambda_{k+1})$ that

$$p_2(\lambda_k) = p_1(\lambda_k, x_k) - p_1(\lambda_{k+1}, x_k) \geq a_k(\lambda_k - \lambda_{k+1}). \tag{8.36}$$

Hence by relations (8.35) and (8.36) we obtain $-a_k(\lambda_{k+1} - \lambda_k) \leq p_2(\lambda_k) \leq c_k(\lambda_k - \lambda_*)$. Since $a_k < 0$, this implies

$$\lambda_{k+1} - \lambda_k \leq -c_k a_k^{-1}(\lambda_k - \lambda_*). \tag{8.37}$$

Using relation (8.37) it follows that

$$\lambda_{k+1} - \lambda_* = \lambda_k - \lambda_* + \lambda_{k+1} - \lambda_k \leq (1 - c_k a_k^{-1})(\lambda_k - \lambda_*),$$

and this completes the proof. □

In case of a single-ratio fractional program the function $\lambda \to p_{1,x}(\lambda)$ reduces to $p_{1,x}(\lambda) = f(x) - \lambda g(x)$, and so for every $\lambda \in \mathbb{R}$ it follows that $\partial p_{1,x}(\lambda) = \{-g(x)\}$. Hence we obtain that the inequality in Theorem 8.5 reduces to

$$0 \leq \frac{\lambda_{k+1} - \lambda_*}{\lambda_k - \lambda_*} \leq (1 - \frac{g(x_0)}{g(x_k)}) \tag{8.38}$$

for any optimal solution x_0 of the optimization problem

$$\inf_{x \in B} f(x)(g(x))^{-1}$$

(cf. [67]).

Before introducing convergence results for the primal Dinkelbach-type algorithm, we need the following definition (cf. [54]).

Definition 8.2 *A sequence* $\{s_k : k \in \mathbb{N}\} \subseteq \mathbb{R}^n$ *with limit* s_∞ *converges Q-linearly if there exists some* $0 < r < 1$ *such that*

$$\limsup_{k\uparrow\infty} \frac{\|s_{k+1} - s_\infty\|}{\|s_k - s_\infty\|} \leq r.$$

The sequence $\{s_k : k \in \mathbb{N}\}$ *converges Q-superlinearly if*

$$\lim_{k\uparrow\infty} \frac{\|s_{k+1} - s_\infty\|}{\|s_k - s_\infty\|} = 0.$$

If a slightly stronger condition as used in Lemma 8.10 holds, then one can show that the sequence $\{\lambda_k : k \in \mathbb{N}\}$ generated by the primal Dinkelbach-type algorithm converges Q-linearly. The same result was shown for a generalized fractional program in [22].

Theorem 8.6 *If Condition 8.5 holds,* λ_* *is finite and the sequence* $\{b_k : k \in \mathbb{N}\}$ *satisfies* $\inf_{k\in\mathbb{N}} b_k > -\infty$, *then* $\lim_{k\uparrow\infty} \lambda_k = \lambda_*$ *and* $\{\lambda_k : k \in \mathbb{N}\}$ *converges Q-linearly.*

Proof. By Lemma 8.10 we obtain $p_2(\lambda_*) = 0$. Since Condition 8.5 holds, one can find some $x \in B$ satisfying $0 = p_2(\lambda_*) = p_1(\lambda_*, x)$, and this shows by Lemma 8.6 that $\lambda_* = \lambda_*(x)$. Hence the set $\{x \in B : \lambda_* = \lambda_*(x)\}$ is nonempty, and for every x belonging to this set it follows by Theorem 8.5 that

$$0 \leq \frac{\lambda_{k+1} - \lambda_*}{\lambda_k - \lambda_*} \leq (1 - c_k a_k^{-1}) \tag{8.39}$$

with $c_k \in \partial p_{1,x}(\lambda_k)$ and $a_k \in \partial p_{1,x_k}(\lambda_{k+1})$. Since $\{\lambda_k : k \in \mathbb{N}\}$ is strictly decreasing and $\lambda_k > \lambda_*$, it follows by Lemma 8.5 that the sequence $\{c_k : k \in \mathbb{N}\}$ is decreasing and satisfies $0 > c_k \geq \sigma$ with $\sigma := \max\{t : t \in \partial p_{1,x}(\lambda_*)\}$ This shows that $\lim_{k\uparrow\infty} c_k = c_\infty$ exists. To identify c_∞ we observe in view of $c_k \in \partial p_{1,x}(\lambda_k)$ that

$$p_1(\lambda, x) \geq p_1(\lambda_k, x) + c_k(\lambda - \lambda_k)$$

for every $\lambda \in \mathbb{R}$. Since the function $p_{1,x}$ is continuous, this yields using $\lambda_k \downarrow \lambda_*$ and $\lim_{k\uparrow\infty} c_k = c_\infty$ that

$$p_1(\lambda, x) \geq p_1(\lambda_*, x) + c_\infty(\lambda - \lambda_*)$$

for every $\lambda \in \mathbb{R}$, and so $c_\infty \in \partial p_{1,x}(\lambda_*)$. Therefore $c_\infty = \sigma$, and we have identified this limit. Also by our assumption we obtain that there exists some $-\infty < M \leq b_k \leq a_k$, and this shows

$$\limsup_{k\uparrow\infty}(1 - c_k a_k^{-1}) \leq 1 - \frac{\sigma}{M} < 1.$$

Applying now relation (8.39) yields the desired result. □

If the conditions of Theorem 8.5 hold and additionally we assume that $\inf_{k\in\mathbb{N}} a_k > -\infty$, then it can be shown in view of the proof of Theorem 8.6 that the sequence λ_k converges Q-linearly to λ_*. This condition is slightly weaker than the one used in Theorem 8.6. Observe the condition $\inf_{k\in\mathbb{N}} b_k > -\infty$ was used in the proof of Lemma 8.10 to show that $p_2(\lambda_*) = 0$, and this implies as shown in the first part of the proof of Theorem 8.6 that $\lambda_* = \lambda_*(x)$ for some $x \in B$. Therefore, if there exists some $x \in B$ satisfying $\lambda_*(x) = \lambda_*$ and $\inf_{k\in\mathbb{N}} a_k > -\infty$, then assuming Condition 8.5 holds the sequence λ_k converges Q-linearly to λ_*. A disadvantage of the first part of the previous assumption is that in general we do not know looking at a min-max problem whether there exists some $x \in B$ satisfying $\lambda_* = \lambda_*(x)$. Hence we imposed some stronger algorithmic condition on the sequence b_k, $k \in \mathbb{N}$ implying this result. In case the (primal) min-max fractional program (P) has a unique optimal solution and some additional topological properties are satisfied, then one can show that the sequence $\{\lambda_k : k \in \mathbb{N}\}$ converges superlinearly.

Theorem 8.7 *If Condition 8.5 holds, the functions f and g are continuous on some open set $W \subseteq \mathbb{R}^{m+n}$ containing the compact set $A \times B$ and the min-max fractional program (P) has a unique optimal solution (y_*, x_*), then $\lim_{k\uparrow\infty} x_k = x_*$, $\lim_{k\uparrow\infty} y_k = y_*$ and $\lim_{k\uparrow\infty} \lambda_k = \lambda_*$, and the sequence λ_k converges Q-superlinearly.*

Proof. Using Lemmas 8.9 and 8.11 the first part follows, and so we only have to show that λ_k converges superlinearly. Considering the proof of Theorem 8.6 it follows that

$$\limsup_{k\uparrow\infty}(1 - c_k a_k^{-1}) = 1 - \sigma(\limsup_{k\uparrow\infty} a_k)^{-1}$$

with $\sigma := \max\{t : t \in \partial p_{1,x_*}(\lambda_*)\}$ and $a_k \in \partial p_{1,x_k}(\lambda_{k+1}), k \in \mathbb{N}$. Since a_k is uniformly bounded by the compactness of $A \times B$ and the function g is continuous, there exists a converging subsequence a_{n_k} satisfying $a_\infty = \lim_{k\uparrow\infty} a_{n_k} = \limsup_{k\uparrow\infty} a_k$. To identify a_∞ we observe for every $k \in \mathbb{N}$ that

$$p_1(\lambda, x_k) \geq p_1(\lambda, x_k) - p_1(\lambda_{k+1}, x_k) \geq a_k(\lambda - \lambda_{k+1}) \tag{8.40}$$

with $a_k \in \partial p_{1,x_k}(\lambda_{k+1})$. Since B is compact and p continuous, it follows by Proposition 1.7 of [3] that $x \to p_1(\lambda, x)$ is upper semicontinuous, and this implies by relation (8.40) that

$$p_1(\lambda, x_*) \geq \limsup_{k\uparrow\infty} p_1(\lambda, x_k) \geq a_\infty(\lambda - \lambda_*) \tag{8.41}$$

Since $x_* \in S_{p_2}(\lambda_*)$, we obtain $p_1(\lambda_*, x_*) = p_2(\lambda_*) = 0$, and this shows by relation (8.41) that $a_\infty \in \partial p_{1,x_*}(\lambda_*)$. By the uniqueness of the optimal solution and Lemma 8.5 we obtain $a_\infty = \sigma$. This shows the desired result. □

In case we consider a single-ratio fractional program with B compact and the functions f, g continuous it follows by Lemma 8.11 that

$$\limsup_{k\uparrow\infty} g(x_k) = \lim_{k\uparrow\infty} g(x_{n_k}) = g(x_*)$$

with x_* an optimal solution of this fractional programming problem. Replacing now in relation (8.38) x_0 by x_* we obtain for a single-ratio fractional program with B compact and f, g continuous that the sequence $\{\lambda_k : k \in \mathbb{N}\}$ always converges Q-superlinearly. Clearly in practice the (primal) Dinkelbach-type algorithm stops in a finite number of steps, and so we need to derive a practical stopping rule. Such a rule is constructed in the next lemma. For other practical stopping rules yielding so-called ϵ-optimal solutions the reader may consult [16].

Lemma 8.12 *If Condition 8.5 holds and there exists some subsequence* $\{a_{n_k} : k \in \mathbb{N}\}$ *satisfying* $\sum_{k=1}^{\infty} a_{n_k}^{-1} = -\infty$ *and some* $x \in B$ *satisfying* $\lambda_* = \lambda_*(x)$, *then the sequence* $\{c_k^{-1} p_2(\lambda_k) : c_k \in \partial p_{1,x}(\lambda_k)\}_{k\in\mathbb{N}}$ *is decreasing and its limit equals* 0. *Moreover, it follows for every* $k \in \mathbb{N}$ *that*

$$\lambda_* \leq \lambda_k \leq \lambda_* + c_k^{-1} p_2(\lambda_k).$$

Proof. By Lemma 8.7 the sequence λ_k is strictly decreasing, and this implies by Lemma 8.5 that the negative sequence c_k is decreasing. Also, since p_2 is decreasing, we obtain that the negative sequence $p_2(\lambda_k)$ is increasing and so the positive sequence $c_k^{-1} p_2(\lambda_k)$ is decreasing. Applying now Lemma 8.9 it follows that $\lim_{k\uparrow\infty} c_k^{-1} p_2(\lambda_k) = 0$, while the listed inequality is an immediate consequence of Lemma 8.9 and relation (8.35). □

Using Lemma 8.12 a stopping rule for the (primal) Dinkelbach-type algorithm is given by $c_k^{-1} p_2(\lambda_k) \leq \epsilon$ for some predetermined $\epsilon > 0$. Finally we observe that the (primal) Dinkelbach-type algorithm applied to a generalized fractional program can be regarded as a cutting plane algorithm (cf. [10]). This generalizes a similar observation by Sniedovich (cf. [70]) who showed the result for the (primal) Dinkelbach-type algorithm applied to a single-ratio fractional program.

In the next section we investigate the dual max-min fractional program (D) and its relation to the primal min-max fractional program (P).

6.3 Duality Results for Primal Min-Max Fractional Programs.

In this subsection we first investigate under which conditions the optimal objective function values of the primal min-max fractional program (P) and the dual max-min fractional program (D) coincide. To start with this analysis, we introduce the following class of bifunctions.

Definition 8.3 *The function $h : \mathbb{R}^m \times \mathbb{R}^n \to [-\infty, \infty]$ is called a concave/convex bifunction on the convex set $C_1 \times C_2$ with $C_1 \subseteq \mathbb{R}^m$ and $C_2 \subseteq \mathbb{R}^n$if for every $x \in C_2$ the function $y \to h(y,x)$ is concave on C_1 and for every $y \in C_1$ the function $x \to h(y,x)$ is convex on C_2. Moreover, a function $h : \mathbb{R}^m \times \mathbb{R}^n \to [-\infty, \infty]$ is called a convex/concave bifunction on $C_1 \times C_2$ if $-h$ is a concave/convex bifunction on the same set. It is called an affine/affine bifunction if it is both a concave/convex and a convex/concave bifunction.*

To guarantee that μ_* equals λ_*, we introduce the following sufficient condition.

Condition 8.6 *The set $B \subseteq R^n$ is a closed convex set and $A \subseteq R^m$ is a compact convex set. Moreover, there exists some open convex set $A_1 \times B_1$ containing $A \times B$ such that g is a positive finite-valued convex/concave bifunction and f a positive finite-valued concave/convex bifunction on $A_1 \times B_1$. If the function g is a positive affine/affine bifunction, then f is a finite-valued concave/convex bifunction.*

If the set B is given by relation (8.1), one can also introduce another dual max-min fractional program. To guarantee that for this problem strong duality holds, we need the following slightly stronger condition.

Condition 8.7 *The set $B \subseteq R^n$ is a closed convex set and $A \subseteq R^m$ is a compact convex set. Moreover, there exists some open convex set $A_1 \times C_1$ containing $A \times C$ such that g is a positive finite-valued convex/concave bifunction and f a positive finite-valued concave/convex bifunction on $A_1 \times C_1$. If the function g is a positive affine/affine bifunction, then f is a finite-valued concave/convex bifunction*

If Condition 8.6 holds, then by Theorem 1.15 of [29] we obtain that the function $y \to f(y,x)$ is continuous on A_1 for every $x \in B$ and $x \to f(x,y)$ is continuous on B_1 for every $y \in A$. The same property also holds for the function g. By the compactness of A this implies

$$0 < g_{\inf}(x) \leq g_{\sup}(x) < \infty$$

for every $x \in B$, and so Condition 8.6 implies Condition 8.1. Also, since for every $x \in B$ the function $y \to f(y,x)(g(y,x))^{-1}$ is continuous on A and the set A is compact, we obtain that $\lambda_*(x)$ is finite for every $x \in B$ implying $\lambda_* < \infty$. For $\lambda_* < \infty$ we derive in Theorem 8.8 that the optimal objective function value of the (primal) min-max fractional program (P) equals the optimal objective function value of the (dual) max-min fractional program (D). Contrary to the proof of the same result in [5] for generalized fractional programs based on Sion's minimax result (cf. [31, 69]) the present proof is an easy consequence of the easier-to-prove minimax result by Ky Fan (cf. [26, 27, 32]) and Theorem 8.1. Note that we do not assume that there exists some $x \in B$ satisfying $\lambda_* = \lambda_*(x)$.

Theorem 8.8 *If Condition 8.6 holds, then there exists some $y_0 \in A$ satisfying*

$$\lambda_* = \mu_* = \mu_*(y_0).$$

Proof. Since we know that $\mu_* \leq \lambda_* < \infty$, it follows for $\lambda_* = -\infty$ that $-\infty = \lambda_* = \mu_* \geq \mu_*(y)$ for every $y \in A$. This shows the desired result for $\lambda_* = -\infty$. If λ_* is finite, then we need to verify that $\lambda_* \leq \mu_*$. Since λ_* is finite, we obtain by Condition 8.6 that the function $(y,x) \to p(\lambda_*, y, x)$ is a concave/convex bifunction on $A \times B$ and for every $x \in B$ the function $y \to p(\lambda_*, y, x)$ is continuous on A_1. Applying now Theorem 3.2 of [32] (see also [27]) we obtain

$$p_2(\lambda_*) = \inf_{x\in B} \sup_{y\in A} p(\lambda_*, y, x) = \max_{y\in A} \inf_{x\in B} p(\lambda_*, y, x).$$

This shows by Theorem 8.1 and the remark after Condition 8.6 that

$$0 \leq p_2(\lambda_*) = \max_{y\in A} \inf_{x\in B} p(\lambda_*, y, x) = \inf_{x\in B} p(\lambda_*, y_0, x) \qquad (8.42)$$

for some $y_0 \in A$. Since $g(y_0, x) > 0$ for every $x \in B$, we obtain

$$\frac{f(y_0,x)}{g(y_0,x)} \geq \lambda_*$$

for every $x \in B$. Hence

$$\mu_* = \sup_{y\in A} \inf_{x\in B} \frac{f(y,x)}{g(y,x)} \geq \inf_{x\in B} \frac{f(y_0,x)}{g(y_0,x)} \geq \lambda_*. \qquad (8.43)$$

Using now relation (8.43) the desired result follows. □

Since there are rather general necessary and sufficient conditions on the bifunctions such that for those functions min-max equals max–min (cf. [28, 30]), the above result holds for a much larger class than the class of concave/convex bifunctions. However, since the class of concave/convex bifunctions is most known, we have restricted ourselves to this well-known class. An easy consequence of Theorem 8.8 is given by the next result.

Lemma 8.13 *If Condition 8.6 holds and there exists some $x_0 \in B$ satisfying $\lambda_* = \lambda_*(x_0)$ and some $y_0 \in A$ satisfying $\mu_* = \mu_*(y_0)$, then the vector (y_0, x_0) is an optimal solution of the (primal) min-max fractional program (P) and an optimal solution of the (dual) max-min fractional program (D).*

Proof. By the definition of $\mu_*(y)$ and $\lambda_*(x)$ it is clear that for every vector $(y, x) \in A \times B$ that

$$\mu_*(y) \leq \frac{f(y,x)}{g(y,x)} \leq \lambda_*(x).$$

This implies by Theorem 8.8 for the given vector $(y_0, x_0) \in A \times B$ that

$$\mu_* = \mu_*(y_0) = \frac{f(y_0,x_0)}{g(y_0,x_0)} = \lambda_*(x_0) = \lambda_*.$$

Hence (y_0, x_0) is an optimal solution of the (primal) min-max fractional program (P) and an optimal solution of the (dual) max-min fractional program (D). □

If the (dual) max-min fractional program (D) has a unique optimal solution and the optimal solution set of the (primal) min-max fractional program (P) is nonempty, then by Lemma 8.13 the unique optimal solution of (D) is an optimal solution of (P). If Condition 8.6 holds and we use the so-called dual Dinkelbach-type algorithm to be discussed in subsection 6.4 for identifying λ_*, this observation will be useful. To analyze the properties of the optimization problem (D) and at the same time construct some generic algorithm to solve problem (D), we introduce similar parametric optimization problems as done for problem (P) at the beginning of subsection 6.1. For every $(\lambda, y) \in \mathbb{R} \times A$ consider the parametric optimization problem

$$d_1(\lambda, y) := \inf_{x \in B} p(\lambda, y, x). \qquad (D_\lambda^y)$$

For every $y \in A$ the function $d_{1,y} : \mathbb{R} \to (-\infty, \infty]$ is now given by

$$d_{1,y}(\lambda) := d_1(\lambda, y).$$

Since $g > 0$ on $A \times B$ and $d_{1,y}$ is the infimum of affine functions, it is obvious that $d_{1,y}$ is a decreasing upper semicontinuous concave function. The so-called effective domain $dom(d_{1,y})$ is defined by

$$dom(d_{1,y}) := \{\lambda \in \mathbb{R} : d_{1,y}(\lambda) > -\infty\} \subseteq \mathbb{R}.$$

By the finiteness of p on $\mathbb{R} \times A \times B$ it is obvious for every $y \in A$ that actually $dom(d_{1,y}) = \{\lambda \in \mathbb{R} : d_{1,y}(\lambda) \text{ finite}\}$. A more difficult optimization problem than problem (D_λ^y) is now given by the parametric optimization problem

$$d_2(\lambda) = \sup_{y \in A} d_1(\lambda, y). \tag{D_λ}$$

As for the concave function $d_{1,y}$ we also introduce the effective domain $dom(d_2)$ of the function d_2 given by

$$dom(d_2) := \{\lambda \in \mathbb{R} : d_2(\lambda) > -\infty\}.$$

It should be clear to the reader that we actually apply the Dinkelbach-type approach to the (dual) max-min fractional program (D) while at the beginning of subsection 6.1 we applied the same approach to the (primal) min-max fractional program (P). It is easy to show that

$$\sup_{y \in A} \inf_{x \in B} p(\lambda, y, x) \leq \inf_{x \in B} \sup_{y \in A} p(\lambda, y, x), \tag{8.44}$$

and so we obtain $d_2(\lambda) \leq p_2(\lambda)$ for every $\lambda \in \mathbb{R}$. If the optimization problem (P) is a single-ratio fractional program, then the set A consists of one element, and as already observed there is no difference in the representation of the (primal) min-max fractional program (P) and the (dual) max-min fractional program (D). Hence for A consisting of one element it is not surprising that also the functional representation of the functions d_2 and p_2 are the same. If the set A consists of more than one element, then we want to know, despite the different functional representations of the functions d_2 and p_2, under which conditions $d_2(\lambda) = p_2(\lambda)$ for some λ. It should come as no surprise that this equality holds under the same conditions as used in Theorem 8.8. Note that in the next result we do not assume that the set $S_{p_2}(\lambda)$ is nonempty.

Theorem 8.9 *Assume Condition 8.6 holds where g is a convex/concave bifunction on $A \times B$. Then it follows for every $\lambda \geq 0$ that there exists some $y_\lambda \in A$ satisfying*

$$p_2(\lambda) = d_2(\lambda) = d_1(\lambda, y_\lambda).$$

Moreover, if g is an affine/affine bifunction, the same result holds for every $\lambda \in \mathbb{R}$.

Proof. Since $\lambda_* < \infty$, we obtain by Lemma 8.1 that $p_2(\lambda) < \infty$ for every $\lambda \in \mathbb{R}$. Also, for a convex/concave bifunction g it follows by Condition 8.6 and $\lambda \geq 0$ that the function $(y, x) \to p(\lambda, y, x)$ is a concave/convex bifunction on $A \times B$ and $y \to p(\lambda, y, x)$ is continuous on A_1 for every $(\lambda, x) \in \mathbb{R}_+ \times B$. A similar observation holds for $\lambda \in \mathbb{R}$, if g is an affine/affine bifunction. Since A is compact, we can now apply Theorem 3.2 of [32]. This shows

$$\begin{aligned} p_2(\lambda) &= \inf_{x\in B} \sup_{y\in A} p(\lambda, y, x) \\ &= \max_{y\in A} \inf_{x\in B} p(\lambda, y, x) = d_2(\lambda). \end{aligned} \tag{8.45}$$

Hence by relation (8.45) there exists for $\lambda \geq 0$ and a convex/concave bifunction g or $\lambda \in \mathbb{R}$ and an affine/affine bifunction g some $y_\lambda \in A$ satisfying $d_1(\lambda, y_\lambda) = d_2(\lambda)$. This completes the proof. □

Applying similar proofs as in Lemma 8.1 and Theorem 8.1 one can verify the following results.

Lemma 8.14 *Assume Condition 8.2 holds. Then $\mu_* > -\infty$ if and only if $dom(d_2) = \mathbb{R}$, and $\mu_*(y) > -\infty$ if and only if $dom(d_{1,y}) = \mathbb{R}$.*

Clearly Lemma 8.14 can be compared with Lemma 8.1 while the next result is the counterpart of Theorem 8.1.

Theorem 8.10 *Assume Condition 8.2 holds and $\mu_* > -\infty$. Then $\lambda < \mu_*$ if and only if $d_2(\lambda) > 0$. Moreover, if $\mu_*(y) > -\infty$, then $\lambda < \mu_*(y)$ if and only if $d_1(\lambda, y) > 0$.*

A direct consequence of the above results is given by the following.

Theorem 8.11 *Assume Condition 8.6 holds where g is a positive convex/concave bifunction on $A \times B$. Then it follows that $0 \leq \lambda_* = \mu_* < \infty$, $p_2(\lambda) = d_2(\lambda)$ for every $\lambda \geq 0$, and these functions are finite-valued on $(-\infty, \lambda_*]$. Moreover, if g is a positive affine/affine bifunction on $A \times B$ and λ_* is finite, then $\mu_* = \lambda_*$, $p_2(\lambda) = d_2(\lambda)$ for every $\lambda \in \mathbb{R}$, and these functions are finite-valued on $(-\infty, \lambda_*]$.*

Proof. If g is a positive convex/concave bifunction on $A \times B$, then by Condition 8.6 the function f must be a positive concave/convex bifunction on $A \times B$. Then automatically $0 \leq \lambda_* < \infty$. Also, by Theorem 8.8 and 8.9 we obtain $\mu_* = \lambda_*$ and $p_2(\lambda) = d_2(\lambda)$ for every $\lambda \geq 0$. Since Condition 8.6 implies Condition 8.1, it follows by the remark after Theorem 8.1 that $p_2(\lambda)$ is finite for every $\lambda \leq \lambda_*$. This yields $d_2(\lambda) = p_2(\lambda)$ is finite-valued on $[0, \lambda_*]$. Using the monotonicity of d_2, we see

$$\infty > p_2(\lambda) \geq d_2(\lambda) \geq d_2(0) = p_2(0) \geq 0$$

for every $\lambda \leq 0$. Hence the first part follows. The second part can be proved similarly, and its proof is therefore omitted. □

If Condition 8.6 holds and hence also Condition 8.1 and λ_* is finite, then it might happen (as shown in Example 8.1) that the value $p_2(\lambda_*)$ is not equal to zero. If additionally there exists some $x_0 \in B$ satisfying $\lambda_* = \lambda_*(x_0)$, then by Theorems 8.3 and 8.11 we know that $d_2(\mu_*) = d_2(\lambda_*) = p_2(\lambda_*) = 0$, and we need this assumption in combination with Condition 8.6 to identify λ_* by the so-called dual Dinkelbach-type algorithm to be discussed in the next subsection. The next result is the counterpart of Theorem 8.2. It can be proved by similar techniques.

Theorem 8.12 *Assume Condition 8.2 holds. Then the decreasing function $d_2 : \mathbb{R} \to [-\infty, \infty]$ is lower semicontinuous.*

Similarly as in Section 6.1 it follows by Theorem 8.12 that $\lim_{s \uparrow \lambda} d_2(s) = d_2(\lambda)$, and the function d_2 is right-continuous with left-hand limits.

As in Section 6.1 we now introduce the following set-valued mappings $S_{d_1} : \mathbb{R} \times A \to 2^B$ and $S_{d_2} : \mathbb{R} \to 2^A$ given by

$$S_{d_1}(\lambda, y) := \{x \in B : d_1(\lambda, y) = p(\lambda, y, x)\} \tag{8.46}$$

and

$$S_{d_2}(\lambda) := \{y \in A : d_2(\lambda) = d_1(\lambda, y)\}. \tag{8.47}$$

The set $S_{d_1}(\lambda, y)$ represents the set of optimal solutions of optimization problem (D_λ^y), while the set $S_{d_2}(\lambda)$ denotes the set of optimal solutions in A of optimization problem (D_λ). Also we consider the set-valued mapping $S_d : \mathbb{R} \to 2^{A \times B}$ given by

$$S_d := \{(y, x) \in A \times B : d_2(\lambda) = d_1(\lambda, y) = p(\lambda, y, x)\}. \tag{8.48}$$

This set represents the set of optimal solutions in $A \times B$ of optimization problem (D_λ). In the next result it is assumed that the sets $S_{d_1}(\lambda, y), S_{d_2}(\lambda)$ and $S_d(\lambda)$ are nonempty on their domain. Applying Theorem 8.12 and using a similar proof as in Lemma 8.3 we obtain the following counterpart of Lemma 8.3.

Lemma 8.15 *Assume Condition 8.2 holds and the functions f and g are finite-valued and continuous on some open set $W \subseteq \mathbb{R}^{m+n}$ containing $A \times B$. Then the set-valued mappings S_{d_1}, S_{d_2} and S_d are closed.*

Considering now the function $d_{1,y} : \mathbb{R} \to [-\infty, \infty)$ given by

$$d_{1,y}(\lambda) := d_1(\lambda, y)$$

one can show as in Lemma 8.4 the following result.

Lemma 8.16 *Assume Condition 8.2 holds and $\mu_*(y)$ is finite for $y \in A$. Then the function $d_{1,y} : \mathbb{R} \to (-\infty, \infty)$ is strictly decreasing and Lipschitz continuous with Lipschitz constant $\overline{g}_{\text{sup}}(y)$ and the function satisfies $\lim_{\lambda \uparrow \infty} d_{1,y}(\lambda) = -\infty$ and $\lim_{\lambda \downarrow -\infty} d_{1,y}(\lambda) = \infty$.*

As in Section 6.1 with respect to the function $p_{1,x}$ it follows in case Condition 8.2 holds that the subgradient set of the convex strictly increasing function $-d_{1,y}$ is nonempty for every $\lambda \in \mathbb{R}$ and this set satisfies

$$\partial(-d_{1,y})(\lambda) \subseteq [\underline{g}_{\text{inf}}(y), \overline{g}_{\text{sup}}(y)]. \tag{8.49}$$

Moreover, the subgradient inequality is given by

$$-d_{1,y}(\mu) \geq -d_{1,y}(\lambda) + a(\mu - \lambda) \tag{8.50}$$

for every $a \in \partial(-d_{1,y})(\lambda)$. Also one can show the following counterpart of Lemma 8.5.

Lemma 8.17 *Assume Condition 8.4 holds. Then it follows for every $y \in A$ that $\mu_*(y)$ is finite, $S_{d_1}(\lambda, y)$ is a nonempty compact set for every $(\lambda, y) \in \mathbb{R} \times A$ and*

$$\partial(-d_{1,y})(\lambda) = [\min_{x \in S_{d_1}(\lambda,y)} g(y,x), \max_{x \in S_{d_1}(\lambda,y)} g(y,x)].$$

Also for every $a_\lambda \in \partial(-d_{1,y})(\lambda)$ and $a_\mu \in \partial(-d_{1,y})(\mu)$ and $\lambda > \mu$ it holds that $a_\lambda \geq a_\mu > 0$.

The next result can be compared with Lemma 8.6.

Lemma 8.18 *Assume Condition 8.2 holds. Then the set $\{\lambda \in \mathbb{R} : d_1(\lambda, y) = 0\}$ is nonempty if and only if $\mu_*(y) > -\infty$. Moreover, if this set is nonempty, then it only contains the finite value $\mu_*(y)$.*

Up to now we did not assume that there exists some $y \in A$ satisfying $\mu_* = \mu_*(y) > -\infty$ or equivalently the dual max-min fractional program (D) has an optimal solution in B. In the next lemma the implications of this assumption are discussed. To do so, consider the (possibly empty) set $D_2 \subseteq \mathbb{R}$ given by

$$D_2 := \{\lambda \in \mathbb{R} : d_2(\lambda) = 0 \text{ and } S_{d_2}(\lambda) \text{ is nonempty}\}.$$

The counterpart of Theorem 8.3 is given by the following result.

Theorem 8.13 *Assume Condition 8.2 holds. Then* $\mu_* = \mu_*(y_0) > -\infty$ *for some* $y_0 \in A$ *if and only if* $D_2 = \{\mu_*\}$. *Moreover, if* $\mu_* = \mu_*(y_0) > -\infty$ *for some* $y_0 \in A$, *then the set* $S_{d_2}(\lambda_*)$ *is nonempty and*

$$S_{d_2}(\lambda_*) = \{y \in A : \mu_* = \mu_*(y)\}.$$

If we introduce the (possibly) empty set $D_3 \subseteq \mathbb{R}$ given by

$$D_3 := \{\lambda \in \mathbb{R} : d_2(\lambda) = 0 \text{ and } (D) \text{ has an optimal solution}\},$$

then without Condition 8.2 one can show the following counterpart of Theorem 8.4. Remember a vector (y, x) is an optimal solution of (D) if and only if $(y, x) \in A \times B$ and $\mu_* = \mu_*(y) = f(y, x)(g(y, x))^{-1}$.

Theorem 8.14 *The (dual) max-min fractional program* (D) *has an optimal solution if and only if* $D_3 = \{\mu_*\}$. *Moreover, if* (D) *has an optimal solution, then the set* $S_d(\mu_*)$ *is nonempty and*

$$S_d(\mu_*) = \{(y, x) \in A \times B : \mu_* = \mu_*(y) = \frac{f(y, x)}{g(y, x)}\}.$$

Finally we will consider in this section another dual max-min fractional program if the nonempty set B is given by (see also relation (8.1))

$$B = \{x \in C : h_k(x) \leq 0, \ k = 1, ..., l\}. \tag{8.51}$$

In case the set B is specified as in relation (8.51) we always assume for the corresponding primal min-max fractional program (P) that the function g is positive on $A \times C$. Introducing now the vector-valued function $h : \mathbb{R}^n \to \mathbb{R}^l$ given by $h(x)^\top = (h_1(x), ..., h_l(x))$, we consider for every $(y, z) \in A \times \mathbb{R}^l_+$ the single-ratio fractional program

$$\mu_*^p(y, z) := \inf_{x \in C} \frac{f(y, x) + z^\top h(x)}{g(y, x)}. \tag{$D_p^{(y,z)}$}$$

A more complicated optimization problem is now introduced by the so-called partial dual of the (primal) min-max fractional program given by

$$\mu_*^p := \sup_{y \in A, z \geq 0} \inf_{x \in C} \frac{f(y, x) + z^\top h(x)}{g(y, x)}. \tag{D_p}$$

Again this is a max-min fractional program, and using only $g > 0$ on $A \times C$ it is easy to show the following result.

Lemma 8.19 *If* g *is positive on* $A \times C$, *then it follows that* $\mu_*^p \leq \mu_* \leq \lambda_*$.

Proof. Since $B \subseteq C$ and $z^\top h(x) \leq 0$ for every $x \in B$ and $z \geq 0$, we obtain by the positivity of g on $A \times C$ that

$$\mu_*^p(y,z) \leq \inf_{x\in B} \frac{f(y,x) + z^\top h(x)}{g(y,x)} \leq \inf_{x\in B} \frac{f(y,x)}{g(y,x)}$$

for every $z \geq 0$ and $y \in A$. This shows

$$\mu_*^p = \sup_{y\in A,\ z\geq 0} \mu_*^p(y,z) \leq \sup_{y\in A} \inf_{x\in B} \frac{f(y,x)}{g(y,x)} = \mu_*,$$

and so the first inequality is verified. We already showed that $\mu_* \leq \lambda_*$. Hence the proof is complete. □

To verify that $\mu_*^p = \lambda_*$, it is clear from Lemma 8.19 that

$$\lambda_* = -\infty \Rightarrow \lambda_* = \mu_* = \mu_*^p = \mu_*^p(y,z) = -\infty$$

for every $(y,z) \in A \times \mathbb{R}_+^l$. If λ_* is finite and we want to ensure that $\mu_*^p = \lambda_*$, then the following so-called Slater-type condition on the nonempty set B should be considered. Before introducing this condition, we assume throughout the remainder of this section that the (possibly empty) set $I \subseteq \{1,...,l\}$ denotes the set of indices for which $h_k : \mathbb{R}^n \to \mathbb{R}$ is affine. Note that $ri(C)$ denotes the relative interior of the set C (cf. [29, 58]).

Condition 8.8 *There exists some $x \in ri(C)$ where C is a closed convex set satisfying $h_k(x) < 0$ for every $k \notin I$ and $h_k(x) \leq 0$ for every $k \in I$. Moreover, for every $k \notin I$ the functions $h_k : \mathbb{R}^n \to \mathbb{R}$ are convex.*

To show under which conditions the equality $\mu_*^p = \lambda_*$ and the finiteness of λ_* holds, we first need to prove the following Lagrangean duality result.

Lemma 8.20 *Assume Condition 8.8 holds and for a given $y \in A$ the function $x \to f(y,x)$ is convex on C and $x \to g(y,x)$ is concave on C. Then it follows for every $\lambda \geq 0$ that there exists some $z_{\lambda,y} \geq 0$ satisfying*

$$\inf_{x\in B} p(\lambda,y,x) = \inf_{x\in C}\{f(y,x) - \lambda g(y,x) + z_{\lambda,y}^\top h(x)\}$$

with B defined in relation (8.51). Moreover, the same result holds for every $\lambda \in \mathbb{R}$ if $x \to f(y,x)$ is convex and $x \to g(y,x)$ is affine.

Proof. Using the definition of the set B and $z \geq 0$, it is easy to see that

$$\inf_{x\in B} p(\lambda,y,x) \geq \inf_{x\in C}\{f(y,x) - \lambda g(y,x) + z^\top h(x)\}.$$

Moreover, for either $\lambda \geq 0$ and $x \to g(y,x)$ concave or $\lambda \in \mathbb{R}$ and $y \to g(y,x)$ affine we see that the function $x \to p(\lambda,y,x)$ is convex on C. Applying now Theorem 28.2 of [58] or Theorem 1.25 of [29] we obtain that there exists some dual solution $z_{\lambda,y} \geq 0$ such that the above inequality is actually an equality. □

Using Lemma 8.20 it is now possible to show that the optimal objective function value of the partial dual equals λ_*.

Theorem 8.15 *Assume Conditions 8.7 and 8.8 hold. Then there exists some* $(y_0, z_0) \in A \times \mathbb{R}^l_+$ *satisfying*

$$\lambda_* = \mu^p_* = \mu^p_*(y_0, z_0).$$

Proof. For $\lambda_* = -\infty$ we know by the remark after Lemma 8.19 that the result holds. Hence we only need to verify the result for λ_* finite. To start we observe by relation (8.42) that

$$0 \leq p_2(\lambda_*) = \inf_{x\in B} p(\lambda_*, y_0, x)$$

for some $y_0 \in A$. Applying now Lemma 8.20 one can find some $z_0 \geq 0$ satisfying

$$\inf_{x\in B} p(\lambda_*, y_0, x) = \inf_{x\in C}\{f(y_0,x) - \lambda_* g(y_0,x) + z_0^\top h(x)\}.$$

This shows

$$0 \leq p_2(\lambda_*) = \inf_{x\in C}\{f(y_0,x) - \lambda_* g(y_0,x) + z_0^\top h(x)\}. \tag{8.52}$$

By relation (8.52) and $g(y_0,x) > 0$ for every $x \in C$ we obtain $\mu^p_*(y_0,z_0) \geq \lambda_*$ which completes the proof. □

In case we use the partial dual (D_p) it follows that the partial dual of the single-ratio fractional program

$$\inf_{x\in B} \frac{f(x)}{g(x)}$$

with B given by relation (8.51) is given by

$$\sup_{z\geq 0} \inf_{x\in C} \frac{f(x) + z^\top h(x)}{g(x)}.$$

Thus for this (Lagrangean) dual (cf. [66, 68]) the single-ratio fractional program and its dual have a different representation. If Theorem 8.15 holds, one can always apply a Dinkelbach-type algorithm to the partial dual (D_p) to find λ_*. This is discussed in detail in [6] and [9]. In the next subsection we will introduce a similar Dinkelbach-type algorithm applied to the (dual) max-min problem (D).

6.4 The Dual Dinkelbach-Type Algorithm.

In this section we apply the Dinkelbach-type approach to the (dual) max-min fractional program (D). Parallel to subsection 6.2 we assume that the next condition holds. Note that this condition is the counterpart of Condition 8.5 used in the primal Dinkelbach-type algorithm which was applied to the (primal) min-max fractional program (P).

Condition 8.9

- Condition 8.2 holds and $\mu_*(y)$ is finite for every $y \in A$.

- If μ_* is finite, then for every $\lambda \leq \mu_*$ the set $S_{d_2}(\lambda)$ is nonempty while for $\mu_* = -\infty$ the set $S_{d_2}(\lambda)$ is nonempty for every $\lambda \in \mathbb{R}$.

If condition 8.9 holds, then one can execute the following so-called dual Dinkelbach-type algorithm. As for the (primal) Dinkelbach-type algorithm introduced in Section 6.2 one can give a similar geometrical interpretation of the next algorithm.

Dual Dinkelbach-type algorithm.

1 Select $y_0 \in A$ and $k := 1$ and compute

$$\mu_k := \mu_*(y_0).$$

2 Determine $y_k \in S_{d_2}(\lambda_k)$. If $d_1(\mu_k, y_k) \leq 0$ stop and return μ_k and y_k. Otherwise compute

$$\mu_{k+1} := \mu_*(y_k),$$

let $k := k + 1$ and go to 1.

Observe in Step 1 and 2 one has to solve a single-ratio fractional program. If B is a finite set, then solving such a problem is easy. Moreover, by Lemma 8.18 it is sufficient to find in step 2 of the primal Dinkelbach-type algorithm the solution of the equation $d_1(\lambda, y_k) = 0$. As already observed, this yields an easy geometrical interpretation of the above algorithm (see also [5]). The next result shows that the sequence μ_k generated by the dual Dinkelbach-type algorithm is strictly increasing. The proof of this result is similar to the proof of the corresponding result for the primal Dinkelbach-type algorithm in Lemma 8.7. This also shows that the primal Dinkelbach-type algorithm approaches the optimal objective function value from above while the dual Dinkelbach-type algorithm approaches it from below.

Lemma 8.21 *If Condition 8.9 holds, then the sequence μ_k generated by the dual Dinkelbach-type algorithm is strictly increasing and satisfies $\mu_k \leq \mu_* \leq \infty$ for every $k \in \mathbb{N}$.*

By Lemma 8.21 we obtain that the sequence μ_k generated by the dual Dinkelbach-type algorithm converges to some limit $v \leq \infty$. Using a similar proof as in Lemma 8.8 one can show the following result in case the generated sequence is finite. If strong duality holds and so $\mu_* = \lambda_*$, one can also use this algorithm to approximate λ_*.

Lemma 8.22 *If Condition 8.9 holds and the dual Dinkelbach-type algorithm stops at μ_k, then $\mu_* = \mu_k = \mu_{k+1}$ and $d_2(\mu_k) = 0$.*

In the remainder of this subsection we only consider the case where the dual Dinkelbach-type algorithm generates an infinite sequence $\mu_k, k \in \mathbb{N}$. By Lemma 8.21 it follows that $\lim_{k\uparrow\infty} \mu_k = v \leq \infty$ exists. Imposing some additional condition it will be shown in Lemma 8.23 that this limit equals μ_*. To simplify the notation in the following lemmas, we introduce for the sequence $\{(\mu_k, y_k) \in \mathbb{R} \times A : y_k \in S_{d_2}(\mu_k)\}$ generated by the primal Dinkelbach-type algorithm the sequence $\{\underline{a}_k : k \in \mathbb{N}\}$ given by

$$-\underline{a}_k \in \partial(-d_{1,x_k})(\mu_{k+1}) \tag{8.53}$$

and for μ_* finite the sequence $\{\underline{b}_k : k \in \mathbb{N}\}$ given by

$$-\underline{b}_k \in \partial(-d_{1,x_k})(\mu_*). \tag{8.54}$$

By the observation after Lemma 8.16 these subgradient sets are nonempty. Using a similar proof as in Lemma 8.9 it is possible to verify the next result.

Lemma 8.23 *If Condition 8.9 holds and there exists a subsequence $\{\underline{a}_{n_k} : k \in \mathbb{N}\}$ satisfying $\sum_{k=1}^{\infty} \underline{a}_{n_k}^{-1} = -\infty$, then $\lim_{k\uparrow\infty} \mu_k = \mu_*$. Moreover for μ_* finite it follows that $\lim_{k\uparrow\infty} d_2(\mu_k) = 0 \geq d_2(\mu_*)$.*

By relation (8.49) it follows that

$$0 > \underline{a}_k \geq -\overline{g}_{\sup}(y_k) \tag{8.55}$$

for every $-\underline{a}_k \in \partial(-d_{1,y_k})(\mu_{k+1})$. Hence one can apply Lemma 8.23 in case $\sum_{k=1}^{\infty} \overline{g}_{\sup}(y_{n_k})^{-1} = \infty$. To show that $d_2(\mu_*) = 0$, we can follow the proof of Lemma 8.10 and obtain the following result.

Lemma 8.24 *If Condition 8.9 holds, μ_* is finite and there exists a subsequence $\{\underline{b}_{n_k} : k \in \mathbb{N}\}$ satisfying $\inf_{k\in\mathbb{N}} \underline{b}_{n_k} > -\infty$, then $\lim_{k\uparrow\infty} \mu_k = \mu_*$ and $\lim_{k\uparrow\infty} d_2(\mu_k) = 0 = d_2(\mu_*)$.*

By relation (8.55) it follows in case $\sup_{k\in\mathbb{N}} \overline{g}_{\text{sup}}(y_k) < \infty$ that the condition of Lemma 8.24 is satisfied. The next result should be contrasted with Lemma 8.11.

Lemma 8.25 *If Condition 8.9 holds, the functions f and g are finite-valued and continuous on some open set $W \subseteq \mathbb{R}^{m+n}$ containing $A \times B$, the set A is compact and there exists a subsequence $\{\underline{a}_{n_k} : k \in \mathbb{N}\}$ satisfying $\sum_{k=1}^{\infty} \underline{a}_{n_k}^{-1} = -\infty$, then the sequence $\{y_k : y_k \in S_{d_2}(\mu_k)\}_{k\in\mathbb{N}}$ has a converging subsequence and every limit point y_∞ of the sequence $\{y_k : k \in \mathbb{N}\}$ satisfies $\mu_* = \mu_*(y_\infty)$ with μ_* finite. Additionally, if there exist a unique $y_* \in A$ satisfying $\mu_* = \mu_*(y_*)$, then $\lim_{k\uparrow\infty} y_k = y_*$. Moreover, for $A \times B$ compact the generated sequence $\{(y_k, x_k) : (y_k, x_k) \in S_d(\mu_k)\}_{k\in\mathbb{N}}$ has a converging subsequence and every limit point of the sequence $\{(y_k, x_k) : k \in \mathbb{N}\}$ is an optimal solution of problem (D). If the optimization problem (D) has a unique optimal solution (y_*, x_*), then $\lim_{k\uparrow\infty} x_k = x_*$ and $\lim_{k\uparrow\infty} y_k = y_*$.*

We now want to investigate how fast the sequence μ_k converges to μ_*. Before discussing this in detail, we list for μ_* finite the following inequality for the sequence $\{\mu_k : k \in \mathbb{N}\}$ generated by the dual Dinkelbach-type algorithm. The proof is similar to the proof of the corresponding result listed in Theorem 8.5 for the primal Dinkelbach-type algorithm.

Theorem 8.16 *If Condition 8.9 holds and there exists some $y \in A$ satisfying $\mu_* = \mu_*(y)$, then it follows for every $-\underline{c}_k \in \partial(-d_{1,y})(\mu_k)$ and $-\underline{a}_k \in \partial(-d_{1,y_k})(\mu_{k+1})$ that*

$$0 \leq \frac{\mu_* - \mu_{k+1}}{\mu_* - \mu_k} \leq (1 - \underline{c}_k \underline{a}_k^{-1}).$$

If a slightly stronger condition as used in Lemma 8.24 holds, then one can show that the sequence $\{\mu_k : k \in \mathbb{N}\}$ generated by the primal Dinkelbach-type algorithm converges Q-linearly. The same result was shown for the dual generalized fractional program in [5] and [8]. The proof of the next result is similar as the proof of the corresponding result for the primal Dinkelbach-type algorithm given in Theorem 8.6

Theorem 8.17 *If Condition 8.9 holds, μ_* is finite and the sequence $\{\underline{b}_k : k \in \mathbb{N}\}$ satisfies $\inf_{k\in\mathbb{N}} \underline{b}_k > -\infty$, then $\lim_{k\uparrow\infty} \mu_k = \mu_*$ and the sequence μ_k converges Q-linearly.*

Finally we show in case the dual (max-min) fractional program (D) has a unique optimal solution and some other topological conditions hold that the sequence $\{\mu_k : k \in \mathbb{N}\}$ converges Q-superlinearly. In case also strong duality holds, then we know by the remark after Lemma 8.13 that this unique optimal solution of (D) is also an optimal solution of the primal min-max fractional program P assuming this set is nonempty. By the compactness of $A \times B$ in the next result the set of optimal solutions of (P) is nonempty.

Theorem 8.18 *If Condition 8.9 holds, the functions f and g are continuous on some open set W containing the compact set $A \times B$ and the max-min fractional program (D) has a unique optimal solution (y_*, x_*), then $\lim_{k\uparrow\infty} x_k = x_*$, $\lim_{k\uparrow\infty} y_k = y_*$ and $\lim_{k\uparrow\infty} \lambda_k = \lambda_*$ and the sequence μ_k converges Q-superlinearly.*

If strong duality holds, then it is clear that one can also use the dual Dinkelbach-type algorithm to determine the value λ_*. This is the main use of this algorithm in the literature (cf. [8, 9]). Also one could combine the dual and primal approach in case strong duality holds and use simultaneously both. An example of such an approach applied to a generalized fractional program with an easy geometrical interpretation is discussed by Gugat (cf. [39, 41]). In [39] it is shown under slightly stronger conditions that always a Q-superlinear convergence rate holds. This concludes our discussion of the parametric approach used in min-max fractional programming which was a major emphasis in this chapter on fractional programming.

References

[1] Almogy, Y. and O. Levin, *Parametric analysis of a multi-stage stochastic shipping problem*, in Operational Research '69, J. Lawrence, ed., Tavistock Publications, London, 1970, 359-370.

[2] Asmussen, S., *Applied Probability and Queues*, Wiley, New York, 1987.

[3] Aubin, J.B., *Optima and Equilibria (An Introduction to Nonlinear Analysis)*, Graduate Texts in Mathematics Vol. 140, Springer Verlag, Berlin, 1993.

[4] Avriel, M., Diewert, W.E., Schaible, S. and I. Zang, *Generalized Concavity*, Plenum Press, New York, 1988.

[5] Barros, A., *Discrete and Fractional Programming Techniques for Location Models*, Kluwer Academic Publishers, Dordrecht, 1998.

[6] Barros, A.I., Frenk, J.B.G. and J. Gromicho, *Fractional location problems*, Location Science 5, 1997, 47-58.

[7] Barros, A.I., Dekker, R., Frenk, J.B.G. and S. van Weeren, *Optimizing a general optimal replacement model by fractional programming techniques*, J. Global Optim. 10, 1997, 405-423.

[8] Barros, A.I., Frenk, J.B.G., Schaible, S. and S. Zhang, *A new algorithm for generalized fractional programming*, Math. Program. 72, 1996, 147-175.

[9] Barros, A.I., Frenk, J.B.G., Schaible, S. and S. Zhang, *Using duality to solve generalized fractional programming problems*, J. Global Optim. 8, 1996, 139-170.

[10] Barros, A.I. and J.B.G. Frenk, *Generalized fractional programming and cutting plane algorithms*, J. Optim. Theory Appl. 87, 1995, 103-120.

[11] Bazaraa, M.S., Sherali, H.D. and C.M. Shetty, *Nonlinear Programming (Theory and Applications)*, Wiley, New York, 1993.

[12] Bázsa, E., *Decision Support for Inventory Systems with Complete Backlogging*, PhD Thesis, Tinbergen Institute Research Series No. 282, Econometric Institute, Erasmus University, Rotterdam, 2002.

[13] Bázsa, E., den Iseger, P.W. and J.B.G. Frenk, *Modeling of inventory control with regenerative processes*, International Journal of Production Economics 71, 2001, 263-276.

[14] Bereanu, B., *Decision regions and minimum risk solutions in linear programming*, in: A. Prekopa (ed.), Colloquium on Applications of Mathematics to Economics, Budapest 1963 (Publication House of the Hungarian Academy of Sciences, Budapest 1965), 37-42.

[15] Birbil, Ş.İ., Frenk, J.B.G. and S. Zhang, *A progressive finite representation approach to minimax optimization*, 2004, in preparation.

[16] Birbil, Ş.İ., Frenk, J.B.G. and S. Zhang, *Generalized fractional programming with user interaction*, 2004, submitted.

[17] Charnes, A., Cooper, W.W., Levin, A.Y and L.M. Seiford, eds., *Data Envelopment Analysis: Theory, Methodology and Applications*, Kluwer Academic Publishers, Dordrecht, 1994.

[18] Charnes, A. and W.W. Cooper, *Programming with linear fractional functionals*, Naval Research Logistics Quarterly 9, 1962, 181-186.

[19] Charnes, A. and W.W. Cooper, *Deterministic equivalents for optimizing and satisficing under chance constraints*, Operations Research 11, 1963, 18-39.

[20] Chen, D.Z., Daescu, O., Dai, Y., Katoh, N., Wu, X. and J. Xu, *Optimizing the sum of linear fractional functions and applications*, in Proceedings of the Eleventh Annual ACM-SIAM Symposium on Discrete Algorithms, ACM, New York, 2000, 707-716.

[21] Craven, B.D., *Fractional Programming*, Heldermann Verlag, Berlin, 1988.

[22] Crouzeix, J.P., Ferland, J.A. and S. Schaible, *An algorithm for generalized fractional programs*, J. Optim. Theory Appl. 47, 1985, 35-49.

[23] Drezner, Z., Schaible, S. and D. Simchi-Levi, *Queueing-location problems on the plane*, Naval Research Logistics 37, 1990, 929-935.

[24] Falk, J.E., Polacsay, S.W., Sacco, W.J., Copes, W.S., and H.R. Champion, *Bounds on a trauma outcome function via optimization*, Oper. Res. 20 (Supp.1), 1992, 86-95.

[25] Falk, J.E., *Maximization of signal-to-noise ratio in an optical filter*, SIAM J. Appl. Math. 7, 1969, 582-592.

[26] Fan, K., *Minimax theorems*, Proc. Nat. Acad. Sc. U.S.A. 39, 1953, 42-47.

[27] Frenk, J.B.G., Kassay, G. and J. Kolumbán, *On equivalent results in minimax theory*, Eur. J. Oper. Res. 157, 2004, 46-58.

[28] Frenk, J.B.G., Protassov,V. and G. Kassay, *On Borel probability measures and noncooperative game theory*, Optimization, 2004, to appear.

[29] Frenk, J.B.G. and G. Kassay, *Introduction to convex and quasiconvex analysis*, Chapter 1 of this volume.

[30] Frenk, J.B.G., Kas, P. and G. Kassay, *On linear programming duality and necessary and sufficient conditions in minmax theory*, Econometric Institute Report E.I 2004-14, Econometric Institute, Erasmus University, 2004, submitted.

[31] Frenk, J.B.G. and G. Kassay, *The level set method of Joó and its use in minimax theory*, Econometric Institute Report E.I 2003-03, Econometric Institute, Erasmus University, 2003, submitted.

[32] Frenk, J.B.G. and G. Kassay, *Minimax results and finite dimensional minimization*, J. Optim. Theory Appl. 113, 2002, 409-421.

[33] Frenk, J.B.G. and S. Schaible, *Fractional programming: introduction and applications*, in Encyclopedia of Optimization, Vol II, E-Integer, C.A. Floudas and P.M. Pardalos, eds., Kluwer Academic Publishers, Dordrecht, 2001, 234-240.

[34] Frenk, J.B.G. and M.J. Kleijn, *On regenerative processes and inventory control*, Pure Appl. Math. 9, 1998, 61-94.

[35] Frenk, J.B.G., Dekker, R. and M.J. Kleijn, *On the marginal cost approach in maintenance*, J. Optim. Theory Appl. 94, 1998, 771-781.

[36] Freund, R.W. and F. Jarre, *Solving the sum-of-ratios problem by an interior point method*, J. Global Optim. 19, 2001, 83-102.

[37] Gilmore, P.C. and R.E. Gomory, *A linear programming approach to the cutting stock problem - Part II*, Operations Research 11, 1963, 863-888.

[38] Goedhart, M.H and J. Spronk, *Financial planning with fractional goals*, Eur. J. Oper. Res. 82, 1995, 111-124.

[39] Gugat, M., *A fast algorithm for a class of generalized fractional programs*, Management Science 42, 1996, 1493-1499.

[40] Gugat, M., *Computation of lower bounds for spectra via fractional semi-infinite programming*, Approximation Optimization 8, 1995, 379-391.

[41] Gugat, M., *Fractional Semi-infinite Programming*, PhD Thesis, Mathematical Institute, University of Trier, Trier, 1994.

[42] Heinen, E., *Grundlagen betriebswirtschaftlicher Entscheidungen, Das Zielsystem der Unternehmung, 2.Auflage*, Gabler, 1971.

[43] Hoskins, J.A. and R. Blom, *Optimal allocation of warehouse personnel: a case study using fractional programming*, FOCUS (U.K.) 3 (2), 1984, 13-21.

[44] Konno, H., *Minimization of the sum of several linear fractional functions*, in Generalized Convexity and Generalized Monotonicity, N. Hadjisavvas, J.E. Martínez-Legaz and J.P. Penot, eds., Lecture Notes in Economics and Mathematical Systems Vol. 502, Springer Verlag, Berlin, 2001, 3-20.

[45] Konno, H. and K. Fukaishi, *A brand and bound algorithm for solving low rank linear multiplicative and fractional programming problems*, J. Global Optim. 18, 2000, 283-299.

[46] Konno, M. and M. Inori, *Bond portfolio optimization by bilinear fractional programming*, J. Oper. Res. Soc. Japan 32, 1989, 143-158.

[47] Kuno, T., *A branch-and-bound algorithm for maximizing the sum of several linear ratios*, J. Global Optim. 22, 2002, 155-174.

[48] Lo, A. and C. MacKinlay, *Maximizing predictability in the stock and bond markets*, Macroeconomic Dynamics 1, 1997, 102-134.

[49] Martos, B., *Hyperbolic programming*, Naval Research Logistics Quarterly 11, 1964, 135-155; originally published in Math. Institute of Hungarian Academy of Sciences 5, 1960, 383-406 (Hungarian).

[50] Martos, B., *Nonlinear Programming, Theory and Applications*, North Holland, Amsterdam, 1975.

[51] Matthis, F.H. and L.J. Matthis, *A nonlinear programming algorithm for hospital management*, SIAM Review 37, 1995, 230-234.

[52] Meszaros, Cs. and T. Rapcsák, *On sensitivity analysis for a class of decison systems*, Decision Support Systems 16, 1996, 231-240.

[53] Mjelde, K.M., *Methods of the Allocation of Limited Resources*, Wiley, New York, 1983.

[54] Nocedal, J. and S.J. Wright, *Numerical Optimization*, Springer Series in Operations Research, Springer Verlag, Berlin, 1999.

[55] Polak, E., *Optimization: Algorithms and Consistent Approximations*, Applied Mathematical Sciences Vol. 124, Springer Verlag, Berlin, 1997.

[56] Radzik, T., *Fractional combinatorial optimization*, in Handbook of Combinatorial Optimization, Vol. 1, D.Z. Du and P.M. Pardalos, eds., Kluwer Academic Publishers, Dordrecht, 1998, 429-478.

[57] Rao, M.R., *Cluster analysis and mathematical programming*, J. Amer. Statist. Assoc. 66, 1971, 622-626.

[58] Rockafellar, R.T., *Convex Analysis*, Princeton Mathematical Series Vol. 28, Princeton University Press, Princeton, New Jersey, 1970.

[59] Schaible, S. and J. Shi., *Recent developments in fractional programming: single-ratio and max-min case*, in Proceedings of the 3rd International Conference in Nonlinear Analysis and Convex Analysis, Aug.25-29, 2003, Takahashi, W. and T. Tanaka, eds., Yokohama Publishing, 2004, to appear.

[60] Schaible, S. and J. Shi, *Fractional programming: the sum-of-ratios case*, Optimization Methods and Software 18, 2003, 219-229.

[61] Schaible, S., *Fractional programming*, in Handbook of Global Optimization, R. Horst and P.M. Pardalos, eds., Kluwer Academic Publishers, Dordrecht, 1995, 495-608.

[62] Schaible, S. and T. Ibaraki, *Fractional programming (invited review)*, Eur. J. Oper. Res. 12, 1983, 325-338.

[63] Schaible, S. and T. Lowe, *A note an a material control problem*, IIE Transactions 15, 1983, 177-179.

[64] Schaible, S., *Fractional programming: applications and algorithms*, Eur. J. Oper. Res. 7, 1981, 111-120.

[65] Schaible, S., *Analyse und Anwendungen von Quotientenprogrammen, Ein Beitrag zur Planung mit Hilfe der nichtlinearen Programmierung*, Mathematical Systems in Economics Vol. 42, Hain Verlag, Meisenheim, 1978.

[66] Schaible, S., *Fractional programming, I, Duality*, Management Science 22, 1976, 858-867.

[67] Schaible, S., *Fractional programming, II, On Dinkelbach's algorithm*, Management Science 22, 1976, 868-873.

[68] Schaible, S., *Duality in fractional programming: a unified approach*, Oper. Res. 24, 1976, 452-461.

[69] Sion, M., *On general minimax theorems*, Pacific J. Math. 8, 1958, 171-176.

[70] Sniedovich, M., *Fractional programming revisited*, Eur. J. Oper. Res. 33, 1988, 334-341.

[71] Stancu-Minasian, I.M., *Fractional Programming: Theory, Methods and Applications*, Kluwer Academic Publishers, Dordrecht, 1997.

[72] Ţigan, S., *On some procedures for solving fractional max-min problems*, Mathematica-Revue D'Analyse Numérique et de Théorie de L'Approximation 17, 1988, 73-91.

[73] Ţigan, S., *A parametrical method for max-min nonlinear fractional problems*, Seminarul Itinerant de Ecuaţii Funcţionale, Aproximare şi Convexitate, Cluj-Napoca, 1983, 175-184.

[74] Von Neumann, J., *Über ein ökonomisches Gleichungssystem und eine Verallgemeinerung des Brouwerschen Fixpuntsatzes*, in Ergebnisse eines mathematischen Kolloquiums (8), Menger, K., ed., Leipzig und Wien, 1937, 73-83.

[75] Zhang, S., *Stochastic Queue Location Problems*, Tinbergen Institute Research Series No. 14, Econometric Institute, Erasmus University, Rotterdam, 1991.

II

GENERALIZED MONOTONICITY

Chapter 9

GENERALIZED MONOTONE MAPS

Nicolas Hadjisavvas
Department of Product and Systems Design Engineering
University of the Aegean
84100 Syros, GREECE
nhad@aegean.gr

Siegfried Schaible
A.G. Anderson Graduate School of Management
University of California
Riverside, CA 92521-0203, USA
siegfried.schaible@ucr.edu

Abstract We first present nine kinds of (generalized) monotone maps and in case of gradient maps their counterpart of nine kinds of (generalized) convex functions. In addition we present topologically pseudomonotone maps. We then derive sufficient and/or necessary conditions for various kinds of generalized monotonicity for several subclasses of maps. We study differentiable maps, locally Lipschitz maps, general continuous maps and affine maps.

Keywords: Generalized convexity, generalized convex functions, generalized monotonicity, generalized monotone maps, gradient maps.

1. Introduction

In many classical models generalized monotonicity can replace the rigid assumption of monotonicity without the loss of valuable properties for the analysis and solution of such models, for instance in complementarity problems and variational inequalities or more generally in equilibrium models in the sense of Blum and Oettli [24]; for example see [34, 46] in this volume.

In working with generalized monotonicity concepts one quickly realizes the need for sufficient and/or necessary conditions which are easier to verify than the original definitions, much like in generalized convexity [1]. In this chapter a major focus is the derivation of criteria for generalized monotonicity of certain subclasses of maps [44].

As a basis for such a study we first present nine kinds of (generalized) monotone maps and their counterpart, nine kinds of generalized convex functions in Section 2. This is followed in Section 3 by the presentation of topologically pseudomonotone maps. Then in Sections 4, 5 and 6 criteria for generalized monotonicity of subclasses of maps are derived. Section 4 deals with differentiable maps. In Section 5 we consider the nondifferentiable case, namely locally Lipschitz and general continuous maps. Finally in Section 6 we obtain criteria for generalized monotone affine maps. These results can be viewed as extensions of characterizations of generalized convex quadratic functions which are presented here as well.

Let us fix some notation. Given $x, y \in \mathbb{R}^n$, we denote by $\langle x, y \rangle$ their scalar product; $[x, y]$ will be the closed line segment

$$[x, y] = \{(1-t)\,x + ty : t \in [0,1]\}\,.$$

The line segments (x, y), $(x, y]$ and $[x, y)$ are defined analogously. For any $C \subseteq \mathbb{R}^n$ we denote by $\operatorname{int} C$ its interior.

Let $\mathbb{R}^n_+$ be the nonnegative orthant of $\mathbb{R}^n$.

The set of all linear maps from $\mathbb{R}^n$ to $\mathbb{R}^m$ will be denoted by $L(\mathbb{R}^n, \mathbb{R}^m)$. For any $M \in L(\mathbb{R}^n, \mathbb{R}^m)$ the corresponding matrix will be denoted by the same symbol.

We denote by F' the Jacobian of a differentiable map F.

2. Various kinds of generalized monotonicity

Let C be a subset of $\mathbb{R}^n$ and $F : C \to \mathbb{R}^n$ be a map. The classical definition of monotonicity and its stronger variants is as follows:

Definition 9.1 *(i) F is monotone, if for all $x, y \in C$,*

$$\langle F(y) - F(x), y - x \rangle \geq 0.$$

(ii) F is strictly monotone, if for all $x, y \in C, x \neq y$,

$$\langle F(y) - F(x), y - x \rangle > 0.$$

(iii) F is strongly monotone, if there exists $\beta > 0$ such that for all $x, y \in C$,

$$\langle F(y) - F(x), y - x \rangle \geq \beta \left\| y - x \right\|^2.$$

It is clear that every strongly monotone map is strictly monotone, and every strictly monotone map is monotone.

The pseudomonotone counterparts of the above are defined as follows:

Definition 9.2 *(i) F is pseudomonotone [28], if for all $x, y \in C$,*

$$\langle F(x), y-x\rangle \geq 0 \Rightarrow \langle F(y), y-x\rangle \geq 0,$$

or, equivalently,

$$\langle F(x), y-x\rangle > 0 \Rightarrow \langle F(y), y-x\rangle > 0.$$

(ii) F is strictly pseudomonotone [29], if for all $x, y \in C$, $x \neq y$,

$$\langle F(x), y-x\rangle \geq 0 \Rightarrow \langle F(y), y-x\rangle > 0.$$

(iii) Let C be open. Then F is strongly pseudomonotone [21], if for every $x \in C$ and $v \in \mathbb{R}^n$ such that $\|v\| = 1$, $\langle F(x), v\rangle = 0$, there exist positive ε, β such that for all $t \in [0, \varepsilon)$

$$x + tv \in C \text{ and } \langle F(x+tv), v\rangle \geq \beta t. \tag{9.1}$$

In the definition of strong pseudomonotonicity we can equivalently require that $v \in \mathbb{R}^n \setminus \{0\}$ instead of $\|v\| = 1$.

It is evident that a monotone (strictly monotone, strongly monotone) map is pseudomonotone (strictly pseudomonotone, strongly pseudomonotone). Also, a strictly pseudomonotone map is pseudomonotone. In order to compare strictly pseudomonotone with strongly pseudomonotone maps, we first show:

Proposition 9.1 *Let C be open and convex and F be continuous. Then F is pseudomonotone (respectively, strictly pseudomonotone) if and only if, for every $x \in C$ and $v \in \mathbb{R}^n \setminus \{0\}$ such that $\langle F(x), v\rangle = 0$, there exists $\varepsilon > 0$ such that for all $t \in (0, \varepsilon)$,*

$$x + tv \in C \text{ and } \langle F(x+tv), v\rangle \geq 0$$

(respectively, $x + tv \in C$ and $\langle F(x+tv), v\rangle > 0$).

Proof. Let F be pseudomonotone and $x \in C, v \in \mathbb{R}^n \setminus \{0\}$ be such that $\langle F(x), v\rangle = 0$. Since C is open, there exists $\varepsilon > 0$ such that $x + tv \in C$ for all $t \in (0, \varepsilon)$. If we set $y = x + tv$, then obviously $\langle F(x), y-x\rangle = 0$ from which we deduce that $\langle F(x+tv), v\rangle = \frac{1}{t}\langle F(y), y-x\rangle \geq 0$.

We now show the converse. Let $x, y \in C$, $x \neq y$ be such that $\langle F(x), y-x \rangle \geq 0$. Set

$$v = y - x, \quad g(t) = \langle F(x + tv), v \rangle$$

and

$$t_0 = \sup \{t \in [0,1] : g(t) \geq 0\}.$$

Note that $g(0) \geq 0$ and that, by continuity, $g(t_0) \geq 0$. Suppose that $t_0 < 1$. If $g(t_0) > 0$, then by continuity of F, $g(t) \geq 0$ for all $t > t_0$ sufficiently close to t_0. On the other hand, if $g(t_0) = 0$, then we set $z = x + t_0 v$. By assumption, $\langle F(z + tv), v \rangle \geq 0$ for all $t > 0$ sufficiently small, thus we find again that $g(t) \geq 0$ for all $t > t_0$ sufficiently close to t_0. But this contradicts the definition of t_0. Thus, $t_0 = 1$ and $\langle F(y), y-x \rangle = g(1) \geq 0$.

For the second case, one direction is again easy, so we prove only the converse. Let $x, y \in C$, $x \neq y$ be such that $\langle F(x), y-x \rangle \geq 0$. By the previous part, we know that F is pseudomonotone, thus $\langle F(y), y-x \rangle \geq 0$. Suppose that $\langle F(y), y-x \rangle = 0$. Then setting $v = x - y$, we deduce from our assumption that for every $t > 0$ sufficiently small $\langle F(y + tv), v \rangle > 0$, thus $\langle F(y+tv), x-(y+tv) \rangle > 0$. By pseudomonotonicity, $\langle F(x), x-(y+tv) \rangle > 0$, i.e., $\langle F(x), y-x \rangle < 0$, a contradiction. Thus, $\langle F(y), y-x \rangle > 0$ and F is strictly pseudomonotone. □

The following corollary now becomes obvious.

Corollary 9.1 *Let C be open and convex. If F is strongly pseudomonotone and continuous, then F is strictly pseudomonotone.*

We now recall various quasimonotonicity notions.

Definition 9.3 *(i) F is quasimonotone [25, 29], if for all $x, y \in C$,*

$$\langle F(x), y-x \rangle > 0 \Rightarrow \langle F(y), y-x \rangle \geq 0.$$

(ii) Let C be convex. F is strictly quasimonotone [21], if F is quasimonotone and for all $x, y \in C$ there exists $z \in (x, y)$ such that

$$\langle F(z), y-x \rangle \neq 0.$$

(iii) Let C be convex. F is semistrictly quasimonotone [21], if F is quasimonotone and for all $x, y \in C$ with $x \neq y$, the following implication holds:

$$\langle F(x), y-x \rangle > 0 \Rightarrow \exists z \in \left(\tfrac{x+y}{2}, y\right) \text{ such that } \langle F(z), y-x \rangle > 0. \tag{9.2}$$

The following implications are obvious from the definitions: a pseudomonotone map is semistrictly quasimonotone; a strictly pseudomonotone map is strictly quasimonotone. Let us now show:

Proposition 9.2 *[21] A strictly quasimonotone map F is semistrictly quasimonotone.*

Proof. If $\langle F(x), y - x\rangle > 0$ for some $x, y \in C$, then obviously $\langle F(x), z - x\rangle > 0$ holds for every $z \in (x, y)$. Since F is quasimonotone, we deduce that $\langle F(z), z - x\rangle \geq 0$ or, equivalently, $\langle F(z), y - x\rangle \geq 0$. By assumption of strict quasimonotonicity there exists a point $z \in \left(\frac{x+y}{2}, y\right)$ such that $\langle F(z), y - x\rangle \neq 0$. Thus, $\langle F(z), y - x\rangle > 0$ and F is semistrictly quasimonotone. □

Whenever F is continuous, the assumption of quasimonotonicity is superfluous in the definition of semistrict quasimonotonicity.

Proposition 9.3 *[21] Let C be convex. If $F : C \to \mathbb{R}^n$ is continuous and such that for every $x, y \in C$ with $x \neq y$ implication (9.2) holds, then F is semistrictly quasimonotone.*

Proof. We have to show that F is quasimonotone. If $x, y \in C$ are such that $\langle F(x), y - x\rangle > 0$, then we can find $z_1 \in \left(\frac{x+y}{2}, y\right)$ such that $\langle F(z_1), y - x\rangle > 0$ or, equivalently, $\langle F(z_1), y - z_1\rangle > 0$. Using the same argument, we can construct inductively a sequence $(z_i)_{i\in\mathbb{N}}$ such that $z_{i+1} \in \left(\frac{z_i+y}{2}, y\right)$ and $\langle F(z_i), y - x\rangle > 0$. Obviously, z_i converges to y as $i \to +\infty$. By continuity, $\langle F(y), y - x\rangle \geq 0$, i.e., F is quasimonotone. □

The following table gathers all implications between the nine (generalized) monotonicity notions we encountered so far. We use the abbreviations m, pm, qm, str., ss. and s. for monotone, pseudomonotone, quasimonotone, strongly, semistrictly and strictly, respectively. Some implications hold under the assumption of continuity of F and/or convexity of C.

$$
\begin{array}{ccccc}
 & & & & \text{qm} \\
 & & & & \Uparrow \\
\text{m} & \Longrightarrow & \text{pm} & \Longrightarrow & \text{ss.qm} \\
\Uparrow & & \Uparrow & & \Uparrow \\
\text{s.m} & \Longrightarrow & \text{s.pm} & \Longrightarrow & \text{s.qm} \\
\Uparrow & & \Uparrow & & \\
\text{str.m} & \Longrightarrow & \text{str.pm} & &
\end{array}
$$

We note that all implications are proper for general maps.

It is useful to have in mind what happens in the special case $n = 1$. Actually, the following proposition holds.

Proposition 9.4 *A function $f : \mathbb{R} \to \mathbb{R}$ is:*

(i) monotone if and only if it is increasing;

(ii) pseudomonotone if and only if there exist disjoint consecutive intervals (possibly empty) I_1, I_2 and I_3 such that $I_1 \cup I_2 \cup I_3 = \mathbb{R}$ and f is negative on I_1, zero on I_2 and positive on I_3;

(iii) quasimonotone if and only if there exist disjoint consecutive intervals I_1, I_2, one of which may be empty, such that $I_1 \cup I_2 = \mathbb{R}$ and f is nonpositive on I_1 and nonnegative on I_2.

Proof. Let us show (ii) as an example. Let $I_3 = \{x \in \mathbb{R} : f(x) > 0\}$. If $I_3 \neq \emptyset$, choose any $x \in I_3$ and $y > x$. Then $f(x)(y - x) > 0$. By pseudomonotonicity, $f(y)(y - x) > 0$, hence $f(y) > 0$. Thus, I_3 is an interval that has $+\infty$ as right endpoint. By the same argument, $I_1 = \{x \in \mathbb{R} : f(x) < 0\}$ is empty or an interval with $-\infty$ as left endpoint. Consequently, $I_2 = \{x \in \mathbb{R} : f(x) = 0\} = \mathbb{R} \setminus (I_1 \cup I_3)$ is also an interval. □

The case $n = 1$ is not so special as it seems. In fact even in the general n-dimensional case all generalized monotonicity properties that we encountered so far are in some sense one-dimensional. Indeed, $\langle F(x), y - x\rangle$ and $\langle F(y), y - x\rangle$ are the projections of $F(x)$ and $F(y)$ on the straight line passing through x and y, respectively. Thus F is monotone, pseudomonotone or quasimonotone if and only if, roughly speaking, the projection of F on any straight line in C is monotone, pseudomonotone or quasimonotone, respectively. See also [7] for a more precise formulation of this fact, and [33] for a general result on bifunctions and set-valued maps.

To each generalized monotonicity notion defined so far, there exists a corresponding notion of generalized convexity. The reader may find the definitions of convex, strictly convex, quasiconvex, strictly quasiconvex, pseudoconvex, strictly pseudoconvex functions elsewhere in this volume (see for instance [7]). In addition to these six generalized convexity notions we recall three less known ones:

Definition 9.4 *Let C be convex and $f : C \to \mathbb{R}$ be a function.*

(i) f is strongly convex [38] if there exists $\beta > 0$ such that for every $x, y \in C$ and $t \in [0, 1]$,

$$tf(x) + (1 - t)f(y) - f(tx + (1 - t)y) \geq \tfrac{\beta}{2}t(1 - t)\|x - y\|^2. \quad (9.3)$$

(ii) f is semistrictly quasiconvex [1, 27] if for every $x, y \in C$, $x \neq y$, and every $z \in (x, y)$,

$$f(x) < f(y) \Rightarrow f(z) < f(y).$$

Definition 9.5 *Let C be open and convex. A differentiable function $f : C \to \mathbb{R}$ is strongly pseudoconvex [1] if for every $x \in C$ and $v \in \mathbb{R}^n$ such that $\|v\| = 1$ and $\langle \nabla f(x), v \rangle = 0$ there exist positive numbers ε and α such that for all $t \in [0, \varepsilon)$,*

$$x + tv \in C \text{ and } f(x + tv) \geq f(x) + \tfrac{\alpha}{2} t^2.$$

These notions are interrelated as follows:

Proposition 9.5 *Let C be open and convex and $f : C \to \mathbb{R}$ be differentiable.*

(i) If f is strongly convex, then it is strongly pseudoconvex.

(ii) If f is strongly pseudoconvex, then it is strictly pseudoconvex.

Proof. (i) For each $x, y \in C$ and $t \in [0, 1]$ we have, setting $x = x_2$ and $y = x_1$ in (9.3):

$$\frac{f((1-t)x_1 + tx_2) - f(x_1)}{t} \leq -f(x_1) + f(x_2) - \tfrac{\beta}{2}(1-t)\|x_2 - x_1\|^2.$$

Taking the limit as $t \to 0$ we obtain

$$\langle \nabla f(x_1), x_2 - x_1 \rangle \leq f(x_2) - f(x_1) - \tfrac{\beta}{2}\|x_2 - x_1\|^2. \qquad (9.4)$$

Now let $x \in C$ and $v \in \mathbb{R}^n$, $\|v\| = 1$ be such that $\langle \nabla f(x), v \rangle = 0$. Choose $\varepsilon > 0$ such that $x + tv \in C$ for all $t \in [0, \varepsilon)$. Setting $x_1 = x$ and $x_2 = x + tv$ in (9.4) we obtain

$$0 \leq f(x + tv) - f(x) - \tfrac{\beta}{2} t^2,$$

i.e., f is strongly pseudoconvex with $\alpha = \frac{\beta}{2}$.

(ii) We recall that f is strictly pseudoconvex [1] if for every $x, y \in C$, $x \neq y$,

$$\langle \nabla f(x), y - x \rangle \geq 0 \Rightarrow f(y) > f(x).$$

Suppose that $x, y \in C$ are such that $\langle \nabla f(x), y - x \rangle \geq 0$ holds. We set $v = y - x$, $g(t) = f(x + tv) - f(x)$ and

$$t_0 = \sup\{t \in [0, 1] : g(t) \geq 0\}.$$

Using the same argument as in the proof of Proposition 9.1 we deduce that $t_0 = 1$ and $g(1) \geq 0$, i.e., $f(y) \geq f(x)$. Since we just showed that

$$\langle \nabla f(x), y - x \rangle \geq 0 \Rightarrow f(y) \geq f(x),$$

f is pseudoconvex. In particular, it is quasiconvex. The definition of strong pseudoconvexity implies that there exists some $z = x + tv \in (x, y)$ such that

$$f(z) \geq f(x) + \tfrac{\alpha}{2} t^2 > f(x). \tag{9.5}$$

By quasiconvexity,

$$f(z) \leq \max\{f(x), f(y)\} = f(y).$$

Combining with relation (9.5), we deduce that $f(x) < f(y)$ as desired. □

Other implications between the nine notions of (generalized) convexity we mentioned are given in [1]. The following table shows all these implications where we use the abbreviations cx, pcx, qcx, s., ss. and str. to stand for convex, pseudoconvex, quasiconvex, strictly, semistrictly and strongly, respectively. Some implications hold under special assumptions of continuity or differentiabilty.

$$\begin{array}{ccccc}
 & & & & \text{qcx} \\
 & & & & \Uparrow \\
\text{cx} & \Longrightarrow & \text{pcx} & \Longrightarrow & \text{ss.qcx} \\
\Uparrow & & \Uparrow & & \Uparrow \\
\text{s.cx} & \Longrightarrow & \text{s.pcx} & \Longrightarrow & \text{s.qcx} \\
\Uparrow & & \Uparrow & & \\
\text{str.cx} & \Longrightarrow & \text{str.pcx} & &
\end{array}$$

We note that all implications are proper for general functions.

The analogy between the table for generalized convex functions and the table for generalized monotone maps is not accidental. In fact, if $F = \nabla f$, then generalized monotonicity of F is equivalent to generalized convexity of f.

Theorem 9.1 *Let C be open and convex and $f : C \to \mathbb{R}$ be differentiable. Then f is convex (respectively, strictly convex, strongly convex, pseudoconvex, strictly pseudoconvex, quasiconvex, semistrictly quasiconvex, strictly quasiconvex) if and only if ∇f is monotone (respectively, strictly monotone, strongly monotone, pseudomonotone, strictly pseudomonotone, quasimonotone, semistrictly quasimonotone, strictly quasimonotone).*

If f is twice differentiable, then f is strongly pseudoconvex if and only if ∇f is strongly pseudomonotone.

Proof. The equivalence for the convex, strictly convex and strongly convex case is standard knowledge; see for example [1]. The pseudoconvex, strictly pseudoconvex and quasiconvex case was treated in [7].

We show that f is strictly quasiconvex if and only if ∇f is strictly quasimonotone. Let ∇f be strictly quasimonotone and $x, y \in C$. Then ∇f is quasimonotone, hence f is quasiconvex [7, 29]. It follows that for every $w \in (x, y)$, $f(w) \leq \max\{f(x), f(y)\}$ holds. If, say $\max\{f(x), f(y)\} = f(y)$, we have to show that $f(w) < f(y)$. Indeed, if $f(w) = f(y)$, then it is easy to verify that f is constant on $[w, y]$. This means that the function $g(t) = f(tw + (1-t)y)$ is constant on $[0,1]$, hence

$$g'(t) = \langle \nabla f(tw + (1-t)y), y - w \rangle = 0, \quad t \in [0,1].$$

Thus $\langle \nabla f(z), y - w \rangle = 0$ for all $z \in [w, y]$ which contradicts the strict quasimonotonicity of ∇f.

Conversely, if f is strictly quasiconvex, then it is quasiconvex, hence ∇f is quasimonotone. Further, if $x, y \in C$, then strict quasiconvexity of f implies that the function $g(t) = f(tx + (1-t)y)$ is not constant on $[0,1]$, hence $g'(t) \neq 0$ for at least one $t \in [0,1]$. It follows immediately that $\langle \nabla f(z), y - x \rangle \neq 0$ for some $z \in (x, y)$, i.e., ∇f is strictly quasimonotone.

Now we show that f is semistrictly quasiconvex if and only if ∇f is semistrictly quasimonotone. If ∇f is semistrictly quasimonotone, then it is quasimonotone, hence f is quasiconvex. Suppose that f is not semistrictly quasiconvex. Then there exist $x, y \in C$ and $z = sx + (1-s)y$, $s \in (0,1)$ such that $f(x) < f(z) = f(y)$. It is easy to see that f is constant on $[z, y]$. Set $g(t) = f((1-t)x + ty)$. Then $g(0) < g(1)$. Hence by the mean value theorem there exists $t_1 \in (0,1)$ such that $g'(t_1) > 0$. By the strict quasimonotonicity of ∇f there exists $t_2 \in \left(\frac{1+t_1}{2}, 1\right)$ such that $g'(t_2) > 0$. We construct inductively a sequence $(t_i)_{i \in \mathrm{N}}$ such that $g'(t_i) > 0$ and $t_{i+1} \in \left(\frac{1+t_i}{2}, 1\right)$. Obviously $t_i \to 1$ hence $t_i > s$ for some i. Since g is constant on $[s, 1]$ we should have $g'(t_i) = 0$, a contradiction.

The proof that semistrict quasiconvexity of f implies semistrict quasimonotonicity of ∇f is similar to the one above and is left to the reader.

We now prove the last assertion, i.e., that f strongly pseudoconvex is equivalent to ∇f strongly pseudomonotone, provided that f is twice differentiable.

Assume that ∇f is strongly pseudomonotone. If $x \in C$ and $v \in \mathbb{R}^n$ are such that $\|v\| = 1$ and $\langle \nabla f(x), v \rangle = 0$, then we can find $\varepsilon > 0$ and $\beta > 0$ such that $\langle \nabla f(x + tv), v \rangle \geq \beta t$ for all $t \in [0, \varepsilon)$. Set $g(t) = f(x + tv)$. Then

$$g'(t) = \langle \nabla f(x + tv), v \rangle \geq \beta t,$$

hence $g(t) \geq g(0) + \frac{\beta}{2}t^2$, i.e.,

$$f(x+tv) \geq f(x) + \frac{\beta}{2}t^2,$$

and f is strongly pseudoconvex.

Assume that f is strongly pseudoconvex. Take x, v, α and ε as in Definition 9.5. We set $g(t) = f(x+tv)$ and write the Taylor expansion

$$g(t) = g(0) + g'(0)\,t + g''(0)\,\tfrac{t^2}{2} + o\left(t^2\right). \tag{9.6}$$

By definition of v, we know that $g'(0) = \langle \nabla f(x), v\rangle = 0$; also, by definition of ε, $g(t) \geq g(0) + \frac{\alpha}{2}t^2$ holds for all $t \in [0, \varepsilon)$. Thus, (9.6) entails

$$g''(0)\,\tfrac{t^2}{2} + o\left(t^2\right) \geq \tfrac{\alpha}{2}t^2, \qquad t \in [0, \varepsilon).$$

It follows that $g''(0) \geq \alpha$. Hence for every $\beta \in (0, \alpha)$ there exists $\varepsilon' > 0$ such that for all $t \in (0, \varepsilon')$, $g'(t) - g'(0) \geq \beta t$ or, equivalently, $\langle \nabla f(x+tv), v\rangle \geq \beta t$ holds. Thus, ∇f is strongly pseudomonotone. □

Note that without the assumption that f is twice differentiable, the gradient of a strongly pseudoconvex function may not be strongly pseudomonotone [22].

We finish this section by a note on quasimonotone maps which are not pseudomonotone. It is easy to construct such maps. For instance, the function $f : \mathbb{R} \to \mathbb{R}$ defined by $f(x) = \sin x + 1$ is quasimonotone, but not pseudomonotone. This function has zeros, and this plays an important role. In fact, it is known that a quasimonotone map F defined on an open convex subset of $\mathbb{R}^n$ that has no zeros, is actually pseudomonotone [7]. This is no longer true when C is not open. For example, we can take $n = 2$, $C = \mathbb{R} \times \{0\} \subseteq \mathbb{R}^2$ and $F(x_1, 0) = (\sin x_1 + 1, 1)$. Then F is quasimonotone, but not pseudomonotone. But here again we can consider equivalently the projection of F onto C, and this map has zeros. More precisely, there are points $x \in C$ such that $F(x)$ is perpendicular to C in the sense that $\langle F(x), x - y\rangle = 0$, for all $y \in C$. Can we construct an example of a convex set C and a continuous quasimonotone, but not pseudomonotone map F on C such that $F(x)$ is never perpendicular to C? The answer is yes, as shown by the following example, taken from [23]: Set $C = [0, 1] \times [0, 1]$ and

$$f(x_1, x_2) = \left(\frac{-t}{t+1}, \frac{-1}{t+1}\right) \text{ where } t = \frac{x_1 + \sqrt{x_1^2 + 4x_2}}{2}.$$

This map is quasimonotone, but not pseudomonotone. It can be shown that such a situation cannot occur if F is affine (see the last section).

The following proposition is useful when dealing with quasimonotone maps that are not pseudomonotone:

Proposition 9.6 *Suppose that $C \subseteq \mathbb{R}^n$ is convex and $F : C \to \mathbb{R}^n$ is quasimonotone and continuous (or, more generally, that its restriction on every straight line segment in C is continuous). Let $x, y \in C$ be such that $\langle F(x), y - x \rangle \geq 0$ and $\langle F(y), y - x \rangle < 0$. Then*

$$\langle F(x), z - x \rangle \leq 0, \forall z \in C. \tag{9.7}$$

Proof. If (9.7) does not hold, then there exists $z \in C$ such that $\langle F(x), z - x \rangle > 0$. If we set $y_t = (1-t)y + tz$, $t \in [0,1]$ we infer easily from the assumptions that $\langle F(x), y_t - x \rangle > 0$. By quasimonotonicity, $\langle F(y_t), y_t - x \rangle \geq 0$. Taking the limit as $t \to 0^+$ we deduce that $\langle F(y), y - x \rangle \geq 0$, a contradiction. □

If $F(x) \neq 0$ in the above proposition, then (9.7) means that

$$\{z \in \mathbb{R}^n : \langle F(x), z - x \rangle = 0\}$$

is a supporting hyperplane of C at x. Thus, if F has no zeros, then x lies on the boundary of C. We thus recover the result that whenever F is quasimonotone on an open convex set and F has no zeros, then F is pseudomonotone (see [7, 30]).

3. Other kinds of generalized monotonicity

As we noted in the previous section, all generalized monotonicity notions presented so far are in some sense one-dimensional; besides, the careful reader should have noticed that most proofs contain a transcription of the problem to one dimension through the introduction of a function $g : \mathbb{R} \to \mathbb{R}$. In recent years, new generalized monotonicity notions were investigated, which involve n instead of two points and are in essence multi-dimensional. These are cyclic pseudomonotonicity, cyclic quasimonotonicity and proper quasimonotonicity which were found to be useful, especially for the study of variational inequalities. We refer to Chapter 11 of this volume for details.

Another recent advance is the unification of the study of pseudomonotone maps in the sense introduced in the previous section, and the pseudomonotone maps in the sense of Brezis [4] which we will call topologically pseudomonotone to avoid confusion. We give the definition in Banach spaces, since the weak topology is involved. Let X be a Banach space, X^* be its dual, $\langle \cdot, \cdot \rangle$ denote the duality product and $C \subseteq X$ be closed and convex. In such a setting, pseudomonotonicity of a map $F : C \to X^*$ is defined again by (9.2). We recall [4, 46]:

Definition 9.6 *A map $F : C \to X^*$ is topologically pseudomonotone if for each net $(x_i)_{i\in I} \subset C$ and for all $x, y \in C$ the assumptions*

$$x_i \to x \text{ weakly and } \liminf \langle F(x_i), x - x_i \rangle \geq 0$$

imply that

$$\limsup \langle F(x_i), y - x_i \rangle \leq \langle F(x), y - x \rangle .$$

Note that in finite-dimensional spaces, any continuous map is topologically pseudomonotone. This shows clearly the difference between pseudomonotonicity as discussed in the previous section and topological pseudomonotonicity. In view of this fact, it is rather astonishing that Domokos and Kolumban [15] succeeded in treating both cases simultaneously, in relation to variational inequalities. Let us first recall that a map $F : C \to X^*$ is called *hemicontinuous* if for every $x, y \in C$ the function

$$g(t) = \langle F(x + t(y - x), y - x) \rangle$$

is continuous at 0^+. Note that hemicontinuity is almost always assumed in order to show the existence of a solution of variational inequality problems with pseudomonotone maps, see for instance [46].

Definition 9.7 *A map $F : C \to X^*$ is B-pseudomonotone if for each net $(x_i)_{i\in I} \subset C$ and for every $x, y \in C$ the assumptions*

$$x_i \to x \text{ weakly and } \langle F(x_i), x + t(y - x) - x_i \rangle \geq 0, \forall t \in [0,1], i \in I$$

imply that

$$\langle F(x), y - x \rangle \geq 0.$$

The new notion is related to the older ones as follows:

Theorem 9.2 *[15] (i) Every topologically pseudomonotone map is B-pseudomonotone,*

(ii) Every pseudomonotone, hemicontinuous map is B-pseudomonotone.

We also mention a simplified version of [15, Theorem 2]:

Theorem 9.3 *Let $C \subseteq X$ be nonempty, weakly compact and convex. If $F : C \to X^*$ is B-pseudomonotone and continuous on the intersection of C with every finite-dimensional subspace, then there exists $x_0 \in C$ such that*

$$\langle F(x_0), x - x_0 \rangle \geq 0, \forall x \in C.$$

As a consequence of Theorem 9.2, the above theorem generalizes some theorems on the existence of a solution of variational inequalities, for pseudomonotone and for topologically pseudomonotone maps.

Recently some other generalized monotonicity notions, closely related to generalized convexity, were introduced. These include *hypomonotone* and *submonotone* maps, related to *d-weakly convex* and *d-approximately convex* functions, respectively. We refer the reader to [11, 12] for details.

4. Differentiable generalized monotone maps

One of the most interesting problems in the study of generalized monotonicity is to find characterizations of generalized monotone maps [44]. We will present here some of the results existing in the literature, for three special cases: (a) the differentiable case, (b) the nondifferentiable case and (c) the affine case.

For pseudomonotone and quasimonotone differentiable maps, necessary and sufficient conditions have recently been found; these results are presented in another chapter of this volume [7]. For later purposes, we recall from there the main characterization theorem. Consider the conditions:

$$x \in C, v \in \mathbb{R}^n, \quad \langle F(x), v\rangle = 0 \Rightarrow \langle F'(x)v, v\rangle \geq 0. \tag{Sdp}$$

$$\left.\begin{array}{c} x \in C, x - v \in C \\ F(x) = 0, F'(x)v = 0 \\ \langle F(x-v), v\rangle > 0 \end{array}\right\} \Rightarrow \exists t_1 > 0, \forall t \in (0, t_1] : \langle F(x + tv, v)\rangle \geq 0. \tag{Bqr}$$

$$\left.\begin{array}{c} x \in C, F(x) = 0 \\ F'(x)v = 0 \end{array}\right\} \Rightarrow \exists t_1 > 0, \forall t \in (0, t_1] : \langle F(x + tv, v)\rangle \geq 0. \tag{Bpr}$$

Theorem 9.4 *Let C be open and convex and $F : C \to \mathbb{R}^n$ be continuously differentiable.*

(i) F is quasimonotone if and only if conditions (Sdp) and (Bqr) hold;

(ii) F is pseudomonotone if and only if conditions (Sdp) and (Bpr) hold.

We now present a characterization of differentiable strongly pseudomonotone maps.

Proposition 9.7 *[21] Let C be open and convex and $F : C \to \mathbb{R}^n$ be differentiable. Then F is strongly pseudomonotone if and only if for every $x \in C$ there exists $\beta > 0$ such that*

$$\|v\| = 1 \text{ and } \langle F(x), v\rangle = 0 \text{ imply that } \langle F'(x)v, v\rangle \geq \beta. \tag{9.8}$$

Proof. For every $h \in \mathbb{R}^n$ sufficiently small

$$F(x+h) = F(x) + F'(x)h + o(h),$$

thus

$$\langle F(x+h), h\rangle = \langle F(x), h\rangle + \langle F'(x)h, h\rangle + \langle o(h), h\rangle . \tag{9.9}$$

Now suppose that F is strongly pseudomonotone. For each $v \in \mathbb{R}^n$ such that $\|v\| = 1$ and $\langle F(x), v\rangle = 0$, choose $\beta_v > 0$ (depending in general on v) such that (9.1) holds. Then (9.9) entails that for sufficiently small $t > 0$,

$$t^2 \langle F'(x)v, v\rangle + t\langle o(tv), v\rangle \geq \beta_v t^2.$$

Dividing by t^2 and taking $t \to 0^+$, we obtain $\langle F'(x)v, v\rangle \geq \beta_v > 0$.

Since the set

$$D = \{v \in \mathbb{R}^n : \|v\| = 1 \text{ and } \langle F(x), v\rangle = 0\}$$

is compact, we obtain that

$$\min\left\{\langle F'(x)v, v\rangle : v \in D\right\} > 0,$$

i.e., (9.8) holds.

Conversely, suppose that (9.8) holds. We choose again $v \in \mathbb{R}^n$ such that $\|v\| = 1$ and $\langle F(x), v\rangle = 0$ and set $h = tv$ in (9.9). Taking account of (9.8), we get

$$\langle F(x+tv), v\rangle \geq \beta t + \langle o(tv), v\rangle .$$

For any $\beta' \in (0, \beta)$, there exists $\varepsilon > 0$ such that for every $t \in (0, \varepsilon)$,

$$|\langle o(tv), v\rangle| < (\beta - \beta')\, t.$$

We obtain immediately that $\langle F(x+tv), v\rangle \geq \beta' t$, i.e., F is strongly pseudomonotone. □

5. Nondifferentiable maps

We now present some necessary and/or sufficient conditions for a nondifferentiable map to be generalized monotone. We shall consider two cases: (a) the map is locally Lipschitz and (b) the map is continuous. The latter case is of course more general, but we begin with the locally Lipschitz case because it will serve as an introduction to what follows.

5.1 Locally Lipschitz maps

Let $C \subseteq \mathbb{R}^n$ be open and convex, and $F : C \to \mathbb{R}^n$ be a locally Lipschitz map. Then Rademacher's Theorem asserts that F is differentiable almost everywhere on C, i.e., the set of points where F is not differentiable has Lebesgue measure zero. The *generalized Jacobian* at any point $x \in C$ is defined as the set of matrices

$$\partial F(x) = co\left\{\lim F'(x_i) : x_i \to x,\ F \text{ is differentiable at } x_i\right\} \tag{9.10}$$

where co denotes the convex hull. Note that at points x where F is differentiable $\partial F(x)$ contains $F'(x)$, but may contain other elements too, unless the components of F are regular at x [5]. The elements of F are called subgradients. Likewise, if $f : C \to \mathbb{R}$ is a locally Lipschitz function, the subdifferential of f at $x \in C$ is defined by

$$\partial f(x) = co\left\{\lim f'(x_i) : x_i \to x,\ f \text{ is differentiable at } x_i\right\}.$$

The following mean value theorem holds. Part (i) is the well-known Lebourg Mean Value Theorem.

Theorem 9.5 *[5] (i) Let $f : C \to \mathbb{R}$ be locally Lipschitz. Then for every $x, y \in C$ there exists $z \in (x, y)$ and $w \in \partial f(z)$ such that*

$$f(y) - f(x) = \langle w, y - x \rangle .$$

(ii) Let $F : C \to \mathbb{R}^n$ be locally Lipschitz. Then for every $x, y \in C$ there exists

$$A \in co\{\partial F(z) : z \in [x, y]\}$$

such that

$$F(y) - F(x) = A(y - x).$$

One can now show:

Theorem 9.6 *[35] Let $F : C \to \mathbb{R}^n$ be locally Lipschitz. Then the following are equivalent:*

(i) F is monotone.

(ii) At each point $x \in C$ at which F is differentiable, $F'(x)$ is positive semidefinite.

(iii) For every $x \in C$ the subgradients $A \in \partial F(x)$ are positive semidefinite.

Proof. (i) $\Rightarrow$ (ii): Let F be differentiable at x. For every $v \in \mathbb{R}^n$ and $t > 0$ sufficiently small,

$$\langle F(x + tv) - F(x), (x + tv) - x \rangle \geq 0.$$

It follows that

$$\langle F'(x)\, v, v\rangle = \lim_{t\to 0^+} \left\langle \frac{F(x+tv)-F(x)}{t}, v\right\rangle \geq 0.$$

(ii) $\Rightarrow$ (iii): This is evident in view of relation (9.10).

(iii) $\Rightarrow$ (i): Suppose that all elements of $\partial F(x)$ are positive semidefinite for each $x \in C$. Then for each $x, y \in C$ the mean value theorem implies that

$$\langle F(y) - F(x), y-x\rangle \in co\left\{\langle A(y-x), y-x\rangle : A \in \bigcup\nolimits_{z\in[x,y]} \partial F(z)\right\}. \tag{9.11}$$

By our assumption, each element of the convex hull is nonnegative, thus $\langle F(y) - F(x), y-x\rangle \geq 0$, i.e., F is monotone. □

In the same manner, one can show:

Theorem 9.7 *[35] A locally Lipschitz map F is strictly monotone if for every $x \in C$ the subgradients $A \in \partial F(x)$ are positive definite.*

Note that the converse is not true.

We now turn our attention to generalized monotone maps. For each $x \in C$, $v \in \mathbb{R}^n$ define

$$D_+F(x;v) = \sup\{\langle Av, v\rangle : A \in \partial F(x)\}$$
$$D_-F(x;v) = \inf\{\langle Av, v\rangle : A \in \partial F(x)\}.$$

The following proposition holds:

Proposition 9.8 *[35] Let F be locally Lipschitz.*

(i) If F is strongly pseudomonotone, then for each $x \in C$ and $v \in \mathbb{R}^n$ such that $\|v\| = 1$, $\langle F(x), v\rangle = 0$, one has $D_+F(x;v) > 0$.

(ii) If for each $x \in C$ and $v \in \mathbb{R}^n$ such that $\|v\| = 1$, $\langle F(x), v\rangle = 0$, one has $D_-F(x;v) > 0$, then F is strongly pseudomonotone.

Proof. (i) Since F is strongly pseudomonotone, there exist $\beta > 0$ and $\varepsilon > 0$ such that for all $t \in [0, \varepsilon)$, $\langle F(x+tv), v\rangle \geq \beta t$ holds. By the mean value theorem, for each $t \in [0, \varepsilon)$ there exists

$$A \in co\{\partial F(z) : z \in [x, x+tv]\} \tag{9.12}$$

such that $F(x+tv) - F(x) = tAv$, hence $\langle F(x+tv), v\rangle = t\langle Av, v\rangle$. This implies that $\langle Av, v\rangle \geq \beta$. Bearing in mind (9.12), we deduce that there exists $t' \in (0,t)$ and $B \in \partial F(x+t'v)$ such that $\langle Bv, v\rangle \geq \beta$. Taking a sequence $t_i \to 0^+$, we obtain $t'_i \to 0^+$ and a sequence $(B_i)_{i\in\mathbb{N}}$,

$B_i \in \partial F(x + t_i' v)$ such that $\langle B_i v, v\rangle \geq \beta$. Since ∂F is upper semicontinuous at x [5], there exists $B \in \partial F(x)$ such that B *is* a cluster point of (B_i). Hence $\langle Bv, v\rangle \geq \beta$ which implies that $D_+F(x; v) \geq \beta > 0$.

(ii) If F is not strongly pseudomonotone, then there exist $x \in C, v \in \mathbb{R}^n$ with $\|v\| = 1$, $\langle F(x), v\rangle = 0$ such that for each positive ε, β one can find $t \in [0, \varepsilon)$ such that $\langle F(x + tv), v\rangle < \beta t$. Applying again the mean value theorem, we can find A given by (9.12) such that $\langle Av, v\rangle < \beta$. Working as in the first part of the proof, we obtain $B \in \partial F(x)$ such that $\langle Bv, v\rangle \leq \beta$. Since β is an arbitrary positive number, it follows that $D_-F(x; v) \leq 0$, a contradiction. □

We further provide necessary and sufficient conditions for a map to be quasimonotone or pseudomonotone. The interested reader may find the proofs in [35].

Proposition 9.9 *The map F is quasimonotone if and only if the following conditions hold for every $x \in C$ and $v \in \mathbb{R}^n$:*

$$\langle F(x), v\rangle = 0 \Rightarrow D_+F(x; v) \geq 0;$$

$$\left.\begin{array}{c} \langle F(x), v\rangle = 0 \\ 0 \in \{\langle Av, v\rangle : A \in \partial F(x)\} \\ \langle F(x + \bar{t}v), v\rangle > 0 \text{ for some } \bar{t} < 0 \end{array}\right\} \Rightarrow \left\{\begin{array}{c} \exists t_1 > 0 \text{ such that } \forall t \in [0, t_1], \\ \langle F(x + tv), v\rangle \geq 0. \end{array}\right.$$

Proposition 9.10 *The map F is pseudomonotone if and only if the following conditions hold for every $x \in C$ and $v \in \mathbb{R}^n$:*

$$\langle F(x), v\rangle = 0 \Rightarrow D_+F(x; v) \geq 0;$$

$$\left.\begin{array}{c} \langle F(x), v\rangle = 0 \\ 0 \in \{\langle Av, v\rangle : A \in \partial F(x)\} \end{array}\right\} \Rightarrow \exists t_1 > 0, \forall t \in [0, t_1] : \langle F(x + tv), v\rangle \geq 0.$$

The sufficient conditions may be improved [35]:

Proposition 9.11 *Suppose that the following conditions hold for every $x \in C, v \in \mathbb{R}^n \setminus \{0\}$:*

$$\langle F(x), v\rangle = 0 \Rightarrow D_-F(x; v) \geq 0;$$

$$\left.\begin{array}{c} F(x) = 0, D_-F(x; v) = 0 \\ \langle F(x + \bar{t}v), v\rangle > 0 \text{ for some } \bar{t} < 0 \end{array}\right\} \Rightarrow \left\{\begin{array}{c} \exists t_1 > 0 \text{ such that } \forall t \in [0, t_1], \\ \langle F(x + tv), v\rangle \geq 0. \end{array}\right.$$

Then F is quasimonotone.

Proposition 9.12 *Suppose that the following conditions hold for every $x \in C, v \in \mathbb{R}^n$:*

$$\langle F(x), v\rangle = 0 \Rightarrow D_-F(x; v) \geq 0;$$

$$F(x) = 0, D_-F(x; v) = 0 \Rightarrow \exists t_1 > 0, \forall t \in [0, t_1] : \langle F(x + tv), v\rangle \geq 0.$$

Then F is pseudomonotone.

5.2 Continuous maps

In the more general case where F is continuous, but not necessarily locally Lipschitz, necessary and sufficient conditions can still be found with the help of *approximate Jacobians.* Let $C \subseteq \mathbb{R}^n$ be open and convex, and $F : C \to \mathbb{R}^m$ be continuous. Given $v \in \mathbb{R}^m$, let $vF : C \to \mathbb{R}$ be the function $vF(x) = \langle v, F(x)\rangle$. The two Dini directional derivatives of vF are defined as follows:

$$(vF)^- (x,u) = \liminf_{t\to 0^+} \frac{(vF)(x+tu) - (vF)(x)}{t}$$
$$(vF)^+ (x,u) = \limsup_{t\to 0^+} \frac{(vF)(x+tu) - (vF)(x)}{t}.$$

Definition 9.8 *A closed subset S of $L(\mathbb{R}^n, \mathbb{R}^m)$ is an approximate Jacobian of the map $F : C \to \mathbb{R}^m$ at $x \in C$ if*

$$(vF)^- (x,u) \leq \sup_{M\in S} \langle Mv, u\rangle, \ \forall u \in \mathbb{R}^n, \ \forall v \in \mathbb{R}^m. \tag{9.13}$$

Definition 9.9 *A closed subset S of $L(\mathbb{R}^n, \mathbb{R}^m)$ is a regular approximate Jacobian of the map $F : C \to \mathbb{R}^m$ at $x \in C$ if*

$$(vF)^+ (x,u) = \sup_{M\in S} \langle Mv, u\rangle, \ \forall u \in \mathbb{R}^n, \ \forall v \in \mathbb{R}^m. \tag{9.14}$$

Note that (9.14) implies (9.13). Also, note that

$$\begin{aligned}((-v)F)^- (x,u) &= \liminf_{t\to 0^+} \frac{(-vF)(x+tu) - (-vF)(x)}{t} \\ &= -\limsup_{t\to 0^+} \frac{(vF)(x+tu) - (vF)(x)}{t} = -(vF)^+ (x,u)\end{aligned}$$

and likewise $\sup_{M\in S} \langle M(-v), u\rangle = -\inf_{M\in S} \langle Mv, u\rangle$. Since (9.13) and (9.14) must hold for all v, it is thus clear that they can also be written

$$(vF)^+ (x,u) \geq \inf_{M\in S} \langle Mv, u\rangle, \ \forall u \in \mathbb{R}^n, \ \forall v \in \mathbb{R}^m$$

and

$$(vF)^- (x,u) = \inf_{M\in S} \langle Mv, u\rangle, \ \forall u \in \mathbb{R}^n, \ \forall v \in \mathbb{R}^m,$$

respectively.

Approximate Jacobians are by no means unique. For instance, the whole space $L(\mathbb{R}^n, \mathbb{R}^m)$ is an approximate Jacobian of every function F

at every point $x \in C$. Whenever F is locally Lipschitz, the Clarke generalized Jacobian $\partial F(x)$ is one approximate Jacobian. In what follows, the symbol $\partial^* F(x)$ is used to denote an arbitrary approximate Jacobian of F at x.

The following mean value theorem holds:

Theorem 9.8 *[26] Let $F : C \to \mathbb{R}^n$ be continuous and $x, y \in C$. Suppose that $\partial^* F(z)$ is an approximate Jacobian of F at each $z \in [x, y]$. Then*

$$F(y) - F(x) = \overline{co}\left\{A(y-x) : A \in \bigcup\nolimits_{z \in [x,y]} \partial^* F(z)\right\}$$

where $\overline{co}$ denotes the closed convex hull. If each $\partial^* F(z)$, $z \in [x, y]$ is bounded, then co may replace $\overline{co}$ in the above relation.

Note that the theorem becomes trivial, and uninteresting, if we choose $\partial^* F(z) = L(\mathbb{R}^n, \mathbb{R}^m)$ for all $z \in [x, y]$. The theorem becomes more interesting if $\partial^* F(z)$ is small. In case F is locally Lipschitz, then the Clarke generalized Jacobian $\partial F(x)$ is a bounded approximate Jacobian, hence the above theorem generalizes Theorem 9.5 of the previous subsection.

Remark 9.1 Let S be an arbitrary subset of $L(\mathbb{R}^n, \mathbb{R}^m)$. If we set $S_1 = \overline{co}S$, then it is easy to show that the following equality holds for every $v \in \mathbb{R}^m$, $u \in \mathbb{R}^n$:

$$\sup_{M \in S} \langle Mv, u \rangle = \sup_{M \in S_1} \langle Mv, u \rangle . \tag{9.15}$$

Note also that in the mean value theorem only convex hulls appear. Thus we could suppose, with no harm of generality, that all approximate Jacobians considered are convex sets. However we will follow the usual practice and will not do so.

In the remaining part of this paragraph we will suppose that $m = n$, and study the generalized monotonicity of F by using approximate Jacobians. Let us begin with the monotone case:

Proposition 9.13 *[26] Let $F : C \to \mathbb{R}^n$ be continuous and $\partial^* F(x)$ be an approximate Jacobian of F at each $x \in C$.*

(i) If for each $x \in C$ the matrices $M \in \partial^* F(x)$ are positive semidefinite, then F is monotone.

(ii) If for each $x \in C$ the set $\partial^* F(x)$ is bounded, and its elements are positive definite, then F is strictly monotone.

Proof. (i) For every $x, y \in C$ by the Mean Value Theorem,

$$\langle F(y) - F(x), y - x \rangle = \overline{co} \left\{ \langle A(y - x), y - x \rangle : A \in \bigcup_{z \in [x,y]} \partial^* F(z) \right\}. \tag{9.16}$$

Note that by our assumption,

$$\langle A(y - x), y - x \rangle \geq 0, \quad \forall A \in \bigcup_{z \in [x,y]} \partial^* F(z).$$

Hence all elements of the set at the right-hand side of (9.16) are non-negative. This implies that $\langle F(y) - F(x), y - x \rangle \geq 0$, so F is monotone.

(ii) The assumption of boundedness implies that $\overline{co}$ in (9.16) may be replaced by *co*. Since every $A \in \bigcup_{z \in [x,y]} \partial^* F(z)$ is positive definite, the expression $\langle A(y - x), y - x \rangle$ is positive for every $x \neq y$ in C. The convex combinations of such expressions are also positive, thus $\langle F(y) - F(x), y - x \rangle > 0$, and F is strictly monotone. □

Because of the arbitrariness in the choice of $\partial^* F(x)$, one cannot hope that the converse of the above proposition holds without any restriction on $\partial^* F(x)$. As the following result shows, it is sufficient to assume that $\partial^* F(x)$ is everywhere regular, or even less than that:

Proposition 9.14 *[26] Let $F : C \to \mathbb{R}^n$ be continuous and $\partial^* F(x)$ be an approximate Jacobian at every $x \in C$. Suppose that $\partial^* F(x)$ is regular at a dense subset K of C and that at each $x \in C \backslash K$,*

$$\partial^* F(x) \subseteq \left\{ \lim_{i \to \infty} M_i : (x_i)_{i \in \mathbb{N}} \subset K,\ x_i \to x,\ M_i \in \partial^* F(x_i) \right\}.$$

If F is monotone, then for each $x \in C$ the matrices $M \in \partial^* F(x)$ are positive definite.

Proof. For every $x \in K$, $u \in \mathbb{R}^n$, and for all $t > 0$ sufficiently small, monotonicity of F implies that $\langle F(x + tu) - F(x), u \rangle \geq 0$. It follows that $(uF)^-(x, u) \geq 0$. By regularity,

$$\inf_{M \in \partial^* F(x)} \langle Mu, u \rangle = (uF)^-(x, u) \geq 0.$$

Hence for each $M \in \partial^* F(x)$, $\langle Mu, u \rangle \geq 0$ holds, i.e., M is positive semidefinite.

If $x \in C \backslash K$ and $u \in \mathbb{R}^n$, then for every $M \in \partial^* F(x)$ there exist by assumption $(x_i) \subset K$, $M_i \in \partial^* F(x_i)$ such that $x_i \to x$ and $M_i \to M$. We just showed that $\langle M_i u, u \rangle \geq 0$; hence $\langle Mu, u \rangle = \lim_{i \to +\infty} \langle M_i u, u \rangle \geq 0$ and M is positive semidefinite. □

The following results have been established for quasimonotone maps [26].

Proposition 9.15 *Let $F : C \to \mathbb{R}^n$ be continuous and $\partial^* F(x)$ be an approximate Jacobian at every $x \in C$. If F is quasimonotone, then*

(i) $\langle F(x), u\rangle = 0 \Rightarrow \sup_{M\in\partial^* F(x)} \langle Mu, u\rangle \geq 0$,

(ii) $\langle F(x), u\rangle = 0$ *and* $\langle F(x+\bar{t}u), u\rangle > 0$ *for some* $\bar{t} < 0$ *imply the existence of* $t_1 > 0$ *such that* $\langle F(x+tu), u\rangle \geq 0$ *for all* $t \in [0, t_1]$.

There is also a kind of converse:

Proposition 9.16 *[26] Let $F : C \to \mathbb{R}^n$ be continuous and $\partial^* F(x)$ be a bounded approximate Jacobian at every $x \in C$. Suppose that $\partial^* F(x)$ is regular at a dense subset K of C and that at each $x \in C\backslash K$,*

$$\partial^* F(x) \subseteq \left\{ \lim_{i\to\infty} M_i : (x_i)_{i\in\mathbb{N}} \subset K,\ x_i \to x,\ M_i \in \partial^* F(x_i) \right\}.$$

Assume further that the following conditions hold for each $x \in C$, $u \in \mathbb{R}^n$:

(i) $\langle F(x), u\rangle = 0 \Rightarrow \max_{M\in\partial^* F(x)} \langle Mu, u\rangle \geq 0$,

(ii) $\langle F(x), u\rangle = 0$, $0 \in \{\langle Mu, u\rangle : M \in \partial^* F(x)\}$ *and* $\langle F(x+\bar{t}u), u\rangle > 0$ *for some* $\bar{t} < 0$ *imply the existence of* $t_1 > 0$ *such that* $\langle F(x+tu), u\rangle \geq 0$ *for all* $t \in [0, t_1]$.

Then F is quasimonotone.

Finally, we mention the analogous results for pseudomonotone maps [26].

Proposition 9.17 *Let $F : C \to \mathbb{R}^n$ be continuous and $\partial^* F(x)$ be an approximate Jacobian at every $x \in C$. If F is pseudomonotone, then*

(i) $\langle F(x), u\rangle = 0 \Rightarrow \sup_{M\in\partial^* F(x)} \langle Mu, u\rangle \geq 0$,

(ii) $\langle F(x), u\rangle = 0$ *implies the existence of* $t_1 > 0$ *such that for all* $t \in [0, t_1]$, $\langle F(x+tu), u\rangle \geq 0$.

Proposition 9.18 *Let $F : C \to \mathbb{R}^n$ be continuous and $\partial^* F(x)$ be a bounded approximate Jacobian at every $x \in C$. Suppose that $\partial^* F(x)$ is regular at a dense subset K of C and that at each $x \in C\backslash K$,*

$$\partial^* F(x) \subseteq \left\{ \lim_{i\to\infty} M_i : (x_i)_{i\in\mathbb{N}} \subset K,\ x_i \to x,\ M_i \in \partial^* F(x_i) \right\}.$$

Assume further that the following conditions hold for each $x \in C$, $u \in \mathbb{R}^n$:

(i) $\langle F(x), u\rangle = 0 \Rightarrow \max_{M\in\partial^* F(x)} \langle Mu, u\rangle \geq 0$,

(ii) $\langle F(x), u\rangle = 0$, *and* $0 \in \{\langle Mu, u\rangle : M \in \partial^* F(x)\}$ *imply the existence of* $t_1 > 0$ *such that* $\langle F(x+tu), u\rangle \geq 0$ *for all* $t \in [0, t_1]$.

Then F is pseudomonotone.

The last four propositions generalize the previously mentioned results for locally Lipschitz maps and, of course, corresponding results for differentiable maps [9, 30].

6. Affine maps

Generalized monotone affine maps of the form

$$F(x) = Mx + q, x \in C$$

where M is an $n \times n$ matrix, $q \in \mathbb{R}^n$ and $C \subseteq \mathbb{R}^n$ is convex, have been studied by various authors [8, 10, 19, 20, 30, 37][1]. We distinguish between two cases, depending on whether M is symmetric or not necessarily symmetric.

6.1 Generalized convex quadratic functions (M symmetric)

We assume that $C \subseteq \mathbb{R}^n$ is a solid convex set, i.e., $\operatorname{int} C$ is nonempty. It is well known that the affine map F is the gradient of a function f if and only if M is symmetric. In this case, f is quadratic and is given (up to a constant) by

$$f(x) = \frac{1}{2}\langle Mx, x\rangle + \langle q, x\rangle .$$

A complete characterization of quasiconvex functions f on C is as follows (for a related result obtained independently, see [16, 17])

Theorem 9.9 *[40] A nonconvex function $f(x) = \frac{1}{2}\langle Mx, x\rangle + \langle q, x\rangle$ is quasiconvex on a solid convex set $C \subseteq \mathbb{R}^n$ if and only if*

(a) M has exactly one (simple) negative eigenvalue,

(b) $\operatorname{rank}(M, q) = \operatorname{rank} M$,

(c) $C \subseteq C_1$ or $C \subseteq C_2$ where

$$C_1 = \left\{x \in \mathbb{R}^n : f(x) \le f(s), \left\langle t^1, x\right\rangle \le \left\langle t^1, s\right\rangle\right\},$$
$$C_2 = \left\{x \in \mathbb{R}^n : f(x) \le f(s), \left\langle t^1, x\right\rangle \ge \left\langle t^1, s\right\rangle\right\},$$

s solves $\nabla f(s) = Ms + q = 0$ and t^1 is an eigenvector associated with the one negative eigenvalue of M.

[1] Very recently a relationship between arbitrary pseudomonotone maps T on $\mathbb{R}^n$ such that $-T$ is also pseudomonotone on $\mathbb{R}^n$ and certain affine maps has been derived in [3].

The three conditions (a), (b) and (c) in Theorem 9.9 hold also under the weaker assumption that f is locally quasiconvex[2] at least at one point x_0 for which $\nabla f(x_0) \neq 0$; see [31, 32] for details.

Pseudoconvexity holds if $C \subseteq C_1^0$ or $C \subseteq C_2^0$ where C_1^0, C_2^0 are obtained from C_1, C_2 by requiring the second inequality to be strict. Hence on open convex sets C quasiconvex functions are pseudoconvex.

The function f is strictly pseudoconvex if and only if f is pseudoconvex and M is nonsingular.

For quadratic functions on open convex sets the diagram in Section 2 is simplified as follows [43]:

$$\begin{array}{ccccc} & & & & \text{qcx} \\ & & & & \Updownarrow \\ \text{cx} & \Longrightarrow & \text{pcx} & \Longleftrightarrow & \text{ss. qcx} \\ \Uparrow & & \Uparrow & & \Uparrow \\ \text{s. cx} & \Longrightarrow & \text{s. pcx} & \Longleftrightarrow & \text{s. qcx} \\ \Updownarrow & & \Updownarrow & & \\ \text{str. cx} & \Longrightarrow & \text{str. px} & & \end{array}$$

Like in the convex case, there are only two types of generalized convex quadratic functions (on open convex sets), and they are distinguished by whether M is nonsingular or singular.

An essential step in the proof of Theorem 9.9 is to show:

Lemma 9.1 *[40] If f is nonconvex quasiconvex on a solid convex set $C \subseteq \mathbb{R}^n$, then*

$$f(x) \leq f(s) \text{ for all } x \in C$$
$$\text{and } g(x) = -(f(s) - f(x))^{\frac{1}{2}} \text{ is convex on } C.$$

Hence quasiconvex quadratic functions are convexifiable through scaling.

Lemma 9.1 also leads to a characterization of quasiconvex functions in terms of an augmented Hessian [41] and a bordered Hessian [2, 6, 43, 45]; see also [42].

In the special case of $C = \mathbb{R}_+^n$ finite criteria for quasiconvexity have been derived. First these were obtained by Martos [36] with help of so-called positive subdefinite matrices. These can also be derived by specializing Theorem 9.9.

[2]A function f is called locally quasiconvex at x_0 if there exists a neighborhood $U(x_0)$ of x_0 such that for all $x \in U(x_0)$, $f(x) \leq f(x_0)$ implies $\langle \nabla f(x_0), x - x_0 \rangle \leq 0$.

Theorem 9.10 *[42] A nonconvex quadratic function $f(x) = \frac{1}{2}\langle Mx, x\rangle + \langle q, x\rangle$ is quasiconvex on $\mathbb{R}^n_+$ if and only if*
(a) M has exactly one (simple) negative eigenvalue,
(b) there exists $s \in \mathbb{R}^n$ such that $\nabla f(s) = Ms + q = 0$ and $\langle q, s\rangle \geq 0$,
(c) $M \leq 0$ and $q \leq 0$.

Finite criteria in terms of the principal minors of $\begin{pmatrix} M & q \\ q^t & 0 \end{pmatrix}$ can also be given [42].

For an extensive treatment of the various characterizations of generalized convex quadratic functions we refer the reader to [1, Ch. 6]

6.2 Generalized monotone affine maps (M not necessarily symmetric)

An affine map F is continuously differentiable with $F' = M$. It turns out that pseudomonotonicity of an affine map F requires only condition (Sdp) which, written for affine maps, becomes

$$x \in C, v \in \mathbb{R}^n, \quad \langle Mx + q, v\rangle = 0 \Rightarrow \langle Mv, v\rangle \geq 0 \qquad \text{(Sdp')}$$

Proposition 9.19 *[30] If condition (Sdp') holds for every $x \in C$, then F is pseudomonotone. The converse also holds, provided that C is open.*

Proof. Suppose that (Sdp') holds for every $x \in C$, but F is not pseudomonotone. Then there exist $x, y \in C$ such that $\langle Mx + q, y - x\rangle \geq 0$ and $\langle My + q, y - x\rangle < 0$. Set $v = y - x$ and consider the function

$$\psi(t) = t\langle Mv, v\rangle + \langle Mx + q, v\rangle .$$

Note that $\psi(0) \geq 0$ and $\psi(1) < 0$. Hence, there exists $t' \in [0, 1)$ such that $\psi(t') = 0$. If we set $z = x + t'v$, then $z \in [x, y] \subseteq C$, $\langle Mz + q, v\rangle = \psi(t') = 0$ and $\langle Mv, v\rangle < 0$, thus contradicting (Sdp').

Conversely, let C be open and F be pseudomonotone. Suppose that for some $x \in C, v \in \mathbb{R}^n$, $\langle Mx + q, v\rangle = 0$ holds. By pseudomonotonicity of F,

$$\langle Mx + q, (x + tv) - x\rangle = 0 \Rightarrow \langle M(x + tv) + q, (x + tv) - x\rangle \geq 0.$$

If we subtract the left from the right hand-side, we obtain $\langle Mv, v\rangle \geq 0$. □

In fact, whenever C is open and convex, the above proposition can be derived directly from Theorem 9.4. Indeed, it is easy to see that conditions (Bpr) and (Bqr) are automatically satisfied if F is affine.

This observation leads to the following immediate corollary of Theorem 9.4:

Proposition 9.20 *[30] An affine map defined on an open convex subset of $\mathbb{R}^n$ is pseudomonotone if and only if it is quasimonotone.*

Propositions 9.19 and 9.20 have another easy consequence.

Proposition 9.21 *[30] Let C be open and convex and $F : C \to \mathbb{R}^n$ be affine and quasimonotone. Then F is monotone if one of the following holds:*
(i) $F(x_0) = 0$ for some $x_0 \in C$;
(ii) $C = \mathbb{R}^n$.

Proof. Since F is quasimonotone, it is also pseudomonotone. In case $F(x_0) = 0$ for some $x_0 \in C$, Proposition 9.19 implies that $\langle Mv, v \rangle \geq 0$ for all $v \in \mathbb{R}^n$, i.e., M is positive semidefinite. This is obviously equivalent to monotonicity of F.

Now suppose that $C = \mathbb{R}^n$. Given any $v \in \mathbb{R}^n$, choose any $x \in \mathbb{R}^n$ and set $x_t = x + tv$ and

$$\psi(t) = \langle Mx_t + q, v \rangle = \langle Mv, v \rangle t + \langle Mx + q, v \rangle .$$

Note that either $\langle Mv, v \rangle = 0$ or ψ is a nonconstant affine function. In the latter case, there exists $t \in \mathbb{R}$ such that $\psi(t) = 0$. Then Proposition 9.19 implies that $\langle Mv, v \rangle \geq 0$. Thus M is again positive semidefinite and F is monotone. □

In many cases of interest, the set C is not open. This happens for instance in linear complementarity problems where C is the nonnegative orthant $\mathbb{R}^n_+$. In such a case the following result holds:

Proposition 9.22 *Let C be convex with nonempty interior and F be quasimonotone.*
(a) If $x, y \in C$ are such that $\langle F(x), y - x \rangle = 0$ and $\langle F(y), y - x \rangle < 0$, then x belongs to the boundary of C and $F(x) = 0$.
(b) If F has no zeros on the boundary of C, then F is pseudomonotone.

Proof. (a) Suppose that $\langle F(x), y - x \rangle = 0$ and $\langle F(y), y - x \rangle < 0$. Then F is not pseudomonotone. If we suppose that $F(x) = 0$, then it is not possible that $x \in \operatorname{int} C$ since Proposition 9.21 would entail that F is monotone. Thus x belongs to the boundary of C, and we are done.

Now suppose that $F(x) \neq 0$. If we set $v = y - x$, we find that $\langle Mx + q, v \rangle = 0$ and $\langle M(x + v) + q, v \rangle < 0$, from which follows that $\langle Mv, v \rangle < 0$.

Choose any $z \in \operatorname{int} C$ and set $z_t = (1-t)x + tz$, $t \in (0,1)$. Then $z_t \in \operatorname{int} C$ and $\lim_{t\to 0^+} F(z_t) = F(x) \neq 0$. Hence $F(z_t) \neq 0$ for t sufficiently small. Thus we may define

$$v_t = v - \langle F(z_t), v\rangle \frac{F(z_t)}{\|F(z_t)\|^2}.$$

Note that $\langle F(z_t), v_t\rangle = 0$. Since F is quasimonotone on $\operatorname{int} C$, in virtue of Proposition 9.20 it is also pseudomonotone on $\operatorname{int} C$. Hence, applying Proposition 9.19 to z_t, v_t we deduce that $\langle Mv_t, v_t\rangle \geq 0$. Taking the limit as $t \to 0^+$ and using $\lim_{t\to 0^+} v_t = v$ we obtain $\langle Mv, v\rangle \geq 0$, a contradiction.

(b) Follows from (a). □

We now mention some results from [8, 10] that link generalized monotonicity of F to intrinsic properties of the matrix M. We first give some definitions and notation.

Given a $n\times n$ matrix D, its Moore-Penrose pseudoinverse is a uniquely defined $n \times n$ matrix $D^\dagger$ which satisfies

$$DD^\dagger D = D, \quad D^\dagger DD^\dagger = D^\dagger, \quad \left(DD^\dagger\right)^t = DD^\dagger, \quad \left(D^\dagger D\right)^t = D^\dagger D.$$

Let $F(x) = Mx + q$ be defined on a convex subset C of $\mathbb{R}^n$ with nonempty interior. We set:

$B = \frac{1}{2}\left(M + M^t\right)$,

n_+, n_- and n_0 is the number of positive, negative and zero eigenvalues of B, respectively,

$r = \dim(Kern(M))$,

$D = M^t B^\dagger M$,

$f(x) = \left\langle Mx + q, B^\dagger (Mx+q)\right\rangle$,

$S = \{x : f(x) \leq 0\}$,

$T = \{x : \langle Dx, x\rangle \leq 0\}$.

The following result holds:

Proposition 9.23 *[10] F is quasimonotone on C if and only if one of the following conditions holds:*

(i) $n_- = 0$ (in which case B is positive semidefinite and F is monotone on $\mathbb{R}^n$);

(ii) $n_- = 1$, $-q \notin M(\operatorname{int} C)$, $q \in B(\mathbb{R}^n) \supseteq M(\mathbb{R}^n)$ and $C \subseteq S$.

A detailed analysis of case (ii) results in the following theorem:

Theorem 9.11 *[10] F is quasimonotone on C (and pseudomonotone on $\operatorname{int} C$) if and only if one of the following conditions holds:*

(i) $n_- = 0$, *i.e., B is positive semidefinite and F is monotone on* $\mathbb{R}^n$;

(ii) $n_- = 1$, $r = n_0 + 1$, $-q \notin M(\operatorname{int} C)$, $q \in B(\mathbb{R}^n) \supseteq M(\mathbb{R}^n)$, *D is positive semidefinite, S is a closed convex set, and* $C \subseteq S$;

(iii) $n_- = 1$, $r = n_0$, $-q \notin M(\operatorname{int} C)$, $q \in B(\mathbb{R}^n) = M(\mathbb{R}^n)$, *and* $T = T_+ \cup -T_+$ *where* T_+ *is a closed convex cone with nonempty interior; and if* $\overline{x}$ *is such that* $M\overline{x} = q$, *then either* $C + \overline{x} \subseteq T_+$ *or* $C + \overline{x} \subseteq -T_+$.

If M is symmetric, then $D = B = M$ and and case (ii) in Theorem 9.11 cannot occur since $r = n_0$. In the symmetric case a nonmonotone affine map is characterized by (iii) which recovers the earlier results for generalized convex quadratic functions, in particular Theorem 9.9 and its implications; see [40, 41, 42].

If M is nonsingular, case (ii) in Theorem 9.11 cannot occur either since $n_0 = r - 1$ would imply $n_0 = -1$.

Of particular interest in Theorem 9.11 is the case $C = \mathbb{R}^n_+$ because of its relevance to complementarity problems. We need some additional notation and definitions. Given $z = (z_1, z_2, \ldots, z_n) \in \mathbb{R}^n$, we write $z \geq 0$ if $z \in \mathbb{R}^n_+$ and denote by z^+ and z^- the vectors such that $z_i^+ = \max\{z_i, 0\}$ and $z_i^- = \max\{-z_i, 0\}$.

A matrix M is called:

positive subdefinite if $\langle Mz, z\rangle < 0$ implies either $M^t z \leq 0$ or $M^t z \geq 0$;

copositive if $\langle Mx, x\rangle \geq 0$ for all $x \geq 0$.

One can show the following characterization of generalized monotone affine maps of the form $F(x) = Mx + q$:

Theorem 9.12 *[8] Let* $F(x) = Mx + q$ *be an affine map.*

(i) F is pseudomonotone on $\mathbb{R}^n_+$ *if and only if*

$$z \in \mathbb{R}^n,\ \langle Mz, z\rangle < 0 \Rightarrow \begin{cases} M^t z \geq 0 \text{ and } \langle z, q\rangle \geq 0 \\ \text{or} \\ M^t z \leq 0,\ \langle z, q\rangle \leq 0 \text{ and } \langle Mz^- + q, z\rangle < 0. \end{cases}$$

(ii) F is quasimonotone on $\mathbb{R}^n_+$ *if and only if*

$$z \in \mathbb{R}^n,\ \langle Mz, z\rangle < 0 \Rightarrow \begin{cases} M^t z \geq 0 \text{ and } \langle z, q\rangle \geq 0 \\ \text{or} \\ M^t z \leq 0, \text{ and } \langle z, q\rangle \leq 0. \end{cases}$$

We mention a result that characterizes pseudomonotone *linear* maps on $\mathbb{R}^n_+$.

Proposition 9.24 *[8] A linear map defined by an $n \times n$ matrix M is pseudomonotone on $\mathbb{R}^n_+$ if and only if it is positive subdefinite and copositive, with the additional assumption in case $M = ab^t$ that $a_i = 0$ whenever $b_i = 0$.*

The following result was shown in [8] as a consequence of the characterization contained in Theorem 9.12. We present here a more direct proof.

Proposition 9.25 *Let $F(x) = Mx + q$ be affine and quasimonotone on $\mathbb{R}^n_+$. If $q \neq 0$, then F is pseudomonotone.*

Proof. Suppose that F is not pseudomonotone. Then there exist $x, y \in C$ such that $\langle F(x), y - x\rangle = 0$ and $\langle F(y), y - x\rangle < 0$. From Proposition 9.22, it follows that $F(x) = 0$, i.e., $Mx = -q$.

For every $\lambda > 0$ we obtain $F(\lambda x) = \lambda M x + q = (1 - \lambda) q$. Since F is continuous, we can choose $w \in \operatorname{int} \mathbb{R}^n_+$, sufficiently close to y, such that $\langle F(w), w - x\rangle < 0$. Then for any λ sufficiently close to 1, $\langle F(w), \lambda x - w\rangle > 0$. By quasimonotonicity $\langle F(\lambda x), \lambda x - w\rangle \geq 0$, hence $(1 - \lambda)\langle q, \lambda x - w\rangle \geq 0$. For $\lambda < 1$ we obtain $\langle q, \lambda x - w\rangle \geq 0$, thus $\langle q, x - w\rangle \geq 0$. Likewise, the case $\lambda > 1$ gives $\langle q, x - w\rangle \leq 0$. Thus we obtain finally $\langle q, x - w\rangle = 0$.

We deduce that $\langle q, x - w\rangle = 0$ for any w in a small ball near y. It is now easy to infer that $q = 0$, a contradiction. □

The assumption $q \neq 0$ cannot be omitted in the above proposition, as one can check by considering $F(x) = Mx$ defined on $\mathbb{R}^2_+$ by

$$M = \begin{pmatrix} 0 & 0 \\ 0 & -1 \end{pmatrix}.$$

We finally present another characterization of quasimonotone affine maps. Set $M^s = 2M^t(M + M^t)^{\dagger} M$.

Theorem 9.13 *[8] Assume that M is not positive semidefinite. Then $F(x) = Mx + q$ is quasimonotone on $\mathbb{R}^n_+$ if and only if exactly one of the following conditions holds:*

(i) $M = ab^t$, $b \neq ta$ for all $t > 0$, and either ($b \geq 0$ and $q = \lambda a - \mu b$ with $\lambda, \mu \geq 0$) or ($b \leq 0$ and $q = \mu b - \lambda a$ with $\lambda, \mu \geq 0$).

(ii) $\operatorname{rank}(M) \geq 2$, $q \in M(\mathbb{R}^n) = M^t(\mathbb{R}^n) = (M + M^t)(\mathbb{R}^n)$, $n_- = 1$, and $M^s \leq 0$. Let $\overline{x}$ be such that $q = M\overline{x}$. Then $\langle M^s \overline{x}, \overline{x}\rangle \leq 0$ and $M^s \overline{x} \leq 0$.

In [8] also the question of feasibility and solvability of linear complementarity problems is studied using the characterizations of pseudomonotone and quasimonotone affine maps derived therein. The results extend earlier ones by Gowda on pseudomonotone affine maps and matrices [18, 19, 20].

References

[1] Avriel, M., Diewert, W.E., Schaible, S. and I. Zang, *Generalized Concavity*, Plenum Publishing Corporation, New York, 1988.

[2] Avriel, M. and S. Schaible, *Second order characterizations of pseudoconvex functions*, Math. Programming 14, 1978, 170–185.

[3] Bianchi, M., Hadjisavvas, N. and S. Schaible, *On pseudomonotone maps T for which $-T$ is also pseudomonotone*, J. Convex Anal. 10, 2003, 149-168.

[4] Brezis, H., *Equations et inéquations nonlinéaires dans les espaces vectoriels en dualité*, Ann. Inst. Fourier (Grenoble) 18, 1968, 115–175.

[5] Clarke, F.H., *Optimization and Nonsmooth Analysis*, Wiley Interscience, New York, New York, 1983.

[6] Cottle, R.W. and J.A. Ferland, *Matrix-theoretic criteria for the quasiconvexity and pseudoconvexity of quadratic functions*, Linear Algebra Appl. 5, 1972, 123–136.

[7] Crouzeix, J.P., Chapter 2, this volume.

[8] Crouzeix, J.P., Hassouni, A., Lahlou, A. and S. Schaible, *Positive sub-definite matrices, generalized monotonicity and linear complementarity problems*, SIAM J. Matrix Anal. Appl. 22, 2000, 66–85.

[9] Crouzeix, J.P. and J.A. Ferland, *Criteria for differentiable generalized monotone maps*, Math. Program. 75, 1996, 399–406.

[10] Crouzeix, J.P. and S. Schaible, *Generalized monotone affine maps*, SIAM J. Matrix Anal. Appl. 17, 1996, 992–997.

[11] Daniilidis, A. and P. Georgiev, *Approximate convexity and submononicity*, Preprint, 2002.

[12] Daniilidis, A., Georgiev, P. and J.P. Penot, *Integration of multivalued operators and cyclic submonotonicity*, Trans. Amer. Math. Soc. 355, 2003, 177–195.

[13] Daniilidis, A. and N. Hadjisavvas, *On the subdifferentials of quasiconvex and pseudoconvex functions and cyclic monotonicity*, J. Math. Anal. Appl. 237, 1999, 30–42.

[14] Daniilidis, A. and N. Hadjisavvas, *On generalized cyclically monotone operators and proper quasimonotonicity*, Optimization 47, 2000, 123–135.

[15] Domokos, A. and J. Kolumbán, *Comparison of two different types of pseudomonotone mappings*, in Seminaire de la Théorie de la Meilleure Approximation, Convexity and Optimization, E.Popoviciu, Ed., Cluj-Napoca 2000, 95–103.

[16] Ferland, J.A., *Quasi-convex and pseudo-convex functions on solid convex sets*, Ph.D. dissertation, Technical Report 71-4, Department of Operations Research, Stanford University, 1971.

[17] Ferland, J.A., *Maximal domains of quasiconvexity and pseudoconvexity for quadratic functions*, Math. Programming 3, 1972, 178–192.

[18] Gowda, M.S., *Pseudomonotone and copositive star matrices*, Linear Algebra Appl. 113, 1989, 107–118.

[19] Gowda, M.S., *Affine pseudomonotone mappings and the linear complementarity problem*, SIAM J. Matrix Anal. Appl. 11, 1990, 373–380.

[20] Gowda, M.S., *On the transpose of a pseudomonotone matrix and the linear complementarity problem*, Linear Algebra Appl. 140, 1990, 129–137.

[21] Hadjisavvas, N. and S. Schaible, *On strong pseudomonotonicity and (semi)strict quasimonotonicity,* J. Optim. Theory Appl. 79, 1993, 139–155.

[22] Hadjisavvas, N. and S. Schaible, *Errata corrige*, J. Optim. Theory Appl. 85, 1995, 741–742.

[23] Hadjisavvas, N. and S. Schaible, *Quasimonotone variational inequalities in Banach spaces*, J. Optim. Theory Appl. 90, 1996, 95–111.

[24] Hadjisavvas, N. and S. Schaible, *Quasimonotonicity and pseudomonotonicity in variational inequalities and equilibrium problems*, in Generalized Convexity, Generalized Monotonicity, J.-P. Crouzeix, J.-E. Martinez-Legaz and M. Volle, eds., Kluwer Academic Publishers, Dordrecht, pp. 257–275 (1998).

[25] Hassouni A., *Sous-differentiels des fonctions quasiconvexes*, Thèse de 3ème Cycle, Université Paul Sabatier, Toulouse, 1983.

[26] Jeyakumar, V., Luc, D.T. and S. Schaible, *Characterizations of generalized monotone nonsmooth continuous maps using approximate Jacobians*, J. Convex Anal. 5, 1998, 119–132.

[27] Karamardian, S., *Duality in mathematical programming*, J. Math. Anal. Appl. 20, 1967, 344–357.

[28] Karamardian, S., *Complementarity problems over cones with monotone or pseudomonotone maps*, J. Optim. Theory Appl. 18, 1976, 445–454.

[29] Karamardian, S. and S. Schaible, *Seven kinds of monotone maps*, J. Optim. Theory Appl. 66, 1990, 37–46.

[30] Karamardian, S., Schaible, S. and J. P. Crouzeix, *Characterizations of generalized monotone maps*, J. Optim. Theory Appl. 76, 1993, 399–413.

[31] Komlósi, S., *Contributions to the theory and methods of quasiconvex programming*, unpublished C.Sc. Thesis, Baku, 1984, pp. 99 (in Russian).

[32] Komlósi, S., *Generalized convexity of quadratic functions*, Izv. Vyssh. Uchebn. Zaved. Mat. 9, 1984, 38–43 (in Russian).

[33] Komlósi, S., *Generalized monotonicity in nonsmooth analysis*, in Generalized Convexity, S. Komlosi, T. Rapcsak and S. Schaible, eds., Springer, Heidelberg, 1994, 263–276.

[34] Konnov, I.V., Chapter 13, this volume.

[35] Luc, D.T. and S. Schaible, *On generalized monotone nonsmooth maps*, J. Convex Anal. 3, 1996, 195–205.

[36] Martos, B., *Sub-definite matrices and quadratic forms*, SIAM J. Appl. Math. 17, 1969, 1215–1223.

[37] Pini, R. and S. Schaible, *Invariance properties of generalized monotonicity*, Optimization 28, 1994, 211–222.

[38] Poljak, B.T., *Existence theorems and convergence of minimizing sequences in extremum problems with restrictions*, Sov. Math. 7, 1966, 72–75.

[39] Rockafellar, T., *Convex Analysis*, Princeton University Press, Princeton, N.J., 1970.

[40] Schaible, S., *Beiträge zur quasikonvexen Programmierung*, Doctoral Dissertation, University of Köln 1971.

[41] Schaible, S., *Second-order characterizations of pseudoconvex quadratic functions*, J. Optim. Theory Appl. 21, 1977, 15–26.

[42] Schaible, S., *Quasiconvex, pseudoconvex, and strictly pseudoconvex quadratic functions*, J. Optim. Theory Appl. 35, 1981, 303–338.

[43] Schaible, S., *Generalized convexity of quadratic functions*, in: Generalized Concavity in Optimization and Economics, eds. Schaible, S. and W.T. Ziemba, Academic Press, New York, 1981, 183–198.

[44] Schaible, S., *Criteria for generalized monotonicity*, in: New Trends in Mathematical Programming, eds. Komlosi, S., Giannessi, F. and

T. Rapcsak, Kluwer Academic Publishers, Dordrecht - Boston - London, 1998, 277–288.

[45] Schaible, S. and R.W. Cottle, *On pseudoconvex quadratic forms*, in: General Inequalities 2, ed. E.F. Beckenbach, Birkhäuser-Verlag, Basel, 1980, 81–88.

[46] Yao, J.-C. and O. Chadli, Chapter 12, this volume.

Chapter 10

GENERALIZED CONVEXITY AND GENERALIZED DERIVATIVES

Sándor Komlósi*
Faculty of Business and Economics
University of Pécs, 7622 Pécs, Hungary
komlosi@ktk.pte.hu

Abstract This chapter is devoted to the study of nonsmooth generalized convex functions with the help of special classes of generalized derivatives. Several results are presented on the links between generalized monotonicity of the generalized derivatives and generalized convexity of the functions under discussion. The abundance of the different notions of generalized derivatives has motivated an axiomatic treatment resulting, among others, in the concept of first order approximation. The usefulness of quasiconvex first order approximations in optimization theory is investigated, in particular, generalized upper quasidifferentiable functions are studied, quasiconvex Farkas Theorems and KKT-type optimality conditions are elaborated.

Keywords: Generalized convexity, generalized derivatives, first order quasiconvex approximations, generalized upper quasidifferentiability, quasiconvex Farkas theorems, KKT-type optimality conditions.

1. Introduction

The core of the different topics discussed in this chapter is the nonlinear programming problem:

$$\begin{gathered} f(x) \to \min \\ x \in F = \{x \in X : g_i(x) \le 0, \quad i = 1, 2, \dots, m\}, \end{gathered} \qquad (NLP)$$

*Financial supports from the Hungarian Scientific Research Fund (OTKA T 025442) are gratefully acknowledged.

where X denotes a locally convex Hausdorff topological vector space the dual of which is denoted by X^*. For the concepts we study in this chapter this general setting is quite appropriate, for the results, however, we have to impose sometimes additional assumptions on the space X. For simplicity, readers may assume $X = X^* = R^n$.

The central problem with (NLP) is to find local or global minimizers of $f(x)$ over the feasible set F. The analysis of (NLP) is twofold. On the one hand there is a need of tractable characterizations of the solutions, while on the other hand there is a need of working algorithms by the help of which (the) solutions can be found.

In the analysis of (NLP) the use of derivatives is inevitable. In smooth problems the classical concepts of derivative (Fréchet-, Gâteaux- and directional) has been proved one of the most useful tools.

Besides treating classical smooth problems, the mathematicians got many impulses from other sciences and important real life problems in the last decades in order to treat nonsmooth, nondifferentiable problems.

The analysis of such problems has definitely required some generalizations of the derivative. Since the early 1960's much effort has gone into the development of a generalized kind of differentiation that can be useful in the analysis of optimization problems. The subject has grown rapidly since then.

One may distinguish two prominent approaches: the one works with generalized directional derivatives using special combinations of the limsup, liminf, sup and inf operations, while the other works with generalized gradients, subgradients, subdifferentials. It should be mentioned that the two approaches are linked very strongly, especially in those cases when the generalized derivatives are either convex or quasiconvex.

We concentrate in this chapter only to generalized directional derivatives, because the "subdifferential approach" is the topic of Chapter 11 of the present volume. The "subdifferential-approach" is discussed in details in [2], [3], [44] and [54], too.

Some of the attempts have gone back to classical sources: to Dini and Hadamard, whereas others provided quite new kinds of generalized derivatives [7], [53], [54], [58], or abstract derivatives [17], [20], [27], [37].

The success of the above approach depends heavily on specific properties of the problem functions. The role of convexity and generalized convexity in this respect is well known. The classical analysis of these properties utilizes the Lagrange Mean Value Theorem, by the help of which the increment $f(x)-f(a)$ is substituted by a suitable total derivative $\langle f'(z), x-a\rangle$, $z \in (a,x)$. Fortunately in nonlinear analysis we still have powerful Mean Value Theorems, by the help of which we can find "*good approximates*" for the increments under discussion.

This chapter is organized as follows: In several investigations on nonsmooth generalized convex functions the Dini derivatives proved to be the best replacements for the gradient. It is the reason that first we present results based on the Dini derivatives.

Section 2 gives a short account on the most important tools on Dini derivatives serving as technical basis for the subsequent subsections. Subsections 2.1 and 2.2 are devoted to the study of nonsmooth generalized convex functions and their characterizations with the help of their Dini derivatives. Generalized upper quasidifferentiability is the central theme of Subsection 2.3. This class of functions are defined via the quasiconvexity of their upper Dini derivative and shares some nice properties.

In nonsmooth analysis several other generalized derivatives proved to be very useful: the Clarke derivative for locally Lipschitz functions, the Rockafellar derivative for lower semicontinuous functions, the incident derivative for lower semicontinuous quasiconvex functions, etc.

Section 3 provides a limited list on typical generalized derivatives with the Rockafellar derivative in the focus and provides a way of using them in characterizing different kinds of generalized convexity.

The abundance of the notions of generalized derivatives puts forward the question of axiomatization of the derivative concept. The concept of first order approximation was introduced with this aim and is discussed in Section 4. In Subsection 4.2.1 a Generalized Farkas Lemma is presented for quasiconvex first order approximations. The section ends with deriving Karush-Kuhn-Tucker type optimality conditions using first order approximations and some results on steepest descent directions, paving the way towards algorithmic design.

2. Dini derivatives

This section is devoted to Dini derivatives. The importance of them in quasiconvex optimization was shown in several works of the two pioneers: J.-P- Crouzeix [8]-[10], and W.E. Diewert [15]. Further results on the topic can be found in [13], [14] and [21]-[23].

We recall that the directional upper and lower Dini derivatives of $f(x)$ at a are defined as follows:

$$f^D(a;d) := \limsup_{t \to 0+} \frac{f(a+td) - f(a)}{t},$$

$$f_D(a;d) := \liminf_{t \to 0+} \frac{f(a+td) - f(a)}{t}.$$

A simple but very important consequence of the definition is the following inequality: for any $a, d \in X$ one has

$$f_D(a;d) \leq f^D(a;d). \tag{10.1}$$

It should also be noted that the Dini derivatives are positively homogeneous in their second arguments, namely for all $\lambda > 0$ we have

$$f_D(a;\lambda d) = \lambda f_D(a;d) \quad \text{and} \quad f^D(a;\lambda d) = \lambda f^D(a;d).$$

It is well-known from the classical mathematical analysis, that convexity of a differentiable function can be completely characterized via the monotonicity of its derivative or gradient. The use of the famous Lagrange Mean Value Theorem ($LMVT$) cannot be ignored in this theory. In nondifferentiable context it is no longer possible to apply the classic $LMVT$. Fortunately we may use instead its extension, the Diewert's Mean Value Theorem ($DMVT$) ([15], Theorem 1, Corollary 1).

Theorem 10.1 (Diewert's Mean Value Theorem) *Let the function $f(x)$ be defined on the line segment $[a,b] \subset X$ and let the function $f_{a,b}(t) = f(a+t(b-a))$ be lower semicontinuous on $[0,1]$. Then there exists $\tau \in [0,1)$ such that*

$$f_D(z; b-a) \geq f(b) - f(a),$$

where $z = a + \tau(b-a)$.

By the help of $DMVT$ one can prove several important theorems for nondifferentiable functions. The following implicit-function theorem was proved by the author in [30] for Euclidean spaces. (See also [21], Theorem 2.8.)

Let us consider R^n and a direction $d \in R^n, d \neq 0$. Denote by L the $n-1$ dimensional subspace of R^n orthogonal to d. Then every $x \in R^n$ has a unique representation as $x = u + vd$, where $u \in M$ and $v \in R$. We may regard u as an element of R^{n-1} and identify x with (u,v).

Theorem 10.2 (Implicit Function Theorem) *Let $f(x)$ be continuous on the open set $S \subset R^n$. Assume that any of the two conditions holds.*

(a) *Let $f_D(a;d) < 0$ for a certain $a = (u_0, v_0) \in S$ and let $f_D(x;d)$ be upper semicontinuous in x at a,*

(b) Let $f^D(a;d) < 0$ for a certain $a = (u_0, v_0) \in S$ and let $f^D(x;d)$ be upper semicontinuous in x at a.

Then there exist convex neighborhoods N and U of a and u_0 respectively and a unique function $h_{a,d}(u)$ defined on N such that:

(i) $x = (u,v) \in N$ *and* $f(x) = f(a) \Longleftrightarrow u \in U$ *and* $h_{a,d}(u) = v$,

(ii) $x = (u,v) \in N$ *and* $f(x) < f(a) \Longleftrightarrow u \in U$ *and* $h_{a,d}(u) < v$,

(iii) $x = (u,v) \in N$ *and* $f(x) > f(a) \Longleftrightarrow u \in U$ *and* $h_{a,d}(u) > v$.

Another simple consequence of the *DMVT*, useful in studying quasiconvexity, is given in the following "*Three Point Lemma*". It uses the frequently used *nonconstancy property.*

Definition 10.1 *We say that $f(x)$ satisfies the nonconstancy property (in short: NC-property) on the convex set $C \subset X$, if there is no line segment $[a,b]$ in C along which $f(x)$ is constant.*

Lemma 10.1 (Three Point Lemma) *Let $f(x)$ be defined on the line segment $[a,b] \subset X$ and let the function $f_{a,b}(t) = f(a + t(b-a))$ be lower semicontinuous on $[0,1]$. If there exists $r = a + \rho(b-a), 0 < \rho < 1$ such that*

$$f(r) > M := \max\{f(a), f(b)\},$$

then there exist $u, v \in [a,b]$ for which one has

$$f(u) \leq f(v) \quad \text{and} \quad f_D(v; u-v) > 0. \tag{10.2}$$

If $f(x)$ satisfies the NC-property on $[a,b]$, then (10.2) holds in a stronger form

$$f(u) < f(v) \quad \text{and} \quad f_D(v; u-v) > 0. \tag{10.3}$$

Proof. Let

$$\gamma = \max\{t : 0 \leq t \leq \rho,\ f_{a,b}(t) \leq M\},$$

and

$$\delta = \min\{t : \rho \leq t \leq 1,\ f_{a,b}(t) \leq M\}.$$

Set

$$c = a + \gamma(b-a) \quad \text{and} \quad d = a + \delta(b-a).$$

Since $f_{a,b}(t)$ is lower semicontinuous, therefore $\alpha < \rho < \beta$; $f(c), f(d) \leq M$ and

$$f(x) > M \text{ for all } x \in (c,d).$$

Case 1: Assume first that $f(x)$ is constant over (c,d). If K denotes that constant, then $K = f(r) > M$. Since $f(x) \equiv K$ for every $x \in (c,d)$ it follows that

$$f_D(c; d-c) = f_D(d; c-d) = +\infty.$$

Put $u = c$ and $v = d$ if $f(c) \leq f(d)$ and $u = d$ and $v = c$ if $f(c) \geq f(d)$. It is easy to check that (10.2) holds with these choices.
Case 2: Assume now that $f(x)$ is not constant over (c,d). Then there exist $w, y \in (c,d)$ such that $f(w) < f(y)$. By *DMVT* there exists $z \in [w,y)$ such that

$$f_D(z; y-w) \geq f(y) - f(w) > 0.$$

Since $z \in (c,d)$ it follows that $f(z) > M \geq f(c), f(d)$. If $y \in (w,d)$ then put $u = d$ and $v = z$. If $y \in (c,w)$ then put $u = c$ and $v = z$. Taking into account the positive homogeneity of the Dini derivative in its second argument it is easy to check that (10.3) holds with these choices. □

2.1 Dini derivatives and Generalized Convexity

In this subsection quasiconvexity and pseudoconvexity will be characterized in terms of directional Dini derivatives.

Definition 10.2 *[4] The function $f(x)$ is called convex (CX), strictly convex (SCX), quasiconvex (QCX), strictly quasiconvex $(SQCX)$, semistrictly quasiconvex $(SSQCX)$ on the convex set $C \subset X$ if for all $x, a \in C$, $x \neq a$ and $t \in (0,1)$ the following conditions below hold, respectively:*

$$f(tx + (1-t)a) \leq tf(x) + (1-t)f(a), \qquad (CX)$$

$$f(tx + (1-t)a) < tf(x) + (1-t)f(a), \qquad (SCX)$$

$$f(x) \leq f(a) \Longrightarrow f(a + t(x-a)) \leq f(a), \qquad (QCX)$$

$$f(x) \leq f(a) \Longrightarrow f(a + t(x-a)) < f(a), \qquad (SQCX)$$

$$f(x) < f(a) \Longrightarrow f(a + t(x-a)) < f(a). \qquad (SSQCX)$$

The following interrelations are immediate consequences of the definitions:

$$(SCX) \Rightarrow (CX) \Rightarrow (QCX),$$

$$(SQCX) \Rightarrow (QCX), \qquad (SQCX) \Rightarrow (SSQCX).$$

The following statement, due to Crouzeix [8], supplements the above interrelations and, in particular, will be very useful in further investigations.

Lemma 10.2 *The function $f(x)$ is (strictly) convex on the convex set $C \subset X$, if and only if for all $g \in X^*$ the function*

$$g(x) = f(x) + \langle g, x \rangle$$

is (strictly) quasiconvex.

Proof. Necessity: Obvious, since the sum of a (strictly) convex and a linear function is always (strictly) convex and thus (strictly) quasiconvex.

Sufficiency: Assume that $g(x) = f(x) + \langle g, x \rangle$ is quasiconvex on C for all $g \in X^*$. Suppose for contradiction that $f(x)$ fails to be convex on C. Then there exist two distinct points $a, b \in C$ and a third point $z = ta + (1-t)b$ in the open line segment (a, b) such that

$$tf(a) + (1-t)f(b) < f(z).$$

By virtue of the Hahn-Banach Extension Theorem you can always find an appropriate $g^* \in X^*$ such that

$$g^*(a) = g^*(b) < g^*(z),$$

where $g^*(x) = f(x) + \langle g^*, x \rangle$. This condition, however, contradicts to the quasiconvexity of $g^*(x)$. This contradiction proves the thesis. The strict variant can be handled completely in the same way. □

We recall, in view of further developments, the following classical result due to Arrow and Enthoven [1].

Theorem 10.3 *Let $f(x)$ be differentiable on the convex set $C \subset X$. Then $f(x)$ is quasiconvex if and only if for all $x, a \in C$*

$$f(x) \leq f(a) \Longrightarrow \langle f'(a), x - a \rangle \leq 0.$$

It is easy to prove that if the function $f(x)$ is quasiconvex on C, then for all $x, a \in C$ any of the following implications holds:

$$f(x) \leq f(a) \Longrightarrow f^D(a; x - a) \leq 0, \qquad QCX(UDini)$$

$$f(x) \leq f(a) \Longrightarrow f_D(a; x - a) \leq 0. \qquad QCX(LDini)$$

Furthermore it is obvious that

$$QCX \Longrightarrow QCX(UDini) \Longrightarrow QCX(LDini).$$

The following simple example shows that, in general, the above implications can not be reversed.

Example 10.1 Let

$$f(t) = \begin{cases} 0 & \text{if} \quad 0 < |t| \leq 1, \\ 1 & \text{if} \quad t = 0. \end{cases}$$

This function possesses both $QCX(UDini)$ and $QCX(LDini)$ but fails to have property QCX. It can be proved that for radially lower semicontinuous functions the three properties under consideration are equivalent.

We recall that $f(x)$ is called *radially lower semicontinuous* on the convex set C, if for every $a, b \in C$ the function $f_{a,b}(t) = f(a+t(b-a)), t \in [0,1]$ is lower semicontinuous.

Theorem 10.4 *([15], Theorem 4) Let $f(x)$ be radially lower semicontinuous on the convex set $C \subset X$. Then conditions QCX, $QCX(UDini)$ and $QCX(LDini)$ are equivalent.*

Proof. Since the implications $QCX \Rightarrow QCX(UDini) \Rightarrow QCX(LDini)$ are rather obvious it is sufficient to prove that $QCX(LDini)$ implies QCX.

Assume for contradiction that $f(x)$ enjoys property $QCX(LDini)$ but fails to be quasiconvex. Then there exist $a, b \in C$ and $r \in (a, b)$ such that

$$f(r) > M = \max\{f(a), f(b)\}.$$

By the Three Point Lemma it follows the existence of $u, v \in (a, b) \subset C$ such that

$$f(u) \leq f(v) \quad \text{and} \quad f_D(v; u - v) > 0.$$

It contradicts, however, the $QCX(LDini)$ property. The thesis follows. □

Modifying the Arrow-Enthoven's characterization of differentiable quasiconvex functions given in Theorem 10.3, O.L. Mangasarian introduced the property called pseudoconvexity.

Definition 10.3 *[50] The differentiable function $f(x)$ is called pseudoconvex, strictly pseudoconvex on the convex set $C \subset X$ if for all $x, a \in C$, $x \neq a$ condition (PCX), $(SPCX)$ holds, respectively:*

$$f(x) < f(a) \Longrightarrow \langle f'(a), x - a \rangle < 0. \qquad (PCX)$$

$$f(x) \leq f(a) \Longrightarrow \langle f'(a), x - a \rangle < 0. \qquad (SPCX)$$

There are two possibilities to extend this notion for nondifferentiable functions by means of the Dini derivatives. Diewert defined the lower-Dini pseudoconvexity by replacing implications (PCX) and $(SPCX)$ with the following ones [15]: for all $x, a \in C$, $x \neq a$

$$f(x) < f(a) \Longrightarrow f_D(a; x-a) < 0, \qquad PCX(LDini)$$

$$f(x) \leq f(a) \Longrightarrow f_D(a; x-a) < 0, \qquad SPCX(LDini)$$

whereas pseudoconvexity is defined in, [6, 10, 30, 33] by the following implications: for all $x, a \in C$, $x \neq a$

$$f(x) < f(a) \Longrightarrow f^D(a; x-a) < 0, \qquad PCX(UDini)$$

$$f(x) \leq f(a) \Longrightarrow f^D(a; x-a) < 0, \qquad SPCX(UDini)$$

Sometimes it is more convenient to use the following equivalent forms:

$$f_D(a; x-a) \geq 0 \Longrightarrow \ f(x) \geq f(a), \qquad PCX(LDini)$$

$$f^D(a; x-a) \geq 0 \longrightarrow \ f(x) \geq f(a). \qquad PCX(UDini)$$

Taking into account inequality $f_D(a; d) \leq f^D(a; d)$ it is obvious that $PCX(UDini) \Longrightarrow PCX(LDini)$. A tricky example, due to Diewert demonstrates that the reverse implication does not hold in general. (See [15] Example 1 or [22] Example 8.)

Diewert's example shows that his definition of pseudoconvexity embraces a wider class of functions then the other one.

It is well-known that for differentiable functions pseudoconvexity implies quasiconvexity. Example 10.1 shows that it is no longer the case for arbitrary functions. Function $f(t)$ from this example is $PCX(LDini)$ but fails to be QCX. If, however, we restrict ourselves only to lower semicontinuous functions, we shall have the desired result.

Theorem 10.5 *Let $f(x)$ be radially lower semicontinuous on the convex set $C \subset X$. If $f(x)$ is $PCX(LDini)$ on C then $f(x)$ is quasiconvex on C. Since $PCX(UDini)$ implies $PCX(LDini)$ the statement of the theorem holds for $PCX(UDini)$ functions, as well.*

Proof. Assume for contradiction that $f(x)$ is pseudoconvex but fails to be quasiconvex. Then there exist $a, b \in C$ and $r \in (a, b)$ such that

$$f(r) > M = \max\{f(a), f(b)\}.$$

Let us consider $c, d \in (a, b)$ defined in the proof of Lemma 10.1. It is easy to check that $f(c), f(d) \leq M$ and $f(x) > M$ for every $x \in (c, d)$. Obviously $r \in (c, d)$. If $f(x)$ were constant over (c, d) we would have

$f(c) < f(r)$ and $f_D(r; c - r) = 0$, which is impossible by $PCX(LDini)$. By the Three Point Lemma it follows the existence of $u, v \in (a, b) \subset C$ such that

$$f(u) < f(v) \quad \text{and} \quad f_D(v; u - v) > 0,$$

but it contradicts, again, the $PCX(LDini)$ property. It proves the thesis. □

Remark 10.1 Let us observe that strict pseudoconvexity always implies quasiconvexity.

Pseudoconvexity, as it is well-known, is a remarkable property from optimization point of view. Necessary optimality condition with respect to minimization provides the concept of inf-stationarity.

Definition 10.4 *A point $a \in C$ is called an inf-stationary point of $f(x)$ over C if $f_D(a; x - a) \geq 0$ holds for every $x \in C$.*

It is obvious that if $f(x)$ attains a local minimum at $a \in C$, then a is an inf-stationary point of $f(x)$ over C.

Theorem 10.6 *Let $f(x)$ be $PCX(LDini)$ over the convex set $C \subset X$ and let $a \in C$ be inf-stationary point of $f(x)$ over C. Then a is a global minimizer of $f(x)$ over C.*

Proof. The proof is an immediate consequence of the definitions. □

The following theorem (which generalizes a result of Crouzeix and Ferland ([12], Theorem 2.2]), proved by the author in [30], provides sufficient condition for quasiconvex functions to be pseudoconvex.

Theorem 10.7 *Let $f(x)$ be radially upper semicontinuous quasiconvex function defined over an open convex set $C \subset X$. If $f(x)$ attains a global minimum at $a \in C$, whenever a is an inf-stationary point of $f(x)$, then $f(x)$ is $PCX(LDini)$.*

Proof. Let $z, a \in C$. Assume that $f(z) < f(a)$. We shall prove that $f^D(a; z - a) < 0$.

From the hypothesis of the present theorem it follows that a cannot be an inf-stationary point of $f(x)$ over C, consequently there exists a (feasible) direction $d \in X$, such that $f_D(a; d) < 0$.

By hypothesis the function $\varphi(t) = f(z - td)$ is upper semicontinuous, $\varphi(0) = f(z) < f(a)$ and thus there exists $T > 0$ such that for every

$t \in (0,T)$ we have $f(z - td) < f(a)$. Without loss of generality we may assume that

$$u := a + d \in C, v := z - d \in C \quad \text{and} \quad f(v) < f(a).$$

For $0 < t < 1$, define

$$x(t) = a + t(z - a) \quad \text{and} \quad y(t) = a + \frac{t}{1-t}d.$$

An easy calculation shows that $x(t)$ belongs to the line segment $[v, y(t)]$. Since $f(x)$ is quasiconvex, we therefore have for every $t \in (0,1)$ that

$$f(x(t)) \leq \max\{f(v), f(y(t))\},$$

and hence for $t > 0$

$$\frac{f(x(t)) - f(a)}{t} \leq \max\left\{\frac{f(v) - f(a)}{t}, \frac{f(y(t)) - f(a)}{t}\right\}.$$

Since $f(v) < f(a)$, if $t \to 0^+$, the first term in the above maximum tends to $-\infty$ and thus

$$f_D(a; z - a) = \liminf_{t \to 0+} \frac{f(x(t)) - f(a)}{t} \leq \liminf_{t \to 0+} \frac{f(y(t)) - f(a)}{t}. \tag{10.4}$$

Since $y(t) = a + \tau d$, where $\tau = t/(1-t)$, we have

$$\frac{f(y(t)) - f(a)}{t} = \frac{f(a + \tau d) - f(a)}{\tau} \frac{1}{1-t}.$$

Taking limits as $t \to 0^+$ yields

$$\liminf_{t \to 0+} \frac{f(y(t)) - f(a)}{t} = f_D(a; d).$$

By assumption $f_D(a; d) < 0$. This inequality along with (10.4) yields the desired results. □

From Theorems 10.5, 10.6 and 10.7 one can deduce the following characterization of continuous pseudoconvex functions.

Theorem 10.8 *Let $f(x)$ be radially continuous on the open convex set C. Then $f(x)$ is PCX(LDini) if and only if $f(x)$ is quasiconvex on C and has a global minimum at $a \in C$ whenever a is an inf-stationary point of $f(x)$ over C.*

Remark 10.2 The notion of inf-stationarity given in Definition 10.4 uses the lower Dini derivative. If one replaces it with the upper Dini derivative you will obtain the "upper" version of inf-stationarity, which is somewhat stronger then the "lower" one. If you replace $PCX(LDini)$ with $PCX(UDini)$ and "lower" inf-stationarity with its "upper" version in Theorems 10.6, 10.7 and 10.8, "upper" versions of the mentioned theorems are obtained.

The characterization of functions via implicit functions is a useful tool in several fields of mathematics, economics, etc. Such a characterization, via the Hessian of the implicit function was elaborated by the author in [31], [32], [38] and [39] for twice differentiable pseudoconvex functions. Certain elements of this characterization can be extended for nonsmooth functions by weakening the notion of smoothness.

Definition 10.5 *A function $f(x)$ defined on an open set $S \subset X$ is called upper semismooth if it is continuous and whenever $a \in S$, $f_D(a;d) < 0$, either $f_D(x;d)$ or $f^D(x;d)$ is upper semicontinuous in x at a.*

Theorem 10.9 *Let $f(x)$ be defined and upper semismooth on the open convex set $C \subset R^n$. Then $f(x)$ is $PCX(LDini)$ on C if and only if*

(i) $f(x)$ attains a global minimum at $a \in C$ whenever a is an inf-stationary point of $f(x)$, and

(ii) $f_D(a;d) < 0$ implies that $h(u) = h_{a,d}(u)$ is a convex function of u, where $h_{a,d}(u)$ is the implicit function defined in Theorem 10.2.

Proof. Throughout the proof we use the notation of Theorem 10.2.
Necessity: Let $f(x)$ be $PCX(LDini)$. Then (i) follows immediately from Theorem 10.6. To prove (ii) let $u_1, u_2 \in U$ and $t \in (0,1)$. Define

$$x_1 = (u_1, h(u_1)) \quad \text{and} \quad x_2 = (u_2, h(u_2))$$

and

$$x(t) = tx_1 + (1-t)x_2 = (tu_1 + (1-t)u_2, th(u_1) + (1-t)h(u_2)) .$$

By Theorem 10.5 $f(x)$ is quasiconvex on C and thus

$$f(x(t)) \leq \max\{f(x_1), f(x_2)\} = f(a).$$

From statements (i) and (ii) in Theorem 10.2, it follows that

$$h(tu_1 + (1-t)u_2) \leq th(u_1) + (1-t)h(u_2),$$

which proves the convexity of $h(u)$.
Sufficiency. By Theorem 10.7 we need only to prove the quasiconvexity of $f(x)$. Let $x_1, x_2 \in C$ and assume that $f(x_1) \leq f(x_2)$. Set

$$x(t) = tx_1 + (1-t)x_2 \quad \text{and} \quad \varphi(t) = f(x(t))$$

for every $t \in [0,1]$. We have to prove that $\varphi(t) \leq \varphi(0)$ for every $t \in [0,1]$. Assume for contradiction that

$$\varphi_{\max} = \max\{\varphi(t) : t \in [0,1]\} > 0.$$

Define t_0 to be the largest maximizer, i.e.

$$t_0 = \sup\{t \in [0,1] : \varphi(t) = \varphi_{\max}\}.$$

It is obvious that $0 < t_0 < 1, \varphi(t_0) = \varphi_{\max}$ and $\varphi(t_0) > \varphi(t)$ for every $t \in [t_0, 1]$. Set $a = x(t_0)$. According to (i) of the present theorem a is not an inf-stationary point of $f(x)$ and thus there exists a direction d such that $f_D(a; d) < 0$.

Consider the function $h(u) = h_{a,d}(u)$, which is convex by hypothesis. Without loss of the generality, we may assume that the points $x_1 = (u_1, v_1)$ and $x_2 = (u_2, v_2)$ are appropriately close to $a = (u_0, v_0)$, that is $u_1, u_2 \in U$. Since $f(x_1) \leq f(x_2) < f(a)$, we have by Theorem 10.2 that $v_1 > h(u_1)$ and $v_2 > h(u_2)$. Thus

$$v_0 = t_0 v_1 + (1-t_0)v_2 >$$

$$> t_0 h(u_1) + (1-t_0)h(u_2) \geq h(t_0 u_1 + (1-t_0)u_2) = v_0.$$

This contradiction proves the thesis. □

2.2 Dini Derivatives and Generalized Monotonicity

Monotonicity plays an important role also in functional analysis, operator theory, nonlinear analysis, optimization theory and related fields, such as variational inequalities and general equilibrium problems. A classical theorem from mathematical analysis claims that a differentiable function is convex if and only if its gradient map is monotone. This statement can be extended for nondifferentiable functions: a directionally differentiable function is convex if and only if its directional derivative is monotone, or a lower semicontinuous function is convex if and only if its convex subdifferential is a nonvoid monotone multifunction.

Since the convexity assumption on the function can be weakened to a certain kind of generalized convexity assumption without "destroying" the nice results valid for the convex case, therefore there have been several attempts to weaken the monotonicity assumption to some kind of generalized monotonicity concept. In 1976 Karamardian introduced the notion of *pseudomonotonicity* for maps in [28]. Some years later in 1983, Hassouni introduced the concept of *quasimonotonicity* for multifunctions [25]. The seminal paper of Karamardian and Schaible [29] from 1990, in which several kinds of generalized monotonicity concepts were introduced for gradient maps, might be considered as the one opening a new theory, the theory of *generalized monotonicity*.

These concepts were extended in nondifferentiable setting by the author in [34]-[36] and Bianchi in [5]. This subsection is devoted to these investigations.

From an abstract standpoint any of the Dini derivatives is a bifunction $\varphi(x;d)$, where x refers to a place and d refers to a direction. Since the domain of d is always the whole space, therefore it is sufficient to specify only the domain of x. We introduce now generalized monotonicity concepts for bifunctions, which can also be used for any generalized derivatives.

Definition 10.6 *Let $\varphi(x;d)$ be a bifunction defined on the convex set $C \subset X$. $\varphi(x;d)$ is called monotone, strictly monotone, quasimonotone, strictly quasimonotone, semistrictly quasimonotone, pseudomonotone, strictly pseudomonotone on C, if for every $y, z \in C$, $y \neq z$, condition (M), (SM), (QM), (SQM), (SSQM), (PM), (SPM) holds, respectively.*

$$\varphi(y; z-y) + \varphi(z; y-z) \leq 0, \tag{M}$$

$$\varphi(y; z-y) + \varphi(z; y-z) < 0, \tag{SM}$$

$$\varphi(y; z-y) > 0 \quad \textit{implies} \quad \varphi(z; y-z) \leq 0, \tag{QM}$$

$$\begin{gathered} \textit{there exists } u \in (y,z) \textit{ such that} \\ \textit{either } \varphi(u; z-y) > 0 \textit{ or } \varphi(u; y-z) > 0, \\ \textit{moreover } \varphi(y; z-y) > 0 \quad \textit{implies} \quad \varphi(z; y-z) \leq 0, \end{gathered} \tag{SQM}$$

$$\begin{gathered} \varphi(y; z-y) > 0 \quad \textit{implies} \quad \varphi(z; y-z) \leq 0, \\ \textit{and the existence of } u \textit{ such that} \\ u \in \left(\tfrac{y+z}{2}, z\right) \textit{ and } \varphi(u; z-y) > 0, \end{gathered} \tag{SSQM}$$

$$\varphi(y; z-y) \geq 0 \quad \textit{implies} \quad \varphi(z; y-z) \leq 0, \tag{PM}$$

$$\varphi(y; z-y) \geq 0 \quad \textit{implies} \quad \varphi(z; y-z) < 0. \tag{SPM}$$

Remark 10.3 In the definition of (M) and (SM) we adopt the following rule:

$$(+\infty) + (-\infty) = 0.$$

The following chains of implications are immediate consequences of the above definitions:

$$(SM) \Rightarrow (M) \Rightarrow (PM) \Rightarrow (QM),$$

$$(SM) \Rightarrow (SPM) \Rightarrow (PM) \Rightarrow (QM),$$

$$(SQM) \text{ and } (SSQM) \Rightarrow (QM).$$

Remark 10.4 It should be mentioned that if $f(x)$ is differentiable and $\varphi(x; d) = \langle \nabla f(x), d \rangle$, then the concepts introduced above coincide with the ones introduced by Karamardian and Schaible [29] and Hadjisavvas and Schaible [24]. (Semi)strict quasimonotonicity for bifunctions were introduced and investigated by Bianchi [5].

The following lemma provides a very useful supplement to the above mentioned interrelations.

Lemma 10.3 *[44] The bifunction $\varphi(x; d)$ is monotone on C if and only if the bifunction*

$$\psi(x; d) = \varphi(x; d) + \langle g, d \rangle$$

is quasimonotone on C for all $g \in X^*$.

Proof. Necessity: Assume that $\varphi(x; d)$ is monotone on C. Let $g \in X^*$ be arbitrary and set $\psi(x; d) = \varphi(x; d) + \langle g, d \rangle$. Since

$$\psi(x; y - x) + \psi(y; x - y) = \varphi(x; y - x) + \varphi(y; x - y),$$

therefore $\psi(x; d)$ is also monotone and thus quasimonotone on C, as well.

Sufficiency: Let $\psi(x; d)$ be quasimonotone for all $g \in X^*$. This claim is true for $g = 0$, therefore $\varphi(x; d)$ is quasimonotone on C, as well. Assume for contradiction that $\varphi(x; d)$ fails to be monotone. This means that there exist two distinct points x and y in C such that

$$\varphi(x; y - x) + \varphi(y; x - y) > 0. \tag{10.5}$$

without loss of the generality we may assume that $\varphi(x; y - x) > 0$. From the quasimonotonicity of $\varphi(x; d)$ it follows that $\varphi(y; x - y) \leq 0$. Taking inequality (10.5) into account we get

$$0 \leq -\varphi(y; x - y) < \varphi(x; y - x).$$

Let us assume first that $\varphi(x; y - x)$ is finite. Then in virtue of the Hahn-Banach Extension Theorem we can find a $g^* \in X^*$ such that

$$\langle g^*, x - y\rangle = \frac{\varphi(x; y - x) - \varphi(y; x - y)}{2}.$$

It follows immediately that

$$-\varphi(y; x - y) < \langle g^*, x - y\rangle < \varphi(x; y - x).$$

In case of $\varphi(x; y - x) = +\infty$. we can also find $g^* \in X^*$ such that

$$-\varphi(y; x - y) < \langle g^*, x - y\rangle < \varphi(x; y - x) = +\infty.$$

If we consider $\psi^*(x; d) = \varphi(x; d) + \langle g^*, d\rangle$, then we have

$$\psi^*(x; y - x) > 0 \quad \text{and} \quad \psi^*(y; x - y) > 0,$$

which contradicts the assumption that $\psi^*(x; d)$ is quasimonotone on C. This contradiction proves the thesis. □

Let us study first quasiconvex functions.

Theorem 10.10 *Let $f(x)$ be defined and radially lower semicontinuous on the convex set C. Then $f(x)$ is quasiconvex on C, if and only if any of the following conditions holds:*

(i) $f_D(x; d)$ is quasimonotone on C,

(ii) $f^D(x; d)$ is quasimonotone on C.

Proof. Sufficiency. (i): Let $f_D(x; d)$ be quasimonotone on C. For contradiction suppose that $f(x)$ fails to be quasiconvex. Then it means that there exist a line segment $[a, b]$ and a point w in its interior, $w \in (a, b)$ such that

$$f(w) > \max\{f(a), f(b)\}.$$

By *DMVT* there exist $y \in [a, w)$ and $z \in (w, b]$ satisfying conditions

$$f_D(y; w - a) \geq f(w) - f(a) > 0,$$

and

$$f_D(z; w - b) \geq f(w) - f(b) > 0.$$

Taking into account the positive homogeneity of the Dini derivative with respect to its direction argument, the last two inequalities provide the following ones:

$$f_D(y; z - y) > 0 \quad \text{and} \quad f_D(z; y - z) > 0$$

contradicting to the quasimonotonicity assumption.

The sufficiency of statement (*ii*) can be proved by the same way applying the relation $f_D(x;d) \le f^D(x;d)$.

Necessity. (i): Assume that $f(x)$ is quasiconvex. By Theorem 10.4, $f(x)$ possesses property $QCX(LDini)$. Assume now that $f_D(a;x-a) > 0$. By $QCX(LDini)$ $f(x) \le f(a)$ cannot hold. Hence we have $f(a) < f(x)$. Applying again implication $QCX(LDini)$ it follows that $f_D(x;a-x) \le 0$, which proves quasimonotonicity of the lower Dini derivative in this case. The same reasoning can also be applied for the upper Dini derivative.

□

One can prove similar results with the same technique for other classes of generalized convex functions: for (strictly) convex/pseudoconvex functions see [40] and for (semi)strictly quasiconvex functions see [5], [23].

Theorem 10.11 *Let $f(x)$ be defined and radially lower semicontinuous on the convex set C. Then $f(x)$ is (strictly) convex, (semi)strictly quasiconvex, (strictly) lower Dini pseudoconvex on C, if and only if $f_D(x;d)$ is (strictly) monotone, (semi)strictly quasimonotone, (strictly) pseudomonotone on C, respectively.*

Theorem 10.12 *Let $f(x)$ be defined and radially lower semicontinuous on the convex set C. Then $f(x)$ is (strictly) convex, (semi)strictly quasiconvex, (strictly) upper Dini pseudoconvex on C, if and only if $f^D(x;d)$ is (strictly) monotone, (semi)strictly quasimonotone, (strictly) pseudomonotone on C, respectively.*

It might be surprising that characterization for convex functions can be directly derived from characterization of quasiconvex functions. Thanks to Lemmas 10.2, 10.3 and Theorem 10.10, it can easily be done.

Theorem 10.13 *Let $f(x)$ be defined and radially lower semicontinuous on the convex set C. Then $f(x)$ is convex on C if and only if any of the following conditions holds:*

(i) $f_D(x;d)$ is monotone on C,

(ii) $f^D(x;d)$ is monotone on C.

Proof. Let $g \in X^*$ be arbitrary and let $g(x) = f(x) + \langle g, x\rangle$. It is obvious that $g_D(x;d) = f_D(x;d) + \langle g, d\rangle$.

Necessity (*i*): Assume that $f(x)$ is convex. Then, by Lemma 10.2, for any $g \in X^*$ the function $g(x) = f(x) + \langle g, x\rangle$ is quasiconvex. It follows by Theorem 10.10, that $g_D(x;d) = f_D(x;d) + \langle g, d\rangle$ is quasimonotone. By applying Lemma 10.3 we obtain that $f_D(x;d)$ is monotone.

Sufficiency (i): Assume that $f_D(x;d)$ is monotone. By Lemma 10.3 $f_D(x;d) + \langle g, d\rangle$ is quasimonotone. It follows from Theorem 10.10 that the function $g(x) = f(x) + \langle g, x\rangle$ is quasiconvex. Since it holds for any $g \in X^*$, therefore, by considering Lemma 10.2 we obtain that $f(x)$ is convex.

The proof for condition (ii) is completely the same. □

2.3 Generalized Upper Quasidifferentiable Functions

The original notion of quasidifferentiability was introduced by B.N. Pshenichnyi [55] in order to generalize the concept of convexity. We recall that a function $f(x)$ is said to be quasidifferentiable at $a \in X$ by Pshenichnyi if $f(x)$ is directionally differentiable at a and its directional derivative $f'(a;d)$ is a convex function of the direction argument d. It is well-known that convex functions, maximum of smooth functions are quasidifferentiable.

Under the same name the concept of quasidifferentiability was generalized by V.F. Demyanov and A.M. Rubinov [16]. They consider the class of directionally differentiable functions $f(x)$ for which $f'(a;d)$ can be written as a difference of two positively homogeneous convex functions.

The author went out from Pshenichnyi's starting point along another path [33], [35].

Definition 10.7 *The function $f(x)$ is said to be* generalized upper quasidifferentiable *at $a \in X$ if its upper Dini derivative $f^D(a;d)$ is finite for each $d \in X$ and quasiconvex in d.*

The above concept is a generalization of Pshenichnyi's definition inasmuch as this one is weakened at two points: the directional derivative $f'(a;d)$ is replaced with the upper Dini derivative $f^D(a;d)$ and instead of convexity, quasiconvexity is required. It should be noted that the above definition is, however, not a generalization of the Demyanov-Rubinov's concept.

Example 10.2 Consider the following Dirichlet-type function: $f(x) = |\sin x|$ if $x \in R$ is rational and $f(x) = 0$ if x is irrational. It is easy to check that $f^D(0;d) = |d|$, hence $f(x)$ is generalized upper quasidifferentiable at $a = 0$. You may observe that $f'(a;d)$ does not exist.

Quasiconvexity is a generalization of convexity; by this generalization, however, many good properties, ensured by convexity, are lost. Continuity is such a property. In order to avoid the drawbacks due to the lack of continuity it is useful to impose some additional properties on the Dini derivative $f^D(a;d)$.

Definition 10.8 *The generalized upper quasidifferentiable function $f(x)$ is called regular at a if it possesses the following additional properties:*

(i) $f^D(a;d)$ is lower semicontinuous in d,

(ii) the convex cone $\Delta f^D(a) = \{d : f^D(a;d) < 0\}$ is open.

Remark 10.5 Replacing $f^D(a;d)$ with $f_D(a;d)$ in the above definitions the concept of (regular) generalized lower quasidifferentiability at a point can be introduced.

One of the basic motivations to introduce the generalized upper quasidifferentiability is the following result of J.-P. Crouzeix ([9], Proposition 16).

Theorem 10.14 *Let the function $f(x)$ be defined and quasiconvex in a neighborhood of $a \in X$. Then the upper Dini derivative $f^D(a;d)$ is quasiconvex in d.*

It follows that if $f(x)$ is quasiconvex and $f^D(a;d)$ is finite for every d, then $f(x)$ is generalized upper quasidifferentiable at a.

Another strong motivation for creating the concept of generalized upper quasidifferentiability is the nice geometric background of the extremal properties of this class of functions. The central notion of it is the *quasisubdifferential*, whose upper version will now be defined.

Let $f(x)$ be generalized upper quasidifferentiable at $a \in X$ and define the functions $\varphi_-(d)$ and $\varphi_+(d)$ as follows:

$$\varphi_-(d) = \begin{cases} f^D(a;d) & , \text{ if } f^D(a;d) < 0, \\ +\infty & , \text{ otherwise,} \end{cases}$$

$$\varphi_+(d) = \begin{cases} 0 & , \text{ if } f^D(a;d) < 0, \\ f^D(a;d) & , \text{ otherwise.} \end{cases}$$

By a result due to J.-P. Crouzeix ([9], Proposition 12 and Corollary 13) the functions $\varphi_-(d)$ and $\varphi_+(d)$ are positively homogeneous convex functions. From optimization point of view $\varphi_-(d)$, the "negative part" of $f^D(a;d)$ may deserve more interest.

Unfortunately the function $\varphi_-(d)$ fails to be lower semicontinuous in general, therefore, we deal only with regular upper quasidifferentiable functions.

In the regular case we have a representation formula

$$\varphi_-(d) = \sup\{\langle g, d\rangle : g \in \partial^q f(a)\},$$

where $\partial^q f(a) \subset X^*$ denotes a closed convex set, called the *upper quasisubdifferential* of $f(x)$ at a. The subdifferential,upper quasisubdifferential has the following remarkable properties.

Theorem 10.15 *[11] Let $f(x)$ be generalized upper quasidifferentiable at $a \in X$. Then the following statements hold true:*

(i) $0 \in \partial^q f(a)$ is a necessary condition for a to be a local minimizer of $f(x)$.

(ii) If in addition $f(x)$ is locally $PCX(UDini)$ at a, then $0 \in \partial^q f(a)$ is a sufficient condition for a to be a local minimizer of $f(x)$.

We recall that $f(x)$ is called locally $PCX(UDini)$ at a, when the following implication holds on a neighborhood N of a

$$x \in N,\ f(x) < f(a) \quad \Longrightarrow \quad f^D(a; x-a) < 0.$$

Theorem 10.16 *[33] Let the function $f(x)$ be regular generalized upper quasidifferentiable at $a \in X$, where X is a real Hilbert space. If $0 \notin \partial^q f(a)$ and $s_0 \in \partial^q f(a)$ is of minimal norm then the direction*

$$d_0 = -\frac{s_0}{\|s_0\|}$$

is the steepest normalized descent direction of $f(x)$ at a, that is for every $d \in X$, $d \neq d_0$, $\|d\| = 1$ one has

$$f^D(a; d) > f^D(a; d_0) = -\|s_0\|.$$

Optimality conditions in terms of upper quasisubdifferential can be developed for the (NLP) problem, as well. The derivation of these conditions will be discussed in section 10.25.

3. Other Classes of Generalized Derivatives

In nonsmooth analysis several other generalized derivatives proved to be very useful: the Clarke derivative for locally Lipschitz functions, the Rockafellar derivative for lower semicontinuous functions, the incident derivative for quasiconvex functions, etc.

The present section provides a limited list on typical generalized derivatives with the Rockafellar derivative in the focus and provides a way of using them in characterizing different kinds of generalized convexity.

The space X is specified in this section as real Banach space.

Besides the Dini derivatives, the Dini-Hadamard version of them can still be considered as classic generalized derivatives.

Dini-Hadamard derivatives (upper and lower) are defined as:

$$f^{DH}(a;d) := \limsup_{\substack{u\to d\\ t\to 0+}} \frac{f(a+tu)-f(a)}{t},$$

$$f_{DH}(a;d) := \liminf_{\substack{u\to d\\ t\to 0+}} \frac{f(a+tu)-f(a)}{t}.$$

The introduction of the Clarke and the Rockafellar derivatives opened a new chapter in Nonsmooth Optimization and Nonlinear Analysis (see [7], [59]). In their definition we use the following notation:

$$(z,y)\downarrow a \iff (z,y)\longrightarrow (a,f(a)) \quad \text{and} \quad y\geq f(a),$$

$$(z,y)\uparrow a \iff (z,y)\longrightarrow (a,f(a)) \quad \text{and} \quad y\leq f(a),$$

Clarke derivatives (upper and lower) are defined as:

$$f^{C}(a;d) := \limsup_{\substack{(z,y)\downarrow a\\ t\to 0+}} \frac{f(z+td)-y}{t},$$

$$f_{C}(a;d) := \liminf_{\substack{(z,y)\uparrow a\\ t\to 0+}} \frac{f(z+td)-y}{t}.$$

The "*limsup inf*" and "*liminf sup*" operations were introduced by Rockafellar (see [58, 59]). The meaning of

$$\limsup_{\substack{(z,y)\downarrow a\\ t\to 0+}} \inf_{u\to d}$$

operation, for instance, is the following

$$\sup_{\varepsilon>0} \limsup_{\substack{(z,y)\downarrow a\\ t\to 0+}} \inf_{\|u-d\|<\varepsilon}.$$

The meaning of the other operation is similar.

Rockafellar derivatives (upper and lower):

$$f^{R}(a;d) := \limsup_{\substack{(z,y)\downarrow a\\ t\to 0+}} \inf_{u\to d} \frac{f(z+tu)-y}{t},$$

$$f_R(a;d) := \liminf_{\substack{(z,y)\uparrow a \\ t\to 0+}} \sup_{u\to d} \frac{f(z+tu)-y}{t}.$$

weak Rockafellar derivatives:

$$f^{wR}(a;d) := \limsup_{t\to 0+} \inf_{u\to d} \frac{f(a+tu)-f(a)}{t},$$

$$f_{wR}(a;d) := \liminf_{t\to 0+} \sup_{u\to d} \frac{f(a+tu)-f(a)}{t}.$$

Remark 10.6 It should be mentioned that for lower semicontinuous function the convergence $(z,y) \downarrow a$ is equivalent to the simpler convergence $z \to_f a$, whose meaning is

$$(z,y) \to_f a \quad \Longleftrightarrow \quad (z, f(z)) \to (a, f(a))\,.$$

The following theorem is very useful in further studies on the generalized derivatives considered. For the proof consult [17].

Lemma 10.4 *For all $a \in C$ and $d \in X$ one has*

$$\begin{array}{ccccccc}
f^L(a;d) & \geq & f^{DH}(a;d) & \geq & f_{wR}(a;d) & \geq & f_R(a;d) \\
IV & & IV & & IV & & IV \\
f^C(a;d) & \geq & f^D(a;d) & \geq & f_D(x;d) & \geq & f_C(a;d) \\
IV & & IV & & IV & & IV \\
f^R(a;d) & \geq & f^{wR}(a;d) & \geq & f_{DH}(a;d) & \geq & f_S(a;d)
\end{array}$$

where the "largest" and "smallest" derivatives are defined as follows:

$$f^L(a;d) := \limsup_{\substack{(z,y)\downarrow a \\ t\to 0+}} \sup_{u\to d} \frac{f(z+tu)-y}{t},$$

$$f_S(a;d) := \liminf_{\substack{(z,y)\downarrow a \\ t\to 0+}} \inf_{u\to d} \frac{f(z+tu)-y}{t}.$$

$f^L(a;d)$ was introduced in [53] and called the *dag derivative*. $f^R(a;d)$ is frequently called as the Clarke-Rockafellar or circa derivative, $f^{wR}(a;d)$ can also be found under the name *incident derivative*, whereas $f_{DH}(a;d)$ can be referred to as *contingent derivative* [54].

3.1 The Upper Rockafellar Derivative and Generalized Convexity

The Clarke and Rockafellar derivatives play central role in Nonsmooth Analysis, since they enjoy some very remarkable properties. A) The upper Rockafellar derivative is lower semicontinuous sublinear in its direction argument for lower semicontinuous functions. B) The upper Clarke derivative is lower semicontinuous sublinear in its direction argument for locally Lipschitzian functions. C) For locally Lipschitzian functions they coincide.

Generalized convexity can be characterized by the upper Rockafellar derivative, as well.

Theorem 10.17 *Let $f(x)$ be lower semicontinuous on the open convex set $C \subset X$. Then each of the following statements are true:*

(i) $f(x)$ is convex on C iff $f^R(x; d)$ is monotone on C,

(ii) $f(x)$ is quasiconvex on C iff $f^R(x; d)$ is quasimonotone on C.

The proof of (i) of the above theorem was given in [47], whereas statement (ii) was proved in [48]. The proofs are strongly based on the Zagrodny's Approximate Mean Value Theorem [63].

3.2 Generalized Monotonicity and Generalized Convexity

Generalized convexity can be characterized not only via Dini derivatives and the Rockafellar derivatives, but also via other generalized derivatives. In order to extend the statements of Theorems 10.13, 10.10 and 10.17 to other generalized derivatives, the following "*majorizing lemma*" proves to be very useful.

Lemma 10.5 *Let the bifunctions $\varphi(x; d)$ and $\psi(x; d)$ be defined on the set $C \subset X$. Let us have for all $x \in C$ and $d \in X$*

$$\psi(x; d) \leq \varphi(x; d).$$

If $\varphi(x; d)$ is (strictly) monotone, quasimonotone, or (strictly) pseudomonotone on C, then $\psi(x; d)$ is (strictly) monotone, quasimonotone, or (strictly) pseudomonotone on C, respectively, as well.

The following result is a simple consequence of this lemma and Theorem 10.17.

Theorem 10.18 *Let the bifunction $\varphi(x;d)$ be defined on the open convex set C, on which the function $f(x)$ is lower semicontinuous. Let for all $x \in C$ and $d \in X$ have*

$$\varphi(x;d) \leq f^R(x;d).$$

Then any of the following statements holds true:

(i) $\varphi(x;d)$ is monotone on C if $f(x)$ is convex on C,

(ii) $\varphi(x;d)$ is quasimonotone on C if $f(x)$ is quasiconvex on C.

The above theorem can be applied to any members of the following collection:

$$f^{wR}(a;d) \;\; \geq \;\; f_{DH}(a;d) \;\; \geq \;\; f_S(a;d) \;\; .$$

The following result will also be used in the sequel [44].

Lemma 10.6 *Let the bifunction $\varphi(x;d)$ be defined on the open set C. Set*

$$\psi(x;d) = \limsup_{z \to x} \inf_{u \to d} \varphi(z;u).$$

Then

(i) the monotonicity of $\varphi(x;d)$ on C implies the monotonicity of $\psi(x;d)$ on C,

(ii) the quasimonotonicity of $\varphi(x;d)$ on C implies the quasimonotonicity of $\psi(x;d)$ on C.

We may also use any generalized derivatives to extend the pseudoconvexity concept of Mangasarian. In Subsection 2.1, the Diewert's concept was investigated. For locally Lipschitzian functions, pseudoconvexity was defined in [26] via the following implication

$$x,y \in C, f(x) < f(y) \quad \text{implies} \quad f^C(y;x-y) < 0. \qquad PCX(Clarke)$$

These attempts motivate the following general treatment.

Definition 10.9 *Let $\varphi(x;d)$ be a bifunction defined on a convex set C. A function $f(x)$ is called φ-quasiconvex, φ-pseudoconvex, strictly φ-pseudoconvex on C, if for any $x,y \in C$, $x \neq y$ implications $QCX(\varphi)$, $PCX(\varphi)$, $SPCX(\varphi)$ hold, respectively:*

$$f(x) \leq f(y) \quad \textit{implies} \quad \varphi(y;x-y) \leq 0, \qquad QCX(\varphi)$$

$$f(x) < f(y) \quad \textit{implies} \quad \varphi(y;x-y) < 0, \qquad PCX(\varphi)$$

$$f(x) \leq f(y) \quad implies \quad \varphi(y; x - y) < 0. \qquad SPCX(\varphi)$$

We have another very useful "*majorizing lemma*".

Lemma 10.7 *Let $f(x)$ and the bifunctions $\varphi(x; d)$ and $\psi(x; d)$ be defined on the set $C \subset X$. Let us have for all $x \in C$ and $d \in X$*

$$\psi(x; d) \leq \varphi(x; d).$$

Then φ-quasiconvexity, (strict) φ-pseudoconvexity of $f(x)$ implies ψ-quasiconvexity, (strict) ψ-pseudoconvexity of $f(x)$, respectively.

The following result is an immediate consequence of Lemmas 10.5, 10.7 and Theorems 10.10, 10.11.

Theorem 10.19 *Let the bifunction $\varphi(x; d)$ be defined on the convex set C, on which the function $f(x)$ is radially lower semicontinuous. Let for all $x \in C$ and $d \in X$ have*

$$f_D(x; d) \leq \varphi(x; d).$$

Then each of the following statements are true:

(i) $f(x)$ is (strictly) convex on C if $\varphi(x; d)$ is (strictly) monotone on C,

(ii) $f(x)$ is quasiconvex on C if $\varphi(x; d)$ is quasimonotone on C,

(iii) $f(x)$ is (strictly) h-pseudoconvex on C if $\varphi(x; d)$ is (strictly) pseudomonotone on C.

The above theorem can be applied to any members of the following collection:

$$\begin{array}{ccccc} f^L(x; d) & \geq & f^{DH}(x; d) & \geq & f_{wR}(x; d) \\ \text{IV} & & \text{IV} & & \text{IV} \\ f^C(x; d) & \geq & f^D(x; d) & \geq & f_D(x; d) \end{array}.$$

Since for lower semicontinuous functions we have

$$f_{DH}(x; d) \leq f^R(x; d) \leq \limsup_{z \to x} \inf_{u \to d} f_{DH}(z; u),$$

(see [64]), therefore from Theorem 10.17 and Lemmas 10.5, 10.6 one can infer the following results.

Theorem 10.20 *Let $f(x)$ be lower semicontinuous on the open convex set $C \subset X$. Then the upper Rockafellar derivative $f^R(x; d)$ is monotone,* *quasimonotone on C, if and only if the lower Dini-Hadamard derivative $f_{DH}(x;d)$ is monotone, quasimonotone on C, respectively.*

Theorem 10.21 *Let the bifunction $\varphi(x;d)$ be defined on the convex set C, on which the function $f(x)$ be lower semicontinuous and let us have for all $x \in C$ and $d \in X$*

$$f_{DH}(x;d) \leq \varphi(x;d).$$

Then each of the following statements are true:

(i) $f(x)$ is convex on C if $\varphi(x;d)$ is monotone on C,

(ii) $f(x)$ is quasiconvex on C if $\varphi(x;d)$ is quasimonotone on C.

The above theorem can be applied for any members of the following collection:

$$\begin{array}{ccccc} f^{L}(x;d) & \geq & f^{DH}(x;d) & \geq & f_{wR}(x;d) \\ \text{IV} & & \text{IV} & & \text{IV} \\ f^{C}(x;d) & \geq & f^{D}(x;d) & \geq & f_{D}(x;d) \\ \text{IV} & & \text{IV} & & \text{IV} \\ f^{R}(x;d) & \geq & f^{wR}(x;d) & \geq & f_{DH}(x;d) \end{array} .$$

Since the incident derivative $f^{wR}(x;d)$ lies between $f^{R}(x;d)$ and $f_{DH}(x;d)$, therefore, taking into account Theorems 10.18 and 10.21 we have the following corollary.

Theorem 10.22 *Let $f(x)$ be lower semicontinuous on the open convex set $C \subset X$. Then each of the following statements are true:*

(i) $f(x)$ is convex on C iff $f^{wR}(x;d)$ is monotone on C,

(ii) $f(x)$ is quasiconvex on C iff $f^{wR}(x;d)$ is quasimonotone on C.

4. First order approximations

The abundance of the notions of generalized derivatives puts forward the question of axiomatization of the derivative concept. The concept of first order approximation was introduced with this aim and shall be discussed in this part. In subsection 4.2.1 a Generalized Farkas Lemma is presented for quasiconvex first order approximations. The section ends with deriving Karush-Kuhn-Tucker type optimality conditions using first order approximations and some results on steepest descent paving the way for algorithmic design.

We are mainly concerned in this part with the derivation of necessary optimality conditions for the following problem:

$$\begin{gathered} \min \quad f(x) \\ x \in F,\ F = \{x \in X : g_i(x) \leq 0, i = 1, ..., m\}. \end{gathered} \tag{NLP}$$

A frequently used scheme of deriving optimality conditions for (NLP) is the following: Let a be a local optimal solution of the problem and N be a suitable neighborhood of it. Assume for sake of simplicity that the first M constraints are active, whereas all of the others are inactive in a. We assume furthermore that the inactive constraint functions are upper semicontinuous at a. Then, in this case, the definition of local optimality can be given as an implication:

$$x \in N,\ g_i(x) \leq 0,\ i = 1, 2, ..., M \implies f(x) - f(a) \geq 0.$$

Taking into account that $g_i(a) = 0$ for all $i = 1, 2, ..., M$, this implication can be written in a more suitable form:

$$x \in N,\ g_i(x) - g_i(a) \leq 0,\ i = 1, 2, ..., M \Rightarrow f(x) - f(a) \geq 0. \tag{10.6}$$

According to a general treatment one approximates the increments $g_i(x) - g_i(a), i = 1, ..., M$ and $f(x) - f(a)$ with appropriate quantities

$$\gamma_i(a; x - a), \quad i = 1, 2, ..., M, \quad \text{and} \quad \varphi(a; x - a)$$

depending on a and $d = x - a$.

Using these "*approximations*", implication (10.6) is replaced by the following:

$$\gamma_i(d) \leq 0,\ i = 1, 2, ..., M \Rightarrow \varphi(d) \geq 0, \tag{10.7}$$

where, for the sake of simplicity the reference to a is omitted. It is obvious that whenever implication (10.6) $\Rightarrow$ (10.7) holds, then condition (10.7) is a necessary optimality condition for (NLP). The problem of finding conditions ensuring implication (10.6) $\Rightarrow$ (10.7) belongs to the realm of regularity conditions, whereas "*dual*" characterizations of condition (10.7) leads to Generalized Farkas Theorems.

The results obtained for these problems crucially depend on the properties imposed on the problem-functions; however, there is a common idea in most of the different approaches, namely the approximation of the problem-functions by an appropriate derivative at a given point.

Derivative means in this context a certain kind of generalized directional derivatives and appropriate means that the derivative has some "nice" properties, such as linearity, convexity, continuity, etc.

There is a property that the generalized directional derivatives have in common: they are positively homogeneous functions of the direction.

For differentiable functions the directional derivative is a linear function of the direction. For convex (or in more general for quasidifferentiable) functions the directional derivative is a convex function of the direction [55]. For locally Lipschitz functions the Clarke-derivative is convex, as well [7].

These examples with many others show that the most developed classes of the optimization problems are those which admit a certain kind of convex generalized directional derivatives. The reason of this seems to be very simple: Convex Analysis provides a very stable (geometric) background for these classes.

The aim of the present section is to generalize certain "*convex*" results along two lines.

(i) : The one is to treat the directional derivatives in an abstract way, namely to consider them as special positively homogeneous functions of the direction. This approach is not new (see, for instance [17], [20], [27]), but the details are different.

(ii) : The other line is to extend the framework of the "*convexity-realm*": to replace the convexity assumption with the requirement of quasiconvexity. Such an attempt has been made in [33] for the directional Dini derivative.

The abstract treatment of Ioffe [27] or Giannessi [20] has mainly been concerned with the convex case. The results of the present paper obtained for the quasiconvex case could partly be applied for their treatment, too.

4.1 Optimality Conditions

Let a be a feasible solution of the (NLP) problem, so let $a \in F$. For the sake of simplicity suppose that the set F is locally convex at a which means that there exists a neighborhood G of a such that the set $F \cap G$ is convex.

In this case for any $d \in cone(F \cap G - a)$ the half line $\{a + td : t \geq 0\}$ has initial section belonging entirely to F. ($cone(M)$ denotes the cone generated by the set M.) Since in our case $cone(F \cap G - a)$ does not depend on the choice of the neighborhood G, simplifying the notation, let us set

$$F(a) = cone(F \cap G - a).$$

The elements of the convex cone $F(a)$ are called *feasible directions* at a with respect to problem (NLP).

For the sake of convenience let us recall the definition of some well known concepts having a crucial role in the sequel. The closed convex cone

$$T_F(a) = clF(a),$$

is called the tangent cone to F at a. ("cl" is a sign for the closure operation.)

Deriving optimality conditions the notion of descent directions is very useful.

Definition 10.10 *A vector $d \in X$ is called a* descent direction *of the function $f(x)$ at a if there exists $\tau > 0$, such that for any $0 < t < \tau$ one has $f(a + td) < f(a)$.*

Definition 10.11 *A vector $d \in X$ is called a* weak descent direction *of the function $f(x)$ at a if for all $\tau > 0$ there exists $0 < t < \tau$, such that $f(a + td) < f(a)$.*

The descent directions, weak descent directions of $f(x)$ at a form a cone denoted by the symbol $D_f(a)$, $WD_f(a)$, respectively. It is easy to see that

$$D_f(a) \subset WD_f(a).$$

It is quite obvious that if $a \in F$ is a local optimal solution of problem (NLP) then the following condition must hold:

$$WD_f(a) \cap F(a) = \emptyset. \tag{10.8}$$

Although this condition seems simple enough, the applicable determination or convenient description of the cones $WD_f(a)$ and $F(a)$ is, however, not an easy task in general. In order to get more tractable optimality conditions a commonly used method is to approximate the cone $WD_f(a)$, thus approximating the objective function $f(x)$ at a by some kinds of its generalized derivatives.

The abstract scheme of this is the following: let $\varphi(a; d)$ denote a function of the directions $d \in X$.

Definition 10.12 *The function $\varphi(a; d)$, $d \in X$ is called first order approximation of the function $f(x)$ at a if it possesses the following two properties:*

(Ai) $\varphi(a; d)$ is positively homogeneous in d,

(Aii) $\varphi(a; d) < 0$ implies $d \in WD_f(a)$.

Generalized derivatives minorized by the lower Dini derivative are all first order approximations in the above sense.

The concept of the first order approximation was introduced by Ioffe [27] in a slightly different manner. He requires in his definition instead of (Aii) the fulfillment of the following property:

(Aii^*) $\quad f^D(a;d) \leq \varphi(a;d)$ for every $d \in X$,

where $f^D(a;d)$ denotes the upper Dini derivative. It is clear that (Aii^*) => (Aii), so our definition is a bit more general than the one of Ioffe.

φ-descent directions. Let us associate with the first order approximation $\varphi(a;d)$ the cone

$$D_\varphi(a) = \{d : \varphi(a;d) < 0\},$$

called *the cone of the φ-descent directions of $f(x)$ at a*. The term '*φ-descent direction*' is justified by the inclusion

$$D_\varphi(a) \subset WD_f(a) \tag{10.9}$$

coming from property (Aii) and expressing that each φ-descent direction is a weak descent direction, as well. The inclusion (10.9) means that the cone $D_\varphi(a)$ constitutes an "*inner*" approximation for $WD_f(a)$, thus condition

$$D_\varphi(a) \cap F(a) = \emptyset \tag{10.10}$$

is a necessary condition for $a \in F$ to be a local minimizer of problem (NLP).

In most of the cases investigated so far the first order approximation $\varphi(a;d)$ is supposed or proved to be continuous in d. This is the case when X is finite dimensional and $\varphi(a;d)$ is convex in d. Out of the consequences of the continuity we pick up the following two.

Definition 10.13 *The first order approximation $\varphi(a;d)$ is called* regular *if it possesses the following two properties:*

(Ri) $\varphi(a;d)$ is lower semicontinuous in d,

(Rii) $D_\varphi(a)$ is open.

By virtue of (Rii) for regular first order approximations the optimality condition (10.10) is equivalent to condition

$$D_\varphi(a) \cap T_F(a) = \emptyset. \tag{10.11}$$

4.1.1 Constraint Qualification. Consider again the nonlinear programming problem:

$$\begin{aligned} f(x) &\to \min \\ x \in F = \{x \in X : g_i(x) &\leq 0, \quad i = 1, 2, \ldots, m\}. \end{aligned} \tag{NLP}$$

We use the notation introduced at the beginning of this section. Consider the two important implications

$$g_i(x) - g_i(a) \leq 0, \; i = 1, 2, ..., M \;\Rightarrow\; f(x) - f(a) \geq 0. \tag{10.12}$$

and

$$\gamma_i(a; d) \leq 0, \; i = 1, 2, ..., M \;\Rightarrow\; \varphi(a; d) \geq 0, \tag{10.13}$$

where $d = x - a$. It is obvious that whenever implication (10.12) $\Rightarrow$ (10.13) holds, then condition (10.13) is a necessary optimality condition for (NLP). The problem of finding conditions ensuring implication (10.12) $\Rightarrow$ (10.13) belongs to the realm of regularity conditions.

We shall now provide a general regularity condition, which fits well to our abstract treatment. Put

$$\gamma(a; d) = \max\{\gamma_i(a; d) : i = 1, 2, ..., M\}$$

and

$$G(a) = \{d : \gamma(a; d) < 0\}, \qquad T(a) = \{d : \gamma(a; d) \leq 0\}.$$

Definition 10.14 *The constraints $g_i(x) \leq 0$, $i = 1, 2, ..., M$ are said to satisfy constraint qualification (X) at $a \in F$ with respect to $\gamma_i(a; d)$, $i = 1, 2, ..., M$ iff: $\gamma_i(a; d)$, $i = 1, 2, ..., M$ are lower semicontinuous first order approximations and*

$$clG(a) = clT(a). \tag{CQ}$$

Lemma 10.8 *Assume that $F(a)$, the cone of the feasible directions, is convex. Then if the constraint qualification (X) holds, then (10.12) implies (10.13).*

Proof. Since $\gamma_i(a; d)$, $i = 1, 2, ..., M$ are first order approximations we have $G(a) \subset F(a)$ and thus $clG(a) \subset clF(a) = T_F(a)$. In virtue of condition (CQ) it follows that $clG(a) = clT(a) = T(a) \subset T_F(a)$. On the other hand (10.12) implies $T_F(a) \cap D_\varphi(a) = \emptyset$, from which it follows that $T(a) \cap D_\varphi(a) = \emptyset$, which is equivalent to implication (10.13). $\square$

4.2 Quasiconvex first order approximations

First order approximations may provide suitable setting for discussing and deriving results important in the analysis of (NLP). It is also obvious that if we do not impose special properties on the first order approximations then this abstract framework remains rather poor. Convexity is a celebrated property in this context, too.

Our aim with the subsequent parts to show that quasiconvex first order approximations still provide a "*rich structure*". In quasiconvex analysis the *incident derivative* and the *upper Dini derivative* may serve as quasiconvex first order approximations.

In the next part we derive a dual representation for implication (10.13). To elaborate "*dual*" characterization for condition (10.13) is the realm of Farkas Lemmas. (For historical details consult [45], [52].)

4.2.1 A Quasiconvex Farkas-Lemma.

In the sequel we consider only *quasiconvex*, *lower semicontinuous* and *positively homogeneous* $\gamma_i(d)$, $i = 1, 2, ..., M$ and $\varphi(d)$ "*approximations*".

Consider now the set T defined as:

$$T = \{d \in X : \gamma_i(d) \leq 0, \quad i = 1, 2, \ldots, M\},$$

which is a closed convex cone. In order to obtain dual characterization for implication (10.13) we need the concept of the *negative polar* of a cone and two important *representation-theorems* of Convex Analysis.

Definition 10.15 *T^0, the negative polar of T, is defined as follows:*

$$T^0 = \{p \in X^* : \langle p, t\rangle \leq 0, \quad \textit{for all } t \in T\}.$$

Lemma 10.9 *([57], Theorem 13.2) Let $\varphi(d)$ be a positively homogeneous lower semicontinuous convex function. Then there exists a convex compact set $\partial_\varphi \subset X^*$, such that*

$$\varphi(d) = max\{\langle g, d\rangle : g \in \partial_\varphi\}. \tag{10.14}$$

Lemma 10.10 (Crouzeix' Representation Theorem) *([9]) Let $\varphi(d)$, $d \in X$ be a positively homogeneous lower semicontinuous quasiconvex function. Then there exist a convex compact set $\partial_\varphi^+ \subset X^*$ and a nonbounded closed convex set $\partial_\varphi^q \subset X^*$, such that*

$$\varphi(d) = \begin{cases} \sup\{\langle g, d\rangle : g \in \partial_\varphi^q\}, & \textit{if } \varphi(d) \leq 0, \\ \max\{\langle g, d\rangle : g \in \partial_\varphi^+\}, & \textit{if } \varphi(d) > 0. \end{cases} \tag{10.15}$$

and

$$\sup\left\{\langle g, d\rangle : g \in \partial_\varphi^q\right\} = +\infty, \quad \text{if } \varphi(d) > 0, \tag{10.16}$$

$$\max\left\{\langle g, d\rangle : g \in \partial_\varphi^+\right\} = 0, \quad \text{if } \varphi(d) \le 0. \tag{10.17}$$

Based on these representations we prove the following theorems.

Theorem 10.23 *Let $\gamma_i(d)$, $i = 1, 2, ..., M$ be positively homogeneous lower semicontinuous quasiconvex functions and let*

$$T = \{d \in X : \gamma_i(d) \le 0, \quad i = 1, 2, \ldots, M\}.$$

(i) If $\varphi(d)$ is positively homogeneous convex function, then implication (10.13) is equivalent to the following inclusion:

$$0 \in \partial_\varphi + T^0. \tag{10.18}$$

(ii) If $\varphi(d)$ is lower semicontinuous positively homogeneous quasiconvex function, then implication (10.13) is equivalent to the following inclusion:

$$0 \in cl(\partial_\varphi^q + T^0). \tag{10.19}$$

Proof. Due to the hypothesis of the present theorem T is a closed convex cone and coincides with its bipolar cone $(T^0)^0$.

(i): First we prove implication (10.18) $\Rightarrow$ (10.13). According to (10.18) there exists $g \in X^*$, such that

$$g \in \partial_\varphi \quad \text{and} \quad -g \in T^0.$$

In virtue of Lemma 10.9 it is obvious that for all $d \in X$ we have $\varphi(d) \ge \langle g, d\rangle$ and due to the definition of T^0 we have that $d \in T \Rightarrow \langle g, d\rangle \ge 0$. These two assertions prove implication (10.13).

We prove now that (10.13) $\Rightarrow$ (10.18), but we shall do it in an indirect way by proving that $\neg$(10.18) $\Rightarrow$ $\neg$(10.13). Assume that (10.18) does not hold, that is

$$0 \notin \partial_\varphi + T^0. \tag{10.20}$$

Since ∂_φ is compact (and convex), therefore it is closed convex not containing the origin. Let z be the minimal norm element of this set. From the extremal property of z it follows that for any $g \in \partial_\varphi$ and $p \in T^0$ the following inequality holds (cf. [51], Theorem 4.7]):

$$\langle z, g + p\rangle \ge \langle z, z\rangle > 0. \tag{10.21}$$

For $p = 0$ we obtain from (10.21) that

$$\langle z, g\rangle \ge \langle z, z\rangle > 0 \quad \text{for all} \quad g \in \partial_\varphi. \tag{10.22}$$

By Lemma 10.9 it immediately follows that

$$\varphi(-z) < 0. \tag{10.23}$$

We have another important consequence of (10.21): since T^0 is a cone (10.21) can be satisfied only if for all $p \in T^0$ $\langle z, p\rangle \geq 0$ holds. Since $(T^0)^0 = T$, therefore it follows that

$$-z \in T. \tag{10.24}$$

(10.23) and (10.24) ensure that in the case considered (10.13) does not hold.

(ii): We start with proving implication (10.19) $\Rightarrow$ (10.13). Assume for contradiction that (10.19) holds but there exists $d \in T$ such that $\varphi(d) < 0$. In virtue of (10.19) there exist two infinite sequences $\{g_j\}$ and $\{p_j\}$ such that

$$g_j \in \partial^q_\varphi,\ p_j \in T^0 \quad \text{and} \quad g_j + p_j \to 0.$$

Taking into account Lemma 10.10 it is obvious that for the given $d \in T$ we have

$$0 > \varphi(d) \geq \langle g_j, d\rangle\,, \quad \text{for all } j,$$

whereas, from the definition of T^0, it is clear that for the given $d \in T$ we also have

$$\langle p_j, d\rangle \leq 0, \quad \text{for all } j.$$

The two assertions above yield the following condition:

$$\langle g_j + p_j, d\rangle \leq \varphi(d) < 0, \quad \text{for all } j.$$

Since we have $g_j + p_j \to 0$ it is impossible. This contradiction proves implication (10.19) $\Rightarrow$ (10.13).

Now we prove implication (10.13) $\Rightarrow$ (10.19) in the form that $\neg$(10.19) $\Rightarrow$ $\neg$(10.13). Assume that (10.19) does not hold, that is

$$0 \notin cl(\partial^q_\varphi + T^0). \tag{10.25}$$

Let z denote the minimal norm element of $cl(\partial^q_\varphi + T^0)$. As a consequence of the extremal property of z for all $g \in \partial^q_\varphi$ and $p \in T^0$ the following condition holds (cf. [51], Theorem 4.7]):

$$\langle z, g+p\rangle \geq \langle z, z\rangle > 0. \tag{10.26}$$

For $p = 0$ (10.26) gives that:

$$\langle z, g\rangle \geq \langle z, z\rangle > 0 \quad \text{for all} \quad g \in \partial^q_\varphi. \tag{10.27}$$

By Lemma 10.10 it follows that

$$\varphi(-z) \leq -\langle z, z\rangle < 0. \tag{10.28}$$

(10.26) has one more important consequence: since T^0 is a cone, condition (10.26) can hold if for all $p \in T^0$ we have $\langle z, p\rangle \geq 0$. Since $(T^0)^0 = T$, therefore it is equivalent to the following containment:

$$-z \in T. \tag{10.29}$$

(10.28) and (10.29) together prove that in the case considered implication (10.13) does not hold. □

The next result gives a tool for transforming Theorem 10.23 into generalized Farkas Theorems.

Lemma 10.11 *(i): Let $\gamma_i(d)$, $i = 1, 2, ..., M$ be positively homogeneous convex functions. Then*

$$T^0 = cl\, cone\, conv \left\{\bigcup_i \partial_{\gamma_i}\right\}. \tag{10.30}$$

(ii): Let $\gamma_i(d)$, $i = 1, 2, ..., M$ be positively homogeneous lower semicontinuous quasiconvex functions. Then

$$T^0 = cl\, cone\, conv \left\{\bigcup_i \partial^q_{\gamma_i}\right\}. \tag{10.31}$$

Proof. Let $\gamma(d) = \max\{\gamma_i(d) : i = 1, 2, \ldots, M\}$. Then $T = \{d \in X : \gamma(d) \leq 0\}$.

(i): In this case $\gamma(d)$ is positively homogeneous and convex, as well. Moreover

$$\partial_\gamma = conv \left\{\bigcup_i \partial_{\gamma_i}\right\}. \tag{10.32}$$

(10.32) is a well-known formula from Convex Analysis (cf. [55], Theorem 3.6). It can be proved very easily that

$$T = (\partial_\gamma)^0 \quad \text{and} \quad T^0 = cl\, cone\, \partial_\gamma,$$

from which, taking into account (10.32), (10.30) is obtained.

If the compact convex set ∂_γ is polyhedral or does not contain the origin, then the cone generated by itself is closed and thus, in this case, the closure operation in (10.30) can be omitted.

(ii): In this case the max-function $\gamma(d)$ is also positively homogeneous lower semicontinuous and quasiconvex and

$$\partial^q_\gamma = cl\, conv \left\{\bigcup_i \partial^q_{\gamma_i}\right\}. \tag{10.33}$$

This formula can be proved as follows: by the Crouzeix' Representation Theorem we have that

$$\sup\left\{\langle g, d\rangle : g \in \partial^q_{\gamma_i}\right\} = \begin{cases} \gamma_i(d), & \text{if } \gamma_i(d) \leq 0, \\ +\infty, & \text{otherwise,} \end{cases}$$

and

$$\sup\left\{\langle g, d\rangle : g \in \partial^q_{\gamma}\right\} = \begin{cases} \gamma(d), & \text{if } \gamma(d) \leq 0, \\ +\infty, & \text{otherwise.} \end{cases}$$

Since $\gamma_i(d) \leq \gamma(d)$ holds for all d and i therefore, as a consequence of the above formulas, we obtain that for all $d \in X$ and i

$$\sup\left\{\langle g, d\rangle : g \in \partial^q_{\gamma_i}\right\} \leq \sup\left\{\langle g, d\rangle : g \in \partial^q_{\gamma}\right\}.$$

From the separation theorems of Convex Analysis it follows that this inequality is equivalent to the following inclusion:

$$\partial^q_{\gamma_i} \subseteq \partial^q_{\gamma},$$

from which we immediately obtain that

$$G = cl\, conv\left\{\bigcup_i \partial^q_{\gamma_i}\right\} \subseteq \partial^q_{\gamma}.$$

We prove now the coincidence $G = \partial^q_{\gamma}$. Assume for contradiction that there exists $f \in X^*$ such that $f \in \partial^q_{\gamma}$ and $f \notin G$. Since G is closed and convex therefore f can be strictly separated from G and thus there exists $d \in X$ such that

$$\sup\left\{\langle g, d\rangle : g \in G\right\} < \langle f, d\rangle.$$

Since $\partial^q_{\gamma_i} \subseteq G$ for all i, therefore we have

$$\sup\left\{\langle g, d\rangle : g \in \partial^q_{\gamma_i}\right\} \leq \sup\left\{\langle g, d\rangle : g \in G\right\} < \langle f, d\rangle.$$

In virtue of the Crouzeix' Representation Theorem it follows that $\gamma_i(d) \leq 0$ and $\gamma_i(d) < \langle f, d\rangle$, from which we obtain that

$$\gamma(d) \leq 0 \quad \text{and} \quad \gamma(d) < \langle f, d\rangle. \tag{10.34}$$

On the other hand, the Crouzeix' representation formula shows that

$$\gamma(d) = \sup\left\{\langle g, d\rangle : g \in \partial^q_{\gamma}\right\} \geq \langle f, d\rangle,$$

which is in contradiction to (10.34).

The relationship $T = (\partial^q_{\gamma})^0$ can be proved by the following sequence of equivalences:

$$d \in \left(\partial^q_{\gamma}\right)^0 \Leftrightarrow \langle g, d\rangle \leq 0, \forall g \in \partial^q_{\gamma} \Leftrightarrow \gamma(d) \leq 0 \Leftrightarrow d \in T.$$

It follows that $T^0 = cl\ cone\ \partial_\gamma^q$ from which, by the help of (10.33), (10.31) can be obtained. □

Now we are in the position to formulate some generalizations of the Farkas Theorem, which are immediate consequences of Theorem 10.23 and Lemma 10.11. To ease the comparison with the "*original*" Farkas Theorem let us recall an equivalent form of it in case of smooth (NLP) problem ([18], [19], [52]):

$$0 \in \nabla f(a) + cone\ conv \{\nabla g_1(a), \nabla g_2(a), \ldots, \nabla g_M(a)\}.$$

Theorem 10.24 (Generalized Farkas Theorems)
A) Let $\gamma_i(d)$, $i = 1, 2, ..., M$ be positively homogeneous convex functions.

(i) If $\varphi(d)$ is a positively homogeneous convex function, then (10.13) is equivalent to the following condition:

$$0 \in \partial_\varphi + cl\ cone\ conv \{\textstyle\bigcup_i \partial_{\gamma_i}\}.$$

(ii) If $\varphi(d)$ is a positively homogeneous lower semicontinuous quasiconvex function, then (10.13) is equivalent to the following condition:

$$0 \in cl\left(\partial_\varphi^q + cone\ conv \{\textstyle\bigcup_i \partial_{\gamma_i}\}\right).$$

B) Let $\gamma_i(d)$, $i = 1, 2, ..., M$ be positively homogeneous lower semicontinuous quasiconvex functions.

(iii) If $\varphi(d)$ is a positively homogeneous convex function, then (10.13) is equivalent to the following condition:

$$0 \in \partial_\varphi + cl\ cone\ conv \{\textstyle\bigcup_i \partial_{\gamma_i}^q\}.$$

(iv) If $\varphi(d)$ is a positively homogeneous lower semicontinuous quasiconvex function, then (10.13) is equivalent to the following condition:

$$0 \in cl\left(\partial_\varphi^q + cone\ conv \{\textstyle\bigcup_i \partial_{\gamma_i}^q\}\right).$$

Remark 10.7 It should be mentioned that Farkas Theorem for positively homogeneous lower semicontinuous quasiconvex functions has recently been elaborated in [60] and [61] by using a general "abstract convexity" approach.

The application of the above results to obtain necessary optimality conditions for certain classes of nonsmooth programming problems is discussed in the next subsection and in more details in [35], [37].

4.2.2 KKT-Type Theorems.

Theorem 10.25 *Let $a \in F$ be a local optimal solution of the (NLP) problem and let the feasible set F be locally convex at a. Assume that the active constraint and the objective functions admit quasiconvex regular first order approximations $\gamma_i(a;d)$, $i = 1, 2, ..., M$ and $\varphi(a;d)$ at a, respectively, whereas the constraint functions inactive at a are upper semicontinuous at a. Assume furthermore that the active constraints satisfy constraint qualification (CQ) with respect to $\gamma_i(a;d)$, $i = 1, 2, ..., M$. Then*

$$0 \in cl\left(\partial^q_\varphi + cone\,conv\left\{\bigcup_i \partial^q_{\gamma_i}\right\}\right).$$

Proof. From Lemma 10.8 it follows that in our case implication

$$\gamma_i(a;d) \leq 0, \; i = 1, 2, ..., M \; \Rightarrow \; \varphi(a;d) \geq 0$$

is a necessary optimality condition. By virtue of $B)$ $ii)$ of the Generalized Farkas Theorem we have the thesis. □

Corollary 10.1 *[35] Let $a \in F$ be a local optimal solution of the (NLP) problem and let the feasible set F be locally convex at a. Assume that the active constraint and the objective functions are regular generalized upper quasidifferentiable at a, while the constraint functions, inactive at a, are upper semicontinuous at a. Assume furthermore that the active constraints satisfy constraint qualification (CQ) with respect to $g_i^D(a;d)$, $i = 1, 2, ...M$. Then*

$$0 \in cl\left(\partial^q f(a) + cone\,conv\left\{\bigcup_i \partial^q g_i(a)\right\}\right).$$

Remark 10.8 By the help of the Generalized Farkas Theorem one can derive further variants of the KKT-type Theorem (see [45]).

4.2.3 Steepest Descent Directions.

Theorem 10.23 provides a dual characterization of implication

$$d \in T_F(a) \Rightarrow \varphi(d) \geq 0$$

in form of $D_\varphi(a) \cap T_F(a) = \emptyset$. It is well known that for convex generalized derivatives condition,

$$D_\varphi(a) \cap T_F(a) \neq \emptyset \tag{10.35}$$

has very interesting geometrical background based on the subdifferential and the normal cone.

We assume that X is a real Hilbert space and start with recalling a well-known theorem on the steepest descent direction (c.f. [42], [62]).

Theorem 10.26 *Let $\varphi(a;d)$ be a convex first order approximation of $f(x)$ at a. Then $D_\varphi(a) \cap T_F(a) \neq \emptyset$ if and only if $0 \notin N_F(a) + \partial_\varphi f(a)$. If $0 \notin N_F(a) + \partial_\varphi f(a)$ and s_0 is the minimal norm element of the closed convex set $N_F(a) + \partial_\varphi f(a)$, then $d_0 = -\frac{s_0}{\|s_0\|} \in D_\varphi(a) \cap T_F(a)$ is the normalized steepest descent direction of $\varphi(a,d)$ in the following sense: for all $d \in D_\varphi(a) \cap T_F(a)$, $d \neq d_0$, $\|d\| = 1$ one has*

$$\varphi(a;d) > \varphi(a;d_0) = -\|s_0\|.$$

This theorem was extended for quasiconvex regular first order approximation by the author in [37].

Theorem 10.27 *Let $\varphi(a;d)$ be regular quasiconvex first order approximation of $f(x)$ at a. Then $D_\varphi(a) \cap T_F(a) \neq \emptyset$ if and only if $0 \notin N_F(a) + \partial_\varphi f(a)$. If $0 \notin N_F(a) + \partial_\varphi f(a)$ and s_0 is the minimal norm element of the closed convex set $N_F(a) + \partial_\varphi f(a)$, then $d_0 = -\frac{s_0}{\|s_0\|} \in D_\varphi(a) \cap T_F(a)$ is the normalized steepest descent direction of $\varphi(a,d)$ in the following sense: for all $d \in D_\varphi(a) \cap T_F(a)$, $d \neq d_0$, $\|d\| = 1$ one has*

$$\varphi(a;d) > \varphi(a;d_0) = -\|s_0\|.$$

References

[1] Arrow, K.J. and A.C. Enthoven, *Quasiconcave programming*, Econometrica, 29 (1961) 779-800.

[2] Aussel, D., J.-N. Corvellec and M. Lassonde, *Subdifferential characterization of quasiconvexity and convexity*, J. Convex Analysis, 1 (1994) 195-202.

[3] Aussel, D., *Subdifferential properties of quasiconvex and pseudoconvex functions: a unified approach*,

[4] Avriel, M., W.E. Diewert, S. Schaible and W.T. Ziemba, *Generalized Concavity*, Plenum Press, New York, 1988.

[5] Bianchi, M., *Generalized Quasimonotonicity and strong pseudomonotonicity of bifunction*, Optimization, 36 (1996) 1-10.

[6] Borde, J.- J.-P. Crouzeix, *Continuity properties of the normal cone to the level sets of a quasiconvex function*, Journal of Optimization Theory and Applications, 97 (1998) 29-45.

[7] Clarke, F.H., *Optimization and nonsmooth analysis*, John Wiley, New York, 1983.

[8] Crouzeix, J-P., *Contributions à l'étude des fonctiones quasiconvexes*, Thèse d'Etat, U.E.R. des Sciences Exactes et Naturelles, Université de Clermont-Ferrand II, 1977.

[9] Crouzeix, J.-P., *Continuity and differentiability properties of quasiconvex functions on* R^n, in: Generalized concavity in optimization and economics, S. Schaible and W.T. Ziemba (eds.), Academic Press, New York, 1981, pp. 109-130.

[10] Crouzeix, J.-P., *Some differentiability properties of quasiconvex functions on* R^n, In: Optimization and Optimal Control, Proceedings of a conference held at Oberwolfach, March 16-22, 1980., A. Auslender, W. Oettli and J. Stoer (eds.), Springer Verlag, Berlin, 1981, pp. 9-20.

[11] Crouzeix, J.-P., *About differentiability of order one of quasiconvex functions on* R^n, Journal of Optimization Theory and Applications 36 (1982) 367-385.

[12] Crouzeix, J.-P. and J..A. Ferland, *Criteria for quasiconvexity and pseudoconvexity: relationships and comparisons*, Mathematical Programming 23 (1982) 193-205.

[13] Crouzeix, J.-P., *Some properties of Dini-derivatives of quasiconvex and pseudoconvex functions*, New Trends in Mathematical Programming, Giannessi, F., S. Komlósi and T. Rapcsák (eds.), Kluwer Academic Publishers, Dordrecht, 1998, pp. 41-57.

[14] Crouzeix, J.-P., *Continuity and differentiability of quasiconvex functions*, Chapter 3, this volume.

[15] Diewert, W.E., *Alternative characterizations of six kinds of quasiconcavity in the nondifferentiable case with applications to nonsmooth programming*, in: Generalized Concavity in Optimization and Economics, S. Schaible and W.T. Ziemba (eds.), Academic Press, New York, 1981.

[16] Demyanov, V.F. and A.M. Rubinov, *On quasidifferentiable functionals*, Dokl. Akad. Nauk. SSR, 250 (1980) 21-25.

[17] Elster, K.-H. and J. Thierfelder, *Abstract cone approximations and generalized differentiability in nonsmooth optimization*, Optimization 19 (1988) 315-341.

[18] Farkas, J, *Über die Anwendungen des mechanischen Princips von Fourier*, Mathematische und Naturwissenschaftliche Bericht aus Ungarn 12 (1895) 263-281.

[19] Farkas, J., *Theorie der einfachen Ungleichungen*, Journal für die Reine und Angewandte Mathematik 124 (1901) 1-27.

[20] Giannessi, F., *Semidifferentiable functions and necessary optimality conditions*, Journal of Optimization Theory and Applications 60 (1989) 191-241.

[21] Giorgi, G. and S. Komlósi, *Dini Derivatives in Optimization - Part I*, Rivista di Matematica per le Scienze Economiche e Sociali, 15 (1992) No1., 3-30.

[22] Giorgi, G. and S. Komlósi, *Dini Derivatives in Optimization - Part II*, Rivista di Matematica per le Scienze Economiche e Sociali, 15 (1992) No2., 3-24.

[23] Giorgi, G. and S. Komlósi, *Dini Derivatives in Optimization - Part III*, Rivista di Matematica per le Scienze Economiche e Sociali, 18 (1995) No3., 46-63.

[24] Hadjisavvas, N. and S. Schaible, *On Strong Pseudomonotonicity and (Semi)strict Quasimonotonicity*, Journal of Optimization Theory and Applications, 79 (1993) 139-155.

[25] Hassouni, A., *Sous-differentiels des fonctions quasi-convexes*, Thèse de 3ème cycle de l'Université Paul Sabatier, Toulouse, 1983.

[26] Hiriart-Urruty, J.B., *New concepts in Nondifferentiable Programming*, Bulletin de la Sociète Mathèmatique de France,. Mémoires, 60 (1979) 57-85.

[27] Ioffe, A.D., *Necessary and sufficient conditions for a local minimum:I*, SIAM J.Control and Optimization 17 (1979) 245-250.

[28] Karamardian, S., *Complementarity Over Cones with Monotone and Pseudomonotone Maps*, Journal of Optimization Theory and Applications 18 (1976) 445-454.

[29] Karamardian, S. and S. Schaible, *Seven kinds of monotone maps*, Journal of Optimization Theory and Applications 66 (1990) 37-46.

[30] Komlósi, S., *Some properties of nondifferentiable pseudoconvex functions*, Mathematical Programming 26 (1983) 232-237.

[31] Komlósi, S., *Contribution to the theory and methods of quasiconvex programming*, Ph.D. Thesis, Baku State University, 1984.

[32] Komlósi, S., *Generalized convexity of quadratic functions*, Izv. Vyssh. Uchebn. Zaved. Math., 9 (1984) 38-43.

[33] Komlósi, S., *On a possible generalization of Pshenichnyi's quasidifferentiability*, Optimization 21 (1990) 3-11.

[34] Komlósi, S., *Generalized Monotonicity of Generalized Derivatives*, Working Paper, 1991, pp. 8.

[35] Komlósi, S., *On generalized upper quasidifferentiability* in: Nonsmooth Optimization: Methods and Applications, Giannessi, F., (ed.), Gordon and Breach, London, (1992) 189-201.

[36] Komlósi, S., *Generalized Monotonicity of Generalized Derivatives*, in: Proceedings of the Workshop on Generalized Concavity for Economic Application, Mazzoleni, P., (ed.), Verona, 1992, 1-7.

[37] Komlósi, S., *Quasiconvex first order approximations and Kuhn-Tucker type optimality conditions*, European Journal of Operational Research 65 (1993) 327- 335.

[38] Komlósi, S., *First and second order characterizations of pseudolinear functions*, European Journal of Operational Research 67 (1993) 278-286.

[39] Komlósi, S., *On pseudoconvex functions*, Acta Scientiarum Mathemathicarum (Szeged), 57 (1993) 569-586.

[40] Komlósi, S., *Generalized Monotonicity in Nonsmooth Analysis*, in: Generalized Convexity, Komlósi, S., T. Rapcsák and S. Schaible (eds.), Springer Verlag, Heidelberg, (1994) 263-275.

[41] Komlósi, S., *Generalized Global Monotonicity of Generalized Derivatives*, International Transactions in Operational Research 1 (1994) 259-264.

[42] Komlósi, S. and M. Pappalardo, *A General Scheme for First Order Approximations in Optimization*, Optimization Methods and Software, 3 (1994) 143.152.

[43] Komlósi, S., *Generalized Monotonicity and Generalized Convexity*, Journal of Optimization Theory and Application, 84 (1995) 361-376.

[44] Komlósi, S., *Monotonicity and Quasimonotonicity in Nonsmooth Analysis*, in: Recent Advances in Nonsmooth Optimization, D.-Z. Du, L.Qi, R.S.Womersely (eds.), World Scientific Publishers, Singapore, 1995., pp.193-214.

[45] Komlósi, S., *Farkas Theorems for positively homogeneous quasiconvex functions*, Journal of Statistics and Management Systems, 5 (2002) Nos 1-3, pp. 107-123.

[46] Kuhn, H.W. and A.W. Tucker, *Nonlinear Programming*, in: Proceedings of the Second Berkeley Symposium on Mathematical Statistics and Probability, Neyman, J. (ed.), University of California Press, Berkeley, 1951, pp. 481-492.

[47] Luc, D.T., and S. Swaminathan, *A characterization of convex functions*, Nonlinear Analysis, Theory, Methods and Applications, 20 (1993) 697-701.

[48] Luc, D.T., *Characterization of quasiconvex functions*, Bulletin of the Australian Mathematical Society, 48 (1993) 697-701.

[49] Luc, D.T., *On Generalized Convex Nonsmooth Functions*, Bulletin of the Australian Mathematical Society, 49 (1994) 139-149.

[50] Mangasarian, O.L., *Pseudoconvex functions*, SIAM Journal on Control 3 (1965) 281-290.

[51] Martos, B., *Nonlinear programming: theory and methods*, Akadémiai Kiadó, Budapest, 1975.

[52] Prékopa, A., *On the development of optimization theory*, Amer. Math. Monthly 87 (1980) 527-542.

[53] Penot, J.-P., *Generalized Convexity in the Light of Nonsmooth Analysis*, in: Recent Developments in Optimization, Durier, R. and C. Michelot (eds.), Lecture Notes in Economics and Mathematical Systems, Springer Verlag, Heidelberg-Berlin, Vol. 429, 1995, pp. 269-290.

[54] Penot, J.-P., *Are generalized derivatives useful for generalized convex functions?* in: Generalized convexity, generalized monotonicity: recent results, Crouzeix, J.-P., J.-E. Martínez-Legaz and M. Volle (eds.), Kluwer, Dordrecht, 1998, pp. 3-59.

[55] Pshenichnyi, B.T., *Necessary conditions for an extremum*, Nauka, Moscow, 1969.

[56] Pshenichnyi, B.T., *Convex analysis and extremum problems*, Nauka, Moscow, 1980.

[57] Rockafellar, R.T., *Convex analysis*, Princeton University Press, Princeton, 1970.

[58] Rockafellar, R.T., *Generalized directional derivatives and subgradients of nonconvex functions*, Canadian Journal on Mathematics, 32 (1980) 257-280.

[59] Rockafellar, R.T., *The Theory of Subgradients and its Applications to Problems of Optimization: Convex and Nonconvex Functions*, Heldermann Verlag, Berlin, 1981.

[60] Rubinov, A.M. and Glover, B.M., *Quasiconvexity via two step functions*, in: Generalized Convexity, Generalized Monotonicity, Crouzeix, J.-P., J.-E. Martínez-Legaz and M. Volle (eds.) Kluwer Academic Publishers, Dordrecht, 1998, pp. 159-184.

[61] Rubinov, A.M., *Abstract convexity: examples and applications*, Optimization 47 (2000) 1-33.

[62] Zalinescu, C., *Solvability Results for Sublinear Functions and Operators*, Zeitschrift für Operations Research, 31 (1987) A79-A101.

[63] Zagrodny, D., *Approximate mean value theorem for upper subderivatives*, Nonlinear Analysis, Theory, Methods and Applications, 12 (1988) 1413-1438.

[64] Zagrodny, D., *A note on the equivalence between the mean value theorem for the Dini derivative and the Clarke-Rockafellar derivative*, Optimization, 21 (1990) 179-183.

Chapter 11

GENERALIZED CONVEXITY, GENERALIZED MONOTONICITY AND NONSMOOTH ANALYSIS

Nicolas Hadjisavvas

Department of Product and Systems Design Engineering

University of the Aegean

84100 Hermoupolis, Syros, Greece

nhad@aegean.gr

Abstract This chapter is an introduction to generalized monotone multivalued maps and their relation to generalized convex functions through subdifferential theory. In particular, it contains the characterization of various types of generalized convex functions through properties of their subdifferentials. Also, some recent results on properly quasimonotone maps, maximal pseudomonotone maps, and a new "quasiconvex" subdifferential are presented.

Keywords: Generalized convexity, generalized monotonicity, subdifferential, maximal monotone operators.

1. Introduction

One of the cornerstones of Convex Analysis is the use of the subdifferential of convex functions. The reason is of course that in most problems arising in practice, convex functions are not differentiable, and in this case the subdifferential (also called Fenchel-Moreau subdifferential) provides an elegant and powerful tool for their study. In spite of its importance, the class of convex functions is a very small one. Since the advances of B.N. Pshenichnyi in Nonsmooth Optimization [64] and F. Clarke [14] who extended the notion of subdifferential to the class of locally Lipschitz functions, considerable effort has been devoted to create subdifferentials that can be used in nonconvex problems. As is usual in cases of generalization, the result was the definition of a large number of subdifferentials, each of which was introduced having in mind a particular application, or a property of the Fenchel-Moreau subdifferential that one wishes to preserve. Of course, none of these subdifferentials is best in all situations.

Given the variety of subdifferentials, it is remarkable that most of them are suitable for characterizing generalized convex functions. In fact, at first one has to introduce appropriate multivalued generalizations of the single valued generalized monotone maps. Having done so, one has then to show that a function is generalized convex in some sense, if and only if its subdifferential is generalized monotone in a corresponding sense. The literature on this subject is vast, and not easy to follow. The complexity of the field is mainly the result of the multitude of generalized monotonicity notions, the large number of subdifferentials, and the different assumptions of continuity etc., used by each author. In an effort to show the underlying unity, several authors proposed axiomatic schemes that cover a large number of subdifferentials and showed that many of the results are valid for all subdifferentials that fall into their axiomatic schemes. Let us mention in this respect the work of Penot [55, 57, 60], Aussel, Corvellec and Lassonde [2, 3] and Thibault and Zagrodny [68].

A large part of this chapter is devoted to an exposition of the main results on the relation of generalized convexity and generalized monotonicity via subdifferential theory. We exclude from our exposition all contributions not directly related to all three of these topics, such as: results involving generalized derivatives rather than subdifferentials (see Chapter 10 of this volume), the application of nonsmooth analytical tools and generalized convexity to duality methods, Mathematical Economics, Hamilton Jacobi-equations etc. (see [58] and Chapter 6 of this volume, and the references therein), the calculus of subdifferentials adapted to Generalized Convexity [62]. Even so, we will have to leave aside some important contributions on the integration of generalized monotone maps [7], the characterization of quasiconvexity via the normal cones [4, 8] etc.

The plan of the chapter is as follows. The next section introduces the various kinds of subdifferentials we are going to use. Section 3 studies the relation of convexity to monotonicity, and some variants. Section 4 is devoted to quasiconvexity and quasimonotonicity, and Section 5 concerns pseudoconvexity and pseudomonotonicity. Section 6 studies the recent concept of proper quasimonotonicity, its relation to variational inequalities, and explores the various connections between the notions of generalized monotonicity. Section 7 presents a recently introduced "quasiconvex" subdifferential and some of its properties. The last section contains some recent results on the maximality of pseudomonotone maps.

Let us fix the notation. In what follows, X is a Banach space and X^* its dual. For simplicity, readers may assume that $X = X^* = \mathbb{R}^n$ with no loss of an important feature (besides, functional analytic arguments are almost entirely missing in this chapter, except in the last section). Given $x, y \in X$, we denote by $[x, y]$ the segment $\{tx + (1-t)\,y : t \in [0,1]\}$. The open and half-open segments $]x, y[$, $]x, y]$ and $[x, y[$ are defined analogously. Given $S \subseteq X$, we denote by $\operatorname{int} S$, $\operatorname{core} S$, $\operatorname{co} S$ respectively, the interior, the algebraic interior and the convex hull of S. The set S is called radially open if $\operatorname{core} S = S$. For $x \in X$ and $\varepsilon > 0$, $B(x, \varepsilon)$ will denote the open ball of center x and radius ε. A *cycle* is a finite sequence $x_1, x_2, \ldots x_{n+1}$ of elements of X such that $x_{n+1} = x_1$.

Given a set $C \subseteq X$ and $x \in X$, we will denote by $N_C(x)$ the normal cone to C at x i.e., the set

$$N_C(x) = \{x^* \in X^* : \forall y \in C,\ \langle x^*, y - x\rangle \leq 0\}.$$

We will mainly consider *multivalued* maps T from X into X^*, i.e., maps that associate to each element $x \in X$ a subset $T(x)$ (possibly empty) of X^*. We will denote such maps as $T : X \rightrightarrows X^*$. The domain of T is the set $D(T) = \{x \in X : T(x) \neq \emptyset\}$.

Given an extended real-valued function $f : X \to \mathbb{R} \cup \{+\infty\}$ and $a \in \mathbb{R}$, we denote by $S_a f$ (resp. $S_a^< f$) the sublevel set (strict sublevel set):

$$S_a f = \{x \in X : f(x) \leq a\}$$
$$S_a^< f = \{x \in X : f(x) < a\}.$$

The function is called radially continuous if its restriction to line segments is continuous.

2. Subdifferentials

We recall the definition of the subdifferential of a convex function (here called Fenchel-Moreau subdifferential, to distinguish it from other subdifferentials to be introduced later).

Let $f : X \to \mathbb{R} \cup \{+\infty\}$ be a proper convex function with effective domain $\operatorname{dom} f := \{x \in X : f(x) < +\infty\}$ (see Chapter 1 for the relevant definitions). The Fenchel-Moreau subdifferential of f at $x_0 \in \operatorname{dom} f$ is the set

$$\partial^{FM} f(x_0) = \{x^* \in X^* : \langle x^*, x - x_0\rangle \leq f(x) - f(x_0), \forall x \in X\}, \quad (11.1)$$

while $\partial^{FM} f(x_0) = \emptyset$ if $x_0 \notin \operatorname{dom} f$.

The above definition makes use of the values of f on the whole space. However, it can be seen that the subdifferential depends only on the values of f at a neighborhood of x_0; indeed, the following result is standard in Convex Analysis [66, Th. 23.2]:

Proposition 11.1 *For any proper convex function f and any $x_0 \in \operatorname{dom} f$, $d \in X$, the directional derivative*

$$f'(x_0, d) := \lim_{t \searrow 0} \frac{f(x_0 + td) - f(x_0)}{t} \tag{11.2}$$

exists in $[-\infty, +\infty]$. In addition,

$$\partial^{FM} f(x_0) = \left\{x^* \in X^* : \langle x^*, d\rangle \leq f'(x_0, d), \forall d \in X\right\}. \tag{11.3}$$

Many attempts to define a subdifferential for more general classes of functions are inspired by relation (11.3). However, since the directional derivative may not exist if the function f is not convex, new kinds of generalized derivatives are introduced. The generalization follows two directions: first, instead of ordinary limits as in (11.2), one makes use of upper or lower limits. Second, besides t which approaches 0 from above, also x_0 and d may be replaced by vectors x and d' which approach (in some sense) x_0 and d, respectively. This creates a large number of possible generalized directional derivatives and an equal number of subdifferentials. For instance, given a lower semicontinuous (lsc) function f, for each $x_0 \in \operatorname{dom} f$, the upper Dini derivative and the Clarke-Rockafellar derivative are defined, respectively, by

$$f^{D^+}(x_0, d) = \limsup_{t \searrow 0} \frac{1}{t}\left(f(x_0 + td) - f(x_0)\right) \tag{11.4}$$

and

$$f^{CR}(x_0, d) = \sup_{\varepsilon > 0} \limsup_{x \to_f x_0, t \searrow 0} \inf_{d' \in B_\varepsilon(d)} \frac{f(x + td') - f(x)}{t}. \tag{11.5}$$

Here, $x \to_f x_0$ means that both $x \to x_0$ and $f(x) \to f(x_0)$. Both derivatives are majorized by the "dag" derivative [57]

$$f^{\dagger}(x_0, d) = \limsup_{t \searrow 0, y \to_f x_0} \frac{1}{t}\left(f(y + t(d + x_0 - y)) - f(y)\right). \tag{11.6}$$

Given any generalized derivative $f^{\Diamond}(x, d)$, $x \in \operatorname{dom} f$, $d \in X$, the corresponding subdifferential is the multivalued map $\partial^{\Diamond} f : X \rightrightarrows X^*$ defined by

$$\partial^{\Diamond} f(x_0) = \left\{x^* \in X^* : \langle x^*, v\rangle \leq f^{\Diamond}(x_0, v), \forall v \in X\right\}$$

for all $x_0 \in \operatorname{dom} f$, and $\partial^{\Diamond} f(x_0) = \emptyset$ for all $x_0 \notin \operatorname{dom} f$.

Accordingly, one can define the upper Dini, the Clarke-Rockafellar, and the dag subdifferentials of f at x_0, thus denoted $\partial^{D^+} f(x_0)$, $\partial^{CR} f(x_0)$ and $\partial^{\dagger} f(x_0)$, respectively. Note that since $f^{\dagger}$ majorizes both f^{D^+} and f^{CR}, $\partial^{\dagger}$ is larger than ∂^{D^+} and ∂^{CR} in the sense that $\partial^{CR} f \subseteq \partial^{\dagger} f$ and $\partial^{D^+} f \subseteq \partial^{\dagger} f$. Actually, $\partial^{\dagger}$ is one of the largest subdifferentials.

Whenever f is locally Lipschitz, the Clarke-Rockafellar derivative equals the somewhat simpler Clarke derivative

$$f^o(x_0; d) = \limsup_{x \to x_0, t \searrow 0} \frac{f(x + td) - f(x)}{t}$$

and then $\partial^{CR} f(x_0)$ is the Clarke subdifferential $\partial^o f(x_0)$ [15].

In many cases where the subdifferential is defined through a generalized directional derivative, the latter can be recovered from the subdifferential. For instance, the Clarke subdifferential $\partial^{CR} f(x_0)$ of a locally Lipschitz function is always w*-compact and the Clarke derivative is given by

$$f^o(x_0; d) = \max\{\langle x^*, d\rangle : x^* \in \partial^{CR} f(x_0)\rangle.$$

An analogous statement holds for the directional derivative and the subdifferential of convex functions. In such cases, many results in terms of subdifferentials are equivalent to results in terms of the corresponding generalized derivatives, see for instance [41, 42].

A subdifferential whose definition does not use a directional derivative, is the Fréchet subdifferential (also called regular subdifferential [67]) $\partial^F f(x)$, defined by

$$\partial^F f(x) := \{x^* \in X^* : f(y) \geq f(x) + \langle x^*, y - x\rangle + o(y - x), \forall y \in X\}$$

where $o : X \to R$ is some real valued function satisfying

$$\lim_{x \to 0} \frac{o(x)}{\|x\|} = 0.$$

One of the most basic tools of smooth or nonsmooth Analysis is the Mean Value Theorem (MVT). If, say, $f : \mathbb{R}^n \to \mathbb{R}$ is a differentiable function, then the MVT asserts that for any $a, b \in \mathbb{R}^n$ there exists $c \in]a, b[$ such that $\langle \nabla f(c), b - a\rangle = f(b) - f(a)$. The same result holds for functions defined in more general spaces. It would be desirable that something analogous holds for subdifferentials; for instance, it would be nice if one could recover, at least in some sense, the value $f(b) - f(a)$ by using the subdifferential of f at some point $c \in]a, b[$. Unfortunately, this is patently false since for most subdifferentials it is possible that

$\partial f(c) = \emptyset$ for all $c \in [a, b]$. However, it was shown by Zagrodny in one of the most far-reaching developments of Nonsmooth Analysis, that a version of the Mean Value Theorem holds for the Clarke-Rockafellar subdifferential, the main difference being the fact that instead of using a point $c \in [a, b]$ one has a sequence of points approaching some $c \in [a, b]$. Various versions of this theorem were shown later to be true also for other subdifferentials. In fact, axiomatic schemes have been invented [3, 55, 68], that incorporate a large number of the existing subdifferentials, and show the theorem to be true under suitable assumptions on the space X.

In order to simplify our discussion, let us define an *abstract subdifferential* as follows:

Definition 11.1 *An abstract subdifferential is any map ∂ which associates to each lsc function $f : X \to \mathbb{R} \cup \{+\infty\}$ and each $x \in X$ a subset $\partial f(x)$ of X^*, and has the properties:*

(a) If f is convex, then $\partial f = \partial^{FM} f$;

(b) If x is a local minimum of f, then $0 \in \partial f(x)$;

(c) For every $v^* \in X^*$ one has $\partial(f + v^*)(x) = \partial f(x) + v^*$.

This definition is quite general and includes most known subdifferentials, and in particular $\partial^{CR}, \partial^{D^+}, \partial^{\dagger}$ and ∂^F. We restrict the class of subdifferentials we are going to use, to those satisfying something like the Mean Value Theorem:

Definition 11.2 *An abstract subdifferential is called* MVT subdifferential *if for all $x, y \in X$ such that $f(y) > f(x)$, there exists $z \in [x, y[$ and sequences $(x_n) \subseteq \operatorname{dom} f$, $(x_n^*) \subseteq X^*$, such that $x_n \to z, x_n^* \in \partial f(x_n)$ and*

$$\langle x_n^*, z + t(y - z) - x_n \rangle > 0, \textit{ for all } t > 0. \tag{11.7}$$

Note that in particular (11.7) implies that

$$\langle x_n^*, v - x_n \rangle > 0, \text{ for all } v \in]z, y] . \tag{11.8}$$

A lot of the subdifferentials encountered in the literature are MVT subdifferentials, provided that the Banach space has some standard regularity property [3]. In particular, ∂^{CR} is a MVT subdifferential in any Banach space X, while ∂^F and ∂^{D^+} are MVT subdifferentials if the space has an equivalent norm which is Fréchet (respectively, Gâteaux) differentiable on $X \setminus \{0\}$.

The class of MVT subdifferentials is extremely broad to be of real use if we do not impose additional assumptions. If we define $\partial^{\max}$ by $\partial^{\max} f(x) = \partial^{FM} f$ if f is convex and $\partial^{\max} f = X^*$ if f is not convex, then it is obvious that $\partial^{\max}$ is a MVT subdifferential, and in fact the largest one. Such a subdifferential cannot be used to study any non-convex function since it does not depend on the function. Small MVT subdifferentials are more interesting [56] as they provide elements x^* that satisfy (11.7), while belonging to a relatively small class. Our usual assumption will be that our subdifferential will be smaller than one of the known subdifferentials, usually $\partial^{\dagger}$.

3. Monotonicity and Convexity

Monotone maps (or "operators") were used in partial differential equations since 70 years at least. However, the modern treatment of such maps started around 1960 by Kachurovskii [38] and Minty [53] who were probably the first to link monotone maps to convex functions. Soon afterwards, Browder [11] studied multivalued monotone maps and made decisive steps in that direction. We refer the interested reader to the treatise of Hu-Papageorgiou [36] for historical details and a modern presentation of the theory.

If $f : X \to \mathbb{R} \cup \{+\infty\}$ is a lsc, proper convex function and ∂ an abstract subdifferential then, according to Definition 11.1, $\partial f = \partial f^{FM}$. It is well known that in this case ∂f^{FM} is a monotone map; in fact, it is also cyclically monotone [65, 66]. Let us recall the appropriate definitions.

A multivalued map $T : X \rightrightarrows X^*$ is called:

1. *Monotone*, if for all $x, y \in X$ and $x^* \in T(x), y^* \in T(y)$,

$$\langle y^* - x^*, y - x \rangle \geq 0.$$

It is called *maximal monotone* if there is no monotone extension of T other than itself, i.e., whenever $T_1 : X \rightrightarrows X^*$ is a monotone map such that for all $x \in X$, $T(x) \subseteq T_1(x)$, then $T = T_1$.

2. *Cyclically monotone*, if for any cycle $x_1, x_2, \ldots, x_{n+1} = x_1 \in X$ and any $x_i^* \in T(x_i), i = 1, 2, \ldots n$,

$$\sum_{i=1}^{n} \langle x_i^*, x_{i+1} - x_i \rangle \leq 0. \tag{11.9}$$

It is called *maximal cyclically monotone* if it has no cyclically monotone extension other than itself.

3. *Strictly monotone*, if for all distinct $x, y \in X$ and all $x^* \in T(x)$, $y^* \in T(y)$,

$$\langle y^* - x^*, y - x \rangle > 0.$$

4. *Cyclically strictly monotone*, if for any cycle $x_1, x_2, \ldots, x_{n+1} \in X$ where at least two x_i's are distinct, and any $x_i^* \in T(x_i)$,

$$\sum_{i=1}^{n} \langle x_i^*, x_{i+1} - x_i \rangle < 0.$$

There are some obvious relations between these notions: Both strict and cyclic monotonicity imply monotonicity. Also, cyclic strict monotonicity implies strict monotonicity and cyclic monotonicity.

The convexity of a function is equivalent to the monotonicity of its subdifferential, as long as the latter is a MVT subdifferential. This is shown by the following theorem, which is more or less classical:

Theorem 11.1 *Let $f : X \to \mathbb{R} \cup \{+\infty\}$ be a lsc function and ∂ be a MVT subdifferential. Then the following assertions are equivalent:*
(i) f is convex.
(ii) For all $x, y \in \operatorname{dom} f$ and all $x^* \in \partial f(x)$ one has:

$$\langle x^*, y - x \rangle \leq f(y) - f(x). \tag{11.10}$$

(iii) The subdifferential ∂f is a cyclically monotone map.
(iv) ∂f is a monotone map.

Proof. Implication (i)⇒(ii) is a consequence of (11.1), since $\partial f = \partial^{FM} f$ whenever f is convex. Implications (ii)⇒(iii)⇒(iv) can be found in most textbooks on Convex Analysis, but we present a proof for the sake of completeness: If (ii) holds, then for any cycle $x_1, x_2, \ldots, x_n, x_{n+1} = x_1$ in X, we have

$$\langle x_i^*, x_{i+1} - x_i \rangle \leq f(x_{i+1}) - f(x_i).$$

Adding these inequalities we get relation (11.9).

If (iii) holds, then by considering a cycle $x_1, x_2, x_3 = x_1$ consisting of two points, we infer immediately (iv). We postpone the proof of implication (iv)⇒(i) until the next section. □

Likewise, we can characterize strictly convex functions (see Chapter 2 of this book) by strict monotonicity:

Theorem 11.2 *Let $f : X \to \mathbb{R} \cup \{+\infty\}$ be a lsc function and ∂ a MVT subdifferential. Consider the following assertions:*
(i) The function f is strictly convex.
(ii) For all distinct $x, y \in \operatorname{dom} f$ and $x^* \in \partial f(x)$, one has

$$\langle x^*, y - x \rangle < f(y) - f(x).$$

(iii) ∂f *is a cyclically strictly monotone map.*
(iv) ∂f *is a strictly monotone map.*

Then (i)⇒(ii)⇒(iii)⇒(iv). If, in addition, for all $x, y \in \operatorname{dom} f$ *there exists* $z \in]x, y[$ *such that* $\partial f(z) \neq \emptyset$, *then (iv)⇒(i).*

Proof. (i) ⇒ (ii): If f is strictly convex, then it is convex. Hence if $x, y \in X$ are distinct, we can apply relation (11.10) to x and $z = (x+y)/2$:

$$\langle x^*, y - x\rangle = 2\langle x^*, z - x\rangle \leq 2(f(z) - f(x)) < f(y) - f(x).$$

(ii)⇒(iii): Let $x_1, x_2, \ldots, x_{n+1}$ be a cycle in $\operatorname{dom} f$ such that at least two of the points are distinct; then $x_{i_0} \neq x_{i_0+1}$ for some i_0. If $x_i^* \in \partial f(x_i)$, $i = 1, 2, \ldots, n$, then by condition (ii), $\langle x_i^*, x_{i+1} - x_i\rangle \leq f(x_{i+1}) - f(x_i)$ where at least one of the inequalities is strict. Adding these inequalities, we deduce that (iii) holds.

(iii)⇒(iv): This is again obvious.

(iv)⇒(i): If ∂f is strictly monotone, then it is monotone. By Theorem 11.1 and Definition 11.1, f is convex and ∂f is the Fenchel-Moreau subdifferential. Suppose that f is not strictly convex; then there exist distinct $x, y \in \operatorname{dom} f$ such that $f(z) = tf(x) + (1-t)f(y)$ for every $z = tx + (1-t)y$, $t \in]0, 1[$. By assumption, we can find $z \in]x, y[$ such that $\partial f(z) \neq \emptyset$. Choose $z^* \in \partial f(z)$, then by Theorem 11.1, $\langle z^*, x - z\rangle \leq f(x) - f(z)$ holds. It follows easily that $\langle z^*, x - y\rangle \leq f(x) - f(y)$. Likewise we deduce that $\langle z^*, y - x\rangle \leq f(y) - f(x)$, hence $\langle z^*, y - x\rangle = f(y) - f(x)$.

Now we repeat the argument and find a point $z_1 \in]z, y[$ such that for some $z_1^* \in \partial f(z_1)$, $\langle z_1^*, y - x\rangle = f(y) - f(x)$ holds. It follows that $\langle z^* - z_1^*, y - x\rangle = 0$ and then $\langle z^* - z_1^*, z - z_1\rangle = 0$, thus contradicting the strict monotonicity of ∂f. □

The equivalence (i)⟺(iv) was established in [49] under the slightly less general assumption $\partial f(x) \neq \emptyset$ for all $x \in X$; the same reference contains a counterexample showing that the implication (iv)⇒(i) does not hold without some assumption on the domain of ∂f.

Note that there is no relation between strict monotonicity and cyclic monotonicity, as the two notions strengthen monotonicity in two different ways. This can be seen by an easy example:

Example 11.1 Let the map $T : \mathbb{R}^2 \to \mathbb{R}^2$ be defined by

$$T(a, b) = \left(\frac{a}{2} - b, a + \frac{b}{2}\right)$$

is strictly monotone since for all $x, y \in \mathbb{R}^2$, $(T(y) - T(x), y - x) = \frac{1}{2}\|y - x\|^2$. However, it is not cyclically monotone as one can see by

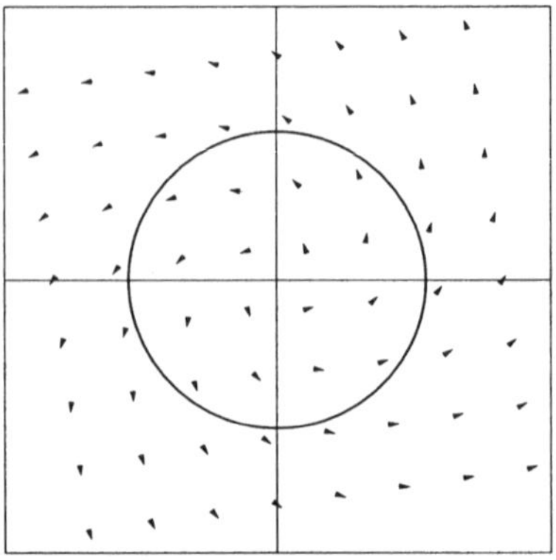

Figure 11.1. A strictly monotone map that is not cyclically monotone. The contour integral around a circle centered at the origin is not zero.

considering the points $x_1 = (1,0)$, $x_2 = (0,1)$, $x_3 = (-1,0)$ and $x_4 = (0,-1)$.

We can understand better the situation if we combine facts from Convex Analysis with Advanced Calculus: According to a classical theorem of Rockafellar [66], a cyclically monotone map is the subdifferential of a convex function if and only if it is maximal. Further, a continuous (cyclically) monotone map from a reflexive Banach space to its dual is always maximal (cyclically) monotone [72, Proposition 32.7, Vol. IIA]. Thus, if a continuous vector field $T : \mathbb{R}^n \to \mathbb{R}^n$ is cyclically monotone, then it is the gradient of a proper convex function $f : \mathbb{R}^n \to \mathbb{R}$. This corresponds to well-known facts from vector Calculus: indeed, if T is continuous and cyclically monotone, then it is an immediate consequence of the definition of the line integral that the contour integral $\oint_C T \cdot d\mathrm{s}$ along any continuous closed curve C is nonpositive; by considering the same curve with the opposite direction we deduce that $\oint_C T \cdot d\mathrm{s} = 0$. This is exactly the condition that guarantees that T is a gradient of a function. In the example above, the contour integral along a circle centered at the origin is not zero, as one can see by a look at the shape of the vector field shown in Figure 11.1. For a detailed analysis of the relation of line integrals to maximal monotonicity we refer the reader to [9, 69].

4. Quasimonotonicity and quasiconvexity

Multivalued quasimonotone maps were first introduced by Hassouni [32]. Also, Hassouni and Ellaia [34] generalized a result of Karamardian and Schaible [39] and established the connection between quasimonotonicity of a locally Lipschitz function and quasiconvexity of its Clarke subdifferential; an analogous connection was shown to hold for lower semicontinuous functions by Luc [48]. Later on, multivalued quasimonotone maps were studied in connection with variational inequalities and equilibrium problems, see [30, 31] and the references therein. Very recently, it has been shown that quasimonotonicity, together with continuity along straight lines and a coercivity condition, is enough for showing the existence of a solution to the variational inequality problem [5]. Also, cyclically quasimonotone maps were also introduced and studied (see [45] for the single valued and [21, 23] for the multivalued case).

We recall the relevant definitions. A multivalued map $T : X \rightrightarrows X^*$ is called:

1. *Quasimonotone*, if for all $x, y \in X$ and $x^* \in T(x), y^* \in T(y)$ the following implication holds:

$$\langle x^*, y - x \rangle > 0 \Rightarrow \langle y^*, y - x \rangle \geq 0. \tag{11.11}$$

2. Cyclically quasimonotone, if for any cycle $x_1, x_2, \ldots, x_{n+1}$ in X and any $x_i^* \in T(x_i)$, $i = 1, 2, \ldots, n$, the following implication holds:

$$\langle x_i^*, x_{i+1} - x_i \rangle > 0 \text{ for all } i = 1, 2, \ldots n-1 \Rightarrow \langle x_n^*, x_1 - x_n \rangle \leq 0. \tag{11.12}$$

Equivalently, T is cyclically quasimonotone if for any cycle $x_1, x_2, \ldots, x_{n+1} \in X$ and any $x_i^* \in T(x_i)$, $i = 1, 2, \ldots, n$, there exists some i in $\{1, 2, \ldots, n\}$ such that

$$\langle x_i^*, x_{i+1} - x_i \rangle \leq 0. \tag{11.13}$$

It is obvious that a monotone map is quasimonotone, and a cyclically monotone map is cyclically quasimonotone. Also, a cyclically quasimonotone map is quasimonotone. However, Example 11.1 shows that a monotone or even strictly monotone map need not be cyclically quasimonotone.

Quasiconvexity of a function is equivalent to quasimonotonicity of the subdifferential, under very mild conditions:

Theorem 11.3 *Let $f : X \to \mathbb{R} \cup \{+\infty\}$ be a lsc function and ∂ a MVT subdifferential. Consider the following assertions:*
(i) f is quasiconvex;
(ii) If $\langle x^*, y - x \rangle > 0$ for some $x^* \in \partial f(x)$, then $f(z) \leq f(y)$ for all $z \in [x, y]$;
(iii) ∂f is a quasimonotone map;
(iv) ∂f is a cyclically quasimonotone map.

Then (iv)⇒(iii)⇒(ii)⇒(i). If $\partial \subseteq \partial^\dagger$, then all four conditions are equivalent.

Proof. Implication (iv)⇒(iii) is trivial. Implications (iii)⇒(ii)⇒(i) are shown in [1]. Implication (i)⇒(ii) under the assumption $\partial \subseteq \partial^\dagger$ is shown in [60].

We show the remaining implication (ii)⇒(iv). It is sufficient to show that $\partial^\dagger f$ is cyclically quasimonotone. If not, then there is a cycle $x_1, x_2, \ldots, x_{k+1} = x_1$ and $x_i^* \in \partial^\dagger f(x_i)$, $i = 1, 2, \cdots, k$ such that $\langle x_i^*, x_{i+1} - x_i \rangle > 0$ for all $i = 1, 2, \ldots, k$. From (ii) we deduce that $f(x_{i+1}) \geq f(x_i)$ for all i's. It follows that $f(x_1) = f(x_2) = \ldots = f(x_k)$.

Since $\langle x_1^*, x_2 - x_1 \rangle > 0$, it follows from the definition of $\partial^\dagger$ that $f^\dagger(x_1, x_2 - x_1) > 0$. Using the definition of $f^\dagger$ we infer that there exist sequences $v_n \to x_1$ and $t_n \to 0^+$ such that $f(v_n + t_n(x_2 - v_n)) - f(v_n) > 0$. Since (ii)⇒(i), f is quasiconvex. From $v_n + t_n(x_2 - v_n) \in [v_n, x_2]$, we deduce that

$$f(v_n) < f(v_n + t_n(x_2 - v_n)) \leq \max\{f(v_n), f(x_2)\} = f(x_2) = f(x_1).$$

However, since $\langle x_k^*, x_1 - x_k \rangle > 0$, for n sufficiently large we have $\langle x_k^*, v_n - x_k \rangle > 0$. Hence, (ii) implies that $f(v_n) \geq f(x_k) = f(x_1)$, a contradiction. □

Implication (i)⇔(iii) for the case $\partial = \partial^{CR}$ was shown in [48], see also [2].

Note that some restriction to the size of the subdifferential is necessary in order to have (i)⇒(iv) or (i)⇒(iii) in Theorem 11.3. For instance, if $\partial = \partial^{\max}$, then obviously the implication does not hold. Further, we note that property (ii), which roughly corresponds to property (ii) in Theorem 11.1 for convex functions, means that if $x^* \in \partial f(x) \setminus \{0\}$ then for all d in the open half-space $\{d \in X : \langle x^*, d \rangle > 0\}$, f is non-decreasing along the half-line stemming from x in the direction d, i.e., the function $g(t) = f(x + td), t \geq 0$ is non-decreasing.

The following continuity result from [26] (see also [24]) will be useful in the sequel:

Proposition 11.2 *If f is quasiconvex, lsc and radially continuous, then it is continuous.*

Proof. Since f is lsc, it suffices to show that it is also upper semicontinuous, i.e., that for every $a \in \mathbb{R}$, the strict sublevel set $S_a^< = \{x \in X : f(x) < a\}$ is open. For any $x \in S_a^<$, let b be such that $f(x) < b < a$. Since f is radially continuous, for any $y \in X$ we can find $\varepsilon > 0$ such that $]x - \varepsilon y, x + \varepsilon y[$ is contained in the sublevel set S_b.

Hence x belongs to the algebraic interior of the closed convex set S_b. For closed convex sets in Banach spaces the algebraic and the topological interior coincide [35, pg 139]. It follows that $x \in \operatorname{int} S_b \subseteq \operatorname{int} S_a^<$. Hence $S_a^<$ is open. □

Convex functions can be characterized through quasiconvexity, as the following result of Crouzeix [20] shows:

Proposition 11.3 *A function $f : X \to \mathbb{R} \cup \{+\infty\}$ is convex if and only if for every $x^* \in X^*$, $f + x^*$ is quasiconvex.*

Proof. If f is convex, then obviously $f + x^*$ is convex, hence quasiconvex. Conversely, let $f + x^*$ be quasiconvex for each $x^* \in X^*$. For any $x, y \in X$ and $z = tx + (1-t)\,y$, $t \in [0,1]$, choose $x^* \in X^*$ such that $\langle x^*, y - x \rangle = f(x) - f(y)$. If, say, $(f + x^*)(y) \geq (f + x^*)(x)$ then by quasiconvexity of $f + x^*$ we have:

$$f(y) + \langle x^*, y \rangle = (f + x^*)(y) \geq (f + x^*)(z) = f(z) + \langle x^*, z \rangle .$$

Hence, $f(z) \leq f(y) + \langle z^*, y - z \rangle = f(y) + t \langle x^*, y - x \rangle = tf(x) + (1-t)\,f(y)$. □

We are now in position to present the missing part of the proof of Theorem 11.1:

Proof of (iv)⇒(i) in Theorem 11.1. If ∂f is monotone then obviously for any $x^* \in X^*$, $\partial(f + x^*) = \partial f + x^*$ is monotone. Hence, by Theorem 11.3, $f + x^*$ is quasiconvex. By Proposition 11.3, f is convex. □

The above proof is contained in [2]. The equivalence of convexity of a function to monotonicity of its Clarke-Rockafellar subdifferential was established in [17] for reflexive Banach spaces and in [46] for general Banach spaces; see also [18].

Semistrictly or strictly quasiconvex functions can also be characterized by means of their subdifferential. We first define the corresponding generalized monotonicity notions. A multivalued map $T : X \rightrightarrows X^*$ is called [50]:

1. *Semistrictly quasimonotone* if it is quasimonotone and for any distinct $x, y \in D(T)$, the following implication holds:

$$\exists x^* \in T(x) : \langle x^*, y - x \rangle > 0 \Rightarrow$$
$$\exists z \in \left] \frac{x+y}{2}, y \right[, \exists z^* \in T(z) : \langle z^*, y - x \rangle > 0. \tag{11.14}$$

2. *Strictly quasimonotone* if it is quasimonotone and for any distinct $x, y \in D(T)$, there exist $z \in]x, y[$ and $z^* \in T(z)$ such that $\langle z^*, y - x\rangle \neq 0$.

An equivalent formulation of (11.14) is the following: If $\langle x^*, y-x\rangle > 0$, $x^* \in T(x)$, then

$$\{z \in]x, y[: \langle z^*, y - x\rangle > 0 \text{ for some } z^* \in T(z)\} \text{ is dense in } [x, y]. \tag{11.15}$$

If $D(T)$ is convex (or more generally, if for all $x, y \in D(T)$, $[x, y] \cap D(T)$ is dense in $[x, y]$), it can be easily seen that a strictly quasimonotone map is semistrictly quasimonotone; also, a strictly monotone map is strictly quasimonotone.

The following characterizations were shown in [27] and in slightly less general form in [22]. Also, some of the implications were shown previously in [50].

Theorem 11.4 *Let $f : X \to \mathbb{R} \cup \{+\infty\}$ be a lsc function and ∂ a MVT subdifferential such that $\partial \subseteq \partial^\dagger$. Consider the following assertions:*
(a) f is semistrictly quasiconvex.
(b) For all $x, y \in X$ the following implication holds:

$$\exists x^* \in \partial f(x) : \langle x^*, y - x\rangle > 0 \Rightarrow \forall z \in [x, y[: f(z) < f(y) \tag{11.16}$$

i.e., f is strictly increasing along the segment $[x, y]$.
(c) ∂f is semistrictly quasimonotone.

Then (a)$\Rightarrow$(b). If f is radially continuous, then (c)$\Rightarrow$(b)$\Rightarrow$(a). If f is locally Lipschitz and $\partial = \partial^o$, then all three assertions are equivalent.

Theorem 11.5 *Let f be lsc and radially continuous, and ∂ be a MVT subdifferential such that $\partial \subseteq \partial^\dagger$. If ∂f is strictly quasimonotone, then f is strictly quasiconvex. Conversely, if $\partial = \partial^o$ and f is locally Lipschitz and strictly quasiconvex, then ∂f is strictly quasimonotone.*

5. Pseudomonotonicity and pseudoconvexity

The usual definition of pseudoconvexity presupposes differentiability of the function involved (see Chapter 2). For non-differentiable functions, one can find some non-equivalent definitions in the literature [54, 59]. Here we will use the definition from [23] (see also [1]):

Definition 11.3 *Given a MVT subdifferential ∂, a lsc function f is called pseudoconvex (with respect to ∂), if for every $x, y \in X$, the following implication holds:*

$$\exists x^* \in \partial f(x) : \langle x^*, y - x\rangle \geq 0 \Longrightarrow \forall z \in [x, y] : f(z) \leq f(y). \tag{11.17}$$

An equivalent formulation of pseudoconvexity is given below:

Proposition 11.4 *Let ∂ be a MVT subdifferential and f a lsc function f. Consider the following assertions:*
(a) f is pseudoconvex;
(b) $\operatorname{dom} f$ *is convex and for every $x, y \in X$ the following implication holds:*

$$\exists x^* \in \partial f(x) : \langle x^*, y - x \rangle \geq 0 \Longrightarrow f(x) \leq f(y); \tag{11.18}$$

(c) f is radially continuous and for every $x, y \in X$, (11.18) holds.
Then (c)⇒(b)⇔(a).

Proof. (a)⇒(b): We have only to prove that $\operatorname{dom} f$ is convex. If not, there exist $x, y \in \operatorname{dom} f$ and $z \in [x, y]$ such that $f(z) = +\infty$. By lower semicontinuity of f, there exists $\varepsilon > 0$ such that $f(z') > f(y)$ for all $z' \in B_\varepsilon(z)$. According to Definition 11.2, there exist $c \in [x, z)$ and sequences $x_n \to c$ and $x_n^* \in \partial f(x_n)$ such that $\langle x_n^*, y - x_n \rangle > 0$. For n sufficiently large, $B_\varepsilon(z)$ intersects $(x_n, y]$. Choose $z' \in B_\varepsilon(z) \cap [x_n, y]$. Using the pseudoconvexity of f, we deduce that $f(y) \geq f(z')$, a contradiction. Hence, $\operatorname{dom} f$ is convex.

(b)⇒(a): We first show that f is quasiconvex. If not, then there exist $x_1, x_2 \in \operatorname{dom} f$ and $z \in]x_1, x_2[$ such that $+\infty > f(z) > m := \max\{f(x_1), f(x_2)\}$. Since f is lower semicontinuous, there exists $\varepsilon > 0$ such that $f(z') > m$ for all $z' \in B(z, \varepsilon)$. Note that $0 \notin \partial f(z)$ since otherwise (11.18) would imply that z is a global minimum, thus contradicting $f(z) > f(x_1)$. Hence, by Definition 11.1, z is not a local minimum. We deduce that there exists $w \in B(z, \varepsilon)$ such that $m < f(w) < f(z)$.

Since ∂ is a MVT subdifferential, there exist a sequence (u_n), converging to some $u \in [w, z[$, and $u_n^* \in \partial f(u_n)$, such that $\langle u_n^*, z - u_n \rangle > 0$. Since $z \in [x_1, x_2]$, it follows that $\langle u_n^*, x_1 - u_n \rangle > 0$ or $\langle u_n^*, x_2 - u_n \rangle > 0$. Using (11.18) we infer that $f(u_n) < m$ and by lower semicontinuity, $f(u) \leq m$. But this contradicts with the fact that $u \in B(z, \varepsilon)$.

Hence f is quasiconvex. For x, y as in (11.18), $f(x) \leq f(y)$ clearly implies that for all $z \in [x, y]$, $f(z) \leq f(y)$; hence, (11.17) holds.

(c)⇒(a): According to Proposition 3.1 in [59], f is quasiconvex; hence, as in the previous part, we infer that f is pseudoconvex. □

Note that (a)⇒(c) does not hold in the proposition above. For instance, assume that $\partial = \partial^{D^+}$. Then for the lsc function

$$f(x) = \begin{cases} 0, & x \leq 0 \\ 1 + x, & x > 0 \end{cases}$$

the subdifferential is

$$\partial^{D^+} f(x) = \begin{cases} 0, & x < 0 \\ [0, +\infty[& x = 0 \\ 1, & x > 0. \end{cases}$$

Thus, f is pseudoconvex and has convex domain without being continuous.

Let us see how pseudoconvexity compares with other generalized convexity notions. First, a lsc convex function is pseudoconvex: to see this, combine Theorem 11.1 with Proposition 11.4. We also have:

Proposition 11.5 *Let ∂ be a MVT subdifferential and f a lsc pseudoconvex function. Then*
(i) f is quasiconvex,
(ii) every local minimum is a global minimum,
(iii) if f is radially continuous, then it is semistrictly quasiconvex.

Proof. (i) This is an immediate consequence of Theorem 11.3.

(ii) If x_0 is a local minimum, then by Definition 11.1, $0 \in \partial f(x_0)$. Setting $x^* = 0$ in (11.17) we infer that x_0 is a global minimum.

(iii) By part (i) of this proof, f is quasiconvex. Also, by Proposition 11.2, f is continuous. Suppose that f is not semistrictly quasiconvex. Then there exist $x, y \in \operatorname{dom} f$ and $z = \lambda x + (1-\lambda)y$, $\lambda \in]0,1[$, such that $f(x) < f(z) = f(y)$.
By continuity, there exists $\varepsilon > 0$ such that $f(x') < f(z)$ for all x' with $\|x - x'\| < \varepsilon$. For any $\overline{y}$ with $\|y - \overline{y}\| < \frac{\lambda}{1-\lambda}\varepsilon$, set $\overline{x} = x + \frac{1-\lambda}{\lambda}(y - \overline{y})$. Then one has $z = \lambda\overline{x} + (1-\lambda)\overline{y}$ and $\|x - \overline{x}\| < \varepsilon$. Hence $f(\overline{x}) < f(z) \leq f(\overline{y})$. It follows that $f(\overline{y}) \geq f(y)$, i.e., y is a local minimum of f. By part (ii), y is also a global minimum, thus contradicting $f(x) < f(y)$.

□

As expected, pseudoconvex functions can be characterized by a corresponding property (pseudomonotonicity) of their subdifferential. There are mainly two notions of pseudomonotonicity that appear in the literature. One is the pseudomonotonicity of maps in the sense of Brezis and Browder [10, 11] whose definition relies mainly on topological properties; see Chapter 12 of this book, where these maps are called "topologically pseudomonotone". The other is the notion defined by Karamardian in the single valued, and by Yao in the multivalued case. Let us recall some definitions.

A multivalued map $T : X \rightrightarrows X^*$ is called:

1. *Pseudomonotone* [41], if for all $x, y \in X$ and $x^* \in T(x), y^* \in T(y)$, the following implication holds:

$$\langle x^*, y - x \rangle \geq 0 \Rightarrow \langle y^*, y - x \rangle \geq 0. \tag{11.19}$$

2. *Cyclically pseudomonotone* [21] if for any cycle $x_1, x_2, \ldots, x_{n+1}$ in X and any $x_i^* \in T(x_i)$, $i = 1, 2, \ldots, n$, the following implication holds:

$$\langle x_i^*, x_{i+1} - x_i \rangle \geq 0 \text{ for all } i = 1, 2, \ldots, n-1 \Rightarrow \langle x_n^*, x_1 - x_n \rangle \leq 0. \tag{11.20}$$

Equivalently, T is cyclically pseudomonotone if for any cycle $x_1, x_2, \ldots$ x_{n+1}, the following implication holds:

$$\exists i \in \{1, 2, \ldots, n\}, \exists x_i^* \in T(x_i) : \langle x_i^*, x_{i+1} - x_i \rangle > 0 \Longrightarrow$$
$$\exists j \in \{1, 2, \ldots, n\}, \forall x_j^* \in T(x_j) : \left\langle x_j^*, x_{j+1} - x_j \right\rangle < 0.$$

Considering a cycle consisting of two points, one can see that cyclically pseudomonotone maps are pseudomonotone. Also, it is obvious that a pseudomonotone map is quasimonotone; actually, under an additional assumption on the domain, we have more:

Proposition 11.6 *Let $T : X \rightrightarrows X^*$ be a pseudomonotone map. If $D(T)$ is convex (or more generally, if for all $x, y \in D(T)$, $[x, y] \cap D(T)$ is dense in $[x, y]$), then T is semistrictly quasimonotone.*

Proof. T is obviously quasimonotone. For any distinct $x, y \in D(T)$ and $x^* \in T(x)$ such that $\langle x^*, y - x \rangle > 0$, choose $z \in \left[\frac{x+y}{2}, y\right] \cap D(T)$. Obviously, $\langle x^*, z - x \rangle > 0$. If $z^* \in T(z)$ then by pseudomonotonicity, $\langle z^*, z - x \rangle > 0$. Hence, $\langle z^*, y - x \rangle > 0$, i.e., T satisfies (11.14). □

A theorem analogous to Theorem 11.3 holds:

Theorem 11.6 *Suppose that f is a lsc, radially continuous function and ∂ a MVT subdifferential. Then the following assertions are equivalent:*
(i) f is pseudoconvex,
(ii) ∂f is cyclically pseudomonotone,
(iii) ∂f is pseudomonotone.

Proof. (i)⇒(ii): If ∂f is not cyclically pseudomonotone, then there exist points $x_1, x_2, \ldots, x_n$ and $x_i^* \in \partial f(x_i)$ such that $\langle x_i^*, x_{i+1} - x_i \rangle \geq 0$ for all $i = 1, 2, \ldots, n-1$, and $\langle x_n^*, x_1 - x_n \rangle > 0$. By definition of pseudoconvexity, the former gives $f(x_{i+1}) \geq f(x_i)$, $i = 1, 2, \ldots, n-1$, hence $f(x_n) \geq f(x_1)$. By Proposition 11.5, f is semistrictly quasiconvex. Applying now Theorem 11.4 we get from $\langle x_n^*, x_1 - x_n \rangle > 0$ that $f(x_1) > f(x_n)$, a contradiction.

(ii)⇒(iii): Obvious.
(iii)⇒(i): This is shown in Theorem 4.1 of [59]. □

Remark. Note that implications (i)⇒(ii)⇒(iii) can also be derived without radial continuity, provided that $\partial \subseteq \partial^{\dagger}$. Indeed, to show (i)⇒(ii), suppose that $x_1, x_2, \ldots, x_n$ are as in the first part of the above proof. As before, we get $f(x_n) \geq f(x_1)$.

From $\langle x_n^*, x_1 - x_n \rangle > 0$ we deduce that there exists $\varepsilon > 0$ such that $\langle x_n^*, y - x_n \rangle > 0$ for all $y \in B(x_1, \varepsilon)$. Since $\partial \subseteq \partial^{\dagger}$, we deduce that $f^{\dagger}(x_n, y - x_n) > 0$. Using the definition of $f^{\dagger}$ (cf. relation (11.6)) we infer that there exist sequences $t_k \to 0^+$ and $y_k \to x_n$ such that

$$f(y_k + t_k(y - y_k)) > f(y_k).$$

Since f is also quasiconvex, we deduce that $f(y) > f(y_k)$ for all k. In particular, no $y \in B(x_1, \varepsilon)$ can be a global minimum. On the other hand, by lower semicontinuity, $f(y) \geq f(x_n)$ and *a fortiori* $f(y) \geq f(x_1)$. This is true for all y near x_1; hence x_1 is a local minimum. By Proposition 11.5, x_1 is also a global minimum. This contradicts the fact that no member of $B(x_1, \varepsilon)$ is a global minimum. Hence, (i)⇒(ii) is true, while (ii)⇒(iii) is obvious. □

Theorem 11.6 was shown under various settings and forms in several papers; see for instance [1, 23, 43, 50, 59].

Pseudoconvexity and pseudomonotonicity have their "strict" counterparts. A function f will be called *strictly pseudoconvex* if for every $x, y \in X$, the following implication holds:

$$\exists x^* \in \partial f(x) : \langle x^*, y - x \rangle \geq 0 \Longrightarrow \forall z \in [x, y[: f(z) < f(y). \quad (11.21)$$

A result similar to Proposition 11.4 holds. Condition (b) below is a slightly different definition of strict pseudoconvexity, introduced by Penot and Quang [59].

Proposition 11.7 *Let ∂ be a MVT subdifferential and f a lsc, radially continuous function f. The following assertions are equivalent:*
(a) f is strictly pseudoconvex;
(b) For every $x, y \in X$ the following implication holds:

$$\exists x^* \in \partial f(x) : \langle x^*, y - x \rangle \geq 0 \Longrightarrow f(x) < f(y); \quad (11.22)$$

Proof. We have only to prove that (b)⇒(a). If relation (11.22) holds, then by Proposition 11.4, f is pseudoconvex; further, by Proposition 11.5 it is semistrictly quasiconvex. Consequently, if $\langle x^*, y - x \rangle \geq 0$ for some $x^* \in \partial f(x)$, then $f(x) < f(y)$ implies that $f(z) < f(y)$ for all $z \in [x, y[$, i.e. (11.21) holds. □

We now compare with other generalized convexity notions. For locally Lipschitz functions, part (ii) of the following proposition is contained in [59].

Proposition 11.8 *Let f be a continuous function and ∂ a MVT subdifferential.*
(i) If f is strictly convex, then it is strictly pseudoconvex.
(ii) If f is strictly pseudoconvex and for all $x, y \in \operatorname{dom} f$ there exists $z \in \,]x, y[$ such that $\partial f(z) \neq \emptyset$, then f is strictly quasiconvex.

Proof. (i) This is a consequence of Theorem 11.2 and Proposition 11.7.

(ii) We know already by Proposition 11.5 that f is semistrictly quasiconvex. Thus, we have only to prove that for every $x, y \in \operatorname{dom} f$ with $f(x) = f(y)$ and every $w \in \,]x, y[$ one has $f(w) < f(x)$. Suppose that $f(w) = f(x)$ for some w; then from semistrict quasiconvexity it follows immediately that f is constant on $[x, y]$. Now choose $z \in \,]x, y[$ such that $\partial f(z) \neq \emptyset$, and $z^* \in \partial f(z)$. It is obvious that $\langle z^*, x - z\rangle$ and $\langle z^*, y - z\rangle$ have opposite sign. If, say, $\langle z^*, x - z\rangle \geq 0$, then by strict pseudoconvexity we get $f(x) > f(z)$, a contradiction. □

Let us compare the various notions of generalized convexity introduced so far. The following implications hold (some of them need extra continuity assumptions) and none other. We use the abbreviations "s." and "ss." for "strict" and "semistrict" respectively:

$$
\begin{array}{ccc}
\text{s. convex} & \rightarrow & \text{convex} \\
\downarrow & & \downarrow \\
\text{s. pseudoconvex} & \rightarrow & \text{pseudoconvex} \\
\downarrow & & \downarrow \\
\text{s. quasiconvex} & \rightarrow & \text{ss. quasiconvex} \\
 & & \downarrow \\
 & & \text{quasiconvex}
\end{array}
$$

A multivalued map T is called *strictly pseudomonotone* [70] if for all distinct $x, y \in X$ and all $x^* \in T(x), y^* \in T(y)$, the following implication holds:

$$\langle x^*, y - x\rangle \geq 0 \Rightarrow \langle y^*, y - x\rangle > 0. \tag{11.23}$$

It is clear that a strictly pseudomonotone map is pseudomonotone and that a strictly monotone map is strictly pseudomonotone. We also have:

Proposition 11.9 *If T is strictly pseudomonotone with convex domain, then T is strictly quasimonotone.*

Proof. Obviously, a strictly pseudomonotone map T is quasimonotone. If T is not strictly quasimonotone then there would exist $x, y \in D(T)$ such that for all $w \in]x, y[$ and $w^* \in T(w)$, $\langle w^*, y - x\rangle = 0$. Hence, if we choose arbitrary $w_1, w_2 \in]x, y[\cap D(T)$ and $w_1^* \in T(w_1)$, $w_2^* \in T(w_2)$ then $\langle w_1^*, w_1 - w_2\rangle = \langle w_2, w_1 - w_2\rangle = 0$. This contradicts (11.23). □

In order to prove a characterization of strictly pseudomonotone functions, we need a lemma:

Lemma 11.1 *Let f be lsc and assume that its subdifferential ∂f is a pseudomonotone map. If $x^* \in \partial f(x)$ and $\langle x^*, y - x\rangle > 0$, then $f(y) > f(x)$.*

Proof. Choose $\varepsilon > 0$ such that $\langle x^*, y' - x\rangle > 0$ for all $y' \in B(\varepsilon, y)$. Since ∂f is also quasimonotone, by Theorem 11.3 one has $f(y') \geq f(x)$ for all $y' \in B(\varepsilon, y)$. In particular $f(y) \geq f(x)$. Suppose that $f(y) = f(x)$. Then y would be a local minimum of f, hence $0 \in \partial f(y)$ by Definition 11.1. However, pseudomonotonicity and $\langle x^*, y - x\rangle > 0$ imply that $\langle y^*, y - x\rangle > 0$ for all $y^* \in \partial f(y)$, a contradiction. □

Proposition 11.10 *Let f be lsc and ∂ a MVT subdifferential. If f is strictly pseudoconvex, then ∂f is strictly pseudomonotone. Conversely, if ∂f is strictly pseudomonotone and for all $x, y \in \operatorname{dom} f$ there exists $z \in]x, y[$ such that $\partial f(z) \neq \emptyset$, then f is strictly pseudoconvex.*

Proof. The first part of the theorem is an easy consequence of the definitions (see also [59]). For the second part, suppose that for some $x, y \in \operatorname{dom} f$ and $x^* \in \partial f(x)$ one has $\langle x^*, y - x\rangle \geq 0$. For any $z \in [x, y[$ choose $w \in]z, y[$ such that $\partial f(w) \neq \emptyset$, and $w^* \in \partial f(w)$. By strict pseudomonotonicity we have $\langle w^*, w - x\rangle > 0$, hence $\langle w^*, y - w\rangle > 0$. On the other hand, ∂f is pseudomonotone. From Lemma 11.1 we deduce that $f(w) < f(y)$. Finally, we note that ∂f is quasimonotone, hence by Theorem 11.3, $\langle x^*, w - x\rangle \geq 0$ implies that $f(z) \leq f(w)$. Thus, $f(z) < f(y)$, i.e. f is strictly pseudoconvex. □

In case f is locally Lipschitz the above Proposition was established in [59].

6. Proper quasimonotonicity

Cyclic generalized monotonicity of a map T is defined by considering the quantity $\langle x_i^*, x_{i+1} - x_i\rangle$ where x_i, x_{i+1} are subsequent elements of a cycle and $x_i^* \in T(x_i)$. If we consider instead the quantity $\langle x_i^*, x - x_i \rangle$ where x is a point in the interior of the cycle, we get a notion which is in general weaker, but sometimes equivalent to the corresponding generalized monotonicity notion:

Definition 11.4 *A map $T : X \rightrightarrows X^*$ is called:*

1. Properly monotone, *if for every finite sequence $x_1, x_2, \ldots, x_n$ in X, every $x_i^* \in T(x_i)$ and every $x = \sum_{i=1}^n \lambda_i x_i$, with $\sum_{i=1}^n \lambda_i = 1$ and $\lambda_i > 0$, the following inequality holds:*

$$\sum_{i=1}^n \lambda_i \langle x_i^*, x - x_i \rangle \leq 0. \tag{11.24}$$

2. Properly pseudomonotone *if for every $x_1, x_2, \ldots, x_n$, x_i^* and x as before, the following implication holds:*

$$\langle x_i^*, x - x_i \rangle \geq 0 \text{ for all } i = 1, 2, \ldots, n-1 \Longrightarrow \langle x_n^*, x - x_n \rangle \leq 0. \tag{11.25}$$

3. Properly quasimonotone, *if for every $x_1, x_2, \ldots, x_n$, x_i^* and x as before, the following implication holds:*

$$\langle x_i^*, x - x_i \rangle > 0 \text{ for all } i = 1, 2, \ldots n-1 \Longrightarrow \langle x_n^*, x - x_n \rangle \leq 0$$

or equivalently, if for every $x_1, x_2, \ldots, x_n$ and x as before, there exists $i \in \{1, 2, \ldots n\}$ such that

$$\forall x_i^* \in T(x_i) : \langle x_i^*, x - x_i \rangle \leq 0. \tag{11.26}$$

Note that in the definitions above, the points $x_1, x_2, \ldots, x_n$ can be equivalently be assumed to belong to $D(T)$; this of course is everywhere obvious, but needs a moment's thought for relation (11.26).

It is evident that proper monotonicity implies proper pseudomonotonicity, which in turn implies proper quasimonotonicity. We also have:

Proposition 11.11 *Let $T : X \rightrightarrows X^*$ be a map.*

1. T is properly monotone if and only if it is monotone.

2. If $D(T)$ is convex, then T is properly pseudomonotone if and only if it is pseudomonotone.

Proof. If T is properly monotone (respectively, properly pseudomonotone), then if we use two points x_1, x_2 and $x = \frac{x_1+x_2}{2}$ we conclude immediately that T is monotone (respectively, pseudomonotone).

1. Suppose that T is monotone. Then for any $x_1, x_2, \ldots, x_n \in D(T)$, any $x_i^* \in T(x_i)$ $(i = 1, 2, \ldots, n)$ and any $x = \sum_{j=1}^n \lambda_i x_i$, with $\sum_{i=1}^n \lambda_i = 1$ and $\lambda_i > 0$, we calculate:

$$\sum_{i=1}^n \lambda_i \langle x_i^*, x - x_i \rangle = \sum_{i=1}^n \sum_{j=1}^n \lambda_i \lambda_j \langle x_i^*, x_j - x_i \rangle =$$

$$\sum_{i>j} \lambda_i \lambda_j [\langle x_i^*, x_j - x_i \rangle + \langle x_j^*, x_i - x_j \rangle] = \sum_{i>j} \lambda_i \lambda_j \langle x_i^* - x_j^*, x_j - x_i \rangle .$$

By monotonicity of T, the last expression is nonpositive. Hence, T is properly monotone.

2. Suppose that T is pseudomonotone and $D(T)$ is convex. If T is not properly pseudomonotone, then there exist $x_1, x_2, \ldots, x_n \in D(T)$, $x_i^* \in T(x_i)$ $(i = 1, 2, \ldots, n)$, and $x = \sum_{i=1}^n \lambda_i x_i$ with $\sum_{i=1}^n \lambda_i = 1$ and $\lambda_i > 0$, such that

$$\langle x_i^*, x - x_i \rangle \geq 0,\ i = 1, \ldots, n-1 \text{ and } \langle x_n^*, x - x_n \rangle > 0. \qquad (11.27)$$

By assumption, $x \in D(T)$, hence we may choose $x^* \in T(x)$. Since T is pseudomonotone, relations (11.27) imply

$$\langle x^*, x - x_i \rangle \geq 0, i = 1, \ldots n-1 \text{ and } \langle x^*, x - x_n \rangle > 0. \qquad (11.28)$$

Thus, $0 = \langle x^*, x - x \rangle = \sum_{i=1}^n \lambda_i \langle x^*, x - x_i \rangle > 0$, a contradiction. □

If $D(T)$ is not convex, T may be pseudomonotone without being properly pseudomonotone (see Chapter 14 of this book).

In contrast to proper monotonicity and proper pseudomonotonicity, proper quasimonotonicity is distinct from quasimonotonicity. Properly quasimonotone maps are obviously quasimonotone, but the converse is not necessarily true. To see this, consider the map $T : \mathbb{R}^2 \to \mathbb{R}^2$ defined as follows. Let $x_1 = (0, 1)$, $x_2 = (0, 0)$, $x_3 = (1, 0)$ and set $T(x_1) = (-1, -1)$, $T(x_2) = (1, 0)$, $T(x_3) = (0, 1)$ and $T(x) = 0$ on $\mathbb{R}^2 \setminus \{x_1, x_2, x_3\}$. Then T is quasimonotone but not properly quasimonotone (take $x = \frac{x_1 + x_2 + x_3}{3}$). On may modify slightly this example to make T continuous, so this behavior is not related to discontinuity.

On the other hand, proper quasimonotonicity is weaker than many other generalized monotonicity notions:

Proposition 11.12 *If T is cyclically quasimonotone, or semistrictly quasimonotone, then it is properly quasimonotone.*

Proof. Suppose that T is not properly quasimonotone. Then there exist $x_1, x_2, \ldots, x_n \in X$, $x_i^* \in T(x_i)$ $(i = 1, 2, \ldots n)$ and $x = \sum_{i=1}^n \lambda_i x_i$ with

$\sum_{i=1}^{n} \lambda_i = 1$ and $\lambda_i > 0$, such that

$$\langle x_i^*, x - x_i \rangle > 0, i = 1, 2, \ldots, n. \tag{11.29}$$

We show that T is neither cyclically quasimonotone nor semistrictly quasimonotone. Set $x_{i(1)} = x_1$. Then $\left\langle x_{i(1)}^*, x - x_{i(1)} \right\rangle > 0$ implies that $\sum_j \lambda_j \left\langle x_{i(1)}^*, x_j - x_{i(1)} \right\rangle > 0$. It follows that for some $x_j \neq x_{i(1)}$, $\left\langle x_{i(1)}^*, x_j - x_{i(1)} \right\rangle > 0$ must hold. We set $x_{i(2)} = x_j$. Using the same argument, we define inductively a sequence $x_{i(1)}, x_{i(2)}, \ldots$ such that

$$(x_{i(k)}^*, x_{i(k+1)} - x_{i(k)}) > 0 \tag{11.30}$$

for all $k \in \mathbb{N}$. Since the set $\{x_1, x_2, \ldots, x_n\}$ is finite, there exist $m, k \in \mathbb{N}$, $m < k$ such that $x_{i(k+1)} = x_{i(m)}$. If we consider the finite sequence $x_{i(m)}, x_{i(m+1)}, \ldots, x_{i(k)}$, relation (11.30) means that T is not cyclically quasimonotone.

To show that T is not semistrictly quasimonotone, note that (11.29) implies that there exists $\varepsilon > 0$ such that for all $y \in B(x, \varepsilon)$ one has:

$$\langle x_i^*, y - x_i \rangle > 0, i = 1, 2, \ldots, n. \tag{11.31}$$

If T is semistrictly quasimonotone, then $\langle x_1^*, x - x_1 \rangle > 0$ implies that there exist $z \in]x_1, x[\cap B(x, \varepsilon)$ and $z^* \in T(z)$ such that $\langle z^*, x - x_1 \rangle > 0$, hence $\langle z^*, x - z \rangle > 0$. Since $\sum_{i=1}^{n} \lambda_i \langle z^*, x_i - z \rangle = \langle z^*, x - z \rangle > 0$, it follows that for some $i = 1, 2, \ldots, n$ we must have $\langle z^*, x_i - z \rangle > 0$. Since T is quasimonotone, we infer that $\langle x_i^*, x_i - z \rangle \geq 0$, for all $x_i^* \in T(x_i)$, in contradiction to (11.31). □

Since proper quasimonotonicity implies quasimonotonicity and is implied by cyclic quasimonotonicity, we infer immediately from Theorem 11.3 the following characterization of quasiconvex functions:

Theorem 11.7 *Let ∂ be a MVT subdifferential such that $\partial \subseteq \partial^\dagger$. A lsc function $f : X \to \mathbb{R} \cup \{+\infty\}$ is quasiconvex if and only if ∂f is properly quasimonotone.*

Thus, quasiconvexity can be characterized either by quasimonotonicity or by proper quasimonotonicity of its subdifferential. Historically, quasimonotonicity was the first to be found, certainly because it has a simpler, basically one-dimensional definition. However, proper quasimonotonicity is often more useful, at least in relation to Minty variational inequalities. Given a subset K of X and a multivalued map T, an element $x_0 \in K$ is called a solution of the *Minty Variational Inequality Problem* (MVIP) for K if

$$\forall x \in K, \forall x^* \in T(x), \langle x^*, x - x_0 \rangle \geq 0. \tag{11.32}$$

Given a set $A \subseteq X$, we recall that a map $G : A \rightrightarrows X$ is called a KKM map if for every $x_1, x_2, \ldots, x_k \in X$, $\operatorname{co}\{x_1, x_2, \ldots, x_k\} \subseteq \bigcup_{i=1}^{n} G(x_i)$. It is straightforward to check that a multivalued map $T : X \rightrightarrows X^*$ is properly quasimonotone if and only if the multivalued map $G_T : X \rightrightarrows X$ defined by

$$G_T(x) = \{y \in X : \forall x^* \in T(x) : \langle x^*, x - y \rangle \geq 0\}$$

is KKM. This observation has important consequences, as shown by the following theorem, due essentially to John [37].

Theorem 11.8 *Let $T : X \rightrightarrows X^*$ be a map with convex domain. The following are equivalent:*
(i) T is properly quasimonotone,
(ii) For every nonempty, weakly compact, convex subset S of X, the MVIP for S has a solution,
(iii) For every $x_1, x_2, \ldots, x_k \in X$, there exists $x_0 \in \operatorname{co}\{x_1, x_2, \ldots, x_k\}$ such that $\forall i = 1, 2, \ldots k$, $\forall x_i^* \in T(x_i)$, $\langle x_i^*, x_i - x_0 \rangle \geq 0$.

Proof. (i)⇒(ii). Since T is properly quasimonotone, the map G_T is KKM. Consequently, the map $G : S \to S$ defined by $G(x) = G_T(x) \cap S$, $x \in S$ is also KKM. The values of G are nonempty and weakly compact; hence, by the Ky Fan Lemma (see [44] or next chapter for instance), $\bigcap_{x \in S} G(x) \neq \emptyset$. It is obvious that any element of $\bigcap_{x \in S} G(x)$ is a solution of the MVIP for S.

(ii)⇒(iii). It is sufficient to set $S = \operatorname{co}\{x_1, x_2, \ldots, x_k\}$.

(iii)⇒(i). We reproduce the proof of [37]. We show that for any $x_1, x_2, \ldots, x_k \in X$ and $x \in \operatorname{co}\{x_1, x_2, \ldots x_k\}$ there exists i such that (11.26) holds, by using induction with respect to k. For $k = 1$, this is evident. Assume that this is true for $k - 1$ and let $x_1, x_2, \ldots, x_k$ and x be as above. Let x_0 be the element whose existence is asserted by (iii). If $x = x_0$, then x obviously satisfies (11.26). If $x \neq x_0$ then the straight line through x and x_0 intersects the border of $\operatorname{co}\{x_1, x_2, \ldots, x_k\}$ at two points at least; one of them, which we will denote by y, is such that $x \in [x_0, y]$. Since y belongs to the convex hull of $k - 1$ of the points $x_1, x_2, \ldots, x_k$, by the induction hypothesis there exists $j \in \{1, 2, \ldots, k\}$ such that

$$\forall x_j^* \in T(x_j), \langle x_j^*, x_j - y \rangle \geq 0.$$

Since we have also

$$\forall x_j^* \in T(x_j), \langle x_j^*, x_j - x_0 \rangle \geq 0$$

we immediately infer that (11.26) holds. □

Since $D(T)$ is convex, the set S and the points $x_1, x_2, \ldots, x_k$ in the above theorem can be equivalently assumed to be contained in $D(T)$.

The following table summarizes the relations between the various kinds of generalized monotonicity.

s. mon.	$\to$	mon. (= p. mon.)	$\overset{\sim}{\leftarrow}$	c. mon.
$\downarrow$		$\downarrow$		$\downarrow$
s. pseudom.	$\to$	pseudom. ($\overset{*}{=}$ p. pseudomon.)	$\overset{\sim}{\leftarrow}$	c. pseudomon.
$\downarrow *$		$\downarrow *$		
s. quasimon.	$\overset{*}{\to}$	ss. quasimon.		
		$\downarrow$		$\downarrow$
		p. quasimon.		
		$\downarrow\sim$		
		quasimon.	$\overset{\sim}{\leftarrow}$	c. quasimon.

We used the abbreviations "c.", "p.", "s." , "ss." and "mon." for "cyclically", "properly", "strictly", "semistrictly" and "monotone", respectively. Some of the implications require the assumption that $D(T)$ is convex (or more generally, that for any $x, y \in D(T)$, $[x, y] \cap D(T)$ is dense in $[x, y]$), in which case we mark the implication by an asterisk. Finally, some of the implications become equivalence relations in case T is the subdifferential of a lsc function; these are marked by $\sim$.

7. A quasiconvex subdifferential

Besides the "all-purpose" subdifferentials defined before, there are some subdifferentials that were designed specifically for certain classes of functions and are not included in our definition of an abstract subdifferential. In particular, many subdifferentials have been defined for the class of quasiconvex functions. Such is the subdifferential of Greenberg-Pierskalla [25], the tangential of Crouzeix [20], the weak lower subdifferential of Martinez-Legaz [51], and the Q-subdifferential of Martinez-Legaz and Sach [52]. All these subdifferentials have some nice properties. For instance, the nonemptyness of the subdifferential on a dense subset of the domain of the function implies that the function is quasiconvex, and this partially justifies their denomination as quasiconvex subdifferentials. But they also have some drawbacks. For instance, the Greenberg-Pierskalla, tangential, and weak lower subdifferentials are too big; the Q-subdifferential is small, but it is probably too small since when the function is quasiconvex, one has to impose severe restrictions to infer that the subdifferential is nonempty on a dense subset of the function's domain. In this section we will define a subdifferential which remedies these drawbacks.

Given a lsc proper function $f : X \to \mathbb{R} \cup \{+\infty\}$, we define $N_f(x) := N_{S_{f(x)}}(x)$ (resp. $N_f^<(x) := N_{S_{f(x)}^<}(x)$) to be the normal cone to the sublevel (resp. strict sublevel) set corresponding to the value $f(x)$. It is easy to check the following equivalences:

$$x^* \in N_f(x) \Longleftrightarrow (\forall y \in X, \langle x^*, y - x\rangle > 0 \Rightarrow f(y) > f(x)) \quad (11.33)$$

$$x^* \in N_f^<(x) \Longleftrightarrow (\forall y \in X, \langle x^*, y - x\rangle > 0 \Rightarrow f(y) \geq f(x)) \quad (11.34)$$

Let us fix a MVT subdifferential ∂ such that $\partial \subseteq \partial^{D^+}$ (actually, $\partial = \partial^{D^+}$ is a good choice).

Definition 11.5 *[24] The* quasiconvex subdifferential $\partial^q f$ *of* f *at* $x \in \operatorname{dom} f$ *is the following subset of* X^**:*

$$\partial^q f(x) = \begin{cases} \partial f(x) \cap N_f(x), & \text{if } N_f^<(x) \neq \{0\} \\ \emptyset, & \text{if } N_f^<(x) = \{0\} \end{cases}$$

If $x \notin \operatorname{dom} f$*, then we set* $\partial^q f(x) = \emptyset$*.*

Let us explore the properties of ∂^q. First, note that N_f is a cyclically quasimonotone map; indeed, if not, then there exists a cycle $x_1, x_2, \ldots,$ $x_{n+1} = x_1$ and $x_i^* \in N_f(x_i)$, $i = 1, 2, \ldots n$, such that $\langle x_i^*, x_{i+1} - x_i\rangle > 0$ for all i's. From (11.33) we infer that $f(x_{i+1}) > f(x_i)$, which leads to the contradiction $f(x_1) > f(x_1)$. Since $\partial^q f \subseteq N_f$, it follows that $\partial^q f$ is a cyclically quasimonotone map for *any* lsc proper function f (compare with Theorem (11.3)). But what distinguishes this subdifferential, is the following result:

Theorem 11.9 *[24] Let f be a lsc radially continuous function (respectively, f is a lsc function with convex domain and its sublevel sets have nonempty interior). Then the following are equivalent:*
(i) f is quasiconvex;
(ii) $\partial^q f = \partial f$;
(iii) $\partial^q f$ has the property detailed in Definition 11.2 (Mean Value Theorem);
(iv) The domain of $\partial^q f$ is dense in $\operatorname{dom} f$*.*

The equivalence (i)$\Leftrightarrow$(ii) means that whenever f is a continuous quasiconvex function, then $\partial^q f$ is simply the subdifferential ∂f; further, this happens only for quasiconvex functions, provided they are continuous. The equivalence (i)$\Leftrightarrow$(iv) is a characterization of quasiconvexity based solely on the domain of $\partial^q f$.

If x_0 is a global minimum of a proper lsc function f, then $N_f^{<}(x_0) = X^*$ and in particular $N_f^{<}(x_0) \neq \{0\}$. Since $0 \in N_f(x_0)$ always holds, it follows that $0 \in \partial^q f(x_0)$. If f is continuous and quasiconvex, then equivalence (i)$\Leftrightarrow$(ii) in the above theorem and Definition 11.1 imply that $0 \in \partial^q f(x_0)$ also holds for local minima.

The quasiconvex subdifferential obeys some nice calculus rules for the supremum of a family of functions and the composition of functions; see [24] for details.

8. Maximal pseudomonotone maps

As already said, the subdifferential of a convex function is not only monotone, but also maximal monotone. This is a classical result of Rockafellar [66], with many applications, since maximal monotone maps enjoy important continuity properties and play a preponderant role in variational inequalities. At first look, it seems that maximality is not a relevant property of generalized monotone maps. For instance, let us tentatively define maximal pseudomonotone maps as those pseudomonotone maps that admit no pseudomonotone extension other than themselves. Let $f : \mathbb{R} \to \mathbb{R}$ be a continuously differentiable pseudoconvex function such that $f'(x) = 0$ for $x \in [a,b]$, $f'(x) > 0$ for $x > b$ and $f'(x) < 0$ for $x < a$. The subdifferential of f is the single valued pseudomonotone map $T(x) = \{f'(x)\}$ and it is not maximal pseudomonotone since the following map is a pseudomonotone extension:

$$\widetilde{T}(x) = \begin{cases}]-\infty, 0[\,, & x < a \\ ]-\infty, 0]\,, & x = a \\ 0, & x \in (a,b) \\ [0, +\infty[\,, & x = b \\ ]0, +\infty[\,, & x > b. \end{cases}$$

Note that $\widetilde{T}(x)$ has no pseudomonotone extension, and contains all positive multiples of $f'(x)$. Note also that an essential feature of single valued pseudomonotone maps is that if we multiply them by a positive function, then the resulting map is also pseudomonotone. This motivates a new definition of maximality, given below. But first let us recall some other definitions and introduce some additional notation for this section.

For any $K \subseteq X^*$, we set $\mathbb{R}_+K = \bigcup_{t\geq 0} tK$ and $\mathbb{R}_{++}K = \bigcup_{t>0} tK$. A map T is called radially upper semicontinuous (rusc) if for all $x \in D(T)$ and $y \in X$, the map $t \to \langle T(x+ty), y\rangle$ is upper semicontinuous (as a multivalued map) at 0. For instance, if the restriction of T to straight lines is upper semicontinuous with respect to the weak* topology in X^*, then T is rusc.

We now introduce an equivalence relation on the set of all pseudomonotone maps from X to X^* as follows: $T \sim S$ if:

- T and S have the same domain,
- For all $x \in X$, $0 \in T(x) \Leftrightarrow 0 \in S(x)$, and
- For all $x \in X$ with $0 \notin T(x)$, $\mathbb{R}_{++}T(x) = \mathbb{R}_{++}S(x)$.

The use of the term "equivalent" is due to the relation to variational inequalities. Given a convex subset K of X, we denote by $S(T,K)$ the set of solutions of the *Stampacchia variational inequality*:

$$x \in S(T,K) \Longleftrightarrow x \in K \text{ and } \forall\, y \in K,\ \exists x^* \in T(x) : \langle x^*, y-x\rangle \geq 0.$$

It is easy to check that, if T_1 and T_2 are equivalent pseudomonotone maps, then $S(T_1,K) = S(T_2,K)$. In fact, the converse is also true [28]:

Theorem 11.10 *Let T_1, T_2 be pseudomonotone maps. If $T_1 \sim T_2$ then $S(T_1,K) = S(T_2,K)$ for every convex subset K of X. Conversely, if $S(T_1,K) = S(T_2,K)$ for every convex subset K of X and the maps T_1, T_2 have w^*-closed convex values, then $T_1 \sim T_2$.*

Based on equivalence, we define maximality as follows.

Definition 11.6 *A pseudomonotone map T will be called D-maximal pseudomonotone if there exists an equivalent pseudomonotone map S whose graph is not strictly contained in the graph of another pseudomonotone map with the same domain.*

The above means that if S' is a pseudomonotone map such that $D(S') = D(S)$ and for all $x \in X, Sx \subseteq S'x$, then $S = S'$.

Given a pseudomonotone map $T : X \rightrightarrows X^*$, if $x \in X$ is such that $0 \in T(x)$, then for all $(y, y^*) \in Gr(T)$, $\langle y^*, y-x\rangle \geq 0$. Set

$$D_x = \{y \in X : \exists y^* \in T(y), \langle y^*, y-x\rangle = 0\}.$$

It can be easily shown that the map $\widetilde{T}$ defined by

$$\widetilde{T}(x) = \begin{cases} \mathbb{R}_{++}T(x), \text{ if } x \in D(T), 0 \notin T(x) \\ N_{D_x}(x), \text{ if } 0 \in T(x) \\ \emptyset, \text{ if } x \notin D(T) \end{cases}$$

(where $N_{D_x}(x)$ is the normal cone of D_x at x) is equivalent to T. In fact, $\widetilde{T}$ is the largest element in the equivalence class of T with respect to domain inclusion. Thus, T is D-maximal pseudomonotone if and only if $\widetilde{T}$ has no pseudomonotone extension with the same domain, apart from itself.

In case of a convex domain, the following proposition gives a very simple characterization of maximality [28].

Proposition 11.13 *Let T be pseudomonotone, with convex domain. Then T is D-maximal pseudomonotone if and only if every pseudomonotone extension of T with the same domain is equivalent to T.*

The following result from [29] is central for the considerations that follow:

Theorem 11.11 *Let T be pseudomonotone and rusc, $D(T)$ be radially open and assume that $T(x)$ is w^*-compact and convex for all $x \in D(T)$. Then T is $D-$maximal pseudomonotone.*

An immediate result of this theorem is the following:

Corollary 11.1 *Let $f : X \to \mathbb{R} \cup \{+\infty\}$ be a pseudoconvex, locally Lipschitz function. Then $\partial^o f$ is a $D-$maximal pseudomonotone map.*

Proof. Since f is locally Lipschitz, it is known that $\operatorname{dom} f$ is open, $D(\partial^o f) = \operatorname{dom} f$, $\partial^o f(x)$ is w^*-compact and convex for every $x \in \operatorname{dom} f$, and finally that $\partial^o f$ is usc in the strong$\times$weak* topology of $X \times X^*$ [15]. Also, according to Theorem 11.6, $\partial^o f$ is pseudomonotone. Hence, by Theorem 11.11, $\partial^o f$ is $D-$maximal. □

A well-known feature of maximal monotone maps is that they are upper semicontinuous in the interior of their domain, with respect to the w^*-topology in X^*. One may expect that something analogous holds for $D-$maximal pseudomonotone maps. However, it is not true that every maximal pseudomonotone map has an equivalent map which is even rusc. To see this, consider the single valued map $T : \mathbb{R}^2 \to \mathbb{R}^2$ defined by

$$T(x,y) = \begin{cases} (0,1), & \text{if } y \geq 0 \\ (1,1), & \text{if } y < 0. \end{cases}$$

Then T is maximal pseudomonotone but has no equivalent rusc map. Nevertheless, the following proposition shows that pseudomonotonicity "helps" continuity:

Proposition 11.14 *[29] Let $T : \mathbb{R}^k \rightrightarrows \mathbb{R}^k$ be a pseudomonotone map with compact convex values, rusc on $D(T)$. Suppose that $D(T)$ is an open convex set. Then there exists a pseudomonotone map T_1, usc on $D(T_1)$ with compact convex values, such that $T_1 \sim T$. If T is single valued, then T_1 can also be chosen single valued.*

The following corollary is an immediate result of the proposition:

Corollary 11.2 *If $T : D(T) \to \mathbb{R}^k$ is a single valued, radially continuous pseudomonotone map with open, convex domain $D(T) \subseteq \mathbb{R}^k$, then there exists a function $f : D(T) \to \mathbb{R}_{++}$ and a continuous map $T_1 : D(T) \to \mathbb{R}^k$ such that $T(x) = f(x) T_1(x)$ for all $x \in D(T)$.*

Thus, in case we have to solve a Stampacchia Variational Inequality Problem (VIP) with a pseudomonotone, single valued map T, the assumption that T is continuous along straight lines is not really weaker than the assumption that it is continuous on finite dimensional subspaces; indeed, given any such subspace E we may replace the projection E^*TE of T on E by an equivalent continuous map T'_E and solve the problem for T'_E; then a limit procedure as in Theorem 2.1 of [19] can ensure that the original VIP has a solution. Analogous considerations hold in the multivalued case.

We close with a note on lower semicontinuity. Lower semicontinuity of multivalued maps is almost never used for monotone maps; the reason is a theorem of Rockafellar, which states that a lower semicontinuous monotone map is actually single valued on the interior of its domain. Something analogous holds for pseudomonotone maps. Let us call T radially lower semicontinuous at $x \in D(T)$ if for all $y \in X$, the map $t \to \langle T(x+ty), y\rangle$ is lower semicontinuous (as a multivalued map) at 0.

Proposition 11.15 *Let T be a pseudomonotone map, which is radially lower semicontinuous at $x \in \operatorname{core} D(T)$. Then $Tx = \{0\}$ or there exists $x^* \in X^*$ such that $T(x) \subseteq \mathbb{R}_{++}x^*$.*

Proof. If not, then there exist $x^* \in T(x)$, $y^* \in T(x) \setminus \{0\}$ such that there exists no $\lambda > 0$ with $x^* = \lambda y^*$. Hence there exists $z \in X$ such that $\langle x^*, z\rangle \geq 0 > \langle y^*, z\rangle$. Since T is radially lower semicontinuous, there exists a sequence $x_n = x + t_n z$, $t_n \searrow 0$ and $x_n^* \in T(x_n)$ such that $\langle x_n^*, z\rangle \to \langle y^*, z\rangle$. Hence, for n sufficiently large, $\langle x_n^*, z\rangle < 0$ i.e., $\langle x_n^*, x - x_n\rangle > 0$. Using pseudomonotonicity we infer that $\langle x^*, x - x_n\rangle > 0$, i.e., $\langle x^*, z\rangle < 0$, a contradiction. □

An analogous result holds for quasimonotone maps.

By applying the same proof as for monotone maps [36], we obtain that, topologically, the set of points at which T cannot be made single valued by equivalence, is very "thin".

Corollary 11.3 *[29] Let T be pseudomonotone, usc with w^*-compact values. Suppose that X is separable and* $\operatorname{int}(D(T)) \neq \emptyset$. *Then the set*

$$\{x \in X : T(x) \text{ is contained in } \{0\} \text{ or } \mathbb{R}_{++}x^* \text{ for some } x^*\}$$

is of the first category.

References

[1] Aussel, D., *Subdifferential properties of quasiconvex and pseudoconvex functions: unified approach*, J. Optim. Theory Appl. 97, 1998, 29-45.

[2] Aussel, D., Corvellec, J.N. and M. Lassonde, *Subdifferential characterization of quasiconvexity and convexity*, J. Convex Anal. 1, 1994, 195-201.

[3] Aussel, D., Corvellec, J.N. and M. Lassonde, *Mean Value Property and Subdifferential Criteria for Lower Semicontinuous Functions*, Trans. Amer. Math. Soc. 347, 1995, 4147-4161.

[4] Aussel, D. and A. Daniilidis, *Normal characterization of the main classes of quasiconvex functions*, Set-Valued Anal. 8, 2000, 219-236.

[5] Aussel, D. and N. Hadjisavvas, *Quasimonotone variational inequalities*, J. Optim. Theory Appl. 121 no 2, 2004 (to appear).

[6] Avriel, M., Diewert, W.E., Schaible, S. and I. Zang, *Generalized Concavity*, Plenum Press, New York, 1988.

[7] Bachir, M., Daniilidis, A. and J.P. Penot, *Lower subdifferentiability and integration*, Set-Valued Anal. 10, 2002, 89-108.

[8] Borde, J., and J.P. Crouzeix, *Continuity of the normal cones to the level sets of quasi convex functions*, J. Optim. Theory Appl. 66, 1990, 415-429.

[9] Borwein, J.M., Moors, W.B., and Y. Shao, *Subgradient representation of multifunctions*, J. Austral. Math. Soc. ser. B 4, 1998, 1-13.

[10] Brezis, H., *Equations et inéquations nonlinéaires dans les espaces vectoriels en dualité*, Ann. Inst. Fourier 18, 1968, 115-175.

[11] Browder, F.E., *Multivalued monotone nonlinear mappings and duality mappings in Banach spaces*, Trans. Amer. Math. Soc. 118, 1965, 338-351.

[12] Browder F.E., *Nonlinear operators and nonlinear equations of evolution in Banach spaces*, Proc. Symposia in Pure Math., vol. XVIII, part 2, American Mathematical Society, Providence, 1976.

[13] Cambini, A., Marchi, A. and L. Martein, *On nonlinear scalarization in vector optimization*, Technical Report n. 75, Dipartimento di Statistica e Matematica Applicata all' Economia, Universitá di Pisa, 1994.

[14] Clarke, F., *Necessary conditions for nonsmooth problems in optimal control and the calculus of variations*, PhD Thesis, University of Washington, 1973.

[15] Clarke, F.H., *Optimization and Nonsmooth Analysis*, Wiley Interscience, New York, New York, 1983.

[16] Clarke, F., Stern, R.J. and P.R. Wolenski, *Subgradient criteria for monotonicity, the Lipschitz condition and convexity*, Can. J. Math. 45, 1993, 1167-1183.

[17] Correa, R., Jofré, A. and L. Thibault, *Characterization of lower semicontinuous convex functions*, Proc. Amer. Math. Soc. 116, 1992, 67-72.

[18] Correa, R., Jofré, A. and L. Thibault, *Subdifferential monotonicity as characterization of convex functions*, Num. Funct. Anal. Optim. 15, 1994, 531-535.

[19] Cottle, R.W. and J.C. Yao, *Pseudo-monotone complementarity problems in Hilbert space*, J. Optim. Theory Appl. 75, 1992, 281-295.

[20] Crouzeix, J.P., *Contributions à l'étude des fonctions quasiconvexes*, Thèse d'Etat, Université de Clermont-Ferrant, 1977.

[21] Daniilidis, A. and N. Hadjisavvas, *On the Subdifferentials of Quasiconvex and Pseudoconvex Functions and Cyclic Monotonicity*, J. Math. Anal. Appl. 237, 1999, 30-42.

[22] Daniilidis, A. and N. Hadjisavvas, *Characterization of nonsmooth semistrictly quasiconvex and strictly quasiconvex functions*, J. Optim. Theory Appl. 102, 1999, 525-536.

[23] Daniilidis, A. and N. Hadjisavvas, *On generalized cyclically monotone operators and proper quasimonotonicity*, Optimization 47, 2000, 123-135.

[24] Daniilidis, A., Hadjisavvas, N. and J.E. Martínez-Legaz, *An Appropriate Subdifferential for Quasiconvex Functions*, SIAM. J. Optim. 12, 2002, 407-420.

[25] Greenberg, H.P. and W.P. Pierskalla, *Quasi-conjugate functions and surrogate duality*, Cahiers Centre Etudes Recherche Opér. 15, 1973, 437-448.

[26] Hadjisavvas, N., *Continuity properties of quasiconvex functions in infinite dimensional spaces*, Working Paper 94-3, A.G. Anderson Graduate School of Management, University of California, Riverside, 1994.

[27] Hadjisavvas, N., *The use of subdifferentials for studying generalized convex functions*, J. Stat. Manag. Syst. 5, 2002, 125-139.

[28] Hadjisavvas, N., *Maximal pseudomonotone operators*, in Recent advances in Optimization, G.P. Crespi, A. Guerraggio, E. Miglierina, M. Rocca (eds), Proceedings of a workshop held in Varese, 13-14/6/2002, Datanova Editrice, 2003, pp. 87-101.

[29] Hadjisavvas, N., *Continuity and maximality properties of pseudomonotone operators*, J. Conv. Anal. 10, 2003, 465-475.

[30] Hadjisavvas, N. and S. Schaible, *Quasimonotonicity and pseudomonotonicity in variational inequalities and equilibrium problems*, in Generalized Convexity, Generalized Monotonicity, J.-P. Crouzeix, J.-E. Martinez-Legaz and M. Volle, eds., Kluwer Academic Publishers, Dordrecht, 1998, pp. 257-275.

[31] Hadjisavvas, N. and S. Schaible, *From scalar to vector equilibrium problems in the quasimonotone case*, J. Optim. Theory Appl. 96, 1998, 297-309.

[32] Hassouni, A., *Sous-differentiels des fonctions quasiconvexes*, Thèse de troisième cycle, Université Paul Sabatier, Toulouse, 1983.

[33] Hassouni A., *Opérateurs quasimonotones; applications à certains problèmes variationnels*, Thèse d'Etat, Université de Toulouse, 1993.

[34] Hassouni, A. and R. Ellaia, *Characterization of nonsmooth functions through their generalized gradients*, Optimization 22, 1991, 401-416.

[35] Holmes R., *Geometric functional analysis and its applications*, Springer, Berlin, 1975.

[36] Hu, S. and N.S. Papageorgiou, *Handbook of Multivalued Analysis*, Vols I,II, Kluwer Academic Publishers, 1997.

[37] John, R., *A Note on Minty Variational Inequalities and Generalized Monotonicity,* Proceedings of the 6th International Symposium on Generalized Convexity/Monotonicity, N. Hadjisavvas, J.E. Martínez-Legaz and J.P. Penot (eds), Springer-Verlag, 2001.

[38] Kachurovskii R.I., *On monotone operators and convex functionals*, Uspekhi Mat. Nauk., 15, 1960, 213-215.

[39] Karamardian, S. and S. Schaible, *Seven Kinds of Monotone Maps*, J. Optim. Theory Appl. 66, 1990, 37-46.

[40] Komlósi S., *Some properties of nondifferentiable pseudoconvex functions*, Math. Prog. 26, 1983, 232-237.

[41] Komlósi S., *Generalized monotonicity in Nonsmooth Analysis*, in Generalized Convexity, S. Komlosi, T. Rapcsak, and S. Schaible, eds., Springer-Verlag, 1994, 263-275.

[42] Komlósi S., *Monotonicity and Quasimonotonicity in Nonsmooth Analysis*, in Recent Advances in Nonsmooth Optimization, D.-Z. Du, L. Qi and R.S. Womersely, eds., World Scientific Publishers, Singapor, 1994, 101-124.

[43] Komlósi S., *Generalized monotonicity and generalized convexity*, J. Optim. Theory Appl. 84, 1995, 361-376.

[44] Fan, K., *A generalization of Tychonoff's fixed-point Theorem*, Math. Ann. 142, 1961, 305-310.

[45] Levin, V.L., *Quasi-convex functions and Quasi-monotone operators*, J. Conv. Anal. 2, 1995, 167-172.

[46] Luc, D.T., *On the maximal monotonicity of subdifferentials*, Acta Math. Vietnamica, 18, 1993, 99-106.

[47] Luc, D.T. and S. Swaminathan, *A characterization of convex functions*, Nonlin. Anal. 20, 1993, 667-701.

[48] Luc, D.T., *Characterizations of quasiconvex functions*, Bull. Austral. Math. Soc. 48, 1993, 393-405.

[49] Luc, D.T., *On generalized convex nonsmooth functions,* Bull. Austr. Math. Soc. 49, 1994, 139-149.'

[50] Luc, D.T., *Generalized monotone set valued maps and support bifunctions*, Acta Math. Vietnamica 21, 1996, 213-252.

[51] Martínez–Legaz, J.E., *Weak lower subdifferentials and applications*, Optimization 21, 1990, 321-341.

[52] Martínez–Legaz, J.E. and P.H. Sach, *A new subdifferential in quasiconvex analysis*, J. Convex Anal., 6, 1999, 1-12.

[53] Minty, G., *Monotone (nonlinear) operators in a Hilbert space*, Duke Math. J. 29, 1962, 341-346.

[54] Ortega, J.M. and W.C. Rheinboldt, *Iterative solution of nonlinear equations in several variables*, Academic Press, New York, 1970.

[55] Penot, J.P., *Generalized Convexity in the light on Nonsmooth Analysis,* in Recent developments in Optimization, R. Durier and C. Michelot (eds), Springer-Verlag, Berlin, 1995.

[56] Penot, J.P., *A mean value theorem with small subdifferentials*, J. Optim. Theory Appl. 94, 1997, 209-221.

[57] Penot, J.P., *Are generalized derivatives useful for generalized convex functions?*, in Generalized Convexity, Generalized Monotonicity, J.-P. Crouzeix, J.-E. Martínez-Legaz and M. Volle, eds., Kluwer Academic Publishers, Dordrecht, 1998, 3-59.

[58] Penot, J.P., *What is quasiconvex Analysis?*, Optimization 47, 2000, 35-110.

[59] Penot, J.P. and P.H. Quang, *Generalized Convexity of Functions and Generalized Monotonicity of Set-Valued Maps*, J. Optim. Theory Appl. 92, 1997, 343-356.

[60] Penot, J.P. and P.H. Sach, *Generalized Monotonicity of Subdifferentials and Generalized Convexity*, J. Optim. Theory Appl. 64, 1997, 251-262.

[61] Penot, J.-P. and M. Volle, *Convexity and generalized convexity methods for the study of Hamilton-Jacobi equations*, in Generalized Convexity and Generalized Monotonicity, N. Hadjisavvas, J.E. Martínez-Legaz and J.P. Penot, eds, Springer-Verlag, Berlin, 2001.

[62] Penot, J.P. and C. Zalinescu, *Elements of quasiconvex subdifferential calculus*, J. Conv. Anal., 7, 2000, 243-270.

[63] Popovici, N., *Contribution à l' optimization vectorielle*, Ph.D. Thesis, Faculté des Sciences, Université de Limoges, 1995.

[64] Pshenichnyi, B.N., *Necessary Conditions for an Extremum*, Marcel Dekker, New York 1971.

[65] Rockafellar, T., *On the maximal monotonicity of subdifferential mappings*, Pacific J. Math. 33, 1970, 209-216.

[66] Rockafellar, T., *Convex Analysis*, Princeton University Press, 1972.

[67] Rockafellar, R.T. and J.B. Wets, *Variational Analysis*, Springer-Verlag, Berlin, 1*998.*

[68] Thibault, L. and D. Zagrodny, *Integration of subdifferentials of lower semicontinuous functions in Banach spaces*, J. Math. Anal. Appl. 189, 1995, 33-58.

[69] Wang, S.X., *Fine and pathological properties of subdifferentials*, PhD Thesis, Department of Mathematics and Statistics, Simon Fraser University, 1999.

[70] Yao, J.-C., *Multi-valued variational inequalities with K-generalized monotone operators*, J. Optim. Theory Appl. 83, 1994, 399-401.

[71] Zagrodny, D., *Approximate mean value theorem for upper subderivatives*, Nonlin. Anal. 12, 1988, 1413-1428.

[72] Zeidler, E., *Nonlinear Functional Analysis and its applications*, Vols.I-IV, Springer-Verlag, 1990.

Chapter 12

PSEUDOMONOTONE COMPLEMENTARITY PROBLEMS AND VARIATIONAL INEQUALITIES*

Jen-Chih Yao

Department of Applied Mathematics

National Sun Yat-Sen University

Kaohsiung, Taiwan 80424, ROC.

yaojc@math.nsysu.edu.tw

Ouayl Chadli

Department of Economics

Faculty of Economics and Social Sciences, University Ibn Zohr

Agadir B.P. 8658, Hay Dakhla, Agadir, Morocco.

ochadli@esta.ac.ma

Abstract In this chapter, we report recent results mainly on existence for complementarity problems and variational inequalities in infinite-dimensional spaces under generalized monotonicity, especially (algebraic) pseudomonotonicity. Variational inequalities associated to a topological pseudomonotone operator have been also considered and some possible extensions of complementarity problems and variational inequalities have been included. Finally some discussions on the equivalence of complementarity problems for pseudomonotone operators are given.

Keywords: generalized monotonicity, pseudomonotonicity, complementarity problems, variational inequalities.

*This research was supported by a grant from the National Science Council of the Republic of China.

1. Introduction

Given a mapping f from $\mathbb{R}^n$ into itself, the *complementarity problem* relative to f, denoted $CP(f)$, is the following system

$$x \geq 0, \quad f(x) \geq 0, \quad \langle x, f(x) \rangle = 0 \tag{12.1}$$

where $x \geq 0$ means that all the components of x are nonnegative and $\langle \cdot, \cdot \rangle$ is the usual scalar product in $\mathbb{R}^n$. Geometrically, the complementarity problem involves finding a nonnegative vector x such that the image $f(x)$ is also nonnegative and is orthogonal to x. When f is nonlinear, $CP(f)$ is called a *nonlinear complementarity problem.* In the case where f is an affine transformation, i.e. $f(x) = Mx + q$ for some $q \in \mathbb{R}^n$ and $M \in \mathbb{R}^{n \times n}$, $CP(f)$ is said to be a *linear complementarity problem* and is denoted by the pair (q, M).

It is interesting to see how the complementarity problem gets its name by the following remarks. For any mapping $f : \mathbb{R}^n \to \mathbb{R}^n$, there is a pairing between the coordinates of a vector x and its image $f(x)$. If $f_1, ..., f_n$ are the component functions of f, then we call x_j and $f_j(x)$ complementary variables or complements of each other. Let

$$y = f(x). \tag{12.2}$$

Then (12.1) becomes the following equivalent system

$$y = f(x), \quad x \geq 0, \quad y \geq 0, \quad \langle x, y \rangle = 0. \tag{12.3}$$

We define $N_n = \{1, 2, .., n\}$. A solution $\overline{x}, \overline{y}$ of (12.3) is said to be *nondegenerate* if at most n of its $2n$ components equal zero. If $\overline{x}, \overline{y}$ is a nondegenerate solution of (12.3), then the sets

$$A = \{i \in N_n : \overline{x}_i > 0\} \text{ and } B = \{i \in N_n : \overline{y}_i > 0\} \tag{12.4}$$

are complementary subsets of N_n. Also, regardless of whether $\overline{x}, \overline{y}$ is a nondegenerate solution of (12.2), if it is a solution of (12.3), the sets A and B defined in (12.4) are disjoint subsets of N_n. For details, see [26] and [24].

We remark that behind the identification of the complementarity problem in the early 1960's was the Kuhn-Tucker theorem for nonlinear programming problem which gives the necessary conditions of optimality when certain conditions of differentiability and regularity are met. We also observe that in 1961 Dorn [35] showed that if M is a positive-definite but not necessarily symmetric matrix, then the minimum value of the following quadratic programming problem

$$\min_{x \in K} \ \langle x, Mx + q \rangle$$
$$K = \{x \in \mathbb{R}^n : x \geq 0, Mx + q \geq 0\}$$
$$q \in \mathbb{R}^n$$

is zero. Dorn's paper was the first paper which treats the complementarity problem as an independent problem.

Dorn's result was generalized by Dantzig and Cottle [31] in 1963 to the case where all the principal minors of the matrix M are positive. The result obtained by Dantzig and Cottle [31] in 1963 was further generalized by Cottle [22, 23] in 1964 and 1966, respectively, to a certain class of nonlinear functions.

The complementarity problem $C(f)$ can be further generalized to more general spaces other than the finite-dimensional space $\mathbb{R}^n$. Such idea was initiated by Habetler and Price [42] and was later refined by Karamardian [55] whose development is the following.

Let X be a locally convex Hausdorff topological vector space over $\mathbb{R}$, and let Y be another vector space over $\mathbb{R}$. Let K be a convex cone in X with polar which is also a convex cone

$$K^* = \{y \in Y : \langle x, y \rangle \geq 0 \text{ for all } x \in K\}$$

where $\langle \cdot, \cdot \rangle$ denotes a bilinear form on $X \times Y$. If f is a mapping from X into Y, then the *complementarity problem relative to f and K*, denoted $CP(f, K)$, is the following system

$$x \in K, \quad f(x) \in K^*, \quad \langle x, f(x) \rangle = 0. \tag{12.5}$$

If we introduce preorders $\geq^K$ and $\geq^{K^*}$ in X and Y, respectively by the following definitions

$$x \geq^K y \quad \text{if and only if} \quad x - y \in K \text{ for all } x, y \in X, \tag{12.6}$$

and

$$z \geq^{K^*} w \quad \text{if and only if} \quad z - w \in K^* \text{ for all } x, y \in Y, \tag{12.7}$$

then we can rewrite the system (12.5) alternately as the following system

$$x \geq^K 0, \quad f(x) \geq^{K^*} 0, \quad \langle x, f(x) \rangle = 0. \tag{12.8}$$

It should be noted that the preorder $\geq^K$ defined by (12.6) is a partial order, that is, it is a reflexive, antisymmetric and transitive relation if and only if the convex cone K is pointed, that is,

$$K \cap (-K) = \{0\};$$

and the preorder $\geq^{K^*}$ defined by (12.7) is a partial order if the convex cone K has nonempty interior, i.e. $intK \neq \emptyset$.

Clearly, in $\mathbb{R}^n$, the nonnegative orthant $\mathbb{R}^n_+$ defined by

$$\mathbb{R}^n_+ = \{x \in \mathbb{R}^n : x \geq 0\}$$

is a closed convex cone and $(\mathbb{R}^n_+)^* = \mathbb{R}^n_+$ in the ordinary case where $\langle x, y\rangle = \sum_{i=1}^{n} x_i y_i$. Therefore, the notation $CP(f)$ as defined in (12.1) is an abbreviation for $CP(f, \mathbb{R}^n_+)$.

It is worth noting that the complementarity problem has very useful applications in various fields such as optimization, economics, game theory, mechanics, engineering, elasticity, free boundary problems and equilibrium problems etc. See, for example [24], [25]-[49] and the references therein. In particular, Lemke [64] employed the complementarity problem as a method to solve matrix games in 1965 and Ingleton [48] gave applications of the complementarity problem in engineering.

We now turn to another problem which is very closely related to the complementarity problem. Again, we start with the finite-dimensional setting. Let K denote a (closed convex) subset of $\mathbb{R}^n$ and suppose $f : K \to \mathbb{R}^n$ is a given function. The *variational inequality*, denoted $VI(f, K)$, is to satisfy the following conditions

$$x \in K, \quad \langle y - x, f(x)\rangle \geq 0 \quad \text{for all } y \in K. \tag{12.9}$$

A vector x for which (12.9) holds is said to be a solution to the variational inequality $VI(f, K)$. Roughly speaking, the variational inequality (12.9) states that the vector $f(x)$ must be at an acute angle with all feasible vectors emanating from x. It is easy to see that $x \in K$ is a solution to $VI(f, K)$ if and only if the vector $f(x)$ is inward normal to K at x, that is, $-f(x)$ belongs to the normal cone $N_K(x)$ where

$$N_K(x) = \begin{cases} \{w : \langle y - x, w\rangle \geq 0 \text{ for all } y \in K\} & \text{if } x \in K \\ \emptyset & \text{otherwise.} \end{cases}$$

The variational inequality was introduced by Hartmann and Stampacchia [46] in 1966, and was later expanded by Stampacchia in several important papers [65, 67, 82]. It is interesting to remark that in 1966 Karamardian also obtained existence results for variational inequalities in his Ph.D dissertation [54] where he used the fixed-point theory which is totally different from that used by Hartmann and Stampacchia [46].

Like complementarity problems, we can also consider the variational inequality in a more general space setting. Let X denote a locally convex

Hausdorff topological vector space over $I\!R$, and X^* denote the topological dual of X. Let $\langle x, u\rangle$ be the pairing between X and X^*. If K is a closed convex subset of X and $T : K \to X^*$ is a given operator, then the variational inequality, denoted $VI(T, K)$, is to find x in K such that

$$\langle y - x, Tx\rangle \geq 0 \quad \text{for all } y \in K. \tag{12.10}$$

Since $(I\!R^n)^* = I\!R^n$, it is easy to see that when $X = I\!R^n$, then the problem (12.10) reduces to the finite-dimensional variational inequality (12.9).

It is interesting to observe that most of early studies of the variational inequality were set in the context of calculus of variations as well as optimal control theory and in connection with the solution of free boundary value problems which are posed in the form of partial differential equations. Readers who are interested in these materials are referred to the books by Kinderlehrer and Stampacchia [60], Baiocchi and Capelo [3] as well as Isac [49], respectively, where a complete and thorough introduction to these applications of variational inequalities in infinite-dimensional metric spaces is given. On the other hand the survey paper by Harker and Pang [45] provides an excellent article for the theory, algorithms and applications of variational inequalities and complementarity problems in finite-dimensional spaces setting.

It should be pointed out that the complementarity problem (12.5) and the variational inequality (12.10) are closely related. Karamardian [55] was the first to establish the relationship between these problems which states that if K is a convex cone, then the complementarity problem $CP(T, K)$ with $Y = X^*$ and the variational inequality (12.10) $VI(T, K)$ have precisely the same solutions, that is $x \in K$ solves the problem $CP(T, K)$ if and only if x solves the problems $VI(T, K)$. Despite the above fact that the complementarity problem is a special case of variational inequality, the early developments of these two problems followed quite different paths. The main distinction is the space setting in which these two problems were studied. Variational inequality is usually in infinite-dimensional metric spaces, e.g., Hilbert spaces or reflexive Banach spaces, and the space setting of the complementarity problem is finite-dimensional Euclidean space.

Due to Karamardian's result on the equivalence of the complementarity problem and the variational inequality, most of the existence results for complementarity problem are based on the corresponding existence results for variational inequality. But it should be observed that the domain of variational inequality is not necessarily unbounded whereas the domain of the complementarity problem is always unbounded since it is a cone. Consequently, it is usually a difficult task to obtain existence results for complementarity problem without the aid of variational inequality.

The first existence result of finite-dimensional variational inequality was obtained independently by Hartmann and Stampacchia [46] and by Karamardian [54] in 1966, respectively, which states that if the set K is compact, convex and the mapping f is continuous, then the variational inequality $VI(f, K)$ has a solution. For variational inequality in infinite-dimensional spaces, monotonicity assumption on the operators is usually needed to derive existence results. Recent research shows that monotonicity assumption can be relaxed to generalized monotonicity assumption. It is our goal of this chapter to report recent results mainly on existence for complementarity problems and variational inequalities in infinite-dimensional spaces under generalized monotonicity, especially, (algebraic) pseudomonotonicity.

The organization of the remainder of this chapter is as follows. In Section 2, we give the definition of pseudomonotone mappings from which several existence results of variational and complementarity problems are derived. Uniqueness of solutions is discussed and applications to optimization problems and to the post-critical equilibrium state of a thin elastic plate are given. In Section 3, instead of algebraic pseudomonotonicity considered in Sections 2 and 3, we consider topological pseudomonotonicity and the variational inequality for topologically pseudomonotone operators. Again, existence results are obtained and some applications are given. In Section 4, we discuss several possible extensions of complementarity problems and variational inequalities. Several existence and uniqueness results are derived. Finally in Section 5, we discuss the equivalence of complementarity problems like nonlinear programs, least element problems, etc, for pseudomonotone operators.

2. Complementarity Problems and Variational Inequalities for Pseudomonotone Mappings

In this section, we shall consider complementarity problems and variational inequalities for pseudomonotone mappings in the setting of Banach spaces. More precisely, let B be a real Banach space with norm $\|\cdot\|$, B^* its topological conjugate space endowed with weak* topology, $\langle u, v\rangle$ the pairing between $u \in B$ and $v \in B^*$. Let K be a nonempty subset of B and T be an operator from K into B^*. The variational inequality, denoted $VI(T, K)$, is the following system

$$x \in K, \ \ \langle y - x, Tx\rangle \geq 0 \quad \text{for all } y \in K.$$

In the case where K is a convex cone of B, with

$$K^* = \{y \in B^* : \langle x, y\rangle \geq 0 \text{ for all } x \in K\}.$$

The complementarity problem relative to T and K, denoted $CP(T, K)$, is the following system

$$x \in K, \quad Tx \in K^*, \quad \langle x, Tx\rangle = 0.$$

In the sequel, for any real number α, $|\alpha|$ denotes the absolute value of α. For $D \subset B$, $int(D)$, $\overline{D}$ and D^c denote the interior, the closure and complement of D, respectively. For $E, F \subset B$, $int_F(E)$ denotes the relative interior of E in F. A subset of a Banach space is said to be *solid* if it has a nonempty interior.

For the relationship between complementarity problems and variational inequalities, we have the following result

Proposition 12.1 *Let K be a convex cone of a Banach space B, and let $T : K \to B^*$ be an operator. Then the following statements are equivalent.*

(i) x^* is a solution of $CP(T, K)$,

(ii) x^* is a solution of $VI(T, K)$.

Proof. If (i) holds, then it is obvious to see that x^* is a solution of $VI(T, K)$. Conversely if (ii) holds, then

$$\langle y - x^*, Tx^*\rangle \geq 0 \quad \text{for all } y \in K. \tag{12.11}$$

Setting in turn $y = 2x^*$ and $y = 0$ in (12.11), we obtain that $\langle x^*, Tx^*\rangle = 0$, and therefore $\langle y, Tx^*\rangle \geq 0$ for all $y \in K$. Hence (i) is established. □

Therefore, one approach to study the complementarity problem is by studying the variational inequality over closed convex cones due to Proposition 12.1.

We remark that the first existence result for the problem $VI(T, K)$ was established by Hartmann and Stampacchia [46] in 1966 which asserts that the $VI(T, K)$ will possess a solution given that T is continuous, B is a finite-dimensional Euclidean space and K is a nonempty compact convex set. Since then, many extensions of this result have been obtained by many authors, e.g., Kinderlehrer and Stampacchia [60, Theorem 1-4, p.84] for the corresponding result in a reflexive Banach space assuming the monotonicity of operators; Holmes [47, Corollary, p.187] for a general variational inequality result in a locally convex space; Théra [84, Theorem] for an existence result without the monotonicity assumption in a Banach space; Barbu and Precupanu [4, Theorem 2-6, p.133] for an existence result under the pseudomonotonicity assumption (in the sense of Brézis); and Yao [91, Theorem 2.1] for an existence result without monotonicity assumption and without using the Ky Fan minimax inequality theorem. It is worth noting that most of the existence results for the problem $VI(T,K)$ in infinite-dimensional spaces require the monotonicity assumption of the operator under consideration.

The purpose of this section is to present some existence results for the problems $VI(T,K)$ and $CP(T,K)$ for pseudomonotone operators (in the sense of Karamardian) in Banach spaces. Uniqueness results will be obtained and some applications to optimization problems and the study of the post-critical equilibrium state of a thin elastic plate subject to unilateral conditions are given.

Recall that $T : K \subset B \to B^*$ is said to be monotone if

$$\langle x - y, Tx - Ty \rangle \geq 0 \quad \text{for all } (x, y) \in K \times K, \tag{12.12}$$

and strictly monotone if equality in (12.12) implies $x = y$. Let us use symbols "$\to$", "$\rightharpoonup$" and "$\overset{*}{\rightharpoonup}$" to denote the norm convergence, the weak convergence and the weak* convergence, respectively, i.e. $x_n \rightharpoonup x$ if $x^*(x_n) \to x^*(x)$ for all $x^* \in B^*$ and $x_n^* \overset{*}{\rightharpoonup} x^*$ if $x_n^*(x) \to x^*(x)$ for all $x \in B$.

Definition 12.1 *Let K be a convex subset of B and $T : K \to B^*$. Then*

(i) The operator T is hemicontinuous on K if T is continuous on line segments in K, i.e., for every pair of points $x, y \in K$, the following function is continuous

$$t \mapsto \langle x - y, T(tx + (1-t)y) \rangle, \quad 0 \leq t \leq 1.$$

(ii) The operator T is continuous on finite-dimensional subspaces if for any finite-dimensional subspace M of B with $K \cap M \neq \emptyset$, the restricted operator $T : K \cap M \to B^*$ is continuous from the norm topology of $K \cap M$ to the weak topology of B^*.*

(iii) The operator T is demicontinuous on K if $x_n \to x$ implies $Tx_n \overset{*}{\rightharpoonup} Tx$.

It is easy to see that the following implications hold.

$$\begin{array}{c}\text{Demicontinuity}\\ \Downarrow \\ \text{Continuity on finite-dimensional subspaces}\\ \Downarrow \\ \text{Hemicontinuity}\end{array}$$

We also note that if T is monotone and hemicontinuous on K, then it is continuous on finite-dimensional subspaces on K. See, e.g., [82]. Consequently if B is finite-dimensional and $T : B \to B^*$ is monotone and hemicontinuous on B then T is itself continuous.

Definition 12.2 *Let K be a subset of B and $T : K \to B^*$. Then T is pseudomonotone if for every pair of points $x, y \in K$, we have*

$$\langle x - y, Ty \rangle \geq 0 \quad \textit{implies} \quad \langle x - y, Tx \rangle \geq 0.$$

Clearly, monotone operators are pseudomonotone but not conversely as the following example shows [58] :

$$T(x) = \frac{1}{1+x}, \qquad K = \{x \in I\!R : x \geq 0\}.$$

It should be also remarked that the notion of pseudomonotonicity was essentially initiated by Brézis, Nirenberg and Stampacchia [13] in 1972 without the name of pseudomonotonicity which was coined later by Karamardian [56] in 1976.

It should also be observed that there exists operator which is both pseudomonotone and hemicontinuous and yet it is not continuous on finite-dimensional subspaces. See, e.g., [85].

2.1 Existence

Before we state and prove existence results for problems $VI(T, K)$ and $CP(T, K)$, we need the following lemma.

Lemma 12.1 *Let K be a closed convex subset in the real Banach space B, and let $T : K \to B^*$ be pseudomonotone and hemicontinuous.*

Then $x \in K$ is a solution of $VI(T, K)$, i.e.,

$$\langle y - x, Tx \rangle \geq 0 \quad \textit{for all } y \in K$$

if and only if

$$\langle y - x, Ty \rangle \geq 0 \quad \textit{for all } y \in K.$$

Proof. If $x \in K$ is a solution of $VI(T, K)$, then it follows by pseudomonotonicity of T that $\langle y - x, Ty \rangle \geq 0$ for all $y \in K$. Conversely, for $y \in K$ and $t \in]0, 1]$ let us set $y_t = (1 - t)x + ty \in K$. Since $\langle y_t - x, Ty_t \rangle \geq 0$, it follows

$$\langle y - x, Ty_t \rangle \geq 0. \tag{12.13}$$

Letting $t \to 0$ in (12.13) and by hemicontinuity of T, one deduces that $\langle y - x, Tx \rangle \geq 0$. Thus x is a solution to $VI(T, K)$. □

Lemma 12.1 was originally proved by Minty [68] for the case of variational inequality for monotone operators in Hilbert space, and was later generalized by Karamardian [57] to the case of pseudomonotone mappings.

Now we can state and prove the first existence result of this section.

Theorem 12.1 *Let K be a weakly compact and convex subset in the real Banach space B, and $T : K \to B^*$ be a pseudomonotone and hemicontinuous operator. Then the problem $VI(T, K)$ has a solution, i.e., there exists $x \in K$ such that*

$$\langle y - x, Tx \rangle \geq 0 \quad \textit{for all } y \in K. \tag{12.14}$$

Moreover, the set of solutions to (12.14) is nonempty weakly compact and convex.

Proof. Consider for $y \in K$ the following nonempty sets

$$\Sigma(y) = \{x \in K : \langle y - x, Tx \rangle \geq 0\}$$
$$H(y) = \{x \in K : \langle y - x, Ty \rangle \geq 0\}.$$

We have to show as a first step that $\bigcap_{y \in K} H(y) \neq \emptyset$. Since T is pseudomonotone, then $\Sigma(y) \subset H(y)$. Also since $H(y)$ is a closed set, it follows that

$$\overline{\Sigma(y)} \subset H(y). \tag{12.15}$$

Let $\{y_1, \cdots, y_n\}$ be a finite subset of K and let $y \in conv(\{y_1, \cdots, y_n\})$ be arbitrary. Then $y = \sum_{i=1}^{n} \alpha_i y_i$ with $\alpha_i \geq 0$ and $\sum_{i=1}^{n} \alpha_i = 1$. Assume for contradiction that $y \notin \bigcup_{i=1}^{n} \overline{\Sigma(y_i)}$. Then

$$\langle y_i - y, Ty \rangle < 0 \quad \forall i = 1, \cdots, n.$$

From this it follows, $\sum_{i=1}^{n} \alpha_i \langle y_i - y, Ty \rangle = \langle (\sum_{i=1}^{n} \alpha_i y_i) - y, Ty \rangle < 0$, a contradiction. Thus, $conv(\{y_1, \cdots, y_n\}) \subset \bigcup_{i=1}^{n} \overline{\Sigma(y_i)}$; and since for arbitrary $y_0 \in K$ $\overline{\Sigma(y_0)}$ is compact, it follows from the Ky Fan Lemma that $\bigcap_{y \in K} \overline{\Sigma(y)} \neq \emptyset$. Therefore, from relation (12.15), $\bigcap_{y \in K} H(y) \neq \emptyset$. Now, let $\overline{x} \in \bigcap_{y \in K} H(y)$. Then $\langle y - \overline{x}, Ty \rangle \geq 0 \ \forall y \in K$. Consequently, from Lemma 12.1, one deduces that $\langle y - \overline{x}, T\overline{x} \rangle \geq 0 \ \forall y \in K$ and hence $\overline{x}$ is a solution to $VI(T, K)$. Furthermore, the solution set is $\bigcap_{y \in K} H(y)$ which is weakly compact and convex. □

As a consequence of Theorem 12.1, we have the following corollary.

Corollary 12.1 *Let K be a weakly compact and convex subset in the real reflexive Banach space B and $T : B \to B^*$ be monotone. Suppose that for every pair of points $y, z \in B$, we have*

$$\liminf_{t \to 0^+} \langle z, T(y + tz) \rangle \leq \langle z, Ty \rangle. \tag{12.16}$$

Then the problem $VI(T, K)$ has a solution and the solution set is nonempty weakly compact and convex.

Proof. We shall show that the operator T satisfying (12.16) is demicontinuous from which the result follows from Theorem 12.1. To this end, let $x \in K$ and $x_n \to x$. Since T is monotone, it is locally bounded at x. (See, e.g., [11, Theorem 2]). Hence we may assume that $\{Tx_n\}$ is bounded. Since B is reflexive, we may assume without loss of generality that $Tx_n \stackrel{*}{\rightharpoonup} x^*$ for some $x^* \in B^*$. For any $z \in B$, since T is monotone, we have

$$0 \leq \langle x_n - z, Tx_n - Tz \rangle \to \langle x - z, x^* - Tz \rangle \text{ as } n \to \infty. \tag{12.17}$$

For any $v \in B$, let $z = x + tv$ with $t > 0$. By substituting z into (12.17) and dividing by t, we have

$$\langle v, x^* - T(x + tv) \rangle \leq 0$$

from which and (12.16) it follows that

$$\begin{aligned} \langle v, x^* \rangle &\leq \liminf_{t \to 0^+} \langle v, T(x + tv) \rangle \\ &\leq \langle v, Tx \rangle. \end{aligned}$$

Consequently, $\langle v, Tx - x^* \rangle \geq 0$ for all $v \in B$. Therefore, $Tx = x^*$ and hence T is demicontinuous on K. □

For $VI(T, K)$ where K may not be weakly compact, we have the following result.

Theorem 12.2 *Let K be a closed convex subset in the real Banach space B and $T : K \to B^*$ which is pseudomonotone and hemicontinuous. Suppose T satisfies the following coercivity condition : there exists a nonempty weakly compact and convex subset C of K satisfying for each $x \in K \setminus C$ there exists $u \in C$ such that $\langle x - u, Tx \rangle > 0$. Then $VI(T, K)$ has at least one solution.*

Proof. Let $A = \{y_1, \cdots, y_n\}$ be a finite subset of K and consider the convex subset $C_1 = conv(A \cup C)$ which is compact since C a is convex and compact subset of B, see [8, Theorem 15]. Then, from Theorem 12.1, one has

$$\exists \overline{x} \in C_1 \text{ such that } \langle y - \overline{x}, T\overline{x} \rangle \geq 0 \quad \forall y \in C_1. \tag{12.18}$$

We argue that $\overline{x} \in C$. Otherwise $\overline{x} \in K \setminus C_1$; and by the coercivity assumption, there exists $u \in C$ such that $\langle \overline{x} - u, T\overline{x} \rangle > 0$ which contradicts relation (12.18).

Now, consider for $y \in K$ the following nonempty sets

$$S(y) = \{x \in C : \langle y - x, Tx \rangle \geq 0\}$$
$$W(y) = \{x \in C : \langle y - x, Ty \rangle \geq 0\}.$$

It follows from (12.18) that $\bigcap_{i=1}^{n} S(y_i) \neq \emptyset$ and therefore $\bigcap_{i=1}^{n} \overline{S(y_i)} \neq \emptyset$. Since T is pseudomonotone,

$$\emptyset \neq \bigcap_{i=1}^{n} \overline{S(y_i)} \subset \bigcap_{i=1}^{n} W(y_i).$$

Thus, the family of closed subsets $\{W(y)\}_{y \in K}$ has the finite intersection property. Since C is compact, $\bigcap_{y \in K} W(y) \neq \emptyset$.

Let $x^* \in \bigcap_{y \in K} W(y)$. Then $\langle y - x^*, Ty \rangle \geq 0 \ \forall y \in K$. Hence from Lemma 12.1, one deduces that x^* is a solution to $VI(T, K)$. □

It is interesting to note that if the real Banach space B is reflexive, then the sufficient condition for the existence of solution to the variational inequality in Theorem 12.2 turns out to be also necessary as the following result shows.

Theorem 12.3 *Let K be a closed convex subset in the real reflexive Banach space B, and $T : K \to B^*$ which is pseudomonotone and continuous on finite-dimensional subspaces. Then the following statements are equivalent :*

(i) The problem $VI(T, K)$ has a solution.

(ii) There exists a weakly compact and convex subset C of K with $int_K(C) \neq \emptyset$ satisfying the following condition : for each $x \in \partial_K(C)$, there exists $u \in int_K(C)$ such that $\langle x - u, Tx\rangle \geq 0$.

The proof of Theorem 12.3 can be found in [87].

More existence results for the problem $VI(T, K)$ can be derived from Theorem 12.2. Let us recall some definitions.

Definition 12.3 *Let K be a subset of a Banach space B and $T : K \to B^*$.*

(i) T is coercive if there exists $x_0 \in K$ such that

$$\frac{\langle x - x_0, Tx\rangle}{\|x\|} \to +\infty \text{ as } \|x\| \to +\infty \text{ and } x \in K.$$

(ii) T is weakly coercive if there exists $x_0 \in K$ such that

$$\langle x - x_0, Tx\rangle \to +\infty \text{ as } \|x\| \to +\infty \text{ and } x \in K.$$

(iii) T is β-monotone if there exists an increasing function $\beta\colon [0,\infty) \to [0,\infty)$ with $\beta(0) = 0$ and $\beta(r) \to +\infty$ as $r \to +\infty$, such that

$$\langle x - y, Tx - Ty\rangle \geq \|x - y\|\beta(\|x - y\|) \text{ for all } x, y \in K.$$

(iv) T is strongly monotone if $\beta(r) = kr$ for some $k > 0$ in (iii).

Obviously, we have the following implications

Strong Monotonicity

$\Downarrow$

β-Monotonicity

$\Downarrow$

Coercivity

$\Downarrow$

Weak coercivity

Proposition 12.2 *(i) If T is weakly coercive, then*

$$\liminf_{\|x\| \to +\infty, x \in K} \langle x - x_0, Tx\rangle > 0.$$

(ii) If B is a reflexive Banach space and T is weakly coercive then there exists a weakly compact convex subset C of K such that $\langle x - x_0, Tx\rangle > 0$ for all $x \in K \setminus C$.

Proof. (*i*) The proof is direct from the definition of T being weakly coercive.

(*ii*) Since T is weakly coercive, then by (*i*) one has

$$\liminf_{\|x\|\to+\infty, x\in K} \langle x - x_0, Tx\rangle = \sup_{\rho>0} \inf_{\|x\|\geq\rho, x\in K} \langle x - x_0, Tx\rangle > 0.$$

Thus, $\exists \rho > 0$ such that $\forall x \in K$ with $\|x\| > \rho$ one has $\langle x - x_0, Tx\rangle > 0$. Set $C = B(0,\rho) \cap K$ which is a weakly compact subset of B. Then for $x \in K \backslash C$ one has $\langle x - x_0, Tx\rangle > 0$. □

The following result is then an easy consequence of Theorem 12.2 and Proposition 12.2.

Corollary 12.2 *Let K be a closed convex subset in the real reflexive Banach space B and $T : K \to B^*$ which is pseudomonotone and hemicontinuous. If T is coercive or weakly coercive, then the problem $VI(T, K)$ has a solution.*

Corollary 12.2 can be employed to derive existence results of zeros of operators as the following result indicates.

Corollary 12.3 *Let B be a real reflexive Banach space and $T : B \to B^*$ be pseudomonotone and hemicontinuous. If T is also coercive or weakly coercive, then T has a zero, i.e., there exists $x \in B$ such that $Tx = 0$.*

Proof. By Corollary 12.2, we know that there exists $x \in B$ such that

$$\langle y - x, Tx\rangle \geq 0 \quad \text{for all } y \in B.$$

Since $y \in B$ is arbitrary, the standard trick gives $Tx = 0$. □

Now we derive some existence results of solutions of complementarity problems. The following result is an easy consequence of Theorem 12.2 and Proposition 12.1.

Theorem 12.4 *Let K be a closed convex cone in the real Banach space B. Let $T : K \to B^*$ be pseudomonotone and hemicontinuous. Suppose T satisfies the following coercivity condition : there exists a nonempty weakly compact and convex subset C of K satisfying for each $x \in K \setminus C$ there exists $u \in C$ such that $\langle x - u, Tx\rangle > 0$. Then the complementarity problem $CP(f, K)$ has a solution.*

The next result is a generalization of a result of Karamardian [56, Theorem 4-1] to infinite-dimensional spaces.

Theorem 12.5 *Let K be a pointed solid closed convex cone in the real Banach space B. Let $T : K \to B^*$ be pseudomonotone and hemicontinuous. Suppose that there exists $u \in K$ such that $T(u) \in int(K^*)$. Then the complementarity problem $CP(f, K)$ has a solution.*

The proof of this theorem can be found in [43], [27].

We remark that results similar to Theorem 12.5 in topological vector space setting have been proved by Allen [1, Theorem 4] and Borwein [9, Thoerem 10], respectively.

2.2 Uniqueness

Uniqueness results for variational inequalities and complementarity problems can be obtained by imposing some more restrictive assumption, e.g., strict pseudomonotonicity, on the operator under consideration. In 1960, Stampacchia [82] showed that the variational inequality has a unique solution if the mapping under consideration is strictly monotone. In Banach space, strict monotonicity requires that

$$\langle x - y, T(x) - T(y) \rangle > 0$$

for any two distinct elements in the domain of the mapping T. To motivate the suitable assumptions under which the variational inequality will possess a unique solution, let us suppose that the problem $VI(T, K)$ has two solutions, say, x and y. Then

$$\langle u - x, Tx \rangle \geq 0 \quad \text{for all } u \in K$$

and

$$\langle u - y, Ty \rangle \geq 0 \quad \text{for all } u \in K.$$

In particular, we have

$$\langle x - y, Ty \rangle \geq 0 \tag{12.19}$$

and

$$\langle y - x, Tx \rangle \geq 0. \tag{12.20}$$

Adding (12.19) and (12.20) together, we obtain

$$\langle x - y, Tx - Ty \rangle \leq 0. \tag{12.21}$$

Therefore, if we assume T to be strictly monotone, then (12.21) can not hold unless $x = y$ and in this case the solution is then unique.

With the above observation, we can state the following result.

Lemma 12.2 *If $T : K \to B^*$ is strictly monotone, then the problem $VI(T, K)$ has at most one solution.*

Lemma 12.2 can be extended to the peudomonotone case. Before we state some uniqueness results for pseudomonotone operators, let us recall some more definitions.

Definition 12.4 *Let K be a nonempty subset of the real Banach space B and $T : K \to B^*$.*

(i) The operator T is strictly pseudomonotone on K if for every distinct pair of points $x, y \in K$ we have

$$\langle x - y, Ty \rangle \geq 0 \implies \langle x - y, Tx \rangle > 0$$

(ii) The operator T is β-pseudomonotone on K if, there exist $x_0 \in K$ and $\beta : [0, +\infty) \to [0, +\infty)$ with $\beta(0) = 0$, $\beta(t) > 0$ for $t > 0$ and $\liminf_{t \to +\infty} \beta(t) > \|Tx_0\|$ such that for every pair of points $x, y \in K$, we have

$$\langle x - y, Ty \rangle \geq 0 \implies \langle x - y, Tx \rangle \geq \|x - y\| \beta(\|x - y\|).$$

(iii) The operator T is strongly pseudomonotone if $\beta(t) = \alpha t$ for some $\alpha > 0$ in (ii).

Clearly, we have the following implications.

Strong monotonicity	$\Longrightarrow$	Strong pseudomonotonicity
$\Downarrow$		$\Downarrow$
β-Monotonicity	$\Longrightarrow$	β-Pseudomonotonicity
$\Downarrow$		$\Downarrow$
Strict monotonicity	$\Longrightarrow$	Strict pseudomonotonicity
$\Downarrow$		$\Downarrow$
Monotonicity	$\Longrightarrow$	Pseudomonotonicity

The concept of strict and strong pseudomonotonicity were introduced by Karamardian and Schaible [58] in 1990.

Now we can extend Lemma 12.2 to pseudomonotone operators.

Lemma 12.3 *Let K be a nonempty subset of the real Banach space B and $T : K \to B^*$. If T is strictly pseudomonotone, then the problem $VI(T, K)$ has at most one solution.*

Proof. Suppose that $VI(T, K)$ has two distinct solutions, say x and y.

Then

$$\langle x - y, T(y) \rangle \geq 0 \tag{12.22}$$

and

$$\langle y - x, T(x) \rangle \geq 0. \tag{12.23}$$

Since T is strictly pseudomonotone, we have from (12.22) that

$$\langle x - y, T(x) \rangle > 0$$

and hence $\langle y - x, T(x) \rangle < 0$ which contradicts (12.23). Therefore $VI(T, K)$ can have at most one solution. □

From Theorem 12.1, and Lemma 12.3, we have the following uniqueness result for the problem $VI(T, K)$.

Theorem 12.6 *Let K be a weakly compact convex subset of a real Banach space B and $T : K \to B^*$ which is strictly pseudomonotone and hemicontinuous. Then the problem $VI(T, K)$ has a unique solution.*

It is interesting to observe that the variational inequality problem has a unique solution for α-pseudomnotone mappings as the following result shows.

Theorem 12.7 *Let K be a closed convex set in a reflexive Banach space B and $T : K \to B$. Suppose T is β-pseudomonotone and hemicontinuous. Then the variational inequality problem $VI(T, K)$ has a unique solution.*

Proof. Without loss of generality, we may assume $0 \in K$. For each positive integer n, let B_n be the closed ball with center at 0 and radius n. Let $K_n = K \cap B_n$. Then K_n is closed bounded and convex for each n. By Theorem 12.1, since β-pseudomonotonicity implies pseudomonotonicity, there exists $x_n \in K_n$ such that

$$\langle x - x_n, Tx_n \rangle \geq 0 \text{ for all } x \in K_n. \tag{12.24}$$

Since $0 \in K_n$ for each n, by β-pseudomonotonicity of T, we have

$$\langle -x_n, T(0) \rangle \geq \|x_n\| \beta(\|x_n\|) \qquad \text{for all } n$$

from which it follows that $\beta(\|x_n\|) \leq \|T(0)\|$ for all n. Consequently, the sequence $\{x_n\}$ is bounded. Without loss of generality, we may assume that $\{x_n\}$ is weakly convergent to some $x^* \in K$. For any $x \in K$, let m be such that $x \in K_m$. Then from (12.24) we have

$$\langle x - x_n, Tx_n \rangle \geq 0 \text{ for all } n \geq m$$

and hence by Lemma 12.1

$$\langle x - x_n, Tx \rangle \geq 0 \text{ for all } n \geq m. \tag{12.25}$$

Letting $n \to +\infty$ in (12.25), we have

$$\langle x - x^*, Tx \rangle \geq 0$$

and by Lemma 12.1 again

$$\langle x - x^*, Tx^* \rangle \geq 0. \tag{12.26}$$

As x in (12.26) is arbitrary, we conclude that x^* is a solution to $VI(T, K)$. Since β-pseudomonotonicity implies strict pseudomonotonicity, the result then follows from Lemma 12.3. □

Now from the above results, one can derive some uniqueness results of solutions to complementarity problems.

Lemma 12.4 *Let K be a closed convex cone of a real Banach space B and $T : K \to B^*$. If T is strictly pseudomonotone, then the complementarity problem $CP(T, K)$ has at most one solution.*

Lemma 12.4 is a direct consequence of Proposition 12.1 and Lemma 12.3.

The following uniqueness result can be proved by the same argument as that in Theorem 12.7.

Theorem 12.8 *Let K be a closed convex cone in a real reflexive Banach space B and $T : K \to B^*$ which is hemicontinuous. If T is either β-pseudomonotone or strongly pseudomonotone, then the problem $CP(T, K)$ has a unique solution.*

The following result is a consequence of Theorem 12.8.

Corollary 12.4 *Let K be a closed convex cone in a real reflexive Banach space B and $T : K \to B^*$. Suppose T is strongly monotone and hemicontinuous. Then the complementarity problem $CP(T, K)$ has a unique solution.*

More uniqueness result can be obtained by combining Lemma 12.4 and Theorem 12.4.

Theorem 12.9 *Let K be a closed convex cone in the real Banach space B. Let $T : K \to B^*$ be strictly pseudomonotone and hemicontinuous. Suppose T satisfies the following coercivity condition : there exists a nonempty weakly compact and convex subset C of K satisfying for each*

$x \in K \setminus C$ there exists $u \in C$ such that $\langle x - u, Tx \rangle > 0$. Then the complementarity problem $CP(f, K)$ has a unique solution.

We remark that in finite-dimensional spaces, more uniqueness results can be obtained. Instead of strict monotonicity or strict pseudomonotonicity, one can employ the concepts of P-functions and uniform P-functions. The notion of P-function was introduced by Moré and Rheinboldt [71] in 1973 and the notion of uniform P-functions was introduced by Megiddo and Kojima [70] in 1977. Readers who are interested in this topic are referred to [71], [70] and the references therein.

2.3 Applications to Optimization Problems

Here, we are interested in the existence of a solution to the following problem :

$$(P) \qquad \underset{x \in K}{\text{Minimize}} f(x)$$

where K is a closed convex subset of a Banach space and f is a real function defined on K. We shall employ results derived in previous sections to obtain existence and uniqueness results of problem (P) for generalized convex functions.

Let us first recall the following definitions. Let Ω be an open subset of a real Banach space B and $f : \Omega \to I\!R$. The function f is said to be *Fréchet-differentiable*, *F-differentiable* for short, at $x \in \Omega$ if there is an $f'(x) \in B^*$ such that

$$f(x+h) = f(x) + \langle h, f'(x) \rangle + w(x,h)$$

where $w(x,h) = o(\|h\|)$ as $h \to 0$. The functional $f'(x)$ is called the *F-derivative* of f at the point x. The function f is *F-differentiable* on Ω if f is F-differentiable at each $x \in \Omega$. The function f is *continuously differentiable* on Ω if f is F-differentiable on Ω and $f' : \Omega \to B^*$ is continuous. Also the function f is said to be *Gâteaux-differentiable*, *G-differentiable* for short, at $x \in \Omega$ if there exists $\nabla f(x) \in B^*$ such that

$$\lim_{t \to 0^+} \frac{f(x+th) - f(x)}{t} = \langle h, \nabla f(x) \rangle \quad \text{for all } h \in B.$$

The functional $\nabla f(x)$ is called the *G-derivative* of f at the point x, and f is *G-differentiable* on Ω if f is G-differentiable at each $x \in \Omega$.

It should be observed that F-differentiability implies G-differentiability and in this case $f' = \nabla f$. Conversely, if f is G-differentiable in a neighborhood of $x \in \Omega$ and ∇f is continuous at x, then f is F-differentiable. For details, see, e.g., [33, Proposition 7.5].

Again, let Ω be a nonempty convex subset of a real Banach space B and $f : \Omega \to I\!R$. The function f is said to be *quasiconvex* on Ω if for every pair of points $x, y \in \Omega$, we have

$$f(tx + (1-t)y) \leq \max\{f(x), f(y)\} \quad \text{for each } t \in [0,1].$$

It is easy to see that f is quasiconvex if and only if for each real number α the following level set of f is convex

$$\{x \in \Omega : f(x) \leq \alpha\}.$$

For G-differentiable functions, we have the following characterization of quasiconvexity in terms of G-derivatives.

Theorem 12.10 *([87, Theorem 6.3]) Let Ω be an open convex subset of a real Banach space B and $f : \Omega \to I\!R$ be G-differentiable. Then the following statements are equivalent :*
(i) f is quasiconvex on Ω.
(ii) If $x, y \in \Omega$ and $f(y) \leq f(x)$ then $\langle y - x, \nabla f(x)\rangle \leq 0$.

We remark that in the case that B is finite-dimensional, Karamardian and Schaible used the quasimonotonicity of the F-derivative to characterize F-differentiable quasiconvex functions. See [58].

Definition 12.5 *Let Ω be an open subset of a real Banach space B and $f : \Omega \to I\!R$ be G-differentiable.*
(i) The function f is pseudoconvex on Ω if for every pair of points $x, y \in \Omega$, we have

$$\langle y - x, \nabla f(x)\rangle \geq 0 \Longrightarrow f(y) \geq f(x).$$

(ii) f is strictly pseudoconvex if for every pair of distinct points $x, y \in \Omega$,

$$\langle y - x, \nabla f(x)\rangle \geq 0 \Longrightarrow f(y) > f(x).$$

(iii) f is strongly pseudoconvex if there exists $\alpha > 0$ such that for every pair of points $x, y \in \Omega$, we have

$$\langle y - x, \nabla f(x)\rangle \geq 0 \Longrightarrow f(y) \geq f(x) + \alpha\|x - y\|^2.$$

The concepts of strict and strong pseudoconvexity were introduced by Karamardian and Schaible [58] in 1990. Also, we have the following implication.

$$\text{Pseudonconvexity} \Longrightarrow \text{Quasiconvexity}$$

For details of proof, see, e.g., [87, Lemma 6.5]. It is interesting to observe that the class of pseudoconvex functions is significant in the following sense. Consider the problem (P) again. If the function f is pseudoconvex in an open set containing K, then any solutions of the problem $VI(\nabla f, K)$ is a global minimum of the problem (P) and in particular every point with zero G-derivative is a global minimum.

It is well known that a G-differentiable function f from an open convex subset Ω of a real Banach space B into $I\!R$ is convex if and only if the operator $\nabla f : \Omega \to B^*$ is monotone. See, e.g., [36, Proposition 5.5, p.25]. We can also use pseudomonotonicity of G-derivatives to characterize pseudoconvex functions as the following result illustrates.

Theorem 12.11 *[87, Theorem 6.6] Let Ω be an open convex subset of a real Banach space B and $f : \Omega \to I\!R$ be G-differentiable. Then the following statements are equivalent :*
(i) The function f is pseudoconvex on Ω.
(ii) The operator $\nabla f : \Omega \to B^*$ is pseudomonotone.

Now we have the first existence result for the problem (P) by Corollary 12.2 and Theorem 12.11.

Theorem 12.12 *Let K be a closed convex subset of a real reflexive Banach space B and let f be a G-differentiable function from an open convex set Ω containing K into $I\!R$. Suppose that $\nabla f : \Omega \to B^*$ is weakly coercive, pseudomonotone and hemicontinuous. Then the problem (P) has a solution.*

The next result is immediate from Theorem 12.12.

Corollary 12.5 *Let K be a closed convex subset of a real reflexive Banach space B and f be a function from an open convex set Ω containing K into $I\!R$. Suppose that f is continuously differentiable on Ω and f' is pseudomonotone and weakly coercive on Ω. Then the problem (P) has a solution.*

Finally, let us consider the uniqueness question of solutions of the problem (P). First we have the following results characterizing strictly and strongly pseudoconvex functions.

Theorem 12.13 *([87, Theorems 6.9, 6.10]) Let Ω be an open convex subset of a real Banach space B and f be a G-differentiable function on Ω. Then, f is strictly pseudoconvex if and only if ∇f is strictly pseudomonotone.*

We then state the following uniqueness result for the problem (P).

Theorem 12.14 *Let K be a closed convex subset of a real reflexive Banach space B and f be a G-differentiable function from an open convex set Ω containing K into $\mathbb{R}$. Then the problem (P) possesses a unique solution under each of the following conditions.*

(i) ∇f is strictly pseudomonotone, weakly coercive and hemicontinuous.

(ii) ∇f is strongly pseudomonotone and hemicontinuous.

Proof. The assertion of (i) follows from Corollary 12.2, Lemma 12.3 and Theorem 12.13 (i).

The assertion of (ii) follows from Theorem 12.8 and Theorem 12.13 (ii). □

2.4 Application to the study of a mathematical model for mechanical problems

In this subsection, we shall consider some applications of existence results established in the previous section to a class of nonlinear complementarity problems and in particular to the study of the post-critical equilibrium state of a thin elastic plate subject to unilateral conditions.

Let K be a closed convex cone in a real Hilbert space H and let $L_1, L_2 : K \to H$ be two mappings. The *nonlinear complementarity problem*, denoted $NCP(T, K)$, is the following system :

$$x^* \in K, \;\; T(x^*) \in K^* \text{ and } \langle x^*, T(x^*) \rangle = 0$$

where $T(x) = x - L_1(x) + L_2(x)$ for each $x \in K$. Such problem was studied by Isac and Théra [52] and was used as a mathematical model for mechanical problems.

The following result is then a consequence of Theorem 12.4.

Theorem 12.15 *Suppose that the mapping T is pseudomonotone and hemicontinuous. If there exists $\rho > 0$ such that for $x \in K$ with $\|x\| > \rho$, there exists $u \in K$ satisfying $\|u\| \leq \rho$ and*

$$\langle x - u, Tx \rangle > 0$$

then the $NCP(T, K)$ has a solution.

It is worth nothing that the closed convex cone K in Theorem 12.15 need not be necessarily pointed, and the mapping T need not be the one-sided Gâteaux directional derivative of a functional defined on K (See [52, Theorem 3.1]).

Another application is as follows. Let Ω be a thin elastic plate whose thickness is supposed to be constant and which rests without friction on a flat rigid support. The material is also assumed to be homogeneous and isotropic. Mathematically, Ω may be identified as a bounded domain (i.e. an open and connected set) in $\mathbb{R}^2$. The plate Ω is assumed to be clamped on $\gamma_1 \subset \Omega$ and simply supported on $\gamma_2 = \gamma \backslash \gamma_1$ where γ is the boundary of Ω which is supposed to be sufficiently regular.

We assume that a lateral variable load λL, where λ is positive and increasing which represents the magnitude of lateral loading, is applied to the boundary of Ω. Consider the following Sobolev space

$$H^2(\Omega) = \{v \in L^2(\Omega) : \frac{\partial v}{\partial x_i}, \frac{\partial^2 v}{\partial x_i \partial x_j} \in L^2(\Omega), \forall i, j = 1, 2\}$$

equipped with the norm $\|\cdot\|_{H^2(\Omega)}$, and let E be the closed subspace of $H^2(\Omega)$ defined as follows :

$$E = \{z \in H^2(\Omega) : z|_\gamma = 0 \text{ and } \frac{\partial z}{\partial n}|_{\gamma_1} = 0, \text{ a.e.}\}$$

where n denotes the normal to γ exterior to Ω and $\frac{\partial(\cdot)}{\partial n}$ denotes the normal exterior derivative. We can equip E with the inner product defined by a continuous bilinear form on $E \times E$ such that the associated norm is equivalent to the initial norm $\|\cdot\|_{H^2(\Omega)}$. See, e.g., [52]. For a fixed λ, the post-critical equilibrium state of the plate subject to unilateral conditions is governed by the following variational inequality:

$$w \in K, \quad \langle z - w, w - \lambda L(w) + G(w) \rangle \geq 0 \quad \text{for all } z \in K, \qquad (12.27)$$

where the set

$$K = \{z \in E : z \geq 0, \text{ a.e. on } \Omega\}$$

represents all the admissible vertical displacements of the plate; $\lambda \geq 0$ is exactly the intensity of the lateral load λL; L is a self-adjoint linear compact operator defined by the nature of the load applied on the boundary of Ω; and G is a bounded nonlinear continuous operator connected with the expansive properties of the plate. In practice, the mapping G is also supposed to be a Gâteaux derivative of a nonlinear functional Φ such that $\Phi(0) = 0$. See, e.g., [34]. In this case, $w = 0$ is a trivial solution of the problem (12.27).

Since K is a closed convex cone, by Proposition 12.1, problem (12.27) is equivalent to the nonlinear complementarity problem $NCP(T, K)$ where

$$T(x) = x - \lambda L(x) + G(x) \quad \text{for all } x \in K. \qquad (12.28)$$

Hence by Theorem 12.4, we have the following existence result for problem (12.27) and therefore the existence of the post-critical equilibrium state of a thin elastic plate subjected to unilateral conditions.

Theorem 12.16 *Suppose that the mapping T defined (12.28) is pseudomonotone and there exists $\rho > 0$ such that for $x \in K$ with $\|x\| > \rho$ there exists $u \in K$ satisfying $\|u\| \leq \rho$ and $\langle x - u, Tx \rangle > 0$. Then there exists a solution to the problem (12.27).*

It should be remarked that if the mapping G is monotone and the mapping L is dissipative (i.e. $-L$ is monotone), then the problem (12.27) has a unique solution because in this case the mapping T is strongly monotone.

We close this section by pointing out that results derived in this section can be employed to get more existence and uniqueness results for the generalized complementarity problem in Banach spaces. We will leave this work to readers who are interested in this part and refer readers to references [45, 58, 43, 44, 87, 5, 10, 32, 50, 83, 72] and the references therein.

3. Variational Inequalities for Topologically Pseudomonotone Mappings

This section will be concerned with another notion of pseudomonotonicity for operators, which is a topological notion, much older than the one cited in the previous sections. In order to make a difference we will call it *topological pseudomonotonicity*, in short *t-pseudomonotonicity*. We will reserve the pseudomonotonicity appellation to the algebraic notion introduced before.

The notion of topological pseudomonotonicity was introduced by Brézis [12] in 1968 for nonlinear operators and it contains many monotone-like operators which were used by Minty [68, 69] and Browder [14] in order to obtain existence theorems for quasi-linear elliptic and parabolic partial differential equations.

Definition 12.6 *An operator $T : K \subset B \to B^*$ is called topologically pseudomonotone on the subset K of the real Banach space B if, for each $u \in K$ and each sequence $\{u_n\}$ in K,*

$$u_n \rightharpoonup u \text{ and } \limsup_{n \to +\infty} \langle u_n - u, Tu_n \rangle \leq 0$$

imply

$$\langle u - w, Tu \rangle \leq \liminf_{n \to +\infty} \langle u_n - w, Tu_n \rangle \text{ for all } w \in B.$$

This definition seems to be obscure. However, we shall see that it is a quite natural notion due to Proposition 12.4 which states that if $T : B \to B^*$ is a nonlinear operator on a real reflexive Banach space such that

$$T = T_1 + T_2$$

where $T_1 : B \to B^*$ is monotone and hemicontinuous, and $T_2 : B \to B^*$ is strongly continuous, i.e., $u_n \rightharpoonup u$ implies $T_2 u_n \to T_2 u$ as $n \to +\infty$, then T is topologically pseudomonotone.

Hence, one can say that: *the theory of topologically pseudomonotone operators unifies both monotonicity arguments and compactness arguments.*

Another definition of topological pseudomonotonicity which has been used by Aubin [2] in a more general setting and involves the previous one is the following.

Definition 12.7 *[2] Let K be a closed subset of a topological vector space X and f a real function defined on $K \times K$. We shall say that f is topologically pseudomonotone if for any generalized sequence (i.e. net) $\{u_\mu\} \subset K$ satisfying*

$$\{u_\mu\} \textit{ stays in a compact set and converges to } \overline{u}$$

and

$$\limsup_\mu f(u_\mu, \overline{u}) \leq 0,$$

its limit $\overline{u}$ satisfies

$$\textit{for all } v \in K, \quad f(\overline{u}, v) \leq \liminf_\mu f(u_\mu, v).$$

When X is a reflexive Banach space and $f(u, v) = \langle u - v, Tu \rangle$ where T is an operator mapping K into X^*, then T is topologically pseudomonotone iff f is topologically pseudomonotone. To see this we use the following lemma, see e.g. [15, Proposition 7.2]

Lemma 12.5 (Kaplansky) *If D is a bounded subset of a reflexive Banach space, then any element in the weak closure of D is a limit of a weakly convergent sequence of D.*

It is also obvious that in Definition 12.7, if f is lower semicontinuous with respect to its first argument then it is topologically pseudomonotone.

In Section 3.2, we will present some existence results on existence of solutions of variational inequalities associated to topologically pseudomonotone operators. The approach used is based on the Ky Fan lemma and its extension as shown by Brézis, Nirenberg and Stampacchia [13]. An other approach based on the Galerkin method was used by Browder [15] and Brézis [12].

3.1 Properties of Topologically Peudomonotone Operators

Let us recall the following definitions

Definition 12.8 *Let B be a real Banach space and $T : B \to B^*$ a nonlinear operator.*

(i) T satisfies condition (S_+) if for each sequence $\{x_n\} \subset B$,

$$x_n \rightharpoonup x \text{ and } \limsup_{n\to\infty} \langle x_n - x, Tx_n - Tx\rangle \le 0 \Rightarrow x_n \to x$$

(ii) T satisfies condition (S) if for each sequence $\{x_n\} \subset B$,

$$x_n \rightharpoonup x \text{ and } \lim_{n\to\infty} \langle x_n - x, Tx_n - Tx\rangle = 0 \Rightarrow x_n \to x.$$

We remark that the notion of operator of (S_+) was introduced by Browder [15]. Typical examples of operators satisfying the (S_+) condition are uniformly monotone operators, i.e.,

$$\langle u - v, Tu - Tv\rangle \ge \alpha(\|v-u\|)\|v-u\|$$

where α is a positive real function with $\alpha(0) = 0$ and $\alpha(t) \to +\infty$ as $t \to +\infty$. Indeed, It follows from the uniform monotonicity of T that

$$0 \le \limsup_{n\to\infty} \alpha(\|u_n - u\|)\|u_n - u\| \le \limsup_{n\to\infty} \langle u_n - u, Tu_n - Tu\rangle \le 0.$$

Hence

$$\lim_{n\to\infty} \alpha(\|u_n - u\|)\|u_n - u\| = 0,$$

and from the properties of α one deduces that $\lim_{n\to\infty} \alpha(\|u_n - u\|) = 0$.

The following result gives some necessary conditions for an operator to be topologically pseudomonotone.

Proposition 12.3 *Let $T : B \to B^*$ be an operator on the real Banach space B. The following properties hold :*

(i) If T is monotone and hemicontinuous, then T is topologically pseudomonotone.

(ii) If T is strongly continuous, then T is topologically pseudomonotone.

(iii) If T is demicontinuous and satisfies the condition (S_+), then T is topologically pseudomonotone.

(iv) If T is continuous and $dim(B) < \infty$, then T is topologically pseudomonotone.

Proof.

(i) Let $\{u_n\}$ be a sequence in B such that $u_n \rightharpoonup u$ and

$$\limsup_{n\to\infty}\langle u_n - u, Tu_n\rangle \leq 0. \tag{12.29}$$

Since the operator T is monotone, it follows that

$$\langle u_n - u, Tu_n\rangle \geq \langle u_n - u, Tu\rangle. \tag{12.30}$$

From the fact that $\langle u_n - u, Tu\rangle \to 0$ as $n \to \infty$; and relations (12.29) and (12.30) one deduces that

$$\langle u_n - u, Tu_n\rangle \to 0 \quad \text{as } n \to \infty.$$

Set $u_t = u + t(w - u)$ for $t \in [0,1]$ and $w \in B$ arbitrary. The monotonicity of T yields

$$\langle u_n - u_t, Tu_n - Tu_t\rangle \geq 0$$

and hence

$$t\langle u - w, Tu_n\rangle \geq \langle u - u_n, Tu_n\rangle + \langle u_n - u, Tu_t\rangle + t\langle u - w, Tu_t\rangle.$$

This implies

$$\begin{aligned}\langle u - w, Tu_t\rangle &\leq \liminf_{n\to\infty}\langle u - w, Tu_n\rangle \\ &= \liminf_{n\to\infty}[\langle u - u_n, Tu_n\rangle + \langle u_n - w, Tu_n\rangle] \\ &= \liminf_{n\to\infty}\langle u_n - w, Tu_n\rangle.\end{aligned}$$

From the hemicontinuity of T, by letting $t \to 0$ one obtains

$$\langle u - w, Tu\rangle \leq \liminf_{n\to\infty}\langle u_n - w, Tu_n\rangle \quad \text{for all } w \in B,$$

and hence T is topologically pseudomonotone.

(ii) From the strong continuity of T, one has that if $u_n \rightharpoonup u$, then $Tu_n \to Tu$. Hence

$$\langle u - w, Tu\rangle = \lim_{n\to\infty}\langle u_n - w, Tu_n\rangle \quad \text{for all } w \in B.$$

(iii) Let $\{u_n\}$ be a sequence in B such that $u_n \rightharpoonup u$ and

$$\limsup_{n\to\infty}\langle u_n - u, Tu_n\rangle \le 0.$$

This implies that

$$\limsup_{n\to\infty}\langle u_n - u, Tu_n - Tu\rangle \le 0.$$

Since the operator T satisfies the (S_+) condition, $u_n \to u$ as $n \to \infty$. From the demicontinuity of T, one deduces that $Tu_n \rightharpoonup Tu$ and hence

$$\langle u - w, Tu\rangle = \lim_{n\to\infty}\langle u_n - w, Tu_n,\rangle \quad \text{for all } w \in B.$$

(iv) This follows from (ii) since on finite-dimensional Banach spaces the weak convergence and the strong convergence coincide. □

We have the following result on the additivity of topologically pseudomonotone operators, which leads us also to have a characterization of the topological pseudomonotonicity notion.

Proposition 12.4 *Let $T_1, T_2 : B \to B^*$ be operators on the real Banach space B.*

(i) If T_1 and T_2 are topologically pseudomonotone, then T_1+T_2 is also topologically pseudomonotone.

(ii) If T_1 is monotone and hemicontinuous and T_2 is strongly continuous, then $T_1 + T_2$ is topologically pseudomonotone.

Proof.

(i) Let $\{u_n\}$ be a sequence in B such that $u_n \rightharpoonup u$ and

$$\limsup_{n\to\infty}\langle u_n - u, T_1u_n + T_2u_n\rangle \le 0. \tag{12.31}$$

This implies that

$$\limsup_{n\to\infty}\langle u_n - u_1, Tu_n\rangle \le 0 \quad \text{and} \quad \limsup_{n\to\infty}\langle u_n - u, T_2u_n\rangle \le 0. \tag{12.32}$$

Indeed, if $\limsup_{n\to\infty}\langle u_n - u_1, Tu_n\rangle > 0$, then there exists a subsequence also denoted by $\{u_n\}$ such that

$$\lim_{n\to\infty}\langle u_n - u, T_1u_n\rangle = a > 0.$$

Therefore, from (12.31) one deduces $\limsup_{n\to\infty}\langle u_n-u, T_2u_n\rangle \le -a$. Since T_2 is topologically pseudomonotone,

$$\langle u - w, T_2u\rangle \le \liminf_{n\to\infty}\langle u_n - w, T_2u_n\rangle \quad \text{for all } w \in B.$$

Letting $w = u$, we obtain the contradiction $0 \leq -a$. Similarly we must have

$$\limsup_{n \to \infty} \langle u_n - u, T_2 u_n \rangle \leq 0.$$

Since T_1 and T_2 are topologically pseudomonotone, from (12.32) we conclude that, for $w \in B$

$$\langle u - w, T_1 u \rangle \leq \liminf_{n \to \infty} \langle u_n - w, T_1 u_n \rangle,$$

$$\langle u - w, T_2 u \rangle \leq \liminf_{n \to \infty} \langle u_n - w, T_2 u_n \rangle,$$

and consequently

$$\langle u - w, T_1 u + T_2 u \rangle \leq \liminf_{n \to \infty} \langle u_n - w, T_1 u + T_2 u_n \rangle \quad \text{for all } w \in X.$$

Therefore $T_1 + T_2$ is topologically pseudomonotone.

(ii) This follows directly from (i) and from Proposition 12.3 (i) and (ii). □

The previous proposition leads us to have a characterization of the notion of topological pseudomonotonicity. More precisely, this notion can be seen as a combination between monotonicity and compactness arguments. Note also that this notion is reserved by addition which is not the case for the pseudomonotonicity in the algebraic sense introduced in Section 3. This makes the topological pseudomonotonicity in some cases more useful for studying some mixed variational inequalities.

We can summarize the above properties by the following implications.

$$\begin{array}{rcl} & & \text{Monotonicity} + \text{Hemicontinuity} \\ & & \Downarrow \\ (\text{Demicontinuity} + \text{Condition } (S_+)) & \Longrightarrow & \text{T-pseudomonotonicity} \\ & & \Uparrow \\ & & \text{T-pseudomonotonicity} + \text{T-pseudomonotonicity} \\ & & \Uparrow \\ & & (\text{Monotonicity, Hemicontinuity}) + \text{Strong continuity} \end{array}$$

3.2 Existence Results for Topologically Pseudomonotone Variational Inequalities

In this section, we will present the approach given by Brézis, Nirenberg and Stampacchia [13] to solve variational inequalities associated to a topologically pseudomonotone operator. Their approach is based on Lemma 12.6 which is a slight extension of the Ky Fan Lemma. An other approach based on the Galerkin method was used by Browder [15] and Brézis [12].

Lemma 12.6 *[13] Let K be an arbitrary set in X a Hausdorff topological vector space. To each $x \in K$ let a set $F(x)$ in E be given satisfying*

(i) $\overline{F(x_0)}$ is compact for some x_0 in K.

(ii) The convex hull of every finite subset $\{x_1, x_2, ..., x_n\}$ of K is contained in the corresponding union $\bigcup_{i=1}^{n} F(x_i)$.

(iii) For every $x \in K$, the intersection of $F(x)$ with any finite dimensional subspace is closed.

(iv) For every convex subset D of X we have

$$\overline{\Big(\bigcap_{x \in K \cap D} F(x)\Big)} \cap D = \Big(\bigcap_{x \in K \cap D} F(x)\Big) \cap D.$$

Then $\bigcap_{x \in K} F(x) \neq \emptyset$.

Assumptions (iii) and (iv) of Lemma 12.6 are satisfied if $F(x)$ is closed for each $x \in K$. In this case we obtain the Ky Fan Lemma [37].

Now we present the following result due to Brézis, Nirenberg and Stampacchia [13] which includes both the Ky Fan minimax principle and general variational inequalities.

Theorem 12.17 *[13] Let K be a closed convex subset of a Hausdorff topological vector space X and let f be a real valued function defined on $K \times K$ such that*

(i) $f(x, x) \geq 0$ for all $x \in K$.

(ii) For every fixed $x \in K$, the set $\{y \in K : f(x, y) < 0\}$ is convex.

(iii) For every fixed $y \in K$, the function $x \in K \mapsto f(x, y)$ is upper semicontinuous on the intersection of K with any finite-dimensional subspace of X.

(iv) For each $y \in K$ and any convergent net $\{x_\alpha\}$ on K with $x_\alpha \to x$, if $f(x_\alpha, ty + (1-t)x) \geq 0$ for each $t \in [0, 1]$ then $f(x, y) \geq 0$.

(v) There is a compact subset L of X and $y_0 \in L \cap K$ such that $f(x, y_0) < 0$ for $x \in K \backslash L$.

Then there exists $x_0 \in L \cap K$ such that

$$f(x_0, y) \geq 0 \quad \text{for all } y \in K.$$

We note that if $-f$ is topologically pseudomonotone in the sense of Definition 12.7 and K is compact, then condition (iv) of Theorem 12.17 is satisfied.

In the case where f is topologically pseudomonotone, Aubin [2, Theorem 1, p.413] obtained a similar result to Theorem 12.17 by using techniques related to minimax type inequalities.

When X is a reflexive Banach space endowed with its weak topology, Theorem 12.17 holds if condition (iv) is replaced by :

$(iv)'$ for each $y \in K$ and any convergent sequence $\{x_n\}$ on K with $x_n \to x \in K$, if $f(x_n, ty + (1-t)x) \geq 0$ for each $t \in [0,1]$ then $f(x, y) \geq 0$.

Now we will employ Theorem 12.17 to solve the following mixed variational inequality : Find $x \in K$ such that

$$\langle y - x, Tx \rangle + \varphi(y) - \varphi(x) \geq 0 \quad \text{for all } y \in K. \tag{12.33}$$

Theorem 12.18 *[13] Let K be a closed convex subset of a Hausdorff topological vector space X, $T : K \to X^*$ be such that whenever $\{x_\alpha\}$ is a net converging to x with* $\limsup \langle Tx_\alpha, x_\alpha - x \rangle \leq 0$ *then* $\liminf \langle Tx_\alpha, x_\alpha - y \rangle \geq \langle Tx, x - y \rangle$ *for all $y \in K$. Assume that T is continuous on finite-dimensional subspaces and that φ is a convex and lower semicontinuous function. Furthermore, if there exists a compact subset L of X and $y_0 \in L \cap K$ such that*

$$\langle y_0 - x, Tx \rangle + \varphi(y_0) - \varphi(x) < 0 \qquad \textit{for all } y \in K \backslash L.$$

Then the solution set of (12.33) is nonempty.

4. Extension

There are several extensions of the variational inequality $VI(T, K)$ and the complementarity problem $CP(T, K)$. These extensions basically involve replacing either the fixed set K or the mapping T by multi-valued mappings. See, e.g., [3, 39, 20, 17, 18, 19, 79, 92, 75, 88, 40, 53, 89, 73]. In this section, we shall consider variational inequality and complementarity problems associated to a multi-valued mapping. Some main existence results will be presented when the multi-valued mapping is pseudomonotone.

To begin, let B be a real Banach space, K a nonempty subset of B and $T : K \to 2^{B^*}$, where 2^{B^*} denotes the family of all nonempty subsets of B^*. The *generalized variational inequality*, denoted $GVI(T, K)$, is to find $x \in K$ such that there exists $u \in Tx$ satisfying the following property :

$$\langle y - x, u \rangle \geq 0 \quad \text{for all } y \in K.$$

Several existence results for the $GVI(T, K)$ have been derived in the literature. For example, Aubin [2], Brézis [12], Browder [15, 16], Shih and Tan [81] gave an existence result of $GVI(T, K)$ for monotone operators; Guo and Yao [41] gave an existence result for nonmonotone operators; Cubiotti and Yao [29] gave some more existence results for operators satisfying $(S)^1_+$ conditions. In the following, we shall employ the celebrated Ky Fan Lemma [37] which is an infinite-dimensional generalization of the KKM Theorem [62], to derive a very general existence result for the $GVI(T, K)$ involving multi-valued pseudomonotone operators.

4.1 Existence and Uniqueness for *GVI(T,K)*

Let K be a nonempty subset of the real Banach space B and $T : K \to 2^{B^*}$. Recall that the operator T is monotone if

$$\langle x - y, u - w \rangle \geq 0 \quad \text{for any } x, y \in K, \ \ u \in Tx, \ \ w \in Ty.$$

Also the operator T is said to be pseudomonotone [90] if for every pair of distinct points $x, y \in K$ and any $u \in Tx$, $w \in Ty$, we have

$$\langle x - y, w \rangle \geq 0 \Longrightarrow \langle x - y, u \rangle \geq 0.$$

The concept of multi-valued pseudomonotone operators was introduced by Saigal [79] in a finite-dimensional space setting and it is clear that if the multi-valued operator is monotone then it is pseudomonotone as the case for single-valued mappings.

Definition 12.9 *Let X, Y be topological spaces and $T : X \to 2^Y$. The operator T is upper semicontinuous at $x_0 \in X$ if for any open set V in Y containing Tx_0, there is an open neighborhood U of x_0 in X such that $Tx \subset V$ for all $x \in U$. The operator T is upper semicontinuous on X if it is upper semicontinuous at each point of X.*

The following celebrated Ky Fan Lemma will play a crucial role in proving the existence result of the $GVI(T, K)$.

Theorem 12.19 *([37, Lemma 6.1]) In a Hausdorff topological vector space, let Y be a convex set and $\emptyset \neq X \subset Y$. For each $x \in X$, let $F(x)$ be a closed subset of Y such that the convex hull of every finite subset*

$\{x_1, .., x_n\}$ of X is contained in the corresponding union $\bigcup_{i=1}^{n} F(x_i)$. If there is a point x_0 such that $F(x_0)$ is compact, then $\bigcap_{x \in X} F(x) \neq \emptyset$.

We also need the following Kneser minimax theorem.

Theorem 12.20 *([63]) Let X be a nonempty convex set in a vector space, and let Y be a nonempty compact convex subset of a Hausdorff topological vector space. Suppose that f is a real-valued function on $X \times X$ such that for each fixed $x \in X$, $f(x, y)$ is lower semicontinuous and convex on Y, and for each fixed $y \in Y$, $f(x, y)$ is concave on X. Then*

$$\min_{y \in Y} \sup_{x \in X} f(x, y) = \sup_{x \in X} \min_{y \in Y} f(x, y).$$

Now we can state and prove the following existence result for the generalized variational inequality.

Theorem 12.21 *Let K be a nonempty closed convex subset of the real reflexive Banach space B. Suppose that $T : K \to 2^{B^*}$ is pseudomonotone such that Tx is a nonempty weakly compact subset of B^* and T is upper semicontinuous from the line segments in K to the weak topology of B^*. Assume in addition that either K is bounded or there exists bounded subset C of K and $x_0 \in C$ such that for any $x \in K \backslash C$,*

$$\inf_{u \in Tx} \langle x - x_0, u \rangle > 0. \tag{12.34}$$

Then there exists $\overline{x} \in K$ such that

$$\sup_{x \in K} \inf_{u \in T\overline{x}} \langle \overline{x} - x, u \rangle \leq 0. \tag{12.35}$$

If in addition the set $T\overline{x}$ is convex, then the GVI(T,K) has $\overline{x}$ as a solution, i.e., there exists $\overline{u} \in T\overline{x}$ such that

$$\langle x - \overline{x}, \overline{u} \rangle \geq 0 \quad \textit{for all } x \in K.$$

Proof. Define two multi-valued operators $F, G : K \to 2^K$ by

$$F(x) = \{y \in K : \inf_{w \in Ty} \langle y - x, w \rangle \leq 0\}$$
$$G(x) = \{y \in K : \sup_{u \in Tx} \langle y - x, u \rangle \leq 0\}$$

for each $x \in K$. The proof of inequality (12.35) is divided into the following five steps.

(i) The convex hull of every finite subset $\{x_1, .., x_n\}$ of K is contained in the corresponding union $\bigcup_{i=1}^{n} F(x_i)$. Indeed, let y be in the convex hull of $\{x_1, .., x_n\}$. Then

$$y = \sum_{i=1}^{n} \lambda_i x_i$$

for some $\lambda_i \geq 0$, $i = 1, ..., n$ and $\sum_{i=1}^{n} \lambda_i = 1$. If $y \notin \bigcup_{i=1}^{n} F(x_i)$, then

$$\inf_{w \in Ty} \langle y - x_i, w \rangle > 0 \quad \text{for each } i = 1, 2, .., n,$$

from which it follows that

$$\begin{aligned} 0 &= \inf_{w \in Ty} \langle y - y, w \rangle \\ &= \inf_{w \in Ty} \sum_{i=1}^{n} \lambda_i \langle y - x_i, w \rangle \\ &\geq \lambda_1 \inf_{w \in Ty} \langle y - x_1, w \rangle + ... + \lambda_n \inf_{w \in Ty} \langle y - x_n, w \rangle \\ &> 0, \end{aligned}$$

which is a contradiction.

(ii) $F(x) \subset G(x)$ for all $x \in K$. Indeed, for each $x \in K$, let $y \in F(x)$. Since the set Ty is weakly compact, there exists $\overline{w} \in Ty$ such that

$$\langle y - x, \overline{w} \rangle = \inf_{w \in Ty} \langle y - x, w \rangle \leq 0$$

and hence

$$\langle x - y, \overline{w} \rangle \geq 0.$$

By the pseudomonotonicity of T, we have

$$\langle x - y, u \rangle \geq 0 \quad \text{for all } u \in Tx.$$

Therefore

$$\sup_{u \in Tx} \langle y - x, u \rangle \leq 0$$

and consequently, $y \in G(x)$. Therefore

$$F(x) \subset G(x) \quad \text{for all } x \in K$$

as claimed.

(iii) $\bigcap_{x \in K} F(x) = \bigcap_{x \in K} G(x)$. We note that, since B is reflexive, the weak and weak* topologies of B^* coincide. Therefore by [81, Lemma 2],

$$\bigcap_{x \in K} G(x) \subset \bigcap_{x \in K} F(x).$$

The inclusion

$$\bigcap_{x \in K} F(x) \subset \bigcap_{x \in K} G(x)$$

follows from Step (ii).

(iv) The weak closure $\overline{F(x_0)}^w$ of $F(x_0)$ is a weakly compact subset of K. If K is bounded, the conclusion follows from the fact that since B is reflexive, K is itself weakly compact. Suppose that K is not bounded. If $F(x_0)$ is not bounded, then there exists a point $x \in F(x_0)$ such that $x \notin C$ and it follows from (12.34) that

$$0 < \inf_{u \in Tx} \langle x - x_0, u \rangle \leq 0$$

which is a contradiction. Therefore, $F(x_0)$ is bounded and hence $\overline{F(x_0)}^w$ is a weakly compact subset of K as claimed.

(v) There exists $\overline{x} \in K$ such that

$$\sup_{x \in K} \inf_{u \in Tx} \langle u, \overline{x} - x \rangle \leq 0.$$

By Steps (i) and (iv) and Theorem 12.19, we have that

$$\bigcap_{x \in K} \overline{F(x)}^w \neq \emptyset.$$

For fixed $x \in K$, since the real function

$$x \mapsto \sup_{u \in Tx} \langle y - x, u \rangle$$

is convex and lower semicontinuous on B, it is also convex and weakly lower semicontinuous. Therefore, $G(x)$ is weakly closed in K, and hence

$$\overline{F(x)}^w \subset G(x), \quad \text{for all } x \in K,$$

from which it follows that

$$\bigcap_{x \in K} G(x) \neq \emptyset.$$

Therefore, by Step (iii), we also have that

$$\bigcap_{x \in K} F(x) \neq \emptyset.$$

Hence, there exists $\overline{x} \in K$ such that

$$\sup_{x \in K} \inf_{u \in T\overline{x}} \langle \overline{x} - x, u \rangle \leq 0.$$

The proof of inequality (12.35) is now completed.
If, in addition, the set $T\overline{x}$ is also convex, by Theorem 12.20, we have

$$\inf_{u \in T\overline{x}} \sup_{x \in K} \langle \overline{x} - x, u \rangle = \sup_{x \in K} \inf_{u \in T\overline{x}} \langle \overline{x} - x, u \rangle \tag{12.36}$$

We note that the real function

$$u \mapsto \sup_{x \in K} \langle \overline{x} - x, u \rangle$$

is convex and weakly lower semicontinuous on B^*. Since $T\overline{x}$ is weakly compact, it follows from (12.36) that there exists $\overline{u} \in T\overline{x}$ such that

$$\langle \overline{x} - x, \overline{u} \rangle \leq 0 \quad \text{for all } x \in K$$

or

$$\langle x - \overline{x}, \overline{u} \rangle \geq 0 \quad \text{for all } x \in K.$$

Hence $\overline{x}$ is a solution of the problem $GVI(T, K)$. □

Remark. By a similar argument to one in the proof of Proposition 12.2, one can easily show that if there exists $x_0 \in K$ such that

$$\liminf_{\|x\| \to \infty, x \in K} \inf_{u \in Tx} \langle x - x_0, u \rangle > 0,$$

then there exists a bounded subset C of K such that for any $x \in K \backslash C$,

$$\inf_{u \in Tx} \langle x - x_0, u \rangle > 0.$$

The following corollary is an immediate consequence of Theorem 12.21.

Corollary 12.6 *Let K be a nonempty closed convex subset of the real reflexive Banach space B. Suppose that $T : K \to 2^{B^*}$ is pseudomonotone such that Tx is weakly compact and convex for each $x \in K$ and T is upper* *semicontinuous from the line segments in K into the weak topology of B^*. If there exists $x_0 \in K$ such that*

$$\liminf_{\|x\|\to\infty, x\in K} \inf_{u\in Tx} \langle x - x_0, u\rangle > 0,$$

then the GVI(T,K) has a solution.

Recall that a single-valued operator $T : K \to B^*$ is called hemicontinuous if it is continuous from the line segments in K into the weak* topology of B^*. The following existence result for the $VI(T, K)$ is then again a consequence of Theorem 12.21.

Corollary 12.7 *Let K be a nonempty closed convex subset of the real reflexive Banach space B. Suppose that $T : K \to B^*$ is hemicontinuous and pseudomonotone. Assume in addition that either K is bounded or there exists $x_0 \in K$ such that*

$$\liminf_{\|x\|\to\infty, x\in K} \langle x - x_0, Tx\rangle > 0.$$

Then the variational inequality $VI(T, K)$ has a solution.

Now we turn to the uniqueness of the solution of the generalized variational inequality. Let us first recall the concepts of strict, α, and strong pseudomonotonicity of multi-valued operators, respectively, introduced by Yao [90].

Definition 12.10 *Let K be a nonempty subset of the real Banach space B and $T : K \to 2^{B^*}$.*

(i) The operator T is strictly pseudomonotone on K if, for every pair of distinct points $x \in K$, $y \in K$, and any $u \in Tx$, $w \in Ty$, we have

$$\langle x - y, w\rangle \geq 0 \Longrightarrow \langle x - y, u\rangle > 0.$$

(ii) The operator T is α-pseudomonotone on K if there exist $x_0 \in K$, $u_0 \in Tx_0$ and $\alpha : [0, +\infty) \to [0, +\infty)$ with

$$\alpha(0) = 0, \quad \alpha(t) > 0 \text{ for } t > 0 \text{ and } \liminf_{r\to+\infty} \alpha(t) > \|u_0\|,$$

such that, for every pair of distinct points $x \in K$, $y \in K$, and any $u \in Tx$, $w \in Ty$, we have

$$\langle x - y, w\rangle \geq 0 \Longrightarrow \langle x - y, u\rangle \geq \|x - y\|\alpha(\|x - y\|).$$

(iii) The operator T is strongly pseudomonotone if $\alpha(t) = kt$ for some $k > 0$ in (ii).

The concept of strict and strong pseudomonotonicity of multi-valued operators in Definition 12.10 are generalizations of the concepts of strict and strong pseudomonotonicity of single-valued mappings, respectively, introduced by Karamardian and Schaible [58].

Lemma 12.7 *Let K be a nonempty subset of the real Banach space B and $T : K \to 2^{B^*}$. If T is strictly pseudomonotone, then the $GVI(T, K)$ can have at most one solution.*

Proof. Suppose on the contrary that the $GVI(T, K)$ has two distinct solutions, say, $\overline{x}$ and $\overline{y}$. Then there exists $\overline{u} \in T\overline{x}$ and $\overline{w} \in T\overline{y}$ such that

$$\langle x - \overline{x}, \overline{u} \rangle \geq 0 \quad \text{for all } x \in K$$

$$\langle x - \overline{y}, \overline{w} \rangle \geq 0 \quad \text{for all } x \in K$$

In particular, we have

$$\langle \overline{y} - \overline{x}, \overline{u} \rangle \geq 0 \tag{12.37}$$

$$\langle \overline{x} - \overline{y}, \overline{w} \rangle \geq 0. \tag{12.38}$$

Since T is strictly pseudomonotone, it follows from (12.37) that

$$\langle \overline{y} - \overline{x}, \overline{w} \rangle > 0$$

from which it follows that

$$\langle \overline{x} - \overline{y}, \overline{w} \rangle < 0$$

which is a contradiction to (12.38). Hence the $GVI(T, K)$ can have at most one solution. □

The following uniqueness result follows from Lemma 12.7 and Theorem 12.21.

Theorem 12.22 *Let K be a nonempty closed convex subset of the real reflexive Banach space B. Suppose that $T : K \to 2^{B^*}$ is strictly pseudomonotone such that Tx is weakly compact and convex for each $x \in K$ and T is upper semicontinuous from the line segments in K into the weak topology of B^*. Assume that either K is bounded or there exists $x_0 \in K$ such that*

$$\liminf_{\|x\| \to \infty, x \in K} \inf_{u \in Tx} \langle x - x_0, u \rangle > 0.$$

Then the $GVI(T,K)$ has a unique solution.

We have more uniqueness results for α-pseudomonotone and strongly pseudomonotone operators.

Theorem 12.23 *([90, Theorem 3.2, Corollary 3.1]) Let K be a nonempty closed convex subset of the real reflexive Banach space B and $T : K \to 2^{B^*}$ such that Tx is weakly compact and convex for each $x \in K$ and T is upper semicontinuous from the line segments in K into the weak topology of B^*. If T is either α-pseudomonotone or strongly pseudomonotone, then the $GVI(T,K)$ has a unique solution.*

We observe that Theorem 12.23 is not a consequence of Theorem 12.22. We also note that few uniqueness results for the $GVI(T,K)$ have been derived in the literature. The uniqueness results presented in this section to our best knowledge are the most recent ones. More efforts should be put in this direction.

4.2 *GVI(T,K)* for Multi-valued Topologically Pseudomonotone Operators

In this section, we first introduce the topological pseudomonotonicity notion for multi-valued operators (See, e.g., Browder [15], Zeidler [93, p.913]). Some existence results of $GVI(T,K)$ are then presented.

Definition 12.11 *Let K be a closed convex subset of a real reflexive Banach space B. The multi-valued operator $T : K \to 2^{B^*}$ is said to be topologically pseudomonotone if the following holds:*

Let $(u_n, u_n^*) \in G(T)$ for all $n \in I\!N$ and suppose that $u_n \rightharpoonup u$ such that

$$\limsup_{n\to\infty} \langle u_n - u, u_n^* \rangle \le 0.$$

Then, for each $v \in K$, there exists $u_v^* \in B^*$ such that $(v, u_v^*) \in G(T)$ and

$$\langle u - v, u_v^* \rangle \le \liminf_{n\to\infty} \langle u_n - v, u_n^* \rangle.$$

We have the following property of multi-valued topologically pseudomonotone operators which represents an extension to the multi-valued case of the same property in the single-valued case.

Proposition 12.5 *Let $T : K \to 2^{B^*}$ be monotone and upper semicontuous on line segments in the nonempty closed convex subset K of the real reflexive Banach space B. If for each $u \in K$, the set Tu is nonempty, convex and closed in B^*, then T is topologically pseudomonotone.*

Proof. We refer to Kluge [61, p.93] by using a similar argument as in the proof of Proposition 12.3. □

The property of the sum of two single-valued topologically pseudomonotone operators is also reserved for multi-valued topologically pseudomonotone operators as shown by the following proposition.

Proposition 12.6 *Let $T_1, T_2 : B \to 2^{B^*}$ be topologically pseudomonotone on the real reflexive Banach space B. Then, the sum*

$$T_1 + T_2 : B \to 2^{B^*}$$

is also topologically pseudomonotone.

Proof. The proof is similar to the one of Proposition 12.4 (i). □

Now, we again consider the following generalized variational inequality $GVI(T, K)$:

$$\text{Find } \overline{x} \in K,\ \overline{\xi} \in T(\overline{x}) \text{ such that } \langle y - \overline{x}, \overline{\xi} \rangle \geq 0 \text{ for all } y \in K, \quad (12.39)$$

where K is a subset of a real Banach space B and $T : K \to 2^{B^*}$.

We have the following existence result for (12.39) where T is a topologically pseudomonotone operator.

Theorem 12.24 *Let B be a real reflexive Banach space and $T : K \to 2^{B^*}$ a multi-valued operator with nonempty bounded closed convex values, where K is a closed convex subset of B.*

Suppose that :

(i) Either K is bounded or there exist $y_0 \in K$ and $r > 0$ such that

$$\sup_{\xi \in T(x)} \langle y_0 - x, \xi \rangle < 0 \text{ for all } x \in K \text{ with } \|x\| > r.$$

(ii) T is topologically pseudomonotone and upper semicontinuous on the intersection of K with any finite-dimensional subspaces of B.

Then the problem GVI(T,K) has a solution.

Proof. We shall apply Theorem 12.17 with B endowed with its weak topology $\sigma(X, X^*)$ and f defined by

$$f(x, y) = \sup_{\xi \in T(x)} \langle y - x, \xi \rangle.$$

Assumptions (ii) and (v) of Theorem 12.17 are immediate. On the other hand, from the topological pseudomonotonicity of T one deduces that f is topologically pseudomonotone and hence one can easily verify that f satisfies assumption (iv) of Theorem 12.17. So we need only to verify that $f(.,y)$ is upper semicontinuous on the intersection of K with finite-dimensional subspaces of X. This follows from Berge Theorem [6, p.122].

Therefore, from Theorem 12.17, there exists $\overline{x} \in K$ such that

$$\sup_{\xi \in T(\overline{x})} \langle y - \overline{x}, \xi \rangle \geq 0 \text{ for all } y \in K.$$

Since $T(\overline{x})$ is convex compact, from [7, Lemma1], we conclude that there exists $\overline{\xi} \in T(\overline{x})$ such that $\langle y - \overline{x}, \overline{\xi} \rangle \geq 0$ for all $y \in K$. Hence the $GVI(T, K)$ has a solution. □

4.3 Generalized Quasi-variational Inequality

Recently Chowdhury and Tan [21] gave an extension of the definition of topological pseudomonotonicity related to a multi-valued operator. They introduced this notion specially to study the following *generalized quasi-variational inequality* on a paracompact set K in a locally convex Hausdorff topological vector space X :

$$\begin{cases} \text{Find } \overline{y} \in S(\overline{y}) \text{ and } \overline{\xi} \in T(\overline{y}) \text{ such that} \\ \langle \overline{y} - x, \overline{\xi} \rangle \leq h(x) - h(\overline{y}) \text{ for all } x \in S(\overline{y}), \end{cases} \tag{12.40}$$

where $S : K \to 2^K$, $T : K \to 2^{X^*}$ and $h : K \to I\!R$.

As we can see that problem (12.40) reduces to the generalized variational inequality problem when $S(y) = K$ for all $y \in K$.

Definition 12.12 *Let X be a topological vector space, K a nonempty subset of X and $T : K \to 2^{X^*}$. If $h : K \to I\!R$, then*

(i) T is said to be topologically h-pseudomonotone if for each $y \in K$ and each net $\{y_\alpha\}_{\alpha \in \Gamma}$ in K converging to y with

$$\limsup_\alpha [\inf_{u \in T(y_\alpha)} \langle y_\alpha - y, u \rangle + h(y_\alpha) - h(y)] \leq 0,$$

we have

$$\liminf_\alpha [\inf_{u \in T(y_\alpha)} \langle y_\alpha - u, u \rangle + h(y_\alpha) - h(x)] \geq \inf_{w \in T(y)} \langle y - x, w \rangle + h(y) - h(x)$$

for all $x \in K$.

(ii) T is said to be strongly topologically h-pseudomonotone if for each continuous function $\theta : K \to [0,1]$, for each $y \in K$ and every net $\{y_\alpha\}_{\alpha \in \Gamma}$ in K converging to y with

$$\limsup_\alpha [\theta(y_\alpha)(\inf_{u \in T(y_\alpha)} \langle y_\alpha - y, u \rangle + h(y_\alpha) - h(y))] \leq 0,$$

we have

$$\liminf_{\alpha}[\theta(y_\alpha)(\inf_{u\in T(y_\alpha)}\langle y_\alpha - u, u\rangle + h(y_\alpha) - h(x))] \geq$$

$$[\theta(y)(\inf_{w\in T(y)}\langle y - x, w\rangle + h(y) - h(x))] \quad \textit{for all } x \in K.$$

(iii) T is said to be topologically strongly pseudomonotone, if $h = 0$ in (ii).

Clearly, T is topologically 0-pseudomonotone if and only if the function f defined on $K\times K$ by $f(x,y) = \inf_{u\in T(x)}\langle x-y, u\rangle$ is topologically pseudomonotone in the sense of Definition 12.7. Also, it is easy to see that if T is strongly topologically pseudomonotone then $f(x,y) = \inf_{u\in T(x)}\langle x-y, u\rangle$ is topologically pseudomonotone.

Proposition 12.7 *Let K be a nonempty subset of a topological vector space X. If $T : K \to X^*$ is monotone and continuous from the relative weak topology on K to the weak* topology on X^*, then T is strongly topologically pseudomonotone.*

Proof. Let us consider any arbitrary continuous function $\theta : K \to [0,1]$. Suppose $\{y_\alpha\}_{\alpha\in\Gamma}$ is a net in K converging to y and

$$\limsup_{\alpha}[\theta(y_\alpha)(\langle y_\alpha - y, Ty_\alpha\rangle)] \leq 0.$$

Then for any $x \in K$ and $\varepsilon > 0$, there are $\beta_1, \beta_2 \in \Gamma$ with

$$|\theta(y_\alpha)\langle y_\alpha - y, Ty_\alpha\rangle| < \frac{\varepsilon}{2} \text{ for all } \alpha \geq \beta_1$$

and

$$|\theta(y_\alpha)\langle y - x, Ty_\alpha - Ty\rangle| < \frac{\varepsilon}{2} \text{ for all } \alpha \geq \beta_2.$$

Choose $\beta_0 \in \Gamma$ with $\beta_0 \geq \beta_1, \beta_2$. Then, since T is monotone, we have

$$\begin{aligned}\theta(y_\alpha)\langle y_\alpha - x, Ty_\alpha\rangle &= \theta(y_\alpha)\langle y_\alpha - y, Ty_\alpha\rangle + \theta(y_\alpha)\langle y - x, Ty_\alpha\rangle\\ &\geq \theta(y_\alpha)\langle y_\alpha - y, Ty\rangle + \theta(y_\alpha)\langle y - x, Ty_\alpha\rangle\\ &= \theta(y_\alpha)\langle y_\alpha - y, Ty\rangle + \theta(y_\alpha)\langle y - x, Ty_\alpha - Ty\rangle +\\ &\quad\ \theta(y_\alpha)\langle y - x, Ty\rangle\\ &> -\tfrac{\varepsilon}{2} - \tfrac{\varepsilon}{2} + \theta(y_\alpha)\langle y - x, Ty\rangle \quad \text{for all } \alpha \geq \beta_0,\end{aligned}$$

so that

$$\inf_{\alpha\geq\beta_0}\theta(y_\alpha)\langle y_\alpha - x, Ty_\alpha\rangle \geq -\varepsilon + \inf_{\alpha\geq\beta_0}\theta(y_\alpha)\langle y - x, Ty\rangle.$$

It follows that

$$\begin{aligned}\limsup_{\alpha} \theta(y_\alpha)\langle y_\alpha - x, Ty_\alpha\rangle &\geq \liminf_{\alpha} \theta(y_\alpha)\langle y_\alpha - x, Ty_\alpha\rangle \\ &\geq -\varepsilon + \theta(y)\langle y - x, Ty\rangle.\end{aligned}$$

As $\varepsilon > 0$ is arbitrary, we now have

$$\limsup_{\alpha} \theta(y_\alpha)\langle y_\alpha - x, Ty_\alpha\rangle \geq \theta(y)\langle y - x, Ty\rangle.$$

Hence T is strongly topologically pseudomonotone. □

The following result due to Chowdhury and Tan [21] is related to the existence of solutions of (12.40) associated to a strongly topologically h-pseudomonotone operator.

Theorem 12.25 *Let X be a locally convex Hausdorff topological vector space, K be a nonempty paracompact convex subset of X and $h : X \to I\!R$ be convex. Let $S : K \to 2^K$ be upper semicontinuous such that each $S(x)$ is compact convex and $T : K \to 2^{X^*}$ be stronly topologically h-pseudomonotone and upper semicontinous from $co(A)$ to the weak* topology on X^* for each finite set A in K such that $T(x)$ is weak* compact and convex. Suppose that the set*

$$\Sigma = \{y \in K : \sup_{x\in S(x)} [\inf_{w\in T(w)} \langle y - x, w\rangle + h(y) - h(x)] > 0\}$$

is open in K. Suppose further that there exist a nonempty compact subset B of K and a point $x_0 \in K$ such that

$$x_0 \in B \cap S(y) \text{ and } \inf_{w\in T(y)} \langle y - x_0, w\rangle + h(y) - h(x_0) > 0 \text{ for all } y \in K\backslash B.$$

Then there exist $\overline{y} \in S(\overline{y})$ and $\overline{\xi} \in T(\overline{y})$ such that $\langle \overline{y} - x, \overline{\xi}\rangle \leq h(x) - h(\overline{y})$ for all $x \in S(\overline{y})$.

For the proof of this theorem we refer to [21].

4.4 The Generalized Complementarity Problem

In this section, we shall consider a generalization of the complementarity problem for multi-valued operators. Existence and uniqueness results will be obtained by employing the results presented in Section 4.1. More specifically, let K be a closed convex cone in a real Banach space B and $T : K \to 2^{B^*}$. The *generalized complementarity problem*, denoted $GCP(T, K)$, is to find $x \in K$ such that there is $y \in Tx$ satisfying the following properties :

$$y \in K^*, \quad \langle x, y\rangle = 0.$$

The problem $GCP(T,K)$ was originally introduced by Saigal [79] in the finite-dimensional space setting. We remark that there are very few results either on existence or the algorithmic aspect for $GCP(T,K)$ in the literature. The results presented in this section are efforts in this direction.

As in the single-valued case, one can expect that both problems $GVI(T,K)$ and $GCP(T,K)$ have the same solution set if the underlying set K is a closed convex cone as the following result shows. Since the proof is very similar to the proof of single-valued case, it will be omitted.

Lemma 12.8 *Let K be a closed convex cone in the real Banach space B and $T : K \to 2^{B^*}$. Then the problems GVI(T,K) and GCP(T,K) have the same solution set.*

Combining Corollary 12.7 and Lemma 12.8, we have the following existence result for the $GCP(T,K)$.

Theorem 12.26 *Let K be a closed convex cone in the real reflexive Banach space B and $T : K \to 2^{B*}$. Suppose that T is pseudomonotone such that Tx is weakly compact and convex for each $x \in K$ and T is upper semicontinuous from the line segments in K into the weak topology of B^*. If there exists $x_0 \in K$ such that*

$$\liminf_{\|x\|\to\infty, x\in K} \inf_{u\in Tx} \langle x - x_0, u\rangle > 0,$$

then the GCP(T,K) has a solution.

We also have the following uniqueness result for $GCP(T,K)$ which is a consequence of Theorem 12.23 and Lemma 12.8.

Theorem 12.27 *Let K be a closed convex cone in the real reflexive Banach space B and $T : K \to 2^{B^*}$. Suppose that T is upper semicontinuous from the line segments in K into the weak topology of B^* and Tx is weakly compact and convex for each $x \in K$. If T is either α-pseudomonotone or strongly pseudomonotone, then the $GCP(T,K)$ has a unique solution.*

We refer readers to [39, 20, 79, 92, 73, 66, 74] for more results on various types of generalized complementarity problems.

5. Equivalence of Complementarity Problems and Least Element Problems

This final section devotes to the equivalence of the complementarity problem, the least element problem and related problems under suitable assumptions. These equivalences are important in the sense that we can use new methods to study complementarity problems by these connections. Also, problems like least element problems and other related problems can be studied by the complementarity problems.

To start our discussion, let K be a closed convex cone in a real Banach space B. As we mentioned in the Introduction, one can define preorders induced by K on B and K^* on B^*, respectively, which are denoted by the same symbol "$\leq$", that is

$$\begin{array}{lcl} x \leq y \in B & \Leftrightarrow & y - x \in K \\ u \leq v \in B^* & \Leftrightarrow & v - u \in K^*. \end{array}$$

Nonzero elements of K and K^* are said to be *positive*, and $u \in K^*$ is said to be *strictly positive* if $\langle x, u\rangle > 0$ for all $x \in K$, $x \neq 0$. The space B is said to be a *vector lattice* with respect to $\leq$ if each pair of point $x, y \in B$ has a unique infimum characterized by the following properties

$$x \wedge y \leq x, \ \ x \wedge y \leq y; \ \ z \leq x, \ \ z \leq y \Longrightarrow z \leq x \wedge y.$$

We note that if B is a vector lattice with respect to $\leq$ induced by K, so is B^* with respect to $\leq$ induced by K^*. See, e.g., [59, Appendix, p.224-231].

Let a closed convex cone $K \subset B$, $T : K \to B^*$ and $f : K \to I\!\!R$ be given. We denote by F the feasible set of T with respect to K, that is,

$$F = \{x \in B : x \in K \text{ and } Tx \in K^*\}.$$

We shall consider in this section the following five problems.

(I) *Nonlinear program* : For a given $u \in B^*$, find $x \in F$ such that

$$\langle x, u\rangle = \min_{y \in F} \langle y, u\rangle.$$

(II) *Least element problem* : find $x \in F$ such that

$$x \leq y \text{ for all } y \in F.$$

(III) *Complementarity problem* : Find $x \in F$ such that $\langle x, Tx \rangle = 0$

(IV) *Variational inequality* : Find $x \in K$ such that

$$\langle y - x, Tx \rangle \geq 0 \text{ for all } y \in K.$$

(V) *Unilateral minimization problem* : Find $x \in K$ such that

$$f(x) = \min_{y \in K} f(y).$$

In [78], Riddel considered the equivalences of the above five problems for strictly monotone Z-map in Banach lattices. Instead of strict monotonicity, we shall use strict pseudomonotonicity to derive equivalences of the above five problems.

First let us recall the following definitions.

Definition 12.13 *([78]) Let B be a real Banach space which is also a vector lattice with positive cone K, and let $T : K \to B^*$. Then T is said to be a Z-mapping relative to K provided*

$$\langle z, Tx - Ty \rangle \leq 0 \quad \textit{whenever } (x - y) \wedge z = 0.$$

It should be observed that Definition 12.13 reduces to the definition of condition Z introduced by Cryer and Dempster [30] when T is linear. Also when B is finite-dimensional and K is the nonnegative orthant, T is a Z-mapping relative to K if and only if T is off-diagonally antitone in the sense of Rheinboldt [77].

Definition 12.14 *([78]) Let K be a nonempty convex subset of the real Banach space B and $T : K \to B^*$. The operator T is positive at infinity if for any $x \in K$, there exists a positive real number $\rho(x)$ such that $\langle y - x, Ty \rangle > 0$ for every $y \in K$ such that $\|y\| \geq \rho(x)$.*

We have the following translation property of pseudomonotone operators.

Lemma 12.9 *([80]) Let K be a convex cone of the real Banach space B and $T : K \to B^*$. If T is pseudomonotone or strictly pseudomonotone, then the operator T_z defined by*

$$T_z(x) = T(x + z), \;\; x \in K$$

for fixed $z \in K$ is also pseudomonotone or strictly pseudomonotone, respectively.

We have the following existence and uniqueness result of variational inequality which is a consequence of Theorem 12.4 and Lemma 12.3.

Theorem 12.28 *Let K be a closed and convex cone of a real reflexive Banach space and $T : K \to B^*$. Suppose that T is pseudomonotone, hemicontinuous and positive at infinity. Then there exists $x \in K$ such that*

$$\langle y - x, Tx \rangle \geq 0 \quad \textit{for all } y \in K.$$

Furthermore, if in addition T is strictly pseudomonotone, the solution is unique.

Next, we employ Theorem 12.28 to derive the following existence result for perturbed variational inequalities.

Corollary 12.8 *Let K be a closed and convex cone of a real reflexive Banach space and $T : K \to B^*$. Suppose that T is pseudomonotone, hemicontinuous and positive at infinity. Then for each fixed $z \in K$, there exists $x \in K$ such that*

$$VI(T, K, z) \qquad \langle y - x, T(x + z) \rangle \geq 0 \quad \textit{for all } y \in K.$$

If in addition, T is strictly pseudomonotone, then for each fixed $z \in K$, the problem $VI(T, K, z)$ has a unique solution.

Proof. For fixed $z \in K$, let us define $T_z : K \to B^*$ by

$$T_z(x) = T(x + z), \quad \text{for all } x \in K.$$

Clearly, T_z is hemicontinuous and it is also pseudomonotone by Lemma 12.9. Now let us show that T_z is positive at infinity. Let $x \in K$ and define $\rho_1(x) = \rho(x + z) + \|z\|$, then for all $y \in K$ such that $\|y\| > \rho_1(x)$ one has $\|y + z\| \geq \|y\| - \|z\| > \rho(x + z)$. Hence using that T is positive at infinity, one has

$$\langle y + z - (x + z), T(x + z) \rangle > 0$$

Therefore, $\langle y - x, T_z(x) \rangle > 0$. Hence by Theorem 12.28, one deduces that $VI(T, K, z)$ has a solution.

If in addition T is strictly pseudomonotone, then by Lemma 12.9, T_z is also strictly pseudomonotone. Therefore $VI(T, K, z)$ has a unique solution. □

Now we give a sequence of propositions each with its own hypotheses and from these propositions, we will derive the main result of this section.

Proposition 12.8 *Let $T : K \to B^*$ be the G-derivative of the function $f : K \to I\!R$. Then any solution of* (**V**) *is also a solution of* (**IV**). *If, in addition, T is pseudomonotone, then conversely any solution of* (**IV**) *is also a solution of* (**V**).

Proof. The first part of the proposition can be found in [78].

Conversely, suppose in addition that T is pseudomonotone. If x is any solution of (**IV**), then

$$\langle y - x, Tx \rangle \geq 0 \quad \text{for all } y \in K.$$

By Theorem 12.11, f is pseudoconvex. By the pseudoconvexity of f, we then have

$$f(y) \geq f(x) \quad \text{for all } y \in K.$$

Hence x is also a solution of (**V**). □

Proposition 12.9 *Let $T : K \to B^*$. Then x is a solution of* (**III**) *if and only if x is a solution of* (**IV**)

Proof. The result follows from [55, Lemma 3.1]. □

Proposition 12.10 *Suppose that $T : K \to B^*$ is strictly pseudomonotone and a Z-mapping relative to K. Then any solution of* (**IV**) *is also a solution of* (**II**).

Proof. Suppose that x is a solution of (**IV**). Then

$$\langle y - x, Tx \rangle \geq 0 \quad \text{for all } y \in K. \tag{12.41}$$

By Proposition 12.9, x is a solution of (**III**) and hence $x \in F$. For any fixed $y \in F$, we claim that $x \leq y$. Indeed, let $y_0 = x \wedge y$. Since $0 \leq x$, $0 \leq y$ and B is a vector lattice with respect to $\leq$, $x \wedge y \geq 0$ or equivalently, $y_0 \in K$. Substituting y_0 into (12.41), we get

$$\langle y_0 - x, Tx \rangle \geq 0. \tag{12.42}$$

If $x \neq y_0$, then by (12.42) and the strict pseudomonotonicity of T, we have

$$\langle y_0 - x, Ty_0 \rangle > 0. \tag{12.43}$$

Now since $y \in F$, $Ty \in K^*$. As $x - y_0 \in K$,

$$\langle x - y_0, Ty \rangle \geq 0. \tag{12.44}$$

On the other hand, the trio y, y_0 and $z = x - y_0$ satisfy

$$(y - y_0) \wedge z = (y - y_0) \wedge (x - y_0) = (y \wedge x) - y_0 = 0.$$

Consequently, by the fact that T is a Z-mapping, we have

$$\langle z, Ty - Ty_0 \rangle \leq 0$$

from which it follows that

$$\langle x - y_0, Ty_0 - Ty \rangle \geq 0. \tag{12.45}$$

By adding (12.44) and (12.45), we obtain

$$\langle x - y_0, Ty_0 \rangle \geq 0,$$

and hence

$$\langle y_0 - x, Ty_0 \rangle \leq 0,$$

which is a contradiction to (12.43). As a result, we conclude that $x = y_0$ and hence $x \leq y$ for all $y \in F$. Therefore x is a solution of (**II**) and the proof is complete. □

Proposition 12.11 *Let $T : K \to B^*$ and $u \in K^*$ be arbitrary. Then any solution of* (**II**) *is also a solution of* (**I**).

Proof. Let x be a solution of (**II**). Then for any $y \in F$, $x \leq y$. Hence $y - x \in K$. Since $u \in K^*$, we have $\langle y - x, u \rangle \geq 0$ and thus

$$\langle x, u \rangle = \min_{y \in F} \langle y, u \rangle.$$

Consequently, x is a solution of (**I**). □

Proposition 12.12 *Let B be reflexive, $T : K \to B^*$ be a Z-mapping relative to K. Assume that T is strictly pseudomonotone, hemicontinuous and positive at infinity. Then for any pair $x \in F$ and $y \in F$, we have $x \wedge y \in F$, i.e. F is a $\wedge$-sublattice.*

Proof. Suppose that $x \in F$ and $y \in F$ and let $z = x \wedge y$. It remains to show that $Tz \geq 0$ or equivalently, $Tz \in K^*$ since $x \geq 0$, $y \geq 0$ and $x \wedge y \geq 0$. By Proposition 12.8, there exists $\overline{x} \in K$ such that

$$\langle w - \overline{x}, T(\overline{x} + z) \rangle \geq 0 \quad \text{for all } w \in K. \tag{12.46}$$

For any $v \in K$ which is a convex cone, $v + \overline{x} \in K$. Thus by (12.46), we have

$$\langle v, T(\overline{x} + z) \rangle \geq 0 \quad \text{for all } v \in K.$$

Thus $T(\overline{x}+z) \in K^*$. We now claim that $z = \overline{x}+z$. Indeed, let $z_0 = x \wedge (\overline{x}+z)$. Since $\overline{x}+z \geq z$ and $x \geq z$, we have $z_0 \geq z$. So $w = z_0 - z \geq 0$ and thus $z_0 - z \in K$. Using w in (12.46), we get

$$\langle z_0 - (\overline{x}+z), T(\overline{x}+z)\rangle \geq 0. \tag{12.47}$$

Suppose $z_0 \neq \overline{x}+z$. Then by the strict pseudomonotonicity of T and (12.47), we have

$$\langle z_0 - (\overline{x}+z), T(z_0)\rangle > 0. \tag{12.48}$$

Note that

$$(x - z_0) \wedge (\overline{x}+z-z_0) = (x \wedge (\overline{x}+z)) - z_0) = 0.$$

The definition of a $Z-$mapping gives

$$\langle (\overline{x}+z) - z_0, Tx - T(z_0)\rangle \leq 0. \tag{12.49}$$

Since $Tx \in K^*$ and $(\overline{x}+z) - z_0 \in K$, we have

$$\langle (\overline{x}+z) - z_0, Tx\rangle \geq 0$$

or

$$\langle (\overline{x}+z) - z_0, -Tx\rangle \leq 0. \tag{12.50}$$

Adding (12.49) and (12.50), we obtain

$$\langle (\overline{x}+z) - z_0, -Tz_0\rangle \leq 0$$

from which it follows that

$$\langle z_0 - (\overline{x}+z), Tz_0\rangle \leq 0$$

which is in contradiction to (12.48). As a result, we must have $z_0 = \overline{x}+z$ and by the definition of z_0, we conclude that $\overline{x}+z \leq x$ as claimed.

Repeating the above argument with y in place of x, one can show that $\overline{x}+z \leq y$. Thus $\overline{x}+z \leq x \wedge y = z$. But on the other hand, $\overline{x}+z \geq z$, so $\overline{x}+z = z$ and the proof is complete. □

Proposition 12.13 *Let B be reflexive, $T : K \to B^*$ be a Z-mapping relative to K. Assume that T is strictly pseudomonotone, hemicontinuous and positive at infinity. Let $u \in K^*$ be strictly positive. Then problem* (**I**) *corresponding to u has at most one solution, and any solution of* (**I**) *is also a solution of* (**II**).

Proof. Suppose $x \in F$ and $y \in F$ are both solutions of the problem (**I**) :

$$\min_{w \in F} \langle w, u\rangle.$$

By the Proposition 12.13, we know $x \wedge y \in F$. Since u is strictly positive,

$$\langle x \wedge y, u\rangle \leq \langle x, u\rangle$$

with strict inequality if $x \wedge y \neq x$. But since x is a solution, we must have $x \wedge y = x$. Similarly $x \wedge y = y$ and thus $x = y$ by the uniqueness of $x \wedge y$.

Suppose now x is a solution of (**I**) corresponding to u and let $w \in F$. From Proposition 12.12 again, $w \wedge x \in F$. By the optimality of x and the positivity of u, we have

$$\langle x, u\rangle \leq \langle w \wedge x, u\rangle \leq \langle x, u\rangle = \min_{y\in F}\langle y, u\rangle.$$

Consequently, $w \wedge x$ solves (**I**). By the uniqueness, $w \wedge x = x$ and so $x \leq w$ for all $w \in F$. Hence x is a solution of (**II**). □

Now by combining Propositions 12.8-12.13, we can state the main result of this section.

Theorem 12.29 *Let K be a closed convex cone of a real reflexive Banach space B which is also a vector lattice with respect to the order $\leq$ induced by K. Let $T : K \to B^*$ be a Z-mapping relative to K and assume that T is strictly pseudomonotone, hemicontinuous and positive at infinity. If $u \in K^*$ is strictly positive, then there exists a unique $x \in F$ which is a solution of problems* (**I**) – (**IV**). *If in addition, T is the G-derivative of $f : K \to I\!R$, then x is also a unique solution of problem* (**V**).

It is not difficult to check that any strongly pseudomonotone operator is also positive at infinity. See, e.g., [80]. Consequently, the following result follows immediately from Theorem 12.29.

Corollary 12.9 *Let K be a closed convex cone of a real reflexive Banach space B which is also a vector lattice with respect to the order $\leq$ induced by K. Let $T : K \to B^*$ be a Z-mapping relative to K. Assume that T is strongly pseudomonotone and hemicontinuous. If $u \in K^*$ is strictly positive, then there exists a unique $x \in F$ which is a solution of problems* (**I**) – (**IV**). *If in addition, T is the G-derivative of $f : K \to I\!R$, then x is also a unique solution of problem* (**V**).

Example. Let $T : [0, \infty) \to I\!R$ be defined as

$$Tx = 2 + \frac{1}{10}x + \sin x, \quad x \geq 0.$$

Then one can easly see that T is strictly pseudomonotone, hemicontinuous and a Z-mapping relative to $[0, \infty)$. We note that T is not monotone since

$$\langle x - y, Tx - Ty \rangle < 0$$

for $x = \frac{3}{2}\pi$, and $y = 0$.

We remark that in Hilbert space, the complementarity problem is also equivalent to fixed point problem. See, e.g., [49].

References

[1] Allen, G., *Variational inequalities, complementarity problems and duality theorems*, J. Math. Anal. Appl. 58, 1977, 1-10.

[2] Aubin, J. P., *Mathematical Methods of Game and Economic Theory*, revised ed., Stud. Math. Appl., 7, North-Holland, Amsterdam 1982.

[3] Baiocchi, C. and A. Capelo, *Variational and Quasivariational Inequalities : Application to Free Boundary Problems*, Wiley, New York 1984.

[4] Barbu, V. and Th. Precupanu, *Convexity and Optimization in Banach Spaces*, Sijthoff and Nourdhoff, Bucarest, Romania 1986.

[5] Bazarra, M. S., Goode, J. J. and M. Z. Nashed, *A nonlinear complementarity problem in mathematical programming in Banach space*, Proc. Am. Math. Soc. 35, 1972, 165-170.

[6] Berge, C., *Espace Topologiques - Functions Multivoques*, Dunod, Paris 1966.

[7] Blum, E. and W. Oettli, *From Optimization and Variational Inequalities to Equilibrium Problems*, The Mathematics Student 63, 1994, 123-145.

[8] Bourbaki, N., *Espaces Vectoriels Topologiques*, Hermann, Paris, 1966.

[9] Borwein, J. M., *Alterative theorems for general complementarity problems*, Lecture Notes in Economics and Mathematical Systems, Springer-Verlag, New York, 98, 1984, 328-331.

[10] Borwein, J. M., *Generalized linear complementarity problem treated without fixed point theory*, J. Optimization Theory Appl. 22, 1984, 343-356.

[11] Borwein, J. M. and S. Fitzpatrick, *Local boundedness of monotone operators under minimal hypotheses*, Bull. Aust. Math. Soc. 39, 1988, 439-441.

[12] Brézis, H., *Equations et inequations non linéaires dans les espaces vectoriels en dualité*, Ann. Inst. Fourier 18, 1968, 115-175.

[13] Brézis, H., Nirenberg L. and G. Stampacchia, *A remark on Ky Fan's minimax principle*, Boll. Unione Mat. Ital. (4) 6, 1972, 293-300.

[14] Browder, F., *Nonlinear elliptic boundary value problems*, Bull. Am. Math. Soc. 69, 1963, 862-874.

[15] Browder, F., *Nonlinear Operators and Nonlinear Equations of Evolution in Banach Spaces*, Proc. Symp. Pure Math. 18, Part 2 (the entire issue). Amer. Math. Soc., Providence, RI, 1976 (preprint version 1968).

[16] Browder, F., *Coincidence theorems, minimax theorems, and variational inequalities*, Contemporary Math. 26, 1984, 67-80.

[17] Chadli, O., Chbani, Z. and H. Riahi, *Equilibrium Problems with Generalized Monotone Bifunctions and Applications to Variational Inequalities*, J. Optimization Theory Appl. 105, 2000, 299-323.

[18] Chadli, O., Chbani, Z. and H. Riahi, *Recession Methods for Equilibrium Problems and Applications to Variational and Hemivariational Inequalities*, Discrete Contin. Dyn. Syst. 5, 1999, 185-195.

[19] Chadli, O., Chbani, Z. and H. Riahi, *Equilibrium problems and noncoercive variational inequalities*, Optimization 50, 2001, 17-27.

[20] Chan, D. and J. S. Pang, *The generalized quasi-variational inequality problem*, Math. Oper. Res. 7, 1982, 211-222.

[21] Chowdhury, M. and K. K. Tan, *Applications of pseudo-monotone operators with some kind of upper semicontinuity in generalized quasi-variational inequalities on non-compact sets*, Proc. Am. Math. Soc. 10, 1998, 2957-2968.

[22] Cottle, R. W., *Notes on a fundamental theorem in quadratic programming*, SIAM J. Appl. Math. 12, 1964, 663-665.

[23] Cottle, R. W., *Nonlinear programs with positively bounded Jacobians*, SIAM J. Appl. Math. 14, 1966, 147-158.

[24] Cottle, R. W., *Complementarity and variational problems*, Sympos. Math., Instituto Nazionale di Alta Mathematica, Volume XIX, 1976, 177-208.

[25] Cottle, R. W., *Numerical methods for complementarity problems in engineering and applied science*, Technical Report SOL 77-24, Systems Optimization Laboratory, Department of Operations Research, Stanford University, 1977.

[26] Cottle, R. W., Habetler, G. J. and C. E. Lemke, *Quadratic forms semi-definite over convex cones*, in Proceedings of the Princeton Symposium on Mathematical Programming, H. W. Kuhn, ed., Princeton University Press, Princeton, New Jersey, 1970, 551-565.

[27] Cottle, R. W. and J. C. Yao, *Pseudomonotone complementarity problems in Hilbert space*, J. Optimization Theory Appl. 75(2), 1992, 281-295.

[28] Cubiotti, P.,*General nonlinear variational inequalities with $(S)^1_+$ operators*, Appl. Math. Lett. 10 (2), 1997, 11-15.

[29] Cubiotti, P. and J. C. Yao, *Multivalued $(S)^1_+$ operators and generalized variational inequalities*, Comput. Math. Appl. 29 (12), 1995, 49-56.

[30] Cryer, C.W. and M. A. H. Dempster, *Equivalence of linear complementarity problems and linear program in vector lattice Hilbert spaces*, SIAM J. Control Optimization 18, 1980, 76-90.

[31] Dantzig, G. B. and R. W. Cottle, *Positive semi-definite programming*, in Nonlinear Programming, Abadie, ed., A course, North-Holland, Amsterdam, 1967, 55-73.

[32] Dash, A. T. and S. Nanda, *A complementarity problem in mathematical programming in Banach space*, J. Math. Anal. Appl. 98, 1984, 328-331.

[33] Deimling, K., *Nonlinear Functional Analysis*, Springer-Verlag, Berlin, 1985.

[34] Do, C., *Bifurcation theory for elastic plates subjected to unilateral conditions*, J. Math. Anal. Appl. 60, 1977, 435-448.

[35] Dorn, W. S., *Self-dual quadratic programs*, SIAM J. Appl. Math. 9(1), 1961, 51-54.

[36] Ekeland, I. and R. Temam, *Convex Analysis and Variational Problems*, North-Holland, Amsterdam, 1976.

[37] Fan, K., *A generalization of Tychonoff's fixed point theorem*, Math. Ann. 142, 1961, 305-310.

[38] Fang, S.C., *A linearization method for generalized complementarity problems*, IEEE Trans. Autom. Control 38, 1982, 363-383.

[39] Fang, S. C. and E. L. Peterson, *Generalized variational inequalities*, J. Optimization Theory Appl. 38, 1982, 363-383.

[40] Guo, J. S. and J. C. Yao, *Extension of strongly nonlinear quasivariational inequalities*, Appl. Math. Lett. 5(3), 1992, 35-38.

[41] Guo, J. S. and J. C. Yao, *Variational inequalities with nonmonotone operators*, J. Optimization Theory Appl. 80(1), 1994, 63-74.

[42] Habetler, G. J. and A. J. Price, *Existence theory for generalized nonlinear complementarity problems*, J. Optimization Theory Appl. 7, 1971, 223-239.

[43] Hadjisavvas, N. and S. Schaible, *Quasimonotone variational inequalities in Banach spaces*, J. Optimization Theory Appl. 90, 1996, 95-111.

[44] Hadjisavvas, N. and S. Schaible, *On strong pseudomonotonicity and (semi)strict quasimonotonicity*, J. Optimization Theory Appl. 79, 1993, 139-155.

[45] Harker, P. T. and J. S. Pang, *Finite-dimensional variational inequality and nonlinear complementarity problems : A survey of theory, algorithms and applications*, Math. Program., Series B, 48, 1990, 161-220.

[46] Hartmann, P. and G. Stampacchia, *On some nonlinear elliptic differential functional equations*, Acta Math. 115, 1966, 153-188.

[47] Holmes, R. B., *Geometric Functional Analysis and Its Applications*, Springer-Verlag, New York, 1975.

[48] Ingleton, A. W., *A problem in linear inequalities*, Proc. London Math. Soc. 16, 1966, 519-536.

[49] Isac, G., *Complementarity Problems*, Lecture Notes in Mathematics, 1528, Springer-Verlag, Berlin Heidelberg, 1992.

[50] Isac, G., *Nonlinear complementarity problem and Galerkin method*, J. Math. Anal. Appl. 108, 1985, 563-574.

[51] Isac, G. and M. S. Gowda, *Operators of class $(S)^1_+$, Altman's condition and the complementarity problem*, J. Fac. Sci., Univ. Tokyo, Sect. IA, Math. 40, 1993, 1-16.

[52] Isac, G. and M. Théra, *Complementarity problems and the existence of the post-critical equilibrium state of a thing elastic plate*, J. Optimization Theory Appl. 58, 1988, 241-257.

[53] Jou, C. R. and J. C. Yao, *Extension of generalized variational inequalities*, Appl. Math. Lett. 6, 1993, 21-25.

[54] Karamardian, S., *Duality in Mathematical Programming*, P.h.D. Thesis, University of California, Berkley, 1966.

[55] Karamardian, S., *Generalized complementarity problems*, J. Optimization Theory Appl. 8, 1971, 161-168.

[56] Karamardian, S., *Complementarity problems over cones with monotone and pseudomonotone maps*, J. Optimization Theory Appl. 18, 1976, 445-454.

[57] Karamardian, S., *An existence theorem for the complementarity problem*, J. Optimization Theory Appl. 19 (2), 1976, 227-232.

[58] Karamardian, S. and S. Schaible, *Seven kinds of monotone maps*, J. Optimization Theory Appl. 66, 1990, 37-46.

[59] Kelley, J. L. and I. Namioka, *Linear Topolical Spaces*, Springer, New York, 1976.

[60] Kinderlehrer, D. and G. Stampacchia, *An Introduction to Variational Inequalities and Their Applications*, Academic Press, New york, 1980.

[61] Kluge, R., *Nichtlineare Variationsungleichungen und Extremalaufgaben.* Verl. der Wiss., Berlin, 1979.

[62] Knaster, B., Kuratowski, C. and S. Mazukiewicz, *Ein bewies des fixpunktsatzes für n-dimensional simplexe*, Fund. Math. 14, 1929, 132-137.

[63] Kneser, H., *Sur un Théorème fondamental de la théorie des jeux*, C. R. Acad. Sci. Paris Sr. I Math. 234, 1952, 2418-2420.

[64] Lemke, C. E., *Bimatrix equilibrium points and mathematical programming*, Managment Sciences 11(7), 1965, 681-689.

[65] Lions, J. L. and G. Stampacchia, *Variational inequalities*, Comm. Pure Appl. Math. 20, 1967, 493-519.

[66] Luna, G., *A remark on the nonlinear complementarity problem*, Proc. Amer. Math. Soc. 48, 1975, 132-134.

[67] Mancino, O. and G. Stampacchia, *Convex programming and variational inequalities*, J. Optimization Theory Appl. 9, 1972, 3-23.

[68] Minty, G. J., *Monotone (non-linear) operators in Hilbert spaces*, Duke Math. J. 29, 1962, 341-346.

[69] Minty, G., *On a monotonicity method for the solution of non-linear equations in Banach spaces*, Proc. Natl. Acad. Sci. USA 50, 1963, 1038-1041.

[70] Megiddo, N. and M. Kojima, *On the existence and uniqueness of solutions in nonlinear complementarity theory*, Math. Program. 12, 1977, 110-130.

[71] Moré, J. J. and W. C. Rheinboldt, *On P- and S-functions and related classes of n-dimensional nonlinear mappings*, Linear Algebra Appl. 6, 1973, 45-68.

[72] Mosco, U., *A nonlinear complementarity problem in Banach spaces*, J. Math. Anal. Appl. 21, 1980, 351-356.

[73] Mosco, U., *Implicit Variational Problems and Quasi-Variational Inequalities*, Lecture Notes in Mathematics, 543, Springer, Berlin, 1976.

[74] Noor, M. A., *The quasi-complementarity problem*, J. Math. Anal. Appl. 130, 1988, 344-353.

[75] Parida, J. and A. Sen, *A variational-like inequality for multifuctions with applications*, J. Math. Anal. Appl. 924, 1987, 73-81.

[76] Phelps, R. R., *Convex Functions, Monotone Operators and Differentiability*. Lecture Notes in Mathematics, 1364, Springer-Verlag, 1993.

[77] Rheinboldt, W. C., *On M-functions and their application to nonlinear Gauss-Seidel iterations and network flows*, J. Math. Anal. Appl. 32, 1970, 274-307.

[78] Riddel, R. C., *Equivalence of nonlinear complementarity problems and least element problems in Banach lattices*, Math. Oper. Res. 6, 1981, 462-474.

[79] Saigal, R., *Extension of the generalized complementarity problem*, Math. Oper. Res. 1, 1976, 260-266.

[80] Schaible, S. and J. C. Yao, *On the equivalence of nonlinear complementarity problems and least-element problems*, Math. Program. 70, 1995, 191-200.

[81] Shih, M. H. and K. K. Tan, *Browder-Hartmann-Stampacchia variational inequalities for multi-valued monotone operators*, J. Math. Anal. Appl. 134, 1988, 431-440.

[82] Stampacchia, G., *Variational inequalities*, in Theory and Applications of Monotone Operators, A. Ghizzetti ed., Proceedings of the NATO Advanced Study Institute, Venice Italy (Edizioni Oderisi, Gubbio, Italy), 1968, 102-192.

[83] Théra, M., *Existence results for the nonlinear complementarity problem and applications to nonlinear analysis*, J. Math. Anal. Appl. 154, 1991, 572-584.

[84] Théra, M., *A note on the Hartmann-Stampacchia thoerem*, in Nonlinear Analysis and Applications, V. Lakshmikanthan ed., Dekker, New York, 1987, 573-577.

[85] Watson, P. J., *Variational inequalities in nonreflexive Banach spaces*, Appl. Math. Lett. 10 (2), 1997, 45-48.

[86] Watson, L. M., *Solving the nonlinear complementarity problem by a homotopy method*, SIAM J. Control Optim. 17, 1979, 36-46.

[87] Yao, J. C., *Variational inequalities with generalized monotone operators*, Math. Oper. Res. 19 (3), 1994, 691-705.

[88] Yao, J. C., *On the generalized complementarity problem*, J. Austral. Math. Soc. Ser. B 35, 1994, 420-428.

[89] Yao, J. C., *Existence of generalized variational inequalities*, Oper. Res. Lett. 15, 1994, 35-40.

[90] Yao, J. C., *Multi-valued variational inequalities with K-pseudomonotone operators*, J. Optimization Theory Appl. 83(2), 1994, 391-403.

[91] Yao, J. C., *Variational inequality*, Appl. Math. Lett. 5 (1), 1992, 39-42.

[92] Yao, J. C., *The generalized quasi-variational inequality problem with applications*, J. Math. Anal. Appl. 158(1), 1991, 139-160.

[93] Zeidler, E., *Nonlinear Functional Analysis and its Applications*, Vol II/B, Nonlinear Monotone Operators, Springer-Verlag, 1985.

Chapter 13

GENERALIZED MONOTONE EQUILIBRIUM PROBLEMS AND VARIATIONAL INEQUALITIES

Igor Konnov
Department of Applied Mathematics
Kazan State University
Kazan 420008
and
Informatics Problems Institute of AS RT
Kazan 420012
RUSSIA
ikonnov@ksu.ru

Abstract This chapter is devoted to equilibrium problems and variational inequalities under generalized monotonicity assumptions on cost functions. We present basic existence and uniqueness results of solutions both for scalar and for vector problems. Relationships between generalized monotonicity properties of cost functions of these problems are also considered. Moreover, we describe basic approaches to construct iterative solution methods, including their convergence properties.

Keywords: equilibrium problems, variational inequalities, generalized monotonicity.

1. Introduction

Let K be a nonempty subset of a topological space E and let $\Phi : K \times K \to \mathbb{R}$ be an *equilibrium bifunction*, i.e., $\Phi(x, x) = 0$ for all $x \in K$. Then one can define the *equilibrium problem* (EP) which is to find an element $x^* \in K$ such that

$$\Phi(x^*, y) \geq 0 \quad \forall y \in K. \tag{13.1}$$

To our knowledge, this formulation of EP was considered first by Nikaido and Isoda in [155] as a generalization of the Nash equilibrium problem in non-cooperative games. EP of the form (13.1) provides a general and convenient format to write and investigate not only various problems arising in applications, but also many well known general problems in Nonlinear Analysis, such as optimization, saddle point, fixed point, complementarity and variational inequality problems; see e.g. [156, 10, 22]. Among these problems, we should pay special attention to variational inequalities. We recall that the *variational inequality problem* (VI) is to find an element $x^* \in K$ such that

$$\exists g^* \in G(x^*), \langle g^*, y - x^* \rangle \geq 0 \quad \forall y \in K, \tag{13.2}$$

where G is a (multivalued) mapping from K into the dual space E', $\langle \cdot, \cdot \rangle$ denotes the duality pairing between E' and E. This problem (with a single valued mapping) was considered first by Fichera [63] and Stampacchia [178]. They introduced VI in order to study certain problems in Mathematical Physics. Such problems are described in detail in [56, 71]. Afterwards, a great number of direct applications of VIs were revealed in other fields, such as Economics, Transportation and Operations Research; e.g. see [84, 62, 68, 150]. On the one hand, VI seems to be merely a particular case of EP, especially in the single-valued case. In fact, it then suffices to set $\Phi(x, y) = \langle G(x), y - x \rangle$ in (13.1). On the other hand, it is possible to make the reverse transformation if $\Phi(x, \cdot)$ possesses certain (sub) differentiability properties. Generally speaking, VI (13.2) can be viewed as the differential form of EP (13.1). In addition, VI enjoys some additional nice properties, such as the linearity in y which enables one to simplify the derivation of certain results in theory and the construction of iterative solution methods in comparison with those for EP. Thus, we shall consider both problems in this chapter.

Traditionally, most works on various aspects of EPs and VIs were devoted to monotone problems. Note that both EP with monotone Φ and VI with monotone G possess certain additional nice properties which do not hold in the generalized monotone case. For instance, it is well known that the sum of any number of convex functions is convex, but the similar assertion is not true for quasiconvex (or pseudoconvex) functions. Analogously, the sum of quasimonotone (or pseudomonotone) mappings need not be quasimonotone (or, respectively, pseudomonotone) in general. Hence, studying and/or solving generalized monotone EPs and VIs meets certain additional difficulties in comparison with those in the monotone case. However, the monotonicity assumption turned out to be too restrictive for many applied problems, especially in Economics and Operations Research. During the 90's, generalized monotone EPs and VIs were being investigated rather extensively and great progress has been achieved in various aspects related to this field. In particular, various characterizations of generalized monotonicity were studied by many researchers; e.g. see [98, 108, 39, 38, 143, 170, 171, 172] and references therein. This assertion also applies to a great number of extensions of the usual EPs and VIs, in particular, to vector EPs and VIs. In the vector case, the inequalities (13.1) and (13.2) should be considered not in $\mathbb{R}$, but in some ordered vector space. Vector VIs are closely related with vector optimization problems; see e.g. [33, 140, 88, 68, 67]. Giannessi [65] first introduced the vector variational inequality in finite-dimensional spaces. Vector EPs can be also regarded as an extension of vector VIs. However, vector EPs are also originated in non-cooperative games where scalar utility functions are replaced with either preference relations [61] or vector functions [20] and in saddle point theory for multicriteria optimization; see e.g. [184] and [163, Chapter 4]. To our knowledge, the general vector equilibrium problem was considered first by Konnov [113].

In this chapter, we intend to present the basic results in the theory and the construction of solution methods for EPs and VIs under generalized monotonicity assumptions. We do not attempt to describe the most general form of results. We aim to explain the essence of the main approaches, so that any interested reader can evaluate the modern state of this promising field as a whole. For the same reason, we give somewhat weakened versions of several results in comparison with the original ones.

2. Equilibrium Problems

We begin our considerations from the general EP of the form (13.1). Throughout this section, we suppose that the following basic assumptions hold:

(A1) *K is a nonempty, convex and closed subset of a real topological vector space E;*

(A2) *Φ is an equilibrium bifunction.*

We denote by K^e the set of solutions of EP (13.1).

2.1 General existence results for EPs

The existence results of solutions for EPs are traditionally based on suitable fixed point type theorems. Ky Fan [57, Lemma 1] presented a so-called matching theorem, which turned out to be especially helpful and fruitful in deriving various existence results for equilibrium type problems. This result, which is also known as Ky Fan's Lemma, is an infinite-dimensional extension of the classical Knaster-Kuratowski-Mazurkiewicz theorem [106]. We give a somewhat more general variant of Ky Fan's Lemma from [59, Corollary 1]. As usual, for a set A, we denote by conv A its convex hull.

Proposition 13.1 *Let X be an arbitrary set in a topological vector space Y. To each $x \in X$, let a closed set $Q(x)$ in Y be given such that, for every finite subset $\{y^1, \dots, y^n\}$ of X, one has*

$$\operatorname{conv}\{y^1, \dots, y^n\} \subseteq \bigcup_{i=1}^{n} Q(y^i). \tag{13.3}$$

If $Q(x)$ is compact for at least one $x \in X$, then

$$\bigcap_{x \in X} Q(x) \neq \emptyset.$$

We extract the basic property of the mapping Q from Proposition 13.1 in a separate definition.

Definition 13.1 (e.g. see [78]) Let Q be a multivalued mapping from a set X into a set Y, $X \subseteq Y$. The mapping Q is said to be a *KKM-map*, if, for each finite subset $\{y^1, \dots, y^n\}$ of X, (13.3) holds.

Thus, each KKM-map, under the additional compactness assumption, possesses the finite intersection property. In order to obtain existence results for EP we need to recall convexity and continuity properties of scalar functions.

Definition 13.2 (e.g. see [147, 9]) Let $f : K \to \mathbb{R}$ be a function. The function f is said to be

(a) *convex*, if for each pair of points $x', x'' \in K$ and for all $\alpha \in [0, 1]$, we have

$$f(\alpha x' + (1 - \alpha)x'') \leq \alpha f(x') + (1 - \alpha)f(x'');$$

(b) *quasiconvex*, if for each pair of points $x', x'' \in K$ and for all $\alpha \in [0, 1]$, we have

$$f(\alpha x' + (1 - \alpha)x'') \leq \max\{f(x'), f(x'')\};$$

(c) *semistrictly quasiconvex*, if for each pair of points $x', x'' \in K$ such that $f(x') \neq f(x'')$ and for all $\alpha \in (0, 1)$, we have

$$f(\alpha x' + (1 - \alpha)x'') < \max\{f(x'), f(x'')\};$$

(d) *explicitly quasiconvex*, if it is both quasiconvex and semistrictly quasiconvex;

(e) *lower* (respectively, *upper*) *semicontinuous*, if for each sequence $\{x^k\} \to \bar{x}$, we have

$$\liminf_{k\to\infty} f(x^k) \geq f(\bar{x})$$

(respectively, $\limsup_{k\to\infty} f(x^k) \leq f(\bar{x})$).

From the definitions one can conclude that the following implications hold:

$$(a) \Longrightarrow (d) \Longrightarrow (c) \Longrightarrow (b).$$

Moreover, each lower semicontinuous and semistrictly quasiconvex function is clearly explicitly quasiconvex. We are now ready to establish a general existence result for EP (13.1), which is also known as Ky Fan's inequality (see [58]).

Theorem 13.1 *Suppose that the following assumptions hold:*
(a) K *is compact;*
(b) $\Phi(x,\cdot)$ *is quasiconvex for each* $x \in K$*;*
(c) $\Phi(\cdot,y)$ *is upper semicontinuous for each* $y \in K$.
Then EP (13.1) is solvable.

The proof is based on verifying the fact that the set-valued mapping $A : K \to 2^K$, defined by

$$A(y) = \{x \in K \mid \Phi(x,y) \geq 0\}, \tag{13.4}$$

satisfies all the conditions of Proposition 13.1. Clearly, $A(y)$ is compact since $\Phi(\cdot,y)$ is upper semicontinuous and K is compact. Next, from the quasiconvexity of $\Phi(x,\cdot)$ it follows that A is KKM. Hence,

$$K^e = \bigcap_{y\in K} A(y) \neq \emptyset$$

due to Proposition 13.1, as desired.

By using a coercivity condition, we can obtain an existence result on unbounded sets; see [24, Theorem 1] and also [149].

Theorem 13.2 *Suppose that there exist a compact subset* L *of* E *and a point* $\tilde{y} \in L\bigcap K$ *such that*

$$\Phi(x,\tilde{y}) < 0 \quad \forall x \in K\backslash L. \tag{13.5}$$

If, in addition, assumptions (b) and (c) of Theorem 13.1 hold, then EP (13.1) is solvable.

In fact, noticing that $A(\tilde{y}) \subseteq L$ because of (13.4) and (13.5), we conclude that $A(\tilde{y})$ is compact and the assertion of this theorem becomes true due to the same argument as that in the proof of Theorem 13.1.

Coercivity condition (13.5) can be somewhat relaxed; see e.g. [1, Theorem 2]. Nevertheless, the combination of the upper semicontinuity of $\Phi(\cdot, y)$ and the compactness of K (or L) in the same topology seems too restrictive in an infinite-dimensional space setting regardless of the choice of any topology in E. One of the most popular approaches to overcome these difficulties consists in employing monotonicity type properties of Φ.

2.2 Generalized monotonicity and dual equilibrium problems

We first recall the definitions of (generalized) monotonicity properties for bifunctions.

Definition 13.3 [10, 149, 22, 18] A bifunction $\Psi : K \times K \to \mathbb{R}$ is said to be

(a) *monotone*, if for all $x, y \in K$, we have

$$\Psi(x,y) + \Psi(y,x) \leq 0;$$

(b) *pseudomonotone*, if for all $x, y \in K$, we have

$$\Psi(x,y) \geq 0 \Longrightarrow \Psi(y,x) \leq 0;$$

(c) *strictly pseudomonotone*, if for all $x, y \in K, x \neq y$, we have

$$\Psi(x,y) \geq 0 \Longrightarrow \Psi(y,x) < 0;$$

(d) *quasimonotone*, if for all $x, y \in K$, we have

$$\Psi(x,y) > 0 \Longrightarrow \Psi(y,x) \leq 0;$$

(e) *strictly quasimonotone*, if Ψ is quasimonotone and for all $x, y \in K$, $x \neq y$, there exists $z \in (x, y)$ such that

$$\text{either } \Psi(z,y) \neq 0 \quad \text{or } \Psi(z,x) \neq 0.$$

From the definitions one can conclude that the following implications hold:

$$\begin{array}{ccccc} (a) & \Longrightarrow & (b) & \Longrightarrow & (d) \\ & & \Uparrow & & \Uparrow \\ & & (c) & \Longrightarrow & (e) \end{array}$$

Let us consider the so-called *dual equilibrium problem* (DEP) which is to find an element $y^* \in K$ such that

$$\Phi(x, y^*) \leq 0 \quad \forall x \in K \tag{13.6}$$

(e.g. see [125]). We denote by K_d^e the set of solutions of this problem. To our knowledge, the pair of problems (13.1) and (13.6) was considered first in [12, p.88] as an equal two-person game. Taking into account the fact that Φ is an equilibrium bifunction, it is easy to see that points $x^* \in K$ and $y^* \in K$ are solutions to EP (13.1) and DEP (13.6), respectively, if and only if they constitute a saddle point of Φ, i.e.,

$$\Phi(x, y^*) \leq \Phi(x^*, y^*) \leq \Phi(x^*, y) \quad \forall x \in K, \forall y \in K. \tag{13.7}$$

Equivalently, employing the usual duality principle for saddle point problems (see [153]), we conclude that EP (13.1) coincides with the primal problem:

$$\sup_{x \in K} \inf_{y \in K} \Phi(x, y),$$

whereas DEP (13.5) coincides with its dual:

$$\inf_{y \in K} \sup_{x \in K} \Phi(x, y).$$

Note that Φ need not be quasiconcave in x, so that we cannot apply the well-known minimax theorems (see [153, 107]) in order to establish the existence of saddle points in (13.7). Nevertheless, the quasiconcavity of $\Phi(\cdot, y)$ can be replaced by generalized monotonicity of Φ. First we give relationships between the solution sets of the primal and dual problems.

Definition 13.4 A function $f : K \to \mathbb{R}$ is said to be

(a) *u-hemicontinuous*, if its restriction on line segments of K is upper semicontinuous;

(b) *hemicontinuous*, if its restriction on line segments of K is continuous.

Remark 13.1 *Throughout this paper we use the simplest definitions of hemicontinuity type properties (see e.g. [83]). At the same time, there exist somewhat different definitions of the same properties (see [125]). For instance, a function $f : K \to \mathbb{R}$ is called u-hemicontinuous, if for all $x, y \in K$, the function $\tau(\alpha) = f(\alpha x + (1 - \alpha)y)$, defined on $[0, 1]$, is upper semicontinuous at 0^+. It is not too hard to verify that both the definitions present the same class of functions.*

Suppose that $\Phi(x, \cdot)$ is explicitly quasiconvex for each $x \in K$ and $\Phi(\cdot, y)$ is u-hemicontinuous and choose arbitrary $x^* \in K_d^e$ and $y \in K$.

Set $x_\alpha = \alpha y + (1-\alpha)x^*$ for $\alpha \in [0,1]$. If $\Phi(x_\alpha, y) < \Phi(x_\alpha, x^*)$ for some $\alpha \in (0,1)$, then, by explicit quasiconvexity, we have

$$0 = \Phi(x_\alpha, x_\alpha) < \max\{\Phi(x_\alpha, y), \Phi(x_\alpha, x^*)\} = \Phi(x_\alpha, x^*),$$

i.e. $x^* \notin K_d^e$, a contradiction. Hence

$$0 = \Phi(x_\alpha, x_\alpha) \leq \max\{\Phi(x_\alpha, y), \Phi(x_\alpha, x^*)\} = \Phi(x_\alpha, y) \tag{13.8}$$

for each $\alpha \in (0,1)$. Taking the limit $\alpha \to 0$ in this inequality we obtain $\Phi(x^*, y) \geq 0$, i.e. $x^* \in K^e$, which in turn gives $K_d^e \subseteq K^e$. The reverse inclusion is clearly true if Φ is pseudomonotone. Thus, we have proven the basic relationship between EP and DEP, which is also known as Minty's Lemma (see [10, Section 10.1]) and can be formulated precisely as follows; e.g. see [18].

Lemma 13.1 *(i) If Φ is pseudomonotone, then $K^e \subseteq K_d^e$.*

(ii) If $\Phi(x, \cdot)$ is explicitly quasiconvex for each $x \in K$ and $\Phi(\cdot, y)$ is u-hemicontinuous for each $y \in K$, then $K_d^e \subseteq K^e$.

Thus, under the assumptions of Lemma 13.1, the existence of solutions to either EP or DEP implies that there exists a saddle point in (13.7) and that the optimal solutions of both problems coincide.

2.3 Existence and uniqueness results for pseudomonotone EPs

Taking the results above as a basis, we now deduce existence and uniqueness results for EP (13.1) with Φ being pseudomonotone. If we suppose that either K is compact or the coercivity condition (13.5) holds, then, following the proofs of Theorems 13.1 and 13.2, we conclude that

$$\bigcap_{y \in K} \overline{A(y)} \neq \emptyset,$$

if $\Phi(x, \cdot)$ is quasiconvex. Set

$$B(y) = \{x \in K \mid \Phi(y, x) \leq 0\}. \tag{13.9}$$

Then, the pseudomonotonicity of Φ and the lower semicontinuity of $\Phi(x, \cdot)$ imply that

$$\emptyset \neq \bigcap_{y \in K} \overline{A(y)} \subseteq \bigcap_{y \in K} \overline{B(y)} = \bigcap_{y \in K} B(y) = K_d^e,$$

i.e., DEP (13.6) is solvable. Applying now Lemma 13.1 (ii), we obtain the following existence result for EPs.

Theorem 13.3 *[24, 18] Suppose that the following assumptions hold:*

(a) either K is compact or there exist a compact subset L of E and a point $\tilde{y} \in L \bigcap K$ such that (13.5) holds;

(b) $\Phi(x, \cdot)$ is semistrictly quasiconvex and lower semicontinuous for each $x \in K$;

(c) $\Phi(\cdot, y)$ is u-hemicontinuous for each $y \in K$;

(d) Φ is pseudomonotone.

Then, EP (13.1) and DEP (13.6) are solvable.

Note that assumption (b) of Theorem 13.1 is replaced with somewhat stronger one in Theorem 13.3, but assumption (c) of Theorems 13.1 and 13.2 is stronger essentially than that of Theorem 13.3.

The set of solutions of DEP possesses additional useful properties, which turn out to be true for EP in the pseudomonotone case. The following lemma is a slight modification of the corresponding results from [18] and [117, Proposition 5.2].

Lemma 13.2 *(i) If $\Phi(x, \cdot)$ is quasiconvex for each $x \in K$, then K_d^e is convex.*

(ii) If $\Phi(x, \cdot)$ is lower semicontinuous for each $x \in K$, then K_d^e is closed.

Combining Lemmas 13.1, 13.2 and Theorem 13.3 yields the following.

Corollary 13.1 *Suppose that assumptions (a) – (d) of Theorem 13.3 hold. Then the sets of solutions of EP (13.1) and DEP (13.6) coincide and are nonempty, convex and compact.*

Next, it is easy to see that strict pseudomonotonicity implies the uniqueness of solutions for EP, as the following proposition states.

Proposition 13.2 *Suppose Φ is strictly pseudomonotone. Then EP (13.1) has at most one solution.*

Therefore, if assumptions (a) – (c) of Theorem 13.3 hold and Φ is strictly pseudomonotone, then both EP (13.1) and DEP (13.6) have a unique and common solution.

2.4 Existence and uniqueness results for quasimonotone EPs

The approach above can be also applied to quasimonotone EPs. Although the implication $A(y) \subseteq B(y)$ does not hold if Φ is quasimonotone, it was shown in [82] that either EP (13.1) is solvable or the key inclusion

$$\bigcap_{y \in \mathrm{rai} K} \overline{A(y)} \subseteq \bigcap_{y \in \mathrm{rai} K} B(y)$$

holds under the assumptions of Theorem 13.3 with replacing pseudomonotonicity and u-hemicontinuity by quasimonotonicity and hemicontinuity, respectively, if, in addition, the relative interior of K, denoted by rai K, is nonempty. Therefore, we can apply Ky Fan's Lemma in order to establish an existence result in the quasimonotone case as well.

Theorem 13.4 *(see [18, Theorem 3.1] and [82, Theorem 2.1]) Suppose that the following assumptions hold:*

(a) either K is compact or there exist a compact subset L of E and a point $\tilde{y} \in L \bigcap K$ such that (13.5) holds;

(b) $\Phi(x, \cdot)$ is semistrictly quasiconvex and lower semicontinuous for each $x \in K$;

(c) $\Phi(\cdot, y)$ is hemicontinuous for each $y \in K$;

(d) Φ is quasimonotone;

(e) $\mathrm{rai} K \neq \emptyset$.

Then EP (13.1) is solvable.

Although the assertions of Corollary 13.1 are not true in the quasimonotone case, it is possible to obtain their weakened analogues and a uniqueness result for the set of non-trivial solutions

$$K_n^e = \{x^* \in K^e \mid \exists y \in K, \quad \Phi(x^*, y) > 0\}.$$

Proposition 13.3 *[18, Theorem 4.1] Suppose that Φ is quasimonotone, $-\Phi(x, \cdot)$ is semistrictly quasiconvex for each $x \in K$, and $\Phi(\cdot, y)$ is hemicontinuous for $y \in K$. Then:*

(i) $\mathrm{conv} K_n^e \subseteq K^e$*;*

(ii) the closure of K_n^e is contained in K^e;

(iii) if, in addition, Φ is strictly quasimonotone, then K_n^e has at most one element.

2.5 Further generalizations

The results above can be generalized in several directions. In particular, it is possible to consider similar duality relationships between EP (13.1) and another DEP with some different bifunction $\Psi : K \times K \rightarrow \mathbb{R}$. That is, we can define the problem of finding an element $y^* \in K$ such that

$$\Psi(x, y^*) \leq 0 \quad \forall x \in K$$

and introduce analogues of the (generalized) monotonicity properties which involve two, rather than one, bifunctions. Then, following the proofs of the results above, we can obtain their various extensions. This approach turned out to be very fruitful, especially in deriving existence results for vector problems; see [21, 157, 158, 82, 125].

Next, it is possible to replace the usual continuity, quasiconvexity and generalized monotonicity properties with more general formulations in terms of the corresponding level sets. For instance, the assertions of Theorems 13.3 and 13.4 remain valid, if we replace assumptions (b) and (c) with the following:

(b') if $\Psi(x, y) < 0$ and $\Psi(x, z) \leq 0$, then

$$\Phi(x, \alpha y + (1-\alpha)z) < 0 \quad \text{for all } \alpha \in (0,1), \quad \text{and for all } x, y, z \in K;$$

the set $B(x)$, defined by (13.9), is closed for each $x \in K$;

(c') the set $\{u \in [x, z] \mid \Phi(u, y) \leq 0\}$ is closed for all $x, y, z \in K$;

see [82]. This approach was developed by many authors; e.g. see [198, 54, 158, 195, 30, 55, 17] and references therein. Of course, assumptions (b') and (c') are more general, but assumptions (b) and (c) seem to be more convenient for their verification. Nevertheless, we can choose the most suitable set of conditions for each specific problem under consideration.

It should be also noted that there exist a number of various coercivity conditions which modify and specify the condition (13.5) for various settings (e.g. see [1, 21, 22, 138, 195] and references therein), thus allowing to remove the explicit compactness assumption.

3. Variational Inequalities with Single-Valued Mappings

In this section, we consider VI (13.2) in the case where the cost mapping G is single-valued. Then VI (13.2) can be equivalently rewritten as follows: Find an element $x^* \in K$ such that

$$\langle G(x^*), y - x^* \rangle \geq 0 \quad \forall y \in K, \tag{13.10}$$

where $G : K \rightarrow E'$ is a given mapping. Throughout this section, we shall suppose that

(B1) *K is a nonempty, convex and closed subset of a real Banach space E.*

We restrict ourselves to a Banach space setting for the sake of simplicity.

3.1 General existence results and dual VIs

Although the first existence results for single-valued VIs were based on the Brouwer fixed point theorem (see [85, 139]), we shall obtain similar results directly from the existence results for EPs described in Section 2. In fact, letting Φ to be defined by

$$\Phi(x, y) = \langle G(x), y - x \rangle, \tag{13.11}$$

we see that EP (13.1) coincides with VI (13.10), moreover, $\Phi(x, \cdot)$ is now clearly affine and continuous. We need only to specify some continuity properties for mappings.

Definition 13.5 Let Y be a real topological vector space. A mapping $Q : K \to Y$ is said to be

(a) *hemicontinuous*, if Q is continuous on line segments of K;

(b) *w-hemicontinuous*, if $Y = E'$ and the restriction of Q on line segments of K is continuous with respect to the weak* topology on E';

(c) *u-hemicontinuous*, if $Y = E'$ and the restriction of Q on line segments of K is upper semicontinuous with respect to the weak* topology on E'.

Theorem 13.5 *Suppose that the following assumptions hold:*

(a) either K is compact or there exist a compact subset L of E and a point $\tilde{y} \in L \bigcap K$ such that

$$\langle G(x), \tilde{y} - x \rangle < 0 \quad \forall x \in K \backslash L; \tag{13.12}$$

(b) G is a continuous mapping from the strong topology in E to the weak topology in E'.*

Then VI (13.10) is solvable.

The proof follows directly from Theorems 13.1 and 13.2, since $\Phi(\cdot, y)$ in (13.11) is now upper semicontinuous. For further modifications and extensions of this theorem see e.g. [154, 104, 1, 10, 7, 22]. Again, in order to weaken the assumptions of Theorem 13.5, we make use of monotonicity type properties for G.

Definition 13.6 [91, 159, 96, 97, 80] A mapping $Q : K \to E'$ is said to be

(a) *monotone*, if for all $x, y \in K$, we have

$$\langle Q(x) - Q(y), x - y \rangle \geq 0;$$

(b) *pseudomonotone*, if for all $x, y \in K$, we have

$$\langle Q(y), x - y \rangle \geq 0 \Longrightarrow \langle Q(x), x - y \rangle \geq 0;$$

(c) *strictly pseudomonotone*, if for all $x, y \in K$, $x \neq y$, we have

$$\langle Q(y), x - y \rangle \geq 0 \Longrightarrow \langle Q(x), x - y \rangle > 0;$$

(d) *quasimonotone*, if for all $x, y \in K$, we have

$$\langle Q(y), x - y \rangle > 0 \Longrightarrow \langle Q(x), x - y \rangle \geq 0;$$

(e) *strictly quasimonotone*, if it is quasimonotone and for all distinct $x, y \in K$ there exists $z \in (x, y)$ such that

$$\langle Q(z), x - y \rangle \neq 0.$$

From the definitions we obtain the following implications:

$$\begin{array}{ccccc} (a) & \Longrightarrow & (b) & \Longrightarrow & (d) \\ & & \Uparrow & & \Uparrow \\ & & (c) & \Longrightarrow & (e) \end{array}$$

It is easy to see that Definition 13.3 reduces to Definition 13.6 if Φ is defined by (13.11). Now, by analogy with DEP (13.6), we can define the *dual variational inequality problem* (DVI) which is to find an element $x^* \in K$ such that

$$\langle G(y), y - x^* \rangle \geq 0 \quad \forall y \in K. \tag{13.13}$$

Clearly, (13.6) with Φ being defined by (13.11) coincides with (13.13). We denote by K^* and K^*_d the sets of solutions of problems (13.10) and (13.13), respectively. First, from the definition of DVI (13.13) we obtain the following immediately.

Lemma 13.3 *The set of solutions to DVI (13.13) is convex and closed.*

Next, applying (13.11) in Lemma 13.1, we obtain some relationship between VI and DVI, which is also known as Minty's Lemma; see [148, 95].

Lemma 13.4 *(i) If G is pseudomonotone, then $K^* \subseteq K^*_d$.*
(ii) If G is u-hemicontinuous, then $K^*_d \subseteq K^*$.

The proof of part (ii) needs additional explanations. In fact, using the same argument as in the proof of Lemma 13.1 (ii) we obtain (13.8). On account of (13.11), it means that

$$0 \leq \langle G(x_\alpha), y - x_\alpha \rangle = (1 - \alpha) \langle G(x_\alpha), y - x^* \rangle$$

for each $\alpha \in (0, 1)$, where $x_\alpha = \alpha y + (1 - \alpha)x^* \in K$ and $x^* \in K_d^*$. Hence,

$$\langle G(x_\alpha), y - x^* \rangle \geq 0$$

and, by u-hemicontinuity, taking the limit $\alpha \to 0$ in this inequality yields

$$\langle G(x^*), y - x^* \rangle \geq 0,$$

i.e., $x^* \in K$, as desired.

On account of the assertions of Minty's Lemma, the problem (13.13) is also called the *Minty variational inequality*; see e.g. [66, 109].

3.2 Existence and uniqueness results for generalized monotone VIs

By using the representation (13.11) and the assertions of Lemmas 13.3 and 13.4, we can obtain existence results of solutions for VI and DVI as analogues of Theorems 13.3 and 13.4, and Corollary 13.1, the weak topology being chosen in E.

Theorem 13.6 *(see [191]) Suppose that the following assumptions hold:*

(a) either K is weakly compact, or there exists a weakly compact subset L of E and a point $\tilde{y} \in L \bigcap K$ such that (13.12) holds;

(b) G is u-hemicontinuous;

(c) G is pseudomonotone.

Then the sets of solutions to VI (13.10) and DVI (13.13) coincide and are nonempty, convex and weakly compact.

In fact, following the proof of Theorem 13.3 and using (13.11), we see that $K_d^* \neq \emptyset$. Applying now Lemma 13.4 (ii) gives $K_d^* = K^*$, hence $K^* \neq \emptyset$. On account of Lemma 13.3, we now see that both solution sets are convex and weakly compact.

Thus, the pseudomonotonicity assumption of G enables us to replace the continuity assumption (b) of Theorem 13.5 with the u-hemicontinuity of G. Similarly, applying (13.11) in Proposition 13.2 gives a uniqueness result.

Proposition 13.4 *Suppose G is strictly pseudomonotone. Then VI (13.10) has at most one solution.*

We now specify the result of Theorem 13.4 to VI (13.10) with G being quasimonotone.

Theorem 13.7 *Suppose that the following assumptions hold:*

(a) either K is weakly compact, or there exist a weakly compact subset L of E and a point $\tilde{y} \in L \bigcap K$ such that (13.12) holds;
(b) G is w-hemicontinuous;
(c) G is quasimonotone;
(d) $\mathrm{rai} K \neq \emptyset$.
Then VI (13.10) is solvable.

In fact, assumptions (a), (b), (d) and (e) of Theorem 13.4 with Φ being defined by (13.11) are now satisfied. Moreover, w-hemicontinuity of G also implies that $\Phi(\cdot, y)$ is hemicontinuous, and the result follows. The result can be viewed as a modification of those in [81].

Recently, Luc [142] strengthened the existence results from Theorems 13.6 and 13.7. He showed that it is sufficient for G to possess the corresponding generalized monotonicity properties on a dense subset of K.

Generalized monotonicity of a separable product of mappings was studied in [40, 41]. Several specializations of generalized monotonicity concepts for VIs defined on a product sets were suggested in [15, 124]. In [124], the corresponding existence results, which are based on applying Ky Fan's Lemma, are also given.

From the considerations above it is clear that the existence of solutions to the initial VI (13.10) is closely related with that of the dual problem (13.13) even regardless of generalized monotonicity properties of the cost mapping. Besides, the existence of solutions to DVI deserves a separate consideration due to its essential role in the convergence theory of iterative methods (see [117] and references therein). In fact, it is not so easy to verify condition (13.13) directly in many cases. It is more suitable to find generalized monotonicity properties which guarantee for DVI (13.13) to be solvable. From Theorem 13.6 we conclude that pseudomonotonicity is sufficient if assumptions (a) and (b) hold. The problem is to weaken this sufficient condition.

Definition 13.7 (see [80, 49]) A mapping $Q : K \rightarrow E'$ is said to be
(a) *explicitly quasimonotone*, if it is quasimonotone and for any distinct $x, y \in K$ the following implication holds:

$$\langle Q(y), x - y \rangle > 0 \Longrightarrow \exists z \in (0.5(x+y), x), \langle Q(z), x - y \rangle > 0;$$

(b) *properly quasimonotone*, if for every finite set $\{x^1, \dots, x^n\} \subseteq K$ and every $y \in \mathrm{conv}\{x^1, \dots, x^n\}$, there exists i such that

$$\langle Q(x^i), y - x^i \rangle \leq 0.$$

Note that any mapping satisfying property (a) is called semistrictly quasimonotone in [80], but we use the name "explicit quasimonotonicity", since it corresponds to the original definition of explicit quasiconvexity (see (d) and (c) in Definition 13.2).

It is easy to see that each properly quasimonotone mapping is quasimonotone. Moreover, it was shown in [47] that each explicitly quasimonotone mapping is properly quasimonotone. On the other hand, it was shown in [121] that each pseudomonotone mapping is explicitly quasimonotone and that each affine quasimonotone mapping is explicitly quasimonotone. Thus, explicit and proper quasimonotonicity can be viewed as intermediate concepts between pseudo- and quasi- monotonicity ones. Various properties of affine generalized monotone maps were investigated in [76, 77, 98, 170, 172, 42, 44].

We now give an existence result for DVI (13.13) under explicit quasimonotonicity. This result was proved first in [121] under several additional assumptions and afterwards strengthened in [47].

Theorem 13.8 *Suppose that the following assumptions hold:*

(a) either K is weakly compact or there exist a weakly compact subset L of E and a point $\tilde{y} \in L \bigcap K$ such that

$$\langle G(\tilde{y}), \tilde{y} - x \rangle < 0 \quad \forall x \in K \backslash L;$$

(b) G is explicitly quasimonotone.

Then DVI (13.13) is solvable.

Furthermore, it was shown in [47] that the assertion of Theorem 13.8 remains valid if (b) is replaced by

(*b′*) G is properly quasimonotone.

However, this result cannot be extended to the quasimonotone case. The corresponding counter-example was given in [121, Example 6.2]. In addition, it was shown in [90] that given a mapping is properly quasimonotone, if and only if DVI is solvable on *each* nonempty, convex and compact subset K of a finite-dimensional Euclidean space. On the other hand, the cost mapping in solvable DVI need not be even quasimonotone; see e.g. [115, Example 2.1]. Therefore, the class of quasimonotone mappings and the class of mappings which provide for DVI (13.13) to be solvable under certain coercivity assumptions have nonempty intersection, but they do not contain each other.

3.3 Variational inequalities and related problems

We now discuss briefly relationships between VIs with single-valued cost mapping and other general problems of Nonlinear Analysis. We consider ways of transformation of such problems to VI and generalized monotonicity properties of cost mappings of the transformed problem.

For instance, it is well known that VI (13.10) with $K = E$ reduces to the nonlinear (operator) equation:

$$G(x^*) = 0.$$

Next, if K is a cone in E, then VI (13.10) is equivalent to the *nonlinear complementarity problem (NCP)*: Find x^* such that

$$x^* \in K, \quad G(x^*) \in K', \quad \langle G(x^*), x^* \rangle = 0;$$

where $K' = \{q \in E' \mid \langle q, x \rangle \geq 0 \quad \forall x \in K\}$ is the conjugate cone to K; see [94]. Hence, we can use all the above results of this section to study these problems.

We now turn to the well-known *optimization problem* which is to find an element $x^* \in K$ such that

$$f(x^*) \leq f(x) \quad \forall x \in K, \tag{13.14}$$

where $f : E \to \mathbb{R} \bigcup \{\infty\}$ is a given function. First we recall an additional generalized convexity property for differentiable functions.

Definition 13.8 (see [147, 9]) Let $f : E \to \mathbb{R}$ be a differentiable function. The function f is said to be *pseudoconvex* on K, if for each pair of points $x', x'' \in K$, we have

$$\langle \nabla f(x''), x' - x'' \rangle \geq 0 \Longrightarrow f(x') \geq f(x'').$$

Here and below ∇f denotes the gradientmap of f.

From Definitions 13.2 and 13.8 it follows that pseudoconvexity implies quasiconvexity, and that each convex differentiable function is pseudoconvex. The following proposition collects relationships between convexity and monotonicity type properties for functions and their gradientmaps, respectively.

Proposition 13.5 *(see [9, 96, 80, 170]) Let $f : E \to \mathbb{R}$ be a differentiable function. The function f is convex (respectively, pseudoconvex, explicitly quasiconvex, quasiconvex) if and only if its gradientmap ∇f is monotone (respectively, pseudomonotone, explicitly quasimonotone, quasimonotone).*

Note that the gradientmaps of generalized convex functions in fact possess stronger monotonicity type properties, such as proper quasimonotonicity in the quasiconvex case and cyclical monotonicity in the convex case; see [159, 164, 49, 50].

Now we give the well-known differentiable optimality condition for optimization problem (13.14).

Theorem 13.9 *(see [147]) Suppose $f : E \to \mathbb{R}$ is a differentiable function.*

(i) Problem (13.14) implies VI (13.10) with $G = \nabla f$.

(ii) If f is pseudoconvex, then VI (13.10) with $G = \nabla f$ implies problem (13.14).

Thus, the optimization problem (13.14) with f being generalized convex can be regarded as a particular case of VI (13.10) with G being generalized monotone. Some relationships between problem (13.14) and DVI (13.13) were presented in [89, 109].

It has been mentioned that EP (13.1) with Φ being defined by (13.11) becomes equivalent to VI (13.10). We now consider the conditions which admit the reverse transformation. To this end, we suppose that $\Phi(x, \cdot)$ is differentiable for each $x \in K$ and set

$$G(x) = \nabla_y \Phi(x, y)|_{y=x}. \tag{13.15}$$

This mapping was introduced by Rosen in [167]. If Φ is an equilibrium bifunction, each solution x^* of EP (13.1) also solves the minimization problem (13.14), where $f(y) = \Phi(x^*, y)$. Applying now Theorem 13.9, we obtain optimality conditions for EP (13.1).

Theorem 13.10 *(see, e.g. [117]) Suppose that $\Phi : K \times K \to \mathbb{R}$ is an equilibrium bifunction, $\Phi(x, \cdot)$ is differentiable for each $x \in K$. Then:*

(i) EP (13.1) implies VI (13.10) with G being defined by (13.15);

(ii) if, in addition, $\Phi(x, \cdot)$ is pseudoconvex for each $x \in K$, then EP (13.1) is equivalent to VI (13.10) with G being defined by (13.15), and DEP (13.6), (13.15) implies DVI (13.13).

We now also give the relationship between generalized monotonicity properties for bifunctions and mappings (see [117, Proposition 2.2] and [122, Proposition 2.1.17]).

Proposition 13.6 *Suppose that $\Phi : K \times K \to \mathbb{R}$ is an equilibrium bifunction, $\Phi(x, \cdot)$ is convex and differentiable for each $x \in K$. If Φ is monotone (respectively, pseudomonotone, strictly pseudomonotone, quasimonotone, strictly quasimonotone, explicitly quasimonotone), then the mapping G, defined by (13.15), is also monotone (respectively, pseudomonotone, quasimonotone, strictly quasimonotone, explicitly quasimonotone).*

Thus, under the assumptions of Proposition 13.6, EP (13.1) can be replaced by equivalent VI with the cost mapping possessing the same (generalized) monotonicity property as that of Φ. Note that the reverse assertions are not true in general; e.g. see [122, Example 3.2.7].

Hence, it seems reasonable to make use of generalized monotone VIs for investigation and solution of certain classes of EPs. Some additional monotonicity type properties of mappings G of the form (13.15) were investigated in [187].

4. Variational Inequalities with Multi-Valued Mappings

In this section, we consider VI (13.2) also in a Banach space setting. Namely, throughout this section we suppose that Assumption **(B1)** holds and that

(B2) $G : K \to 2^{E'}$ *is a mapping with nonempty, convex and weakly* compact values.*

4.1 Generalized monotonicity and dual variational inequalities

First we note that it is possible to establish existence and uniqueness results of solutions to VI (13.2) by using the corresponding results of Section 2 applied to the following equilibrium bifunction:

$$\Phi(x,y) = \sup_{g \in G(x)} \langle g, y - x \rangle. \tag{13.16}$$

Clearly, (13.16) extends (13.11) to the multivalued case. Note that the supremum in (13.16) is always attainable because of **(B2)**. Obviously, VI (13.2) implies EP (13.1), (13.16). Conversely, if $x^* \in K$ solves EP (13.1), (13.16), then

$$\sup_{g^* \in G(x^*)} \langle g^*, y - x^* \rangle \geq 0 \quad \forall y \in K,$$

or equivalently,

$$\inf_{y \in K} \sup_{g^* \in G(x^*)} \langle g^*, y - x^* \rangle \geq 0.$$

In order to obtain (13.2) we can now apply Kneser's minimax theorem, which is formulated as follows.

Proposition 13.7 *[107] Let X be a nonempty convex subset in a vector space, and let Y be a nonempty compact convex subset of a Hausdorff topological vector space. Suppose that f is a real-valued function on $X \times Y$ such that $f(x, \cdot)$ is convex and lower semicontinuous for each $x \in X$ and $f(\cdot, y)$ is concave for each $y \in Y$. Then,*

$$\min_{y \in Y} \sup_{x \in X} f(x,y) = \sup_{x \in X} \min_{y \in Y} f(x,y).$$

If we set $Y = K$, $X = G(x^*)$, $f(x,y) = \langle x, x^* - y\rangle$, then all the assumptions of Proposition 13.7 will be safisfied and we obtain (13.2). Thus, we have established the following equivalence result.

Lemma 13.5 *VI (13.2) is equivalent to EP (13.1), (13.16).*

In order to apply the results of Section 2 to VI (13.2), we need to provide the corresponding properties of the bifunction Φ in (13.16). Clearly, $\Phi(x,\cdot)$ is convex and lower semicontinuous for each $x \in K$, since it is the supremum of affine functions (see e.g. [51, Chapter 1, Section 2]). We recall several continuity type properties for multivalued mappings.

Definition 13.9 (see [13, 154, 83, 126]) Let Y be a topological vector space. A mapping $Q : K \to 2^Y$ is said to be

(a) *upper semicontinuous*, if for each point $x \in K$ and for each open set Z (in the corresponding topology) such that $Z \supseteq Q(x)$, there is a neighborhood X of x such that $Z \supseteq Q(y)$ whenever $y \in X \bigcap K$;

(b) *upper hemicontinuous*, if its restriction on line segments of K is upper semicontinuous;

(c) *u-hemicontinuous*, if $Y = E'$ and the restriction of Q on line segments of K is upper semicontinuous with respect to the weak* topology on E'.

Since the existence results for VIs without monotonicity type properties require too restrictive assumptions in infinite-dimensional spaces (see Theorem 13.5 and also [7, Theorem 9.9]), we give these results with the help of generalized monotonicity properties for multivalued mappings.

Definition 13.10 (see [25, 169, 192, 141, 47]) A mapping $Q : K \to 2^{E'}$ is said to be

(a) *monotone*, if for all $x, y \in K$ and for all $q' \in Q(x), q'' \in Q(y)$, we have

$$\langle q' - q'', x - y\rangle \geq 0;$$

(b) *pseudomonotone*, if for all $x, y \in K$ and for all $q' \in Q(x), q'' \in Q(y)$, we have

$$\langle q'', x - y\rangle \geq 0 \Longrightarrow \langle q', x - y\rangle \geq 0;$$

(c) *strictly pseudomonotone*, if for all distinct $x, y \in K$ and for all $q' \in Q(x), q'' \in Q(y)$, we have

$$\langle q'', x - y\rangle \geq 0 \Longrightarrow \langle q', x - y\rangle > 0;$$

(d) *quasimonotone*, if for all $x, y \in K$ and for all $q' \in Q(x), q'' \in Q(y)$, we have

$$\langle q'', x - y\rangle > 0 \Longrightarrow \langle q', x - y\rangle \geq 0;$$

(e) *strictly quasimonotone*, if it is quasimonotone and for all distinct $x, y \in K$ there exists $z \in (x, y)$ such that

$$\langle q', x - y \rangle \neq 0 \quad \text{for some} \quad q \in Q(z).$$

From the definitions we obtain the following implications:

$$\begin{array}{ccccc} (a) & \Longrightarrow & (b) & \Longrightarrow & (d) \\ & & \Uparrow & & \Uparrow \\ & & (c) & \Longrightarrow & (e) \end{array}$$

In the single-valued case, these definitions reduce to those in Definition 13.6. Besides, they can be obtained equivalently from those in Definition 13.3 by using (13.16). Combining (13.6) and (13.16), we can also define the dual variational inequality problem (DVI) in the multivalued case, which is to find an element $x^* \in K$ such that

$$\forall x \in K, \quad \forall g \in G(x): \quad \langle g, x - x^* \rangle \geq 0. \tag{13.17}$$

We denote by K^* and K^*_d the sets of solutions of problems (13.2) and (13.17), respectively. By analogy with the proof of Lemma 13.4, applying (13.16) in Lemma 13.1, we obtain the following multi-valued variant of Minty's Lemma; see e.g. [173, 192].

Lemma 13.6 *(i) If G is pseudomonotone, then $K^* \subseteq K^*_d$.*
(ii) If G is u-hemicontinuous, then $K^*_d \subseteq K^*$.

From the definition of DVI (13.17) we obtain the following useful characterization of its solution set.

Lemma 13.7 *The set K^*_d is convex and closed.*

4.2 Existence and uniqueness results for multivalued generalized monotone VIs

First we note that applying (13.16) in Theorem 13.3 with respect to weak topology in E and taking into account Lemma 13.6, we can establish the general existence result for pseudomonotone VIs. In addition, taking into account Lemmas 13.5, 13.6, and 13.7, we obtain the general characterization of solution sets.

Theorem 13.11 *(see [60, 192, 126]) Suppose that the following assumptions hold:*

(a) either K is weakly compact, or there exist a weakly compact subset L of E and a point $\tilde{y} \in L \bigcap K$ such that

$$\langle g, \tilde{y} - x \rangle < 0 \quad \forall g \in G(x), \forall x \in K \backslash L; \tag{13.18}$$

(b) G is u-hemicontinuous;
(c) G is pseudomonotone.
Then the sets of solutions to VI (13.2) and DVI (13.17) coincide and are nonempty, convex and weakly compact.

Moreover, applying (13.16) in Proposition 13.2 yields a uniqueness result.

Proposition 13.8 *Suppose G is strictly pseudomonotone. Then VI (13.2) has at most one solution.*

Observe that the existence result above is due to applying Ky Fan's Lemma to the mappings A and B, defined by (13.4), (13.16) and (13.9), (13.16), respectively. These mappings can be equivalently redefined as follows:

$$A(y) = \{x \in K \mid \exists g \in G(x), \langle g, y - x\rangle \geq 0\} \tag{13.19}$$

and

$$B(y) = \{x \in K \mid \forall g \in G(y), \langle g, y - x\rangle \geq 0\}. \tag{13.20}$$

Replacing $B(y)$ with the set

$$\tilde{B}(y) = \{x \in K \mid \exists g \in G(y), \langle g, y - x\rangle \geq 0\}, \tag{13.21}$$

we can somewhat weaken the generalized monotonicity properties which provide the existence of solutions to VI (13.2). First we need to define such weakened properties.

Definition 13.11 [53, 45] A mapping $Q : K \to 2^{E'}$ is said to be
(a) *weakly pseudomonotone*, if for all $x, y \in K$, the inequality $\langle q'', x - y\rangle \geq 0$ for some $q'' \in Q(y)$ implies $\langle q', x - y\rangle \geq 0$ for some $q' \in Q(x)$;
(b) *weakly quasimonotone*, if for all $x, y \in K$, the relation $\langle q'', x - y\rangle > 0$ for some $q'' \in Q(y)$ implies $\langle q', x - y\rangle \geq 0$ for some $q' \in Q(x)$.

Obviously, each pseudomonotone (respectively, quasimonotone) mapping is weakly pseudomonotone (respectively, quasimonotone), but the reverse assertions are not true in general.

Next, using the same argument as in the proof of Theorems 13.1 and 13.2, we see that A is KKM. Hence, if either K is weakly compact or condition (13.18) holds, then

$$\emptyset \neq \bigcap_{y \in K} \overline{A(y)}^{w},$$

where $\overline{A}^w$ denotes the weak closure of a set A. If G is weakly pseudomonotone, we have $A(y) \subseteq \tilde{B}(y)$, whereas the strong compactness of $G(y)$ implies that $\tilde{B}(y)$ is weakly compact. It now follows that

$$\emptyset \neq \bigcap_{y \in K} \overline{A(y)}^w \subseteq \bigcap_{y \in K} \tilde{B}(y).$$

But u-hemicontinuity of G yields

$$\bigcap_{y \in K} \tilde{B}(y) \subseteq \bigcap_{y \in K} A(y),$$

thus extending the assertion of Lemma 13.6 (ii). Therefore, we have obtained the following existence result, which can be viewed as a modification of Theorem 13.11.

Theorem 13.12 *(see [53, 126, 30]) Suppose that the following assumptions hold:*

(a) either K is weakly compact, or there exist a weakly compact subset L of E and a point $\tilde{y} \in L \bigcap K$ such that (13.18) holds;

(b) G is u-hemicontinuous;

(c) G is weakly pseudomonotone and has strongly compact values.

Then, VI (13.2) is solvable.

There exist a number of various coercivity conditions for VIs which modify and specify conditions (13.12) and (13.18) for many different settings; e.g. see [84, 192, 195, 83]. Crouzeix [37] and Daniilidis and Hadjisavvas [48] studied the situations where coercivity conditions are not only sufficient, but necessary for solvability of VIs. In particular, it was shown in [48] that several different coercivity conditions, including (13.18), are equivalent to each other and also to the fact that K^* is nonempty and bounded, if E is reflexive, G is upper hemicontinuous and pseudomonotone.

Next, it is possible to obtain an existence result for quasimonotone VIs in a reflexive Banach space setting. The corresponding theorem was established in [45] and is based on the direct application of Ky Fan's Lemma to the mappings A and $\tilde{B}$, defined by (13.19) and (13.21), respectively. We first recall that a point $y \in E$ is called an *inner point* of K if for all $q \in E'$ the following implication holds:

$$\langle q, x \rangle \leq \langle q, y \rangle \quad \forall x \in K \Longrightarrow \langle q, x \rangle = \langle q, y \rangle \quad \forall x \in K$$

(see [197]). We denote by inn K the set of all inner points of K. Under Assumption **(B1)**, we must have inn $K \subseteq K$. Moreover, rai $K \subseteq$ inn$K$.

For more detailed description of properties of inner points, see [197, 23, 81].

Theorem 13.13 *Suppose that the following assumptions hold:*

(a) either K is bounded, or there exists $\rho > 0$ such that for all $x \in K$ with $\|x\| \geq \rho$, there exists a point $\tilde{y} \in K$ such that $\|\tilde{y}\| < \rho$ and

$$\langle g, \tilde{y} - x \rangle \leq 0 \quad \forall g \in G(x);$$

(b) G is upper hemicontinuous;

(c) G is weakly quasimonotone and has strongly compact values;

(d) E is reflexive and $\mathrm{inn}K \neq \emptyset$.

Then, VI (13.2) has a solution.

On account of Theorem 13.11, DVI (13.17) is solvable if G is pseudomonotone and u-hemicontinuous and the coercivity condition (13.18) holds. We also can weaken these sufficient conditions for solvability of DVI. First we give the multivalued analogues of the properties from Definition 13.7.

Definition 13.12 (see [47, 49]) A mapping $Q : K \to 2^{E'}$ is said to be

(a) *explicitly quasimonotone*, if it is quasimonotone and for any distinct $x, y \in K$ the following implication holds:

$$\begin{aligned} &\exists q'' \in Q(y), \langle q'', x - y \rangle > 0 \Longrightarrow \\ &\exists z \in (0.5(x + y), x), \exists q' \in Q(z) : \langle q', x - y \rangle > 0; \end{aligned}$$

(b) *properly quasimonotone*, if for every finite set $\{x^1, \ldots, x^n\} \subseteq K$ and every $y \in \mathrm{conv}\{x^1, \ldots, x^n\}$, there exists i such that

$$\langle q^i, y - x^i \rangle \leq 0 \quad \forall q^i \in Q(x^i).$$

Obviously, Definition 13.12 reduces to Definition 13.7 in case G is single-valued. By definition, each properly quasimonotone mapping is quasimonotone. It was shown in [121] and [47] that each pseudomonotone mapping is explicitly quasimonotone and that each explicitly quasimonotone mapping is properly quasimonotone, respectively. The reverse assertions are not true in general, but it follows from [121, Lemma 3.2] that explicit and proper quasimonotonicity become equivalent to the usual quasimonotonicity in the affine case. The existence of solutions for DVI (13.17) under quasimonotonicity type assumptions was investigated in [121] and [47]. The counter-example from [121, Example 6.2] shows that the quasimonotonicity is not sufficient for DVI (13.17) to be solvable. Nevertheless, Theorems 13.11 and 13.12 in [121] show that the explicit quasimonotonicity can be sufficient under certain additional assumptions. We now give the strongest existence result for DVI under proper quasimonotonicity, which was obtained in [47].

Theorem 13.14 *Suppose that the following assumptions hold:*
(a) either K is weakly compact or there exist a weakly compact subset L of E and a point $\tilde{y} \in L \bigcap K$ such that for every $x \in K \backslash L$ there exists $\tilde{g} \in G(y)$ with $\langle \tilde{g}, \tilde{y} - x \rangle < 0$;
(b) G is properly quasimonotone.
Then, DVI (13.17) is solvable.

Moreover, it was shown in [90] that given a mapping is properly quasimonotone, if and only if DVI is solvable on each nonempty, convex and compact subset K of a finite-dimensional Euclidean space.

4.3 Variational inequalities and related problems

In this subsection, we discuss briefly relationships between VIs with multi-valued mappings and other problems of Nonlinear Analysis. For instance, from the definition it follows that VI (13.2) reduces to the multivalued inclusion

$$0 \in G(x^*),$$

in case $K = E$. Also, if K is a cone in E, then VI (13.2) becomes equivalent to multivalued NCP; e.g. see [94, 84]. We now discuss the relationship between VI (13.2) and the optimization problem (13.14) where f is not necessarily differentiable. For the sake of simplicity, we shall consider the case where f is locally Lipschitz.

Definition 13.13 (see [36, 7]) Let $f : K \to \mathbb{R} \bigcup \{\infty\}$ is said to be *locally Lipschitz*, if for each point $x \in K$ there exists a neighborhood X of x such that f is Lipschitz continuous on X.

If f in (13.14) is locally Lipschitz, then the gradientmap ∇f can be replaced by the Clarke subdifferential ∂f (see [36]). We recall that each locally Lipschitz function is *subdifferentiable*, i.e. its Clarke subdifferential is nonempty at each point of its domain. We now recall the definition of pseudoconvexity for nondifferentiable functions.

Definition 13.14 [162] Let $f : E \to \mathbb{R} \bigcup \{+\infty\}$ be a locally Lipschitz function. The function f is said to be *pseudoconvex* on K, if for each pair of points $x', x'' \in K$, we have

$$\exists g \in \partial f(x''), \langle g, x' - x'' \rangle \geq 0 \Longrightarrow f(x') \geq f(x'').$$

From the definition it follows that each locally Lipschitz and pseudoconvex function is explicitly quasiconvex (see [162, 47]). The following proposition presents relationships between convexity and monotonicity type properties in the nonsmooth case.

Proposition 13.9 *(see [36, 86, 141, 8, 162, 47]) Let $f : E \to \mathbb{R} \bigcup \{+\infty\}$ be a locally Lipschitz function. The function f is convex (respectively, pseudoconvex, explicitly quasiconvex, quasiconvex) if and only if its Clarke subdifferential is monotone (respectively, pseudomonotone, explicitly quasimonotone, quasimonotone).*

Note that the subdifferential of a (generalized) convex function possesses certain additional cyclical monotonicity type properties; see e.g. [164, 166, 135, 49, 47] and references therein. The following well known optimality condition for problem (13.14) shows that VI (13.2) is closely related with optimization problems.

Theorem 13.15 *Suppose $f : E \to \mathbb{R} \bigcup \{+\infty\}$ is a locally Lipschitz function. Then:*

(i) [36] Problem (13.14) implies VI (13.2) with $G = \partial f$.

(ii) If f is pseudoconvex, then VI (13.2) with $G = \partial f$ implies (13.14).

Clearly, assertion (ii) follows from the definition of pseudoconvexity. On account of Proposition 13.9, we conclude that optimization problems with (generalized) convex functions can be viewed as instances of VIs with (generalized) monotone mappings.

Lemma 13.5 presents a way to transform VI (13.2) into an equivalent EP. Now we consider the conditions which provide the reverse transformation in the case there $\Phi(x, \cdot)$ is non differentiable. Namely, set

$$G(x) = \partial_y \Phi(x, y)|_{y=x}. \tag{13.22}$$

In the differentiable case (13.22) clearly reduces to (13.15). Using the same argument as that in the proof of Theorem 13.10 and applying now Theorem 13.15, we obtain the relationship between EP (respectively, DEP) and VI (respectively, DVI) in the non-differentiable case.

Theorem 13.16 *(see [117]) Suppose that $\Phi : K \times K \to \mathbb{R}$ is an equilibrium bifunction, $\Phi(x, \cdot)$ is locally Lipschitz for each $x \in K$. Then:*

(i) EP (13.1) implies VI (13.2) with G being defined by (13.22);

(ii) if, an addition, $\Phi(x, \cdot)$ is pseudoconvex for each $x \in K$, then EP (13.1) is equivalent to VI (13.2) with G beind defined by (13.22), and DEP (13.6) implies DVI (13.17).

The following proposition gives relationships between (generalized) monotonicity properties for G and Φ in (13.22), thus extending Proposition 13.6.

Proposition 13.10 *(see [117, Proposition 2.2] and [122, Proposition 2.1.17]) Suppose that* $\Phi : K \times K \to \mathbb{R}$ *is an equilibrium bifunction,* $\Phi(x, \cdot)$ *is convex for each* $x \in K$. *If* Φ *is monotone (respectively, pseudomonotone, strictly pseudomonotone, quasimonotone, strictly quasimonotone, explicitly quasimonotone), the same is true for the mapping* G, *defined by (13.22).*

Thus, convexity of $\Phi(x, \cdot)$ allows one to convert the initial EP with (generalized) monotone bifunction into an equivalent VI, its cost mapping satisfying the same (generalized) monotonicity property. Again, the reverse assertion is not true in general; see [122, Example 3.2.7].

5. Iterative Methods for Variational Inequalities

In this section, we consider several approaches to construct iterative methods which converge to a solution of VI under generalized monotonicity assumptions. Due to the assertions of Theorems 13.10 and 13.16 and, also, Propositions 13.6 and 13.10, such methods can be applied for EPs if their cost bifunctions are convex in the second variable and possess the same monotonicity type properties. To simplify the exposition, we restrict ourselves to a Euclidean space setting. Namely, throughout this section we suppose that

(C1) *K is a nonempty, convex and closed subset of the real n-dimensional space* $\mathbb{R}^n$.

First we note that constructing solution methods for generalized monotone VIs meets certain difficulties since, for instance, the sum of quasimonotone (or pseudomonotone) mappings need not be quasimonotone (or, respectively, pseudomonotone) in general. For this reason, one cannot use the Tikhonov regularization method (see [185, 11]), the usual proximal point method (see [146, 165]) and, also, various Lagrangian-based methods (see e.g. [160, 93]). Next, convergence properties of the averaging method (see [26]) and the descent methods using merit functions (see [64]) are also based essentially on the monotonicity arguments. Thus, there are few general approaches which provide convergence of the corresponding methods for generalized monotone VIs.

5.1 Traditional approaches

First we consider iterative methods which were designed for solving the optimization problem (13.14). It is well known that most of such methods (see e.g. [70, 84, 160]) are convergent to a solution of the corresponding VI which expresses the necessary optimality condition for problem (13.14) without any additional convexity type properties. More precisely, the usual gradient projection method

$$x^{k+1} := \pi_K[x^k - \lambda_k \nabla f(x^k)],$$

where the stepsize $\lambda_k > 0$ is chosen with an Armijo-type rule, generates a sequence whose limit points are solutions of VI (13.10) with $G = \nabla f$ (see e.g. [14, Chapter 1]). Here and below $\pi_K(\cdot)$ denotes the projection mapping onto K. Generalized monotonicity of ∇f, due to Proposition 13.5 and Theorem 13.9, only provides convergence to a solution of the optimization problem (13.14). Hence, each limit point of the projection method

$$x^{k+1} := \pi_K[x^k - \lambda_k G(x^k)] \tag{13.23}$$

with $\lambda_k > 0$ being chosen properly will be a solution of VI (13.10) if G is integrable and possesses certain continuity properties. However, this is not the case if G is not integrable. In fact, the same method (13.23) does not provide convergence even in the non-integrable monotone case, for instance, when $G(x) = Ax + b$ with A being skew-symmetric, regardless of the stepsize choice (e.g. see [122, Example 1.2.1]). Nevertheless, the method (13.23) becomes convergent if we replace monotonicity of G with the following assumption:

(C2) *VI (13.10) is solvable, and for every $x^* \in K^*$, we have*

$$\langle G(x), x - x^* \rangle > 0 \quad \forall x \in K \backslash K^*. \tag{13.24}$$

This property has clear geometric sense: the angle between $-G(x)$ and $x^* - x$ has to be acute at each non-optimal point x. Hence, it is possible to choose $\lambda_k > 0$ so that the distance from $x^k - \lambda_k G(x^k)$ to a solution is less than that from x^k. The corresponding convergence result can be formulated as follows; see [75, Chapter 5].

Theorem 13.17 *Suppose that $G : K \to \mathbb{R}^n$ is a continuous mapping and assumption (C2) holds. If a sequence $\{x^k\}$ is generated by method (13.23) where*

$$\lambda_k > 0, \quad \sum_{k=0}^{\infty} \lambda_k = \infty, \quad \sum_{k=0}^{\infty} \lambda_k^2 < \infty; \tag{13.25}$$

then it converges to a solution of VI (13.10).

Note that the inequality (13.24) is also known in Economics as the *revealed preference property* and was intensively investigated by many authors (see e.g. [6]). It is clear that strict pseudomonotonicity of G implies (13.24). Moreover, each pseudomonotone integrable G also satisfies (13.24). In [29, 43], several classes of mappings which are intermediate between strict monotone (respectively, strict pseudomonotone) and monotone (respectively, pseudomonotone) ones and provide for (13.24) to hold are considered in detail.

The assertion of Theorem 13.17 can be extended to the multivalued case. Namely, (13.23) and (13.24) are then rewritten respectively as

$$x^{k+1} := \pi_K[x^k - \lambda_k g^k], g^k \in G(x^k) \tag{13.26}$$

and

$$\forall g \in G(x), \langle g, x - x^* \rangle > 0 \quad \forall x \in K \backslash K^*. \tag{13.27}$$

A sequence $\{x^k\}$ which is generated by (13.26) and (13.25) will converge to a point of K^* if (13.27) holds and G is an upper semicontinuous mapping with nonempty, convex and compact values; see [75, Chapter 5]. It should be noted that rule (13.25) leads to very slow convergence of the method. Hence, we need to apply other approaches to attain more rapid convergence and to weaken sufficient conditions for convergence. Namely, we intend to describe several methods with the following sufficient condition (see [110]):

(C3) $K^* = K_d^* \neq \emptyset$ *for problem (13.10) or (13.2).*

From Theorems 13.6, 13.8, 13.11, and 13.14 it follows that **(C3)** is weaker essentially than **(C2)**. We will consider the general nonlinear case. The case where G is affine and pseudomonotone was studied by Gowda [76, 77]. In particular, Gowda showed that the Lemke method [133] can be applied directly to solve such VIs.

5.2 The extragradient and modified proximal point methods

Let us consider the single-valued VI (13.10). First we recall some strengthened (pseudo) monotonicity properties.

Definition 13.15 (see [25, 97]) A mapping $Q : K \to \mathbb{R}^n$ is said to be

(a) *strongly monotone* with modulus $\tau > 0$, if for all $x, y \in K$, we have

$$\langle Q(x) - Q(y), x - y \rangle \geq \tau \|x - y\|^2;$$

(b) *strongly pseudomonotone* with modulus $\tau > 0$, if for all $x, y \in K$, we have

$$\langle Q(y), x - y \rangle \geq 0 \Longrightarrow \langle Q(x), x - y \rangle \geq \tau \|x - y\|^2.$$

From the definitions it follows that strong monotonicity implies monotonicity and strong pseudomonotonicity. In turn, strong monotonicity implies strict monotonicity, but the reverse assertions are not true.

The idea of the extragradient method by Korpelevich [128] consists in replacing the usual projection iteration (13.23) with the double projection iteration

$$x^{k+1} := \pi_K[x^k - \lambda_k G(y^k)], \quad y^k := \pi_K[x^k - \lambda_k G(x^k)] \qquad (13.28)$$

with $\lambda_k > 0$. This method was applied first to monotone VIs, and afterwards extended to VIs satisfying **(C3)** (see [5]). In fact, even in the latter case, there exists $\lambda > 0$ such that the angle between $-G[\pi_K(x^k - \lambda G(x^k))]$ and $x^* - x^k$ is acute for every $x^* \in K^*$, hence sequence $\{x^k\}$ generated by process (13.28) will converge to a solution if λ_k was chosen properly. Note that one needs to choose the same stepsize λ_k for two subiterations simultaneously. In [128, 5], the value of λ_k was suggested to be fixed: $\lambda_k = \lambda < L/2$, where L is a Lipschitz constant for G. A linesearch procedure for (13.28) was proposed in [103]. From the results of [5] it follows that the extragradient method attains a linear rate of convergence if G is strongly pseudomonotone (see also [186, 136]).

It has been mentioned that the usual proximal point method, generally speaking, cannot be applied to non-monotone problems. Nevertheless, Billups and Ferris [19] presented a modification of this method for VI (13.10) in the case where K is a box-constrained set and **(C3)** is satisfied. The idea is to replace the initial VI (13.10) with a sequence of perturbed problems which have the cost mapping $G_k^\lambda(x) = G(x) + \lambda(x - x^k)$ and are solved with a prescribed accuracy. However, since G need not be monotone, we cannot choose an arbitrary λ as in the classical proximal point method. It was suggested in [19] to choose λ large enough to provide for G_k^λ to be strongly monotone and solve each auxiliary VI with mapping G_k^λ by a SQP type algorithm. Then the main iteration sequence will converge to a solution under Assumption **(C3)**. It is clear that the efficiency of such a method depends strongly on the choice of the algorithm for inner subproblems and its accuracy, and, also, the strategy of changing the basic parameter λ. The results of numerical experiments on several test problems presented in [19] show that the method is more robust than other SQP type methods.

It would be of interest to extend this approach to more general classes of VIs and verify other algorithms for solving inner subproblems.

5.3 Center type methods

The idea of this class of methods can be explained as follows. We recall that a set C is said to be solid if its interior int$C$ is nonempty.

Suppose that we can associate to each iterate x^k a solid set U_k such that $x^k \in U_k$ and $K^* \bigcap U_k \neq \emptyset$ and, also, a vector a^k such that

$$K^* \subseteq H_k^+, \quad \text{where } H_k^+ = \{x \in K \mid \langle a^k, x^k - x \rangle \geq 0\}. \tag{13.29}$$

Letting

$$U_{k+1} := U_k \bigcap H_k^+, \tag{13.30}$$

we obtain the sequence of sets

$$U_0 \supseteq U_1 \supseteq \ldots \supseteq U_k \supseteq \ldots,$$

whose volumes, under a suitable choice of x^k, will tend to zero, thus providing the convergence of the method. For instance, if we set x^k to be the center of gravity, then the volume of U_k will reduce in a linear rate; see [134]. Let us consider VI (13.10) where the feasible set K is defined as follows:

$$K = \{x \in \mathbb{R}^n \mid h(x) \leq 0\}, \tag{13.31}$$

where $h : \mathbb{R}^n \to \mathbb{R}$ is a convex, but not necessarily differentiable, function and suppose that **(C3)** holds and there exists a point $\tilde{x}$ such that $h(\tilde{x}) < 0$, i.e. the Slater constraint qualification is satisfied. In order to apply the method of centers of gravity to solve this problem it is sufficient to choose

$$a^k \begin{cases} = G(x^k) & \text{if } h(x^k) < 0, \\ \in \partial h(x^k) & \text{if } h(x^k) \geq 0. \end{cases}$$

Then, due to **(C3)** and the subgradient property, the key relation (13.29) holds, thus providing convergence. However, the implementation of this method is very difficult when $n > 2$. There are two main approaches to simplify this method. The first is to choose the sets U_k to be "regular" in the sense of computation of their centers of gravity. For example, one can choose an ellipsoid or a simplex in $\mathbb{R}^n$. The most known is the ellipsoid method, which was proposed first by Yudin and Nemirovskii [196] and Shor [174] for convex programming and afterwards adjusted for saddle point problems and VIs (see [152, 175, 176, 151]). Then each set U_k is an ellipsoid and we have to replace rule (13.30) with the following:

U_{k+1} *is the smallest ellipsoid containing the set* $U_k \bigcap H_k^+$.

It is well known (see [151]) that the volumes of U_k will also tend to zero in a linear rate, but the convergence is slower than that of the method of centers of gravity and depends strongly on the dimensionality of the problem. The second approach consists in replacing the precise "center" of U_k with its approximation in some sense. The general framework involving both the approaches was described in [144]. We now give an example of the method which follows the second approach and is based on the computation of approximate analytic centers; see [72]. This algorithm is applied to VI (13.10) where K is a full-dimensional polyhedron, i.e.,

$$K = \{x \in \mathbb{R}^n \mid Ax \leq b\}.$$

Algorithm 5.1 *Step 0:* Set $k := 0, A^k := A, b^k := b, U^k := \{x \in \mathbb{R}^n \mid A^k x \leq b^k\}$.

Step 1 (Computation of an approximate analytic center): Find x^k as an approximate maximizer of the dual potential $\varphi_0(s) = \sum_j \ln s_j$ over $\{x \in \mathbb{R}^n \mid A^k x + s = b^k, s \geq 0\}$.

Step 2 (Stopping criterion): Compute $g_p(x^k) := \max_{x \in K} \langle G(x^k), x^k - x \rangle$. If $g_p(x^k) = 0$, stop.

Step 3 (Generation of a cutting plane): Set

$$A^{k+1} := \begin{pmatrix} A^k \\ G(x^k)^T \end{pmatrix}, \quad b^{k+1} := \begin{pmatrix} b^k \\ \langle G(x^k), x^k \rangle \end{pmatrix}, \quad k := k+1$$

and go to Step 1.

However, most of such algorithms do not provide convergence of their iteration sequences to a solution of VI (13.10) only under Assumption **(C3)**. It is possible to attain convergence with respect to some merit function (see [151, Theorem 1]). Otherwise, one needs either to introduce additional assumptions such as (13.24) or to incorporate auxiliary linesearch procedures (see [145]) to provide strict cutting planes for non-optimal points.

The idea of various proximal level methods (see [74, 132, 105]) is rather close to that of the center methods. In fact, they are based on sequential updating a polyhedral approximation of a non-smooth merit function for VI and computing the prox-center of the corresponding level sets. These methods possess the same convergence properties with respect to the initial VI.

It should be also noted that all the center and bundle type methods can be easily extended to the multivalued VI (13.2); see e.g. [151, 52, 74, 132, 105]. It is sufficient to replace $G(x^k)$ with an arbitrary element of this set in the above considerations.

5.4 Combined relaxation methods

The idea of combined relaxation (CR) methods consists in defining the next iterate x^{k+1} as the projection of the current iterate x^k onto a hyperplane H_k which separates strictly x^k and the solution set. Clearly, then the distance to all solutions will decrease. This approach to solve VI was proposed first by Konnov [110] where it was also noticed that the parameters of the hyperplane H_k can be found with the help of an iteration of any relaxation method. Namely, several implementable methods whose auxiliary procedures were based on an iteration of the projection method, the Frank-Wolfe method, and the Newton method were presented in [110]. It was also observed in [110] that all these methods ensure convergence to a solution of VI (13.10) under Assumption **(C3)**. Afterwards, CR methods were developed in several directions. Within the general CR framework, different rules for determining the separating hyperplane and auxiliary procedures were presented; see [115, 119, 120, 122] and also [87, 177, 179, 180]. To illustrate this approach, we describe one of the CR methods which was proposed in [180, 177, 87, 120] in somewhat different variants.

Algorithm 5.2 *Step 0:* Choose a point $x^0 \in \mathbb{R}^n$, numbers $\alpha \in (0,1), \beta \in (0,1), \gamma \in (0,2)$, and a sequence of positive definite matrices $\{A_k\}$. Set $k := 0$.

Step 1 (Auxiliary procedure): Compute m as the smallest non-negative integer such that

$$\langle G(x^k) - G(z^{k,m}), x^k - z^{k,m}\rangle \leq (1-\alpha)\beta^{-m}\langle A_k(z^{k,m} - x^k), z^{k,m} - x^k\rangle$$

where $z^{k,m} \in K$ is a (unique) solution to the auxiliary VI:

$$\langle G(x^k) + \beta^{-m} A_k(z^{k,m} - x^k), y - z^{k,m}\rangle \geq 0 \quad \forall y \in K. \tag{13.32}$$

Set $\theta_k := \beta^m, y^k := z^{k,m}$.

Step 2 (Determining the hyperplane). If $x^k = y^k$, stop. Otherwise set

$$\begin{aligned} g^k &:= \; G(y^k) - G(x^k) - \theta_k^{-1} A_k(y^k - x^k), \\ \sigma_k &:= \; \langle g^k, x^k - y^k\rangle / \|g^k\|^2. \end{aligned}$$

Step 3 (Main iteration): Set $x^{k+1} := x^k - \gamma\sigma_k g^k$, $k := k+1$ and go to Step 1.

It was shown in [120], that the point x^{k+1} in this algorithm is the projection of x^k onto the hyperplane

$$H_k(\gamma) = \{x \in \mathbb{R}^n \mid \langle g^k, x - x^k\rangle = -\gamma\sigma_k\|g^k\|^2\},$$

H_k (1) strictly separating x^k and K^* if **(C3)** holds. It follows that $\{x^k\}$ converges to a solution of VI (13.10). Moreover, the rate of convergence is linear under certain strong pseudomonotonicity type assumptions (see [120, 122]). Note that problem (13.32) corresponds to an iteration of the projection method if $A_k \equiv I$ and an iteration of the Newton method if $A_k = \nabla G(x^k)$. Varying the rules of choosing A_k, we obtain various auxiliary procedures within the same Algorithm 5.2.

This approach can be easily extended to the case of multivalued VI (13.2); see [111, 112, 114, 118, 122]. It suffices to replace the auxiliary procedure for computing the parameters of the separating hyperplane H_k, which is now based on an iteration of the relaxation subgradient method. In addition to **(C3)**, we suppose that the feasible set K is defined by (13.31), where $h : \mathbb{R}^n \to \mathbb{R}$ is a convex, but not necessarily differentiable, function; G is an upper semicontinuous mapping with nonempty, convex and compact values; and that the Slater constraint qualification is satisfied, i.e., there exists a point $\tilde{x}$ such that $h(\tilde{x}) < 0$. Set

$$P(x) = \begin{cases} G(x) & \text{if } h(x) < 0, \\ \text{conv}\{G(x) \bigcup \partial h(x)\} & \text{if } h(x) = 0, \\ \partial h(x) & \text{if } h(x) > 0. \end{cases}$$

Then the corresponding CR method can be described as follows.

Algorithm 5.3 *Step 0*: Choose a point $x^0 \in \mathbb{R}^n$, numbers $\alpha \in (0,1)$, $\gamma \in (0,2)$, and sequences $\{\varepsilon_l\} \searrow 0$ and $\{\eta_l\} \searrow 0$. Set $k := 0$, $l := 1$.

Step 1 (Auxiliary procedure):

Step 1.1 : Choose q^0 from $P(x^k)$, set $i := 0$, $p^i := q^i$.

Step 1.2: If $\|p^i\| \le \eta_l$, set $x^{k+1} := x^k$, $k := k+1$, $l := l+1$ and go to Step 1. (*null step*)

Step 1.3: Set $w^{i+1} := x^k - \varepsilon_l p^i/\|p^i\|$, choose $q^{i+1} \in P(w^{i+1})$. If

$$\langle q^{i+1}, p^i \rangle > \theta \|p^i\|^2,$$

then set $y^k := w^{i+1}$, $g^k := q^{i+1}$ and go to Step 2. (*descent step*) Otherwise, set

$$p^{i+1} := \pi_{\text{conv}\{p^i, q^{i+1}\}}(0),$$

$i := i+1$ and go to Step 1.2.

Step 2 (Main iteration): Set $\sigma_k := \langle g^k, x^k - y^k \rangle / \|g^k\|^2$, $x^{k+1} := x^k - \gamma\sigma_k g^k$, $k := k+1$ and go to Step 1.

In contrast to the previous algorithm, Algorithm 5.3 combines the direction finding procedure with tuning the tolerances ε_l and η_l. Also, it does not involve any linesearch. Nevertheless, it was shown in [112, 114] (see also [122, Sections 2.3 and 2.4]) that Algorithm 5.3 converges to a solution of VI (13.2), (13.31) under the assumptions above and attains a linear rate of convergence under certain additional assumptions. Note that Algorithm 5.3 is completely implementable even for nonlinearly constrained problems. Its convergence properties are based on the fact that VI (13.2), (13.31) is now equivalent to the inclusion

$$0 \in P(x^*),$$

which in turn becomes equivalent to the dual problem of finding a point x^* of $\mathbb{R}^n$ such that

$$\forall x \in \mathbb{R}^n, \quad \forall p \in P(x): \quad \langle p, x - x^* \rangle \geq 0$$

(cf. (13.17)) due to **(C3)**.

6. Vector Equilibrium Problems and Vector Variational Inequalities with Single-Valued Mappings

In this section, we consider vector extensions of EP (13.1) and VI (13.10). However, scalar concepts of convexity and (generalized) monotonicity admit several different vector extensions; see e.g. [140, 88, 182, 109, 67]. For this reason, we give our analysis for one of the most popular variants and hope that any interested reader will be able to make the corresponding modifications in dealing with other variants of the same concepts and problems.

6.1 Formulations of problems and generalized monotonicity for vector bifunctions

We begin our considerations in a topological vector space setting. Throughout this section, we suppose that Assumption **(A1)** holds. Also, we suppose that

(D1) *F is a real topological vector space with a partial order $\geq$ induced by a pointed closed convex and solid cone C.*

Therefore, $f' \geq f''$ is equivalent to $f' - f'' \in C$, $f' > f''$ is equivalent to $f' - f'' \in \text{int}C$, and $f' \not> f''$ is equivalent to $f' - f'' \notin \text{int}C$. Next, we suppose that

(D2) *$T : K \times K \to F$ is a bifunction such that $T(x, x) \geq 0$ for every $x \in K$.*

Thus, **(D2)** can be viewed as an extension of the notion of the scalar equilibrium bifunction. Being based on the scalar problems (13.1) and (13.6), we now define their vector analogues. Namely, the *vector equilibrium problem* (VEP) is to find an element $x^* \in K$ such that

$$T(x^*, y) \not< 0 \qquad \forall y \in K. \tag{13.33}$$

The *dual vector equilibrium problem* (DVEP) is to find an element $y^* \in K$ such that

$$T(x, y^*) \not> 0 \qquad \forall x \in K. \tag{13.34}$$

We denote by K^e and K^e_d the sets of solutions to problems (13.33) and (13.34), respectively. We intend to present existence results of solutions for VEPs by employing the same Ky Fan Lemma, hence it is necessary to recall vector analogues of some definitions from Section 2.

Definition 13.16 (e.g. see [140, 88, 182, 16]) Let $f : K \to F$ be a function. The function f is said to be

(a) *convex*, if for each pair of points $x', x'' \in K$ and for all $\alpha \in [0, 1]$, we have

$$f(\alpha x' + (1 - \alpha)x'') \leq \alpha f(x') + (1 - \alpha)f(x'');$$

(b) *quasiconvex*, if for each pair of points $x', x'' \in K$ and for all $\alpha \in [0, 1]$, we have

$$f(\alpha x' + (1 - \alpha)x'') \leq \max\{f(x'), f(x'')\};$$

(c) *explicitly quasiconvex*, if it is quasiconvex and for each pair of points $x', x'' \in K$ such that $f(x') < f(x'')$ and for all $\alpha \in (0, 1)$, we have

$$f(\alpha x' + (1 - \alpha)x'') < f(x'');$$

(d) *lower* (respectively, *upper*) *semicontinuous*, if for all $\varphi \in F$, the set $\{x \in K \mid f(x) \not> \varphi\}$ (respectively, the set $\{x \in K \mid f(x) \not< \varphi\}$) is closed;

(e) *u-hemicontinuous*, if its restriction on line segments of K is upper semicontinuous;

(f) *hemicontinuous*, if its restriction on line segments of K is continuous.

From the definitions one can conclude that the following implications hold:

$$(a) \Longrightarrow (c) \Longrightarrow (b) \quad \text{and} \quad (f) \Longrightarrow (e).$$

For other kinds of convexity and continuity type concepts see e.g. [140, 181, 183, 127]. Now we turn to monotonicity type properties.

Definition 13.17 [16, 83] A bifunction $T : K \times K \to F$ is said to be

(a) *monotone*, if for all $x, y \in K$, we have

$$T(x, y) + T(y, x) \leq 0;$$

(b) *pseudomonotone*, if for all $x, y \in K$, we have

$$T(x, y) \not< 0 \Longrightarrow T(y, x) \not> 0,$$

or equivalently,

$$T(x,y) > 0 \Longrightarrow T(y,x) < 0;$$

(c) *quasimonotone*, if for all $x, y \in K$, we have

$$T(x,y) > 0 \Longrightarrow T(y,x) \leq 0;$$

From the definitions we obviously have the following implications:

$$(a) \Longrightarrow (b) \Longrightarrow (c),$$

but the reverse assertions are not true in general as in the scalar case.

6.2 Existence results for generalized monotone VEPs

Existence results of solutions for vector problems which are based upon employing generalized monotonicity concepts also enable one to essentally weaken continuity and compactness assumptions in comparison with the general results; e.g. see [31, 188, 158, 131, 102]. In order to obtain such existence results of solutions for VEP (13.33) it is sufficient to replace the mappings $A, B : K \to 2^K$, defined by (13.4) and (13.9), with the following:

$$A(y) = \{x \in K \mid T(x,y) \not< 0\} \tag{13.35}$$

and

$$B(x) = \{y \in K \mid T(x,y) \not> 0\}, \tag{13.36}$$

respectively, and verify the conditions of Proposition 13.1 for these mappings. Note that

$$K^e = \bigcap_{y \in K} A(y) \quad \text{and} \quad K^e_d = \bigcap_{x \in K} B(x). \tag{13.37}$$

Hence, using the same argument as in the proof of Lemma 13.1, we obtain its analogue for the vector case.

Lemma 13.8 *[16, 127] (i) If T is pseudomonotone, then $K^e \subseteq K^e_d$.*

(ii) If $T(x,\cdot)$ is explicitly quasiconvex for each $x \in K$ and $T(\cdot,y)$ is u-hemicontinuous for each $y \in K$, then $K^e_d \subseteq K^e$.

Next, taking into account (13.33)–(13.37) and using the argument similar to that in the proof of Theorem 13.5, we obtain an existence result for pseudomonotone VEPs.

Theorem 13.18 *[16, 82, 127] Suppose that the following assumptions hold:*

(a) either K is compact or there exist a compact subset L of E and a point $\tilde{y} \in L \bigcap K$ such that

$$T(x, \tilde{y}) < 0 \quad \forall x \in K \backslash L; \tag{13.38}$$

(b) $T(x, \cdot)$ is explicitly quasiconvex and lower semicontinuous for each $x \in K$;

(c) $T(\cdot, y)$ is u-hemicontinuous for each $y \in K$;

(d) T is pseudomonotone.

Then, VEP (13.33) and DVEP (13.34) are solvable and have the same solution sets.

Note that assumption (b) of Theorem 13.18 implies that $B(x)$ is closed for each $x \in K$, hence the solution sets of VEP (13.33) and DVEP (13.34) are also closed, but they need not be convex even under stronger monotonicity and convexity assumptions on T; e.g. see [16, Remark 3.4].

To obtain existence results in the quasimonotone case, we need a somewhat stronger concept of explicit quasiconvexity; see [16, 82].

Definition 13.18 A function $f : K \to F$ is said to be *s-explicitly quasiconvex*, if it is quasiconvex and for each pair of points $x', x'' \in K$ such that $f(x') < \varphi$ and $f(x'') \not> \varphi$ for some $\varphi \in F$, we have

$$f(\alpha x' + (1 - \alpha)x'') < \varphi \quad \text{for all} \quad \alpha \in (0, 1).$$

Theorem 13.19 *[82] Suppose that the following assumptions hold:*

(a) either K is compact or there exist a compact subset L of E and a point $\tilde{y} \in L \bigcap K$ such that (13.38) holds;

(b) $T(x, \cdot)$ is s-explicitly quasiconvex and lower semicontinuous for each $x \in K$;

(c) $T(\cdot, y)$ is hemicontinuous for each $y \in K$;

(d) T is quasimonotone;

(e) $\text{rai}K \neq \emptyset$.

Then VEP (13.1) is solvable.

Thus, Theorem 13.19 is a vector analogue of Theorem 13.4.

6.3 Existence results for vector variational inequalities with single-valued mappings

In the scalar case, equilibrium problems extend variational inequalities. Similarly, vector variational inequalities can be viewed as some specialization of VEP (13.33). Suppose that Assumptions **(B1)** and **(D1)** hold and that E and F are Banach spaces. Denote by $\mathcal{L}(E,F)$ the space of all linear continuous mappings from E into F and consider a mapping $G : K \to \mathcal{L}(E,F)$. Then, letting T to be defined by

$$T(x,y) = G(x)(y-x) \tag{13.39}$$

in (13.33), we see that (13.33) reduces to the following *vector variational inequality problem* (VVI): Find an element $x^* \in K$ such that

$$G(x^*)(y-x^*) \not< 0 \quad \forall y \in K. \tag{13.40}$$

Analogously, letting (13.39) in (13.34), we obtain the *dual vector variational inequality problem* (DVVI): Find an element $x^* \in K$ such that

$$G(x)(x-x^*) \not< 0 \quad \forall x \in K. \tag{13.41}$$

We denote by K^* and K_d^* the sets of solutions to problems (13.40) and (13.41), respectively. It is clear that VVI (13.40) and DVVI (13.41) are vector extensions of the scalar problems (13.10) and (13.13), respectively. VVI of the form (13.40) is one of the most investigated equilibrium type problems; see e.g. [65, 32, 33, 35, 31, 188, 68, 46, 194, 109, 67]. At the same time, by using the representation (13.39), we can specialize the existence results from Subsection 6.2 for VVI (13.40). First we note that $T(x,\cdot)$ in (13.39) is clearly convex (even affine). Next, applying (13.39) in Definition 13.17, we can obtain the monotonicity type properties for G.

Definition 13.19 [65, 129, 46] A mapping $Q : K \to \mathcal{L}(E,F)$ is said to be

(a) *monotone*, if for all $x, y \in K$, we have

$$[Q(x) - Q(y)](x-y) \geq 0;$$

(b) *pseudomonotone*, if for all $x, y \in K$, we have

$$Q(y)(x-y) \not< 0 \Longrightarrow Q(x)(x-y) \not< 0,$$

or equivalently,

$$Q(y)(x-y) > 0 \Longrightarrow Q(x)(x-y) > 0;$$

(c) *quasimonotone*, if for all $x, y \in K$, we have

$$Q(y)(x-y) > 0 \Longrightarrow Q(x)(x-y) \geq 0.$$

We need also certain continuity properties for mappings.

Definition 13.20 A mapping $Q : K \to \mathcal{L}(E,F)$ is said to be *u-hemicontinuous*, if its restriction on line segments of K is upper semicontinuous with respect to a given topology on $\mathcal{L}(E,F)$.

We shall apply the weak topology in E, the strong topology in F, and the strong operator topology in $\mathcal{L}(E,F)$. For this reason, we need the concept of a completely continuous mapping. We recall that a mapping $A \in \mathcal{L}(E,F)$ is said to be *completely continuous* if it maps each weakly convergent sequence into a strongly convergent sequence. Applying (13.39) in Lemma 13.8 and using the argument similar to that in the proof of Lemma 13.4, we obtain their analogue for VVIs.

Lemma 13.9 *(see [35, 31, 194]) (i) If G is pseudomonotone, then $K^* \subseteq K^*_d$.*

(ii) If G is u-hemicontinuous, then $K^*_d \subseteq K^*$.

Similarly, combining (13.39) with (13.35) and (13.36), we specialize Theorem 13.18 for pseudomonotone VVIs.

Theorem 13.20 *[194, 46] Suppose that the following assumptions hold:*

(a) either K is weakly compact, or there exist a weakly compact subset L of E and a point $\tilde{y} \in L \bigcap K$ such that

$$G(x)(\tilde{y} - x) < 0 \quad \forall x \in K \backslash L; \tag{13.42}$$

(b) G is u-hemicontinuous and has completely continuous values;

(c) G is pseudomonotone.

Then VVI (13.40) and DVVI (13.41) are solvable and have the same solution sets.

In the quasimonotone case, it is possible to strengthen the existence result for VVIs in comparison with that of Theorem 13.19 with using (13.39).

Theorem 13.21 *(see [46]) Suppose that the following assumptions hold:*

(a) either K is weakly compact, or E is reflexive and there exist a number $\rho > 0$ and a point $\tilde{y} \in K$ with $\|\tilde{y}\| < \rho$ such that

$$G(x)(\tilde{y} - x) \leq 0 \quad \forall x \in K, \|\tilde{y}\| \geq \rho; \tag{13.43}$$

(b) G is hemicontinuous with respect to the strong operator topology in $\mathcal{L}(E,F)$ and has completely continuous values;

(c) G is quasimonotone;

(d) $\text{inn}K \neq \emptyset$.
Then VVI (13.40) is solvable.

Note that (13.43) can be viewed as a specialization of coercivity condition (13.42) in a reflexive space setting.

6.4 Further generalizations and modifications

Together with VEPs and VVIs described in this section, there are a great number of their generalizations and modifications, which are also extensively investigated. For example, the cone C can be replaced with its "moving" analogue $C(x)$ or/and the bifunction T and the mapping G can be supposed to be multivalued; see e.g. [31, 194, 46, 137, 126, 30, 127, 129, 82, 83, 131, 3, 67]. The multivalued and moving case for VVIs will be considered in the next section in more detail.

In addition to the direct applications of mathching type theorems to deriving existence results for vector problems, Oettli [157] suggested another approach which is based on preliminary reducing them to scalar ones and considering new generalized monotonicity type concepts which are based on two, rather than one, functions. This approach was further developed in [158, 82]. Existence results for VEPs and VVIs which are based on vector extensions of these new generalized monotonicity concepts were obtained in [158, 127, 123]. Various coercivity conditions for VVIs and VEPs are considered in [194, 137, 126, 4, 3, 102]. Another field of investigations is related to various modifications of vector inequalities in defining the corresponding VVIs and VEPs; see [4, 158, 2, 190, 193, 109, 67, 69].

In the vector-valued case, there are no such simple relationships between generalized convexity of functions and generalized monotonicity of their derivatives, as those in the scalar case presented in Propositions 13.5 and 13.9 (see e.g. [88]). A scalarization approach to overcome these difficulties was developed in [27, 28, 109, 168]. Several relationships between vector optimization problems and primal (dual) VVIs were presented in [66, 109].

In [123], the relationships between monotonicity properties of vector bifunctions and mappings which are similar to those in Propositions 13.6 and 13.10 were established in the case where F is a Kantorovich space (see [92] for definitions). It should be also mentioned that various applications of VVIs and VEPs were considered by many researchers; e.g. see [33, 113, 188, 189, 190, 73, 161, 116, 34, 31, 68, 99, 100, 101].

7. Vector Variational Inequalities with Multi-Valued Mappings

In this section, we consider vector generalizations of VI (13.2). For simplicity, we also restrict our considerations to a Banach space setting. Namely, throughout this section we suppose that Assumption **(B1)** holds and also that

(E1) *C is a multivalued mapping from K into a real Banach space F such that $C(x)$ is a closed convex and solid cone of F for each $x \in K$;*

(E2) *G is a multivalued mapping from K into $\mathcal{L}(E, F)$ such that $G(x)$ is a nonempty set of completely continuous operators.*

Next, we shall apply the weak topology in E, the strong topology in F, and the strong operator topology in $\mathcal{L}(E, F)$.

7.1 Formulations of problems and existence results

In the multivalued vector case, we have to consider two different formulations of a variational inequality. Namely, a point $x^* \in K$ is said to be a *weak solution* to the *vector variational inequality problem* (VVI) if

$$\forall y \in K, \exists P^* \in G(x^*) : P^*(y - x^*) \notin -\text{int}C(x^*). \tag{13.44}$$

Also, a point $x^* \in K$ is said to be a *strong solution* to VVI if

$$\exists P^* \in G(x^*) : P^*(y - x^*) \notin -\text{int}C(x^*) \qquad \forall y \in K. \tag{13.45}$$

In the case where $F = \mathbb{R}$, $C(x) \equiv \mathbb{R}_+ = \{\alpha \in \mathbb{R} \mid \alpha \geq 0\}$, (13.45) reduces to (13.2), whereas (13.44) reduces to (13.1) where Φ is defined by (13.16). Thus, both problems coincide if G has convex and weakly* compact values, as Lemma 13.5 states. However, this assertion is not in general true with respect to problems (13.44) and (13.45). Clearly, (13.45) implies (13.44), but the reverse assertion is not true. We also define the *dual vector variational inequality problem* (DVVI) as follows: Find an element $x^* \in K$ such that

$$\forall y \in K, \exists P \in G(y) : P(y - x^*) \notin -\text{int}C(x^*). \tag{13.46}$$

We denote by K^*, K^w and K_w^* the sets of solutions to problems (13.44), (13.45) and (13.46), respectively. We shall also need definitions of generalized monotonicity properties and continuity properties for vector multivalued mappings.

Definition 13.21 (see [79, 129, 46, 126]) A mapping $Q : K \to 2^{\mathcal{L}(E,F)}$ is said to be

(a) *monotone*, if for all $x, y \in K$ and for all $P' \in Q(x), P'' \in Q(y)$, we have

$$(P' - P'')(x - y) \in C(x);$$

(b) *pseudomonotone*, if for all $x, y \in K$ and for all $P' \in Q(x), P'' \in Q(y)$, we have

$$P''(x - y) \notin -\text{int}C(x) \Longrightarrow P'(x - y) \notin -\text{int}C(x);$$

(c) *weakly pseudomonotone*, if for all $x, y \in K$, the relation

$$P''(x - y) \notin -\text{int}C(x) \qquad \text{for some } P'' \in Q(y)$$

implies

$$P'(x - y) \notin -\text{int}C(x) \qquad \text{for some } P' \in Q(x);$$

(d) *weakly quasimonotone*, if for all $x, y \in K$, the relation

$$P''(x - y) \notin -C(x) \qquad \text{for some } P'' \in Q(y)$$

implies

$$P'(x - y) \notin -\text{int}C(x) \qquad \text{for some } P' \in Q(x).$$

From the definitions it follows that the following implications hold:

$$(a) \Longrightarrow (b) \Longrightarrow (c) \Longrightarrow (d),$$

but the reverse assertion is not true in general.

Definition 13.22 A mapping $Q : K \to 2^{\mathcal{L}(E,F)}$ is said to be *u-hemicontinuous*, if its restriction on line segments of K is upper semicontinuous with respect to a given topology on $\mathcal{L}(E, F)$.

Using the definitions above and argument similar to that in Lemma 13.1, we obtain its vector analogue.

Lemma 13.10 *[126] (i) If G is pseudomonotone, then $K^w \subseteq K^*_w$.*
(ii) If G is u-hemicontinuous, then $K^*_w \subseteq K^w$.

Unfortunately, in the multivalued vector case, it is not possible to obtain existence results directly from the corresponding results for VEP (13.33). Nevertheless, we can redefine the mappings $A, \tilde{B} : K \to 2^K$ from (13.19) and (13.21) as follows:

$$A(y) = \{x \in K \mid \exists P \in G(x), P(y - x) \notin -\text{int}C(x)\}$$

and

$$\tilde{B}(y) = \{x \in K \mid \exists P \in G(y), P(y - x) \notin -\text{int}C(x)\}.$$

Observe that

$$K^w = \bigcap_{y \in K} A(y) \quad \text{and} \quad K_w^* = \bigcap_{y \in K} \tilde{B}(y).$$

Hence, we can obtain existence results of solutions for VVI (13.44) by verifying the conditions of Proposition 13.1 and using the argument similar to those in the proofs of Theorems 13.3 and 13.11.

Theorem 13.22 *[126, 46, 82] Suppose that the following assumptions hold:*

(a) either K is weakly compact, or E is reflexive and there exist a number $\rho > 0$ and a point $\tilde{y} \in K$ with $\|\tilde{y}\| < \rho$ such that

$$P(\tilde{y} - x) \in -\text{int}C(x) \quad \forall P \in G(x),$$

for all $x \in K$, $\|\tilde{y}\| \geq \rho$;

(b) the graph of the mapping $x \mapsto F \backslash (-\text{int}C(x))$ is sequentially closed in the (weak) $\times$ (norm) topology on $E \times F$;

(c) G is u-hemicontinuous;

(d) G is weakly pseudomonotone and has compact values.

Then VVI (13.44) and DVVI (13.46) are solvable and have the same solution sets.

In the pseudomonotone case, the assertion of this theorem can be somewhat strengthened. In fact, it is sufficient to redefine the mapping $B : K \to 2^K$ from (13.20) as follows:

$$B(y) = \{x \in K \mid \forall P \in G(y), P(y - x) \notin -\text{int}C(x)\}$$

and replace $\tilde{B}$ with B in our considerations. This approach yields the following existence result.

Theorem 13.23 *[126] The assertion of Theorem 13.22 remains true if we replace assumption (d) with the following:*

(d′) G is pseudomonotone.

In the quasimonotone case, we have the following vector extension of Theorem 13.13.

Theorem 13.24 *[46, 82] Suppose that assumptions (a) and (b) of Theorem 13.22 hold and also that*

(c′) G is upper hemicontinuous;

(d′) G is weakly quasimonotone and has compact values.

(e) $\text{inn}K \neq \emptyset$.

Then VVI (13.44) is solvable.

7.2 Scalarization of VVIs

In order to derive existence of strong solutions to VVI (13.44), a scalarization technique was proposed in [126]. It is based on maintaining certain generalized monotonicity properties for a scalarized VI.

Given an element $s \in F'$ and a mapping $Q : K \to 2^{\mathcal{L}(E,F)}$, we can define the scalarized mapping $Q_s : K \to 2^{E'}$ by

$$\langle Q_s(x), y\rangle = \langle s, Q(x)y\rangle$$

for all $x \in K$ and $y \in E$. Also, set

$$H_s = \{\ \varphi \in F \mid \langle s, \varphi\rangle \geq 0\}.$$

Definition 13.23 [126] Suppose D is a given set in the space F. A mapping $Q : K \to 2^{\mathcal{L}(E,F)}$ is said to be

(a) (D)-*pseudomonotone*, if for all $x, y \in K$ and for all $P' \in Q(x)$, $P'' \in Q(y)$, we have

$$P'(y - x) \in D \Longrightarrow P''(y - x) \in D;$$

(b) *weakly* (D)-*pseudomonotone*, if for all $x, y \in K$, the relation

$$P''(x - y) \in D \qquad \text{for some } P'' \in Q(y)$$

implies

$$P'(x - y) \in D \qquad \text{for some } P' \in Q(x);$$

(c) *weakly v-coercive*, if there exists $\tilde{y} \in K$ and $s \in F'\backslash\{\mathbf{0}\}$ such that

$$\inf_{q \in Q_s(x)} \langle q, x - \tilde{y}\rangle \to \infty, \quad \text{as } x \in K, \ \ \|x\| \to \infty.$$

Lemma 13.11 *[126] Suppose that* $Q : K \to 2^{\mathcal{L}(E,F)}$ *is* (H_s)-*pseudomonotone (respectively, weakly* (H_s)-*pseudomonotone) for some* $s \in F'\backslash\{\mathbf{0}\}$. *Then the mapping* Q_s *is pseudomonotone (respectively, weakly pseudomonotone).*

Thus, we can reduce VVI (13.44) to a scalar VI and obtain existence results by applying Theorems 13.11 and 13.12. Set

$$C^* = \{\ s \in F' \mid \langle s, \varphi\rangle \geq 0 \quad \forall \varphi \in \mathrm{conv}_{x \in K} C(x)\}.$$

Theorem 13.25 *[126] Suppose that the following assumptions hold:*

(a) either K *is weakly compact or there exists an element* $s \in C^*\backslash\{\mathbf{0}\}$ *such that* G *is weakly v-coercive;*

(b) G *is u-hemicontinuous;*

(c) G is weakly (H_s)-pseudomonotone with respect to the same $s \in C^*\backslash\{\mathbf{0}\}$;

(d) G has compact and convex values.

Then VVI (13.45) is solvable.

Again, in the pseudomonotone case, the assertion of this theorem can be somewhat strenghtened.

Theorem 13.26 *[126] Suppose that assumptions (a) and (b) of Theorem 13.25 hold and also that*

(c′) G is (H_s)-pseudomonotone with respect to the same $s \in C^*\backslash\{\mathbf{0}\}$;

(d′) G has convex values.

Then VVI (13.45) is solvable.

Note that, under the assumptions of Theorems 13.25 or 13.26, elements of $G(x)$ need not be completely continuous operators. Next, replacing pseudomonotonicity with quasimonotonicity and using Theorem 13.24, we can obtain existence results of strong solutions for multivalued quasimonotone VVIs in the same manner.

Other scalarization techniques for VVIs were considered in [109, 88, 66, 193, 73, 130, 69, 168].

Acknowledgments

The author is grateful to Nicolas Hadjisavvas, Sandor Komlósi and Siegfried Schaible for their valuable comments and suggestions. This work was supported in part by RFBR Grant No. 01-01-00070 and by the R.T. Academy of Sciences.

References

[1] Allen, G. *Variational inequalities, complementarity problems, and duality theorems*, J. Math. Anal. and Appl. 58, 1977, 1–10.

[2] Ansari, Q.H. *Vector equilibrium problems and vector variational inequalities*, In: Vector Variational Inequalities and Vector Equilibria. Mathematical Theories, F. Giannessi, ed., Kluwer Academic Publishers, Dordrecht-Boston-London, 2000, 1–15.

[3] Ansari, Q.H., Konnov, I.V., and Yao, J.C. *On generalized vector equilibrium problems*, Nonl. Anal. 47, 2001, 543–554.

[4] Ansari, Q.H., Oettli, W., and Schläger, D. *A generalization of vectorial equilibria*, Math. Meth. Oper. Res. 46, 1997, 147–152.

[5] Antipin, A.S. *On convergence of proximal methods to fixed points of extremal mappings and estimates of their rate of convergence*, Comp. Maths. Math. Phys. 35, 1995, 539–551.

[6] Arrow, K.J. and Hahn, F.H. *General competitive analysis*, Holden Day, San Francisco, 1972.

[7] Aubin, J.-P. *Optima and equilibria*, Springer-Verlag, Berlin, 1998.

[8] Aussel, D., Corvellec, J.-N., and Lassonde, M. *Subdifferential characterization of quasiconvexity and convexity*, J. Convex Analysis 1, 1994, 195–201.

[9] Avriel, M., Diewert, W.E., Schaible, S., and Zang, I. *Generalized convexity*, Plenum Press, New York, 1988.

[10] Baiocchi, C. and Capelo, A. *Variational and quasivariational inequalities. Applications to free boundary problems*, John Wiley and Sons, New York, 1984.

[11] Bakushinskii, A.B. and Goncharskii, A.V. *Ill-posed problems: Theory and applications*, Kluwer Academic Publishers, Dordrecht, 1994.

[12] Belen'kii, V.Z. and Volkonskii, V.A., Editors, *Iterative methods in game theory and programming*, Nauka, Moscow, 1974. (in Russian)

[13] Berge, C. *Topological spaces*, Oliver and Boyd, Edinburgh, 1963.

[14] Bertsekas, D.P. *Constrained optimization and Lagrange multiplier methods*, Academic Press, New York, 1982.

[15] Bianchi, M. *Pseudo P-monotone operators and variational inequalities*, Research Report No.6, Istituto di Econometrica e Matematica per le Decisioni Economiche, Universita Cattolica del Sacro Cuore, Milan, 1993.

[16] Bianchi, M., Hadjisavvas, N., and Schaible, S. *Vector equilibrium problems with generalized monotone bifunctions*, J. Optim. Theory Appl. 92, 1997, 527–542.

[17] Bianchi, M. and Pini, R. *A note on equilibrium problems for KKM-maps*, Quaderno No.37, Istituto di Econometrica e Matematica per le Applicazioni Economiche, Finanziare ed Attuari, Milan, 2000.

[18] Bianchi, M. and Schaible, S. *Generalized monotone bifunctions and equilibrium problems*, J. Optim. Theory Appl. 90, 1996, 31–43.

[19] Billups, S.C. and Ferris, M.C. *QPCOMP: A quadratic programming based solver for mixed complementarity problems*, Math. Progr. 76, 1997, 533–562.

[20] Blackwell, D. *An analogue of the minimax theorem for vector payoffs*, Pacific J. Math. 6, 1956, 1–8.

[21] Blum, E. and Oettli, W. *Variational principles for equilibrium problems*, In: Parametric Optimization and Related Topics III, J. Guddat, H.Th. Jongen, B. Kummer, and F. Nožička, eds., Peter Lang Verlag, Frankfurt am Main, 1993, 79–88.

[22] Blum, E. and Oettli, W. *From optimization and variational inequalities to equilibrium problems*, The Math. Student 63, 1994, 127–149.

[23] Borwein, J.M. and Lewis, A.S. *Partially finite convex programming, Part I: Quasi relative interiors in duality theory*, Math. Progr. 57, 1992, 15–48.

[24] Brézis, H., Nirenberg, L., and Stampacchia, G. *A remark on Ky Fan's minimax principle*, Boll. Unione Mat. Ital. 6, 1972, 293–300.

[25] Browder, F.E. *Multivalued monotone nonlinear mappings and duality mappings in Banach spaces*, Trans. Amer. Math. Soc. 71,1965, 780–785.

[26] Bruck, R. *On weak convergence of an ergodic iteration for the solution of variational inequalities for monotone operators in Hilbert space*, J. Math. Anal. Appl. 61, 1977, 159–164.

[27] Cambini, R. and Komlósi, S. *On the scalarization of pseudoconvexity and pseudomonotonicity concepts for vector-valued functions*, in: Generalized Convexity, Generalized Monotonicity, J.-P. Crouzeix, J.-E. Martinez-Legaz, and M. Volle, eds., Kluwer Academic Publishers, Dordrecht - Boston - London, 1998, 277–290.

[28] Cambini, R. and Komlósi, S. *On polar generalized monotonicity in vector optimization*, Optimization 47, 2000, 111–121.

[29] Censor, Y., Iusem, A.N., and Zenios, S. *An interior point method with Bregman functions for the variational inequality problem with paramonotone operators*, Math. Progr. 81, 1999, 373–400.

[30] Chadli, O., Chbani, Z., and Riahi, H. *Equilibrium problems with generalized monotone bifunctions and applications to variational inequalities*, J. Optim. Theory Appl. 105, 2000, 299–323.

[31] Chen, G.Y. *Existence of solution for a vector variational inequality: An extension of the Hartmann-Stampacchia theorem*, J. Optim. Theory Appl. 74, 1992, 445–456.

[32] Chen, G.Y. and Cheng, G.M. *Vector variational inequality and vector optimization*, Lecture Notes in Economics and Mathematical Systems **258**, Springer-Verlag, Heidelberg, 1987, 408–416.

[33] Chen, G. Y. and Craven, B.D. *A vector variational inequality and optimization over an efficient set*, Zeit. Oper. Res. 3, 1990, 1–12.

[34] Chen, G. Y. and Craven, B.D. *Existence and continuity for vector optimization*, J. Optim. Theory Appl. 81, 1994, 459–468.

[35] Chen, G.Y. and Yang, X.Q., *Vector complementarity problem and its equivalences with weak minimal element in ordered spaces*, J. Math. Anal. Appl. 153, 1990, 136–158.

[36] Clarke, F.H. *Optimization and nonsmooth analysis*, John Wiley and Sons, New York, 1983.

[37] Crouzeix, J.-P. *Pseudomonotone variational inequality problems: Existence of solutions*, Math. Progr. 78, 1997, 305–314.

[38] Crouzeix, J.-P. *Characterization of generalized convexity and generalized monotonicity: A survey*, in: Generalized Convexity, Generalized Monotonicity, J.-P. Crouzeix, J.-E. Martinez-Legaz, and M. Volle, eds., Kluwer Academic Publishers, Dordrecht - Boston - London, 1998, 237–256.

[39] Crouzeix, J.-P. and Ferland, J. A. *Criteria for differentiable generalized monotone maps*, Math. Progr. 75, 1996, 399–406.

[40] Crouzeix, J.-P. and Hassouni, A. *Quasimonotonicity of separable operators and monotonicity indices*, SIAM J. Optim. 4, 1994, 649–658.

[41] Crouzeix, J.-P. and Hassouni, A. *Generalized monotonicity of a separable product of operators: The multivalued case*, Set-Valued Analysis 3, 1995, 351–373.

[42] Crouzeix, J.-P., Hassouni, A., Lahlou, A., and Schaible, S. *Positive definite matrices, generarized monotonicity and linear complementarity problems*, SIAM J. Matrix Anal. Appl. 22, 2000, 66–85.

[43] Crouzeix, J.-P. , Marcotte, P., and Zhu, D.L. *Conditions ensuring the applicability of cutting plane methods for solving variational inequalities*, Math. Progr. 88, 2000, 521–539.

[44] Crouzeix, J.-P., and Schaible, S. *Generalized monotone affine maps*, SIAM J. Matrix Anal. Appl. 17, 1996, 992–997.

[45] Daniilidis, A. and Hadjisavvas, N. *Variational inequalities with quasimonotone multivalued operators*, Working Paper, Department of Mathematics, University of the Aegean, Samos, Greece, March 1995.

[46] Daniilidis, A. and Hadjisavvas, N. *Existence theorems for vector variational inequalities*, Bull. Austral. Math. Soc. 54, 1996, 473–481.

[47] Daniilidis, A. and Hadjisavvas, N. *Characterization of nonsmooth semistrictly quasiconvex and strictly quasiconvex functions*, J. Optim. Theory Appl. 102, 1999, 525–536.

[48] Daniilidis, A. and Hadjisavvas, N. *Coercivity conditions and variational inequalities*, Math. Progr. 86, 1999, 433–438.

[49] Daniilidis, A. and Hadjisavvas, N. *On the subdifferentials of quasiconvex and pseudoconvex functions and cyclic monotonicity*, J. Math. Anal. Appl. 237, 1999, 30–42.

[50] Daniilidis, A. and Hadjisavvas, N. *On generalized cyclically monotone operators and proper quasimonotonicity*, Optimization 47, 2000, 123–135.

[51] Dem'yanov, V.F. and Rubinov, A.M. *Approximate methods for solving extremum problems*, Leningrad University Press, Leningrade, 1968 (in Russian; Engl. transl. in Elsevier Science B.V., Amsterdam, 1970).

[52] Denault, M. and Goffin, J.-L. *On a primal-dual analytic center cutting plane method for variational inequalities*, Comp. Optim. Appl. 12, 1999, 127–155.

[53] Ding, X.P. *Browder-Hartmann-Stampacchia variational inequalities*, J. Math. Anal. Appl. 173, 1993, 577–587.

[54] Ding, X.P. and Tarafdar, E. *Generalized nonlinear variational inequalities with non monotone set-valued mappings*, Appl. Math. Lett. 7, 1994, 5–11.

[55] Ding, X.P. and Tarafdar, E. *Generalized variational-like inequalities with pseudomonotone set-valued mappings*, Arch. Math. 74, 2000, 302–313.

[56] Duvaut, G. and Lions, J.-L. *Les inéquations en mechanique et physique*, Dunod, Paris, 1972.

[57] Fan Ky, *A generalization of Tychonoff's fixed point theorem*, Math. Ann. 142, 1961, 305–310.

[58] Fan Ky, *A minimax inequality and applications*, In: Inequalities. III, O. Shisha, ed., Academic Press, New York, 1972, 103–113.

[59] Fan Ky, *Some properties of convex sets related to fixed point theorems*, Math. Ann. 266, 1984, 519–537.

[60] Fang, S.C. and Petersen, E.L. *Generalized variational inequalities*, J. Optim. Theory Appl. 38, 1982, 363–383.

[61] Farquharson, R. *Sur une généralisation de la notion d' équilibrium*, Compt. Rend. Acad. Sci. Paris 240, 1955, 46–48.

[62] Ferris, M.C. and Pang, J.-S. *Engineering and economic applications of complementarity problems*, SIAM Review 39, 1997, 669–713.

[63] Fichera, G. *Problemi elastostatici con vincoli unilaterali; il problema di Signorini con ambigue al contorno*, Acad. Naz. Lincei. Mem. 8, 1964, 91–140.

[64] Fukushima, M. *Merit functions for variational inequality and complementarity problems*, In: Nonlinear Optimization and Applications, G. Di Pillo and F. Giannessi. Plenum Press, New York, 1996, 155–170.

[65] Giannessi, F. *Theorems of alternative, quadratic programs and complementarity problems*, in: Variational Inequalities and Complementarity Problems, R.W. Cottle, F. Giannessi, and J.-L. Lions, eds., John Wiley and Sons, Chichester, 1980, 151–186.

[66] Giannessi, F. *On Minty variational principle*, in: New Trends in Mathematical Programming, F. Giannessi, S. Komlósi, and T. Rapcsák, eds., Kluwer Academic Publishers, Dordrecht, 1998, 93–99.

[67] Giannessi, F., Editor, *Vector variational inequalities and vector equilibria. Mathematical theories*, Kluwer Academic Publishers, Dordrecht-Boston-London, 2000.

[68] Giannessi, F. and Maugeri, A., Editors, *Variational inequalities and network equilibrium problems*, Plenum Press, New York, 1995.

[69] Giannessi, F. and Pellegrini, L., *Image space analysis for vector optimization and variational inequalities. Scalarization*, In: Proceedings of Combinatorial and Global Optimization Conference, Chania, May 1998, P. Pardalos, ed., Kluwer Academic Publishers, Dordrecht-Boston-London, to appear.

[70] Gill, P.E., Murray, W., and Wright, M.H. *Practical optimization*, Academic Press, New York, 1981.

[71] Glowinski, R., Lions, J.-L., and Trémolières, R. *Analyse numerique des inéquations variationnelles*, Dunod, Paris, 1976.

[72] Goffin, J.-L. , Marcotte, P., and Zhu, D.L. *An analytic center cutting plane method for pseudomonotone variational inequalities*, Oper. Res. Lett. 20, 1999, 1–6.

[73] Goh, C.J. and Yang, X.Q. *Vector equilibrium problems and vector optimization*, Eur. J. Oper. Res. 116, 1999, 615–628.

[74] Gol'shtein, E.G., Nemirovskii, A.S., and Nesterov, Yu.E. *Level method, its extensions and applications*, Ekon. Mat. Met. 31, 1995, 164–181. (in Russian)

[75] Gol'shtein, E.G. and Tret'yakov, N.V. *Modified Lagrangians and monotone maps in optimization*, John Wiley and Sons, New York, 1996.

[76] Gowda, M.S. *Pseudomonotone and copositive star matrices*, Linear Algebra Appl. 113, 1989, 107–118.

[77] Gowda, M.S. *Affine pseudomonotone mappings and the linear complementarity problem*, SIAM J. Matr. Anal. Appl. 11, 1990, 373–380.

[78] Granas, A. *Sur quelques méthodes topologiques en analyse convexe*, in: Méthodes Topologiques en Analyse Convexe, A. Granas, ed., Les Presses de l'Université de Montreal, Montreal, 1990, 151–186.

[79] Hadjisavvas, N., Kravvaritis, D., Pantelidis, G., and Polyrakis, I. *Nonlinear monotone operators with values in $L(X,Y)$*, J. Math. Anal. Appl. 140, 1989, 83–94.

[80] Hadjisavvas, N. and Schaible, S. *On strong pseudomonotonicity and (semi) strict quasimonotonicity*, J. Optim. Theory Appl. 79, 1993, 139–155.

[81] Hadjisavvas, N. and Schaible, S. *Quasimonotone variational inequalities in Banach spaces*, J. Optim. Theory Appl. 90, 1996, 95–111.

[82] Hadjisavvas, N. and Schaible, S. *From scalar to vector equilibrium problems in the quasimonotone case*, J. Optim. Theory Appl. 96, 1998, 297–309.

[83] Hadjisavvas, N. and Schaible, S. *Quasimonotonicity and pseudomonotonicity in variational inequalities and equilibrium problems*, in: Generalized Convexity, Generalized Monotonicity, J.-P. Crouzeix, J.-E. Martinez-Legaz, and M. Volle, eds., Kluwer Academic Publishers, Dordrecht - Boston - London, 1998, 257–275.

[84] Harker, P.T. and Pang, J.-S. *Finite-dimensional variational inequality and nonlinear complementarity problems: A survey of theory, algorithms and applications*, Math. Progr. 48, 1990, 161–220.

[85] Hartmann, P. and Stampacchia, G. *On some nonlinear elliptic differential functional equations*, Acta Mathematica 115, 1966, 153–188.

[86] Hassouni, A. *Quasimonotone multifunctions; applications to optimality conditions in quasiconvex programming*, Num. Funct. Anal. and Optim. 13, 1992, 267–275.

[87] He, B. *A class of projection and contraction methods for monotone variational inequalities*, Appl. Math. and Optim. 35, 1997, 69–76.

[88] Jeyakumar, V., Oettli, W., and Natividad, M. *A solvability theorem for a class of quasiconvex mappings with applications to optimization*, J. Math. Anal. Appl. 179, 1993, 537–546.

[89] John, R. *Variational inequalities and pseudomonotone functions: some characterizations*, in: Generalized Convexity, Generalized Monotonicity, J.-P. Crouzeix, J.-E. Martinez-Legaz, and M. Volle, eds., Kluwer Academic Publishers, Dordrecht - Boston - London, 1998, 291–301.

[90] John, R. *A note on Minty variational inequalities and generalized monotonicity*, in: Generalized Convexity and Generalized Monotonicity, N. Hadjisavvas, J.-E. Martinez-Legaz, and M. Penot, eds., Lecture Notes in Economics and Mathematical Systems **502**, Springer-Verlag, Berlin-Heidelberg- New York, 2001, 240–246.

[91] Kachurovskii, R.I. *On monotone operators and convex functionals*, Usp. Matem. Nauk 15, 1960, 213–215.

[92] Kantorovich, L.V. and Akilov, G.P. *Functional analysis in normed spaces*, Fizmatgiz, Moscow, 1959. (in Russian)

[93] Kanzow, C. and Qi, H.-D. *A QP-free constrained Newton-type method for variational inequality problems*, Math. Progr. 85, 1999, 81–106.

[94] Karamardian, S. *Generalized complementarity problems*, J. Optim. Theory Appl. 8, 1971, 161–167.

[95] Karamardian, S. *An existence theorem for the complementarity problem*, J. Optim. Theory Appl. 19, 1976, 227–232.

[96] Karamardian, S. *Complementarity over cones with monotone and pseudomonotone maps*, J. Optim. Theory Appl. 18, 1976, 445–454.

[97] Karamardian, S. and Schaible, S. *Seven kinds of monotone maps*, J. Optim. Theory Appl. 66, 1990, 37–46.

[98] Karamardian, S., Schaible, S., and Crouzeix, J.-P. *Characterizations of generalized monotone maps*, J. Optim. Theory Appl. 76, 1993, 399–413.

[99] Kazmi, K.R. *Existence of solutions for vector optimization*, Appl. Math. Letters 9, 1996, 19–22.

[100] Kazmi, K.R. *Some remarks on vector optimization problems*, J. Optim. Theory Appl. 96, 1998, 133–138.

[101] Kazmi, K.R. *Existence of solutions for vector saddle points problems*, In: Vector Variational Inequalities and Vector Equilibria. Mathematical Theories, F. Giannessi, ed., Kluwer Academic Publishers, Dordrecht-Boston-London, 2000, 267–275.

[102] Kazmi, K.R. *On vector equilibrium problem*, Proc. Indian Acad. Sci. (Math. Sci.) 110, 2000, 213–223.

[103] Khobotov, E.N. *Modification of the extragradient method for solving variational inequalities and certain optimization problems*, USSR Comp. Maths. Math. Phys. 27, 1987, 120–127.

[104] Kinderlehrer, D. and Stampacchia, G. *An introduction to variational inequalities and their applications*, Academic Press, New York, 1980.

[105] Kiwiel, K.C. *Proximal level bandle methods for convex nondifferentiale optimization, saddle point problems and variational inequalities*, Math. Progr. 69, 1995, 89–109.

[106] Knaster, B., Kuratowski, C., and Mazurkiewicz, S., *Ein Beweis des Fixpunktsatzes für N Dimensionale Simplexe*, Fund. Math. 14, 1929, 132–137.

[107] Kneser, H. *Sur le théorème fondamental de la théorie des jeux*, C. R. Acad. Sci. Paris 234, 1952, 2418–2420.

[108] Komlósi, S. *Generalized monotonicity in nonsmooth analysis*, In: Generalized Convexity, S. Komlósi, T. Rapcsák, and S. Schaible, eds., Springer-Verlag, Heidelberg, 1994, 263–275.

[109] Komlósi, S. *On the Stampacchia and Minty variational inequalities*, in: Generalized Convexity and Optimization in Economic and Financial Decisions, G. Giorgi and F. Rossi, eds., Pitagora Editrice, Bologna, 1999, 231–260.

[110] Konnov, I.V. *Combined relaxation methods for finding equilibrium points and solving related problems*, Russ. Math. (Iz. VUZ) 37, 1993, no.2, 44–51.

[111] Konnov, I.V. *A combined method for variational inequalities*, News-Lett. Math. Progr. Assoc. 4, 1993, 64. (in Russian)

[112] Konnov, I.V. *On combined relaxation method's convergence rates*, Russ. Math. (Iz. VUZ) 37, 1993, no.12, 89–92.

[113] Konnov, I.V. *Combined relaxation methods for solving vector equilibrium problems*, Russ. Math. (Iz. VUZ) 39, 1995, no.12, 51–59.

[114] Konnov, I.V. *A general approach to finding stationary points and the solution of related problems*, Comp. Maths. Math. Phys. 36, 1996, 585–593.

[115] Konnov, I.V. *A class of combined iterative methods for solving variational inequalities*, J. Optim. Theory Appl. 94, 1997, 677–693.

[116] Konnov, I.V. *On systems of variational inequalities*, Russ. Math. (Iz. VUZ) 41, 1997, no.12, 77–86.

[117] Konnov, I.V. *Methods for solving finite-dimensional variational inequalities*, DAS, Kazan, 1998. (in Russian)

[118] Konnov, I.V. *A combined relaxation method for variational inequalities with nonlinear constraints*, Math. Progr. 80, 1998, 239–252.

[119] Konnov, I.V. *Accelerating the convergence rate of a combined relaxational method*, Comp. Maths. Math. Phys. 38, 1998, 49–56.

[120] Konnov, I.V. *On the convergence of combined relaxation methods for variational inequalities*, Optim. Meth. Software 80, 1998, 239–252.

[121] Konnov, I.V. *On quasimonotone variational inequalities*, J. Optim. Theory Appl. 99, 1998, 165–181.

[122] Konnov, I.V. *Combined relaxation methods for variational inequalities*, Lecture Notes in Economics and Mathematical Systems **495**, Springer-Verlag, Berlin-Heidelberg- New York, 2001.

[123] Konnov, I.V. *On vector equilibrium and vector variational inequality problems*, in: Generalized Convexity and Generalized Monotonicity, N. Hadjisavvas, J.-E. Martinez-Legaz, and M. Penot, eds., Lecture Notes in Economics and Mathematical Systems **502**, Springer-Verlag, Berlin-Heidelberg- New York, 2001, 247–263.

[124] Konnov, I.V. *Relatively monotone variational inequalities over product sets*, Oper. Res. Letters 28, 2001, 21–26.

[125] Konnov, I.V. and Schaible, S. *Duality for equilibrium problems under generalized monotonicity*, J. Optim. Theory Appl. 104, 2000, 395–408.

[126] Konnov, I.V. and Yao, J.C. *On the generalized vector variational inequality problem*, J. Math. Anal. Appl. 206, 1997, 42–58.

[127] Konnov, I.V. and Yao, J.C. *Existence of solutions for generalized vector equilibrium problems*, J. Math. Anal. Appl. 233, 1999, 328–335.

[128] Korpelevich, G.M. *Extragradient method for finding saddle points and other problems*, Matecon 12, 1976, 747–756.

[129] Lee, G.M., Kim, D.S., Lee, B.S., and Cho, S.J. *Generalized vector variational inequalities and fuzzy extension*, Appl. Math. Letters 6, 1993, 47–51.

[130] Lee, G.M., Kim, D.S., Lee, B.S., and Yen, N.D. *Vector variational inequalities as a tool for studying vector optimization problems*, Nonlin. Anal. 34, 1998, 745–765.

[131] Lee, G.M. and Kum, S. *Vector variational inequalities in a topological vector space*, In: Vector Variational Inequalities and Vector Equilibria. Mathematical Theories, F. Giannessi, ed., Kluwer Academic Publishers, Dordrecht-Boston-London, 2000, 307–320.

[132] Lemaréchal, C., Nemirovskii, A., Nesterov, Y. *New variants of bundle methods* Math. Progr. 69, 1995, 111–147.

[133] Lemke, C.E. *Bimatrix equilibrium points and mathematical programming*, Manag. Sci. 11, 1965, 681–689.

[134] Levin, A.L. *On an algorithm for the minimization of convex functions*, Soviet Math. Doklady 6, 1965, 286–290.

[135] Levin, V.L. *Quasiconvex functions and quasimonotone operators*, J. Convex Analysis 2, 1995, 167–172.

[136] Lin, G.-H. and Xia, Z.-Q. *Some convergence results for pseudomonotone variational inequalities*, Pure Math. Appl. 10, 1999, 173–182.

[137] Lin, K.L., Yang, D.P., and Yao, J.C. *Generalized vector variational inequalities*, J. Optim. Theory Appl. 92, 1997, 117–125.

[138] Lin, L.-J. and Park, S. *On some generalized quasi-equilibrium problems*, J. Math. Anal. Appl. 224, 1998, 167–181.

[139] Lions, J.-L. and Stampacchia, G. *Variational inequalities*, Comm. Pure Appl. Math. 20, 1967, 493–519.

[140] Luc, D.T. *Theory of vector optimization*, Lecture Notes in Mathematics **543**, Springer-Verlag, Berlin, 1989.

[141] Luc, D.T. *Characterizations of quasiconvex functions*, Bull. Austral. Math. Soc. 48, 1993, 393–406.

[142] Luc, D.T. *Existence results for densely pseudomonotone variational inequalities*, J. Math. Anal. Appl. 254, 2001, 291–308.

[143] Luc, D.T. and Schaible, S. *Generalized monotone nonsmooth maps*, J. Convex Analysis 3, 1996, 195–205.

[144] Magnanti, T.L. and Perakis, G. *A unifying geometric framework and complexity analysis for variational inequalities*, Math. Progr. 71, 1995, 327–352.

[145] Marcotte, P. and Zhu, D.L. *A cutting plane method for quasimonotone variational inequalities*, Comp. Math. Appl. 20, 2001, 317–324.

[146] Martinet, B. *Regularization d'inéquations variationnelles par approximations successives*, Rev. Fr. Inf. Rech. Oper. 4, 1970, 154–159.

[147] Martos, B. *Nonlinear programming. Theory and methods*, Akadémiai Kiado, Budapest, 1975.

[148] Minty, G. *On the generalization of a direct method of the calculus of variations*, Bull. Amer. Math. Soc. 73, 1967, 315–321.

[149] Mosco, U. *Implicit variational problems and quasivariational inequalities*, In: Nonlinear Operators and Calculus of Variations, J.P. Gossez et al., eds., Lecture Notes in Mathematics **543**, Springer-Verlag, Berlin, 1976, 83–156.

[150] Nagurney, A. *Network economics: A variational inequality approach*, Kluwer Academic Publishers, Dordrecht, 1993.

[151] Nemirovskii, A.S. *Effective iterative methods for solving equations with monotone operators*, Ekon. Matem. Met. (Matecon) 17, 1981, 344–359.

[152] Nemirovskii, A.S. and Yudin, D.B. *Problem complexity and method efficiency in optimization*, Nauka, Moscow, 1979. (in Russian; Engl. transl. in John Wiley and Sons, New York,1983).

[153] von Neumann, J. and Morgenstern, O. *Game theory and economic behaviour*, Princeton University Press, Princeton, 1953.

[154] Nikaido, H. *Convex structures and economic theory*, Academic Press, New York, 1968.

[155] Nikaido, H. and Isoda, K. *Note on noncooperative convex games*, Pacific J. Math. 5, 1955, 807–815.

[156] Nirenberg, L. *Topics in nonlinear functional analysis*, New York University Press, New York, 1974.

[157] Oettli, W. *A remark on vector-valued equilibria and generalized monotonicity*, Acta Mathem. Vietnam. 22, 1997, 213–221.

[158] Oettli, W. and Schläger, D. *Generalized vectorial equilibria and generalized monotonicity*, in Functional Analysis with Current Applications in Science, Technology and Industry, M. Brokate and A.H. Siddiqi, eds., Pitman Research Notes in Mathematics Series **377**, Longman Publishing Co., Essex, 1998, 145–154.

[159] Ortega, J.M. and Rheinboldt, W.C. *Iterative solution of nonlinear equations in several variables*, Academic Press, New York, 1970.

[160] Patriksson, M. *Nonlinear programming and variational inequality problems: A unified approach*, Kluwer Academic Publishers, Dordrecht, 1999.

[161] Pellegrini, L. *Vector formulation of some economic equilibrium problems*, in: Generalized Convexity and Optimization in Economic and Financial Decisions, G. Giorgi and F. Rossi, eds., Pitagora Editrice, Bologna, 1999, 323–331.

[162] Penot, J.-P. and Quang, P.H. *Generalized convexity of functions and generalized monotonicity of set-valued maps*, J. Optim. Theory Appl. 92, 1997, 343–356.

[163] Podinovskii, V.V. and Nogin, V.D. *Pareto-optimal solutions of multicriteria problems*, Nauka, Moscow, 1982 (in Russian).

[164] Rockafellar, R.T. *Convex analysis*, Princeton University Press, Princeton, 1970.

[165] Rockafellar, R.T. *Monotone operators and the proximal point algorithm*, SIAM J. Contr. Optim. 14, 1976, 877–898.

[166] Rockafellar, R.T. *The theory of subgradients and its applications to problems of optimization*, Heldermann-Verlag, Berlin, 1981.

[167] Rosen, J.B. *Existence and uniqueness of equilibrium points for concave n-person games*, Econometrica 33, 1965, 520–534.

[168] Sach, P.H. *Characterization of scalar quasiconvexity and convexity of locally Lipschitz vector valued maps*, Optimization 46, 1999, 283–310.

[169] Saigal, R. *Extension of the generalized complementarity problem*, Math. Oper. Res. 1, 1976, 260–266.

[170] Schaible, S. *Generalized monotone maps*, In: Proceedings of a Conference Held at "G. Stampacchia International School of Mathematics", F. Giannessi, ed., Gordon and Breach Science Publishers, Amsterdam, 1992, 392–408.

[171] Schaible, S. *Generalized monotonicity - concepts and uses*, in: Variational Inequalities and Network Equilibrium Problems, F. Giannessi and A. Maugeri, eds., Plenum Press, New York, 1995, 289–299.

[172] Schaible, S. *Criteria for generalized monotonicity*, in: New Trends in Mathematical Programming, F. Giannessi, S. Komlósi, and T. Rapcsák, eds., Kluwer Academic Publishers, Dordrecht, 1998, 277–288.

[173] Shih, M.H. and Tan, K.K. *Browder-Hartmann-Stampacchia variational inequalities for multivalued monotone operators*, J. Math. Anal. Appl. 134, 1988, 431–440.

[174] Shor, N.Z. *Cut-off method with space dilation in convex programming problems*, Cybernetics 13, 1977, 94–96.

[175] Shor, N.Z. *New development trends in nonsmooth optimization methods*, Cybernetics 6, 1977, 87–91.

[176] Shor, N.Z. *Minimization methods for non-differentiable functions*, Naukova Dumka, Kiev, 1979. (in Russian; Engl. transl. in Springer-Verlag, Berlin, 1985).

[177] Solodov, M.V. and Tseng, P. *Modified projection-type methods for monotone variational inequalities*, SIAM J. Contr. Optim. 34, 1996, 1814–1830.

[178] Stampacchia, G. *Formes bilinéaires coercitives sur le ensembles convexes*, C. R. Acad. Sci. Paris 258, 1964, 4413–4416.

[179] Sun, D. *A new step-size skill for solving a class of nonlinear projection equations*, J. Comp. Math. 13, 1995, 357–368.

[180] Sun, D. *A class of iterative methods for solving nonlinear projection equations*, J. Optim. Theory Appl. 91, 1996, 123–140.

[181] Tanaka, T. *Cone convexity of vector-valued functions*, Sci. Rep. Hirosaki Univ. 37, 1990, 170–177.

[182] Tanaka, T. *Generalized quasiconvexities, cone saddle points, and minimax theorems for vector-valued functions*, J. Optim. Theory Appl. 81, 1994, 355–377.

[183] Tanaka, T. *Cone quasiconvexity of vector-valued functions*, Sci. Rep. Hirosaki Univ. 42, 1995, 157–163.

[184] Tanino, T. and Sawaragi, Y. *Duality theory in multiobjective optimization*, J. Optim. Theory Appl. 27, 1979, 509–529.

[185] Tikhonov, A.N. and Arsenin, A.V. *Solution of ill-posed problems*, John Wiley and Sons, New York, 1977.

[186] Tseng, P. *On linear convergence of iterative methods for the variational inequality problem*, J. Comp. Appl. Math. 60, 1995, 237–252.

[187] Uryas'yev, S.P. *Adaptive algorithms of stochastic optimization and game theory*, Nauka, Moscow, 1990. (in Russian)

[188] Yang, X.Q. *Vector variational inequality and its duality*, Nonlin. Anal., Theory, Methods and Appl. 21, 1993, 869–877.

[189] Yang, X.Q. *Vector complementarity and minimal element problems*, J. Optim. Theory Appl. 77, 1993, 483–495.

[190] Yang, X.Q. and Goh, C.J. *On vector variational inequalities: Applications to vector equilibria*, J. Optim. Theory Appl. 95, 1997, 431–443.

[191] Yao, J.C. *Variational inequalities with generalized monotone operators*, Math. Oper. Res. 19, 1994, 691–705.

[192] Yao, J.C. *Multi-valued variational inequalities with K - pseudomonotone operators*, J. Optim. Theory Appl. 83, 1994, 391–403.

[193] Yen, N.D. and Lee, G.M. *On monotone and strongly monotone vector variational inequalities*, in: Generalized Convexity and Optimization in Economic and Financial Decisions, G. Giorgi and F. Rossi, eds., Pitagora Editrice, Bologna, 1999, 347–358.

[194] Yu, S.J. and Yao, J.C. *On vector variational inequalities*, J. Optim. Theory Appl. 89, 1996, 749–769.

[195] Yuan, G.X.Z. *The study of minimax inequalities and applications to economies and variational inequalities*, Memoires of the American Mathematical Society **132**, Number 625, American Mathematical Society, Providence, 1998.

[196] Yudin, D.B. and Nemirovskii, A.S. *Informational complexity and efficient methods for the solution of convex extremal problems, I, II*, Matecon 13, 1977, 25–45; 550–559.

[197] Zarantonello, E.H. *Projections on convex sets in Hilbert space and spectral theory*, in: Contributions to Nonlinear Functional Analysis,

E.H. Zarantonello, ed., Academic Press, New York- London, 1971, 237–424.

[198] Zhou, J.X., and Chen, G. *On diagonal convexity conditions for problems in convex analysis and quasi-variational inequalities*, J. Math. Anal. Appl. 132, 1988, 213–225.

Chapter 14

USES OF GENERALIZED CONVEXITY AND GENERALIZED MONOTONICITY IN ECONOMICS

Reinhard John
Department of Economics
University of Bonn
Adenauerallee 24-42
53113 Bonn, Germany
rjohn@uni-bonn.de

Abstract This chapter presents some uses of generalized concavity and generalized monotonicity in consumer theory and general equilibrium theory. The first part emphasizes the relationship between generalized monotonicity properties of individual demand and axioms of revealed preference theory. The second part points out the relevance of pseudomonotone market excess demand to a well-behaved general equilibrium model. It is shown that this property can be derived from assumptions on the distribution of individual (excess) demands.

Keywords: Consumer, demand, economy, equilibrium, generalized monotonicity, preference, revealed preference, utility, variational inequality.

1. Introduction

Probably the first use of generalized monotonicity in economics was made already 70 years ago. Even more remarkable, it occurred independently and almost at the same time in two seminal articles. The one, by Georgescu-Roegen [13], dealt with the concept of a local preference in consumer theory, the other, by Wald [45], contained the first rigorous proof of the existence of a competitive general equilibrium. We shall outline their contributions in this introduction, borrowing some insights from the illuminating papers by Shafer [42] and Kuhn [32].

1.1 Georgescu-Roegen's Local Preferences

Imagine a consumer who considers alternative consumption bundles in a neighborhood of some given vector $x \in \mathbb{R}^l_+$ that describes the amounts of l consumption goods initially given. The consumer is able to distinguish the possible directions in which he can move away from x according to his taste. More precisely, there are three kinds of directions: Preference, nonpreference, and indifference. This idea had been already formalized in [2] by postulating the existence of a (row) vector $g(x) \in \mathbb{R}^l$ such that a direction $v \in \mathbb{R}^l$ is a preference (resp. nonpreference, indifference) direction if and only if

$$g(x)v > 0 \text{ (resp. } < 0, = 0).$$

It should be emphasized that these comparisons are supposed to be possible only locally. The consumer may not be able to compare x with a bundle y that is very different from x. In addition, the vector $g(x)$ typically varies with x (otherwise, one would obtain the special case of a global preference representing perfect substitutes).

Assume now that x is contained in some given set $X \subseteq \mathbb{R}^l_+$ that represents the consumer's feasible bundles. Georgescu-Roegen calls x an "equilibrium position or a point of saturation" relative to X if no direction away from x to any other alternative y in X is one of preference, i.e. if

$$g(x)(y - x) \leq 0 \quad \text{for all } y \in X.$$

Thus, x is an equilibrium consumption bundle if, in modern terminology, x is a solution to the Stampacchia variational inequality problem with respect to g on X.[1]

By assuming that the consumer always selects such an equilibrium out of a set of alternatives provided it exists, Georgescu-Roegen's concept of a local preference could be a foundation of a demand theory. However, for that he needed existence of an equilibrium bundle, at least in budget sets determined by prices and income.

Since he also wanted the equilibrium to be stable, he formulated another assumption that he called the "*principle of persisting nonpreference*":

If the consumer moves away from an arbitrary bundle x to a bundle $x + \Delta x$ such that Δx is not a preference direction, then Δx is a nonpreference direction at $x + \Delta x$. Formally stated, this principle says

$$g(x)\Delta x \leq 0 \quad \text{implies} \quad g(x + \Delta x)\Delta x < 0,$$

[1] Notice the reverse inequality compared to the usual definition. This convention is employed throughout the chapter.

or, by denoting $x + \Delta x = y$,

$$g(x)(y-x) \leq 0 \quad \text{implies} \quad g(y)(y-x) < 0.$$

Clearly, this property is nothing else than strict pseudomonotonicity of g.[2]

In a sequel contribution, Georgescu-Roegen [14] employed a slightly different formulation of his principle of persisting nonpreference:

An indifference or nonpreference direction remains one of indifference or nonpreference if the consumer extends his consumption according to that direction, or, formally,

$$g(x)(y-x) \leq 0 \quad \text{implies} \quad g(y)(y-x) \leq 0,$$

i.e. g is pseudomonotone.

Georgescu-Roegen has also shown that the set of equilibrium points relative to a feasible set X is convex. In addition, he observed that these equilibria are, in modern terminology, solutions to the Minty variational inequality problem with respect to g on X.

More surprisingly, Georgescu-Roegen [14] proved the existence of a consumption equilibrium in a simplex. Of course, he assumed continuity of g which, as we know, is sufficient for the existence of a solution to the Stampacchia variational inequality problem with respect to g. However, he did not use any advanced argument like Brouwer's fixed point theorem. Instead, he succeeded by induction on the dimension of the simplex. For that, he used pseudomonotonicity of g. Actually, it is the same idea that Wald employed in [45] in order to prove the existence of a competitive equilibrium of a production economy.

1.2 Wald's Equilibrium Existence Proof

Wald considered a general equilibrium model of a production economy where the production sector is characterized by fixed input coefficients. Its origin is the simplification of Walras' famous production equations in [7], [8], [46], and [39]. We shall present here a slightly different version of his model that is due to Kuhn [32]. For a historically closer but nevertheless modern treatment we refer to Hildenbrand [18].

The economic environment is described by m pure factors of production $i = 1, ..., m$ which are available in fixed positive amounts $r = (r_1, ..., r_m)$ and which can be used to produce n final commodities $j = 1, ..., n$. The production of one unit of commodity j requires $a_{ij} \geq 0$ units

[2]In this chapter, all notions of (generalized) monotonicity are defined in the sense of generalizing a *nonincreasing* real valued function of one real variable.

of the primary resource i, i.e. the production sector is characterized by the input coefficient matrix $A = (a_{ij})_{m\times n}$.

Assume that price vectors $p = (p_1, ..., p_n) \geq 0$ for final commodities and $q = (q_1, ..., q_m) \geq 0$ for factors are given. Then the output vector $x = (x_1, ..., x_n) \geq 0$ maximizes profits if unit costs are greater or equal than output price, equality being implied by a nonzero output quantity, i.e. if

$$qA \geq p \quad \text{and} \quad (qA - p)x = 0, \tag{14.1}$$

where prices are represented by row vectors, quantities by column vectors.

All factor markets are in (free disposal) equilibrium if on every market demand is less than or equal to supply, equality being implied by a positive factor price, i.e. if

$$Ax \leq r \quad \text{and} \quad q(Ax - r) = 0. \tag{14.2}$$

By the duality theory of linear programming, it is well known that (14.1) and (14.2) are equivalent to the statement that x and q solve the dual problems

$$\max px \quad \text{subject to} \quad x \geq 0 \text{ and } Ax \leq r$$

and

$$\min qr \quad \text{subject to} \quad q \geq 0 \text{ and } qA \geq p,$$

or, equivalently, that p, q, and x satisfy the conditions

$$Ax \leq r,\ qA \geq p, \quad \text{and} \quad px = qr.$$

It remains to close the model by an equilibrium condition for the product markets. According to Schlesinger's modification of Cassel's equations this is accomplished by assuming the existence of an inverse demand function f, i.e. by

$$p = f(x), \tag{14.3}$$

which means that the vector x of final commodities is demanded if and only if their prices are given by $f(x)$.

It was essentially the system of conditions (14.1), (14.2), and (14.3) for which Wald proved the existence of a solution (p^*, q^*, x^*). His proof is based on the following assumptions:

(W1) For every $j \in \{1, ..., n\}$ there exists $i \in \{1, ..., m\}$ such that $a_{ij} > 0$.

(W2) $f : \mathbb{R}^n_+ \longrightarrow \mathbb{R}^n_{++}$ is continuous.

(W3) For every $x, x' \in \mathbb{R}^n_+$ such that $x \neq x'$,

$$f(x)(x' - x) \leq 0 \quad \text{implies} \quad f(x')(x' - x) < 0.$$

Assumption (W1) guarantees that the production possibilities are bounded, i.e. $X = \{x \geq 0 \,|\, Ax \leq r\}$ is bounded. This implies for any $p \geq 0$ the existence of an element $x^* \in X$ that maximizes px on X, i.e. the primal linear programming problem mentioned above has the optimal solution x^*.

By the duality theorem of linear programming, it follows that the dual problem can be solved by some q^*. Consequently, (p, q^*, x^*) satisfies (14.1) and (14.2) but not necessarily $p = f(x^*)$.

Thus, we have to find a vector x^* that maximizes $f(x^*)x$ on X, i.e. we look for $x^* \in X$ such that for all $x \in X$

$$f(x^*)x \leq f(x^*)x^*,$$

or, equivalently,

$$f(x^*)(x - x^*) \leq 0.$$

In modern terminology, we have to solve a variational inequality problem. Assumption (W2) guarantees the existence of a solution which, as Kuhn [32] has shown, can be proved by applying Kakutani's fixed point theorem. However, this result was not available to Wald.

On the other hand, we have not used (W3) which states that f is strictly pseudomonotone. Obviously, this assumption is not necessary for an existence proof. What was its use for Wald?

The crucial point is that strict pseudomonotonicity of f ensures a unique solution to the variational inequality problem with respect to f on *every* convex subset of X. It is this consequence that enabled Wald to prove existence by induction on the number n of final commodities.

Remember that Georgescu-Roegen proceeded in the same way. An elementary induction proof that resembles their contributions has been given in [27] for the general case of a pseudomonotone variational inequality problem on a compact and convex subset of a finite dimensional Euclidean space.

1.3 An Outline of this Chapter

In the sequel, we shall present some uses of generalized concavity and generalized monotonicity in the same two fields of research that contain the contributions by Georgescu-Roegen and Wald, i.e. consumer theory and general equilibrium theory. Of course, it is not claimed that there are no applications in other fields as, for example, game theory.[3] However, since we want to focus on generalized monotonicity, we have selected those which seem to be more relevant to that part of the subject. The reader who is interested in more applications of generalized concavity is referred to [4].

Part 2 deals with consumer theory. The first section introduces the classical approach which employs the notion of a utility function that represents a transitive preference relation. We stress the role played by quasiconcavity or pseudoconcavity of that function.

Section 2.2 considers the central concept of demand. It is shown that the demand relation derived from a transitive preference displays generalized monotonicity properties that are, in the economic literature, well known as axioms of revealed preference theory.

In the final section of this part, we extend the classical approach by studying nontransitive preferences (a convenient abbreviation for "not necessarily transitive" which would be the correct expression). If these relations are convex (resp. semistrictly or strictly convex), various generalized monotonicity properties of demand arise quite naturally. Like in the case of transitive preferences, these are related to well known revealed preference axioms. However, in order to obtain a fully satisfactory revealed preference theory, some open problems have still to be solved. The section concludes with the first order characterization of pseudomonotone continuously differentiable demand that will be used in the two final sections of this chapter.

Part 3 deals with general equilibrium theory. Its first section points out the relationship to variational inequalities (see e.g. [10]). In particular, the relevance of pseudomonotone excess demand to the stability of an equilibrium is recognized by its representability as a solution to a Minty variational inequality problem. The main insight can be easily visualized by a ball in a pseudoconvex landscape (see Figure 1 below).

If the ball moves according to the force of gravity, it will move away from any point like A or C towards B. As a rest point with respect to the movement of the ball, B represents an equilibrium of the physical system. It is unique if the landscape is strictly pseudoconvex. If there is a flat region around B, the set of equilibria is at least convex.

Notice that these conclusions are not valid if the landscape is only (semistrictly) quasiconvex. Indeed, that case allows a vanishing slope at points like A or C such that these could be equilibria too (although not stable ones if the quasiconvexity is semistrict). On the other hand,

[3] See e.g. the excellent survey by Harker and Pang in [15].

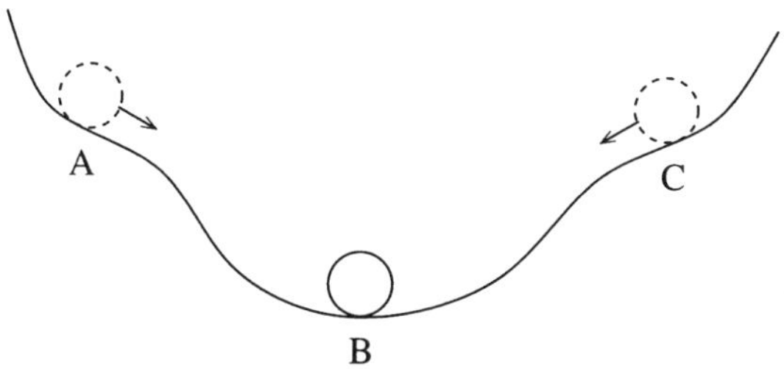

Figure 1

observe that a (strictly) convex landscape is not needed for stability. This observation is important since an economic system may satisfy conditions that ensure a pseudomonotone excess demand but fails to fulfill stronger requirements implying monotonicity.

The final Sections 3.2 and 3.3 provide two standard examples for a general equilibrium model. We shall not present the most general versions. For example, set-valued excess demand as well as production will not be considered.[4] This seems to be acceptable in order to convey the main idea as simply as possible.

While Section 3.1 points out the importance of a pseudomonotone excess demand function for a well-behaved economic model, Sections 3.2 and 3.3 investigate conditions that ensure this property. It will be argued there that distributional assumptions are appropriate in order to reach that goal. In this respect, the presentation will follow closely the contributions by Hildenbrand [17] and Jerison [25].

2. Consumer Theory

2.1 Preference and Utility

We consider an economy with a finite (typically large) number l of consumption goods. A vector $x = (x_i)_{i=1}^{l} \in \mathbb{R}_+^l$ describing the consumption of x_i units of commodity i for $i = 1, ..., l$ is called a *consumption bundle.*

The traditional approach (see e.g. [36]) assumes that a consumer's taste is characterized by a binary relation $\succeq$ on the *consumption set* $\mathbb{R}_+^l$ with the following two basic properties.

Completeness: $\forall x, y \in \mathbb{R}_+^l : x \succeq y \vee y \succeq x.$

[4]For such an exposition of general equilibrium theory, [36] is highly recommended.

Transitivity: $\forall x, y, z \in \mathbb{R}^l_+ : x \succeq y \wedge y \succeq z \Rightarrow x \succeq z.$

The relation $\succeq$ is called the consumer's *preference*. For $x, y \in \mathbb{R}^l_+$, $x \succeq y$ is interpreted as "x is at least as good as y" or "x is weakly preferred to y". While transitivity follows naturally from this interpretation, completeness means that the consumer is able to compare any two commodity bundles. Observe that the latter property implies that $\succeq$ is *reflexive*, i.e. $x \succeq x$ for all $x \in \mathbb{R}^l_+$.

There are two other relations that can be derived from $\succeq$. The *strict preference* $\succ$ is defined by

$$x \succ y : \Leftrightarrow x \succeq y \wedge \neg y \succeq x \tag{14.4}$$

and interpreted as "x is better than y".

If $x \succeq y$ but not $x \succ y$, x and y are called to be *indifferent* which is denoted by $x \sim y$. Obviously, $x \sim y$ if and only if $x \succeq y$ and $y \succeq x$.

Usually, it is also assumed that $\succeq$ is *continuous*, i.e. $\succeq$ (as a subset of $\mathbb{R}^{2l}_+$) is closed in $\mathbb{R}^{2l}_+$.

Continuous preferences are precisely those relations on $\mathbb{R}^l_+$ which can be represented by a continuous real valued function on $\mathbb{R}^l_+$, i.e. we obtain

Proposition 14.1 *A binary relation $\succeq$ is a continuous preference if and only if there is a continuous function $u : \mathbb{R}^l_+ \to \mathbb{R}$ such that for all $x, y \in \mathbb{R}^l_+$*

$$x \succeq y \Leftrightarrow u(x) \geq u(y). \tag{14.5}$$

Proof. It is obvious that a relation $\succeq$ satisfying (14.5) for a continuous function u is complete, transitive, and continuous. A proof of the reverse implication can be found in [12]. □

A real valued function u that satisfies (14.5) is called a *utility representation* (or *utility function*) of the preference $\succeq$. Observe that any monotone transformation of u also represents $\succeq$, i.e. u is far from being unique. In the sequel, we suppose that u is some fixed continuous utility function representing the continuous preference $\succeq$.

In many cases it is natural to assume that the consumption of goods are desirable, at least if they are consumed together. This is made precise by

Definition 14.1 *The preference $\succeq$ (resp. the utility function u) is called*

(1) monotone if for all $x, y \in \mathbb{R}^l_+$, $x \gg y$ implies $x \succ y$ (resp. $u(x) > u(y)$),

(2) *strictly monotone if for all* $x, y \in \mathbb{R}^l_+$, $x > y$ *implies* $x \succ y$ *(resp.* $u(x) > u(y)$*).*[5]

There is a weaker assumption that allows the presence of some "bads".

Definition 14.2 *The preference* $\succeq$ *(resp. the utility function* u*) is called locally nonsatiated with respect to* $X \subseteq \mathbb{R}^l_+$ *at* $x \in X$ *if for every neighborhood* N *(w.r.t.* X*) of* x *there is* $y \in N$ *such that* $y \succ x$ *(resp.* $u(y) > u(x)$*).* $\succeq$ *(resp.* u*) is locally nonsatiated if it is locally nonsatiated with respect to* $\mathbb{R}^l_+$ *at every* $x \in \mathbb{R}^l_+$*.*

The ultimate goal of consumer theory is to derive a useful theory of demand from the hypothesis that a consumer maximizes utility on (convex) budget sets. For that, the following basic notions of convexity turn out to be important.

Definition 14.3 *The preference* $\succeq$ *is called*

(1) *convex, if* $x \succeq y$ *implies* $\lambda x + (1-\lambda)y \succeq y$

(2) *semistrictly convex, if* $x \succ y$ *implies* $\lambda x + (1-\lambda)y \succ y$

(3) *strictly convex, if* $x \succeq y$ *implies* $\lambda x + (1-\lambda)y \succ y$

for all $x, y \in \mathbb{R}^l_+$ *such that* $x \neq y$ *and all* $\lambda \in]0,1[$.

Obviously, (3) implies (2) and, by continuity of $\succeq$, (2) implies (1). The latter implication is proved in [12] where a somewhat different terminology is used. The notation employed here seems to be more appropriate because of the following straightforward relationship(cf.[4]).

Proposition 14.2 *The preference* $\succeq$ *is (strictly, semistrictly) convex if and only if the utility representation* u *of* $\succeq$ *is (strictly, semistrictly) quasiconcave.*

The economic rationale for convexity are "nonincreasing (decreasing) rates of substitution". This property is easily explained for the case of two commodities. Consider a monotone and convex preference on $\mathbb{R}^2_+$. Convexity means that the set X of all bundles that are weakly preferred to some consumption bundle x is convex, i.e. it typically looks like X in Figure 2 below.

The boundary of X consists of all bundles that are indifferent to x. Compare now the amount of good 2 that would compensate the consumer for consuming one unit less of good 1 at the two points x and y.

[5] $x \gg y$ means $x_i > y_i$ for $i = 1, ...l$, and $x > y$ means $x \geq y$ and $x \neq y$.

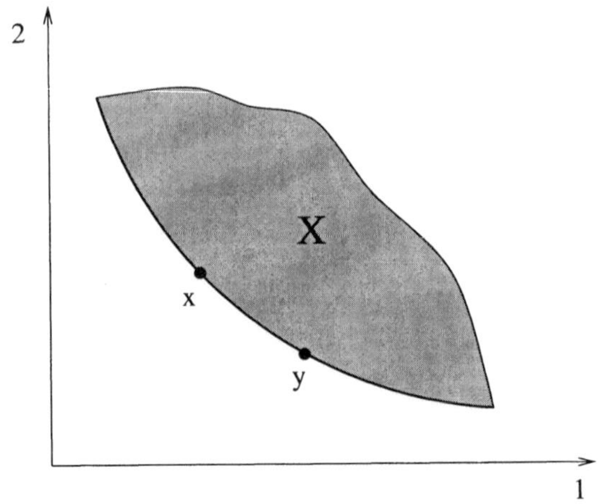

Figure 2

By convexity, it cannot be larger at y than at x (it is actually smaller at y if the preference is strictly convex). The quite intuitive interpretation is that additional consumption is less valuable if the quantity already consumed is larger.

What is added to convexity if the preference is semistrictly convex? An answer is provided by the following characterization.

Proposition 14.3 *Let the preference $\succeq$ be convex. Then $\succeq$ is semistrictly convex if and only if for every convex subset X of $\mathbb{R}^l_+$ the preference $\succeq$ is locally nonsatiated with respect to X at any $x \in X$ that is not a global satiation point of X, i.e. at any $x \in X$ such that there is $y \in X$ with $y \succ x$.*

Proof. Let $\succeq$ be semistrictly convex and let X be a convex subset of $\mathbb{R}^l_+$. Assume that $x, y \in X$ such that $y \succ x$. Since $\lambda y + (1-\lambda)x \succ x$ for arbitrary small $\lambda > 0$ there is $z \in X$ with $z \succ x$ in every neighborhood of x. Thus, $\succeq$ is locally nonsatiated w.r.t. X at x.

In order to prove the converse, assume that $\succeq$ is not semistrictly convex, i.e. there are $x, y \in \mathbb{R}^l_+$ and $\mu \in \,]0,1[$ such that $x \succ y$ and $y \succeq \mu x + (1-\mu)y = z$. It will be shown that $y \succeq z_\lambda = \lambda x + (1-\lambda)y$ for all $\lambda \in [0, \mu]$ (see Fig. 3).

Observe that $x \succ y \succeq z$ implies that $x \succ z_\lambda$ since otherwise, by convexity of $\succeq$, $z \succeq x$. From $x \succ z_\lambda$ it follows, again by convexity of $\succeq$, that $z \succeq z_\lambda$ and, by transitivity, $y \succeq z_\lambda$.

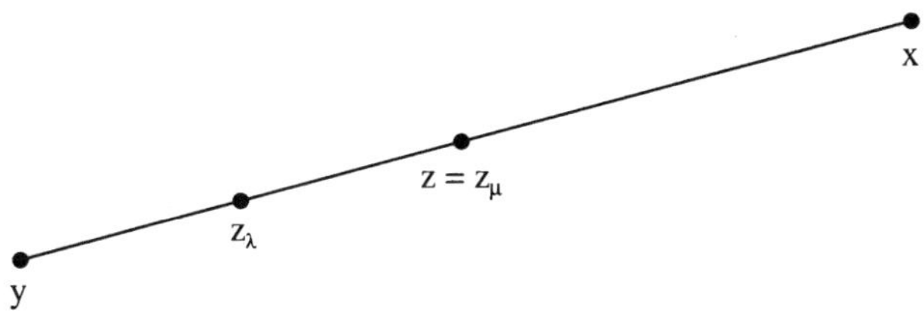

Figure 3

Consequently, $\succeq$ is locally satiated at y with respect to the line segment $[y, x]$ although y is not a global satiation point of that convex set.

□

The most important implication of a convex preference is the convexity (or uniqueness in case of strictly convex preferences) of the set of utility maximizers on a convex set. This property is crucial for the proof of the existence of a market equilibrium. Moreover, it characterizes continuous (strictly) convex preferences as shown by

Proposition 14.4 *Let u be a continuous utility function. Then*

(1) u is quasiconcave if and only if for every convex subset X of $\mathbb{R}^l_+$ the set $X^* = \{x^* \in X \mid \forall x \in X : u(x^*) \geq u(x)\}$ is convex.

(2) u is strictly quasiconcave if and only if for every convex subset X of $\mathbb{R}^l_+$ there is at most one $x^* \in X$ such that $u(x^*) \geq u(x)$ for all $x \in X$.

Proof. (1) Let u be quasiconcave and let X be an arbitrary convex subset of $\mathbb{R}^l_+$. If $x_1, x_2 \in X^*$, then $u(x_1) \geq u(x_2)$ and, by quasiconcavity of u, $u(\lambda x_1 + (1-\lambda)x_2) \geq u(x_2) \geq u(x)$ for all $\lambda \in]0,1[$ and all $x \in X$, i.e. $\lambda x_1 + (1-\lambda)x_2 \in X^*$.

In order to prove the converse, assume that u is not quasiconcave, i.e. there are $x_1, x_2 \in \mathbb{R}^l_+$ and $\mu \in]0,1[$ such that $u(x_1) \geq u(x_2)$ and $u(\mu x_1 + (1-\mu)x_2) < u(x_2)$. Define λ_1, λ_2 by

$$\lambda_1 = \max\{\lambda \in [0,\mu] |\, u(\lambda x_1 + (1-\lambda)x_2) \geq u(x_2)\},$$

$$\lambda_2 = \min\{\lambda \in [\mu,1] |\, u(\lambda x_1 + (1-\lambda)x_2) \geq u(x_2)\}$$

(see Figure 4 below).

By continuity of u, λ_1 and λ_2 exist. Furthermore, $\lambda_1 < \mu < \lambda_2$. Thus, for $X = \{\lambda x_1 + (1-\lambda)x_2 \mid \lambda_1 \leq \lambda \leq \lambda_2\}$ we obtain $X^* =$

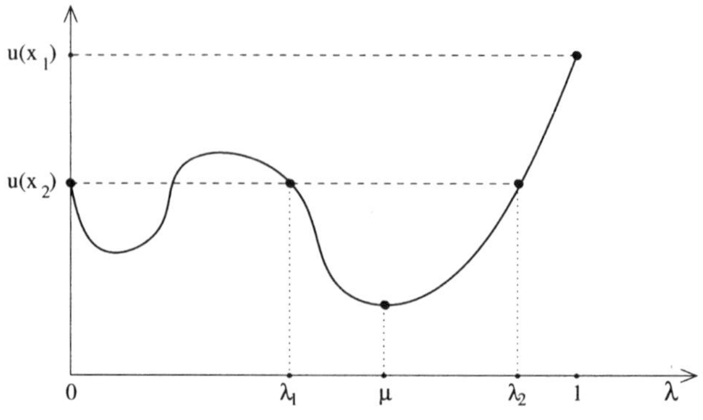

Figure 4

$\{\lambda_1 x_1 + (1-\lambda_1)x_2,\ \lambda_2 x_1 + (1-\lambda_2)x_2\}$. Since X^* is not convex, (1) is proved.

(2) Let u be strictly quasiconcave and let X be an arbitrary convex subset of $\mathbb{R}^l_+$. If $x_1, x_2 \in X$ maximize u on X, then $u(x_1) = u(x_2)$. It follows that $x_1 = x_2$ since otherwise $u(\frac{1}{2}x_1 + \frac{1}{2}x_2) > u(x_2) = u(x_1)$, i.e. x_1 and x_2 would not maximize u on X.

Now, assume that u is not strictly quasiconcave, i.e. there are $x_1, x_2 \in \mathbb{R}^l_+$ and $\mu \in]0,1[$ such that $x_1 \neq x_2$, $u(x_1) \geq u(x_2)$, and $\alpha := u(\mu x_1 + (1-\mu)x_2) \leq u(x_2)$. If there are $\mu_1 \in [0,\mu[$ and $\mu_2 \in]\mu, 1]$ such that $u(\mu_1 x_1 + (1-\mu_1)x_2)$, $u(\mu_2 x_1 + (1-\mu_2)x_2) > \alpha$, then u is not quasiconcave and we can apply part (1). Otherwise, either $u(\lambda x_1 + (1-\lambda)x_2) \leq \alpha$ for all $\lambda \in [0,\mu]$ or $u(\lambda x_1 + (1-\lambda)x_2) \leq \alpha$ for all $\lambda \in [\mu, 1]$. In the first case, x_2 and $\mu x_1 + (1-\mu)x_2$ maximize u on $X = \{\lambda x_1 + (1-\lambda)x_2 \,|\, 0 \leq \lambda \leq \mu\}$. In the second case, $\mu x_1 + (1-\mu)x_2$ and x_1 maximize u on $X = \{\lambda x_1 + (1-\lambda)x_2 \,|\, \mu \leq \lambda \leq 1\}$. Since in both cases there are two maximizers, (2) is proved. □

Let us now assume that the utility function u is differentiable. In that case it is desirable to determine a utility maximizing consumption bundle by a simple first order condition. In general, however, we only obtain the following separate necessary and sufficient conditions.

Proposition 14.5 *Let u be differentiable and let X be a convex subset of $\mathbb{R}^l_+$. Then, for $x^* \in X$ the condition*

$$\text{(N)} \qquad \partial u(x^*)(x - x^*) \leq 0 \quad \text{for all} \quad x \in X$$

is necessary, and the condition

$$\text{(S)} \qquad \partial u(x)(x^* - x) \geq 0 \quad \text{for all} \quad x \in X$$

is sufficient for x^* to maximize u on X.[6]

Proof. If x^* maximizes u on X, then for any $x \in X$ such that $x \neq x^*$ and any $\lambda \in [0,1]$ the inequality $u(x^* + \lambda(x - x^*)) - u(x^*) \leq 0$ holds. Hence,

$$\partial u(x^*)(x - x^*) = \lim_{\lambda \to 0} \frac{1}{\lambda} \left[u(x^* + \lambda(x - x^*)) - u(x^*) \right] \leq 0,$$

which proves that (N) is necessary.

In order to show that (S) is sufficient, assume that x^* does not maximize u on X, i.e. there exists $x \in X$ such that $u(x) > u(x^*)$. By the mean value theorem there exists $\alpha \in]0,1[$ such that

$$\partial u(x + \alpha(x^* - x))(x^* - x) = u(x^*) - u(x) < 0.$$

For $\bar{x} = x + \alpha(x^* - x)$ it follows that $x^* - \bar{x} = (1 - \alpha)(x^* - x)$ and, consequently, $\partial u(\bar{x}) \frac{1}{1-\alpha}(x^* - \bar{x}) < 0$. Thus, we obtain, $\partial u(\bar{x})(x^* - \bar{x}) < 0$ which contradicts (S). □

Condition (N) means that x^* is a solution to the Stampacchia variational inequality problem with respect to ∇u on X and condition (S) means that x^* is a solution to the Minty variational inequality problem with respect to ∇u on X (see [31]). Clearly, (N) is a simpler condition than (S) since it requires knowledge of the gradient of u only at x^*. Therefore, we would be in a nice situation if u would have the property that (N) implies (S).

It is known (see [26]) that (N) implies (S) for any convex subset X of $\mathbb{R}^l_+$ if and only if the gradient map of u is *pseudomonotone*, i.e. if

$$\partial u(x)(y - x) \leq 0 \quad \text{implies} \quad \partial u(y)(y - x) \leq 0$$

for all $x, y \in \mathbb{R}^l_+$.

It is also known that pseudomonotonicity of the gradient map is equivalent to *pseudoconcavity* of u. Put differently, the pseudoconcave utility functions are precisely those functions for which (N) characterizes a utility maximizing element for every convex subset of $\mathbb{R}^l_+$.

Fortunately, the difference between pseudo- and quasiconcavity is not large. Indeed, as shown in [9], if the gradient of u does not vanish on an open and convex domain X, then u is pseudoconcave on X if and only if u is quasiconcave on X.

---

[6] $\partial u(x)$ denotes the Jacobian matrix of u at x, i.e. $\partial u(x) = \nabla u(x)^T$.

2.2 Demand and Revealed Preference

Let $p = (p_1, ..., p_l) \gg 0$ be a market price vector for the various consumption goods and let the consumer's wealth be given by $w > 0$. Then his *budget set* consists of all bundles $x \in \mathbb{R}^l_+$ that are not more expensive than w, i.e.

$$\mathcal{B}(p, w) = \{x \in \mathbb{R}^l_+ \mid px \leq w\},$$

where the price vector p is always considered as a row vector.

The *utility maximization hypothesis* states that the consumer chooses a consumption bundle $x^* \in \mathcal{B}(p, w)$ such that

$$u(x^*) \geq u(x) \quad \text{for all} \quad x \in \mathcal{B}(p, w).$$

The set of all such x^* is called the consumer's *demand* at the price vector p and wealth w. Since $\mathcal{B}(\lambda p, \lambda w) = \mathcal{B}(p, w)$ for all $\lambda > 0$, one can represent all possible budget sets by restricting the wealth to be equal to one. Put differently, p is interpreted as the vector of the price-income ratios. Thus, the *demand relation* $\mathcal{D}_u \subseteq \mathbb{R}^l_{++} \times \mathbb{R}^l_+$ derived from the utility function u is defined by

$$(p, x) \in \mathcal{D}_u :\Leftrightarrow\ x \in \mathcal{B}(p) \wedge \forall y \in \mathcal{B}(p) :\ u(x) \geq u(y)$$

where $\mathcal{B}(p) = \{x \in \mathbb{R}^l_+ \mid px \leq 1\}$.

Clearly, $\mathcal{D}_u$ can be equivalently described as a set-valued mapping from $\mathbb{R}^l_{++}$ into $\mathbb{R}^l_+$, i.e. $\mathcal{D}_u(p) = \{x \in \mathbb{R}^l_+ \mid (p, x) \in \mathcal{D}_u\}$ for each $p \in \mathbb{R}^l_{++}$.

What properties of $\mathcal{D}_u$ can be deduced from u? An answer is provided by

Proposition 14.6 *Let u be continuous. Then the following properties can be derived:*

(1) $\mathcal{D}_u$ is an upper hemicontinuous and compact-valued correspondence, i.e. $\mathcal{D}_u(p) \neq \emptyset$ for all $p \in \mathbb{R}^l_{++}$ and for any sequence $(p_n, x_n) \in \mathcal{D}_u$ such that p_n converges to p there is a subsequence of (x_n) which converges to $x \in \mathcal{D}_u(p)$.

(2) If u is locally nonsatiated, then the budget identity $p\mathcal{D}_u(p) = 1$ holds for every p, i.e. $px = 1$ for all $x \in \mathcal{D}_u(p)$.
Moreover, $\mathcal{D}_u$ satisfies the Generalized Axiom of Revealed Preference (GARP), i.e. for any $(p_1, x_1), ..., (p_n, x_n) \in \mathcal{D}_u$ the inequalities

$$p_i(x_{i+1} - x_i) \leq 0 \quad \text{for} \quad i = 1, ..., n-1$$

imply that $p_n(x_1 - x_n) \geq 0$.

(3) *If* u *is quasiconcave, then* $\mathcal{D}_u$ *is convex-valued.*

(4) *If* u *is strictly quasiconcave, then* $\mathcal{D}_u(p)$ *contains only one element, i.e.* $\mathcal{D}_u$ *can be identified with a continuous function from* $\mathbb{R}^l_{++}$ *into* $\mathbb{R}^l_+$.
Moreover, $\mathcal{D}_u$ *satisfies the Strong Axiom of Revealed Preference (SARP), i.e. for any* $(p_1, x_1), ..., (p_n, x_n) \in \mathcal{D}_u$ *the inequalities*

$$p_i(x_{i+1} - x_i) \leq 0 \quad \text{for} \quad i = 1, ..., n-1$$

and $x_1 \neq x_n$ *imply that* $p_n(x_1 - x_n) > 0$.

Proof. (1) Since the budget sets are compact, continuity of u implies that $\mathcal{D}_u(p)$ is nonempty and compact for each p. The budget correspondence $\mathcal{B}$ from $\mathbb{R}^l_{++}$ into $\mathbb{R}^l_+$ is obviously continuous. Hence, by the maximum theorem (see [5]), $\mathcal{D}_u$ is upper hemicontinuous.

(2) If $px < 1$ there is a neighborhood N of x such that $pN < 1$. Local nonsatiation implies the existence of $y \in N$ with $u(y) > u(x)$. Since $py \leq 1$, x does not maximize u on $\mathcal{B}(p)$.
From $p_i(x_{i+1} - x_i) \leq 0$ it follows that $p_i x_{i+1} \leq p_i x_i = 1$. Hence, $x_{i+1} \in \mathcal{B}(p_i)$ and thus $u(x_i) \geq u(x_{i+1})$. By assumption, this inequality holds for $i = 1, ..., n-1$. Consequently, $u(x_1) \geq u(x_2) \geq ... \geq u(x_{n-1}) \geq u(x_n)$. If one would have $p_n(x_1 - x_n) < 0$ or, equivalently, $p_n x_1 < p_n x_n = 1$, then, by local nonsatiation, $x_1 \notin \mathcal{D}_u(p_n)$ contradicting $u(x_1) \geq u(x_n)$. Therefore, $p_n(x_1 - x_n) \geq 0$.

(3) This follows immediately from Proposition 14.4 (1).

(4) The first part follows immediately from Proposition 14.4 (2).

As in the proof of (2), the inequalities $p_i(x_{i+1} - x_i) \leq 0$ for $i = 1, ..., n-1$ imply that $u(x_1) \geq ... \geq u(x_n)$. By strict quasiconcavity of u, $x_1 \neq x_n$ implies the existence of $\lambda \in]0, 1[$ such that $u(\lambda x_1 + (1 - \lambda)x_n) > u(x_n)$. Since x_n maximizes u on $\mathcal{B}(p_n)$, $\lambda x_1 + (1 - \lambda)x_n$ cannot be a bundle in $\mathcal{B}(p_n)$. Hence, by convexity of $\mathcal{B}(p_n)$, $x_1 \notin \mathcal{B}(p_n)$, i.e. $p_n x_1 > 1$. Since $p_n x_n = 1$, it follows that $p_n x_1 > p_n x_n$, i.e. $p_n(x_1 - x_n) > 0$. □

It is obvious that SARP implies GARP but that the reverse is not true. For example, if p, x_1, x_2 are chosen such that $x_1 \neq x_2$ and $px_1 = px_2$, then $\mathcal{D} = \{(p, x_1), (p, x_2)\}$ satisfies GARP but not SARP.

While SARP was introduced by Houthakker [21], GARP has been formulated by Varian [44] as equivalent to the following condition of *cyclical consistency* which is due to Afriat [1]:

For any $(p_1, x_1), ..., (p_n, x_n) \in \mathcal{D}_u$ the inequalities

$$p_i(x_{i+1} - x_i) \leq 0 \quad \text{for} \quad i = 1, ..., n \ (\text{mod n})$$

imply that $p_i x_{i+1} = p_i x_i$ for $i = 1, ..., n$ (mod n).

Afriat has also shown that any finite demand relation $\mathcal{D} = \{(p_i, x_i) \in \mathbb{R}^l_{++} \times \mathbb{R}^l_+ \mid p_i x_i = 1 \text{ for } i = 1, ..., k\}$ is cyclically consistent if and only if there exist real numbers $\lambda_1, ..., \lambda_k > 0$ such that

$$\sum_{i=1}^{k \,(\mathrm{mod}\, k)} \lambda_i p_i (x_{i+1} - x_i) \geq 0.$$

Put differently, by setting $q_i = \lambda_i p_i$ for $i = 1, ..., k$, the transformed demand relation $\mathcal{D}' = \{(q_i, x_i) \mid i = 1, ...k\}$ is *cyclically monotone.* It is easy to see that any demand relation which is transformable (in the sense above) into a cyclically monotone one satisfies GARP. Indeed, $p_i(x_{i+1} - x_i) \leq 0$ for $i = 1, ...n-1$ implies $\sum_{i=1}^{n-1} \lambda_i p_i(x_{i+1} - x_i) \leq 0 \leq \sum_{i=1}^{n(\mathrm{mod}\, n)} \lambda_i p_i(x_{i+1} - x_i)$. Hence, $\lambda_n p_n(x_1 - x_n) \geq 0$. Since $\lambda_n > 0$, $p_n(x_1 - x_n) \geq 0$.

With regard to the above equivalences, it seems to be quite appropriate to denote the generalized monotonicity property described by GARP as *cyclic pseudomonotonicity.* Actually, this term was first used by Daniilidis and Hadjisavvas [11] for a general set-valued mapping displaying that property.[7]

Analogously, SARP might be called *strict cyclic pseudomonotonicity* since it generalizes the property that

$$\sum_{i=1}^{n(\mathrm{mod}\, n)} p_i(x_{i+1} - x_i) > 0$$

if $x_i \neq x_{i+1}$ for at least one $i = 1, ..., n(\mathrm{mod}\, n)$.

Observe that if the utility function is neither locally nonsatiated nor strictly quasiconcave, no generalized monotonicity property can be deduced. For example, a constant utility function u is even semistrictly quasiconcave but puts no restriction on demand since $\mathcal{D}_u(p) = \mathcal{B}(p)$ for all p.

An important question is now whether the properties of $\mathcal{D}_u$ that were deduced from certain properties of u in Proposition 14.6 actually characterize demand relations derived from utility maximization. For example, if $\mathcal{D} : \mathbb{R}^l_{++} \longrightarrow \mathbb{R}^l_+$ is a continuous demand function satisfying the

[7]To be precise, GARP means that $-\mathcal{D}_u^{-1}$, where $\mathcal{D}_u^{-1}$ denotes the set-valued mapping corresponding to the inverse demand relation, is cyclically pseudomonotone in the sense of Daniilidis and Hadjisavvas.

budget identity and SARP, does there exist a continuous, locally nonsatiated, and strictly quasiconcave utility function u such that $\mathcal{D}_u = \mathcal{D}$?

Such questions are typically posed in the theory of revealed preference that tries to replace conditions on unobservable characteristics as preference and utility by conditions on the (in principle) observable demand relation. The main concepts are introduced by

Definition 14.4 *A demand relation $\mathcal{D} \subseteq \mathbb{R}^l_{++} \times \mathbb{R}^l_+$ is rationalized by a utility function u (or by the corresponding preference relation) if $\mathcal{D} \subseteq \mathcal{D}_u$. If actually $\mathcal{D} = \mathcal{D}_u$, we say that u strongly rationalizes $\mathcal{D}$.*

While the concept of rationalizability should be interpreted as a consistency of the "demand observations" with an underlying utility that is maximized, strong rationalizability corresponds to a characterization of demand relations derived from the utility maximization hypothesis.

Unfortunately, demand functions derived from a continuous and strictly quasiconcave utility function (even if it is locally nonsatiated) cannot be characterized by SARP. This can be seen from the following example due to Hurwicz and Richter [22] and depicted in Figure 5 below.

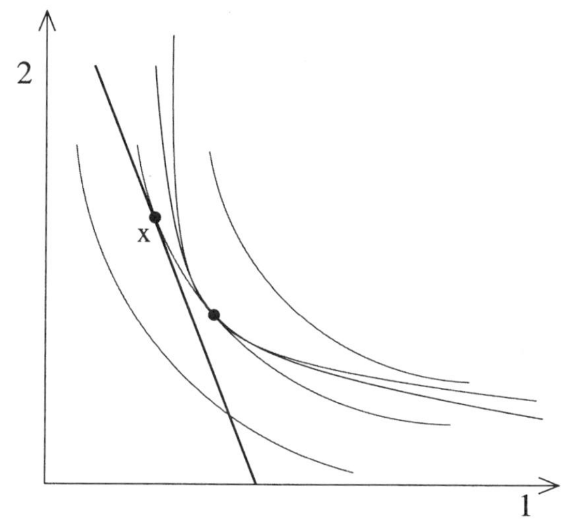

Figure 5

For any budget line given by p, $x = \mathcal{D}(p)$ is chosen as the point for which the budget line is tangent at x to the curve on which x is located. Obviously, $\mathcal{D}$ satisfies the budget identity and is continuous. Since it is not difficult to verify that this demand function can be derived from maximizing an upper semicontinuous utility function it satisfies SARP.

However, as shown by Hurwicz and Richter [22], it cannot be strongly rationalized by a continuous utility function.

Of course, as Houthakker [21] has already observed, if additional assumptions (e.g. $\mathcal{D}$ is "income Lipschitzian" and satisfies a boundary condition) are satisfied, then strict rationalizability by a continuous, monotone, and strictly quasiconcave utility function is possible, see e.g. [34]. On the other hand, these conditions are not necessary. To summarize, there is no way to characterize strong rationalizability of a demand function by a continuous, locally nonsatiated, and strictly quasiconcave utility function.

However, with regard to the weaker requirement of rationalizability, there is a remarkable result due to Afriat [1] and reformulated by Varian [44].

Proposition 14.7 *For a finite demand relation* $\mathcal{D} = \{(p_1, x_1), ..., (p_k, x_k)\} \subseteq \mathbb{R}^l_{++} \times \mathbb{R}^l_+$ *the following properties are equivalent:*

(i) *$\mathcal{D}$ can be rationalized by a locally nonsatiated utility function.*

(ii) *$\mathcal{D}$ satisfies the budget identity and GARP.*

(iii) *$\mathcal{D}$ can be rationalized by a continuous, monotone, and concave utility function.*

Proof. See [44]. □

By the equivalence of (i) and (iii), if a finite demand relation can be rationalized by a nontrivial (i.e. locally nonsatiated) utility function at all, then it can be rationalized by a very nice one (in the sense of the properties given in (iii)). Put differently, continuity, monotonicity, and concavity cannot be revealed by only a finite number of demand observations.

An analogous result characterizing finite demand relations that satisfy SARP is proved in [37].

2.3 Nontransitive Preference

It is well known from empirical studies that consumers often do not behave in accordance with a transitive preference relation. Consequently, one should abandon the transitivity axiom. However, is there still a useful theory of demand in that case? Or does it lead to no behavioral restrictions at all?

First, suppose that the preference is neither assumed to be locally nonsatiated nor strictly convex. Then, as we have seen in the previous section, demand is only restricted by the budget set, even with transitivity, completeness, and continuity. In order to exclude trivial preferences, one has to maintain at least the condition of local nonsatiation. However, without transitivity, demand is only restricted by the budget identity even if completeness, continuity, and monotonicity are assumed. Indeed, an example is given by defining x to be strictly preferred to y iff $x \gg y$.

The conclusion is that some convexity assumption may be needed to obtain nontrivial implications for demand. On the other hand, what kind of restrictions can be expected in that case?

Apart from the budget identity, obvious candidates are generalized monotonicity properties that are weaker than GARP or SARP but are similar in spirit. Actually, this was the basic idea of Samuelson [38] who formulated the following consistency postulate on a demand relation $\mathcal{D}$ which was later called the *Weak Axiom of Revealed Preference* (WARP):

For any two consumption bundles $(p, x), (q, y) \in \mathcal{D}$ such that $x \neq y$,

$$p(y - x) \leq 0 \quad \text{implies} \quad q(y - x) < 0$$

Samuelson justified this condition by a simple argument. If $p(y-x) \leq 0$ and $(p, x) \in \mathcal{D}$ means that x has been chosen at p while y could have been chosen and x is not equal to y, then this should be interpreted as "x is strictly revealed preferred to y". Hence, consistent behaviour should imply that y must not be strictly revealed preferred to x, i.e. $q(x - y) > 0$ for $(q, y) \in \mathcal{D}$.

The problem of this interpretation is the same as for SARP. It does not allow that two different bundles in a budget set are chosen because they are considered to be indifferent.

This objection is taken into account by weakening WARP (which is equivalent to strict pseudomonotonicity of inverse demand) to pseudomonotonicity of inverse demand. We call a demand relation $\mathcal{D}$ *pseudomonotone* if

$$p(y - x) \leq 0 \quad \text{implies} \quad q(y - x) \leq 0$$

for any $(p, x), (q, y) \in \mathcal{D}$.

If $(p, x) \in \mathcal{D}$ and $p(y - x) \leq 0$ is now interpreted as " x is weakly revealed preferred to y" but as "strictly revealed preferred" if the strict inequality holds, then this property means that y must not be strictly revealed preferred to x if x is weakly revealed preferred to y. Although this kind of consistency seems to be quite sensible, there is obviously a slightly weaker requirement. The postulate that x and y cannot be both strictly revealed preferred to each other is formalized by

$$p(y - x) < 0 \quad \text{implies} \quad q(y - x) \leq 0$$

for all $(p,x),(q,y) \in \mathcal{D}$.

Of course, this is equivalent to *quasimonotonicity* of the inverse demand relation.

In the sequel, the relationship between convexity assumptions on the preference relation and generalized monotonicity properties of demand will be studied in detail.

A *preference* is now defined as a complete binary relation R on $\mathbb{R}^l_+$, i.e. we simply drop the transitivity axiom. Like in the transitive case, xRy is interpreted as "x is at least as good as y". The relation "better than" is denoted by P and, as in (14.4), defined by

$$x\,P\,y :\Longleftrightarrow x\,R\,y \wedge \neg y\,R\,x.$$

Of course, completeness of R implies that xPy iff $\neg yRx$. Moreover, we define $R(x) := \{y \in \mathbb{R}^l_+ \mid yRx\}$ and $P(x) := \{y \in \mathbb{R}^l_+ \mid yPx\}$.

Additional possible properties of R are collected in

Definition 14.5 *A preference R is called*

(1) upper continuous, if $R(x)$ is closed for every $x \in \mathbb{R}^l_+$

(2) continuous, if R is closed in $\mathbb{R}^l_+ \times \mathbb{R}^l_+$.

(3) locally nonsatiated, if for every $x \in \mathbb{R}^l_+$ and every neighborhood N of x, there is $y \in N$ such that $y\,P\,x$.

(4) (strictly) monotone, if $\forall x,y \in \mathbb{R}^l_+ : x \gg y (x > y) \Rightarrow x\,P\,y$

(5) convex, if $R(x)$ is convex for every $x \in \mathbb{R}^l_+$.
If, in addition, for all $x \in \mathbb{R}^l_+$ and for all $y,z \in R(x)$,

$$y\,P\,x \vee z\,P\,x \Longrightarrow \forall \lambda \in \,]0,1[:\ \lambda y + (1-\lambda)z\,P\,x,$$

then R is called semistrictly convex.

(6) strictly convex, if $y,z \in R(x)$ and $y \neq z$ imply that $\lambda y+(1-\lambda)z\,P\,x$ for all $\lambda \in \,]0,1[$.

It is easy to show that the convexity conditions (5) and (6) are equivalent to those in Definition 14.3 if R is transitive.

The demand relation induced by R is defined analogously to the transitive case.

Definition 14.6 *Let R be a preference. Then the demand relation $\mathcal{D}_R$ derived from R is given by*

$$(p,x) \in \mathcal{D}_R :\Leftrightarrow x \in \mathcal{B}(p) \wedge \forall y \in \mathcal{B}(p) : x\,R\,y$$

and equivalently considered as the set-valued mapping

$$\mathcal{D}_R : \mathbb{R}^l_{++} \longrightarrow \mathbb{R}^l_+,$$

defined by $\mathcal{D}_R(p) = \{x \in \mathbb{R}^l_+ \mid (p,x) \in \mathcal{D}_R\}$.

In contrast to the transitive case, upper continuity alone is not sufficient for $\mathcal{D}_R$ being a correspondence from $\mathbb{R}^l_{++}$ into $\mathbb{R}^l_+$, i.e. $\mathcal{D}_R(p) \neq \emptyset$ for all p. However, a weak convexity condition on the sets $P(x)$ allows the application of the KKM-Lemma to obtain that property of $\mathcal{D}_R$. In the economic literature, this observation is due to Sonnenschein [43] and presented in the following

Lemma 14.1 *Let R be an upper continuous preference. If R has the KKM-property, i.e.*

(i) the convex hull $co\{x_1, ..., x_n\}$ of finitely many elements in $\mathbb{R}^l_+$ is always contained in $\bigcup_{i=1}^{n} R(x_i)$

or, equivalently,

(ii) $x \notin coP(x)$ for all $x \in \mathbb{R}^l_+$,

then $\mathcal{D}_R(p) \neq \emptyset$ for all $p \in \mathbb{R}^l_{++}$.

Proof. The equivalence of (i) and (ii) is obvious. By definition, $x^* \in \mathcal{D}_R(p)$ iff $x^* \in \bigcap_{x \in \mathcal{B}(p)} R(x) \cap \mathcal{B}(p)$. Since $\mathcal{B}(p)$ is compact and the sets $R(x)$ are closed, the intersection is nonempty if all finite intersections are nonempty. Clearly, the latter property is guaranteed by the KKM-Lemma. □

There is no doubt that an operational definition of a "nontransitive consumer" should at least imply nonempty sets $\mathcal{D}_R(p)$. Observe, however, that the example mentioned above ($x\,R\,y$ iff $\neg y \gg x$) satisfies the conditions of Lemma 14.1 but does only predict $px = 1$ for $(p,x) \in \mathcal{D}_R$.

It will turn out that convex preferences put further restrictions on demand. Moreover, they also ensure that $\mathcal{D}_R$ is a correspondence as shown first by

Lemma 14.2 *If the preference R is upper continuous, locally nonsatiated, and convex, then R has the KKM-property.*

Proof. Assume that the claim is not true, i.e. there are $x_1, ..., x_n \in \mathbb{R}^l_+$ and $\lambda_1, ..., \lambda_n \geq 0$ with $\sum_{i=1}^{n} \lambda_i = 1$ such that $x_i\,P\,x$ for $x = \sum_{i=1}^{n} \lambda_i x_i$ and $i = 1, ..., n$.

By upper continuity of R, the sets $P^{-1}(x_i) = \{y \in \mathbb{R}^l_+ \mid x_i \, P \, y\} = \mathbb{R}^l_+ \setminus R(x_i)$ are open and, consequently, their intersection $W = \bigcap_{i=1}^{n} P^{-1}(x_i)$ is also open.

From $x \in W$ it follows that there is a neighborhood N of x such that $y \in W$ for all $y \in N$. Since, by completeness of R, $y \in W$ implies $x_i \in R(y)$ for $i = 1, ..., n$, we obtain $x = \sum_{i=1}^{n} \lambda_i x_i \in R(y)$ by convexity of R. Hence, R cannot be locally nonsatiated. □

Corollary 14.1 *Let R be an upper continuous, locally nonsatiated, and convex preference. Then $\mathcal{D}_R$ is a compact- and convex-valued correspondence from $\mathbb{R}^l_{++}$ into $\mathbb{R}^l_+$.*

Moreover, $\mathcal{D}_R$ is upper hemicontinuous if R is continuous.

Proof. The first part follows immediately from Lemma 14.1 and Lemma 14.2 since, by upper continuity and convexity of R, $\mathcal{D}_R(p) = \bigcap_{x \in \mathcal{B}(p)} R(x) \cap \mathcal{B}(p)$ is compact and convex.

The second part is implied by the maximum theorem of Berge [5] which is easily adapted to maxima of binary relations. □

Proposition 14.8 *Let R be a locally nonsatiated preference. If, in addition, R is*

(1) convex, then $\mathcal{D}_R$ is properly quasimonotone, i.e. for arbitrary $(p_1, x_1), ..., (p_n, x_n) \in \mathcal{D}_R$ and $\lambda_1, ..., \lambda_n > 0$ with $\sum_{i=1}^{n} \lambda_i = 1$ there exists $j \in \{1, ..., n\}$ such that for $x = \sum_{i=1}^{n} \lambda_i x_i$:

$$p_j(x - x_j) \geq 0.$$

In particular, $\mathcal{D}_R$ is quasimonotone.

(2) semistrictly convex, then $\mathcal{D}_R$ is properly pseudomonotone, i.e. for arbitrary $(p_1, x_1), ..., (p_n, x_n) \in \mathcal{D}_R$ and $\lambda_1, ..., \lambda_n > 0$ with $\sum_{i=1}^{n} \lambda_i = 1$ the following implication holds for $x = \sum_{i=1}^{n} \lambda_i x_i$:

$$\exists i : p_i(x - x_i) < 0 \Rightarrow \exists j : p_j(x - x_j) > 0.$$

In particular, $\mathcal{D}_R$ is pseudomonotone.

(3) *strictly convex, then $\mathcal{D}_R$ is properly strictly pseudomonotone, i.e. for arbitrary $(p_1, x_1), ..., (p_n, x_n) \in \mathcal{D}_R$ and $\lambda_1, ..., \lambda_n > 0$ with $\sum_{i=1}^{n} \lambda_i = 1$ the following implication holds for $x = \sum_{i=1}^{n} \lambda_i x_i$:*

$$(\forall i : p_i(x - x_i) \leq 0) \Rightarrow x_1 = \cdots = x_n .$$

In particular, $\mathcal{D}_R$ satisfies WARP.

Proof. (1) If $\mathcal{D}_R$ is not properly quasimonotone, there exist $(p_1, x_1), ..., (p_n, x_n) \in \mathcal{D}_R$ such that for $x = \sum_{i=1}^{n} \lambda_i x_i$ with $\lambda_1, ..., \lambda_n > 0$ and $\sum_{i=1}^{n} \lambda_i = 1$ the inequality $p_i(x - x_i) < 0$ holds for $i = 1, ..., n$. Hence, there is a neighborhood N of x with $p_i(y - x_i) < 0$ for all $y \in N$ and every i. This implies $x_i \, R \, y$ for all $y \in N$ and every i. From convexity of R, it follows that $x \, R \, y$ for all $y \in N$, contradicting that R is locally nonsatiated.

$\mathcal{D}_R$ is quasimonotone: Assume that (p, x), $(q, y) \in \mathcal{D}_R$. If $p(y - x) < 0$, then $p\left((\frac{1}{2}x + \frac{1}{2}y) - x\right) < 0$. By proper quasimonotonicity, $q\left((\frac{1}{2}x + \frac{1}{2}y) - y\right) \geq 0$ which implies $q(x - y) \geq 0$. Hence, $q(y - x) \leq 0$.

(2) If $\mathcal{D}_R$ is not properly pseudomonotone, there exist $(p_1, x_1), ..., (p_n, x_n) \in \mathcal{D}_R$ such that for $x = \sum_{i=1}^{n} \lambda_i x_i$ with $\lambda_1, ..., \lambda_n > 0$ and $\sum_{i=1}^{n} \lambda_i = 1$ the inequalities $p_j(x - x_j) \leq 0$ hold for $j = 1, ..., n$ and $p_i(x - x_i) < 0$ for some $i \in \{1, ..., n\}$.

Assume that $p_i(x - x_i) < 0$ implies $x_i \, P \, x$ (this will be shown below). Since $p_j(x - x_j) \leq 0$ implies $x_j \, R \, x$, we can conclude, by semistrict convexity of R and induction, that $x = \sum_{j=1}^{n} \lambda_j x_j \, P \, x$ which contradicts reflexivity of R.

It remains to show that $(p, x) \in \mathcal{D}_R$ and $p(y - x) < 0$ implies $x \, P \, y$.

Assume that $x \, P \, y$ does not hold. Since $x \, R \, y$, we obtain, by definition of P, that $y \, R \, x$. Local nonsatiation of R implies the existence of z such that $z \, P \, x$. Since R is semistrictly convex, $\lambda y + (1 - \lambda) z \, P \, x$ for all $\lambda \in \,]0, 1[$. For λ close to 1, $p(y' - x) < 0$, where $y' = \lambda y + (1 - \lambda) z$. However, $y' \, P \, x$ contradicts $(p, x) \in \mathcal{D}_R$.

$\mathcal{D}_R$ is pseudomonotone: Assume that (p, x), $(q, y) \in \mathcal{D}_R$. If $q(x - y) < 0$, then $q\left((\frac{1}{2}x + \frac{1}{2}y) - y\right) < 0$. By proper pseudomonotonicity, $p\left(\frac{1}{2}x + \frac{1}{2}y) - x\right) > 0$ which implies $p(y - x) > 0$, or equivalently, $p(x - y) < 0$.

(3) Assume that $(p_1, x_1), ..., (p_n, x_n) \in \mathcal{D}_R$ and that for $\lambda_1, ..., \lambda_n > 0$ with $\sum_{i=1}^{n} \lambda_i = 1$ and $x = \sum_{i=1}^{n} \lambda_i x_i$ the inequality $p_i(x - x_i) \leq 0$ holds

for $i = 1, ..., n$. By strict convexity of R and induction, it follows that $\sum_{i=1}^{n} \lambda_i x_i \, P \, x$ if not all x_i are equal. This contradicts reflexivity of R.

$\mathcal{D}_R$ satisfies WARP: Assume that $(p, x), (q, y) \in \mathcal{D}_R$. Then $p(y - x) \leq 0$ and $q(x - y) \leq 0$ imply $p\big(\frac{1}{2}x + \frac{1}{2}y) - x\big) \leq 0$ and $q\big(\frac{1}{2}x + \frac{1}{2}y) - y\big) \leq 0$. By proper strict pseudomonotonicity of $\mathcal{D}_R$, $x = y$. □

The notions of proper quasimonotonicity and proper pseudomonotonicity have been introduced by Daniilidis and Hadjisavvas [11]. They also have shown that pseudomonotonicity on a convex domain is equivalent to proper pseudomonotonicity and, in addition, that such an equivalence does not hold for quasimonotonicity.

In general, all three pairwise generalized monotonicity notions are weaker than the corresponding proper ones. This is shown by the following

Example. Let $\mathcal{D} = \{(p_1, x_1), (p_2, x_2), (p_3, x_3)\}$ where $p_1 = (1, 1.2, 0.5)$, $p_2 = (0.5, 1, 1.2)$, $p_3 = (1.2, 0.5, 1)$,

$$x_1 = \begin{pmatrix} 1 \\ 0 \\ 0 \end{pmatrix}, \quad x_2 = \begin{pmatrix} 0 \\ 1 \\ 0 \end{pmatrix}, \quad x_3 = \begin{pmatrix} 0 \\ 0 \\ 1 \end{pmatrix}.$$

Observe first, that $p_i x_i = 1$ for $i = 1, 2, 3$. Furthermore, $p_1 x_3 = p_2 x_1 = p_3 x_2 = 0.5 < 1$ and $p_3 x_1 = p_1 x_2 = p_2 x_3 = 1.2 > 1$. Hence, $\mathcal{D}$ satisfies WARP.

On the other hand, for $x = \sum_{i=1}^{3} \frac{1}{3} x_i$ we obtain $p_i x = 0.9 < 1$ for $i = 1, 2, 3$. Thus, $\mathcal{D}$ is not properly quasimonotone.

This example shows, that even the strongest pairwise generalized monotonicity notion does not imply the weakest proper one. □

The natural question arises whether the necessary properties of a demand relation derived from a locally nonsatiated and (strictly, semistrictly) convex preference R are also sufficient in the sense of revealed preference theory. Analogously to Definition 14.4, a demand relation $\mathcal{D}$ is *rationalized* (resp. *strongly rationalized*) by a (not necessarily transitive) preference R if $\mathcal{D} \subseteq \mathcal{D}_R$ (resp. $\mathcal{D} = \mathcal{D}_R$).

It has been shown in [28] that a demand correspondence $\mathcal{D}$ satisfying the budget identity can be rationalized by an upper continuous, monotone, and convex preference if and only if $\mathcal{D}$ is quasimonotone.

However, according to the spirit of revealed preference theory, it is desirable to have rationalizability results for finite demand relations. Indeed, if $\mathcal{D}$ is interpreted as a set of observed choices, then it is hard to imagine that $\mathcal{D}$ is a demand correspondence which would require demand observations for all price vectors.

From this viewpoint, there is a result in [28] stating that an arbitrary (in particular finite) demand relation satisfying the budget identity is properly quasimonotone if and only if it is rationalized by a reflexive, upper continuous, monotone, and convex relation R.

Unfortunately, this result is not satisfactory. It does not ensure the existence of rationalizing preference since R is not complete.

Put differently, it is desirable to solve the following open

Problem: Let $\mathcal{D} = \{(p_1, x_1), ..., (p_n, x_n)\}$ be a finite demand relation such that $p_i x_i = 1$ for all $i = 1, ..., n$. Does there exist a locally nonsatiated, upper continuous, and convex (resp. semistrictly convex, strictly convex) preference that rationalizes $\mathcal{D}$ if and only if $\mathcal{D}$ is properly quasimonotone (resp. pseudomonotone, strictly pseudomonotone)?

At least, there is a result analogous to Proposition 14.7. In order to present it, we have to introduce the generalization of the utility representation to the nontransitive case which is due to Shafer [40].

Definition 14.7 *A real-valued function r defined on $\mathbb{R}^l_+ \times \mathbb{R}^l_+$ is called a (numerical) representation of the preference R if for all $x, y \in \mathbb{R}^l_+$ the following conditions are satisfied:*

(1) $x\,R\,y \Leftrightarrow r(x,y) \geq 0$

(2) $r(x,y) = -r(y,x)$.

Notice that a utility representation u of a transitive preference $\succeq$ yields a representation in this sense by defining $r_u(x,y) = u(x) - u(y)$. It is also obvious that any preference R can be represented. Defining the indifference relation I by $x\,I\,y$ iff $x\,R\,y$ and $y\,R\,x$, completeness of R implies $\mathbb{R}^l_+ \times \mathbb{R}^l_+ = P \cup I \cup P^{-1}$. Since $R = P \cup I$, R is represented by the function r that takes the values 1 on P, 0 on I, and -1 on P^{-1}.

On the other hand, it is clear that any skew-symmetric function $r : \mathbb{R}^l_+ \times \mathbb{R}^l_+ \longrightarrow \mathbb{R}^l_+$ induces a complete preference R_r which is represented by r (define $x\,R_r\,y \Leftrightarrow r(x,y) \geq 0$). If R is continuous then R_r is also continuous. Conversely, it has been shown in [40] that every continuous preference has a continuous representation. Put differently, describing a "nontransitive consumer" by a continuous preference is equivalent to a description by a continuous and skew-symmetric real-valued bifunction on $\mathbb{R}^l_+$. Of course, the relationship is not one-to-one since any sign-preserving transformation of a representation of R also represents R.

Corresponding to the possible properties of a preference R, we say that the representation r is *nonsatiated*, if for every $x \in \mathbb{R}^l_+$ there exists

$y \in \mathbb{R}^l_+$ such that $r(y,x) > 0$ (observe that this is weaker than local nonsatiation).

Since strict monotonicity means $r(x,y) > 0$ if $x > y$, we call the stronger condition

$$\forall x, y, z \in \mathbb{R}^l_+ : x > y \Rightarrow r(x,z) > r(y,z)$$

strong monotonicity of r.

Finally, a concave utility representation is generalized by a *concave-convex* representation, i.e. for every $x \in \mathbb{R}^l_+$ the function $r(\cdot, x) : \mathbb{R}^l_+ \longrightarrow \mathbb{R}$ is concave (resp. $r(x, \cdot)$ is convex).

Saying that r rationalizes $\mathcal{D}$ if R_r rationalizes $\mathcal{D}$, we can state

Proposition 14.9 *For a finite demand relation* $\mathcal{D} = \{(p_1, x_1), ..., (p_n, x_n)\}$ *the following conditions are equivalent:*

(i) There exists a nonsatiated, concave-convex representation that rationalizes $\mathcal{D}$.

(ii) $\mathcal{D}$ is monotone transformable, i.e. there exist real numbers $\pi_i > 0 \quad (i = 1, ..., n)$ such that

$$(\pi_i p_i - \pi_j p_j)(x_i - x_j) \leq 0$$

for all $i, j \in \{1, ..., n\}$.

(iii) There exists a continuous, strongly monotone, concave-convex representation that rationalizes $\mathcal{D}$.

Proof. See [29]. □

Notice that monotone transformability by the numbers π_i means that the relation $\mathcal{D} = \{(\pi_i p_i, x_i) \mid i = 1, ..., n\}$ is monotone. Thus, this property is completely analogous to the transformability into a cyclically monotone relation which has been observed as equivalent to GARP.

As shown in [29], monotone transformability is a generalized monotonicity property that is weaker than cyclic pseudomonotonicity but stronger than proper pseudomonotonicity. Corresponding to the open problem above, proper pseudomonotone (resp. quasimonotone, strictly pseudomonotone) demand relations might be those which are rationalized by semistrictly quasiconcave-quasiconvex (resp. quasiconcave-quasiconvex, strictly quasiconcave-quasiconvex) representations.

We turn now to the case of continuously differentiable demand functions. Their foundation on properties of the underlying preference will not be considered here. For that, we refer to [35] or, with regard to the nontransitive case, to [3]. Instead, we shall derive some first order implications of pseudomonotonicity that will be used in Section 3.

Let the demand relation $\mathcal{D}$ be represented by the continuous differentiable function $h : \mathbb{R}^l_{++} \longrightarrow \mathbb{R}^l_{+}$, i.e.$(p, x) \in \mathcal{D}$ iff $h(p) = x$. As before, we assume the budget identity $ph(p) = 1$ for all p.

Remember that we have normalized the representation of the budget sets by choosing wealth equal to one. Consequently, demand $f(p, w)$ at the price vector p and arbitrary wealth w can be derived from h by $f(p, w) = h(\frac{p}{w})$. By definition, f is homogeneous of degree 0 in p and w and satisfies the budget identity $pf(p, w) = w$ for all p and w.

Now assume that $\mathcal{D}$ is pseudomonotone as defined above. We claim that this is equivalent to the "weak WARP " (WWA)[8] for the demand function f, i.e.

$$pf(p', w') \le w \quad \text{implies} \quad p'f(p, w) \ge w'$$

for all $(p, w), (p', w') \in \mathbb{R}^{l+1}_{++}$.

Indeed, by setting $q = \frac{p}{w}$ and $q' = \frac{p'}{w'}$, it follows that $pf(p', w') \le w$ is equivalent to $q(h(q') - h(q)) \le 0$ and that $q'(h(q') - h(q)) \le 0$ is equivalent to $p'f(p, w) \ge w'$.

In a straightforward way, WWA implies that, for fixed wealth w, the function $f(\cdot, w)$ is pseudomonotone. By homogeneity of f, it is easy to see that the reverse implication also holds, i.e. if $f(\cdot, w)$ is pseudomonotone for some w then f satisfies WWA.

Since $f(p, w)$ does not vanish, pseudomonotonicity of $f(\cdot, w)$ is equivalent to the following first order condition (see [9]):

For all $p \in \mathbb{R}^l_{++}$, the Jacobian matrix $\partial_p f(p, w)$ is negative semidefinite on $f(p, w)^\perp$, i.e. for all $v \in \mathbb{R}^l$,

$$v^T f(p, w) = 0 \quad \text{implies} \quad v^T \partial_p f(p, w) v \le 0.$$

There is a similar first order condition that involves the *Slutsky matrix* of f at (p, w) which is defined by

$$Sf(p, w) := \partial_p f(p, w) + \partial_w f(p, w) f(p, w)^T. \tag{14.6}$$

This matrix is derived from the *income compensated demand* function g at (p, w) that is defined by

$$g(q) = f\big(q, qf(p, w)\big).$$

[8]In the economic literature, the abbreviation WWA has come into use since it was introduced in [30].

Indeed, by the chain rule of differentiation, it is easy to check that $Sf(p,w)$ is equal to the Jacobian $\partial g(q)$ evaluated at p.

The Slutsky matrix can be interpreted as a measure of substitutability of the commodities if prices change in such a way that real income (given by the initially demanded bundle) stays the same. For example, if income compensated demand does not change there is no substitution at all, i.e. the Slutsky matrix is zero.

The first order characterizations of WWA can now be summarized in

Proposition 14.10 *If f is continuously differentiable then the following conditions are equivalent:*

(i) f satisfies WWA.

(ii) For some (resp. all) $w > 0$, the Jacobian matrices $\partial_p f(p,w)$ of the function $f(\cdot, w)$ are negative semidefinite on $f(p,w)^\perp$.

(iii) For all p and w, the Slutsky matrix $Sf(p,w)$ is negative semidefinite, i.e.

$$v^T Sf(p,w)v \le 0 \quad \text{for all } v \in \mathbb{R}^l.$$

Proof. The equivalence of (i) and (ii) has been already mentioned before and was first proved by Hildenbrand and Jerison [19].

In order to prove that (ii) implies (iii), observe that any $v \in \mathbb{R}^l$ can be written as $v = u + \lambda p^T$, where $\lambda = v^T f(p,w)/w$ such that $u \perp f(p,w)$. It is not difficult to check that the budget identity implies $pSf(p,w) = 0$ and that homogeneity of f implies $Sf(p,w)p^T = 0$. By definition of $Sf(p,w)$, it follows that $v^T Sf(p,w)v = u^T Sf(p,w)u = u^T \partial_p f(p,w)u \le 0$ since $u \perp f(p,w)$.

Assume now (iii) and consider the *Slutsky equation* $\partial_p f(p,w) = Sf(p,w) - \partial_w f(p,w) f(p,w)^T$. Thus, if $v \perp f(p,w)$ then $v^T \partial_p f(p,w)v = v^T Sf(p,w)v \le 0$. □

The first order characterization of WWA by the negative semidefiniteness of the Slutsky matrices is due to Kihlstrom, Mas-Colell, and Sonnenschein [30] and probably the first one obtained for a generalized monotonicity property.

3. General Equilibrium Theory

3.1 Variational Inequalities and Economic Equilibrium

Consider below the graphical illustration of a competitive equilibrium on a market for a certain commodity that is well known from all elementary textbooks (Figure 6).

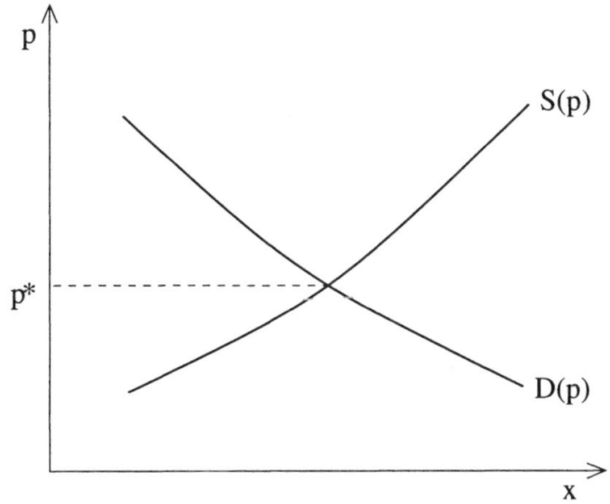

Figure 6

It is assumed that for any given market price p there are quantities $D(p)$ and $S(p)$ of that commodity which describe demand and supply for the commodity at p. An equilibrium on this market is a price p^* for which demand equals supply, i.e. p^* such that $D(p^*) = S(p^*)$.

The situation depicted in Figure 6 is a nice one. There is a unique equilibrium price that is stable with respect to a price adjustment according to the sign of excess demand: If the price increases in the case that demand exceeds supply and decreases in the opposite case, then the price always adjusts towards p^*. Denoting the excess demand by

$$E(p) = D(p) - S(p),$$

we obtain for every $p \neq p^*$ the inequality

$$E(p)(p^* - p) > 0,$$

i.e. p^* is a solution to the Minty variational inequality problem with respect to the excess demand function E.

Of course, this is a very simple example in which a certain market is considered separately. In reality, there are markets for a large number of commodities and demand and supply on each market may depend in general on all prices.

In order to model such an environment, we consider a comprehensive collection of l commodities $i = 1, ..., l$ and an excess demand function

$$E : P \longrightarrow \mathbb{R}^l$$

defined on a convex set P of price vectors $p = (p_1, ..., p_l) \in \mathbb{R}^l$. Of course, $E_i(p)$ is interpreted as the excess demand for commodity i at the market prices $p_1, ..., p_l$. At the moment, we are not concerned about a derivation of these functions from a model of individual behaviour, i.e. they are simply taken as given.

With regard to the domain P, it is often assumed that $P \subseteq \mathbb{R}^l_+$, i.e. negative prices are excluded. However, in general this is too restrictive (see e.g. [41]).

If $P = \mathbb{R}^l_{++}$, an *equilibrium* is given by a price vector $p^* \in P$ such that $E(p^*) = 0$, i.e. demand equals supply on each market . In the case that E is defined on $P = \mathbb{R}^l_+$, the appropriate concept is a *free-disposal equilibrium*, i.e. $p^* \in \mathbb{R}^l_+$ such that $E(p^*) \leq 0$ and $p^*E(p^*) = 0$. Such a p^* is also known as a solution to the nonlinear complementary problem with respect to E and allows a negative excess demand $E_i(p^*)$ for commodity i if the price p_i^* is equal to zero.

In general, an *equilibrium* for the excess demand function $E : P \longrightarrow \mathbb{R}^l$ is a solution to the Stampacchia variational inequality problem for E, i.e. a price vector $p^* \in P$ such that for all $p \in P$

$$(p - p^*)E(p^*) \leq 0.$$

We remark that this definition is easily extended to the case of an excess demand correspondence by using the same solution concept for multi-valued mappings.

The relevance of generalized monotonicity to the theory of general economic equilibrium is demonstrated by the following results.

The first one is of course well known from the theory of variational inequalities (see e.g. [26]).

Proposition 14.11 *Let $E : P \longrightarrow \mathbb{R}^l$ be continuous.*

(1) If E is pseudomonotone then $p^* \in P$ is an equilibrium for E if and only if p^* solves the Minty variational inequality problem with respect to E, i.e. for all $p \in P$

$$(p^* - p)E(p) \geq 0.$$

In particular, the set of equilibria is convex.

(2) If E is strictly pseudomonotone then $p^* \in P$ is an equilibrium for E if and only if p^* solves the strict Minty variational inequality problem with respect to E, i.e. for all $p \in P$ such that $p \neq p^*$

$$(p^* - p)E(p) > 0.$$

In particular, there is at most one equilibrium .

It should be noticed that pseudomonotonicity (resp. strict pseudomonotonicity) of E is also necessary if the set of equilibria are required to be convex (resp. to contain at most one element) for any restriction of E to a (one-dimensional) convex subset of P (see [26]).

Proposition 14.11 implies the next result that was suggested in [26] and proved in [31].

Proposition 14.12 *Let $E : P \longrightarrow \mathbb{R}^l$ be continuously differentiable on the open convex domain P. If E is (strictly) pseudomonotone and p^* is an equilibrium for E, then*

$$p(t) = p^*$$

is a (globally asymptotically) stable solution of the autonomous dynamical system

$$\dot{p} = E(p)$$

on P.

Proof. Define $V(p) := (p^* - p)^2$. Since V is obviously positive definite and

$$\dot{V}(p) = \partial V(p)E(p) = -2(p^* - p)E(p),$$

Proposition 14.11 implies $\dot{V}(p) \leq 0$ (resp. $\dot{V}(p) < 0$) if E is pseudomonotone (resp. strictly pseudomonotone). Thus, V is a (strict) Liapunov function for p^* on P which proves the claim (see Theorem 1 in Chapter 9, §3 in [20]). □

The differential equation $\dot{p} = E(p)$ in Proposition 14.12 has a straightforward economic interpretation. If p is not an equilibrium price vector then prices adjust according to the sign of excess demand, i.e. a certain commodity price increases in case of excess demand and decreases in case of excess supply for that commodity. This price adjustment is known as the *Walrasian tâtonnement process.*

In the following sections we investigate two examples for an excess demand function. They will be derived from specific economic models such that one can ask for conditions that imply excess demand to be (strictly) pseudomonotone. For convenience, production activities are not considered. These are taken into account in a survey by Brighi and John [6].

3.2 Distribution Economies

A simple economic model that nevertheless allows to develop a nontrivial general equilibrium theory had been already suggested by Cassel

[7], [8] and was later explicitly formulated by Malinvaud [33] who called it a *distribution economy.*

Its simplicity is due to the assumption that there is an exogenously given vector $y \in \mathbb{R}^l_{++}$ which represents a fixed supply of l consumption goods. These goods are demanded by a finite set H of consumers (or households) who are described by their wealth $w_h > 0$ and by their demand function f^h.

We assume that for each $h \in H$ the demand function $f^h : \mathbb{R}^l_{++} \times \mathbb{R}_{++} \longrightarrow \mathbb{R}^l_+$ is continuous and satisfies the budget identity

$$pf^h(p, w) = w$$

for all p and w.

Given the individual characteristics $(f^h, w_h)_{h \in H}$, the *aggregate* (or *market*) *demand* function $F : \mathbb{R}^l_{++} \longrightarrow \mathbb{R}^l_+$, defined by

$$F(p) = \sum_{h \in H} f^h(p, w_h)$$

is continuous and satisfies the aggregate budget identity

$$pF(p) = W$$

where $W = \sum\limits_{h \in H} w_h$ denotes aggregate wealth.

The aggregate (or market) excess demand function $E : \mathbb{R}^l_{++} \longrightarrow \mathbb{R}^l$ is then given by $E(p) = F(p) - y$.

As a basic requirement, we have to ensure the existence of an equilibrium.

Proposition 14.13 *Assume that the consumption sector $(f^h, w_h)_{h \in H}$ satisfies the following boundary condition:*

For each sequence $(p^n)_{n \in \mathbb{N}}$ of price vectors in $\mathbb{R}^l_{++}$ that converges to some $\bar{p} \in \mathbb{R}^l_+ \setminus \mathbb{R}^l_{++}$, the sequence $(f^h(p^n, w_h))_{n \in \mathbb{N}}$ is unbounded for at least one consumer h.

Then, for any $y \in \mathbb{R}^l_{++}$, there exists an equilibrium of the distribution economy $\mathcal{E} = ((f^h, w_h)_{h \in H}, y)$.

Proof. For each $n \in \mathbb{N}$, define $S_n = \{p \in \mathbb{R}^l_{++} \mid py = W,\ p_i \geq \frac{1}{n} \cdot \frac{W}{ly_i}$ for $i = 1, ..., l\}$. Since $\hat{p} = \left(\frac{W}{ly_1}, ..., \frac{W}{ly_l}\right) \in S_n$, S_n is a nonempty, convex and compact subset of $\mathbb{R}^n_{++}$.
By the fundamental existence theorem of Hartman and Stampacchia [16], for each $n \in \mathbb{N}$ there is $p^n \in S_n$ such that $(p - p^n)E(p^n) \leq 0$ for all

$p \in S_n$. In particular,

$$(\hat{p} - p^n)E(p^n) = \hat{p}E(p^n) = \hat{p}F(p^n) - W \leq 0.$$

This implies that the sequence $(F(p^n))_{n\in\mathbb{N}}$ and, consequently, all sequences $(f^h(p^n, w_h))_{n\in\mathbb{N}}$ are bounded.

Since $(p^n)_{n\in\mathbb{N}}$ is a sequence in the bounded set $S = \{p \in \mathbb{R}^l_{++} \,|\, py = W\}$, there is a subsequence $(p^k)_{k\in\mathbb{N}}$ converging to some $\bar{p} \in \bar{S} = \{p \in \mathbb{R}^l_+ \mid py = W\}$. By the boundary condition, $\bar{p} \notin \mathbb{R}^l_+ \setminus \mathbb{R}^l_{++}$, i.e. $\bar{p} \in S$.

Consider an arbitrary $p \in S$. Then there is some k_0 such that $p \in S_k$ for $k \geq k_0$. Hence, $(p - p^k)E(p^k) \leq 0$ for $k \geq k_0$ and, by continuity of E, $(p - \bar{p})E(\bar{p}) = pE(\bar{p}) \leq 0$. It follows that $pE(\bar{p}) \leq 0$ for all $p \in \bar{S}$. Defining p^i by $p^i_i = \frac{W}{y_i}$ and $p^i_j = 0$ for $i \neq j$, we obtain $p^i y = p^i_i y_i = W$, i.e. $p^i \in \bar{S}$. Thus $p^i E(\bar{p}) = \frac{W}{y_i} E_i(\bar{p}) \leq 0$ and, consequently, $E_i(\bar{p}) \leq 0$. Since this inequality holds for all $i = 1, ..., l$, $E(\bar{p}) \leq 0$. From $\bar{p} \in \mathbb{R}^l_{++}$ and $\bar{p}E(\bar{p}) = 0$, it follows that $E(\bar{p}) = 0$. □

Obviously, the boundary condition in Proposition 14.13 that ensures the existence of an equilibrium means that each commodity is desirable for at least one consumer. From an economic viewpoint, such an assumption is not strong.

We have seen that an equilibrium $p^* \in \mathbb{R}^l_{++}$ is necessarily an equilibrium with respect to the restricted excess demand function $E' : S \longrightarrow \mathbb{R}^l$, where $S = \{p \in \mathbb{R}^l_{++} \mid py = W\}$. It turns out that E' is (strictly) pseudomonotone if the corresponding property holds for the aggregate demand function F. Applying Proposition 14.11 yields

Proposition 14.14 *If the aggregate demand function F is pseudomonotone, then the set of all equilibria is nonempty, compact and convex. There is a unique equilibrium if F is strictly pseudomonotone.*

Proof. The equilibrium set is compact as a closed subset of a bounded set. The convexity (resp. uniqueness) property follows from Proposition 14.11 if we show that (strict) pseudomonotonicity of F implies (strict) pseudomonotonicity of E'.

In order to prove this, consider $p, q \in S$ such that $p \neq q$ and $(p - q)E(q) = (p - q)(F(q) - y) \leq 0$. Since $(p - q)y = 0$, it follows that $(p-q)F(p) \leq 0$. If F is pseudomonotone, $(p-q)F(p) \leq 0$ and, thus, $(p-q)(F(p) - y) = (p - q)E(p) \leq 0$. Analogously, strict pseudomonotonicity implies that the inequalities are strict. □

What conditions ensure that aggregate demand is pseudomonotone? If there were only one consumer, then this property would be implied by the Weak Axiom of Revealed Preference. However, pseudomonotonicity is not additive. Of course, assuming monotone individual demand functions is even sufficient for monotone aggregate demand. Since this assumption is too strong, one has to look for another solution to the problem.

A promising approach, convincingly presented by Hildenbrand [17], is to take the distribution of the individual characteristics into account. In his book, he has derived various (generalized) monotonicity properties of aggregate demand by postulating different kinds of heterogeneous consumption sectors.

One of these is the *hypothesis of increasing dispersion of households' demands.* It was already studied by Jerison [23], [24] and implies that market demand is (strictly) pseudomonotone.

We choose to present another hypothesis in this section which even implies monotonicity of market demand. The reason is that pseudomonotonicity of market demand, although sufficient for a well-behaved excess demand function on the restricted domain S, does not yield such nice properties on $\mathbb{R}^l_{++}$.

In particular, we obtain the following negative result.

Proposition 14.15 *Assume that $F : \mathbb{R}^l_{++} \longrightarrow \mathbb{R}^l_+$ is not monotone at some p with strictly positive demand $F(p)$, i.e. $(p-q)(F(p)-F(q)) > 0$ for some other q. Then there is a supply vector y such that the distribution economy given by F and y has an equilibrium which is not a solution to the Minty variational inequality problem with respect to the excess demand E on $\mathbb{R}^l_{++}$.*

Proof. If y is chosen equal to $F(p)$ then $F(p) = y$, i.e. p is an equilibrium. On the other hand, $(p-q)E(q) = (p-q)(F(q)-y) = -(p-q)(y-F(q)) = -(p-q)(F(p)-F(q)) < 0$. □

In order to exclude the case described in the proposition above, we need a condition that guarantees monotonicity of market demand. Since we do not want to impose this property on individual demand, from which it would follow trivially for aggregate demand, such a condition also provides an example that aggregation may create properties that are not satisfied at the individual level.

In the sequel, we assume that all demand functions f^h are continuously differentiable. Consequently, the market demand function F is also continuously differentiable.

It is known that F is monotone if the Jacobian matrix $\partial F(p)$ is negative semidefinite for every $p \in \mathbb{R}^l_{++}$. Moreover, if all these matrices are negative definite, then F is strictly monotone.

By definition of aggregate demand,

$$\partial F(p) = \sum_{h \in H} \partial_p f^h(p, w_h),$$

which, by the Slutsky decomposition (see (14.6))

$$\partial_p f^h(p, w_h) = S f^h(p, w_h) - \partial_w f^h(p, w_h)\, f^h(p, w_h)^T,$$

implies that

$$\partial F(p) = S(p) - A(p),$$

where $S(p) = \sum_{h \in H} S f^h(p, w_h)$ and $A(p) = \sum_{h \in H} \partial_w f^h(p, w_h)\, f^h(p, w_h)^T$.

Assuming that all consumers satisfy WWA, the individual Slutsky matrices $S f^h(p, w_h)$ are negative semidefinite and, thus, $S(p)$ is also negative semidefinite.

In order to establish the negative semidefiniteness of $\partial F(p)$, it remains to show that $A(p)$ is positive semidefinite.

The matrix $A(p)$ is positive semidefinite if and only if its symmetrized matrix

$$M(p) := A(p) + A(p)^T$$

is positive semidefinite.

By the product rule of differentiation, we obtain

$$\partial_w f_i^h(p, w_h)\, f_j^h(p, w_h) + \partial_w f_j^h(p, w_h)\, f_i^h(p, w_h) = \partial_w [f_i^h(p, w_h)\, f_j^h(p, w_h)]$$

or, in matrix notation,

$$M(p) = \sum_{h \in H} \partial_w [f^h(p, w_h)\, f^h(p, w_h)^T].$$

Let $v \neq 0$ be an arbitrary vector in $\mathbb{R}^l$. Then

$$\begin{aligned} v^T M(p) v &= \sum_{h \in H} \partial_w [v^T f^h(p, w_h)\, f^h(p, w_h)^T v] \\ &= \partial_w \sum_{h \in H} [v^T f^h(p, w_h)]^2. \end{aligned}$$

Now observe that

$$\sum_{h \in H} [v^T f^h(p, w_h)]^2 = |H| \cdot m^2 \{ v^T f^h(p, w_h) \mid h \in H \},$$

where m^2 denotes the second moment of a set of numbers.

Imagine the hypothetical situation that the wealth w_h of each consumer increases by a small amount $\Delta > 0$. This will lead to the demand vectors $f^h(p, w_h+\Delta)$, $h \in H$. Consider the set of numbers $\{v^T f^h(p, w_h+\Delta) \mid h \in H\}$ and define

$$s(p, v) :=$$
$$\lim_{\Delta\to 0} \tfrac{1}{\Delta}[m^2\{v^T f^h(p, w_h + \Delta) \mid h \in H\} - m^2\{v^T f^h(p, w_h) \mid h \in H\}].$$

By definition, it follows immediately that

$$|H| \cdot s(p, v) = \partial_w \sum_{h\in H} [v^T f^h(p, w_h)]^2.$$

Consequently, $s(p, v) \geq 0$ implies $v^T M(p)v \geq 0$ and $M(p)$ is positive semidefinite if $s(p, v) \geq 0$ for all v.

What is the interpretation of the latter condition? The numbers $v^T f^h(p, w_h + \Delta)$, $h \in H$, represent the orthogonal projections of the demand vectors $f^h(p, w_h + \Delta)$, $h \in H$, onto the straight line through the origin with direction v. The second moment can be interpreted as a measure of spread around zero of this set of numbers. If, for small $\Delta > 0$,

$$m^2\{v^T f^h(p, w_h + \Delta) \mid h \in H\} \geq m^2\{v^T f^h(p, w_h) \mid h \in H\},$$

the spread of the demand vectors $f^h(p, w_h)$ is increasing in the direction v. It implies that $s(p, v) \geq 0$.

Now, we say that the demand sector satisfies the *hypothesis of (strictly) increasing spread* around the origin, if $s(p, v) \geq 0$ (resp. > 0) for every $p \in \mathbb{R}^l_{++}$ and every $v \neq 0$, i.e. if the spread of the demand vectors $f^h(p, w_h)$ is (strictly) increasing in every direction. Thus we have proved

Proposition 14.16 *The aggregate demand function of the consumption sector given by $(f^h, w_h)_{h\in H}$ is (strictly) monotone provided that the hypothesis of (strictly) increasing spread holds.*

For a detailed discussion and justification of the hypothesis of increasing spread we refer to [17].

3.3 Exchange Economies

In contrast to the concept of a distribution economy, an exchange economy is characterized by private ownership of the commodities supplied and by wealth that is dependent on prices. More precisely, the economy is given by a set H of households (or consumers) who are, for each $h \in H$, described by a demand function f^h and an initial endowment $e^h \in \mathbb{R}^l_+$.

For any price vector $p \in P = \mathbb{R}^l_{++}$, the wealth w_h of consumer h is determined by the value of his endowments at that price, i.e. $w_h = pe^h$. Thus, the demand of $h \in H$ at $p \in P$ is given by

$$d^h(p) := f^h(p, pe^h).$$

Assuming that f^h is homogeneous of degree 0 in p and w, i.e., $f^h(\lambda p, \lambda w) = f^h(p, w)$ for $\lambda > 0$, and satisfies the budget identity $pf^h(p, w) = w$, we can conclude that d^h is homogeneous of degree 0 in p, i.e. $d^h(\lambda p) = d^h(p)$ for $\lambda > 0$, and that

$$pd^h(p) = pe^h.$$

Aggregate demand is then defined by

$$D(p) = \sum_{h \in H} d^h(p).$$

Clearly, D is homogeneous of degree 0 and satisfies *Walras' Law*, i.e. $pD(p) = pe$, where $e = \sum_{h \in H} e^h$ is the vector of total endowments.

An equilibrium of the exchange economy is a price vector $p^* \in \mathbb{R}^l_{++}$ such that

$$D(p^*) = e.$$

Equivalently, p^* is a zero of the aggregate excess demand function $Z : \mathbb{R}^l_{++} \longrightarrow \mathbb{R}^l_+$, defined by

$$Z(p) = D(p) - e.$$

Again, Z is homogeneous of degree 0 and satisfies Walras' Law, i.e. $pZ(p) = 0$ for all p.

Observe that Z can be also written as the sum of the individual excess demand functions, i.e.

$$Z(p) = \sum_{h \in H} z^h(p),$$

where $z^h(p) = d^h(p) - e^h$.

By homogeneity of Z, if p^* is an equilibrium then λp^* is also an equilibrium for every $\lambda > 0$. If we say that there is a unique equilibrium, it means uniqueness up to positive scalar multiples. Put differently, only relative prices can be uniquely determined by an equilibrium of an exchange economy.

In order to prove the existence of an equilibrium, prices can be normalized by restricting the domain P to the open unit simplex $S = \{p \in$

$\mathbb{R}^l_{++} \mid \sum_{i=1}^l p_i = 1\}$. Using this restriction, the proof of the next result is essentially analogous to the proof of Proposition 14.13.

Proposition 14.17 *Assume that the individual demand functions f^h are continuous and satisfy the following boundary condition:*

For every sequence $(p^n)_{n\in\mathbb{N}}$ of price vectors in $\mathbb{R}^l_{++}$ that converges to some $\bar{p} \in \mathbb{R}^l_+ \setminus \mathbb{R}^l_{++}$ and every wealth sequence $(w^n)_{n\in\mathbb{N}}$ that converges to some $w > 0$, the sequence $(f^h(p^n, w^n))_{n\in\mathbb{N}}$ is unbounded.

Then there exists an equilibrium of the exchange economy $\mathcal{E} = (f^h, e^h)_{h\in H}$ provided that the aggregate endowment vector is strictly positive.

Proof. For each $n \in \mathbb{N}$, define a subset S_n of S by $S_n := \{p \in S \mid p_i \geq \frac{1}{nl}$ for $i = 1, ..., l\}$. Since $\hat{p} = (\frac{1}{l}, ..., \frac{1}{l}) \in S_n$ for all $n \in \mathbb{N}$, each S_n is nonempty, convex, and compact. By continuity of Z, the existence result by Hartman and Stampacchia [16] implies that there is, for each $n \in \mathbb{N}$, $p^n \in S_n$ such that $(p - p^n)Z(p^n) \leq 0$ for every $p \in S_n$. In particular,

$$(\hat{p} - p^n)Z(p^n) = \hat{p}Z(p^n) = \frac{1}{l}\sum_{i=1}^{l} Z_i(p^n) \leq 0.$$

It follows that $\sum\limits_{i=1}^{l} D_i(p^n) \leq \sum\limits_{i=1}^{l} e_i$, i.e. the sequence $D(p^n)$ is bounded. Consequently, all sequences $d^h(p^n) = f^h(p^n, p^n e^h)$ are bounded.

The sequence $(p^n)_{n\in\mathbb{N}}$ in S has a subsequence $(p^k)_{k\in\mathbb{N}}$ that converges to some $p^* \in \bar{S} = cl\, S$. Hence, for every $h \in H$, $p^k e^h$ converges to $p^* e^h$. However, $p^* e^h > 0$ for some $h \in H$ since otherwise $\sum\limits_{h\in H} p^* e^h = p^* \sum\limits_{h\in H} e^h = p^* e = 0$, contradicting $e \gg 0$.

By the boundary condition, $p^* \notin \mathbb{R}^l_+ \setminus \mathbb{R}^l_{++}$, i.e. $p^* \in S$. From the definition of the sets S_n it follows that an arbitrary $p \in S$ is an element of S_k for all k greater or equal than some k_0. Hence, $(p - p^k)Z(p^k) \leq 0$ for $k \geq k_0$ and, by continuity of Z, $(p - p^*)Z(p^*) = pZ(p^*) \leq 0$.

This implies $qZ(p^*) \leq 0$ for all $q \in \bar{S} = cl\, S$, in particular for q^i defined by $q^i_i = 1$ and $q^i_j = 0$ for $i \neq j$. Thus, $Z_i(p^*) = q^i Z(p^*) \leq 0$ for $i = 1, ..., l$.

Since $p^* \in \mathbb{R}^l_{++}$ and $p^* Z(p^*) = 0$, it follows that $Z(p^*) = 0$. □

Of course, existence is only a minimal requirement. It is desirable that there is a unique equilibrium in S which, in addition, should display some stability property, for example, to be a Minty equilibrium point with respect to Z.

Here, the situation is more difficult than in the case of a distribution economy. Even if one is willing to accept that all consumers have strictly monotone demand functions (which implies a strictly monotone market demand), the problem is that wealth is dependent on prices and endowments. It turns out that the distribution of endowments matters a lot.[9]

Consequently, one has to rely on distributional assumptions in order to obtain a uniqueness result. Such an assumption has been introduced by Jerison [25] and will be presented below.

We suppose that all demand functions f^h are continuously differentiable and satisfy WWA (see Section 2.3). From the Slutsky equation

$$\partial_p f^h(p,w) = S\, f^h(p,w) - \partial_w f^h(p,w) f^h(p,w)^T$$

we obtain by the chain rule of differentiation a similar decomposition of the Jacobian of individual excess demand,

$$\begin{aligned} \partial z^h(p) &= \partial d^h(p) = \partial_p f^h(p, pe^h) \\ &= S\, f^h(p, pe^h) - \partial_w f^h(p, pe^h) z^h(p)^T. \end{aligned} \tag{14.7}$$

Now imagine that, hypothetically, the nominal wealth pe^h increases by $\Delta \geq 0$ for each $h \in H$ and denote by

$$z^h(p,\Delta) = f^h(p, pe^h + \Delta) - e^h$$

the excess demand of consumer h after this increase. The mean excess demand of the wealthier population is then given by

$$Z(p,\Delta) = \frac{1}{|H|} \sum_{h \in H} z^h(p,\Delta).$$

Let $v \neq 0$ be an arbitrary vector in $\mathbb{R}^l$. The *dispersion of excess demand* at (p,Δ) in the direction v is defined as the variance of the real variable $v^T z^h(p,\Delta)$, i.e.

$$\sigma_v^2(p,\Delta) = \frac{1}{|H|} \sum_{h \in H} [v^T z^h(p,\Delta) - v^T Z(p,\Delta)]^2.$$

Clearly, for $\Delta = 0$, $\sigma_v^2(p,0)$ denotes the dispersion of excess demand for the actual situation.

---

[9] The standard example where uniqueness is likely to fail is an economy with two commodities and two consumers such that each consumer owns only one of these commodities and has also a relatively high propensity to consume the same commodity.

Definition 14.8 *The exchange economy satisfies the hypothesis of increasing dispersion of excess demands (IDED), if for all* $p \in \mathbb{R}^l_{++}$ *and all* $v \neq 0$ *such that* $pv = v^T Z(p) = 0$, $\partial_\Delta \sigma_v^2(p, 0) > 0$.

Notice that IDED does not imply particular restrictions on the distribution of the demand functions f^h or the distribution of endowments e^h. It is a hypothesis on the joint distribution of individual characteristics (f^h, e^h). Let us illustrate this point by a simple example with two goods and two consumers.

Example. The utility functions of the agents a and b are respectively $u_a(x,y) = \min\{x, 2y\}$ and $u_b(x,y) = \min\{2x, y\}$. Accordingly, their demands are given by

$$f^a(p,w) = \left(\frac{2w}{2p_1 + p_2}, \frac{w}{2p_1 + p_2}\right), \quad f^b(p,w) = \left(\frac{w}{p_1 + 2p_2}, \frac{2w}{p_1 + 2p_2}\right).$$

(1) We shall consider first the case where the vectors of initial endowments are $e^a = (10, 0)$ and $e^b = (0, 10)$. If $Z(p) \neq 0$ the hypothesis is trivially satisfied in the two goods case. Therefore, let us focus on the price vectors at which Z vanishes, i.e. $p = (1,1)$ in the present example. This case is depicted in Fig. 7a where the dashed line represents the budget line of each agent after a generalized increase in income by the amount $\Delta > 0$. E_a and E_b are the Engel curves of the two agents.

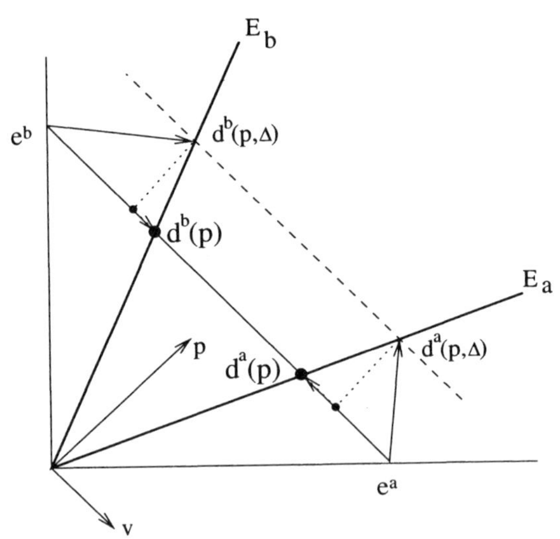

Figure 7a

Fig. 7b translates the above example in terms of excess demand and clearly shows that IDED is violated.

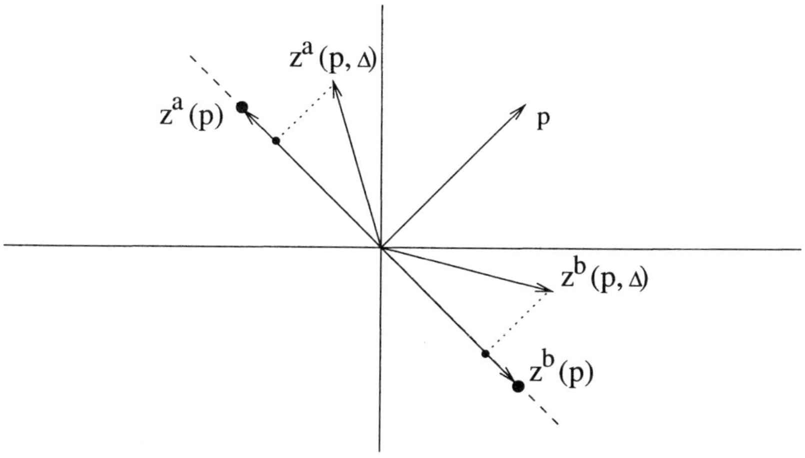

Figure 7b

(2) Now consider the case where the vectors of initial endowments are reversed, i.e. $e^a = (0, 10)$ and $e^b = (10, 0)$. The marginal distributions of e and f are the same as before, but the joint distribution is different. This case is shown in Fig. 8a.

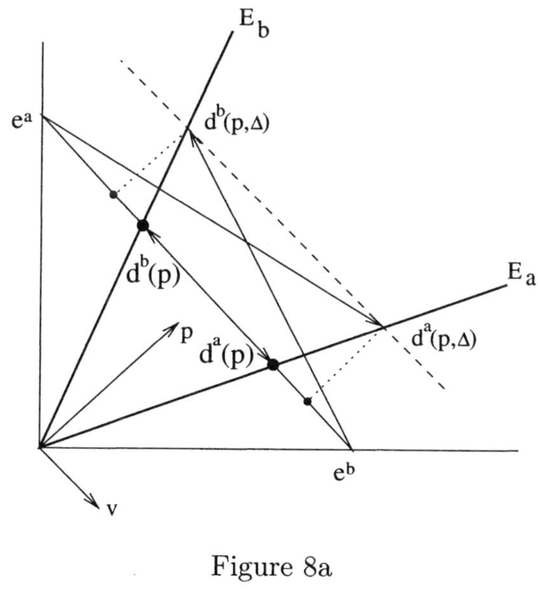

Figure 8a

Fig. 8b shows that the dispersion of excess demand is increasing so that IDED holds. We notice that the individual characteristics are the same as in (1), except for their joint distribution. (1) and (2) thus show that IDED does not depend on the particular shape of the Engel curves.

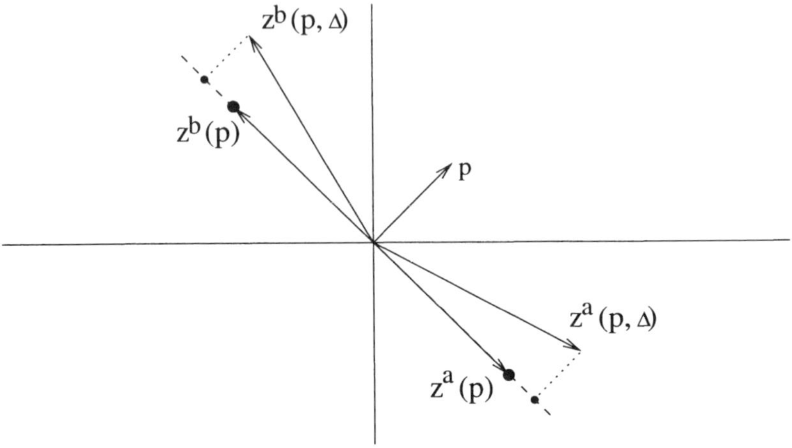

Figure 8b

The following result which is due to Jerison [25] shows that IDED implies a well-behaved aggregate demand of an exchange economy.

Proposition 14.18 *Assume that the exchange economy given by* $(f^h, e^h)_{h\in H}$ *satisfies, in addition to assumptions which ensure existence of an equilibrium, the hypothesis IDED.*

Then, there is (up to positive scalar multiples) a unique equilibrium p^**. Moreover, the aggregate excess demand function satisfies Wald's Axiom, i.e. for all* $p, q \in \mathbb{R}^l_{++}$ *such that* $Z(p) \neq Z(q)$,

$$qZ(p) \leq 0 \quad \text{implies} \quad pZ(q) > 0,$$

in particular, $p^* Z(q) > 0$ *for every price vector* q *which is not an equilibrium.*

Proof. Our first aim is to prove that $\partial Z(p)$ is negative definite on the subspace of $\mathbb{R}^l$ that is orthogonal to p and $Z(p)$, i.e. $v^T \partial Z(p) v < 0$ for all $v \neq 0$ such that $pv = v^T Z(p) = 0$.

From the Slutsky decomposition of the individual excess demands (14.7) it follows that

$$\partial Z(p) = \sum_{h\in H} Sf^h(p, pe^h) - \sum_{h\in H} m^h(p) z^h(p)^T,$$

where we have used the abbreviation $m^h(p) = \partial_w f^h(p, pe^h)$.

Since we have assumed that the individual demand functions satisfy WWA, the sum of negative semidefinite matrices $Sf^h(p, pe^h)$ is negative semidefinite. It remains to show that the matrix

$$A(p) = \sum_{h \in H} m^h(p) z^h(p)^T$$

is positive definite on $p^\perp \cap Z(p)^\perp$ or, equivalently, that $\bar{A}(p) = \frac{1}{|H|} A(p)$ has this property.

In order to prove this, observe first that

$$\begin{aligned}
\sigma_v^2(p, \Delta) &= \frac{1}{|H|} \sum_{h \in H} (v^T [z^h(p, \Delta - Z(p, \Delta)])^2 \\
&= \frac{1}{|H|} \sum_{h \in H} v^T [z^h(p, \Delta) - Z(p, \Delta)] \, [z^h(p, \Delta) - Z(p, \Delta)]^T v \\
&= v^T \left\{ \frac{1}{|H|} \sum_{h \in II} [z^h(p, \Delta) - Z(p, \Delta)] \, [z^h(p, \Delta) - Z(p, \Delta)]^T \right\} v \\
&= v^T \operatorname{cov} z^h(p, \Delta) v,
\end{aligned}$$

where $\operatorname{cov} z^h(p, \Delta)$ denotes the covariance matrix of the individual excess demands $z^h(p, \Delta)$. This implies

$$\begin{aligned}
\partial_\Delta \sigma_v^2(p, 0) &= v^T [\partial_\Delta \operatorname{cov} z^h(p, 0)] v \\
&= v^T \partial_\Delta \left\{ \frac{1}{|H|} \sum_{h \in H} [z^h(p, 0) - Z(p, 0)] \, [z^h(p, 0) - Z(p, 0)]^T \right\} v \\
&= v^T \left\{ \partial_\Delta \left[\frac{1}{|H|} \sum_{h \in H} z^h(p, 0) z^h(p, 0)^T \right] - \partial_\Delta [Z(p, 0) Z(p, 0)^T] \right\} v \\
&= v^T \left\{ \frac{1}{|H|} \sum_{h \in H} m^h(p) z^h(p)^T + \frac{1}{|H|} \sum_{h \in H} z^h(p) m^h(p)^T \right. \\
&\qquad \left. - M(p) Z(p)^T - Z(p) M(p)^T \right\} v,
\end{aligned}$$

where $M(p) = \frac{1}{|H|} \sum\limits_{h \in H} m^h(p)$.

Since the last term is equal to

$$2v^T \left[\frac{1}{|H|} \sum_{h \in H} m^h(p) z^h(p)^T \right] v - 2v^T Z(p) M(p)^T v,$$

it follows from the assumption $v^T Z(p) = 0$ that

$$\partial_\Delta \sigma_v^2(p,0) = 2v^T \bar{A}(p)v.$$

By IDED, $\partial_\Delta \sigma_v^2(p,0) > 0$. Hence, $\bar{A}(p)$ is positive definite on $p^\perp \cap Z(p)^\perp$ and thus we have proved that $\partial Z(p)$ is negative definite on $p^\perp \cap Z(p)^\perp$.

It was shown by Kihlstrom, Mas-Colell, and Sonnenschein [30] (Theorem 4) that this implies Wald's Axiom for Z, i.e. $Z(p) \neq Z(q)$ and $qZ(p) \leq 0$ implies $pZ(q) > 0$.

In particular, it follows by Walras' Law that Z is pseudomonotone, i.e.

$$(q-p)Z(p) \leq 0 \quad \text{implies} \quad (q-p)Z(q) \leq 0$$

for all $p, q \in \mathbb{R}^l_{++}$.

Hence, the set of equilibrium prices is convex by Proposition 14.11. It remains to prove uniqueness.

Assume that p and q are two different equilibria. Setting $v = q - p$, we obtain $Z(p + \lambda v) = 0$ for all $\lambda \in [0,1]$ since the equilibrium set is convex. This implies

$$\partial_\lambda Z(p + \lambda v) \mid_{\lambda=0} = \partial Z(p)v = 0.$$

For $Z(p) = 0$, $\partial Z(p)$ is negative definite on $p^\perp$, i.e. rank $\partial Z(p) = l-1$. By homogeneity of Z, $\partial Z(p)p^T = 0$. Thus, p is collinear with v, resp. p is collinear with q. □

Observe that the conclusions of Proposition 14.18 only need that $\partial Z(p)$ is negative definite on $p^\perp \cap Z(p)^\perp$. This has been shown by using the decomposition

$$\partial Z(p) = S(p) - A(p)$$

and by proving that IDED implies the positive definiteness of $A(p)$ on that subspace (knowing that $S(p)$ is negative semidefinite).

Alternatively, one can assume that $S(p)$ is even negative definite on $p^\perp$ (it cannot be negative definite on the whole space since, by homogeneity, $S(p)p^\top = 0$) and weaken IDED by the hypothesis of *nondecreasing dispersion of excess demands* (NDED) (see [25]), i.e. $\partial_\Delta \sigma_v^2(p,0) \geq 0$ for all p and v such that $pv = v^T Z(p) = 0$.

The stronger assumption on $S(p)$ can be easily justified. It means that there is always substitution between two commodities in case of a relative price change. Although it may be restrictive condition for an individual, it is an acceptable assumption for aggregate demand.

Acknowledgments

I would like to thank Luigi Brighi. This chapter has benefited from my collaboration with him.

References

[1] Afriat, S.N., *The construction of utility functions from expenditure data*, Int. Economic Rev. 8, 1967, 67-77.

[2] Allen, R.G.D., *The foundations of a mathematical theory of exchange*, Economica 12, 1932, 197-226.

[3] Al-Najjar, N., *Non-transitive smooth preferences*, Journal of Economic Theory 60, 1993, 14-41.

[4] Avriel, M., Diewert, W. E., Schaible, S. and I. Zang, *Generalized Concavity*, Plenum Press, New York 1988.

[5] Berge, C., *Topological spaces*, Dover Publications, New York 1997.

[6] Brighi, L. and R. John, *Characterizations of pseudomonotone maps and economic equilibrium*, J. Statistics and Management Systems 5, 2002, 253-273.

[7] Cassel, G., *Grundriss einer elementaren Preislehre*, Zeitschrift f. d. ges, Staatswissenschaft 55, 1899, 395-458.

[8] Cassel, G., *Theoretische Sozialökonomie*, Winter, Leipzig 1918.

[9] Crouzeix, J. P., *Characterizations of generalized convexity and generalized monotonicity, a survey*, in Generalized Convexity, Generalized Monotonicity, J.-P. Crouzeix, J.-E. Martinez-Legaz and M. Volle, eds., Kluwer Academic Publishers, Dordrecht 1998, 237-256.

[10] Dafermos, S., *Exchange price equilibria and variational inequalities*, Math. Program. 46, 1990, 391-402.

[11] Daniilidis, A. and N. Hadjisavvas, *On the subdifferentials of quasiconvex and pseudoconvex functions and cyclic monotonicity*, J. of Math. Analysis and Appl. 237, 1999, 30-42.

[12] Debreu, G., *Theory of value*, Wiley, New York 1959.

[13] Georgescu-Roegen, N., *The pure theory of consumer's behavior*, Quarterly J. of Economics 50, 1936, 545-593.

[14] Georgescu-Roegen, N., *Choice and revealed preference*, Southern Economic Journal 21, 1954, 119-130.

[15] Harker, P.T. and J.-S. Pang, *Finite-dimensional variational inequality and nonlinear complementary problems: A survey of theory, algorithms and applications,* Math. Program. 48, 1990, 161-220.

[16] Hartman, P. and G. Stampacchia, *On some nonlinear elliptic differential functional equations,* Acta Mathematica 115, 1966, 153-188.

[17] Hildenbrand, W., *Market demand: Theory and empirical evidence,* Princeton University Press, Princeton 1994.

[18] Hildenbrand, W., *An exposition of Wald's existence proof,* in Karl Menger: Ergebnisse eines mathematischen Kolloquiums, E. Dierker and K. Sigmund, eds., Springer-Verlag, Wien 1998.

[19] Hildenbrand, W. and M. Jerison, *The demand theory of the weak axioms of revealed preference,* Economics Letters 29, 1989, 209-213.

[20] Hirsch, M. W. and S. Smale, *Differential Equations, Dynamical Systems, and Linear Algebra,* Academic Press, New York 1974.

[21] Houthakker, H. S., *Revealed preference and the utility function,* Economica 17, 1950, 159-174.

[22] Hurwicz, L. and M. K. Richter, *Revealed preference without demand continuity assumptions,* in Preferences, Utility, and Demand, J. S. Chipman, L. Hurwicz, M. K. Richter and H. F. Sonnenschein, eds., Harcourt Brace Jovanovich, New York 1971.

[23] Jerison, M. *The representative consumer and the weak axiom when the distribution of income is fixed,* Report DP 150, SUNY, Albany 1982.

[24] Jerison, M., *Aggregation and pairwise aggregation of demand when the distribution of income is fixed.,* J. Economic Theory 33, 1984, 1-31.

[25] Jerison, M., *Dispersed excess demands, the weak axiom and uniqueness of equilibrium,* J. Math. Economics 31, 1999, 15-48.

[26] John, R., *Variational inequalities and pseudomonotone functions: Some characterizations,* in Generalized Convexity, Generalized Monotonicity, J.-P. Crouzeix, J.-E. Martinez-Legaz and M. Volle, eds., Kluwer Academic Publishers, Dordrecht 1998, 291-301.

[27] John, R., *Abraham Wald's equilibrium existence proof reconsidered,* Economic Theory 13, 1999, 417-428.

[28] John, R., *Quasimonotone individual demand,* Optimization 47, 2000, 201-209.

[29] John, R., *The concave nontransitive consumer,* J. Global Optimization 20, 2001, 297-308.

[30] Kihlstrom, R., Mas-Colell, A. and H. Sonnenschein, *The demand theory of the weak axiom of revealed preference,* Econometrica 44, 1976, 971-978.

[31] Komlósi, S., *On the Stampacchia and Minty variational inequalities*, in Generalized Convexity and Optimization for Economic and Financial Decisions, G. Giorgi and F. A. Rossi, eds. Pitagora Editrice, Bologna 1999.

[32] Kuhn, H. W., *On a theorem of Wald*, in Linear inequalities and related systems, H. W. Kuhn and A. W. Tucker, eds., Annals of Mathematical Studies No. 38, Princeton University Press, Princeton 1956.

[33] Malinvaud, E. *Lectures on microeconomic theory*, North-Holland, Amsterdam 1972.

[34] Mas-Colell, A., *On revealed preference analysis*, Rev. Economic Studies 45, 1978, 121-131.

[35] Mas-Colell, A., *The theory of general economic equilibrium: A differentiable approach*, Cambridge University Press, Cambridge 1985.

[36] Mas-Colell, A., Whinston, M. D. and J. R. Green, *Microeconomic theory*, Oxford University Press, New York 1995.

[37] Matzkin, R. L. and M. K. Richter, *Testing strictly concave rationality*, J. Economic Theory 53, 1991, 287-303.

[38] Samuelson, P. A., *A note on the pure theory of consumer behaviour*, Economica 5, 1938, 61-72.

[39] Schlesinger, K., *Über die Produktionsgleichungen der ökonomischen Wertlehre*, Ergebnisse eines mathematischen Kolloquiums 6, 1935, 10-11.

[40] Shafer, W. J., *The nontransitive consumer*, Econometrica 42, 1974, 913-919.

[41] Shafer, W. J., *Equilibrium in economies without ordered preferences or free disposal,* J. Math. Economics 3, 1976, 135-137.

[42] Shafer, W. J., *Transitivity*, in The New Palgrave: A Dictionary of Economics, J. Eatwell, M. Milgate and P. Newman, eds. Macmillan Press, London 1987.

[43] Sonnenschein, H. F., *Demand theory without transitive preferences, with applications to the theory of competitive equilibrium*, in Preferences, Utility, and Demand, J. S. Chipman, L. Hurwicz, M. K. Richter and H. F. Sonnenschein, eds., Harcourt Brace Jovanovich, New York 1971.

[44] Varian, H. R., *The nonparametric approach to demand analysis*, Econometrica 50, 1982, 945-973.

[45] Wald, A., *Über die Produktionsgleichungen der ökonomischen Wertlehre*, Ergebnisse eines mathematischen Kolloquiums 7, 1936, 1-6.

[46] Zeuthen, F., *Das Prinzip der Knappheit, technische Kombination und ökonomische Qualität,* Zeitschrift f. Nationalökonomie 4, 1933, 1-24.

Index